High Performance Liquid Chromatography in Pesticide Residue Analysis

CHROMATOGRAPHIC SCIENCE SERIES

A Series of Textbooks and Reference Books

Editor:
Nelu Grinberg

Founding Editor:
Jack Cazes

1. Dynamics of Chromatography: Principles and Theory, J. Calvin Giddings
2. Gas Chromatographic Analysis of Drugs and Pesticides, Benjamin J. Gudzinowicz
3. Principles of Adsorption Chromatography: The Separation of Nonionic Organic Compounds, Lloyd R. Snyder
4. Multicomponent Chromatography: Theory of Interference, Friedrich Helfferich and Gerhard Klein
5. Quantitative Analysis by Gas Chromatography, Josef Novák
6. High-Speed Liquid Chromatography, Peter M. Rajcsanyi and Elisabeth Rajcsanyi
7. Fundamentals of Integrated GC-MS (in three parts), Benjamin J. Gudzinowicz, Michael J. Gudzinowicz, and Horace F. Martin
8. Liquid Chromatography of Polymers and Related Materials, Jack Cazes
9. GLC and HPLC Determination of Therapeutic Agents (in three parts), Part 1 edited by Kiyoshi Tsuji and Walter Morozowich, Parts 2 and 3 edited by Kiyoshi Tsuji
10. Biological/Biomedical Applications of Liquid Chromatography, edited by Gerald L. Hawk
11. Chromatography in Petroleum Analysis, edited by Klaus H. Altgelt and T. H. Gouw
12. Biological/Biomedical Applications of Liquid Chromatography II, edited by Gerald L. Hawk
13. Liquid Chromatography of Polymers and Related Materials II, edited by Jack Cazes and Xavier Delamare
14. Introduction to Analytical Gas Chromatography: History, Principles, and Practice, John A. Perry
15. Applications of Glass Capillary Gas Chromatography, edited by Walter G. Jennings
16. Steroid Analysis by HPLC: Recent Applications, edited by Marie P. Kautsky
17. Thin-Layer Chromatography: Techniques and Applications, Bernard Fried and Joseph Sherma
18. Biological/Biomedical Applications of Liquid Chromatography III, edited by Gerald L. Hawk
19. Liquid Chromatography of Polymers and Related Materials III, edited by Jack Cazes

20. Biological/Biomedical Applications of Liquid Chromatography, edited by Gerald L. Hawk
21. Chromatographic Separation and Extraction with Foamed Plastics and Rubbers, G. J. Moody and J. D. R. Thomas
22. Analytical Pyrolysis: A Comprehensive Guide, William J. Irwin
23. Liquid Chromatography Detectors, edited by Thomas M. Vickrey
24. High-Performance Liquid Chromatography in Forensic Chemistry, edited by Ira S. Lurie and John D. Wittwer, Jr.
25. Steric Exclusion Liquid Chromatography of Polymers, edited by Josef Janca
26. HPLC Analysis of Biological Compounds: A Laboratory Guide, William S. Hancock and James T. Sparrow
27. Affinity Chromatography: Template Chromatography of Nucleic Acids and Proteins, Herbert Schott
28. HPLC in Nucleic Acid Research: Methods and Applications, edited by Phyllis R. Brown
29. Pyrolysis and GC in Polymer Analysis, edited by S. A. Liebman and E. J. Levy
30. Modern Chromatographic Analysis of the Vitamins, edited by André P. De Leenheer, Willy E. Lambert, and Marcel G. M. De Ruyter
31. Ion-Pair Chromatography, edited by Milton T. W. Hearn
32. Therapeutic Drug Monitoring and Toxicology by Liquid Chromatography, edited by Steven H. Y. Wong
33. Affinity Chromatography: Practical and Theoretical Aspects, Peter Mohr and Klaus Pommerening
34. Reaction Detection in Liquid Chromatography, edited by Ira S. Krull
35. Thin-Layer Chromatography: Techniques and Applications, Second Edition, Revised and Expanded, Bernard Fried and Joseph Sherma
36. Quantitative Thin-Layer Chromatography and Its Industrial Applications, edited by Laszlo R. Treiber
37. Ion Chromatography, edited by James G. Tarter
38. Chromatographic Theory and Basic Principles, edited by Jan Åke Jönsson
39. Field-Flow Fractionation: Analysis of Macromolecules and Particles, Josef Janca
40. Chromatographic Chiral Separations, edited by Morris Zief and Laura J. Crane
41. Quantitative Analysis by Gas Chromatography, Second Edition, Revised and Expanded, Josef Novák
42. Flow Perturbation Gas Chromatography, N. A. Katsanos
43. Ion-Exchange Chromatography of Proteins, Shuichi Yamamoto, Kazuhiro Naka-nishi, and Ryuichi Matsuno
44. Countercurrent Chromatography: Theory and Practice, edited by N. Bhushan Man-dava and Yoichiro Ito
45. Microbore Column Chromatography: A Unified Approach to Chromatography, edited by Frank J. Yang
46. Preparative-Scale Chromatography, edited by Eli Grushka
47. Packings and Stationary Phases in Chromatographic Techniques, edited by Klaus K. Unger

48. Detection-Oriented Derivatization Techniques in Liquid Chromatography, edited by Henk Lingeman and Willy J. M. Underberg
49. Chromatographic Analysis of Pharmaceuticals, edited by John A. Adamovics
50. Multidimensional Chromatography: Techniques and Applications, edited by Hernan Cortes
51. HPLC of Biological Macromolecules: Methods and Applications, edited by Karen M. Gooding and Fred E. Regnier
52. Modern Thin-Layer Chromatography, edited by Nelu Grinberg
53. Chromatographic Analysis of Alkaloids, Milan Popl, Jan Fähnrich, and Vlastimil Tatar
54. HPLC in Clinical Chemistry, I. N. Papadoyannis
55. Handbook of Thin-Layer Chromatography, edited by Joseph Sherma and Bernard Fried
56. Gas–Liquid–Solid Chromatography, V. G. Berezkin
57. Complexation Chromatography, edited by D. Cagniant
58. Liquid Chromatography–Mass Spectrometry, W. M. A. Niessen and Jan van der Greef
59. Trace Analysis with Microcolumn Liquid Chromatography, Milos Krejcl
60. Modern Chromatographic Analysis of Vitamins: Second Edition, edited by André P. De Leenheer, Willy E. Lambert, and Hans J. Nelis
61. Preparative and Production Scale Chromatography, edited by G. Ganetsos and P. E. Barker
62. Diode Array Detection in HPLC, edited by Ludwig Huber and Stephan A. George
63. Handbook of Affinity Chromatography, edited by Toni Kline
64. Capillary Electrophoresis Technology, edited by Norberto A. Guzman
65. Lipid Chromatographic Analysis, edited by Takayuki Shibamoto
66. Thin-Layer Chromatography: Techniques and Applications: Third Edition, Revised and Expanded, Bernard Fried and Joseph Sherma
67. Liquid Chromatography for the Analyst, Raymond P. W. Scott
68. Centrifugal Partition Chromatography, edited by Alain P. Foucault
69. Handbook of Size Exclusion Chromatography, edited by Chi-San Wu
70. Techniques and Practice of Chromatography, Raymond P. W. Scott
71. Handbook of Thin-Layer Chromatography: Second Edition, Revised and Expanded, edited by Joseph Sherma and Bernard Fried
72. Liquid Chromatography of Oligomers, Constantin V. Uglea
73. Chromatographic Detectors: Design, Function, and Operation, Raymond P. W. Scott
74. Chromatographic Analysis of Pharmaceuticals: Second Edition, Revised and Expanded, edited by John A. Adamovics
75. Supercritical Fluid Chromatography with Packed Columns: Techniques and Applications, edited by Klaus Anton and Claire Berger
76. Introduction to Analytical Gas Chromatography: Second Edition, Revised and Expanded, Raymond P. W. Scott
77. Chromatographic Analysis of Environmental and Food Toxicants, edited by Takayuki Shibamoto

78. Handbook of HPLC, edited by Elena Katz, Roy Eksteen, Peter Schoenmakers, and Neil Miller
79. Liquid Chromatography–Mass Spectrometry: Second Edition, Revised and Expanded, Wilfried Niessen
80. Capillary Electrophoresis of Proteins, Tim Wehr, Roberto Rodríguez-Díaz, and Mingde Zhu
81. Thin-Layer Chromatography: Fourth Edition, Revised and Expanded, Bernard Fried and Joseph Sherma
82. Countercurrent Chromatography, edited by Jean-Michel Menet and Didier Thiébaut
83. Micellar Liquid Chromatography, Alain Berthod and Celia García-Alvarez-Coque
84. Modern Chromatographic Analysis of Vitamins: Third Edition, Revised and Expanded, edited by André P. De Leenheer, Willy E. Lambert, and Jan F. Van Bocxlaer
85. Quantitative Chromatographic Analysis, Thomas E. Beesley, Benjamin Buglio, and Raymond P. W. Scott
86. Current Practice of Gas Chromatography–Mass Spectrometry, edited by W. M. A. Niessen
87. HPLC of Biological Macromolecules: Second Edition, Revised and Expanded, edited by Karen M. Gooding and Fred E. Regnier
88. Scale-Up and Optimization in Preparative Chromatography: Principles and Biopharmaceutical Applications, edited by Anurag S. Rathore and Ajoy Velayudhan
89. Handbook of Thin-Layer Chromatography: Third Edition, Revised and Expanded, edited by Joseph Sherma and Bernard Fried
90. Chiral Separations by Liquid Chromatography and Related Technologies, Hassan Y. Aboul-Enein and Imran Ali
91. Handbook of Size Exclusion Chromatography and Related Techniques: Second Edition, edited by Chi-San Wu
92. Handbook of Affinity Chromatography: Second Edition, edited by David S. Hage
93. Chromatographic Analysis of the Environment: Third Edition, edited by Leo M. L. Nollet
94. Microfluidic Lab-on-a-Chip for Chemical and Biological Analysis and Discovery, Paul C.H. Li
95. Preparative Layer Chromatography, edited by Teresa Kowalska and Joseph Sherma
96. Instrumental Methods in Metal Ion Speciation, Imran Ali and Hassan Y. Aboul-Enein
97. Liquid Chromatography–Mass Spectrometry: Third Edition, Wilfried M. A. Niessen
98. Thin Layer Chromatography in Chiral Separations and Analysis, edited by Teresa Kowalska and Joseph Sherma
99. Thin Layer Chromatography in Phytochemistry, edited by Monika Waksmundzka-Hajnos, Joseph Sherma, and Teresa Kowalska
100. Chiral Separations by Capillary Electrophoresis, edited by Ann Van Eeckhaut and Yvette Michotte
101. Handbook of HPLC: Second Edition, edited by Danilo Corradini and consulting editor Terry M. Phillips
102. High Performance Liquid Chromatography in Phytochemical Analysis, edited by Monika Waksmundzka-Hajnos and Joseph Sherma

103. Hydrophilic Interaction Liquid Chromatography (HILIC) and Advanced Applications, edited by Perry G. Wang and Weixuan He
104. Hyphenated and Alternative Methods of Detection in Chromatography, edited by R. Andrew Shalliker
105. LC-NMR: Expanding the Limits of Structure Elucidation, Nina C. Gonnella
106. Thin Layer Chromatography in Drug Analysis, edited by Łukasz Komsta, Monika Waksmundzka-Hajnos, and Joseph Sherma
107. Pharmaceutical Industry Practices on Genotoxic Impurities, edited by Heewon Lee
108. Advanced Separations by Specialized Sorbents, edited by Ecaterina Stela Dragan
109. High Performance Liquid Chromatography in Pesticide Residue Analysis, edited by Tomasz Tuzimski and Joseph Sherma

High Performance Liquid Chromatography in Pesticide Residue Analysis

Edited by

Tomasz Tuzimski
Joseph Sherma

CRC Press
Taylor & Francis Group
Boca Raton London New York

CRC Press is an imprint of the
Taylor & Francis Group, an **informa** business

CRC Press
Taylor & Francis Group
6000 Broken Sound Parkway NW, Suite 300
Boca Raton, FL 33487-2742

Printed on acid-free paper
Version Date: 20150413

International Standard Book Number-13: 978-1-4665-6881-5 (Hardback)

Visit the Taylor & Francis Web site at
http://www.taylorandfrancis.com

and the CRC Press Web site at
http://www.crcpress.com

Contents

Preface

This book is organized as a monograph that presents, in a properly structured manner, up-to-date, state-of-the-art information on the very important field of high-performance column liquid chromatography (HPLC) applied to analysis of pesticides. It is a well-established fact that HPLC finds broad application in the separation, identification, and quantification of important components, such as pesticides, and yet no monograph on this particular subject has yet been published, especially as a practical guide. Environmental chemistry benefits from the precision, rapidity, versatility, and automation of HPLC, which is a very flexible separation tool well suited for this particular kind of investigation.

WHY READ THIS BOOK?

Albert Einstein has been quoted as saying, "If the bee disappears from the surface of the earth, man would have no more than four years to live. No more bees, no more pollination, no more plants, no more animals, no more men!" Further, the following has been published on the website http://globalclimatechange.wordpress.com/2007/04/20/Einstein-on-bees/: "Bits and pieces of information about farmers' concerns for bee disappearance (or colony collapse disorder) in 24 states around the U.S. have bubbled up to the surface, over the last year and a half, but hardly any large-scale media attention has been drawn to this potentially serious problem. Recently, bees have gone missing from hives around Europe as well. The East Coast of the U.S. is reporting a 70% loss in commercial bee hive habitation, the West Coast 60%; these figures are staggering."

Excessive overuse of pesticides and uncontrolled spraying in agronomy is one of the possible reasons that led to the paralysis of the nervous system of bees on a massive scale. The effect of applications of pesticides may be the cause of irrational behavior of insects, which began en masse to abandon their apiaries. Decreased numbers of apiaries and insects are factors that may be of extreme importance in the process of pollination. This, in turn, led to reduced yields and production of food of plant origin. The consequence may be a reduction in animal breeding and production of food of animal origin.

Does this mean that the mass extinction of bees could lead to the destruction of mankind? To counter this, it is necessary to develop a modern methodology for the analysis of pesticide residues. One of these methods is HPLC.

HIGHLIGHTS

1. This is the first monograph ever published devoted specifically to application of HPLC to pesticide analysis as *a practical guide.*
2. This book should serve as a comprehensive source of critical information on pesticide analysis provided by a team of international experts for an extremely wide variety of scientists with interest in research, industrial analysis, regulatory analysis, or teaching in the fields of toxicology, environmental protection, analytical chemistry, physical chemistry, crop management and production, food protection and production, plant ecology, human and veterinary medicine, herbal drugs and nutritional supplement production, pharmacology, forensics, and others.
3. This book presents HPLC as a flexible and versatile separation and analysis tool with multiple purposes and advantages in investigations of pesticides for food and plant drug standardization, promotion of health, protection of new herbal medicines, etc.
4. This book presents material relative to environmental analytical chemistry, which is one of the important and yet not frequently enough discussed areas of chemistry, biology, and medicine.

This book is divided into three major sections. Section I is devoted to general information concerning the areas of science related to pesticide residue analysis in which HPLC is used. This section of the book is a practical guide to HPLC methods of pesticide residue analysis that shows, step by step, how to achieve a good practice of chromatography. Section I, with nine chapters, is devoted to the general issues related to pesticide residue analysis and to the particular demands imposed on HPLC in applications to environmental analysis. It starts with a chapter giving an overview of the field and a description of the organization of the book, followed by an explanation of the new trends and expectations in environmental issues and modern methods of pesticide residue investigation. Also, classification and properties of pesticides (indicating their activity) are included in the beginning of Section I. Section I is concerned with the choice of optimal conditions of the chromatographic process for separation, identification, and quantitative determination of pesticides. This information is multipurpose and applied to the choice of chromatographic conditions for various analytes.

The two chapters of Section II cover kinetic study of pesticides and their degradation and fragmentation in the environment and HPLC methods applied in pesticide residue analysis.

Section III, with nine chapters, provides coverage of a very wide spectrum of applications of HPLC and ultra-performance liquid chromatography (UPLC or UHPLC) to separation and analysis of pesticides from various classes. The authors address all topics necessary for the complete coverage of HPLC for the following spheres: sample preparation methods applied in pesticide analysis (Chapter 12), quantitative analysis and method validation (Chapter 13), analysis of pesticides by HPLC coupled with other ultraviolet detectors (Chapter 14), the hyphenated techniques of HPLC-mass spectrometry (Chapter 15), multidimensional liquid chromatography (Chapter 16), chiral separation of analytes by HPLC (Chapter 17), application of chemiluminescence detection in analysis of pesticides (Chapter 18), UPLC applied to pesticide analysis (Chapter 19), and comparison of HPLC with other modern analytical techniques for the analysis of pesticides (Chapter 20).

We thank Barbara Glunn, senior editor–chemistry, and Cheryl Wolf, editorial assistant, CRC Press/Taylor & Francis Group, as well as Dr. Nelu Grinberg, editor, Chromatographic Science Series, for their unfailing support during this project. We also thank the expert chapter authors for their valuable contributions to our book.

Tomasz Tuzimski
Joseph Sherma

Editors

Tomasz Tuzimski, PhD, DSc, is an adjunct professor in the Department of Physical Chemistry, Faculty of Pharmacy, Medical Analytics Division, Medical University of Lublin (Lublin, Poland). His scientific interests include the theory and application of liquid chromatography, taking into consideration optimization of chromatographic systems for separation and quantitative analysis of analytes in multicomponent mixtures. Dr. Tuzimski was rewarded for his achievements in the field of study of chromatographic methods in analytical chemistry of pesticides (a series of five publications and monograph, T. Tuzimski and E. Soczewiński, *Retention and Selectivity of Liquid–Solid Chromatographic Systems for the Analysis of Pesticides (Retention Database) in Problems of Science, Teaching and Therapy*, Medical University of Lublin, Poland, October 2002, by the Ministry of Health of Poland (individual prize). Dr. Tuzimski was also rewarded as a coauthor of a handbook for students, *Analytical Chemistry* (edited by R. Kocjan, PZWL, 2000 and 2002, in Polish), by the Ministry of Health of the Polish Republic (team prize).

Dr. Tuzimski was invited by Professor Szabolcs Nyiredy to a 3-month practice in the Research Institute for Medicinal Plants in Budakalász (Hungary). The investigations were financially supported by the Educational Exchange Programme between Hungary and Poland—Hungarian Scholarship Board (No. MÖB 2-13-1-44-3554/2005). He actively participated in numerous scientific symposia, where he presented his research results as oral and poster presentations at 20 international meetings and 18 national scientific symposia.

Dr. Tuzimski has published 47 research papers (including 24 individual papers) in journals with a high level of impact factors (total IF = 64.377). He is the author of articles written at the special invitations of editors of the *Journal of Chromatography A*, *Journal of Liquid Chromatography and Related Technologies*, and *Journal of Planar Chromatography—Modern TLC*. Besides the above-mentioned monograph, Dr. Tuzimski is the author of the following chapters: Use of Planar Chromatography in Pesticide Residue Analysis, in *Handbook of Pesticides: Methods of Pesticide Residues Analysis,* edited by Leo M.L. Nollet and Hamir Singh Rathore, Boca Raton 2010, CRC Press Taylor & Francis Group; Basic Principles of Planar Chromatography and Its Potential for Hyphenated Techniques, in *High-Performance Thin-Layer Chromatography (HPTLC)*, edited by ManMohan Srivastava, Springer, Heidelberg 2011; Multidimensional Chromatography in Pesticides Analysis, in *Pesticides—Strategies for Pesticides Analysis,* edited by Margarita Stoytcheva, InTech, Rijeka 2011; Determination of Pesticides in Complex Samples by One-Dimensional (1D-), Two-Dimensional (2D-) and Multidimensional Chromatography, in *Pesticides in the Modern World—Trends in Pesticide Analysis*, edited by Margarita Stoytcheva, InTech, Rijeka 2011; Pesticide Residues in the Environment, in *Pesticides: Evaluation of Environmental Pollution,* edited by Leo M.L. Nollet and Hamir Singh Rathore, CRC Press Taylor & Francis Group, Boca Raton 2012; and Advanced Spectroscopic Detectors for Identification and Quantification: UV–Visible, Fluorescence, and Infrared Spectroscopy, in *Instrumental Thin-Layer Chromatography*, edited by Colin F. Poole, Elsevier 2015, Amsterdam, Netherlands. He is also a coauthor with Prof. Dr. T. Dzido of the chapter Chambers, Sample Application and Chromatogram Development, in *Thin-Layer Chromatography in Phytochemistry*, edited by M. Waksmundzka-Hajnos, J. Sherma, and T. Kowalska, Boca Raton 2008, CRC Press Taylor & Francis Group. Dr. Tuzimski is a recipient of two grants from the Polish Ministry of Science and Higher Education (2005–2008 and 2009–2011) for the study and procedure implementation of new methods of analysis of pesticides in original samples (e.g., water, medicinal herbs, wines, food) with application of modern analytical methods combined with diode array scanning densitometry.

Dr. Tuzimski has reviewed 150 submitted research manuscripts (*Journal of Chromatography A*, *Journal of Chromatography B*, *Journal of Separation Science*, *Journal of Chromatographic Science*, *Journal of AOAC Int.*, and *Journal of Planar Chromatography—Modern TLC*). He has taught analytical and physical chemistry exercises with second-year students of the Faculty of Pharmacy, and was also an instructor of postgraduate chromatographic courses for scientific research staff from Polish universities and workers from the industry. He was a promoter of research work of three masters of pharmacy and supervised the research work of ten masters of pharmacy. He is a member of the Polish Pharmaceutical Society as well as the editorial board of *The Scientific World Journal/Analytical Chemistry, Advances in Analytical Chemistry, American Journal of Environmental Protection, International Journal of Biotechnology and Food Science (IJBFS)*, and *Advancement in Scientific and Engineering Research (ASER)*. Dr. Tuzimski edited four special sections on pesticide residue analysis of the *Journal of AOAC International* (2010, 2012, 2014, and 2015).

Joseph Sherma earned a BS degree in chemistry from Upsala College, East Orange, New Jersey, in 1955 and a PhD degree in analytical chemistry from Rutgers, the State University, New Brunswick, New Jersey, in 1958 under the supervision of the renowned ion exchange chromatography expert William Rieman III. Professor Sherma is currently the John D. and Francis H. Larkin Professor Emeritus of chemistry at Lafayette College, Easton, PA; he taught courses in analytical chemistry for more than 40 years, was head of the chemistry department for 12 years, and continues to supervise research students at Lafayette. During sabbatical leaves and summers, Professor Sherma did research in the laboratories of the eminent chromatographers Dr. Harold Strain, Dr. Gunter Zweig, Dr. Mel Getz, Professor James Fritz, and Professor Joseph Touchstone.

Professor Sherma has authored, coauthored, edited, or coedited more than 780 publications, including research papers and review articles in approximately 55 different peer-reviewed analytical chemistry, chromatography, and biological journals; approximately 30 invited book chapters; and more than 70 books and U.S. government agency manuals in the areas of analytical chemistry and chromatography.

In addition to his research in the techniques and applications of thin layer chromatography (TLC), including especially drug analysis, Professor Sherma has a very productive interdisciplinary research program in the use of analytical chemistry to study biological systems with Bernard Fried, Kreider Professor Emeritus of Biology at Lafayette College, with whom he has coauthored the book *Thin Layer Chromatography* (first to fourth editions) and edited the *Handbook of Thin Layer Chromatography* (first to third editions), all published by Marcel Dekker, Inc., as well as coediting *Practical Thin Layer Chromatography* for CRC Press. Professor Sherma wrote with Dr. Zweig a book titled *Paper Chromatography* for Academic Press and the first two volumes of the *Handbook of Chromatography* series for CRC Press, and coedited with him 22 more volumes of the chromatography series and 10 volumes of the series *Analytical Methods for Pesticides and Plant Growth Regulators* for Academic Press. After Dr. Zweig's death, Professor Sherma edited five additional volumes of the chromatography handbook series and two volumes in the pesticide series. The pesticide series was completed under the title *Modern Methods of Pesticide Analysis* for CRC Press with two volumes coedited with Dr. Thomas Cairns. Three books on quantitative TLC and advances in TLC were edited jointly with Professor Touchstone for Wiley-Interscience.

For CRC/Taylor & Francis Group, Professor Sherma coedited with Professor Teresa Kowalska *Preparative Layer Chromatography* and *Thin Layer Chromatography in Chiral Separations and Analysis*, coedited with Professor Kowalska and Professor Monika Waksmundska-Hajnos *Thin Layer Chromatography in Phytochemistry*, and coedited with Professor Waksmundska-Hajnos *High Performance Liquid Chromatography in Phytochemical Analysis*. A book titled *Thin Layer Chromatography in Drug Analysis* coedited with Professor Lukasz Komsta and Professor Waksmundzka-Hajnos was published in 2014.

Professor Sherma served for 23 years as the editor for residues and trace elements of the *Journal of AOAC International* and is currently the journal's acquisitions editor. He has guest-edited with Professor Fried 14 annual special issues on TLC of the *Journal of Liquid Chromatography and Related Technologies* and regularly guest-edits special sections of issues of the *Journal of AOAC International* on specific subjects in all areas of analytical chemistry. For 12 years, he also wrote an article on modern analytical instrumentation for each issue of the *Journal of AOAC International.* Professor Sherma has written biennial reviews of planar chromatography that were published in the American Chemical Society journal *Analytical Chemistry* from 1970 to 2010, the *Journal of AOAC International* in 2012, and the *Central European Journal of Chemistry* in 2014. Since 1982, he has also written biennial reviews of pesticide analysis by TLC in the *Journal of Liquid Chromatography and Related Technologies and* the *Journal of Environmental Science and Health, Part B.* He is now on the editorial boards of the *Journal of Planar Chromatography—Modern TLC*; *Acta Chromatographica*; *Journal of Environmental Science and Health, Part B*; and *Journal of Liquid Chromatography and Related Technologies.*

Professor Sherma was a recipient of the 1995 ACS Award for Research at an undergraduate institution sponsored by Research Corporation. The first 2009 issue, volume 12, of *Acta Universitatis Cibiniensis, Seria F, Chemia* was dedicated in honor of Professor Sherma's teaching, research, and publication accomplishments in analytical chemistry and chromatography.

Contributors

María de los Ángeles Herrera Abdo
Department of Chemistry and Physics
University of Almería
Almería, Spain

Hassan Y. Aboul-Enein
Department of Pharmaceutical and Medicinal Chemistry
Pharmaceutical and Drug Industries Research Division
National Research Centre
Cairo, Egypt

Martha Bohrer Adaime
Laboratory of Pesticide Residue Analysis
Chemistry Department
Federal University of Santa Maria
Santa Maria, Brazil

Imran Ali
Department of Chemistry
Jamia Millia Islamia (Central University)
New Delhi, India

Tomasz Bączek
Department of Pharmaceutical Chemistry
Faculty of Pharmacy
Medical University of Gdańsk
Gdańsk, Poland

Szymon Bocian
Chair of Environmental Chemistry and Bioanalytics
Nicholas Copernicus University
Toruń, Poland

Bogusław Buszewski
Chair of Environmental Chemistry and Bioanalytics
Nicholas Copernicus University
Toruń, Poland

Łukasz Cieśla
Department of Inorganic Chemistry
Chair of Chemistry
Faculty of Pharmacy with Medical Analytics Division
Medical University of Lublin
Lublin, Poland

Antonia Garrido Frenich
Department of Analytical Chemistry
University of Almería
Almería, Spain

Tadeusz Górecki
Department of Chemistry
Faculty of Science
University of Waterloo
Waterloo, Ontario, Canada

Iqbal Hussain
Department of Chemistry
Faculty of Science
Universiti Teknologi Malaysia
UTM Johor Bahru, Johor, Malaysia

Pavel Jandera
Department of Analytical Chemistry
Faculty of Chemical Technology
University of Pardubice
Pardubice, Czech Republic

Roman Kaliszan
Department of Biopharmaceutics and Pharmacodynamics
Faculty of Pharmacy
Medical University of Gdańsk
Gdańsk, Poland

Piotr Kawczak
Department of Pharmaceutical Chemistry
Faculty of Pharmacy
Medical University of Gdańsk
Gdańsk, Poland

Manoel Leonardo Martins
Laboratory of Pesticide Residue Analysis
Chemistry Department
Federal University of Santa Maria
Santa Maria, Brazil

Ana Masiá
Food and Environmental Safety Research Group (SAMA-UV)
Faculty of Pharmacy
University of Valencia
Valencia, Spain

Marco Minella
Department of Chemistry
University of Torino
Turin, Italy

Claudio Minero
Department of Chemistry
University of Torino
Turin, Italy

Ahmed Mostafa
Suez Canal Authority
Ismailia, Egypt

and

Department of Chemistry
University of Waterloo
Waterloo, Ontario, Canada

Peipei Pan
School of Life Sciences
University of Nevada
Las Vegas, Nevada

Anna Petruczynik
Department of Inorganic Chemistry
Faculty of Pharmacy with Medical Analytics Division
Medical University of Lublin
Lublin, Poland

Yolanda Picó
Food and Environmental Safety Research Group (SAMA-UV)
Faculty of Pharmacy
University of Valencia
Valencia, Spain

Osmar Damian Prestes
Laboratory of Pesticide Residue Analysis
Chemistry Department
Federal University of Santa Maria
Santa Maria, Brazil

Kristy M. Richards
US FDA
Lenexa, Kansas

Chris Sack
US FDA
Lenexa, Kansas

Mohd Marsin Sanagi
Department of Chemistry
Faculty of Science
Universiti Teknologi Malaysia
UTM Johor Bahru, Johor, Malaysia

Heba Shaaban
Pharmaceutical Analytical Chemistry Department
Faculty of Pharmacy
Suez Canal University
Ismailia, Egypt

Joseph Sherma
Department of Chemistry
Lafayette College
Easton, Pennsylvania

Robert E. Smith
US FDA
Lenexa, Kansas

and

Science Department
Park University
Parkville, Missouri

Edward Soczewiński
Department of Analytical Chemistry
Chair of Chemistry
Faculty of Pharmacy with Medical Analytics Division
Medical University of Lublin
Lublin, Poland

Kevin Tran
US FDA
Lenexa, Kansas

Tomasz Tuzimski
Department of Physical Chemistry
Chair of Chemistry
Faculty of Pharmacy with Medical Analytics Division
Medical University of Lublin
Lublin, Poland

Jeanette M. Van Emon
National Exposure Research Laboratory
U.S. Environmental Protection Agency
Las Vegas, Nevada

José Luis Martínez Vidal
Department of Chemistry and Physics
University of Almería
Almería, Spain

Davide Vione
Department of Chemistry
University of Torino
Turin, Italy

Monika Waksmundzka-Hajnos
Department of Inorganic Chemistry
Faculty of Pharmacy with Medical Analytics Division
Medical University of Lublin
Lublin, Poland

Renato Zanella
Laboratory of Pesticide Residue Analysis
Chemistry Department
Federal University of Santa Maria
Santa Maria, Brazil

Section I

Practical Guide to HPLC Methods of Pesticide Residue Analysis

1 Overview of the Field of Chromatographic Methods of Pesticide Residue Analysis and Organization of the Book

Tomasz Tuzimski and Joseph Sherma

CONTENTS

1.1 TRENDS IN PESTICIDE RESIDUE ANALYSIS

The agricultural production of food and animal feed on an economically competitive basis requires an ever-increasing application of pesticides. "Pesticide" is a general term that includes a variety of chemical and biological products to eliminate or control pests, such as fungi, insects, rodents, and weeds (fungicides, insecticides, rodenticides, and herbicides, respectively). In the European Union (EU), approximately 320,000 tons of active substances are sold every year, which accounts for one quarter of the world market [1]. Residues in fruits and vegetables, cereals, processed baby food, and foodstuffs of animal origin are controlled through a system of statutory maximum residue limits (MRLs), which are defined as "The maximum concentration of pesticide residue (expressed as mg residue/kg commodity likely to occur in or on food commodities and animal feeds after the use of pesticides according to good agricultural practice" [2]. MRLs vary ordinarily within the range 0.0008–50 mg/kg [3], and they are typically between 0.01 and 10 mg/kg for the adult population. Lower values of MRLs are set for baby food: The EU specified a MRL of 0.010 mg/kg [4]; the lowest levels are set for particular special residues [5].

Pesticides are widespread throughout the world. The composition of pesticide mixtures occurring in environmental samples depends on geographical area, season of the year, number of farms, and quantity and intensity of use of plant-protection agents. The variety of their mixtures in different matrices, for example, rivers, is very large. Many sample-preparation techniques are used in pesticide residue analysis; the method selected depends on the complexity of the sample, the natures of the matrix and the analytes, and the analytical techniques available. The most efficient approach to pesticide analysis involves the use of chromatographic methods.

1.2 SURVEY OF CHROMATOGRAPHIC METHODS OF PESTICIDE RESIDUE ANALYSIS

The following chromatographic methods are most frequently applied for pesticide residue analysis: one- and two-dimensional (2-D) thin layer chromatography (TLC); high-performance liquid chromatography

(HPLC); gas chromatography (GC); and multidimensional chromatographic techniques, such as GC × GC, LC–LC, and multidimensional planar chromatography. The acronym "HPLC" is almost universally used to denote column HPLC, and ultra-performance liquid chromatography (UPLC) (or ultra-high performance liquid chromatography [UHPLC] for companies other than Waters, Inc.) is column liquid chromatography (LC) with column stationary phase particles <2 um. High-performance TLC (HPTLC) is technically HPLC because it involves a small particle stationary phase (the layer) and a liquid mobile phase, but HPTLC is never considered as a technique covered by the acronym HPLC. In the literature, LC is reserved usually as an acronym for column liquid chromatography.

HPLC is a chromatographic technique widely used for qualitative and quantitative analysis of organic compounds, for example, pesticides present in multicomponent mixtures. It utilizes a fully automated instrumental system, including a column, mobile phase container, mobile phase pump, and detector. The HPLC system is controlled by a computer program that registers chromatographic profiles and all data of the individual peaks: retention time, peak height, peak width, surface area of a peak, system efficiency, peak symmetry factor, etc. Because the column providing the separation is connected to the detector, HPLC offers wide possibilities of detection and online determination of a wide range of organic and inorganic compounds.

At each stage of the HPLC procedure, the chromatographer should possess basic skills that substantially help in accomplishing the analyses correctly in order to obtain reliable, repeatable, and reproducible results. He or she might meet many pitfalls during work with HPLC systems. This book gives information that will draw the reader's attention to the procedures and equipment that have often been applied and proven in contemporary HPLC practice.

1.3 ORGANIZATION OF THE BOOK

This book covers all topics important in pesticide residue analysis by HPLC. It comprises 20 chapters divided into three parts. Section I is devoted to general information concerning the areas of science related to pesticide residue analysis in which HPLC is used. This section of the book is a practical guide to HPLC methods of pesticide residue analysis, which shows, step by step, how to behave with decorum for good practice of chromatography.

Properties of the analyte are very important for making the correct choice of chromatographic conditions for the identification and quantitative analysis of all types of samples. Chapter 2 is devoted to classification and properties of pesticides (indicating their activity). The choice of the mobile phase depends not only on the properties of the column sorbent and its activity, but also on the structure and type of separated analytes. Chapter 4 is devoted to the choice of the mode of chromatographic analysis of pesticides on the basis of the properties of analytes, for example, lipophilicity, which is an important characteristic of organic compounds in terms of their environmental activity. Quantitative structure–activity relationship and quantitative structure–retention relationship studies have found growing acceptance and application in agrochemical research.

The significance of the relationships between retention and composition of the mobile phase for prediction of separation of sample components has inspired many authors to investigate the characteristics of retention more deeply. The dependence of retention on the composition of the mobile phase can be described using different theoretical models. It is useful to discuss, in more detail, the modes of retention and selectivity optimization that can be applied to obtain appropriate chromatographic resolution. Chapter 3 is devoted to the method development of chromatography: retention–mobile phase composition relationships and their application to analysis of pesticides.

Analytical methods for pesticide analysis were developed in the 1960s, employing an initial extraction with acetone, followed by a partitioning step upon addition of a nonpolar solvent and salt; these methods involved complex and solvent-intensive cleanup steps. Moreover, the instruments available for analysis of the target compounds had relatively low selectivity and sensitivity. The development of technology and robotics in the 1990s to reduce manual methodology and allow sample preparation during nonworking time led to the development of automatic sample preparation techniques, such as supercritical fluid extraction and pressurized liquid extraction. Although initially very promising, these techniques have not succeeded in the field of pesticide analysis for various reasons, namely high price and low reliability of the instruments and inability to extract different pesticide classes in foods with the same efficiency, often requiring separate optimization for different analytes. Later, a successful simplification of "traditional" solvent sample preparation, QuEChERS (Quick, Easy, Cheap, Effective, Rugged, and Safe), was presented by Dr. Steven Lehotay and collaborators. Two similar QuEChERS methods achieved the status of official methods of the AOAC International [6] and the European Committee for Standardization (Standard Method EN 15662). Analysis of environmental samples requires a good extraction method for sample preparation. The great variety of samples and pesticides to be analyzed requires numerous sample-preparation methods. The problem of peak overlapping may occur, and a preseparation of the sample is often necessary. This preseparation aims at reducing the complexity of the original matrix by resolving several simpler fractions of the original matrix. The fractions should contain the same amounts of the analytes as in the whole sample, ready for analysis and free from substances that can interfere during the chromatographic analysis. Chapter 5 is devoted to the choice of the mode of sample preparation for analysis of pesticides on the basis of the properties of the matrix.

Application of appropriate stationary and mobile phases is the key element that influences the resolution of the mixture components and the efficiency of the quantitative and qualitative analysis. Optimization of these elements can be effectively performed on the basis of a good understanding of the theoretical fundamentals and practical knowledge of HPLC. Sophisticated equipment, methods, and software are inherent elements of today's LC and can effectively facilitate optimization of chromatographic separations. Thanks to these features, HPLC is a powerful analytical and separation technique in contemporary analysis, which has gained growing popularity in laboratory practice, especially for separation and analysis of pesticides in environmental, food, and agricultural samples. The next three chapters describe the selection of stationary phases and columns (Chapter 6) and mobile phases for analysis of nonionic analytes (Chapter 7) and ionic analytes (Chapter 8) used in the HPLC of pesticides.

This section of the book also contains Chapter 9, concerning optimization of normal phase and reversed phase (RP) chromatographic systems applied in column chromatography in isocratic- and gradient-elution modes. RP-HPLC [approximately 80% of all chromatographic separations are performed by use of the nonpolar chemically bonded phases, containing mainly octadecyl (C_{18}) chains] can be performed either by isocratic elution (with a mobile phase of constant composition) or by gradient elution (with changes in the mobile phase strength resulting from changes in composition in the course of a chromatographic run). Gradient elution is usually obtained by gradual addition of a high-elution-strength solvent to a low-elution-strength solvent (usually water). Gradient elution lowers the analysis time and improves separation efficiency. The gradient-elution mode is applied to mixtures of components differing considerably in polarity when the "general elution problem" occurs. A linear gradient is usually applied, but the gradient profile can be programmed to have different shapes. Gradient elution suitable for a given separation can be optimized with the aid of special computer programs (Drylab, Chromsword) simulating the separation of a particular mixture that is characterized with known retention parameters. For separation of organic electrolytes, pH gradients or a binary gradient of a mobile phase modifier and pH can be applied. However, gradient elution limits the use of certain detectors and increases the time needed for equilibration of the column.

The next two chapters (Chapters 10 and 11) in Section II cover kinetic studies and degradation of pesticides and their determination by HPLC, respectively. Kinetic studies have revealed several interactions between sorption and degradation. It is commonly accepted that sorbed chemicals are less accessible to microorganisms and that sorption accordingly limits their degradation as well as their transport. Kinetically, the sorption of most organic pesticides is a two-step process: An initial fast step that accounts for the greater part of total sorption is followed by a much slower step tending toward final equilibrium.

Chapter 10 deals with the kinetic changes of pesticides in the environment. After general information focused on different conditions influencing kinetic aspects of pesticides' transformation, the author presents examples of different pathways that common pesticides may undergo after entering the environment. In the chapter, HPLC is presented as an indispensible tool for monitoring the environmental fate of pesticides.

The majority of pesticides exhibit limited stability in the environment and in plants and animals; it is an extremely important property of these compounds that, on the one hand, determines the efficiency of their action and, on the other, permits the safe use of agricultural products by humans. The fate of pesticides in the soil depends on the chemical transformations in which the living organisms participate (biotic, biochemical transformations) and on physical, chemical, and photochemical processes.

Biotic transformations catalyzed by the enzymes of soil microorganisms predominate. The highest ability of efficient degradation of pesticides is exhibited by bacteria, *Actinomycetes*, and mushrooms. The contribution of mushrooms is the greatest, reaching up to 80%. There are at least two causes of this high activity of soil mushrooms: greater resistance to unfavorable vegetation conditions in comparison to bacteria (mushrooms exhibit high liveliness in acidic environments even at pH <5) and greater ability of degradation of pesticides by mushroom enzymes. Soil microorganisms exhibiting the highest activity in degradation of pesticides comprise bacteria from the species *Arthrobacter*, *Bacillus*, *Corynebacterium*, *Flavobacterium*, and *Pseudomonas*; *Actinomycetes* of the *Nocardia* and *Streptomyces* species; and mushrooms of the *Penicillium*, *Aspergillus*, *Fusarium*, and *Trichoderma* species.

Photochemical transformation of pesticides is restricted to soil and plant surfaces and only to compounds sensitive to solar radiation (in the range 290–450 nm). It results in the formation of radicals, which are highly reactive, owing to the presence of a single electron, which leads, for example, in aqueous solutions, to numerous reactions of breaking of bonds as well as to recombination reactions. Many environmental fate processes, including sorption, hydrolysis, volatilization, transport, and accumulation of bound residues, are coupled with degradation; each of these processes may respond differently to environmental conditions, thus making the comprehension of factors controlling degradation challenging. The importance of these processes has been recognized recently, leading to studies in which several key fate processes were investigated in the same experimental system. In Chapter 11, readers will find essential information concerning the causes of transformation of pesticides from various chemical groups, their degradation and fragmentation in the environment, and their identification by HPLC. Also, Chapter 11 includes details of the degradation of pesticides by sunlight in surface waters, soil, and the atmosphere and modeling of transformation kinetics and pathways in surface waters.

Section III contains nine chapters devoted to the main areas in which pesticide analysis is necessary. The first chapter of Section III, Chapter 12, is devoted to determination of pesticides in environmental samples (i.e., methods, problems, and new trends). Sample preparation is an essential part of HPLC analysis, intended to provide a reproducible and homogenous solution that is suitable for injection onto the column. The goal of sample preparation is a sample solution that (i) is free of interferences, (ii) will not damage the column, and (iii) is compatible with an applied HPLC method (i.e., the sample solvent is soluble in the mobile phase, and it does not affect sample retention). Sample pretreatment is usually carried out in a manual offline mode. However, many sample preparation techniques have been automated with the use of appropriate instrumentation. Although automation can be expensive and complicated, it is indispensable when large numbers of samples have to be analyzed and the time or labor per sample is excessive. Column switching—the so-called coupled column chromatography—can also be used as a kind of sample preparation method. It is used not only for the complete resolution of partly separated fractions from the first system, but also for removal of the contaminants (especially "column killers") or the late, strongly sorbed eluates, thus extending the column life and improving performance of the column in subsequent runs. Column switching can be used as an online method of sample pretreatment in the chromatographic run. Chapter 12 presents applications of sample preparation methods to analysis of pesticides in different samples: water, atmosphere, sludge, fruits and vegetables, medicinal plants, animal feed, soil, human blood, urine, and drool.

Chapter 13 presents methods of quantitative analysis of selected mixture components and discusses method validation. Topics such as external and internal standardization, matrix-matched calibration, standard addition, internal calibration and standard normalization, limits of detection and quantification, accuracy, precision, ruggedness and robustness, linearity, specificity, and quality control are discussed.

Detection of compounds is the subject of Chapters 14 and 15. The detectors most often applied in HPLC are ultraviolet (UV)-visible and photodiode array. UV diode array detectors (DADs) allow simultaneous collection of chromatograms at different wavelengths during a single run [$A = f(t, \lambda)$]. Therefore, the DAD provides more information on sample composition than is provided by a single wavelength run. The UV spectrum of each separated peak of samples and standards is also obtained as an important tool for selecting an optimum wavelength for quantification and to verify peak purity and peak identity. The DAD can also be used to examine the chromatograms at different wavelengths, which enables group classification. In environmental analysis, the analytes can be identified on the basis of their retention times and by comparison between the UV spectrum of the reference compound in the library and the UV spectrum of the detected peak in the sample. A match equal to or higher than 990 is fixed to confirm identification between both spectra for the pesticides determined. If the peaks of analyte are pure, then the surface area under the compared spectra of standard and analyte is green. If the peak of the analyte is contaminated, the surface area would be red. Because these peaks are pure, the calculated surface areas of the compared peaks are green. The matrix contains 1024 photodiodes, which corresponds to a nominal difference of 0.9 nm in the UV range. In the visual and near infrared range, the difference is somewhat greater. To correct this optical nonlinearity and transform the discrete diode distances into a linear scale, a linear interpolation algorithm is applied that utilizes a calibration table of wavelengths and real wavelength values obtained from emission lines of a deuterium lamp. The determination of peak purity is carried out using an interpolation algorithm that takes into account a calibration table of wavelengths from the emission lines of the deuterium lamp at 486 and 656 nm. In Chapter 14, readers will find essential information concerning the analysis of pesticides by HPLC coupled with various types of detectors (UV, DAD, photodiode array detector [PAD], light scattering, electrochemical, and others). UV detection is limited to the compounds having chromophore groups (e.g., aromatic rings), and it is not suitable for the compounds that do not absorb in UV range.

The main part of Chapter 15 is devoted to hyphenated LC-mass spectrometry (LC-MS) techniques for identification and quantification of pesticides. The use of a mass spectrometer as an HPLC detector is becoming commonplace for the qualitative and quantitative analysis of mixture components. MS fragmentation patterns can be used to identify each peak. For all MS techniques, the analyte is first ionized in the source because the mass spectrometer can only detect the charged species. Ions having discrete mass/charge ratios are then separated and focused in the mass analyzer.

It is well known that matrix effects are one of the main drawbacks of, for example, LC-MS/MS methods, making quantification in samples problematic in some cases. Co-eluting compounds from the sample matrix can affect the analyte ionization process, leading to signal enhancement or signal suppression. These undesirable effects typically cause a loss of method accuracy, precision, and sensitivity and lead to incorrect quantification and also to problems in correct identity confirmation. In Chapter 15, the author also presents essential information concerning the causes of matrix effects on pesticides analyzed by LC-MS or LC-MS/MS methods.

Sometimes, the resolving power attainable with a single chromatographic system is insufficient for the analysis of complex mixtures. The coupling of chromatographic techniques is clearly attractive for the analysis of multicomponent mixtures of pesticides. Analysis of compounds present at low concentrations in complex mixtures is especially challenging because the number of interfering compounds present at similar or higher concentrations increases exponentially as the concentrations of target compounds decrease. Truly comprehensive 2-D hyphenation is generally achieved by frequent sampling from a first column into a second, which is a very rapid analysis. Multidimensional LC has long been seen as a potential solution to increase resolution and improve the speed of analysis, particularly in the separation of complex mixtures, for example, pesticides in natural samples. Multidimensional LC methods are typically divided into two main groups: comprehensive separations (denoted LC × LC for a 2-D separation) concerned with the separation and quantification of large numbers (ca., tens to thousands) of constituents of a sample, and targeted "heart-cutting" or "coupled-column" methods (LC–LC for a 2-D separation) concerned with the analysis of a few (ca., 1–5) constituents of the sample matrix. In the past decade, research on the development of practically useful LC × LC has been particularly active. Chapter 16 presents different modes of multidimensional LC applied to analysis of pesticides. Multidimensional LC is a good alternative to multidimensional GC for polar or thermolabile compounds. Polar compounds need to be derivatized for GC analysis, and this is not necessary for LC. Multidimensional chromatography has important applications in environmental analysis, especially for pesticides.

Chapter 17 is devoted to selected problems of chiral separations, which can be useful in environmental analysis of pesticides. Approximately 25% of pesticides are chiral compounds, and in most cases, one of the enantiomers presents the pesticide activity, and the other can present different activity toward the target organism. In these cases, the use of enantiomerically pure pesticides would result in the greatest effectiveness in controlling insects or weeds in agriculture and reducing environmental risks. Another reason for using enantiomerically pure pesticides is that, whereas the active enantiomer has the desired effects on the target species, the other enantiomer may have adverse effects on some nontarget organisms. Moreover, biotic processes such as microbiological transformation are commonly enantioselective, and the use of racemic pesticides can result in different environmental fates because one enantiomer is safer than the other. Therefore, the search for new and effective methods for the separation and determination of pesticide enantiomers is necessary in order to optimize enantioselective production processes, assessing the enantiomeric purity of commercial formulations, and monitoring their presence in environmental or other types of matrices. An example of particular interest is the fungicide metalaxyl [(*R*/*S*)-methyl-N-(2-methoxyacetyl)-N-(2,6-xylyl)-DL-alaninate], which is employed to control plant diseases caused by pathogens of the *Oomycota* type. It has been demonstrated that the activity of the *R*-enantiomer (metalaxyl-M) is around 1000 times higher than that of the *S*-enantiomer. In addition, the degradation of metalaxyl enantiomers in the environment is also clearly enantioselective. In fact, the *S*-enantiomer has shown a faster degradation in vegetables because the enantiomer is active for a shorter time whereas, in the case of soils, the first enantiomer being degraded or decomposed is the *R*-enantiomer.

Applications of chemiluminescence detection in HPLC analysis of pesticides are often more cost-effective and sensitive than instrumental procedures. Chapter 18 presents chemiluminescence detection methods for screening analysis of pesticides as well as different couplings of chemiluminescence detection with HPLC for pesticide analysis.

Recently, UPLC has been developed as an innovative and powerful separation technique, leading to higher resolution and sensitivity and shorter analysis time compared to HPLC. Chapter 19 of Section III is devoted to discussion of UPLC techniques applied to analysis of pesticides.

In Chapter 20, readers will find essential information concerning the analysis of pesticides by HPLC as a basic method, pros and cons of this technique, and alternative chromatographic and nonchromatographic techniques used for determination of pesticides.

The authors who contributed chapters to this book are all recognized international experts in their respective fields. We hope that the book will serve as a comprehensive source of information and training on the state-of-the-art pesticide residue analysis methods based on HPLC.

Summing up, HPLC is the principal separation technique in the analysis of all types of samples, including environmental, food, crops, and others. It can be used for identification of pesticides in samples and for their quantitative analysis. HPLC-DAD enables peak purity control and group identification (or class identification) of pesticides. The LC-MS hyphenated techniques enable identification of the known and unknown matrix constituents and pesticides with high sensitivity and selectivity. Hyphenated techniques enable data collection from numerous difficult samples, which proves very useful in correct identification and quantitative analysis of pesticides. We hope that this book will serve as a useful practical guide to HPLC methods of pesticide residue analysis for all readers, ranging from beginners to experts in chromatography.

REFERENCES

1. The Pesticides Safety Directorate, York, United Kingdom.
2. Proposed PAHO/WHO Plan of Action for Technical Cooperation in Food Safety, 2006–2007.
3. The Applicant Guide: Maximum Residue Levels, The Pesticides Safety Directorate, York, United Kingdom.
4. UK National Action Plan for Sustainable Use of Pesticides (Plant Protection Products), February 2013, Department for Environment and Rural Affairs, London, UK.
5. Status of Active Substances under EU Review (doc. 3010). Available at http://ec.europa.eu/food/plant/protection/evaluation/stat_active_subs_3010_en.xls.
6. AOAC Official Method 2007.01, Pesticide Residues in Foods by Acetonitrile Extraction and Partitioning with Magnesium Sulfate—Gas Chromatography/Mass Spectrometry and Liquid Chromatography/Tandem Mass Spectrometry, First Action 2007, AOAC International, Rockville, MD.

2 Pesticide Classification and Properties

Tomasz Tuzimski

CONTENTS

2.1 INTRODUCTION

2.1.1 Definition of a Pesticide

According to the FAO [1], a pesticide is any substance or mixture of substances intended to prevent, destroy, or control any pest, including vectors of human or animal diseases; unwanted species of plants or animals causing harm during or otherwise interfering with the production, processing, storage, or marketing of food, agricultural commodities, wood and wood products, or animal feedstuffs; or which may be administered to animals for the control of insects, arachnids, or other pests in or on their bodies. The term includes chemicals used as growth regulators, defoliants, desiccants, or fruit-thinning agents or agents for preventing the premature ripening of fruits and substances applied to crops either before or after harvest to prevent deterioration during storage or transport. The term "pest" includes insects, weeds, mammals, and microbes among other species [2].

The term "pesticide" is also defined by the FAO in collaboration with the UNEP [3] as a chemical designed to combat the attacks of various pests and vectors on agricultural crops, domestic animals, and human beings.

2.1.2 Short Historical Background of Pesticide Use

Chemical substances have been used by humans to control pests from the beginning of agriculture. The first generation of pesticides involved the use of highly toxic, inorganic compounds. The first known pesticide to be used was sulfur. By the fifteenth century, toxic chemicals, such as arsenic (calcium arsenate, lead arsenate), mercury, and lead, were applied to crops to kill pests. In the seventeenth century, nicotine sulfate was extracted from tobacco leaves for use as an insecticide. The fumigant hydrogen cyanide was used in the 1860s for the control of such pests as fungi, insects, and bacteria. Other compounds included Bordeaux mixture (copper sulfate, lime, and water) and sulfur. Their use was abandoned because of their toxicity and infectiveness. The nineteenth century saw the introduction of two natural pesticides: pyrethrum and rotenone.

The second generation involved the use of synthetic organic compounds. The first important synthetic organic pesticide was dichlorodiphenyltrichloroethane (DDT), first synthesized by the German scientist Othmar Zeidler in 1873 [4]; its insecticidal effect was discovered by the Swiss chemist Paul Hermann Müller in 1939 [5]. The discovery of DDT had a great impact on the control of pests and soon became widely used worldwide. DDT was hailed as a miracle because of its broad-spectrum activity, persistence, insolubility, inexpensiveness, and easy application [6]. At the time, pesticides had a good reputation mainly due to their control of diseases, such as malaria transmitted by mosquitoes and the bubonic plague transmitted by fleas. Nevertheless, this opinion changed after learning of the toxic effects of DDT on birds, particularly after the publication of the book *Silent Spring* by Rachel Carson in 1962 [7].

At present, due to the possible toxic effects of pesticides on human health and the environment, there are strict regulations for their registration and use all over the world. On one hand, the "green revolution," especially in developed countries, has been made possible only with the help of agrochemicals, particularly pesticides. On the other hand, laboratory experiments have indicated that some pesticide residues may cause carcinogenicity upon long exposure. Pesticide residues also exhibit specific effects on species other than the pests for which they are solely intended. It is also known that bodies of water, air, birds, and aquatic animals are constantly moving and transporting poisons from one region of the globe to another. For example, DDT was used in the fields of East Africa, but in a few months, it was found in the water of the Bay of Bengal, that is, at a distance 6000 km away [8].

TABLE 2.1
Chronology of Pesticide Development

Period	Example	Source	Characteristics
1800–1920s	Early organics, nitro-phenols, chlorophenols, creosote, naphthalene, petroleum oils	Organic chemistry, by-products of coal gas production, etc.	Often lack specificity and were toxic to user or nontarget organisms
1945–1955	Chlorinated organics, DDT, HCCH, chlorinated cyclodienes	Organic synthesis	Persistent, good selectivity, good agricultural properties, good public health performance, resistance, harmful ecological effects
1945–1970	Cholinesterase inhibitors, organophosphorus compounds, carbamates	Organic synthesis, good use of structure–activity relationships	Lower persistence, some user toxicity, some environmental problems
1970–1985	Synthetic pyrethroids, avermectins, juvenile hormone mimics, biological pesticides	Refinement of structure–activity relationships, new target systems	Some lack of selectivity, resistance, costs, and variable persistence
1985	Genetically engineered organisms	Transfer of genes for biological pesticides to other organisms and into beneficial plants and animals, genetic alteration of plants to resist nontarget effects of pesticides	Possible problems with mutations and escapes, disruption of microbiological ecology, monopoly on products

Source: Stephenson, G.A., and Solomon, K.R., *Pesticides and Environment.* Department of Environmental Biology, University of Guelph, Guelph, 1993. Available at http://www.fao.org/docrep/w2598e/w2598e07.htm#historical development of pesticides.

Organochlorine pesticides (aldrin, chlordane, DDT, dieldrin, endrin, heptachlor, mirex, toxaphene, and hexachlorobenzene) constitute nine of the 12 chemical substances or groups currently defined under the Stockholm Convention on Persistent Organic Pollutants (POPs) [9]. POPs are stable, fat-soluble, carbon-based compounds that volatilize at warm temperatures and are transported poleward by wind, water, and wildlife. Organochlorine pesticides, which have been classified by the U.S. Environmental Protection Agency as POPs, have the ability to bioaccumulate and biomagnify and can bioconcentrate (i.e., become more concentrated) up to 70,000 times their original concentrations [7,10].

Therefore, it is particularly important to know the properties of the substances used as pesticides. Knowledge of physicochemical properties, for example, the octanol/water partition coefficient (K_{ow} expressed in the logarithmic form as log P) and solubility in water, allows the fate and behavior of such chemicals in the environment to be predicted.

Pesticide use has increased 50-fold since 1950, and 2.5 million tons of industrial pesticides are now used each year. Table 2.1 shows the chronology of pesticide development in the world [11].

2.2 CLASSIFICATION OF PESTICIDES

2.2.1 Classification Based on the Targeted Pest Species

One of the most convenient means for controlling harmful organisms is the use of various pesticides. Depending on the purpose for which they are intended, pesticides are divided into the following basic groups:

- *Acaricides*—for the control of mites or ticks
- *Algicides*—for the destruction of algae and other aquatic vegetation
- *Avicides*—for the control of bird pests
- *Bactericides*—for the control of bacteria and bacterial diseases of plants
- *Fungicides*—for the control of plant diseases and various fungi
- *Herbicides*—for the control of weeds
- *Insecticides*—for the control of harmful insects (individual groups of insecticides also have more specific names, such as *aphicides*, preparations for the control of aphids)
- *Limacides or molluscicides*—for the control of various mollusks, including gastropods
- *Nematicides*—for the control of roundworms (nematodes)
- *Rodenticides*—for the control of rodents

Pesticides include chemical compounds that stimulate or retard the growth of plants; they also include those that remove leaves (*defoliants*) or desiccate plants (*dessicants*) and are used for the purpose of mechanizing the work in harvesting cotton, soybeans, potatoes, and many other crops. The term "pesticides" is also applied to compounds to repel (*repellents*) and attract (*attractants*) pests.

2.2.2 Chemical Classification of Pesticides

Pesticides fall into three main categories: fungicides, herbicides, and insecticides. These three types are used to combat different pests. In addition, three main categories can be classified according to their chemical composition. On different websites, for example, on http://www.alanwood.net/pesticides/, the reader can find a massive amount of information about classifications of pesticides. The compendium contains much more than International Organization for Standardization common names with nomenclature data sheets for more than 1700 different active ingredients and for more than 350 ester and salt derivatives, made accessible by a comprehensive set of indices and a classification.

Under chemical classification, pesticides are organized according to the chemical nature of the active ingredients. Based on chemical classification, pesticides are collected into four main groups, namely, organochlorine, organophosphorus, carbamate, and pyrethroid.

The chemical classification of pesticides is by far the most useful for researchers in the field of pesticides and the environment and those who search for details. Pesticides can be broadly classified according to their general chemical nature into several principal types as shown in Table 2.2.

2.2.2.1 Chemical Classification of Insecticides

Chemical insecticides are usually divided into four major classes: chlorinated hydrocarbons or organochlorines, organophosphates, carbamates, and pyrethroids.

TABLE 2.2
Chemical Classification of Pesticides

Chemical Type	Example (Common Name)	Structure	Typical Action
Organochlorines	p,p′-DDT	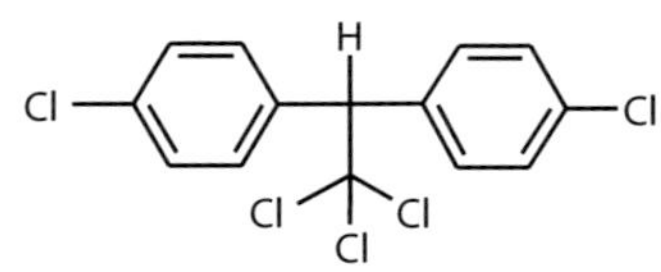	Insecticides
Organophosphates	Chlorpyrifos-methyl	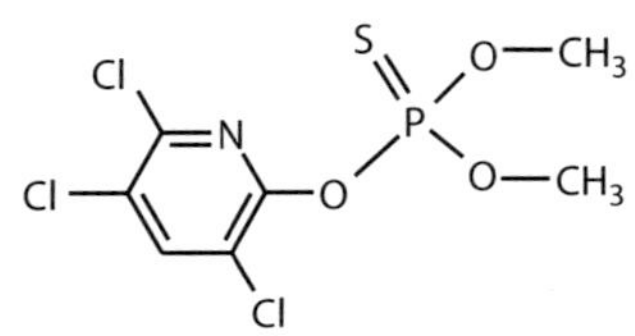	Insecticides
Carbamates	Fenoxycarb	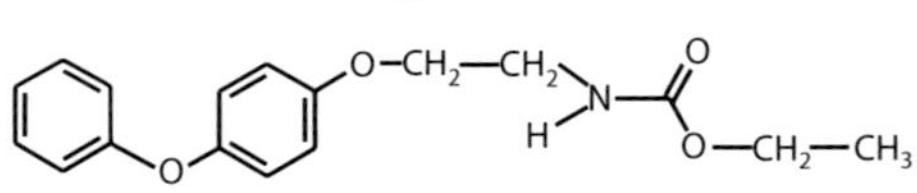	Insecticides
Pyrethroids	Cypermethrin	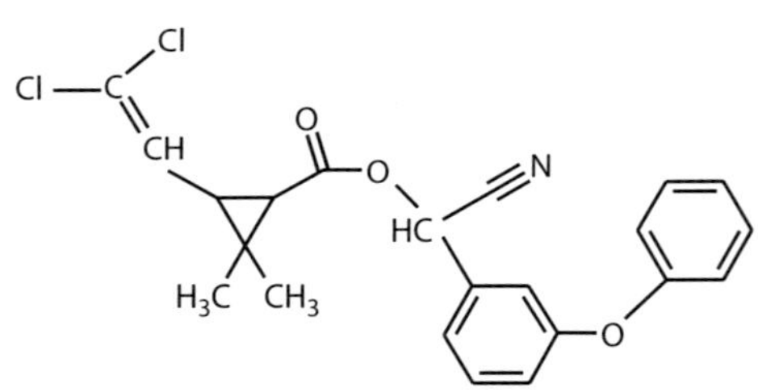	Insecticides
Benzoylphenylureas	Diflubenzuron	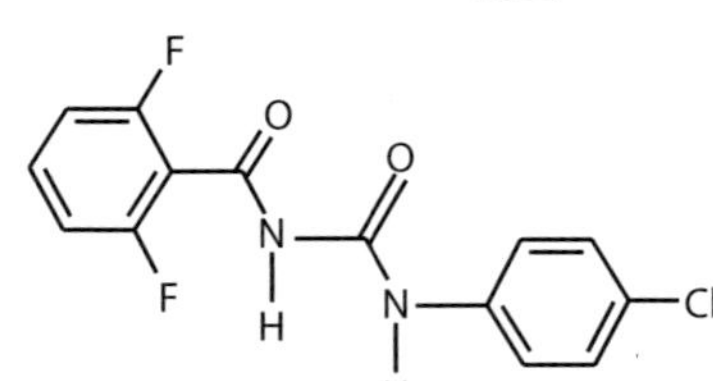	Insecticides
Amides	Beflubutamid	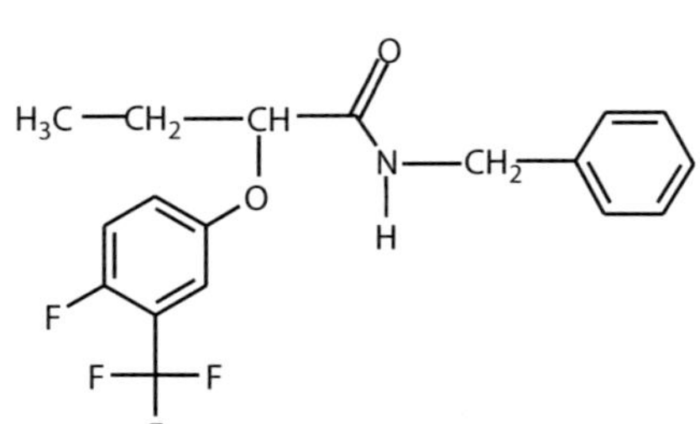	Herbicides

TABLE 2.2 (CONTINUED)
Chemical Classification of Pesticides

Chemical Type	Example (Common Name)	Structure	Typical Action
Chloroacetanilides	Acetochlor	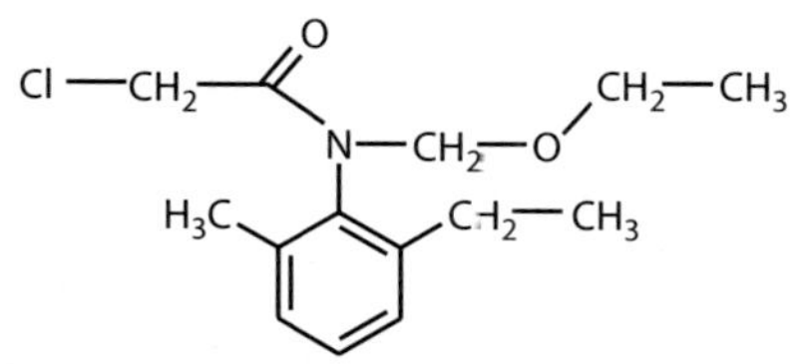	Herbicides
Carbamates	Chlorpropham	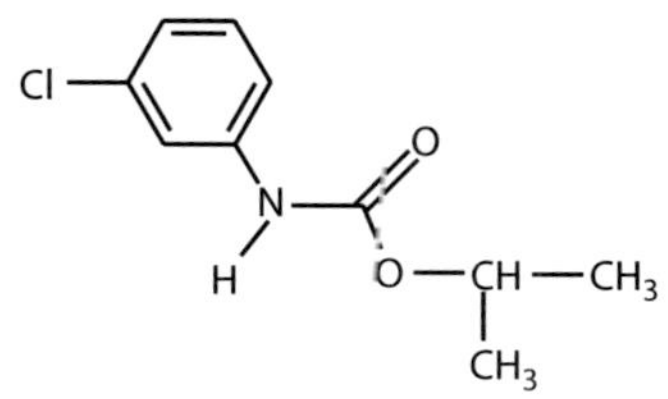	Herbicides
Nitriles	Bromoxynil	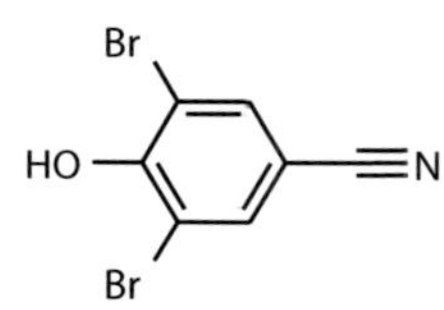	Herbicides
Dinitroanilines	Benfluralin	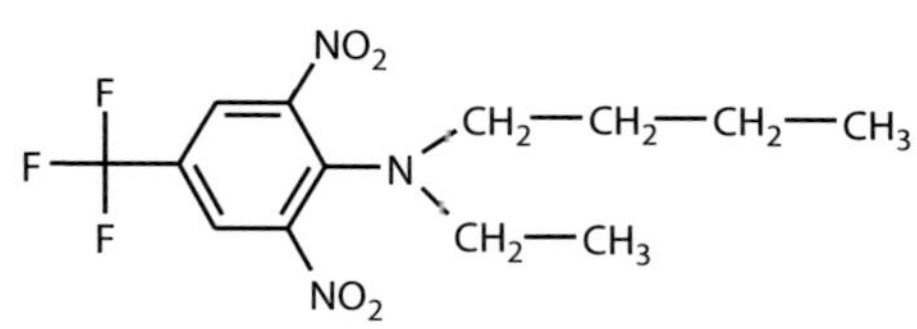	Herbicides
Organophosphates	Glyphosate	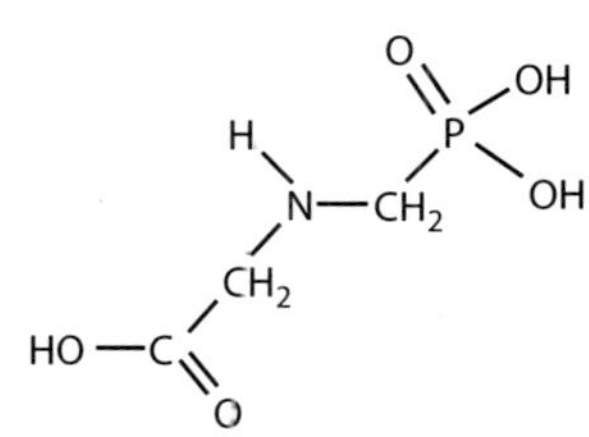	Herbicides
Phenoxy acid derivatives	2,4-D	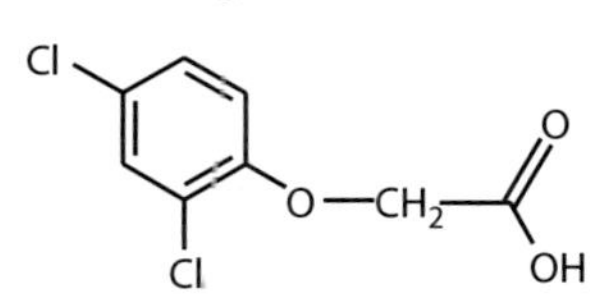	Herbicides
Quaternary ammonium compounds	Chlormequat chloride	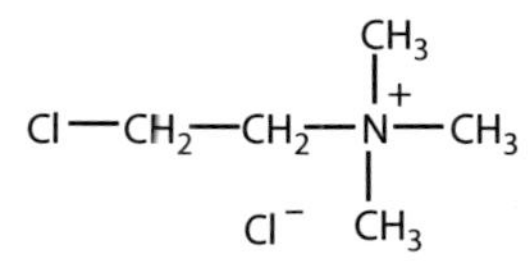	Herbicides

(*Continued*)

TABLE 2.2 (CONTINUED)
Chemical Classification of Pesticides

Chemical Type	Example (Common Name)	Structure	Typical Action
Triazines	Tebuthylazine	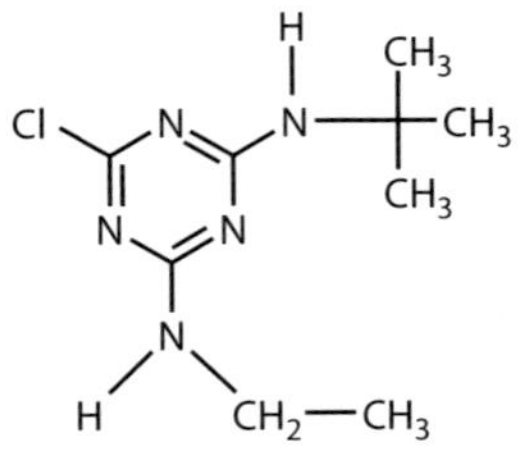	Herbicides
Phenylureas	Chlorotoluron	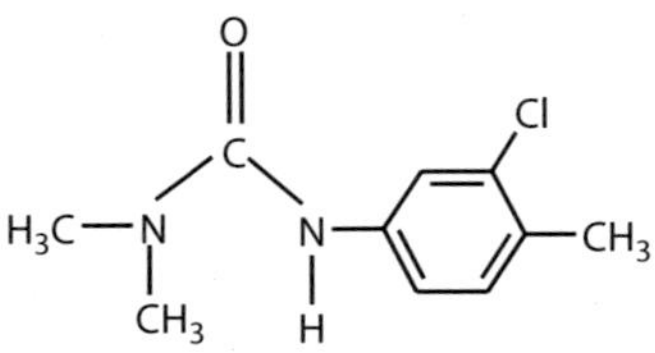	Herbicides
Sulfonylureas	Amidosulfuron	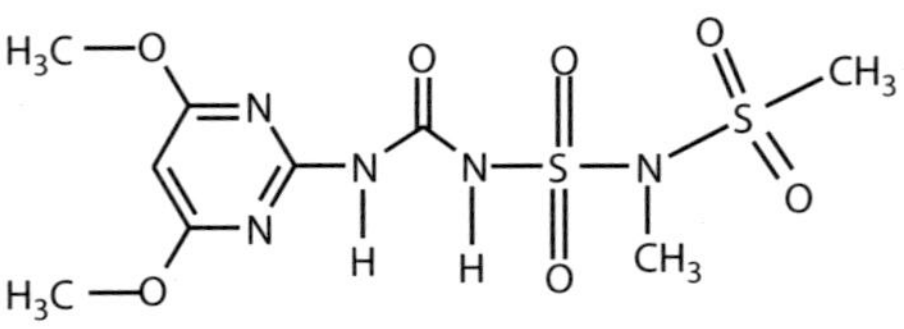	Herbicides
Pyridazinones	Chloridazon	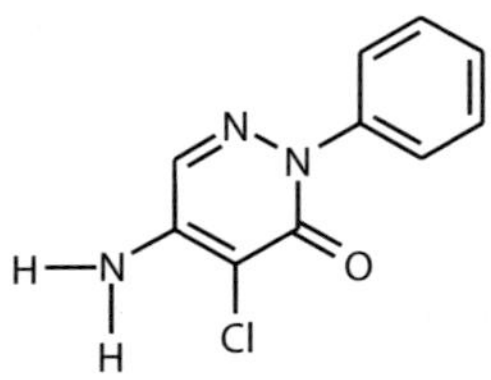	Herbicides
Conazoles	Bromuconazole	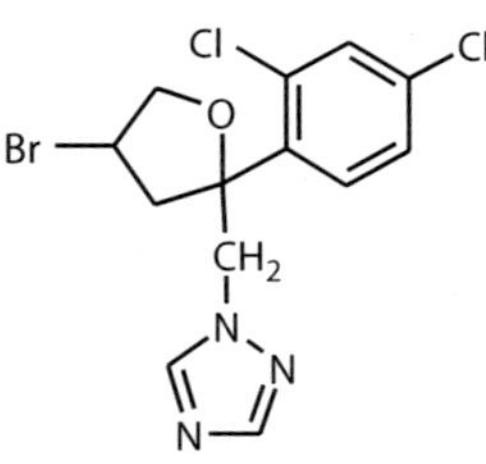	Fungicides
Dithiocarbamates	Maneb	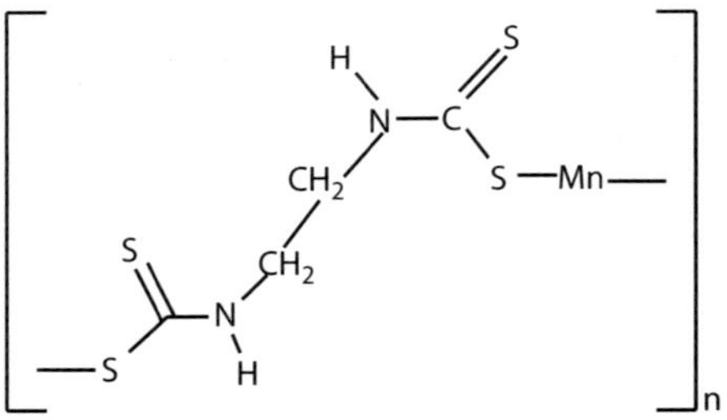	Fungicides

TABLE 2.2 (CONTINUED)
Chemical Classification of Pesticides

Chemical Type	Example (Common Name)	Structure	Typical Action
Morpholines	Dimethomorph	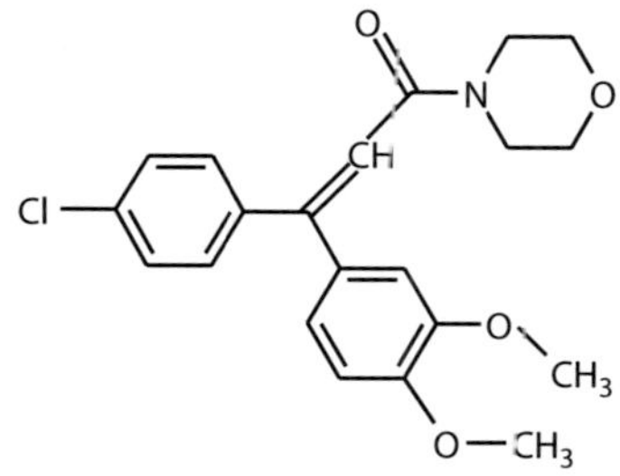	Fungicides
Benzimidazoles	Carbendazim	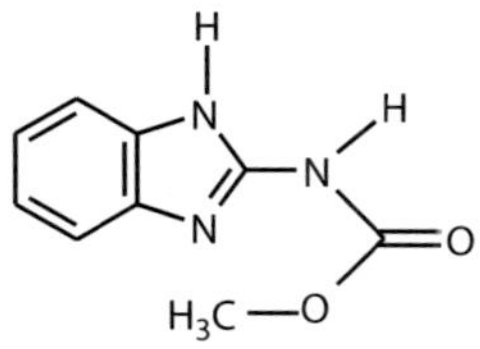	Fungicides
Anilides	Benalaxyl	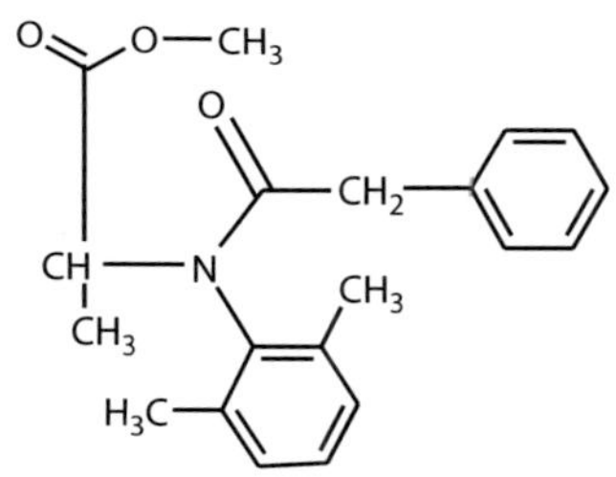	Fungicides

2.2.2.1.1 *Organochlorines*

Chlorinated hydrocarbons or organochlorines are a group of compounds that chemically break down very slowly and can remain in the environment for long periods of time. The main representatives of this group are DDT analogs, benzene hexachloride isomers, and cyclodiene compounds. Some organochlorine compounds are classified as POPs. Due to their persistence and toxicity, most of these organochlorine compounds have been banned, or their use as pesticides has been restricted. Organochlorine pesticides are organic compounds with three, five, or more chlorine atoms. Table 2.3 shows the main representatives of this group [12]. Organochlorine insecticides act as central nervous system stimulants, producing hyperactivity, convulsions, and death of the insect. There are gamma-aminobutyric acid (GABA)-gated chloride channel antagonists and also stomach and contact poisons.

2.2.2.1.2 *Organophosphates*

The first organophosphorus insecticide compound—tetraethyl pyrophosphate (TEPP)—was synthesized by de Clermont in France. It was manufactured in Germany in 1943 and known as Bladan or TEPP.

Organophosphorus insecticides are hydrocarbon compounds that contain one or more phosphorus atoms in their molecules, and they play an important role in the success of modern farming and food production. Because these compounds are characterized by low persistence and high effectiveness, they are widely used as systematic insecticides for plants, animals, and soil treatments. Examples of these insecticides are shown in Table 2.4.

TABLE 2.3
Chemical Names and Properties of Organochlorine Insecticides

Common Name/ Chemical Formula	IUPAC Name	Maximum UV-Vis Absorption (L mol^{-1} cm^{-1})	Vapor Pressure (mPa) at 25°C	Solubility in Water[a] (mg L^{-1})/ Organic Solvents[a] (mg L^{-1})	Dissociation Constant (pK_a) at 25°C	Octanol–Water Partition Coefficient (log P)[a,b]	Soil Degradation (days) (Aerobic) DT_{50}	K_{oc} (Linear)	Bioconcentration Factor BCF/CT_{50} (days)	Mode of Action
Aldrin $C_{12}H_8Cl_6$	(1*R*,4*S*,4a*S*,5*S*,8*R*,8a*R*)-1,2,3,4,10,10-hexachloro-1,4,4a,5,8,8a-hexahydro-1,4:5,8-dimethanonaphthalene	–	3	0.027/ 6,000,000[c] 6,000,000[d] 6,000,000[e]	No dissociation	6.5	365 (field)	17,500	3348/1–5	Central nervous system stimulant, GABA-gated chloride channel antagonist, also stomach and contact toxin
p,p′-DDT $C_{14}H_9Cl_5$	1,1,1-trichloro-2,2-bis(4-chlorophenyl)ethane	–	0.025	0.006/ 1,000,000[f] 850,000[g] 770,000[d] 600,000[e]	–	6.91	2000 (lab)	151,000	3173/not available	Nonsystemic stomach and contact action, sodium channel modulator
Dieldrin $C_{12}H_8Cl_6O$	(1*R*,4*S*,4a*S*,5*R*,6*R*,7*S*,8*S*,8a*R*)-1,2,3,4,10,10-hexachloro-1,4,4a,5,6,7,8,8a-octahydro-6,7-epoxy-1,4:5,8-dimethanonaphthalene	–	0.024	0.14/–	No dissociation	3.7	2000 (lab)	12,000	35,000/1	Central nervous system stimulant, GABA-gated chloride channel antagonist

Endosulfan $C_9H_6Cl_6O_3S$	1,4,5,6,7,7-hexachloro-8,9,10-trinorborn-5-en-2,3-ylenebismethylene sulfite	–	0.83	0.32/ 200,000[h] 65,000[i] 24,000[j] 200,000[k]	–	4.75	86 (field)	11,500	2755/not available	Nonsystemic with contact and stomach action, acts as a noncompetitive GABA antagonist
Methoxychlor $C_{16}H_{15}Cl_3O_2$	1,1,1-trichloro-2,2-bis(4-methoxyphenyl)ethane	–	0.08	0.1/ 440,000[e] 50,000[l] 700,000[m] 1,333,000[g]	–	5.83	120 (lab)	80,000	1622/not available	Contact and stomach action; central nervous stimulant, producing hyperactivity, convulsions and death
Pentachlorophenol C_6HCl_5O	Pentachlorophenol	–	16,000	1000/ 500,000[c] 150,000[d] 1,800,000[l] 12,000,000[i]	4.73/weak acid	3.32	48 (lab)	30	216/not available	Accelerates aerobic metabolism and increases heat production

Source: University of Hertfordshire, The Pesticide Properties DataBase (PPDB) developed by the Agriculture & Environment Research Unit (AERU), University of Hertfordshire, 2006–2013, 2015. Available at http://sitem.herts.ac.uk/aeru/ppdb/en/atoz.htm.

Note: a: at 20°C; b: at pH 7; c: acetone; d: benzene; e: xylene; f: cyclohexanone; g: dichloromethane; h: ethyl acetate; i: ethanol; j: hexane; k: toluene; l: methanol; m: trichloroethane.

TABLE 2.4
Chemical Names and Properties of Organophosphorus Insecticides

Common Name/ Chemical Formula	IUPAC Name	Maximum UV-Vis Absorption (L mol^{-1} cm^{-1})	Vapor Pressure (mPa) at 25°C	Solubility in Water[a] (mg L^{-1})/ Organic Solvents[a] (mg L^{-1})	Dissociation Constant (pK_a) at 25°C	Octanol–Water Partition Coefficient (log *P*)[a,b]	Soil Degradation (days) (Aerobic) DT_{50}	K_{oc} (Linear)	Bioconcentration Factor BCF/CT_{50} (days)	Mode of Action
Azinphos-methyl $C_{10}H_{12}N_3O_3PS_2$	*S*-3,4-dihydro-4-oxo-1,2,3-benzotriazin-3-ylmethyl *O*,*O*-dimethyl phosphorodithioate	–	5.00×10^{-04}	28/250,000[c,h] 170,000[e]	5/weak acid	2.96	31 (lab)	1112 linear	40/not available	Nonsystemic, contact and stomach action, acetylcholinesterase (AChE) inhibitor
Chlorfenvinphos $C_{12}H_{14}Cl_3O_4P$	(*EZ*)-2-chloro-1-(2,4-dichlorophenyl)vinyl diethyl phosphate	–	0.53	145/Miscible[c,e,g,i]	–	3.8	30 (field)	680 linear	250/not available	Contact and stomach action, acetylcholinesterase (AChE) inhibitor
Chlorpyrifos-methyl $C_7H_7Cl_3NO_3PS$	*O*,*O*-dimethyl *O*-3,5,6-trichloro-2-pyridyl phosphorothioate	290 nm > 10	3.0	2.74/250,000[c,h] 154,000[n] 193,000[l]	–	4.00	2.5 (lab)	4645 linear	1800/2.6	Nonsystemic with contact, stomach and respiratory action, Acetylcholinesterase (AChE) inhibitor
Diazinon $C_{12}H_{21}N_2O_3PS$	*O*,*O*-diethyl *O*-2-isopropyl-6-methylpyrimidin-4-yl phosphorothioate	Maxima at 246 nm = 4050, 290 nm = 20.86	11.97	60/ 250,000[h] 9,000,000[c,k,l]	2.6/strong acid	3.69	18.4 (field)	609 linear	500/2	Nonsystemic with respiratory, contact and stomach action, acetylcholinesterase (AChE) inhibitor
Dichlorvos $C_4H_7C_{12}O_4P$	2,2-dichlorovinyl dimethyl phosphate	Maxima at 204 nm	2100	18,000/ Miscible[c,i,j,k]	No dissociation	1.9	2 (lab)	50 linear	Low risk/–	Respiratory, contact and stomach action, acetylcholinesterase (AChE) inhibitor

Dimethoate $C_5H_{12}NO_3PS_2$	2-dimethoxyphosphinothioylthio-*N*-methylacetamide	No maxima observed above 200 nm	0.247	39,800/313,000[e] 1,030,000[k] 1,590,000[l] 295[j]	No dissociation	0.704	7.2 (field)	–	8/not available	Systemic with contact and stomach action, acetylcholinesterase (AChE) inhibitor
Fenitrothion $C_9H_{12}NO_5PS$	*O,O*-dimethyl *O*-4-nitro-m-tolyl phosphorothioate	Neutral solution: 290 nm = 4210; Acidic solution: 290 nm = 4390; Basic solution: 290 nm = 2370	0.676	19/ 25,000[j] 500,000[c,h,l]	No dissociation	3.32	2.7 (lab)	2000 linear	29/0.19	Nonsystemic, broad spectrum with contact and stomach action, acetylcholinesterase (AChE) inhibitor
Fenthion $C_{10}H_{15}O_3PS_2$	*O,O*-dimethyl *O*-4-methylthio-m-tolyl phosphorothioate	–	0.37	4.2/ 100,000[j] 250,000[g,k,o]	–	4.84	34 (lab)	1500 linear	154/5	Contact, stomach and respiratory action, acetylcholinesterase (AChE) inhibitor
Malathion $C_{10}H_{19}O_6PS_2$	Diethyl (dimethoxyphosphinothioylthio) succinate	No absorbance above 290 nm	3.1	148/250,000[c,e,l] 62,000[n]	No dissociation	2.75	1 (field)	1800 linear	103/0.69	Broad-spectrum, nonsystemic with contact, stomach and respiratory action, acetylcholinesterase (AChE) inhibitor
Methamidophos $C_2H_8NO_2PS$	(*RS*)-(*O,S*-dimethyl phosphoramidothioate)	–	2.3	200,000/200,000[c,g] 1000[j] 3500[k]	–	−0.79	4 (field)	1.0 linear	75/not available	Systemic with contact and stomach action, acetylcholinesterase (AChE) inhibitor
Methidathion $C_6H_{11}N_2O_4PS_3$	3-dimethoxyphosphinothioylthio methyl-5-methoxy-1,3,4-thiadiazol-2(3*H*)-one	–	0.25	240/150,000[i] 670,000[c] 11,000[j] 14,000[p]	–	2.57	7 (field)	400 linear	12.6/not available	Nonsystemic with contact and stomach action, acetylcholinesterase (AChE) inhibitor.

(*Continued*)

TABLE 2.4 (CONTINUED)
Chemical Names and Properties of Organophosphorus Insecticides

Common Name/ Chemical Formula	IUPAC Name	Maximum UV-Vis Absorption (L mol^{-1} cm^{-1})	Vapor Pressure (mPa) at 25°C	Solubility in Water[a] (mg L^{-1})/ Organic Solvents[a] (mg L^{-1})	Dissociation Constant (pK_a) at 25°C	Octanol–Water Partition Coefficient (log P)[a,b]	Soil Degradation (days) (Aerobic) DT_{50}	K_{oc} (Linear)	Bioconcentration Factor BCF/CT_{50} (days)	Mode of Action
Oxydemeton-methyl $C_6H_{15}O_4PS_2$	*S*-2-ethylsulfinylethyl *O*,*O*-dimethyl phosphorothioate	Maxima at 213 nm = 1356	2	1,200,000/25[j] 250,000[c,k,p]	Not determinable	−0.74	5 (field)	10 linear	Low risk/–	Systemic with contact and stomach action, rapid knockdown effect, acetylcholinesterase (AChE) inhibitor
Phosmet $C_{11}H_{12}NO_4PS_2$	*O*,*O*-dimethyl *S*-phthalimidomethyl phosphorodithioate	Neutral solution: 221.9 nm = 44,668, 294.6 nm = 1259; Acidic solution: 222.6 nm = 42,658, 295.6 nm = 2399; Basic solution: 218.5 nm = 16,595	0.065	15.2/ 53,000[e] 62,000[h] 29,200[l] 1040[n]	No dissociation	2.96	7 (field)	–	79/not available	Nonsystemic with predominately contact action, acetylcholinesterase (AChE) inhibitor

Pirimiphos-methyl $C_{11}H_{20}N_3O_3PS$	*O*-2-diethylamino-6-methylpyrimidin-4-yl *O,O*-dimethyl phosphorothioate	301 nm = 3690, 270 nm = 142,000, 247 nm = 22,400, 220 nm = 3390	2.00×10^{-03}	11/ 250,000[c,e,h,l]	4.3/weak acid	3.9	39 (field)	1100 linear	741/not available	Broad-spectrum with contact and respiratory action, acetylcholinesterase (AChE) inhibitor
Profenofos $C_{11}H_{15}BrClO_3PS$	(*RS*)-(*O*-4-bromo-2-chlorophenyl *O*-ethyl *S*-propyl phosphorothioate)	–	2.53	28/Miscible[c,e,l]	No dissociation	1.7	7 (field)	2016 linear	1186/not available	Nonsystemic with contact and stomach action, acetylcholinesterase (AChE) inhibitor
Trichlorfon $C_4H_8Cl_3O_4P$	Dimethyl (*RS*)-2,2,2-trichloro-1-hydroxyethylphosphonate	–	0.21	120,000/21,500[e] 363,000[h] 707,000[c] 1,346,000[l]	Not determinable	0.43	18 (lab)	10 linear	0.41/not available	Nonsystemic with contact and stomach action, acetylcholinesterase (AChE) inhibitor

Source: University of Hertfordshire, The Pesticide Properties DataBase (PPDB) developed by the Agriculture & Environment Research Unit (AERU), University of Hertfordshire, 2006–2013, 2015. Available at http://sitem.herts.ac.uk/aeru/ppdb/en/atoz.htm.

Note: a: at 20°C; b: at pH 7; c: acetone; d: benzene; e: xylene; f: cyclohexanone; g: dichloromethane; h: ethyl acetate; i: ethanol; j: *n*-hexane; k: toluene; l: methanol; m: trichloroethane; n: *n*-heptane; o: isopropanol; p: *n*-octanol.

2.2.2.1.3 Carbamates

Carbamates are organic compounds derived from carbamic acid with the general formula as illustrated in Figure 2.1, where R_1 is an alcohol group, R_2 is a methyl group, and R_3 is usually a hydrogen group (Table 2.5).

Some carbamate insecticides contain a sulfur atom in their molecule.

The three subgroups of carbamate insecticides include the following:

- N-methylcarbamates and esters of phenols (e.g., carbaryl [methiocarb, propoxur])
- N-dimethyl carbamate and N-dimethyl carbamate esters of heterocyclic phenols (e.g., carbofuran, pirimicarb)
- Oxime derivatives of aldehydes (e.g., aldicarb, methomyl, and oxamyl)

Carbamate insecticides have a very broad spectrum of action, and they are particularly effective on lepidopterous larvae and on ornamental pests, such as snails, slugs, and household pests.

2.2.2.1.4 Pyrethroids

Pyrethroids are synthetic analogues of the naturally occurring pyrethrins, a product of flowers from certain species of chrysanthemum (*Chrysanthenemum cinerariaefolium*). The naturally occurring forms are esters from (+)-*trans* and (+)-*cis* alcohols. These pesticides were developed by the modification of the pyrethrin structure by introducing a biphenoxy moiety and substituting some hydrogens with halogens in order to confer stability and, at the same time, retain the basic properties of pyrethrins. Pyrethroids are degraded in soil and have no detectable effects on soil microflora. The most widely used examples of pyrethroids are presented in Table 2.6.

2.2.2.1.5 Benzoylphenylureas

Table 2.7 shows the main representatives of this group. Benzoylphenylurea chitin synthesis inhibitors act as an antimolting agent, killing larvae and pupae.

2.2.2.2 Chemical Classification of Herbicides

2.2.2.2.1 Amides

Amides are herbicides with the general formula as illustrated in Figure 2.2 (Table 2.8).

The key subgroups are the N-substituted, for example, chloroacetamides, butyramides, propionamides, and benzamides.

2.2.2.2.2 Chloroacetanilides

The N-substituted chloroacetamides and the substituted anilides are key subgroups of amides (Table 2.9). The half-life in soil in days is the shortest for propachlor (ca., 4 days), and the longest is for acetochlor and alachlor (8–18 and 1–30 days, respectively) [13].

2.2.2.2.3 Carbamates

Carbamates are a broad group of herbicides (Table 2.10). The key subgroup is the carbamic acid, which is shown in Figure 2.3, and thiocarbamate (Figure 2.4):

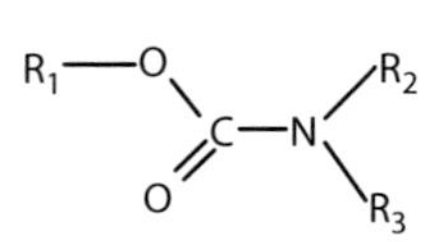

FIGURE 2.1 General formula of carbamates (organic compounds derived of carbamic acid).

TABLE 2.5
Chemical Names and Properties of Carbamate Insecticides

Common Name/ Chemical Formula	IUPAC Name	Maximum UV-Vis Absorption (L mol^{-1} cm^{-1})	Vapor Pressure (mPa) at 25°C	Solubility in Water[a] (mg L^{-1})/ Organic Solvents[a] (mg L^{-1})	Dissociation Constant (pK_a) at 25°C	Octanol–Water Partition Coefficient (log *P*)[a,b]	Soil Degradation (days) (Aerobic) DT_{50}	K_{oc} (Linear)	Bioconcentration Factor BCF/CT_{50} (days)	Mode of Action
Aldicarb $C_7H_{14}N_2O_2S$	(*EZ*)-2-methyl-2-(methylthio) propionaldehyde *O*-methylcarbamoyloxime	–	3.87	4930/ 180,000[d] 110,000[k] 470,000[g] 380,000[c]	No dissociation	1.15	2 (field)	36	42/not available	Systemic with contact and stomach action absorbed through roots, acetylcholinesterase (AChE) inhibitor
Bendiocarb $C_{11}H_{13}NO_4$	2,2-dimethyl-1,3-benzodioxol-4-yl methylcarbamate	–	4.6	280/ 175,000[c] 40,000[i] 225[j]	8.8/very weak acid	1.7	3.5 (field)	385	64.8/not available	Systemic, with contact and stomach action resulting in rapid knockdown, acetylcholinesterase (AChE) inhibitor
Carbaryl $C_{12}H_{11}NO_2$	1-naphthyl methylcarbamate	Neutral solution: 220 nm = 82,696, 270 nm = 5743, 279 nm = 6434, 291 nm = 4211; Acidic solution: 221.5 nm = 18,362, 280.0 nm = 6703, 295 nm < 2743	0.0416	9.1/ 250[n] 9860[e] 87,500[l] 175,000[h]	10.4/very weak acid	2.36	16 (lab)	300	44/32–144	Stomach and contact activity with slight systemic properties, acetylcholinesterase (AChE) inhibitor
Carbofuran $C_{12}H_{15}NO_3$	2,3-dihydro-2,2-dimethylbenzofuran-7-yl methylcarbamate	Neutral solution: 276 nm = 2800, 290 nm = 251; Acidic and basic solution: No significant differences in spectrum	0.08	322/ 110[n] 71,700[l] 61,500[h] 105,200[c]	No dissociation	1.8	14 (field)	–	12/0.5	Systemic with contact and stomach action, acetylcholinesterase (AChE) inhibitor

(*Continued*)

TABLE 2.5 (CONTINUED)
Chemical Names and Properties of Carbamate Insecticides

Common Name/ Chemical Formula	IUPAC Name	Maximum UV-Vis Absorption (L mol^{-1} cm^{-1})	Vapor Pressure (mPa) at 25°C	Solubility in Water[a] (mg L^{-1})/ Organic Solvents[a] (mg L^{-1})	Dissociation Constant (pK_a) at 25°C	Octanol–Water Partition Coefficient (log P)[a,b]	Soil Degradation (days) (Aerobic) DT_{50}	K_{oc} (Linear)	Bioconcentration Factor BCF/CT_{50} (days)	Mode of Action
Carbosulfan $C_{20}H_{32}N_2O_3S$	2,3-dihydro-2,2-dimethylbenzofuran-7-yl (dibutylaminothio) methylcarbamate	In acetonitrile: 200 nm = 43,420, 277.5 nm = 3144; In acetonitrile:water (50:50): 292 nm = 274	0.0359	0.11/ Miscible[c,k] 250,000[h,l]	No dissociation	7.42	21 (field)	–	990/0.09	Systemic with contact and stomach action, acetylcholine esterase inhibitor
Fenoxycarb $C_{17}H_{19}NO_4$	ethyl 2-(4-phenoxyphenoxy) ethylcarbamate	Neutral solution: 228 nm = 15,219, 278 nm = 2453, 300 nm = 745; Acidic solution: 228 nm = 15,062, 278 nm = 2357, 300 nm = 643; Basic solution: 228 nm = 14,879, 278 nm = 2374, 300 nm = 664	8.67×10^{-04}	7.9/ 500,000[c,h,k]	No dissociation	4.07	5.94 (field)	–	215/0.86	Nonneurotoxic with contact and stomach action, acts by mimicking the action of the juvenile hormone keeping the insect in an immature state

Methomyl $C_5H_{10}N_2O_2S$	*S*-methyl (*EZ*)-*N*-(methylcarbamoyloxy) thioacetimidate	Neutral solution: 234 nm = 9010; Acidic solution: 234 nm = 8980; Basic solution: 234 nm = 8890	0.72	55,000/ 97,100[n] 250,000[c] 1,000,000[l] 420,000[i]	No dissociation	0.09	6.97 (lab)	72	Low risk/–	Systemic with contact and stomach action, acetylcholinesterase (AChE) inhibitor
Oxamyl $C_7H_{13}N_3O_3S$	(*EZ*)-*N*,*N*-dimethyl-2-methylcarbamoyloxyimino-2-(methylthio)acetamide	Neutral solution: 290 nm = 80.1; Acidic solution: 290 nm = 60.1; Basic solution: 290 nm = 1154	0.051	148,100/ 10,500[n] 250,000[c,l] 41,300[e]	–2.11/estimated, very strong acid	–0.44	11 (field)	16.6	2/not available	Systemic with contact action, acetylcholinesterase (AChE) inhibitor
Pirimicarb $C_{11}H_{18}N_4O_2$	2-dimethylamino-5,6-dimethylpyrimidin-4-yl dimethylcarbamate	218.8 nm = 5760, 244.7 nm = 20,900, 272.4 nm = 855, 313.5 nm = 3800	0.43	3100/ 235,000[e] 250,000[c,l] 226,000[h]	4.4/weak base	1.7	9 (field)	–	24/not available	Selective, systemic with contact, stomach and respiratory action, acetylcholinesterase (AChE) inhibitor

Source: University of Hertfordshire, The Pesticide Properties DataBase (PPDB) developed by the Agriculture & Environment Research Unit (AERU), University of Hertfordshire, 2006–2013, 2015. Available at http://sitem.herts.ac.uk/aeru/ppdb/en/atoz.htm.

Note: a: at 20°C; b: at pH 7; c: acetone; d: benzene; e: xylene; f: cyclohexanone; g: dichloromethane; h: ethyl acetate; i: ethanol; j: *n*-hexane; k: toluene; l: methanol; m: trichloroethane; n: *n*-heptane; o: isopropanol; p: *n*-octanol.

TABLE 2.6
Chemical Names and Properties of Pyrethroid Insecticides

Common Name/Chemical Formula	IUPAC Name	Maximum UV-Vis Absorption (L mol^{-1} cm^{-1})	Vapor Pressure (mPa) at 25°C	Solubility in Water[a] (mg L^{-1})/ Organic Solvents[a] (mg L^{-1})	Dissociation Constant (pK_a) at 25°C	Octanol–Water Partition Coefficient (log P)[a,b]	Soil Degradation (days) (Aerobic) DT_{50}	K_{oc} (Linear)	Bioconcentration Factor BCF/CT_{50} (days)	Mode of Action
Acrinathrin $C_{26}H_{21}F_6NO_5$	(*S*)-α-cyano-3-phenoxybenzyl (*Z*)-(1*R*)-cis-2,2-dimethyl-3-[2-(2,2,2-trifluoro-1-trifluoromethylethoxycarbonyl) vinyl]cyclopropanecarboxylate	Neutral solution: 291 nm = 466; Acidic solution: 291 nm = 642; Basic solution: 308 nm = 1760	4.40 × 10^{-05}	0.002/ 400,000[e] 17,500[j] 61,400[i] 700,000[c]	Does not dissociate	6.3	22 (field)	48,231	538/3	Contact and stomach action
Cyfluthrin $C_{22}H_{18}Cl_2FNO_3$	(*RS*)-α-cyano-4-fluoro-3-phenoxy benzyl (1*RS*,3*RS*;1*RS*,3*SR*)-3-(2,2-dichlorovinyl)-2,2-dimethylcyclopropanecarboxylate	No absorption above 290 nm	0.0003	0.0066/ 200,000[g,k] 10,000[j]	No dissociation	6	33 (field)	123,930	506/9	Nonsystemic with contact and stomach action and rapid knockdown effect, sodium channel modulator
Cypermethrin $C_{22}H_{19}Cl_2NO_3$	(*RS*)-α-cyano-3-phenoxybenzyl (1*RS*,3*RS*;1*RS*,3*SR*)-3-(2,2-dichlorovinyl)-2,2-dimethylcyclopropanecarboxylate	204 nm = 43,217, 278 nm = 2368	0.00023	0.009/ 2,000,000[h] 450,000[c,l] 142,000[j]	No dissociation	5.3	69 (field)	156,250	1204/not available	Nonsystemic with contact and stomach action, sodium channel modulator
Deltamethrin $C_{22}H_{19}Br_2NO_3$	(*S*)-α-cyano-3-phenoxybenzyl (1*R*,3*R*)-3-(2,2-dibromovinyl)-2,2-dimethylcyclopropane carboxylate	267, 271, and 278 nm, low to very low absorption at 290–300 nm	0.0000124	0.0002/ 450,000[c] 175,000[e] 2470[n]	No dissociation	4.6	21 (field)	10,240,000	1400/not available	Nonsystemic with contact and stomach action, sodium channel modulator

Esfenvalerate $C_{25}H_{22}ClNO_3$	(*S*)-α-cyano-3-phenoxybenzyl (*S*)-2-(4-chlorophenyl)-3-methylbutyrate	–	0.0000012	0.001/500,000[c] 82,000[l] 26,000[j]	No dissociation	6.24	44 (field)	5300	3250/not available	Contact and stomach action, sodium channel modulator
Fluvalinate $C_{26}H_{22}ClF_3N_2O_3$	(*RS*)-α-cyano-3-phenoxybenzyl *N*-(2-chloro-α,α,α-trifluoro-p-tolyl)-*DL*-valinate	–	0.013	0.002/–	–	3.85	–	–	–	Contact and stomach action, sodium channel modulator
Permethrin $C_{21}H_{20}Cl_2O_3$	3-phenoxybenzyl (1*RS*,3*RS*;1*RS*,3*SR*)-3-(2,2-dichlorovinyl)-2,2-dimethylcyclopropanecarboxylate	–	0.007	0.2/ 1,000,000[e,j]	–	6.1	42 (field)	100,000	300/not available	Broad-spectrum with contact and stomach action, slight repellant effect, sodium channel modulator
Resmethrin $C_{22}H_{26}O_3$	5-benzyl-3-furylmethyl (1*RS*,3*RS*;1*RS*,3*SR*)-2, 2-dimethyl-3-(2-methylprop-1-enyl)cyclopropanecarboxylate	–	0.0015	0.01/–	–	5.43	30 (typical)	100,000	68/not available	Nonsystemic with contact action. Sodium channel modulator
Tetramethrin $C_{19}H_{25}NO_4$	cyclohex-1-ene-1,2-dicarboximidomethyl (1*RS*,3*RS*;1*RS*,3*SR*)-2,2-dimethyl-3-(2-methylprop-1-enyl)cyclopropanecarboxylate	–	2.1	1.83/ 20,000[c,j,l,p]	–	4.6	0.32 (lab)	1423	–	Nonsystemic with rapid contact action, sodium channel modulator
Tralomethrin $C_{22}H_{19}Br_4NO_3$	(*S*)-α-cyano-3-phenoxybenzyl (1*R*,3*S*)-2,2-dimethyl-3-[(*RS*)-1,2,2,2-tetrabromoethyl] cyclopropanecarboxylate	–	4.80×10^{-06}	0.08/ 1,000,000[c,e,k]	–	5	27 (lab)	359,732	1200/not available	Nonsystemic with contact and stomach action

Source: University of Hertfordshire, The Pesticide Properties DataBase (PPDB) developed by the Agriculture & Environment Research Unit (AERU), University of Hertfordshire, 2006–2013, 2015. Available at http://sitem.herts.ac.uk/aeru/ppdb/en/atoz.htm.

Note: a: at 20°C; b: at pH 7; c: acetone; d: benzene; e: xylene; f: cyclohexanone; g: dichloromethane; h: ethyl acetate; i: ethanol; j: *n*-hexane; k: toluene; l: methanol; m: trichloroethane; n: *n*-heptane; o: isopropanol; p: *n*-octanol.

TABLE 2.7
Chemical Names and Properties of Benzoylphenylurea Insecticides (Chitin Synthesis Inhibitors)

Common Name/Chemical Formula	IUPAC Name	Maximum UV-Vis Absorption (L mol^{-1} cm^{-1})	Vapor Pressure (mPa) at 25°C	Solubility in Water[a] (mg L^{-1})/ Organic Solvents[a] (mg L^{-1})	Dissociation Constant (pK_a) at 25°C	Octanol–Water Partition Coefficient (log P)[a,b]	Soil Degradation (days) (Aerobic) DT_{50}	K_{oc} (Linear)	Bioconcentration Factor BCF/CT_{50} (days)	Mode of Action
Bistrifluron $C_{16}H_7ClF_8N_2O$	1-[2-chloro-3,5-bis (trifluoromethyl)phenyl]-3-(2,6-difluorobenzoyl)urea	–	0.0027	0.03/33,000[l] 64,000[g] 3500[j]	9.58/weak acid	5.74	–	54,758	–	Chitin synthesis inhibitor
Chlorfluazuron $C_{20}H_9Cl_3F_5N_3O_3$	1-[3,5-dichloro-4-(3-chloro-5-trifluoromethyl-2-pyridyloxy) phenyl]-3-(2,6-difluorobenzoyl) urea	–	1.00×10^{-05}	0.016/6.39[j] 1000[p] 4670[e] 2680[l]	8.1/very weak acid	5.8	90 (typical)	20,800	–	Acts as an antimolting agent, killing larvae and pupae, inhibitor of chitin biosynthesis, type O
Diflubenzuron $C_{14}H_9ClF_2N_2O_2$	1-(4-chlorophenyl)-3-(2,6-difluorobenzoyl)urea	In acetonitrile 257 nm = 15,148, 290 nm = 10,500	0.00012	0.08/6980[c] 1100[l] 290[k] 63[j]	–	3.89	3.2 (lab)	–	320/0.6	Selective, nonsystemic with contact and stomach action, acts by inhibiting chitin synthesis
Flucycloxuron $C_{25}H_{20}ClF_2N_3O_3$	1-{α-[(*EZ*)-4-chloro-α-cyclopropylbenzyl ideneaminooxy]-p-tolyl}-3-(2,6-difluorobenzoyl)urea	–	5.40×10^{-05}	0.001/200[f] 3300[e] 3800[i]	–	6.97	208 (typical)	19,427	–	Nonsystemic, inhibits molting process

Flufenoxuron $C_{21}H_{11}ClF_6N_2O_3$	1-[4-(2-chloro-α,α,α-trifluoro-p-tolyloxy)-2-fluorophenyl]-3-(2,6-difluorobenzoyl)urea	Neutral solution: 206 nm = 31,741, 252 nm = 17,403, 274 nm = 18,589; Acidic solution: 220 nm = 12,073, 258 nm = 9424, 297 nm = 1064; Basic solution: 224 nm = 19,830, 237 nm = 19,571	6.52×10^{-09}	0.0043/83,000[c] 16,000[g] 6000[e] 3500[k]	10.1/very weak acid	5.11	42.9 (field)	157,643	700,500/21	Growth regulator with contact and stomach action, inhibitors of chitin biosynthesis
Hexaflumuron $C_{16}H_8Cl_2F_6N_2O_3$	1-[3,5-dichloro-4-(1,1,2,2-tetrafluoroethoxy)phenyl] -3-(2,6-difluorobenzoyl)urea	–	0.059	0.027/162,000[c] 100,000[h] 5[n]	–	5.68	170 (field)	10,391	4700/not available	Chitin synthesis inhibitor, systemic with stomach action
Lufenuron $C_{17}H_8Cl_2F_8N_2O_3$	(*RS*)-1-[2,5-dichloro-4-(1,1,2,3,3,3-hexafluoropropoxy) phenyl]-3-(2,6-difluorobenzoyl) urea	Neutral solution: 210 nm = 37,293; Acidic solution: 210 nm = 30,588; Basic solution: 295 nm = 4871; No absorption between 295 and 750 nm	4.00×10^{-03}	0.046/460,000[c] 330,000[h] 66,000[k] 100 [j]	10.2/very weak acid	5.12	256 (field)	–	5300/36	Systemic, selective, stomach acting, chitin synthesis inhibitor
Novaluron $C_{17}H_9ClF_8N_2O_4$	(*RS*)-1-[3-chloro-4-(1,1,2-trifluoro-2-trifluoromethoxyethoxy) phenyl]-3-(2,6-difluorobenzoyl) urea	Neutral solution: 253 nm = 15,400; Acidic solution: 253 nm = 9780; Basic solution: 263 nm = 20,500	1.60×10^{-02}	0.003/113,000[h] 1880[e] 8.39[n]	–	4.3	96.5 (field)	9598	2091/7.3	A chitin synthesis inhibitor, with stomach action and some contact activity

(*Continued*)

TABLE 2.7 (CONTINUED)
Chemical Names and Properties of Benzoylphenylurea Insecticides (Chitin Synthesis Inhibitors)

Common Name/Chemical Formula	IUPAC Name	Maximum UV-Vis Absorption (L mol^{-1} cm^{-1})	Vapor Pressure (mPa) at 25°C	Solubility in Water[a] (mg L^{-1})/ Organic Solvents[a] (mg L^{-1})	Dissociation Constant (pK_a) at 25°C	Octanol–Water Partition Coefficient (log *P*)[a,b]	Soil Degradation (days) (Aerobic) DT_{50}	K_{oc} (Linear)	Bioconcentration Factor BCF/CT_{50} (days)	Mode of Action
Teflubenzuron $C_{14}H_6Cl_2F_4N_2O_2$	1-(3,5-dichloro-2, 4-difluorophenyl)-3-(2,6-difluorobenzoyl)urea	Neutral solution: 249 nm = 13,543; Acidic solution: 251 nm = 15,153; Basic solution: 262 nm = 21,115	9.16×10^{-04}	0.01/8850[c] 1060[l] 740[k] 10[n]	9.2/weak acid	4.3	13.7 (field)	26,062	640/0.8	Systemic, chitin synthesis inhibitor
Triflumuron $C_{15}H_{10}ClF_3N_2O_3$	1-(2-chlorobenzoyl)-3-(4-trifluoromethoxyphenyl)urea	249 nm = 14,940,000	0.0002	0.04/26,600[c] 23,300[h] 11,700[g] 100[n]	–	4.9	22 (field)	2967	612/1.36	Chitin synthesis inhibitor, insect growth regulator

Source: University of Hertfordshire, The Pesticide Properties DataBase (PPDB) developed by the Agriculture & Environment Research Unit (AERU), University of Hertfordshire, 2006–2013, 2015. Available at http://sitem.herts.ac.uk/aeru/ppdb/en/atoz.htm.

Note: a: at 20°C; b: at pH 7; c: acetone; d: benzene; e: xylene; f: cyclohexanone; g: dichloromethane; h: ethyl acetate; i: ethanol; j: *n*-hexane; k: toluene; l: methanol; m: trichloroethane; n: *n*-heptane; o: isopropanol; p: *n*-octanol.

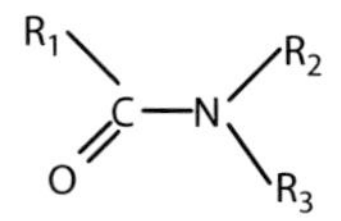

FIGURE 2.2 General formula of amides.

2.2.2.2.4 Nitrile and Dinitroaniline

These compounds are also applied as herbicides (Tables 2.11 and 2.12).

2.2.2.2.5 Organophosphorus

The main component of this group is glyphosate (Table 2.13).

2.2.2.2.6 Phenoxy Acids

The phenoxy acids form a frequently used group of herbicides (Table 2.14). Some components of this group that are formed by stereosomers are commercialized as single enanthiomers or racemic mixtures. The key components of the phenoxy acid herbicides are derivatives of acetic acid, for example, 2,4-D [(2,4-dichlorophenoxy)acetic acid], and propionic acid, for example, mecoprop-P [(R)-2-(4-chloro-o-tolyloxy)propionic acid].

2.2.2.2.7 Pyridines and Quaternary Ammonium Compounds

Pyridate and pyridazines are contact-selective herbicides with foliar activity. The key components of these pesticides are included in Table 2.15.

2.2.2.2.8 Triazines

Triazines are the oldest and the most commonly used herbicides, representing around 30% of the pesticide market in the world (Table 2.16). The main subgroups are chlorotriazine, methoxytriazine, methylthiotriazine, and trazinone herbicides. These compounds have an appreciable persistence in soil. The half-life in soil in days is the shortest for cyanazine (ca., 14 days) and the longest for simazine (27–102 days) [13].

2.2.2.2.9 Phenylureas

Phenylureas are important herbicides and are used worldwide to control weeds in various crops. Phenylureas comprise numerous substituted urea and derivatives of urea (Table 2.17).

2.2.2.2.10 Sulfonylureas

Sulfonylureas are a group of substituted ureas developed recently that have, in general, an herbicidal activity higher than the phenylurea herbicides. Main subgroups are pyrimidinyl sulfonylurea and triazinyl sulfonylurea herbicides (Table 2.18).

2.2.2.2.11 Pyridazines and Pyridazinones

The last chemical group of herbicides is pyridazines and pyridazinones (Table 2.19).

2.2.2.3 Chemical Classification of Fungicides

Fungicides belong to various chemical classes.

2.2.2.3.1 Conazoles

The main subgroups are triazole and imidazole fungicides (Table 2.20).

TABLE 2.8
Chemical Names and Properties of Amide Herbicides

Common Name/Chemical Formula	IUPAC Name	Maximum UV-Vis Absorption (L mol^{-1} cm^{-1})	Vapor Pressure (mPa) at 25°C	Solubility in Water[a] (mg L^{-1})/ Organic Solvents[a] (mg L^{-1})	Dissociation Constant (pK_a) at 25°C	Octanol–Water Partition Coefficient (log *P*)[a,b]	Soil Degradation (days) (Aerobic) DT_{50}	K_{oc} (Linear)	Bioconcentration Factor BCF/CT_{50} (days)	Mode of Action
Allidochlor $C_8H_{12}ClNO$	*N,N*-diallyl-2-chloroacetamide	–	0.0108	197,000/–	4.1/weak acid	1.83	10 (typical)	20	Low risk/–	Translocates and inhibits cell division causing plant death
Beflubutamid $C_{18}H_{17}F_4NO_2$	(*RS*)-*N*-benzyl-2-(α,α,α, 4-tetrafluoro-m-tolyloxy) butyramide	–	1.10 × 10^{-02}	3.29/600,000[c] 571,000[h] 473,000[l] 2180[n]	Unlikely to dissociate	4.28	55 (field)	–	230/not available	Inhibition of carotenoid biosynthesis
Bromobutide $C_{15}H_{22}BrNO$	(*RS*)-2-bromo-3,3-dimethyl-*N*-(1-methyl-1-phenylethyl)butyramide	–	–	3.54/–	–	3.48	–	–	–	Inhibition of 4-hydroxy phenyl-pyruvate dioxygenase, bleaching
Dimethenamid $C_{12}H_{18}ClNO_2S$	(*RS*)-2-chloro-*N*-(2,4-dimethyl-3-thienyl)-*N*-(2-methoxy-1-methylethyl)acetamide	–	0.37	1200/ Miscible[c,h,j,k]	No dissociation	2.2	13 (field)	–	Low risk/–	Absorbed via roots, little translocation, inhibition of mitosis and cell division
Diphenamid $C_{16}H_{17}NO$	*N,N*-dimethyldiphenylacetamide	–	3.04 × 10^{-03}	260/ 189,000[c] 50,000[e]	–	2.17	30 (typical)	210	54/not available	Selective, systemic, absorbed through roots

Napropamide $C_{17}H_{21}NO_2$	(*RS*)-*N*,*N*-diethyl-2-(1-naphthyloxy)propionamide	Neutral solution: 215 nm = 58,800, 282 nm = 10,500; Acidic solution: 215 nm = 58,600, 282 nm = 10,900	2.2×10^{-02}	74.0/440,000[c] 11,100[n] 290,000[h] 692,000[g]	No dissociation	3.3	72 (field)	839	98/not available	Selective, systemic, absorbed through roots and translocated, acts by preventing root cell elongation and so disrupts growth
Pethoxamid $C_{16}H_{22}ClNO_2$	2-chloro-*N*-(2-ethoxyethyl)-*N*-(2-methyl-1-phenylprop-1-enyl) acetamide	240 nm = 12,000, 290 nm = 60	0.34	400/ 117,000[n] 250,000[h,l,n]	No dissociation	2.96	14.2 (field)	–	Low risk/–	Selective, absorbed by roots and to a lesser extent foliage, inhibits auxin transport
Propyzamide $C_{12}H_{11}Cl_2NO$	3,5-dichloro-*N*-(1,1-dimethylprop-2-ynyl)benzamide	206 nm and 284 nm	0.0267	9/139,000[c] 63,800[l] 501[j] 9670[k]	No dissociation	3.3	56 (field)	840	49/42	Selective, systemic absorbed by roots and translocated
Tebutam $C_{15}H_{23}NO$	*N*-benzyl-*N*-isopropyl-2, 2-dimethylpropionamide	–	89	790/–	–	3	60 (typical)	25	–	Selective, microtubule assembly inhibition

Source: University of Hertfordshire, The Pesticide Properties DataBase (PPDB) developed by the Agriculture & Environment Research Unit (AERU), University of Hertfordshire, 2006–2013, 2015. Available at http://sitem.herts.ac.uk/aeru/ppdb/en/atoz.htm.

Note: a: at 20°C; b: at pH 7; c: acetone; d: benzene; e: xylene; f: cyclohexanone; g: dichloromethane; h: ethyl acetate; i: ethanol; j: *n*-hexane; k: toluene; l: methanol; m: trichloroethane; n: *n*-heptane; o: isopropanol; p: *n*-octanol.

TABLE 2.9
Chemical Names and Properties of Chloroacetanilide Herbicides

Common Name/Chemical Formula	IUPAC Name	Maximum UV-Vis Absorption (L mol^{-1} cm^{-1})	Vapor Pressure (mPa) at 25°C	Solubility in Water[a] (mg L^{-1})/ Organic Solvents[a] (mg L^{-1})	Dissociation Constant (pK_a) at 25°C	Octanol–Water Partition Coefficient (log P)[a,b]	Soil Degradation (days) (Aerobic) DT_{50}	K_{oc} (Linear)	Bioconcentration Factor BCF/CT_{50} (days)	Mode of Action
Acetochlor $C_{14}H_{20}ClNO_2$	2-chloro-*N*-ethoxymethyl-6′-ethylacet-o-toluidide	Neutral solution: 273 nm = 448, 265 nm = 538; Acid solution: 273 nm = 466, 265 nm = 552	2.2×10^{-02}	282/ 5,000,000[c] 500,000[h] 756,000[k] 100,000[i]	No dissociation	4.14	12.1 (field)	156	20/not available	Selective, absorbed mainly by shoots and roots of germinating weeds, lipid synthesis inhibitor
Alachlor $C_{14}H_{20}ClNO_2$	2-chloro-2′,6′-diethyl-*N*-methoxymethylacetanilide	–	2.9	240/–	0.62/strong acid	3.09	14 (field)	335	39/not available	Selective, systemic action absorbed by germinating shoots, lipid synthesis inhibitor
Butachlor $C_{17}H_{26}ClNO_2$	*N*-butoxymethyl-2-chloro-2′,6′-diethylacetanilide	–	0.24	20/–	–	4.5	11.5 (field)	700	2.4/1.7	Selective, systemic absorbed primarily via germinating shoots, inhibition of mitosis and cell division

Dimethachlor $C_{13}H_{18}ClNO_2$	2-chloro-*N*-(2-methoxyethyl) acet-2′,6′-xylidide	Neutral solution: 215 nm = 14,461, 265 nm = 486; Acidic solution: 215 nm = 14,768, 265 nm = 572; Basic solution: 215 nm = 9576, 265 nm = 469	0.64	2300/500,000[c,l] 42,000[j] 440,000[p]	No dissociation	3.2	–	–	75/not available	Selective, absorbed by new shoots of seedlings and roots
Metolachlor $C_{15}H_{22}ClNO_2$	2-chloro-*N*-(6-ethyl-o-tolyl)-*N*-[(1*RS*)-2-methoxy-1-methylethyl] acetamide	–	1.7	530/Miscible[c,d,e,j]	No dissociation	3.4	21 (field)	120	68.8/1.5	Selective, reduces seed germination, inhibition of mitosis and cell division
Propachlor $C_{11}H_{14}ClNO$	2-chloro-*N*-isopropylacetanilide	–	30.6	580/353,900[c] 205,500[e] 296,100[k] 655,900[d]	–	1.6	–	80	37/not available	Selective, systemic, effects cell formation and protein synthesis
Propisochlor $C_{15}H_{22}ClNO_2$	2-chloro-6′-ethyl-*N*-isopropoxy methylacet-ortho-toluidide	–	3.1	90.8/483,000[c] 538,000[g] 582,000[n] 598,000[l]	–	3.3	7.63 (lab)	–	72.2/not available	Inhibits protein synthesis and so blocks cell division, selective, absorbed by shoots of germinating plants

Source: University of Hertfordshire, The Pesticide Properties DataBase (PPDB) developed by the Agriculture & Environment Research Unit (AERU), University of Hertfordshire, 2006–2013, 2015. Available at http://sitem.herts.ac.uk/aeru/ppdb/en/atoz.htm.

Note: a: at 20°C; b: at pH 7; c: acetone; d: benzene; e: xylene; f: cyclohexanone; g: dichloromethane; h: ethyl acetate; i: ethanol; j: *n*-hexane; k: toluene; l: methanol; m: trichloroethane; n: *n*-heptane; o: isopropanol; p: *n*-octanol.

TABLE 2.10
Chemical Names and Properties of Carbamate Herbicides

Common Name/ Chemical Formula	IUPAC Name	Maximum UV-Vis Absorption (L mol^{-1} cm^{-1})	Vapor Pressure (mPa) at 25°C	Solubility in Water[a] (mg L^{-1})/ Organic Solvents[a] (mg L^{-1})	Dissociation Constant (pK_a) at 25°C	Octanol–Water Partition Coefficient (log *P*)[a,b]	Soil Degradation (days) (Aerobic) DT_{50}	K_{oc} (Linear)	Bioconcentration Factor BCF/CT_{50} (days)	Mode of Action
Chlorpropham $C_{10}H_{12}ClNO_2$	Isopropyl 3-chlorocar banilate	–	24	110/ 1,000,000[c,e,g,n]	No dissociation	3.76	25 (lab)	–	144/not available	Mitosis inhibitor, absorbed predominately by roots
Desmedipham $C_{16}H_{16}N_2O_4$	Ethyl 3-phenyl carbamoyloxycarbanilate	Neutral solution: 203 nm = 55,726, 236 nm = 43,133, 273 nm = 3363	0.000041	7/ 20[j] 1200[k] 187,000[l] 285,000[c]	No dissociation	3.39	8 (field)	–	157/not available	Selective, systemic, absorbed through leaves, inhibits photosynthesis (photosystem II)
EPTC $C_9H_{19}NOS$	*S*-ethyl dipropyl (thiocarbamate)	–	4500	370/Miscible[c,d,e,i]	No dissociation	3.2	18 (field)	300	60/not available	Selective, systemic absorbed by roots and shoots, lipid synthesis inhibitor
Molinate $C_9H_{17}NOS$	*S*-ethyl azepane -1-carbothioate	200 nm, no absorption >290 nm	500	1100/Miscible[c,h,j,l]	No dissociation	2.86	12.5 (field)	190	72/not available	Selective, systemic absorbed by roots and translocated, inhibition of lipid synthesis

Phenmedipham $C_{16}H_{16}N_2O_4$	Methyl 3-(3-methylcar baniloyloxy)carbanilate	205 nm = 59,646, 237 nm = 37,848, 274 nm = 2761	7.00 × 10^{-07}	1.8/ 970[k] 36,200[l] 165,000[c] 56,300[h]	No dissociation	3.59	18 (field)	888	165/2.7	Selective, systemic, absorbed through leaves and translocated, inhibits photosynthesis (photosystem II)
Propham $C_{10}H_{13}NO_2$	Isopropyl carbanilate	–	1999.5	250/–	–	2.6	11 (lab)	98	Low risk/–	Selective, systemic, mitosis inhibitor
Thiobencarb $C_{12}H_{16}ClNOS$	*S*-4-chlorobenzyl diethyl (thiocarbamate)	Neutral solution: 221 nm = 19,006; Acidic solution: 220.5 nm = 18,888; Basic solution: 222 nm = 18,888	2.39	16.7/500,000[c,j,k,l]	No dissociation	4.23	4 (field)	–	302/0.3	Selective, ACCase inhibitor—inhibition of lipid synthesis
Tri-allate $C_{10}H_{16}Cl_3NOS$	*S*-2,3,3-tricloroallyl di-isopropylthiocarbamate	No absorption between 210 and 900 nm recorded under basic, neutral, or acidic conditions	12	4.1/500,000[c,e,i,l]	No dissociation observed	4.06	46 (field)	3034	1400/not available	Selective

Source: University of Hertfordshire, The Pesticide Properties DataBase (PPDB) developed by the Agriculture & Environment Research Unit (AERU), University of Hertfordshire, 2006–2013, 2015. Available at http://sitem.herts.ac.uk/aeru/ppdb/en/atoz.htm.

Note: a: at 20°C; b: at pH 7; c: acetone; d: benzene; e: xylene; f: cyclohexanone; g: dichloromethane; h: ethyl acetate; i: ethanol; j: *n*-hexane; k: toluene; l: methanol; m: trichloroethane; n: *n*-heptane; o: isopropanol; p: *n*-octanol.

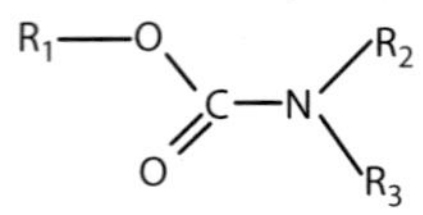

FIGURE 2.3 General formula of carbamate herbicides (for the subgroup of the carbamic acids).

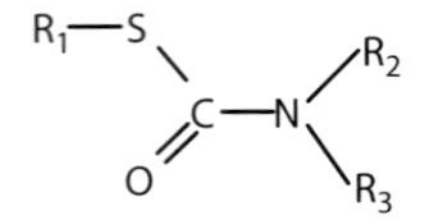

FIGURE 2.4 General formula of thiocarbamate herbicides.

2.2.2.3.2 Dithiocarbamates

Dithiocarbamates are also applied in agriculture as fungicides (Table 2.21). The dithiocarbamate fungicides are the most widely used organic fungicides and have a wide spectrum of activity as foliar spray for fruits, vegetables, and ornamentals and as speed protectants.

2.2.2.3.3 Morpholines

Morpholines are specific fungicides against powdery mildew fungi and are used to control the disease in products such as cereals, cucumbers, and apples. The key components of these fungicides are dimethomorph, dodemorph, and fenpropimorph (Table 2.22).

2.2.2.3.4 Benzimidazoles

Carbendazim and thiabendazole are key components of the benzimidazole fungicide group (Table 2.23).

2.2.2.3.5 Anilides

Table 2.24 contains the main components belonging to the class of anilide fungicides.

2.3 TABULATED PROPERTIES OF PESTICIDES

Tables 2.3 through 2.24 contain parameters describing the physical properties of pesticides, such as the following [12]:

- Absorption maxima and intensities: The wavelength of maximum absorbance is a characteristic value and so can be used for identification purposes.
- Vapor pressure: The pressure at which a liquid is in equilibrium with its vapor at 25°C. It is a measure of the tendency of a material to vaporize (the higher the vapor pressure, the greater the potential).
- Solubility: A measure of how easily a given substance dissolves in a particular solvent (solubility in water allows the fate and behavior of pesticides in the environment to be predicted, for example, pesticides that are very soluble in water will tend not to accumulate in soil or biota because of their strong polar character).
- Acid-base character (pK_a): This allows evaluation of, for example, the sorption process of pesticides. The sorption process is different for nonionic and ionic pesticides and for weakly acidic, weakly basic, and neutral pesticides because the sorption of these pesticides depends on the soil pH, electric charge, and ionic strength.
- Partition coefficient: This is a measured ratio (at equilibrium) of the dissolved mass of the substance between equal layers of n-octanol and water, K_{ow}, expressed in the logarithmic form as log P (log P is considered to be a good indicator of bioaccumulation of pesticides in the environment and food chains and the systematic mode of action of pesticides, for

TABLE 2.11
Chemical Names and Properties of Nitrile Herbicides

Common Name/ Chemical Formula	IUPAC Name	Maximum UV-Vis Absorption (L mol^{-1} cm^{-1})	Vapor Pressure (mPa) at 25°C	Solubility in Water[a] (mg L^{-1})/ Organic Solvents[a] (mg L^{-1})	Dissociation Constant (pK_a) at 25°C	Octanol–Water Partition Coefficient (log P)[a,b]	Soil Degradation (days) (Aerobic) DT_{50}	K_{oc} (Linear)	Bioconcentration Factor BCF/CT_{50} (days)	Mode of Action
Bromoxynil $C_7H_3Br_2NO$	3,5-dibromo-4-hydroxybenzonitrile	221.2 nm = 30,343, 287 nm = 18,302	0.17	90/170,000[c] 90,000[l] 410,000[r]	3.86/weak acid	1.04	8 (field)	302	Low risk/–	Selective, contact action with some systemic activity, inhibits photosynthesis (photosystem II)
Dichlobenil $C_7H_3Cl_2N$	2,6-dichlorobenzonitrile	298.9 nm = 1985.3; 290 nm = 1878.8; 243 nm = 5766.9; 236.9 nm = 6805.9; 211 nm = 38,528.8	0.00014	21.2/86,000[c] 53,000[e] 151,000[g]	No dissociation	2.7	5.4 (field)	257	100/1.0	Systemic, inhibition of cell wall synthesis
Ioxynil $C_7H_3I_2NO$	4-hydroxy-3,5-diiodophenyl cyanide	237.2 nm = 28,030, 287 nm = 17,380	0.00204	3034/73,500[c] 22,200[l] 5800[e] 37,500[n]	4.1/weak acid	2.2	5 (field)	–	29/not available	Selective, systemic with contact action, acts by inhibiting photosynthesis at photosystem II

Source: University of Hertfordshire, The Pesticide Properties DataBase (PPDB) developed by the Agriculture & Environment Research Unit (AERU), University of Hertfordshire, 2006–2013, 2015. Available at http://sitem.herts.ac.uk/aeru/ppdb/en/atoz.htm.

Note: a: at 20°C; b: at pH 7; c: acetone; d: benzene; e: xylene; f: cyclohexanone; g: dichloromethane; h: ethyl acetate; i: ethanol; j: *n*-hexane; k: toluene; l: methanol; m: trichloroethane; n: *n*-heptane; o: isopropanol; p: *n*-octanol; r: tetrahydrofuran.

TABLE 2.12
Chemical Names and Properties of Dinitroaniline Herbicides

Common Name/ Chemical Formula	IUPAC Name	Maximum UV-Vis Absorption (L mol^{-1} cm^{-1})	Vapor Pressure (mPa) at 25°C	Solubility in Water[a] (mg L^{-1})/ Organic Solvents[a] (mg L^{-1})	Dissociation Constant (pK_a) at 25°C	Octanol–Water Partition Coefficient (log P)[a,b]	Soil Degradation (days) (Aerobic) DT$_{50}$	K_{oc} (Linear)	Bioconcentration Factor BCF/CT$_{50}$ (days)	Mode of Action
Benfluralin $C_{13}H_{16}F_3N_3O_4$	*N*-butyl-*N*-ethyl-α,α,α-trifluoro-2,6-dinitro-p-toluidine	Neutral solution: 239 nm = 9180, 283 nm = 8010; Acidic solution: 248 nm = 4390, 298 nm = 4580, 448 nm = 3870; Basic solution: 238 nm = 7550, 283 nm = 6370, 431 nm = 3720	1.73	0.065/ 250,000[e] 39,100[l] 250,000[c,h]	−0.59/very strong acid	5.19	53 (field)	10,777	1572/1.3	Microtubule assembly inhibition
Butralin $C_{14}H_{21}N_3O_4$	(*RS*)-*N*-sec-butyl-4-tert-butyl-2,6-dinitroaniline	–	0.77	0.308/182,800[n] 668,800[e] 68,300[l] 773,300[c]	No dissociation	4.93	105 (field)	46,391	1950/7	Selective, absorbed by germinating seedlings

Ethafluralin $C_{13}H_{14}F_3N_3O_4$	*N*-ethyl-α,α,α-trifluoro-*N*-(2-methylallyl)-2,6-dinitro-p-toluidine	Neutral solution: 269 nm = 8110, 377 nm = 2330; Acidic solution: 269 nm = 8060, 377 nm = 2360; Basic solution: 269 nm = 7860, 377 nm = 2360	12	0.01/1,000,000[c,g] 130,000[j] 124,000[l]	No dissociation	5.11	53 (field)	6364	1330/3	Selective, microtube inhibitor—reducing seedling root growth
Isopropalin $C_{15}H_{23}N_3O_4$	4-isopropyl-2,6-dinitro-*N*,*N*-dipropylaniline	–	1.17	0.11/–	–	5.29	100 (typical)	10,000	17,000/not available	Inhibition of microtubular assembly
Trifluralin $C_{13}H_{16}F_3N_3O_4$	α,α,α-trifluoro-2,6-dinitro-*N*,*N*-dipropyl-p-toluidine	Neutral solution: 209 nm = 19,400, 272 nm = 8460, 385 nm = 2440	9.5	0.221/250,000[c,j,k] 142,000[l]	No dissociation	5.27	170 (field)	15,800	5674/47	Selective, inhibition of mitosis and cell division

Source: University of Hertfordshire, The Pesticide Properties DataBase (PPDB) developed by the Agriculture & Environment Research Unit (AERU), University of Hertfordshire, 2006–2013, 2015. Available at http://sitem.herts.ac.uk/aeru/ppdb/en/atoz.htm.

Note: a: at 20°C; b: at pH 7; c: acetone; d: benzene; e: xylene; f: cyclohexanone; g: dichloromethane; h: ethyl acetate; i: ethanol; j: *n*-hexane; k: toluene; l: methanol; m: trichloroethane; n: *n*-heptane; o: isopropanol; p: *n*-octanol; r: tetrahydrofuran.

TABLE 2.13
Chemical Names and Properties of Organophosphorus Herbicides

Common Name/ Chemical Formula	IUPAC Name	Maximum UV-Vis Absorption (L mol^{-1} cm^{-1})	Vapor Pressure (mPa) at 25°C	Solubility in Water[a] (mg L^{-1})/ Organic Solvents[a] (mg L^{-1})	Dissociation Constant (pK_a) at 25°C	Octanol–Water Partition Coefficient (log P)[a,b]	Soil Degradation (days) (Aerobic) DT_{50}	K_{oc} (Linear)	Bioconcentration Factor BCF/CT_{50} (days)	Mode of Action
Glyphosate $C_3H_7NO_5P$	*N*-(phosphonomethyl) glycine	–	0.0131	10,500/231[l] 78[c] 26[j] 12[h]	2.34/strong acid	−3.2	12 (field)	1435	0.5/not available	Broad-spectrum, systemic, contact action translocated and nonresidual, inhibition of lycopene cyclase
Glufosinate $C_5H_{12}NO_4P$	(2*RS*)-2-amino-4-[hydroxy(methyl) phosphinoyl]butyric acid	–	–	–/–	2/strong acid, pK_a(2) = 2.0; pK_a(3) = 9.8	−3.96	–	–	Low risk/–	Nonselective, contact with some systemic action, glutamine synthetase inhibitor: accumulates ammonium ions, inhibits photosynthesis

Source: University of Hertfordshire, The Pesticide Properties DataBase (PPDB) developed by the Agriculture & Environment Research Unit (AERU), University of Hertfordshire, 2006–2013, 2015. Available at http://sitem.herts.ac.uk/aeru/ppdb/en/atoz.htm.

Note: a: at 20°C; b: at pH 7; c: acetone; d: benzene; e: xylene; f: cyclohexanone; g: dichloromethane; h: ethyl acetate; i: ethanol; j: *n*-hexane; k: toluene; l: methanol; m: trichloroethane; n: *n*-heptane; o: isopropanol; p: *n*-octanol; r: tetrahydrofuran.

TABLE 2.14
Chemical Names and Properties of Phenoxy Acid Herbicides

Common Name/Chemical Formula	IUPAC Name	Maximum UV-Vis Absorption (L mol^{-1} cm^{-1})	Vapor Pressure (mPa) at 25°C	Solubility in Water[a] (mg L^{-1})/ Organic Solvents[a] (mg L^{-1})	Dissociation Constant (pK_a) at 25°C	Octanol–Water Partition Coefficient (log *P*)[a,b]	Soil Degradation (days) (Aerobic) DT_{50}	K_{oc} (Linear)	Bioconcentration Factor BCF/CT_{50} (days)	Mode of Action
2,4-D $C_8H_6Cl_2O_3$	(2,4-dichlorophenoxy) acetic acid	290 nm = 164.97	0.0187	23,180/390,000[c] 810,000[l] 13,000[g] 6400[k]	2.87/strong acid	−0.83	10 (field)	88.4	10/not available	Selective, systemic, absorbed through roots and increases biosynthesis and production of ethylene causing uncontrolled cell division and so damages vascular tissue, synthetic auxin
Diclofop $C_{15}H_{12}Cl_2O_4$	(*RS*)-2-[4-(2,4-dichlorophenoxy) phenoxy]propionic acid	–	0.0000031	122,700/–	3.43/weak acid	1.61	35.2 (field)	–	Low risk/–	Selective, absorbed by leaves and inhibits fatty acid synthesis
Fenoxaprop-P-ethyl $C_{18}H_{16}ClNO_5$	(*R*)-2[4-[(6-chloro-2-benzoxazolyl) oxy]-phenoxy]-propanoic acid	239 nm = 22,862, 278 nm = 7980, 291 nm − 1488	5.30×10^{-04}	0.7/400,000[c] 380,000[h] 43,100[l] 7000[j]	0.18/very strong acid	4.58	0.31 (field)	11,354	338/0.4	Selective, systemic with contact action, inhibits fatty acid synthesis (ACCase)
MCPA $C_9H_9ClO_3$	4-chloro-o-tolyloxyacetic acid	No absorption >290 nm	0.4	29,390/ 775,600[l] 289,300[h] 26,500[k] 323[j]	3.73/weak acid	−0.81	25 (field)	–	1/not available	Selective, systemic with translocation, synthetic auxin
Mecoprop-P $C_{10}H_{11}ClO_3$	(*R*)-2-(4-chloro-o-tolyloxy)propionic acid	203 nm = 21,000, 229 nm = 9800, 280 nm = 1600, 287 nm = 1500, 290 nm = 1200	0.23	860/1,000,000[c,h,l]	3.86/weak acid	0.02	21 (field)	–	3/1.1	Selective, systemic, absorbed through leaves and translocated to roots, synthetic auxin

(Continued)

TABLE 2.14 (CONTINUED)
Chemical Names and Properties of Phenoxy Acid Herbicides

Common Name/Chemical Formula	IUPAC Name	Maximum UV-Vis Absorption (L mol^{-1} cm^{-1})	Vapor Pressure (mPa) at 25°C	Solubility in Water[a] (mg L^{-1})/ Organic Solvents[a] (mg L^{-1})	Dissociation Constant (pK_a) at 25°C	Octanol–Water Partition Coefficient (log *P*)[a,b]	Soil Degradation (days) (Aerobic) DT_{50}	K_{oc} (Linear)	Bioconcentration Factor BCF/CT_{50} (days)	Mode of Action
Quizalofop-P-ethyl $C_{19}H_{17}ClN_2O_4$	ethyl (*R*)-2-[4-(6-chloroquinoxalin-2-yloxy)phenoxy] propionate	Neutral solution: 343 nm = 6342; Acidic solution: 343 nm = 6356; Basic solution: 343 nm = 6242	1.10×10^{-04}	0.61/250,000[c,e,h] 34,870[l]	No dissociation	4.61	1.8 (field)	–	380/not available	Selective, an acetyl CoA carboxylase inhibitor (ACCase)
Triclopyr $C_7H_4Cl_3NO_3$	3,5,6-trichloro-2-pyridyloxyacetic acid	–	0.1	8100/665,000[l] 582,000[c] 19,000[k] 90[j]	3.97/weak acid	4.62	30 (field)	27	0.77/14	Selective, systemic, absorbed through roots and foliage, synthetic auxin

Source: University of Hertfordshire, The Pesticide Properties DataBase (PPDB) developed by the Agriculture & Environment Research Unit (AERU), University of Hertfordshire, 2006–2013, 2015. Available at http://sitem.herts.ac.uk/aeru/ppdb/en/atoz.htm.

Note: a: at 20°C; b: at pH 7; c: acetone; d: benzene; e: xylene; f: cyclohexanone; g: dichloromethane; h: ethyl acetate; i: ethanol; j: *n*-hexane; k: toluene; l: methanol; m: trichloroethane; n: *n*-heptane; o: isopropanol; p: *n*-octanol; r: tetrahydrofuran.

TABLE 2.15
Chemical Names and Properties of Pyridine Herbicides and Quaternary Ammonium Compounds

Common Name/ Chemical Formula	IUPAC Name	Maximum UV-Vis Absorption (L mol^{-1} cm^{-1})	Vapor Pressure (mPa) at 25°C	Solubility in Water[a] (mg L^{-1})/ Organic Solvents[a] (mg L^{-1})	Dissociation Constant (pK_a) at 25°C	Octanol–Water Partition Coefficient (log P)[a,b]	Soil Degradation (days) (Aerobic) DT_{50}	K_{oc} (Linear)	Bioconcentration Factor BCF/CT_{50} (days)	Mode of Action
Diquat dibromide $C_{12}H_{12}N_{22}Br$	9,10-dihydro-8a,10a-diazoniaphenanthrene dibromide	204 nm, 272 nm, 310 nm	0.01	718,000/ 25,000[l] 100[c,h,k]	No dissociation	–4.6	5500 (field)	2,184,750	1/not available	Nonselective, contact absorbed through foliage, some desiccant action, photosystem I (electron transport) inhibitor
Paraquat $C_{12}H_{14}N_2$	1,1′-dimethyl-4,4′-bipyridinium	290 nm	0.01	620,000/143,000[l] 100[c,j,k]	No dissociation	–4.5	2800 (field)	1,000,000	Low risk/–	Broad-spectrum, nonresidual activity with contact and some desiccant action, photosystem I (electron transport) inhibitor
Chlormequat chloride $C_5H_{13}Cl_2N$	2-chloroethyltrimethylammonium chloride	No significant absorption from 200 to 750 nm	1.0×10^{-03}	886,000/365,000[l] 130[c] 10[h,n]	Complete dissociation	−3.47	7 (field)	–	0.01/not available	Inhibits cell elongation
Mepiquat chloride $C_7H_{16}ClN$	1,1-dimethylpiperidinium chloride	No significant absorption between 200 and 750 nm at pH 1, 6, or 13	1.00×10^{-05}	500,000/487,000[l] 20[c] 10[n]	Completely dissociates in aqueous solution	−3.55	18.4 (lab)	–	2/not available	Inhibits biosynthesis of gibberellic acid

Source: University of Hertfordshire, The Pesticide Properties DataBase (PPDB) developed by the Agriculture & Environment Research Unit (AERU), University of Hertfordshire, 2006–2013, 2015. Available at http://sitem.herts.ac.uk/aeru/ppdb/en/atoz.htm.

Note: a: at 20°C; b: at pH 7; c: acetone; d: benzene; e: xylene; f: cyclohexanone; g: dichloromethane; h: ethyl acetate; i: ethanol; j: *n*-hexane; k: toluene; l: methanol; m: trichloroethane; n: *n*-heptane; o: isopropanol; p: *n*-octanol; r: tetrahydrofuran.

TABLE 2.16
Chemical Names and Properties of Triazine Herbicides

Common Name/ Chemical Formula	IUPAC Name	Maximum UV-Vis Absorption (L mol^{-1} cm^{-1})	Vapor Pressure (mPa) at 25°C	Solubility in Water[a] (mg L^{-1})/ Organic Solvents[a] (mg L^{-1})	Dissociation Constant (pK_a) at 25°C	Octanol–Water Partition Coefficient (log *P*)[a,b]	Soil Degradation (days) (Aerobic) DT_{50}	K_{oc} (Linear)	Bioconcentration Factor BCF/CT_{50} (days)	Mode of Action
				Chlorotriazine Herbicides						
Atrazine $C_8H_{14}ClN_5$	6-chloro-*N*2-ethyl-*N*4-isopropyl-1,3,5-triazine-2,4-diamine	–	0.039	35/24,000[h] 28,000[g] 4000[k] 110[j]	1.7/very weak base	2.7	29 (field)	100	4.3/not available	Selective, systemic action with residual and foliar activity, inhibits photosynthesis (photosystem II)
Cyanazine $C_9H_{13}ClN_6$	2-(4-chloro-6-ethylamino-1,3,5-triazin-2-ylamino)-2-methylpropiononitrile	–	0.000213	171/195,000[c] 45,000[i] 15,000[d,j]	12.9/very weak acid	2.1	16 (lab)	190	157/not available	Selective, systemic with contact and residual action, photosystem II inhibitor
Propazine $C_9H_{16}ClN_5$	6-chloro-*N*2,*N*4-diisopropyl-1,3,5-triazine-2,4-diamine	–	0.004	8.6/ 6200[d] 5000[s]	1.7/very weak base	3.95	135 (lab)	154	62/not available	Selective, systemic, absorbed by roots and translocated
Sebuthylazine $C_9H_{16}ClN_5$	(*RS*)-*N*2-sec-butyl-6-chloro-*N*4-ethyl-1,3,5-triazine-2,4-diamine	–	–	–	–	–	–	–	–	Selective, systemic, absorbed by roots and translocated
Simazine $C_7H_{12}ClN_5$	6-chloro-*N*2,*N*4-diethyl-1,3,5-triazine-2,4-diamine	–	0.00081	5/ 570[i] 1500[c] 130[k] 3.1[j]	1.62/very weak base	2.3	90 (field)	130	221/not available	Selective, systemic, absorbed through roots and foliage and translocated, inhibits photosynthesis (photosystem II)

Terbuthylazine $C_9H_{16}ClN_5$	*N*2-tert-butyl-6-chloro-*N*4-ethyl-1,3,5-triazine-2,4-diamine	Neutral solution: 222 nm = 38,696, 262 nm = 3291; Acidic solution: 222 nm = 3291; Basic solution: 222 nm = 38,191, 262 nm = 3241	0.12	6.6/ 41,000[c] 9800[k] 12,000[p] 410[j]	1.9	3.4	22.4 (field)	–	34/0.8	Broad-spectrum with strong and rapid effects
Methoxytriazine Herbicides										
Secbumeton $C_{10}H_{19}N_5O$	(*RS*)-*N*2-sec-butyl-*N*4-ethyl-6-methoxy-1,3,5-triazine-2,4-diamine	–	0.97	600/40,000[c] 60,000[g] 59,000[l] 350,000[k]	4.4/weak acid	3.64	60 (typical)	150	–	Inhibition of photosynthesis at photosystem II, absorbed by roots and leaves with limited translocation
Terbumeton $C_{10}H_{19}N_5O$	*N*2-tert-butyl-*N*4-ethyl-6-methoxy-1,3,5-triazine-2,4-diamine	–	0.27	130/130,000[c] 110,000[k] 220,000[l] 90,000[p]	–	3.04	300 (typical)	295	–	Selective, absorbed through leaves and roots
Methylthiotriazine Herbicides										
Ametryn $C_9H_{17}N_5S$	*N*2-ethyl-*N*4-isopropyl-6-methylthio-1,3,5-triazine-2,4-diamine	–	0.365	200/56,900[c] 4600[k] 1400[j]	10.07/very weak acid	2.63	37 (field)	316	33/not available	Selective, systemic absorbed through foliage and roots, inhibits photosynthesis (photosystem II)
Aziprotryne $C_7H_{11}N_7S$	4-azido-*N*-isopropyl-6-methylthio-1,3,5-triazin-2-amine	–	1.65×10^{-05}	55/–	–	3	5 (typical)	294	–	Selective with residual activity

(Continued)

TABLE 2.16 (CONTINUED)
Chemical Names and Properties of Triazine Herbicides

Common Name/ Chemical Formula	IUPAC Name	Maximum UV-Vis Absorption (L mol^{-1} cm^{-1})	Vapor Pressure (mPa) at 25°C	Solubility in Water[a] (mg L^{-1})/ Organic Solvents[a] (mg L^{-1})	Dissociation Constant (pK_a) at 25°C	Octanol–Water Partition Coefficient (log P)[a,b]	Soil Degradation (days) (Aerobic) DT_{50}	K_{oc} (Linear)	Bioconcentration Factor BCF/CT_{50} (days)	Mode of Action
Desmetryn $C_8H_{15}N_5S$	*N*2-isopropyl-*N*4-methyl-6-methylthio-1,3,5-triazine-2,4-diamine	–	1.30 × 10^{-01}	580/–	4/weak acid	2.38	9 (lab)	150	21/not available	Selective, absorbed by leaves and roots, photosynthetic electron transport inhibitor
Methoprotryne $C_{11}H_{21}N_5OS$	*N*2-isopropyl-*N*4-(3-methoxypropyl)-6-methylthio-1,3,5-triazine-2,4-diamine	–	0.038	320/450,000[c] 650,000[g] 380,000[k] 5000[j]	–	2.82	–	–	Low risk/–	Selective, absorbed through roots and foliage and translocated, interferes with biosynthesis
Prometryn $C_{10}H_{19}N_5S$	*N*2,*N*4-diisopropyl-6-methylthio-1,3,5-triazine-2,4-diamine	–	0.13	33/240,000[c] 160,000[l] 170,000[k] 5500[j]	4.1/weak base, pK_b = 9.95	3.34	41 (lab)	400	85/not available	A selective, systemic, contact and residual triazine, a photosynthetic electron transport inhibitor at the photosystem II receptor site
Terbutryn $C_{10}H_{19}N_5S$	*N*2-tert-butyl-*N*4-ethyl-6-methylthio-1,3,5-triazine-2,4-diamine	–	0.13	25/220,000[c,l] 130,000[p] 9000[j]	4.3/weak base	3.66	52 (field)	2432	72.4/0.5	Selective, absorbed through roots and foliage and translocated

				Triazinone Herbicides						
Hexazinone $C_{12}H_{20}N_4O_2$	3-cyclohexyl-6-dimethylamino-1-methyl-1,3,5-triazine-2,4(1*H*,3*H*)-dione	–	0.03	33,000/626,000[c] 2,146,500[l] 334,000[k] 837,000[d]	2.2/weak base	1.17	90 (lab)	54	7/not available	Nonselective with contact action, absorbed through the roots and foliage of plants, inhibits photosynthesis (photosystem II)
Metamitron $C_{10}H_{10}N_4O$	4-amino-4,5-dihydro-3-methyl-6-phenyl-1,2,4-triazin-5-one	311.1 nm = 11,789	7.44×10^{-04}	1770/37,000[c] 33,000[g] 20,000[h] 2000[e]	No dissociation	0.85	11.1 (field)	77.7	75/not available	Selective, systemic, absorbed mainly by roots and translocated, inhibits photosynthesis (photosystem II)
Metribuzin $C_8H_{14}N_4OS$	4-amino-6-tert-butyl-4,5-dihydro-3-methylthio-1,2,4-triazin-5-one	294 nm = 8175 (mean)	0.121	1165/449,400[c] 250,000[h] 60,000[e] 820[n]	0.99/strong acid, $pK_b = 13$	1.65	19 (field)	–	10/not available	Selective, systemic with contact and residual activity, inhibits photosynthesis (photosystem II)

Source: University of Hertfordshire, The Pesticide Properties DataBase (PPDB) developed by the Agriculture & Environment Research Unit (AERU), University of Hertfordshire, 2006–2013, 2015. Available at http://sitem.herts.ac.uk/aeru/ppdb/en/atoz.htm.

Note: a: at 20°C; b: at pH 7; c: acetone; d: benzene; e: xylene; f: cyclohexanone; g: dichloromethane; h: ethyl acetate; i: ethanol; j: *n*-hexane; k: toluene; l: methanol; m: trichloroethane; n: *n*-heptane; o: isopropanol; p: *n*-octanol; r: tetrahydrofuran; s: diethyl ether.

TABLE 2.17
Chemical Names and Properties of Phenylurea Herbicides

Common Name/ Chemical Formula	IUPAC Name	Maximum UV-Vis Absorption (L mol^{-1} cm^{-1})	Vapor Pressure (mPa) at 25°C	Solubility in Water[a] (mg L^{-1})/ Organic Solvents[a] (mg L^{-1})	Dissociation Constant (pK_a) at 25°C	Octanol–Water Partition Coefficient (log *P*)[a,b]	Soil Degradation (days) (Aerobic) DT_{50}	K_{oc} (Linear)	Bioconcentration Factor BCF/CT_{50} (days)	Mode of Action
Chlorbromuron $C_9H_{10}BrClN_2O_2$	3-(4-bromo-3-chlorophenyl)-1-methoxy-1-methylurea	–	0.053	35/ 460,000[c] 170,000[g] 89,000[j] 72,000[d]	–	3.09	36.5 (field)	470	68/not available	Absorbed via roots and translocated
Chlorotoluron $C_{10}H_{13}ClN_2O$	3-(3-chloro-p-tolyl)-1,1-dimethylurea	Maxima at 241–242 nm = 19,516	0.005	74/ 54,000[c] 48,000[i] 21,000[h] 3000[k]	No dissociation	2.5	34 (field)	196	Low risk/–	Selective, nonsystemic absorbed by roots and foliage, acts by the inhibition of photosynthetic electron transport
Chloroxuron $C_{15}H_{15}ClN_2O_2$	3-[4-(4-chlorophenoxy) phenyl]-1,1-dimethylurea	–	2.30×10^{-04}	3.7/ 106,000[g] 44,000[c]	–	3.4	60 (field)	2820	105/not available	Growth inhibition and chlorotic and necrotic effects on foliage
Diuron $C_9H_{10}Cl_2N_2O$	3-(3,4-dichlorophenyl)-1,1-dimethylurea	Maxima at 250.2 nm, tail at 290 nm	1.15×10^{-03}	35.6/ 53,600[c] 21,200[h] 14,400[t] 1330[e]	No dissociation	2.87	89 (field)	813	9.45/not available	Systemic, absorbed via roots, acts by strongly inhibiting photosynthesis

Fenuron $C_9H_{12}N_2O$	1,1-dimethyl-3-phenylurea	–	5	3850	–	0.98	60 (lab)	42	6/not available	Inhibition of photosynthesis
Fluometuron $C_{10}H_{11}F_3N_2O$	1,1-dimethyl-3-(α,α,α-trifluoro-m-tolyl)urea	Neutral solution: 243 nm = 17,300, 279 nm = 1230; Acidic solution: 243 nm = 17,700, 279 nm = 1390; Basic solution: 243 nm = 17,400, 279 nm = 1360	0.125	111/144,000[c] 109,000[l] 20,600[p] 1980[e]	No dissociation	2.28	89.8 (field)	–	40.4/not available	Selective, inhibiting photosynthesis
Isoproturon $C_{12}H_{18}N_2O$	3-(4-isopropylphenyl)-1,1-dimethylurea	207.8 nm = 32,512, 241.5 nm = 1972, 295 nm = 550	5.50×10^{-03}	70.2/ 46,000[l] 30,000[c] 2000[e] 100[n]	No dissociation	2.5	23 (field)	–	177/not available	Selective, systemic absorbed by roots and leaves, inhibits photosynthesis (photosystem II)
Linuron $C_9H_{10}Cl_2N_2O_2$	3-(3,4-dichlorophenyl)-1-methoxy-1-methylurea	211 nm	5.1	63.8/ 395,000[c] 292,000[h] 170,000[l] 75,000[k]	No dissociation	3	48 (field)	739	49/8	Selective, systemic with contact and residual action, inhibits photosynthesis (photosystem II)

Source: University of Hertfordshire, The Pesticide Properties DataBase (PPDB) developed by the Agriculture & Environment Research Unit (AERU), University of Hertfordshire, 2006–2013, 2015. Available at http://sitem.herts.ac.uk/aeru/ppdb/en/atoz.htm.

Note: a: at 20°C; b: at pH 7; c: acetone; d: benzene; e: xylene; f: cyclohexanone; g: dichloromethane; h: ethyl acetate; i: ethanol; j: *n*-hexane; k: toluene; l: methanol; m: trichloroethane; n: *n*-heptane; o: isopropanol; p: *n*-octanol; r: tetrahydrofuran; s: diethyl ether; t: 1,2-dichloroethane.

TABLE 2.18
Chemical Names and Properties of Sulfonylurea Herbicides

Common Name/ Chemical Formula	IUPAC Name	Maximum UV-Vis Absorption (L mol^{-1} cm^{-1})	Vapor Pressure (mPa) at 25°C	Solubility in Water[a] (mg L^{-1})/ Organic Solvents[a] (mg L^{-1})	Dissociation Constant (pK_a) at 25°C	Octanol–Water Partition Coefficient (log *P*)[a,b]	Soil Degradation (days) (Aerobic) DT_{50}	K_{oc} (Linear)	Bioconcentration Factor BCF/CT_{50} (days)	Mode of Action
			Pyrimidinylsulfonylurea Herbicides							
Amidosulfuron $C_9H_{15}N_5O_7S_2$	1-(4,6-dimethoxypyrimidin-2-yl)-3-mesyl(methyl)sulfamoylurea	Neutral solution: 201 nm = 31,649, 241 nm = 14,938, 291 nm = 10; Acidic solution: 201 nm = 33,226, 241 nm = 13,978, 291 nm = 20; Basic solution: 241 nm = 22,442, 291 nm = 12	0.013	3070/ 3000[h] 256[c,k] 1[j]	3.58/weak acid	−1.56	16.6 (lab)	29.3	4.85/not available	Selective, systemic absorbed through leaves and roots, inhibits plant amino acid synthesis
Azimsulfuron $C_{13}H_{16}N_{10}O_5S$	1-(4,6-dimethoxypyrimidin-2-yl)-3-[1-methyl-4-(2-methyl-2*H*-tetrazol-5-yl) pyrazol-5-ylsulfonyl]urea	242 nm = 23,014, 23,988, 24,099	4.00×10^{-06}	1050/ 26,400[c] 13,000[h] 2100[l] 1800[k]	3.6/weak acid	−1.4	3.5 (field)	73.8	Low risk/–	Selective, absorbed by foliage, ALS (AHAS) inhibitor
Bensulfuron $C_{15}H_{16}N_4O_7S$	α-[(4,6-dimethoxypyrimidin-2-ylcarbamoyl) sulfamoyl]-o-toluic acid	–	–	–/–	–	2.08	26.2 (lab)	–	Low risk/–	Selective, systemic action being absorbed through foliage and roots, inhibits plant amino acid synthesis
Ethoxysulfuron $C_{15}H_{18}N_4O_7S$	1-(4,6-dimethoxypyrimidin-2-yl)-3-(2-ethoxyphenoxysulfonyl)urea	194 nm = 74,000	0.066	5000/ 36,000[c] 7700[l] 2500[k] 6[j]	5.28/weak acid	1.01	17.5 (field)	134	Low risk/–	Selective, inhibits plant cell growth, ALS inhibitor

Flazasulfuron $C_{13}H_{12}F_3N_5O_5S$	1-(4,6-dimethoxypyrimidin-2-yl)-3-(3-trifluoromethyl-2-pyridylsulfonyl)urea	241 nm = 17,300	0.0133	2100/ 22,100[g] 6900[h] 560[k] 0.5[j]	4.37/weak acid	−0.06	10 (field)	46	Low risk/–	Systemic, absorbed through leaves, inhibition of acetolactate synthase ALS
Foramsulfuron $C_{17}H_{20}N_6O_7S$	1-(4,6-dimethoxypyrimidin-2-yl)-3-[2-(dimethylcarbamoyl)-5-formamidophenylsulfonyl]urea	291 nm = 3300, 252 nm = 33,300, 219 nm = 31,900	4.20×10^{-09}	3293/ 1925[c] 1660[l] 362[h] 10[n]	4.6/weak acid	−0.78	5.5 (lab)	–	Low risk/–	Acetolactate synthase (ALS) inhibitor stunting growth and causing death
Halosulfuron-methyl $C_{13}H_{15}ClN_6O_7S$	Methyl 3-chloro-5-(4,6-dimethoxypyrimidin-2-ylcarbamoylsulfamoyl)-1-methylpyrazole-4-carboxylate	Neutral solution: 245 nm = 20,347, 203 nm = 34,896; Acidic solution: 244 nm = 19,570, 203 nm = 39,329; Basic solution: 260 nm = 11,492, 233 nm = 15,312, 215.5 nm = 14,869	3.5×10^{-02}	10.2/ 15,260[h] 3640[k] 1616[l] 127.8[j]	3.44/weak acid	−0.02	14 (field)	109	Low risk/–	Systemic, selective, acts by inhibiting biosynthesis of essential amino acids valine and isoleucine restricting plant growth
Nicosulfuron $C_{15}H_{18}N_6O_6S$	2-[(4,6-dimethoxypyrimidin-2-ylcarbamoyl) sulfamoyl]-*N*,*N*-dimethylnicotinamide	Neutral solution: 244 nm = 23,800; Acidic solution: 241 nm = 19,200; Basic solution: 244 nm = 23,800	8.00×10^{-07}	7500/ 21,300[g] 8900[c] 2400[h] 400[l]	4.78/weak acid; pK_a (2) 7.58	0.61	19.3 (field)	30	Low risk/–	Selective, systemic absorbed by foliage and roots and translocated, acetolactate synthase (ALS) inhibitor
Oxasulfuron $C_{17}H_{18}N_4O_6S$	Oxetan-3-yl 2-[(4,6-dimethylpyrimidin-2-yl)carbamoylsulfamoyl]benzoate	232.7 nm = 23,719, no absorption between 300 nm and 900 nm	0.002	1700/ 9300[c] 1500[l] 320[k] 2.2[n]	5.1/weak acid	−0.81	6 (field)	85	Low risk/–	Absorbed by shoots and roots and translocated, ALS inhibitor
Rimsulfuron $C_{14}H_{17}N_5O_7S_2$	1-(4,6-dimethoxypyrimidin-2-yl)-3-(3-ethylsulfonyl-2-pyridylsulfonyl)urea	pH 5: 240 nm = 22,400, 290 nm = 481; pH 2.1: 290 nm = 203; pH 1.8: 230 nm = 17,800, 290 nm = 181	8.90×10^{-04}	7300/ 14,800[c] 35,500[g] 2850[h] 1550[l]	4/weak acid	−1.46	10.8 (field)	50.3	Low risk/–	Selective, systemic, absorbed through foliage and roots and translocated, acetolactate synthase (ALS) inhibitor

(Continued)

TABLE 2.18 (CONTINUED)
Chemical Names and Properties of Sulfonylurea Herbicides

Common Name/ Chemical Formula	IUPAC Name	Maximum UV-Vis Absorption (L mol^{-1} cm^{-1})	Vapor Pressure (mPa) at 25°C	Solubility in Water[a] (mg L^{-1})/ Organic Solvents[a] (mg L^{-1})	Dissociation Constant (pK_a) at 25°C	Octanol–Water Partition Coefficient (log P)[a,b]	Soil Degradation (days) (Aerobic) DT_{50}	K_{oc} (Linear)	Bioconcentration Factor BCF/CT_{50} (days)	Mode of Action
Sulfosulfuron $C_{16}H_{18}N_6O_7S_2$	1-(4,6-dimethoxypyrimidin-2-yl)-3-(2-ethylsulfonylimidazo[1,2-a]pyridin-3-ylsulfonyl)urea	Maxima at 208 nm, spectrum extends to 320 nm, 300 nm = 4169, 312 nm = 2188	3.05 × 10^{-05}	1627/ 710[c] 330[l] 160[e] 1[n]	3.51/weak acid	−0.77	24 (field)	33.0	Low risk/–	Systemic, absorbed by roots and leaves and translocated, ALS inhibitor
Triazinylsulfonylurea Herbicides										
Chlorsulfuron $C_{12}H_{12}ClN_5O_4S$	1-(2-chlorophenylsulfonyl)-3-(4-methoxy-6-methyl-1,3,5-triazin-2-yl)urea	Neutral solution: 236 nm = 26,399, 204 nm = 31,320, 236 nm = 26,284; Acidic solution pH 2: 202 nm = 42,552, 205 nm = 32,941, 222 nm = 23,968; Basic solution pH 10: 203 nm = 33,752, 236 nm = 26,399	3.07 × 10^{-06}	12,500/ 140,000[g] 37,000[c] 15,000[l] 2800[k]	3.4/weak acid	−0.99	36.2 (field)	–	20/not available	Selective, systemic, absorbed by roots and foliage, acts by inhibiting cell division, ALS inhibitor

Ethametsulfuron-methyl $C_{15}H_{18}N_6O_6S$	Methyl 2-[(4-ethoxy-6-methylamino-1,3,5-triazin-2-yl)carbamoylsulfamoyl]benzoate	Acid solution: 220 nm = 41,432, no significant absorption > 290 nm Neutral solution: 225 nm = 41,432, no significant absorption > 290 nm Basic solution: 225 nm: 41,187, no significant absorption > 290 nm	6.41×10^{-04}	223/ 2066[g] 764[c] 173[h] 3.0[j]	4.2/weak acid	−0.23	22.6 (field)	220.7	Low risk/–	Selective, inhibiting plant growth, ALS inhibitor
Iodosulfuron $C_{13}H_{12}IN_5O_6S$	4-iodo-2-[(4-methoxy-6-methyl-1,3,5-triazin-2-yl)carbamoylsulfamoyl]benzoic acid	–	–	–/–	–	–	–	–	–	Selective to cereals, acetolactate synthase (ALS) inhibitor
Metsulfuron $C_{13}H_{13}N_5O_6S$	2-(4-methoxy-6-methyl-1,3,5-triazin-2-ylcarbamoylsulfamoyl)benzoic acid	–	3.99×10^{-08}	172/–	–	1.7	–	–	Low risk/–	Selective, systemic with contact and residual action, inhibits plant amino acid synthesis
Prosulfuron $C_{15}H_{16}F_3N_5O_4S$	1-(4-methoxy-6-methyl-1,3,5-triazin-2-yl)-3-[2-(3,3,3-trifluoropropyl)phenylsulfonyl] urea	227.5 nm = 21,645, shoulder at 250 nm, no absorption > 290 nm	3.50×10^{-03}	4000/ 160,000[c] 56,000[h] 8400[i] 6100[k]	3.76/weak acid	1.5	16 (field)	–	0.13/not available	Absorbed by leaves and roots, acetolactate synthase (ALS) inhibitor

(Continued)

TABLE 2.18 (CONTINUED)
Chemical Names and Properties of Sulfonylurea Herbicides

Common Name/ Chemical Formula	IUPAC Name	Maximum UV-Vis Absorption (L mol^{-1} cm^{-1})	Vapor Pressure (mPa) at 25°C	Solubility in Water[a] (mg L^{-1})/ Organic Solvents[a] (mg L^{-1})	Dissociation Constant (pK_a) at 25°C	Octanol–Water Partition Coefficient (log P)[a,b]	Soil Degradation (days) (Aerobic) DT_{50}	K_{oc} (Linear)	Bioconcentration Factor BCF/CT_{50} (days)	Mode of Action
Thifensulfuron $C_{11}H_{11}N_5O_6S_2$	3-(4-methoxy-6-methyl-1,3,5-triazin-2-ylcarbamoylsulfamoyl)thiophene-2-carboxylic acid	–	–	–/–	–	–	29 (field)	–	–	Selective, absorbed through foliage, acetolactate synthase (ALS) inhibitor
Triasulfuron $C_{14}H_{16}ClN_5O_5S$	1-[2-(2-chloroethoxy)phenylsulfonyl]-3-(4-methoxy-6-methyl-1,3,5-triazin-2-yl)urea	223.4 nm = 26,051, 282.8 nm = 3415, shoulder between 240 and 250 nm, no absorption after 340 nm	0.0021	815/ 14,000[c] 4300[h] 300[k] 40[j]	4.64/weak acid	−0.59	19 (field)	60	1.3/not available	Selective, absorbed by leaves and roots and translocated, ALS inhibitor
Tribenuron $C_{14}H_{15}N_5O_6S$	2-[4-methoxy-6-methyl-1,3,5-triazin-2-yl(methyl)carbamoylsulfamoyl]benzoic acid	–	–	–/–	–	–	–	–	–	Selective, absorbed through foliage, acetolactate synthase (ALS) inhibitor
Triflusulfuron $C_{16}H_{17}F_3N_6O_6S$	2-[4-dimethylamino-6-(2,2,2-trifluoroethoxy)-1,3,5-triazin-2-ylcarbamoylsulfamoyl]-m-toluic acid	–	–	1.0/–	4.4/weak acid	3.1	–	–	–	Selective, inhibits amino acid synthesis

Source: University of Hertfordshire, The Pesticide Properties DataBase (PPDB) developed by the Agriculture & Environment Research Unit (AERU), University of Hertfordshire, 2006–2013, 2015. Available at http://sitem.herts.ac.uk/aeru/ppdb/en/atoz.htm.

Note: a: at 20°C; b: at pH 7; c acetone; d: benzene; e: xylene; f: cyclohexanone; g: dichloromethane; h: ethyl acetate; i: ethanol; j: *n*-hexane; k: toluene; l: methanol; m: trichloroethane; n: *n*-heptane; o: isopropanol; p: *n*-octanol; r: tetrahydrofuran; s: diethyl ether; t: 1,2-dichloroethane.

TABLE 2.19
Chemical Names and Properties of Pyridazine and Pyridazinone Herbicides

Common Name/Chemical Formula	IUPAC Name	Maximum UV-Vis Absorption (L mol^{-1} cm^{-1})	Vapor Pressure (mPa) at 25°C	Solubility in Water[a] (mg L^{-1})/Organic Solvents[a] (mg L^{-1})	Dissociation Constant (pK_a) at 25°C	Octanol–Water Partition Coefficient (log P)[a,b]	Soil Degradation (days) (Aerobic) DT_{50}	K_{oc} (Linear)	Bioconcentration Factor BCF/CT_{50} (days)	Mode of Action
Chloridazon $C_{10}H_8ClN_3O$	5-amino-4-chloro-2-phenylpyridazin-3(2*H*)-one	210 nm = 18,577, 229 nm = 25,043, 286 nm = 10,088	1.0×10^{-06}	422/ 15,100[l] 3700[h] 190[g] 100[k]	3.38/very weak acid	1.19	34.7 (field)	120	12/not available	Selective, systemic
Pyridate $C_{19}H_{23}ClN_2O_2S$	*O*-6-chloro-3-phenylpyridazin-4-yl *S*-octyl thiocarbonate	–	0.000998	1.49/ 9,000,000[c,e,h]	No dissociation	0.5	5 (field)	–	116/not available	Selective with contact action, absorbed mainly by the leaves

Source: University of Hertfordshire, The Pesticide Properties DataBase (PPDB) developed by the Agriculture & Environment Research Unit (AERU), University of Hertfordshire, 2006–2013, 2015. Available at http://sitem.herts.ac.uk/aeru/ppdb/en/atoz.htm.

Note: a: at 20°C; b: at pH 7; c: acetone; d: benzene; e: xylene; f: cyclohexanone; g: dichloromethane; h: ethyl acetate; i: ethanol; j: *n*-hexane; k: toluene; l: methanol; m: trichloroethane; n: *n*-heptane; o: isopropanol; p: *n*-octanol; r: tetrahydrofuran.

TABLE 2.20
Chemical Names and Properties of Conazole Fungicides

Common Name/ Chemical Formula	IUPAC Name	Maximum UV-Vis Absorption (L mol^{-1} cm^{-1})	Vapor Pressure (mPa) at 25°C	Solubility in Water[a] (mg L^{-1})/ Organic Solvents[a] (mg L^{-1})	Dissociation Constant (pK_a) at 25°C	Octanol–Water Partition Coefficient (log P)[a,b]	Soil Degradation (days) (Aerobic) DT_{50}	K_{oc} (Linear)	Bioconcentration Factor BCF/CT_{50} (days)	Mode of Action
				Triazoles						
Bromuconazole $C_{13}H_{12}BrCl_2N_3O$	1-[(2*RS*,4*RS*;2*RS*,4*SR*)-4-bromo-2-(2,4-dichlorophenyl) tetrahydrofurfuryl]-1*H*-1,2,4-triazole	Neutral solution: 202.5 nm = 43,936, 220 nm = 10,762; Acidic solution: 202.5 nm = 38,527, 220 nm = 11,157; Basic solution: 221.5 nm = 9915	0.004	48.3/ 187,000[k] 50,000[p] 1790[j] 269.2[c]	2.75/very strong acid, pK_a (2) −4.02	3.24	123 (field)	872	131/not available	Systemic, sterol biosynthesis inhibitor
Cyproconazole $C_{15}H_{18}ClN_3O$	(2*RS*,3*RS*;2*RS*,3*SR*)-2-(4-chlorophenyl)-3-cyclopropyl-1-(1*H*-1,2,4-triazol-1-yl)butan-2-ol	Neutral solution: 297 nm = 0.7; Acidic solution: 295 nm = 0.4; Basic solution: 295 nm = 0.8	0.026	93/ 410,000[l] 360,000[c] 240,000[h] 1300[j]	No dissociation	3.09	129 (field)	–	28/1	Systemic with protective, curative and eradicant action, disrupts membrane function, an ergosterol-biosynthesis inhibitor

Difenoconazole $C_{19}H_{17}Cl_2N_3O_3$	3-chloro-4-[(2*RS*,4*RS*;2*RS*,4*SR*)-4-methyl-2-(1*H*-1,2,4-triazol-1-ylmethyl)-1,3-dioxolan-2-yl]phenyl 4-chlorophenyl ether	Acid: 215 nm = 29,306, 235 nm = 17,556, 275 nm = 1743; Neutral: 215 nm = 28,658, 235 nm = 17,392, 275 nm = 1680; Alkaline: 220 nm = 21,210, 235 nm = 17,176, 275 nm = 1542	3.33×10^{-05}	15.0/ 610,000[c] 500,000[k] 330,00[i] 3400[j]	1.07/strong acid	4.36	85 (field)	–	330/1.0	Systemic with preventative and curative action, disrupts membrane function
Epoxiconazole $C_{17}H_{13}ClFN_3O$	(2*RS*,3*SR*)-1-[3-(2-chlorophenyl)-2,3-epoxy-2-(4-fluorophenyl)propyl]-1*H*-1,2,4-triazole	204 nm = 32,000, 263 nm = 390	1.00×10^{-02}	7.1/ 140,000[c] 100,000[h] 40,000[k] 28,800[i]	No dissociation	3.3	120 (field)	–	70/0.72	Preventative and curative action
Fenbuconazole $C_{19}H_{17}ClN_4$	4-(4-chlorophenyl)-2-phenyl-2-(1*H*-1,2,4-triazol-1-ylmethyl)butyronitrile	Neutral solution: 195 nm = 35,600; Acidic solution: 200 nm = 20,600; Basic solution: 217 nm = 12,100	3.4×10^{-04}	2.47/ 250,000[c] 132,000[h] 26,000[e] 680[n]	No dissociation	3.79	61 (field)	–	160/1.4	Systemic protectant and curative, acts by inhibiting sterol biosynthesis in fungi
Flusilazole $C_{16}H_{15}F_2N_3Si$	bis(4-fluorophenyl)(methyl)(1*H*-1,2,4-triazol-1-ylmethyl)silane	Neutral solution: Maxima at 206 nm; Acidic and basic solutions: Maxima at 202 nm	0.0387	41.9/ 200,000[e,h] 85,000[j]	2.5/very weak base	3.87	94 (field)	1664	250/not available	Broad-spectrum, systemic with protective and curative action

(*Continued*)

TABLE 2.20 (CONTINUED)
Chemical Names and Properties of Conazole Fungicides

Common Name/ Chemical Formula	IUPAC Name	Maximum UV-Vis Absorption (L mol^{-1} cm^{-1})	Vapor Pressure (mPa) at 25°C	Solubility in Water[a] (mg L^{-1})/ Organic Solvents[a] (mg L^{-1})	Dissociation Constant (pK_a) at 25°C	Octanol–Water Partition Coefficient (log P)[a,b]	Soil Degradation (days) (Aerobic) DT_{50}	K_{oc} (Linear)	Bioconcentration Factor BCF/CT_{50} (days)	Mode of Action
Flutriafol $C_{16}H_{13}F_2N_3O$	(*RS*)-2,4′-difluoro-α-(1*H*-1,2,4-triazol-1-ylmethyl) benzhydryl alcohol	No absorption above 290 nm	4.0 × 10^{-04}	95.0/ 116,000[c] 115,000[l] 10,000[e] 300[j]	2.3/strong acid	2.3	860 (field)	–	Low risk/–	Broad-spectrum, systemic, contact action with eradicant and protective properties
Ipconazole $C_{18}H_{24}ClN_3O$	(1*RS*,2*SR*,5*RS*;1*RS*,2*SR*,5*SR*)-2-(4-chlorobenzyl)-5-isopropyl-1-(1*H*-1,2,4-triazol-1-ylmethyl)cyclopentanol	Neutral: 276 nm = 315 Acidic: 276 nm = 304 Basic: 276 nm = 312	0.003	9.34/ 679,000[l] 570,000[c] 428,000[h] 425,000[t]	–5.43	4.3	131 (field)	–	350/not available	Systemic, broad-spectrum, inhibits sterol synthesis in fungi
Metconazole $C_{17}H_{22}ClN_3O$	(1*RS*,5*RS*;1*RS*,5*SR*)-5-(4-chlorobenzyl)-2,2-dimethyl-1-(1*H*-1,2,4-triazol-1-ylmethyl)cyclopentanol	196 nm = 17,700, 221 nm = 5900, 226 nm (shoulder) = 4600, 262 nm = 150, 268 nm = 190	2.10 × 10^{-05}	30.4/ 403,000[l] 363,000[c] 103,000[k] 1400[j]	11.38/pK_a(2) 1.08	3.85	265 (field)	–	129/1	Systemic, ergosterol biosynthesis inhibitor

Myclobutanil $C_{15}H_{17}ClN_4$	(*RS*)-2-(4-chlorophenyl)-2-(1*H*-1,2,4-triazol-1-ylmethyl)hexanenitrile	203 nm = 16,400, 219 nm = 17,900, 267 nm = 500, 273 nm = 500, 290 nm = 0	0.198	132/270,000[e] 250,000[i] 1020[n]	2.3/strong acid	2.89	35 (field)	–	Low risk/–	Broad spectrum, systemic with protective, eradicative, and curative action, disrupts membrane function by inhibiting sterol biosynthesis
Penconazole $C_{13}H_{15}Cl_2N_3$	(*RS*)-1-[2-(2,4-dichlorophenyl)pentyl]-1*H*-1,2,4-triazole	Neutral solution: 220 nm = 10,564, 273 nm = 437, 281 nm = 401; Acidic solution: 220 nm = 10,741, 273 nm = 410, 281 nm = 376; Basic solution: 224 nm = 9607, 273 nm = 453, 281 nm = 417; No absorption maximum above 290 nm	0.366	73/ 500,000[c,g,k] 24,000[j]	1.51/very weak base	3.72	90 (field)	–	320/3	Systemic with curative and protective action, acts by interfering with ergosterol biosynthesis
Propiconazole $C_{15}H_{17}Cl_2N_3O_2$	(2*RS*,4*RS*;2*RS*,4*SR*)-1-[2-(2,4-dichlorophenyl)-4-propyl-1,3-dioxolan-2-ylmethyl] 1*H*-1,2,4-triazole	220 nm = 11,666	0.056	150/ 1585[n] Miscible[c,e,l]	1.09/very weak base	3.72	214 (field)	1086	116/8	Systemic with curative and protective action, works via the demethylation of C-14 during ergosterol biosynthesis

(*Continued*)

TABLE 2.20 (CONTINUED)
Chemical Names and Properties of Conazole Fungicides

Common Name/ Chemical Formula	IUPAC Name	Maximum UV-Vis Absorption ($L\ mol^{-1}\ cm^{-1}$)	Vapor Pressure (mPa) at 25°C	Solubility in Water[a] ($mg\ L^{-1}$)/ Organic Solvents[a] ($mg\ L^{-1}$)	Dissociation Constant (pK_a) at 25°C	Octanol–Water Partition Coefficient ($\log P$)[a,b]	Soil Degradation (days) (Aerobic) DT_{50}	K_{oc} (Linear)	Bioconcentration Factor BCF/CT_{50} (days)	Mode of Action
Prothioconazole $C_{14}H_{15}Cl_2N_3OS$	(*RS*)-2-[2-(1-chlorocyclopropyl)-3-(2-chlorophenyl)-2-hydroxypropyl]-2,4-dihydro-1,2,4-triazole-3-thione	Acidic solution: maxima at 196 nm and 244 nm; Basic solution: maxima at 252 nm	0.0004	300/ 250,000[c,h] 8000[e] 100[n]	6.9/weak acid	3.82	1.6 (field)	–	18.8/not available	Systemic with protective, curative, and eradicative action, long-lasting activity
Tebuconazole $C_{16}H_{22}ClN_3O$	(*RS*)-1-p-chlorophenyl-4,4-dimethyl-3-(1*H*-1,2,4-triazol-1-ylmethyl)pentan-3-ol	Neutral solution: 221.4 nm = 11,980, 262.0 nm = 304, 268.5 nm = 408, 276.5 nm = 368, 290.0 nm < 10	1.30×10^{-03}	36/ 200,000[g] 96,000[p] 57,000[k] 80[j]	No data/very weak base	3.7	49.6 (field)	–	78/2	Systemic with protective, curative, and eradicant action, disrupts membrane function
Tetraconazole $C_{13}H_{11}Cl_2F_4N_3O$	(*RS*)-2-(2,4-dichlorophenyl)-3-(1*H*-1,2,4-triazol-1-yl) propyl 1,1,2,2-tetrafluoroethyl ether	281 nm < 310, No significant absorption at >290 nm	0.18	156.6/ 300,000[c,e,h] 15,000[j]	0.65/strong acid pK_a range 0.8–0.5	3.56	430 (field)	–	35.7/0.189	Systemic with protectant, eradicant, and curative properties
Tradimenol $C_{14}H_{18}ClN_3O_2$	(1*RS*,2*RS*;1*RS*,2*SR*)-1-(4-chlorophenoxy)-3,3-dimethyl-1-(1*H*-1,2,4-triazol-1-yl)butan-2-ol	–	0.0005	72/ 250,000[g] 140,000[o] 18,000[e] 450[n]	No dissociation	3.18	64.9 (field)	750	21/0.42	Selective with curative, protective, and eradicant action, disrupts membrane function

Triticonazole $C_{17}H_{20}ClN_3O$	(*RS*)-(*E*)-5-(4-chlorobenzylidene)-2,2-dimethyl-1-(1*H*-1,2,4-triazol-1-ylmethyl)cyclopentanol	Neutral solution: 212 nm = 23,879, 263 nm = 25,731	1.00×10^{-03}	9.3/ 74,500[c] 18,200[l] 12,600[k] 120[j]	No dissociation	3.29	161 (field)	374	94/3	Inhibits sterol demethylation
Imidazoles										
Imazalil $C_{14}H_{14}Cl_2N_2O$	(*RS*)-1-(β-allyloxy-2,4-dichlorophenylethyl)imidazole	At pH 4: 265 nm = 236, 272 nm = 311, 280 nm = 255; At pH 7: 265 nm = 246, 272 nm = 325, 280 nm = 268; At pH 9: 265 nm = 246, 272 nm = 329, 280 nm = 273	0.158	184/ 500,000[h,k,l] 19,000[j]	6.49/weak base	2.56	6.4 (field)	6.4	56.3/35.2	Systemic with curative and protective properties, disrupts membrane function
Triflumizole $C_{15}H_{15}ClF_3N_3O$	(*E*)-4-chloro-α,α,α-trifluoro-*N*-(1-imidazol-1-yl-2-propoxyethylidene)-o-toluidine	201.5 nm = 25,300; 236 nm = 26,400; 301 nm = 4910	0.191	10.5/ 1,486,000[h] 1,440,000[c] 496,000[l] 17,600[j]	3.7/weak base	4.77	–	1373	1417/5.8	Systemic with protective and curative action, inhibitors of chitin biosynthesis

Source: University of Hertfordshire, The Pesticide Properties DataBase (PPDB) developed by the Agriculture & Environment Research Unit (AERU), University of Hertfordshire, 2006–2013, 2015. Available at http://sitem.herts.ac.uk/aeru/ppdb/en/atoz.htm.

Note: a: at 20°C; b; at pH 7; c: acetone; d: benzene; e: xylene; f: cyclohexanone; g: dichloromethane; h: ethyl acetate; i: ethanol; j: *n*-hexane; k: toluene; l: methanol; m: trichloroethane; n: *n*-heptane; o: isopropanol; p: *n*-octanol; r: tetrahydrofuran; s: diethyl ether; t: 1,2-dichloroethane.

TABLE 2.21
Chemical Names and Properties of Dithiocarbamate Fungicides

Common Name/ Chemical Formula	IUPAC Name	Maximum UV-Vis Absorption (L mol^{-1} cm^{-1})	Vapor Pressure (mPa) at 25°C	Solubility in Water[a] (mg L^{-1})/ Organic Solvents[a] (mg L^{-1})	Dissociation Constant (pK_a) at 25°C	Octanol–Water Partition Coefficient (log P)[a,b]	Soil Degradation (days) (Aerobic) DT_{50}	K_{oc} (Linear)	Bioconcentration Factor BCF/CT_{50} (days)	Mode of Action
Mancozeb ($C_4H_6MnN_2S_4$) x(Zn)y	Manganese ethylenebis(dithiocarbamate) (polymeric) complex with zinc salt	–	0.013	6.2/ Insoluble	10.3/very weak acid	1.33	18 (field)	998	3.2/not available	Broad spectrum, nonsystemic, contact with protective action, acts by disrupting lipid metabolism
Maneb $C_4H_6MnN_2S_4$	Manganese ethylenebis(dithiocarbamate) (polymeric)	Maxima at 285 nm	0.014	178/ 10[c,h,n]	No dissociation	–0.45	7 (field)	2000	Low risk/–	Nonspecific with protective action
Metiram ($C_{16}H_{33}N_{11}S_{16}Zn_3$)x	Zinc ammoniate ethylenebis(dithiocarbamate) - poly(ethylenethiuram disulfide)	259 nm = 64,000, 281 nm = 48,000, 29 nm = 30,000	0.01	2/ 100[c,h,k,l]	No dissociation	1.76	7 (field)	500,000	3.2/not available	Broad spectrum, nonsystemic with protective action

Propineb $C_5H_8N_2S_4Zn$	Polymeric zinc propylenebis(dithiocarbamate)	–	0.16	10/ $100^{c,g,k,n}$	Not determinable	−0.26	3 (lab)	–	Low risk/–	Contact action with protective properties and long residual activity
Ziram $C_6H_{12}N_2S_4Zn$	Zinc bis(dimethyldithiocarbamate)	Neutral solution: 251 nm = 39,989, 270 nm = 31,586; Acidic solution: Maxima below UV cutoff point; Basic solution: 251 nm = 52,739	1.8×10^{-02}	$0.967/2300^{c}$ 1010^{h} 900^{e} 110^{l}	–	1.65	6.3 (field)	–	470/not available	Contact action with protective properties

Source: University of Hertfordshire, The Pesticide Properties DataBase (PPDB) developed by the Agriculture & Environment Research Unit (AERU), University of Hertfordshire, 2006–2013, 2015. Available at http://sitem.herts.ac.uk/ppdb/en/atoz.htm.

Note: a: at 20°C; b: at pH 7; c: acetone; d: benzene; e: xylene; f: cyclohexanone; g: dichloromethane; h: ethyl acetate; i: ethanol; j: *n*-hexane; k: toluene; l: methanol; m: trichloroethane; n: *n*-heptane; o: isopropanol; p: *n*-octanol; r: tetrahydrofuran; s: diethyl ether; t: 1,2-dichloroethane.

TABLE 2.22
Chemical Names and Properties of Morpholine Fungicides

Common Name/ Chemical Formula	IUPAC Name	Maximum UV-Vis Absorption (L mol^{-1} cm^{-1})	Vapor Pressure (mPa) at 25°C	Solubility in Water[a] (mg L^{-1})/ Organic Solvents[a] (mg L^{-1})	Dissociation Constant (pK_a) at 25°C	Octanol–Water Partition Coefficient (log P)[a,b]	Soil Degradation (days) (Aerobic) DT_{50}	K_{oc} (Linear)	Bioconcentration Factor BCF/CT_{50} (days)	Mode of Action
Dimethomorph $C_{21}H_{22}ClNO_4$	(*EZ*)-4-[3-(4-chlorophenyl)-3-(3,4-dimethoxyphenyl) acryloyl]morpholine	200 nm = 45,000, 205 nm = 30,000, 221 nm = 16,000, 242 nm = 20,000, 286 nm = 9100, 312 nm = 4500	9.85×10^{-04}	28.95/ 100,400[c] 49,500[k] 39,000[l] 112[j]	−1.3/very strong acid	2.68	44 (field)	–	Low risk/–	Systemic with good protective activity, lipid synthesis inhibitor
Dodemorph $C_{18}H_{35}NO$	4-cyclododecyl-2,6-dimethylmorpholine	–	0.48	100/ 1,000,000[u] 185,000[h] 57,000[c] 50,000[i]	8.08/weak acid	4.6	41 (lab)	–	–	Systemic with protective and curative action
Fenpropimorph $C_{20}H_{33}NO$	cis-4-[(*RS*)-3-(4-tert-butylphenyl)-2-methylpropyl]-2,6-dimethylmorpholine	203 nm = 11,000, 219 nm = 11,000, 242 nm = 210, 464 nm = 420, 270 nm = 320, 272 nm = 420, no absorption above 290 nm	3.9	4.32/ 7,892,000[l] 7,780,000[h] 7,646,000[s] 7,604,000[c]	6.98/weak acid	4.5	25.5 (field)	–	428/3.8	Systemic with protective and curative action, disrupts membrane function

Source: University of Hertfordshire, The Pesticide Properties DataBase (PPDB) developed by the Agriculture & Environment Research Unit (AERU), University of Hertfordshire, 2006–2013, 2015. Available at http://sitem.herts.ac.uk/aeru/ppdb/en/atoz.htm.

Note: a: at 20°C; b: at pH 7; c: acetone; d: benzene; e: xylene; f: cyclohexanone; g: dichloromethane; h: ethyl acetate; i: ethanol; j: *n*-hexane; k: toluene; l: methanol; m: trichloroethane; n: *n*-heptane; o: isopropanol; p: *n*-octanol; r: tetrahydrofuran; s: diethyl ether; t: 1,2-dichloroethane; u: chloroform.

TABLE 2.23
Chemical Names and Properties of Benzimidazole Fungicides

Common Name/ Chemical Formula	IUPAC Name	Maximum UV-Vis Absorption (L mol^{-1} cm^{-1})	Vapor Pressure (mPa) at 25°C	Solubility in Water[a] (mg L^{-1})/ Organic Solvents[a] (mg L^{-1})	Dissociation Constant (pK_a) at 25°C	Octanol–Water Partition Coefficient (log P)[a,b]	Soil Degradation (days) (Aerobic) DT$_{50}$	K_{oc} (Linear)	Bioconcentration Factor BCF/CT$_{50}$ (days)	Mode of Action
Carbendazim $C_9H_9N_3O_2$	Methyl benzimidazol-2-ylcarbamate	242.5–244 nm = 10,410, two smaller peaks at 279–280.5 nm and 285–288 nm = 14,670	0.09	8.0/ 300[i] 100[u] 135[h]	4.2/weak base	1.48	22 (field)	–	25/not available	Systemic with curative and protectant activity, inhibition of mitosis and cell division
Thiabendazole $C_{10}H_7N_3S$	2-(thiazol-4-yl) benzimidazole	254 nm and 302 nm	5.30×10^{-04}	30/ 8230[l] 2430[c] 130[e] 10[n]	4.73/pK_a(2) 12.00	2.39	724 (field)	7344	96.5/not available	Systemic with curative and protective action, acts by compromising the cytoskeleton through a selective interaction with ß-tubulin

Source: University of Hertfordshire, The Pesticide Properties DataBase (PPDB) developed by the Agriculture & Environment Research Unit (AERU), University of Hertfordshire, 2006–2013, 2015. Available at http://sitem.herts.ac.uk/aeru/ppdb/en/atoz.htm.

Note: a: at 20°C; b: at pH 7; c: acetone; d: benzene; e: xylene; f: cyclohexanone; g: dichloromethane; h: ethyl acetate; i: ethanol; j: *n*-hexane; k: toluene; l: methanol; m: trichloroethane; n: *n*-heptane; o: isopropanol; p: *n*-octanol; r: tetrahydrofuran; s: diethyl ether; t: 1,2-dichloroethane; u: chloroform.

TABLE 2.24
Chemical Names and Properties of Anilide Fungicides

Common Name/ Chemical Formula	IUPAC Name	Maximum UV-Vis Absorption (L mol^{-1} cm^{-1})	Vapor Pressure (mPa) at 25°C	Solubility in Water[a] (mg L^{-1})/ Organic Solvents[a] (mg L^{-1})	Dissociation Constant (pK_a) at 25°C	Octanol–Water Partition Coefficient (log P)[a,b]	Soil Segradation (days) (Aerobic) DT_{50}	K_{oc} (Linear)	Bioconcentration Factor BCF/CT_{50} (days)	Mode of Action
Benalaxyl $C_{20}H_{23}NO_3$	Methyl *N*-(phenylacetyl)-*N*-(2,6-xylyl)-*DL*-alaninate	203 nm, No absorption at 290 nm	0.572	28.6/250,000[c,h,l] 19,400[n]	No dissociation	3.54	54 (field)	4998	57/not available	Systemic with protective, curative, and eradicant action, disrupts fungal nucleic acid synthesis
Metalaxyl $C_{15}H_{21}NO_4$	Methyl *N*-(methoxyacetyl)-*N*-(2,6-xylyl)-*DL*-alaninate	–	0.75	7100/750,000[g] 650,000[l] 550,000[d] 9100[j]	0/very strong acid	1.65	46 (field)	–	7/not available	Systemic with curative and protective action, acts by suppressing sporangial formation, mycelial growth, and the establishment of new infections

Source: University of Hertfordshire, The Pesticide Properties DataBase (PPDB) developed by the Agriculture & Environment Research Unit (AERU), University of Hertfordshire, 2006–2013, 2015. Available at http://sitem.herts.ac.uk/aeru/ppdb/en/atoz.htm.

Note: a: at 20°C; b: at pH 7; c: acetone; d: benzene; e: xylene; f: cyclohexanone; g: dichloromethane; h: ethyl acetate; i: ethanol; j: *n*-hexane; k: toluene; l: methanol; m: trichloroethane; n: *n*-heptane; o: isopropanol; p: *n*-octanol; r: tetrahydrofuran; s: diethyl ether; t: 1,2-dichloroethane; u: chloroform.

example, a positive correlation to log K_{ow} values [generally ≤+2] indicates the likely systematic translocation of such pesticides or their metabolites in the plant transvascular system; values of log P +4 or higher are regarded as an indicator that a substance will bioaccumulate).

- Soil–water partition coefficient (adsorption coefficient): K_{oc} is the ratio (at equilibrium) of the mass of a substance that is adsorbed in the soil per unit mass of organic carbon in the soil. Its value is dependent on the organic matter content of soil, polarity of the pesticide, and soil pH. (K_{oc} values greater than 1000 indicate strong adsorption to soil; chemicals with lower K_{oc} values (less than 500) tend to move more with water than be adsorbed into sediment).
- Soil degradation (days): This is an indicator that can have values such as <30 = nonpersistent, 30–100 = moderately persistent, 100–365 = persistent, and >365 = very persistent.

Distribution coefficient (K_d) is an important parameter used to quantify the adsorption of pesticide molecules to soils. It is defined as the ratio of the sorbent phase concentration to the solution phase concentration at equilibrium (Equation 2.1):

$$K_d = \frac{C_a}{C_d} \tag{2.1}$$

where K_d is the distribution coefficient of a pesticide molecule between soil and water (volume/mass), C_a is the amount of pesticide adsorbed per unit of adsorbent mass (mass/mass), and C_d is the concentration of pesticide dissolved (mass/volume).

K_d is directly related to the K_{oc} value (Equations 2.2 and 2.3):

$$K_d = \frac{K_{oc} \times OC}{100} \tag{2.2}$$

where K_{oc} is the soil organic partition coefficient, and OC is the organic carbon content (%) (Equation 2.3):

$$K_{oc} = \frac{K_d \times 100}{OC} \tag{2.3}$$

The speed of degradation of pesticides in soils depends, to an essential degree, on the chemical and biological properties of the soils. Even though soils are very diverse, there exist general regularities and characteristics of degradation and fragmentation of pesticides in soils.

The general regularities are as follows [13]:

- More polar pesticides degrade in soils faster than nonpolar pesticides.
- Anionic pesticides degrade in soils more easily than cationic pesticides.
- Aromatic pesticides are more stable than aliphatic pesticides.
- An increase in temperature usually accelerates the degradation and fragmentation of pesticides.
- A decrease in the moisture content of the soil usually decreases the degradation and fragmentation rate of pesticides (but an excessive increase in the moisture content of the soil causes the formation of oxygen-free soil).
- Increased basicity of soil (an increase of pH values) causes acceleration of chemical processes without enzymes and vice versa; acidification of soil (a decrease of pH values) usually increases the stability of pesticides in soil.
- Fe^{3+}, Cu^{2+}, and other metal cations and aluminium oxide are inorganic catalysts of transformation processes.

Readers may find the details of degradation processes of pesticides in Chapter 11.
Tables 2.3 through 2.24 also contain the following parameters [12]:

- Bioconcentration factor (BCF) is indicative of the accumulation of a chemical in living organisms (biota) compared with the concentration in water (the concentration of the chemical in tissue per concentration of the chemical in water). It is an indicator of how much of a chemical will accumulate in living organisms, such as fish. This describes the accumulation of pollutants through chemical partitioning from the aqueous phase into an organic phase, such as the gill of a fish (<100 = low potential, 100–5000 = threshold for concern, >5000 = high potential). Once adsorbed into an organism, chemicals can move through the food chain, ending up in humans. BCF values are unitless and generally range from 1 to 1,000,000.

  BCF values can be expressed by Equation 2.4:

$$\text{BCF} = \frac{\text{Concentration in fish}}{\text{Concentration in water}} \tag{2.4}$$

- DT_{50} is the time required for the chemical concentration under defined conditions to decline to 50% of the amount at application. In many cases, chemicals show "half-life" behavior, in which subsequent concentrations continue to decline by 50% in the same amount of time. In such cases, several or more half-lives (e.g., in which the concentration declines to one eighth or 1/16) are a measure of the persistence of the chemical (DT_{50} in field studies can be evaluated).
- Mode of action is the mechanism by which the substance performs its main functions.

2.4 MODES OF PESTICIDE ACTION

Pesticides can be classified according to their target and modes or periods of action as presented in Table 2.25 [14]. Details for individual pesticides are shown in the previous Tables 2.3 through 2.24 [12].

2.5 PESTICIDES VERSUS BIOCIDES

Biocides are applied in order to render harmless, otherwise control, or kill harmful and unwanted organisms. They are not used for plant protection. Wood preservatives, disinfectants, rodenticides, textile preservatives, or household insecticides belong to this large family of 23 different product types. Biocides can not only adversely affect harmful organisms, but also humans, the environment, and endangered species. For instance, active substances can be water toxic, carcinogenic, reproductive toxic, or endocrine disruptive. Particularly vulnerable groups, such as children or pregnant women, are threatened by the wide and improper use of hazardous biocides [15].

According to Directive 98/8/EC of the European Parliament and Council of February 16, 1998, biocide products are defined as active substances and preparations containing one or more active substances; put in the form in which they are supplied to the user; and intended to destroy, render harmless, prevent the action of, or otherwise exert a controlling effect on any harmful organism by chemical or biological means.

TABLE 2.25
Classifications of Pesticides

By Target		By Mode or Time of Action		
Type	**Target**	**Type**	**Action**	**By Chemical Structure**
Bactericide (sanitizers or disinfected)	Bacteria	Contact	Kills by contact with pest	Pesticides can be either organic or inorganic chemicals. Most of today's pesticides are organic.
Defoliant	Crop foliage	Eradicant	Effective after infection by pathogen	
Desiccant	Crop foliage	Fumigants	Enters pest as a gas	Commonly used inorganic pesticides include copper-based fungicides limesulfur used to control fungi and mites, boric acid used for cockroach control, and ammonium sulfamate herbicides.
Fungicide	Fungi	Nonselective	Toxic to both crop and weed	
Herbicide	Weeds	Postemergence	Effective when applied after crop or weed emergence	
Insecticide	Insects	Preemergence	Effective when applied after planting and before crop or weed emergence	
Miticide (acaricide)	Mites and ticks	Preplant	Effective when applied prior to planting	Organic insecticides can either be natural (usually extracted from plants or bacteria) or synthetic. Most pesticides used today are synthetic organic chemicals. They can be grouped into chemical families based on their structure.
Molluscicide	Slugs and snails	Protectants	Effective when applied before pathogen infects the plant	
Nematicide	Nematodes	Selective	Toxic only to weed	
Plant growth regulator	Crop growth processes	Soil sterilant	Toxic to all vegetation	
Rodenticide	Rodents	Stomach poison	Kills animal pests after ingestion	
Wood preservative	Wood-destroying organisms	Systemic	Transported through crop or pest following absorption	

Source: Reprinted from *Agric. Ecosyst. Environ.*, 123, Arias-Estévez, M., López-Periago, E., Martínez-Carballo, E., Simal-Gándara, J., Mejuto, J.-C., and García-Río, L. The mobility and degradation of pesticides in soils and the pollution of groundwater resources, 247–260, Copyright 2008, with permission from Elsevier.

The Biocidal Products Directive 98/8/EC classified biocides in four groups and 23 product types [15]:

Main Group 1: Disinfectants and general biocidal products
Product type 1: Human hygiene biocidal products
Product type 2: Private area and public health area disinfectants and other biocidal products
Product type 3: Veterinary hygiene biocidal products
Product type 4: Food and feed area disinfectants
Product type 5: Drinking water disinfectants

Main Group 2: Preservatives
Product type 6: In-can preservatives
Product type 7: Film preservatives
Product type 8: Wood preservatives
Product type 9: Fiber, leather, rubber, and polymerized material preservatives
Product type 10: Masonry preservatives
Product type 11: Preservatives for liquid cooling and processing systems
Product type 12: Slimicides
Product type 13: Metalworking fluid preservatives

Main Group 3: Pest control
Product type 14: Rodenticides
Product type 15: Avicides
Product type 16: Molluscicides
Product type 17: Piscicides
Product type 18: Insecticides, acaricides, and products to control other arthropods
Product type 19: Repellents and attractants

Main Group 4: Other biocidal products
Product type 20: Preservatives for food or feedstocks
Product type 21: Antifouling products
Product type 22: Embalming and taxidermist fluids
Product type 23: Control of other vertebrates

2.6 CONCLUSIONS

An important goal in the field of analytical chemistry is to achieve continuous improvement in the analysis of toxic pollutants, for example, pesticide residues in the environment. Pesticides are widespread throughout the world. The composition of pesticide mixtures occurring in environmental samples depends on geographical area, season of the year, number of farms, and quantity and intensity of use of plant-protection agents. The variety of their mixtures in different matrices, for example, rivers, fruits and vegetables, and medicinal plants, is extraordinarily large.

Knowledge of the physicochemical properties of the pesticides influences their analysis. This knowledge allows choice of the most adequate conditions for optimum analysis of pesticides by high-performance liquid chromatography from the stage of preparing samples through to the determination step.

REFERENCES

1. FAO (Food and Agriculture Organization). International Code of Conduct on the Distribution and Use of Pesticides, Rome, Italy, 1989.
2. EPA (US Environmental Protection Agency). http://www.epa.gov/pesticides/about.
3. WHO (World Health Organization)/UNEP (United Nations Environment Programme). Public Health Impact of Pesticides Use in Agriculture, Geneva, Switzerland, 1990.
4. Zeidler, O.I. Verbindungen von Chloral mit Brom- und Chlorbenzol. *Berichte der deutschen chemischen Gesellschaft* 7 (2), 1180–1181, 1874 (published 27 January, 2006 by *European Journal of Inorganic Chemistry,* doi: 10.1002/cber.18740070278).
5. http://www.britannica.com/EBchecked/topic/396883/Paul-Hermann-Muller.
6. Keneth, M. *The DDT Story*. The British Crop Protection Council, London, UK, 1992.

7. EPA (United States Environmental Protection Agency) *Persistent Organic Pollutants: A Global Issue, A Global Response*. http://www.epa.gov/international/toxics/pop.html.
8. Rathore, H.S., and Nollet, L.M.L. (eds.) *Pesticides Evaluation of Environmental Pollution*, CRC Press Taylor & Francis Group, Boca Raton, 2012 (Preface).
9. Stockholm Convention on Persistent Organic Pollutants. 2001. Secretariat of the Stockholm Convention International Environment House, Geneva, Switzerland. http://chm.pops.int/Portals/0/Repository/convention_text/UNEP-POPS-COP-CONVTEXT-FULL.English.PDF.
10. Mahmoud, M.F., and Loutfy, N. *Uses and Environmental Pollution of Biocides*. In: Rathore, H.S., Nollet, L.M.L. (eds.) *Pesticides Evaluation of Environmental Pollution*, CRC Press Taylor & Francis Group, Boca Raton, 2012, pp. 3–25.
11. FAO (Food and Agriculture Organization of the United Nation). Chapter 4: *Pesticides as water pollutants*. http://www.fao.org/docrep/w2598e/w2598e07.htm#historical development of pesticides.
12. University of Hertfordshire. The Pesticide Properties DataBase (PPDB) developed by the Agriculture & Environment Research Unit (AERU), University of Hertfordshire, 2006–2013, 2015. http://sitem.herts.ac.uk/aeru/ppdb/en/atoz.htm.
13. Tuzimski, T. *Pesticide Residues in the Environment*. In: Rathore, H.S., Nollet, L.M.L. (eds.) *Pesticides Evaluation of Environmental Pollution*, CRC Press Taylor & Francis Group, Boca Raton, 2012, Chapter 6, pp. 147–201.
14. Arias-Estévez, M., López-Periago, E., Martínez-Carballo, E., Simal-Gándara, J., Mejuto, J.-C., and García-Río, L. The mobility and degradation of pesticides in soils and the pollution of groundwater resources, *Agric. Ecosyst. Environ.*, 2008, 123, 247–260.
15. PAN Europe (Pesticide Action Network Europe). http://www.pan-europe.info/Campaigns/biocides.html.

3 Method Development of Chromatography

Retention–Eluent Composition Relationships and Application to Analysis of Pesticides

Tomasz Tuzimski and Edward Soczewiński

CONTENTS

3.1 INTRODUCTION

Chromatography is the science that studies the separation of molecules based on differences in their structures. In this technique, a mixture of compounds is separated over a stationary support due to different interactions with that support. According to these different (stronger or weaker) interactions with the support, the components will move more or less rapidly. In this way, even chemically similar molecules can be separated from each other.

A detailed discussion of the mechanism of chromatographic separations is presented in the next pages on the first part of the chapter.

3.2 MOLECULAR MECHANISM OF CHROMATOGRAPHIC SEPARATION: CHROMATOGRAPHIC PARAMETERS RETENTION FACTOR k, RETENTION TIME t_R, SEPARATION FACTOR α, RESOLUTION R_S, AND EFFICIENCY N

3.2.1 Models of Retention Mechanism in NP Systems

For the characterization of normal-phase (NP) systems, mainly three models of retention can be taken into consideration.

3.2.1.1 Snyder-Soczewiński Retention Model

In 1968, Snyder published an adsorption model in which it was assumed that the adsorbent surface is covered with an adsorbed layer of solvent molecules, and adsorption of a solute molecule must be accompanied by desorption of one or more solvent molecules from the surface [1]. Soczewiński, combining this concept, formulated a simple molecular model of adsorption [2], next known as the Snyder-Soczewiński model of retention [3,4]. The simple molecular model of adsorption was initially elaborated for silica as an adsorbent. Later, it was confirmed for high-performance liquid chromatography (HPLC) nonaqueous NP systems with other polar sorbents: diol, amino, and nitril and even for multifunctional solutes and enantioselective sorbents.

For monofunctional solute Z, adsorption on a silica surface was considered a formation of the molecular complex (AZ) with a silanol group (A) with release of a modifier molecule (S) (Figure 3.1a) [4]:

$$\mathrm{AS} + \mathrm{Z} \rightleftarrows \mathrm{AZ} + \mathrm{S}$$

In accordance with Snyder's model of competitive adsorption, the surface silanol groups (A) are covered with H-bonded molecules of the polar component of the eluent (S). The adsorption of solute molecule (Z) is accompanied with displacement of a modifier molecule in a reversible reaction. For m-point adsorption (for a solute molecule containing m polar groups capable of simultaneous interaction with surface silanol groups) [4]:

$$\begin{aligned} m\mathrm{AS} + \mathrm{Z} &\rightleftarrows \mathrm{A}_m\mathrm{Z} + m\mathrm{S} \\ K &= \frac{\mathrm{X}_{\mathrm{A}_m\mathrm{Z}}\mathrm{X}_\mathrm{S}^m}{\mathrm{X}_\mathrm{AS}^m\mathrm{X}_\mathrm{Z}} \\ K &= \frac{\mathrm{X}_{\mathrm{A}_m\mathrm{Z}}}{\mathrm{X}_\mathrm{Z}} = \frac{\mathrm{const}}{\mathrm{X}_\mathrm{S}^m} \\ R_\mathrm{M} &= \log k = \mathrm{const} - m \log \mathrm{X}_\mathrm{S} \end{aligned} \tag{3.1}$$

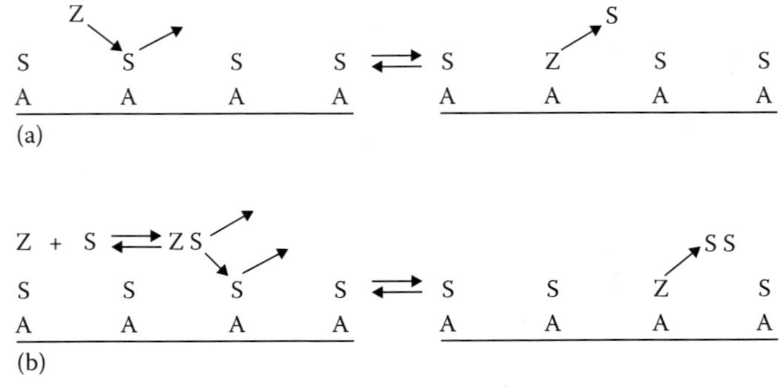

FIGURE 3.1 Molecular mechanism of competitive adsorption (A = silanol group; S = modifier molecule; Z = solute molecule). (a) Simple displacement. (b) Displacement combined with dissociation of solvate ZS. (From Soczewiński, E., *J. Chromatogr. A*, 965, 109–116, 2002. With permission.)

where (Equation 3.1) const is a constant, m is a constant depending on the number of the substituents interacting simultaneously with the adsorbent surface, and X_S is the concentration of polar solvent expressed in molar fractions for monofunctional solute $m = 1$. However, the slope may be different when the H-bonding group is solvated by the modifier; the adsorption of solute Z must then be accompanied by decomposition of the solvate ZS (Figure 3.1b). The solvation of solute Z by the modifier S, that is, formation of the ZS complex, is competitive to adsorption of the solute—formation of the AZ complex. Adsorption of the solute—formation of the A–Z hydrogen bond—results in release of two molecules of modifier S [4]:

$$K = \frac{X_{AZ}X_S^2}{X_{AS}X_{ZS}}$$

$$K = \frac{X_{AZ}}{X_{ZS}} = \frac{\text{const}}{X_S^2} \tag{3.2}$$

$$\log k = \text{const} - 2\log[S]$$

so that (in the case of strong solvation of the solute by the modifier) the slope $m = 2$ in spite of single-point adsorption (Equation 3.2).

3.2.1.2 Other Retention Models

The dependence of retention on the composition of the mobile phase can also be described using different theoretical models:

- Martin–Synge model of partition chromatography [5,6]
- Scott–Kucera model of adsorption chromatography [7,8]
- Kowalska model of adsorption and partition chromatography [9–11]
- Ościk thermodynamic model [12,13]

The model proposed by Scott and Kucera assumes bilayer adsorption of the solvent, sorption of solute molecules without displacement, and also dispersive interactions between eluent components and solute molecules. It leads to the following dependence [7,8]:

$$\frac{1}{k} = A' + B'\phi \tag{3.3}$$

where A′ and B′ are constants, and ϕ is the volume fraction of the polar solvent in eluent.

Kowalska described another model of retention for the characterization of NP systems [14].

3.2.2 Models of Retention Mechanism in RP Systems

For sample analysis, the predominant HPLC mode in use today is reversed-phase (RP) chromatography with a nonpolar column in combination with a (polar) mixture of water plus an organic solvent (e.g., methanol, acetonitrile, dioxane, tetrahydrofuran, acetone) as a mobile phase (RP system). In RP systems of the type octadecyl silica/water + methanol, the retention versus eluent composition relationships of a semilogarithmic type was reported by Snyder et al. [15] in the form (Equation 3.4)

$$\log k = \log k_w - S\phi_{\text{mod}} \tag{3.4}$$

where $\log k_w$ is the retention factor of the solute for pure water as the mobile phase, S is constant, and ϕ_{mod} is the volume fraction of the modifier (e.g., methanol).

In 1962, Soczewiński and Wachtmeister described a model of retention in NP liquid–liquid systems [16] with an analogous linear semilogarithmic retention-modifier concentration equation. For $\phi_{mod} = 1$ (pure modifier), $S = \log k_w - \log k_{mod}$, $S = \log (k_w/k_{mod})$—the logarithm of the hypothetical partition coefficient of solute between water and the modifier (actually miscible) [16]. The constant S increases with decreasing polarity of the organic solvent and is a measure of its elution strength. On the other hand, S rises with an increase in size of the solute molecule. The above equation can be used for prediction of retention and selectivity for a reasonable concentration range. However, for a broad concentration range, this equation does not predict solute retention with good precision. In cases of broad concentration ranges of the mobile phase, the following equation was reported [17] (Equation 3.5):

$$\log k = \log k_w + a\phi + b\phi^2 \quad (3.5)$$

where a and b are constants that are dependent on the solute and the mobile-phase type.

Deviations from this equation occur especially beyond the concentration range $0.1 < \phi < 0.9$, that is, for high and low concentrations of water. These deviations are explained by several reasons. Conformational changes in the alkyl chain structure of the stationary phase at a high water concentration in the mobile phase can influence this effect. When the concentration of water is low, then its participation in the hydrophobic mechanism of retention is eliminated, and additionally, molecular interactions of the solute and unreacted silanols can occur [18].

It seems that, in the RP systems, some displacement mechanism may, in some cases, be operative, as in nonaqueous systems with polar adsorbents as presented in Figure 3.1b [3,4]. The alkyl chains (usually C_8 and C_{18}) bound to the surface are saturated with the modifier molecules, some of which are displaced by the adsorbed solute molecule, which, at the same time, loses some of the solvating modifier molecules (Figure 3.2) [3,4]. Figure 3.2 shows a simplified picture of the molecular retention mechanism, in which the vertical lines are octadecyl chains (C_{18}) on the silica surface. The hydrophobic interactions of the solute Z molecule (solvated with two modifier S molecules) with the hydrocarbon chains is accompanied by displacement of one modifier molecule from the surface film and loss of two solvating modifier molecules. The slope of the log k versus log [S] line is then $m = 3$. In HPLC experiments, we can determine k values only in a narrow range (say, 1–20) corresponding to the narrow range of ϕ_{mod} values, especially in the case of high slopes m [4].

In practice, in RP chromatography, the mechanism is usually mixed, ambiguously, and so far only partly explained. Solvophobic interaction (Figure 3.3a) assumes that the solute molecule is attached to a ligand group (C_8 in this example). Adsorption (Figure 3.3b) implies that the solute molecule does not penetrate into the stationary phase but is retained at the interface between the stationary phase and mobile phases. The partition model (Figure 3.3c) considers the stationary phase to be similar to a liquid phase, into which the solute molecule is dissolved. In both solvophobic interaction and partition, the solute molecule is located within the stationary phase [19].

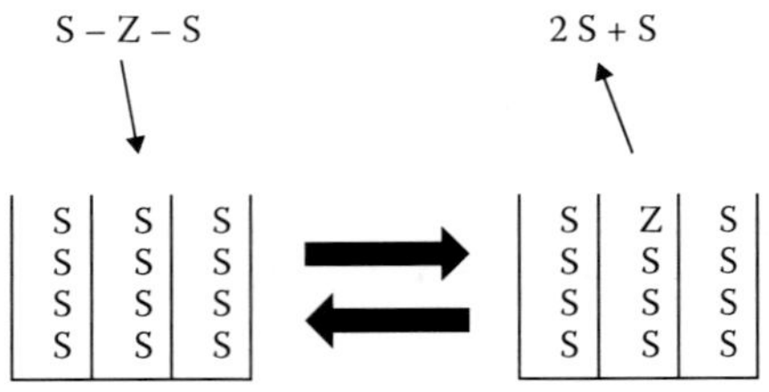

FIGURE 3.2 Presumed mechanism of competitive sorption of solute combined with dissociation of ZS_2 solvate (slope of log k vs. log C_{mod} plot $m = 3$). Octadecyl silica sorbent (RP-18) saturated with modifier (S) molecules. (Adapted from Soczewiński, E., *J. Chromatogr. A*, 965, 109–116, 2002.)

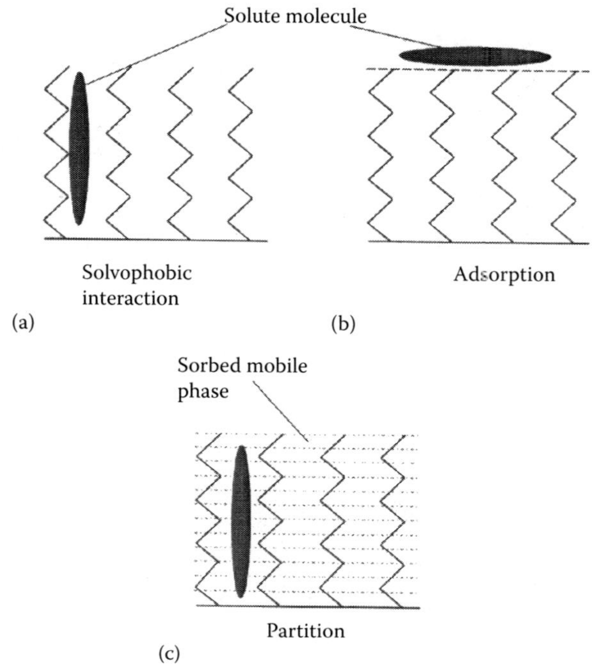

FIGURE 3.3 Different possibilities for the retention of a solute molecule in RP chromatography. (a) Solvophobic interaction, (b) adsorption, (c) partition. (Snyder, L. R., Kirkland, J. J., Dolan, J. W.: *Introduction to Modern Liquid Chromatography*. 2010. Copyright Wiley-VCH Verlag GmbH & Co. KGaA. Reproduced with permission.)

Schoenmakers proposed the following quadratic equation, which describes his model of retention in RP systems [20] (Equation 3.6):

$$\ln k = A\phi^2 + B\phi + C \tag{3.6}$$

where A, B, and C are constants, and ϕ is the volume fraction of the modifier.

Solute retention is determined by various interactions between the solute, mobile phase, and stationary phase (column). The relative importance of different solute (analyte)–column interactions—and column selectivity—depends of the components of the chromatographic system used in HPLC experiments: the composition of the stationary phase and molecular structure of the solute. There are (different) interactions that can influence column selectivity (Figure 3.4a–h) [19]:

- Hydrophobic interaction (Figure 3.4a)
- Steric exclusion of larger solute molecules from the stationary phase (here referred to as "steric interaction") (Figure 3.4b)
- Hydrogen bonding of an acceptor (basic) solute group by a donor (acidic) group within the stationary phase (for silica a silanol –SiOH) (Figure 3.4c)
- Hydrogen bonding of a donor (acidic) solute group by an acceptor (basic) group within the stationary phase (represented here by a group "X") (Figure 3.4d)
- Cation exchange or electrostatic interaction between a cationic solute and ionized silanol ($-SiO^-$) within the stationary phase; also repulsion of ioniozed acid (e.g., R-COO^-) (Figure 3.4e)
- Dipole–dipole interaction between a dipolar solute group (a nitro group in this example) and a dipolar group in the stationary phase (a nitrile group for cyano column) (Figure 3.4f)

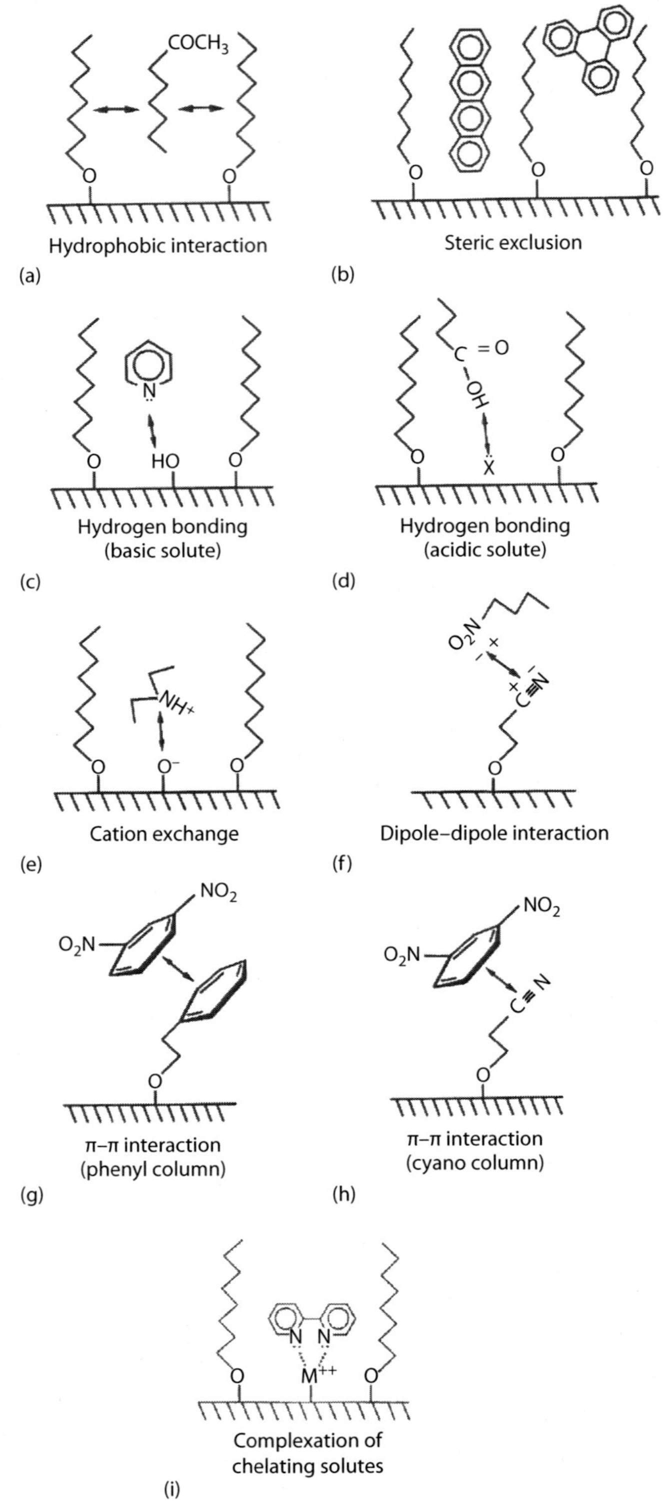

FIGURE 3.4 Solute-column interactions that determine column selectivity (figures omit the connecting silane group [$-Si(CH_3)_2-$]). (Snyder, L. R., Kirkland, J. J., Dolan, J. W.: *Introduction to Modern Liquid Chromatography.* 2010. Copyright Wiley-VCH Verlag GmbH & Co. KGaA. Reproduced with permission.)

- π–π interaction between an aromatic solute and either a phenyl group (phenyl column) (Figure 3.4g) or a nitrile group (cyano column) (Figure 3.4h)
- Complexation between a chelating solute and metal contaminants on the particle surface (Figure 3.4i)

The hydrophobic interactions are by far the most important contribution in RP chromatography. Dipole interactions are only important in the case of a cyano column (Figure 3.4h), and the π–π interaction occurs only for phenyl and cyano columns [19,21]. Both dipole and π–π interactions are inhibited by the use of acetonitrile as a B-solvent, which further minimizes their importance for separation with acetontrile. As shown in Figure 3.4i, complexation with surface metals can result from the use of a less pure, type-A silica, leading to broad, tailing peaks (very undesirable); the chelating solute α,α-bipyridyl has been used to test columns for metal complexation [19].

The attraction between adjacent molecules of solute (analyte) and eluent (solvent) is the result of several different intermolecular interactions as illustrated in Figure 3.5 [19]:

- Dispersion interactions (Figure 3.5a)
- Dipole–dipole interaction (Figure 3.5b)
- Hydrogen bonding interactions (Figure 3.5c)
- Ionic (coulombic) interactions (Figure 3.5d)
- Charge transfer of π–π interaction (Figure 3.5e)

These interactions are also described in Chapter 7. Examples of different interactions between solutes (analytes) and different stationary phases are illustrated in Figure 3.4 [19].

Other details of interactions between solute and stationary phase or solute and components of eluents can be found in other parts of this book in Chapters 6 and 8. The polar interactions of various nonionic aliphatic solvents used in HPLC can be described by the solvent-selectivity triangle (see Section 3.5).

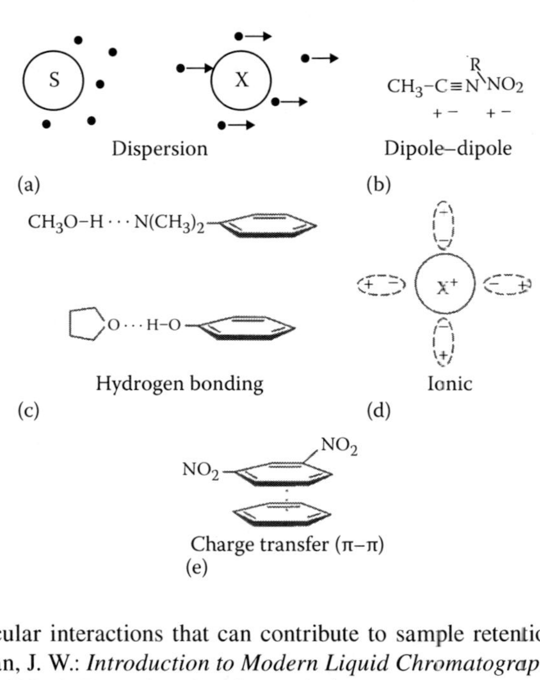

FIGURE 3.5 Intermolecular interactions that can contribute to sample retention and selectivity. (Snyder, L. R., Kirkland, J. J., Dolan, J. W.: *Introduction to Modern Liquid Chromatography*. 2010. Copyright Wiley-VCH Verlag GmbH & Co. KGaA. Reproduced with permission.)

3.2.3 Chromatographic Parameters: Retention Factor k, Retention Time t_R, Separation Factor α, Resolution R_S, and Efficiency N

The successful use of HPLC requires an understanding of how separation is affected by experimental conditions. The success of experiments in HPLC can be expressed by general parameters, such as retention time t_R, retention factor k, efficiency N, resolution R_s, and separation factor α.

The retention time t_R is the most common parameter with which we have to deal with during the HPLC experiments. The retention time t_R is the time from sample injection to the appearance of the top of the peak of solute (analyte) in the chromatogram. The retention factor k is defined as the quantity of solute in the stationary phase (s), divided by the quantity in the mobile phase (m). The quantity of solute (analyte) in each phase is equal to its concentration (in stationary or mobile phases expressed as C_s and C_m, respectively) times the ratio of volumes of the phases (V_s and V_m, respectively). The retention factor k can be expressed as [19] (Equation 3.7):

$$k = \frac{C_s/C_m}{V_s/V_m} = \frac{K}{\psi} \tag{3.7}$$

where k is the retention factor, C_s is the concentration of the solute in the stationary phase, C_m is the concentration of the solute in the mobile phase, V_s is the volume of the stationary phase, V_m is the volume of the mobile phase, $K = (C_s/C_m)$ is the equilibrium constant for Equation 3.5, and $\psi = (V_s/V_m)$ is the phase ratio (the ratio of stationary and mobile-phase volumes within the column).

A solute molecule must be present in either the mobile or stationary phase. The retention factor k can be expressed as [5] (Equations 3.8 and 3.9):

$$k = \frac{1-R}{R} \tag{3.8}$$

or

$$R = \frac{1}{1+k} \tag{3.9}$$

where R is the fraction of molecules in the mobile phase, and $1 - R$ is the fraction of molecules in the stationary phase.

R can be expressed as the ratio (Equation 3.10)

$$R = \mu_x/\mu_0 \tag{3.10}$$

where μ_x is the migration velocity for the solute (analyte), and μ_0 is the migration velocity for solute which is not retained in the stationary phase.

If R is the ratio of μ_x and μ_0, Equation 3.10 can expressed as (Equations 3.11 and 3.12)

$$t_R = t_0(1 + k) \tag{3.11}$$

or

$$V_R = V_m(1 + k) \tag{3.12}$$

where V_R is the total retention volume, the volume of mobile phase needed to elute the solute (analyte) from the column, and V_m is the column dead-volume, the volume needed to elute the solute (analyte) from the column that is not retained by the stationary phase ($k = 0$).

Equation 3.11 can be rearranged to give Equation 3.13:

$$k = \frac{t_R - t_0}{t_0} \tag{3.13}$$

where t_R is retention time, t_0 is retention time of solute that is not retained by the stationary phase.

The relative ability of a column to produce narrow peaks is described as the column efficiency and is defined by the plate number N (measured from the baseline peak width W) (Equation 3.14):

$$N = 16\left(\frac{t_R}{W}\right)^2 \tag{3.14}$$

or as Equation 3.15 by the plate number N (measured in the half-height peak width $W_{1/2}$):

$$N = 5.54\left(\frac{t_R}{W_{1/2}}\right)^2 \tag{3.15}$$

The both methods are illustrated in Figure 3.6.

The separation of two peaks, as in Figure 3.6, is usually described in terms of their resolution R_s (Equation 3.16):

$$R_s = \frac{2\left[t_{R(j)} - t_{R(i)}\right]}{W_i + W_j} \tag{3.16}$$

where $t_{R(i)}$ and $t_{R(j)}$ are retention times for two peaks i and j, respectively, and W_i and W_j are the baseline widths W for peaks i and j, respectively.

The separation factor α (a measure of separation selectivity or relative retention) is defined as (Equation 3.17):

$$\alpha = \frac{k_j}{k_i} \tag{3.17}$$

where k_i and k_j are the values of k for adjacent peaks i and j.

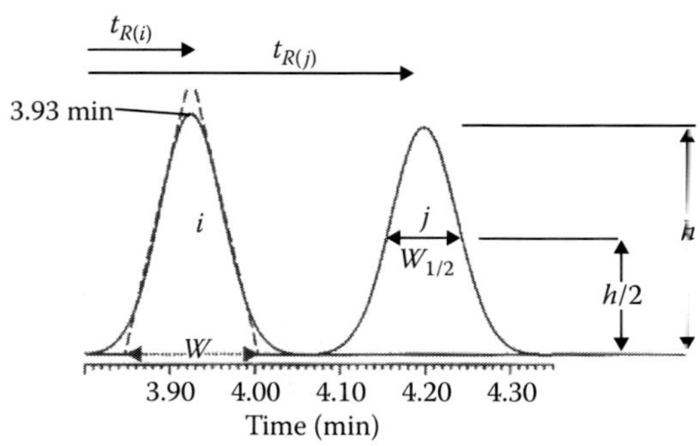

FIGURE 3.6 Measurement of peak width. (Snyder, L. R., Kirkland, J. J., Dolan, J. W.: *Introduction to Modern Liquid Chromatography*. 2010. Copyright Wiley-VCH Verlag GmbH & Co. KGaA. Reproduced with permission.)

TABLE 3.1
Effect of Different Separation Conditions on Retention (*k*), Selectivity (α), and Plate Number (*N*)

Condition	*k*	α	*N*
%B	++	+	–
B-solvent (acetonitrile, methanol, etc.)	+	++	–
Temperature	+	+	+
Column type (C_{18}, phenyl, cyano, etc.)	+	++	–
Mobile phase pH[a]	++	++	+
Buffer concentration[a]	+	+	–
Ion-pair reagent concentration[a]	++	++	+
Column length	0	0	++
Particle size	0	0	++
Flow rate	0	0	+
Pressure	–	–	+[b]

Source: Snyder, L. R., Kirkland, J. J., Dolan, J. W.: *Introduction to Modern Liquid Chromatography*. 2010. Copyright Wiley-VCH Verlag GmbH & Co. KGaA. Reproduced with permission.

Note: ++, major effect; +, minor effect; –, relatively small effect; 0, no effect; bolded quantities denote conditions that are primarily used (and recommended) to control *k*, α, or *N*, respectively (e.g., %B is varied to control *k* or α, column length is varied to control *N*).

[a] For ionizable solutes (acids or bases).

[b] Higher pressures allow larger values of *N* by a proper choice of other conditions; pressure per se, however, has little direct effect on *N*.

The resolution can be expressed (Equation 3.18) as a function of the retention factor *k* for the first peak *i* (term *a*), the separation factor α (term *b*), and column efficiency or the plate number *N* (term *c*):

$$R_s = \left(\frac{1}{4}\right)\underset{(a)}{\left[\frac{k}{(1+k)}\right]}\ \underset{(b)}{(\alpha - 1)}\ \underset{(c)}{N^{0.5}} \tag{3.18}$$

where *k* is the retention factor for the first peak, α is the separation factor, and *N* is the column efficiency (the plate number).

The effect of different separation conditions on retention (*k*), selectivity (α), and plate number (*N*) is summarized in Table 3.1.

3.3 OPTIMAL CHROMATOGRAM

The success of the chromatographic analysis depends on the quality of separation of the peaks in the chromatogram. The separation of two peaks *i* and *j* in HPLC experiments is usually described as R_s (Equation 3.16) and illustrated as in Figure 3.7 [19].

Better separation (increased resolution) results from a larger difference in peak retention times and/or narrower peaks. Accurate quantitative analysis based on a separation as in Figure 3.7 is favored by baseline resolution, in which the valley between the two peaks returns to the baseline. In this case, for two peaks of comparable size (the least well-separated peak pair of solutes/analytes), the value of $R_s > 1.5$. When more than two peaks are to be separated, the goal is usually a resolution $R_s \geq 2$ for the least well-separated peak pair as shown in Figure 3.8 [19,22].

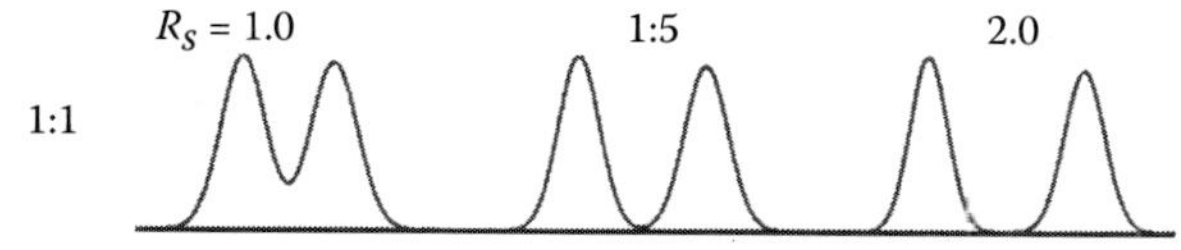

FIGURE 3.7 Separation as a function of resolution for relative peak size 1:1. (Snyder, L. R., Kirkland, J. J., Dolan, J. W.: *Introduction to Modern Liquid Chromatography*. 2010. Copyright Wiley-VCH Verlag GmbH & Co. KGaA. Reproduced with permission.)

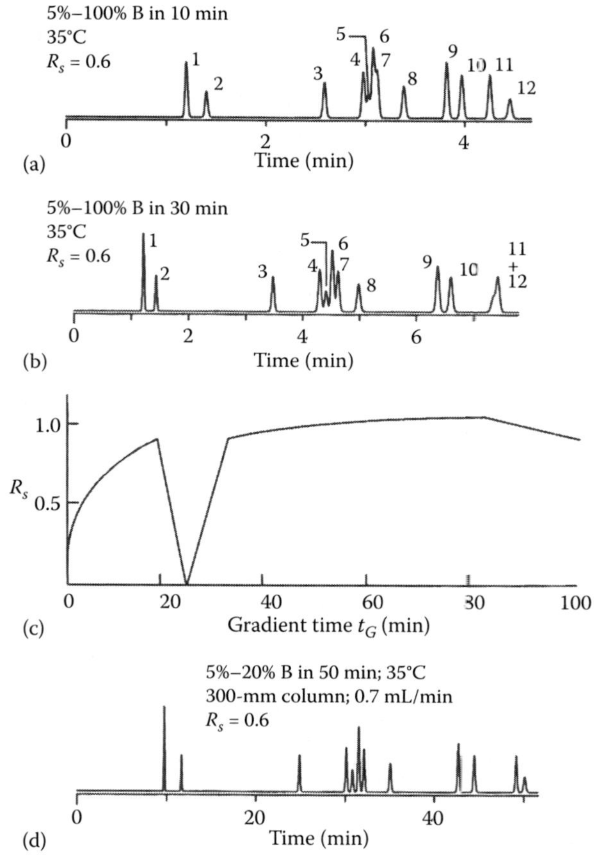

FIGURE 3.8 Illustration of strategy for method development based on the use of computer simulation for selection of final separation conditions (separation of 12-component mixture). (a) Chromatogram—conditions: C_{18} column, acetonitrile/water gradients, other conditions varied. (b) Chromatogram—conditions: same as in (a) except 5%–100% B in 10 minutes, 100 × 4.6-mm (3-μm) C_{18} column. (c) Resolution map. (d) Final separation with condition indicated in figure. (Recreated chromatograms from data of Braumann, T. et al., *J. Chromatogr.*, 261, 329–343, 1983; Snyder, L. R., Kirkland, J. J., Dolan, J. W.: *Introduction to Modern Liquid Chromatography*. 2010. Copyright Wiley-VCH Verlag GmbH & Co. KGaA. Reproduced with permission.)

3.4 CONTROL OF RETENTION: QUANTITATIVE RETENTION–ELUENT COMPOSITION RELATIONSHIPS AND GRADIENT ELUTION

The chromatographic systems are usually illustrated as linear relationships between log k values and concentration or *log* concentration of the modifier (polar solvent). An example is presented in Figure 3.9.

The plots are straight lines and usually not parallel, which indicates variations of selectivity. The slopes are differentiated depending on the difference in eluent strength and number of polar groups

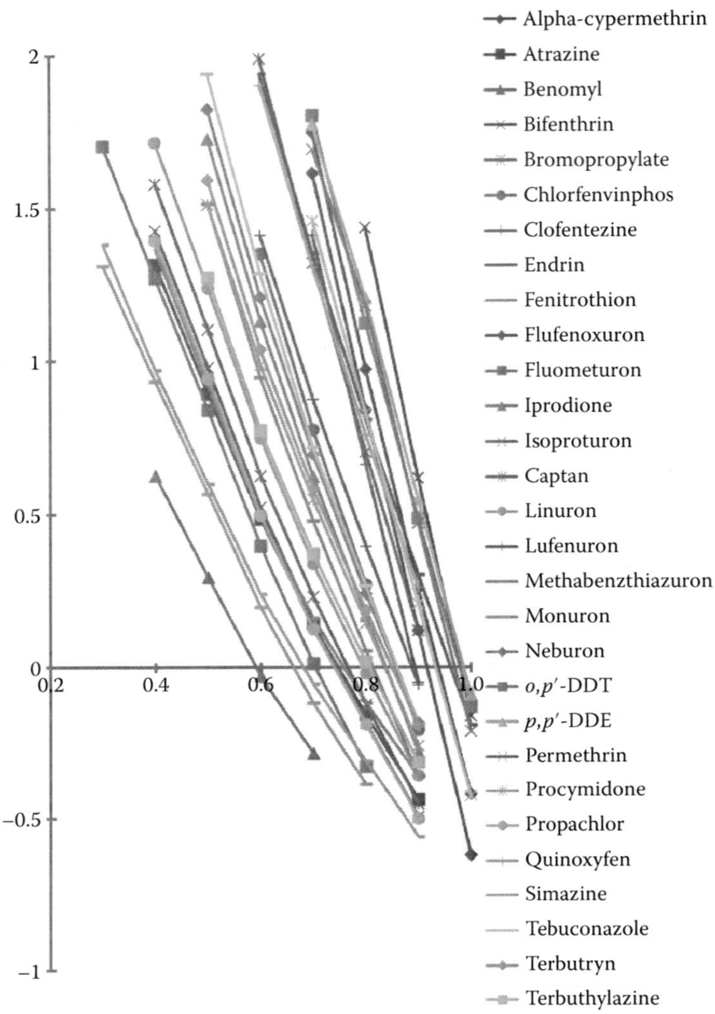

FIGURE 3.9 Relationships log k versus ϕ for mixture of pesticides analyzed in RP-HPLC system (MeOH—water, XDB-C_{18} column, 150 × 4.6 mm, 5 μm, 1.0 mL/min; T = 22°C).

(NP system) in the solute molecule. Also, the size of the molecule, number of substituents, and the position of the substituents in the molecule can affect the slope of these relationships.

Retention and separation of substances can be easily changed by changing the concentration (Figure 3.10) or type of modifier in the mobile phase [19].

More details about the influence of type of modifier on retention and separation of solutes/analytes are described in Chapter 7.

For some samples, however, no single value of percentage of modifier in eluent (%B) can provide a generally satisfactory separation as illustrated in Figure 3.11a and b—an example of separation of a nine-component mixture of herbicides by RP systems. Many samples cannot be successfully separated by the use of isocratic conditions but instead require gradient elution (also solvent programming): changes in mobile phase composition during the separation to progressively reduce the retention of more strongly retained analytes. This example (Figure 3.11a–c) illustrates the general elution problem: the inability of single isocratic separation to provide satisfactory separation within

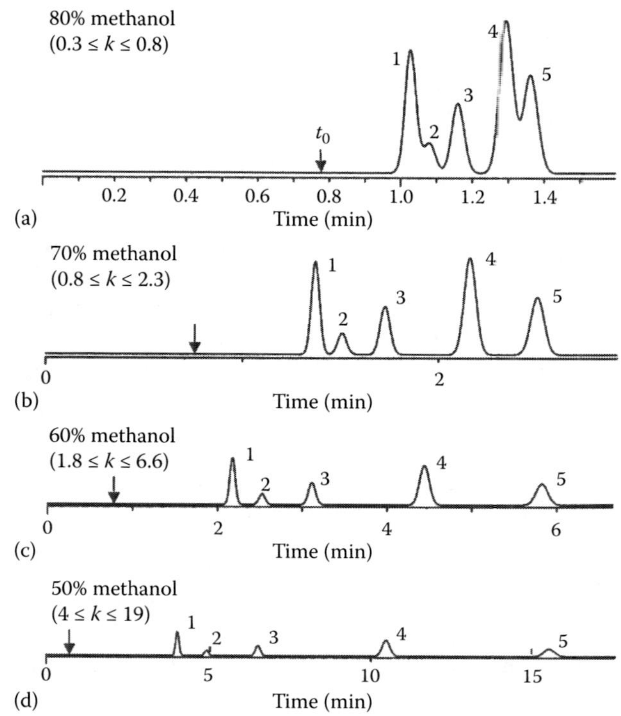

FIGURE 3.10 Separation as a function of mobile phase %B. Herbicide sample: 1, monolinuron; 2, metobromuron; 3, diuron; 4, propazine; 5, chloroxuron. Conditions, 150 × 4.6 mm, 5 μm C_{18} column; methanol–water mixtures as mobile phase; 2.0 mL/min; ambient temperature. (Recreated chromatograms from data of Braumann, T. et al., *J. Chromatogr.*, 261, 329–343, 1983.)

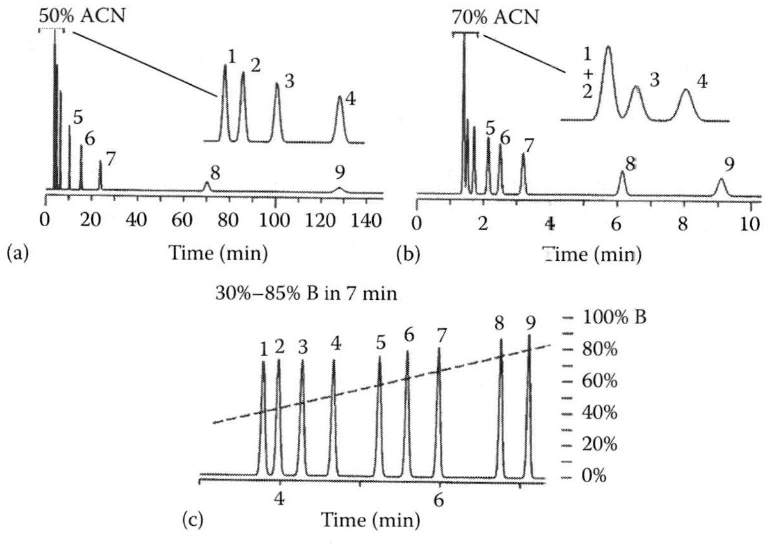

FIGURE 3.11 Illustration of the general elution problem and the need for gradient elution. The sample is a mixture of herbicides. (a) Isocratic elution using 50% acetonitrile (ACN)–water as mobile phase; 150 × 4.6 mm, C_{18} column (5 μm particles), 2.0 mL/min, ambient temperature; (b) same as (a) except 70% ACN–water; (c) same as (a) except gradient elution 30%–85% ACN in seven minutes. (Computer simulations based on data of Quarry, M. A. et al., *J. Chromatogr.*, 285, 19–61, 1984.)

a reasonable run time for samples with a wide range in retention of components (peaks with very different values of k) [19,23].

As demonstrated in Figure 3.11a, the separation of all components of a mixture is impossible because later peaks are very wide and low and have very long retention times. Variation of the percentage of modifier (from 50% to 70%) in the binary mobile phase partly solves the latter two difficulties (Figure 3.11b), but at the same time, it introduces another problem: the poor resolution of peaks (1 to 3). All peaks of a mixture are separated to baseline in a total run time of slightly more than seven minutes with approximately constant peak widths and comparable detection sensitivity for each peak (Figure 3.11c) [19,23].

By gradient shape, we mean the way in which mobile-phase composition (%B) changes with time during a gradient run. Gradient elution can be carried out with different gradient shapes. More details about the influence of the type of gradient on retention and separation of solutes/analytes are described in Chapter 7.

3.5 CONTROL OF SELECTIVITY: EQUIELUOTROPIC SERIES, SNYDER'S TRIANGLE, NYIREDY'S "PRISMA" MODEL

Typical selection of the solvent is based on the eluotropic series, for example, for most popular silica and/or ε^0 parameter. One of the first attempts of the solvent systematization with regard to their elution properties was formulated by Trappe as the eluotropic series [24]. Pure solvents were ordered according to their chromatographic elution strength for various polar adsorbents in terms of the solvent strength parameter ε^0 defined according to Snyder [25,26] and expressed by Equation 3.19:

$$\varepsilon^0 = \Delta G_S^0 / 2.3RTA_S \quad (3.19)$$

where ΔG_S^0 is the adsorption-free energy of the solute molecule, R is the universal gas constant, T is the absolute temperature, and A_S is the area occupied by the solvent molecule on the adsorbent surface expressed in 1/6 benzene molecule area.

The parameter ε^0 represents adsorption energy of the solvent per unit area on the standard activity surface. Solvent strength is the sum of many types of intermolecular interactions.

A simple choice of the mobile phase is possible by a microcircular technique on the basis of the eluotropic series or for binary and/or multicomponent mobile phases [18,27,28].

Neher [29] proposed an equieluotropic series, which gives the possibility of replacing one solvent mixture by another one: composition scales (approximately logarithmic) for solvent pairs are subordinated to give constant elution strengths for vertical scales. The equieluotropic series of mixtures is approximately characterized by constant retention, but these can often show different selectivity [18]. The scales, devised originally for planar chromatography on alumina layers, were later adapted to silica by Saunders [30], who determined accurate retention data by HPLC and subordinated the composition scale to Snyder's elution strength parameter [31].

Snyder [32] proposed the calculation of the elution strength of multicomponent mixtures. The solvent strength ε_{AB} of the binary solvent mobile phase is given by the relationship (Equation 3.20)

$$\varepsilon_{AB} = \varepsilon_A^0 + \frac{\log\left(X_B 10^{\alpha n_b\left(\varepsilon_B^0 - \varepsilon_A^0\right)} + 1 - X_B\right)}{\alpha n_b} \quad (3.20)$$

where ε_A^0 and ε_B^0 are the solvent strengths of two pure solvents A and B, respectively; X_B is the mole fraction of the stronger solvent B in the mixture; α is the adsorbent activity parameter; and n_b is the adsorbent surface area occupied by a molecule of the solvent B.

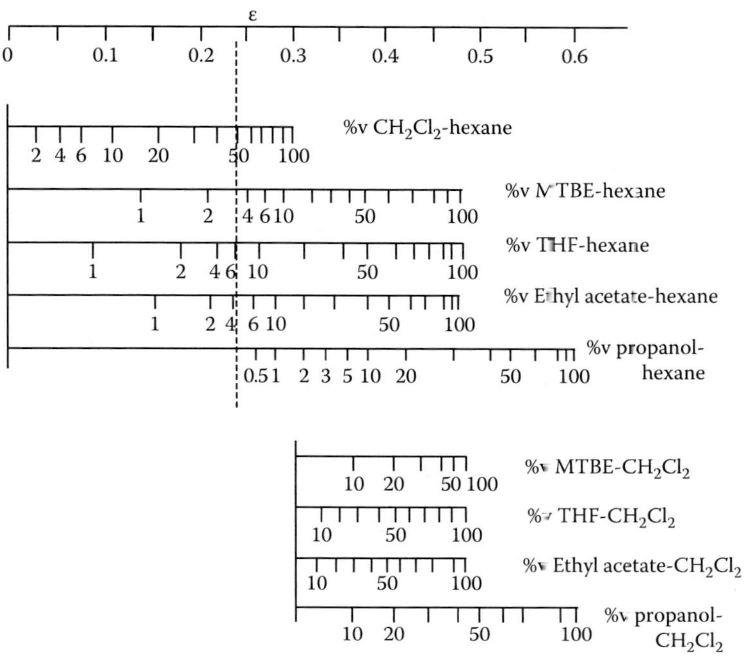

FIGURE 3.12 Solvent nomograph for NP chromatography and silica columns. (Adapted from Meyer, V. R., Palamareva, M. D., *J. Chromatogr.*, 641, 391–395, 1993.)

The example of equieluotropic series for normal-phase chromatography and silica column is presented in Figure 3.12 [19,33].

Some basic physicochemical parameters (viscosity, dipole moment, refractive index, dielectric constant, etc.) are used for the characterization of the solvent ability for molecular interactions, which are of great importance for chromatographic retention, selectivity, and performance. The physical constants mentioned earlier for some common solvents used in chromatography are collected in several monographs [19,34].

Solvent strength (eluent strength, elution strength) refers to the ability of the solvent or solvent mixture to elute the solutes from the stationary phase. This strength rises with an increase in solvent polarity in NP systems. A reversed order of elution strength takes place for RP systems. Solvent polarity is connected with molecular solute–solvent interactions, including dispersion (London), dipole–dipole (Keesom), induction (Debye), and hydrogen-bonding interactions [35].

The first attempts of solvent classifications were performed for characterization of liquid phases applied in gas chromatography (Rohrschneider and McReynolds) [36,37]. Another solvent classification was also described [38–40]. In this classification, the solubility parameter was derived based on values of cohesion energy of pure solvents.

Snyder's polarity scale has gained significance for solvent classification in liquid chromatography practice in which the parameter P′ is used for characterization of solvent polarity [41]. This parameter was calculated based on the distribution constant, K, of test solutes (ethanol, dioxane, nitromethane) in gas–liquid (solvent) systems. Ethanol was chosen for characterization of the solvent with regard to its basic properties (proton-acceptor properties), dioxane to characterize its acidic properties (proton-donor properties), and nitromethane to describe dipolar properties of the solvent. The sum of the log K values of these three test compounds is equal to the parameter P′ of the solvent. In addition, each value of log K of the test solutes was divided by parameter P′; then the relative values of three types of polar interaction were calculated for each solvent: x_d for dioxane (acidic), x_e for ethanol (basic), and x_n for nitromethane (dipolar) [18]. These x_i values were corrected for nonpolar (dispersive) interactions and were demonstrated in a three-component coordinate plot on an equilateral triangle (Figure 3.13).

Snyder characterized more than 80 solvents and obtained eight groups of solvents on the triangle [42,43]. The triangle was named the Snyder-Rohrschneider solvent selectivity triangle (SST). This classification of solvents is useful for selectivity optimization in liquid chromatography. More details about optimization of retention of solutes on the basis of the Snyder-Rohrschneider SST are described in Chapter 7.

As mentioned previously, solvents from each group of the SST show different selectivity, which can lead to changes in the separation order. When the average solvent strengths and selectivity values are calculated for each solvent group of the SST, then linear correlations of these quantities are found for solvent groups I, II, III, IV, and VIII and for solvent groups I, V, and VII [44]. Solvents of group VI do not belong to either correlation due to their different ability for molecular interactions in comparison with solvents of the remaining groups. It was mentioned here that the solvents belonging to the groups in the corners of the SST (groups I, VII, VIII) and from its middle part (group VI) are the most often applied in NP systems of planar chromatography. Nyiredy et al. [45] suggested the selection and testing of 10 solvents with various strengths from eight selectivity groups of SST [diethyl ether (I); 2-propanol and ethanol (II); tetrahydrofuran (III); acetic acid (IV); dichloromethane (V); ethyl acetate, dioxane (VI); toluene (VII); chloroform (VIII)]. All these solvents are miscible with hexane (or heptane) the solvent strength of which is about zero.

It is purposeful to discuss, in more detail, the modes of retention and selectivity optimization that can be applied to obtain appropriate chromatographic resolution. It seems that the strategy of separation optimization based on classification of solvents by Snyder (or solvatochromic parameters) and the PRISMA is the well-known model introduced by Nyiredy [45–48] is the most suitable in laboratory practice for liquid chromatographic separations of sample mixtures. This opinion is expressed, taking into account the simplicity of this procedure and low costs of operations involved (no sophisticated equipment and expensive software are necessary). The PRISMA model has three parts as illustrated in Figure 3.14 [49].

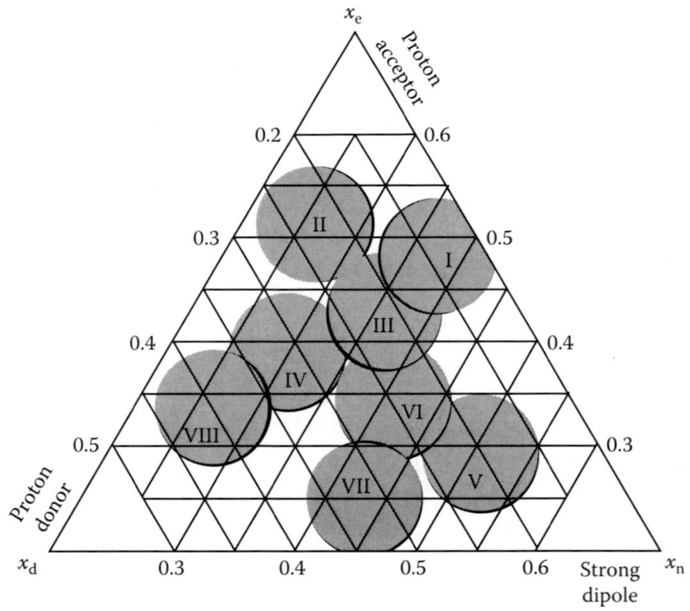

FIGURE 3.13 The solvent selectivity triangle (SST). (Adapted from Snyder, L. R., *J. Chromatogr. Sci.*, 16, 223–234, 1978.)

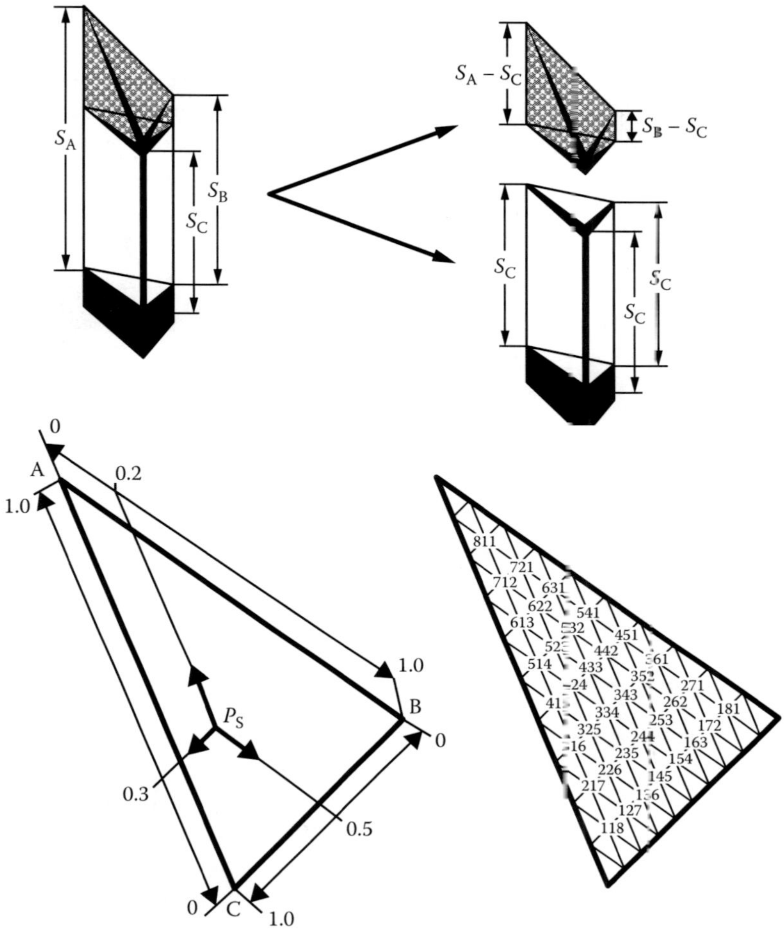

FIGURE 3.14 The PRISMA model. (With kind permission from Springer Science+Business Media: *Planar Chromatography, A Retrospective View for the Third Millennium*, Optimization of the mobile phase, 2001, pp. 47–67, Siouffi, A. M., Abbou, M., Nyiredy, Sz., ed.)

The PRISMA model method was introduced by Nyiredy and coworkers [50] for optimization of the mobile phase in RP HPLC. The details can also be found in Chapter 7.

3.6 CALCULATION OF RETENTION FROM MOLECULAR STRUCTURES OF ANALYTES: COMPUTER-ASSISTED METHOD DEVELOPMENT (DryLab, ChromSword)

It is worthwhile to discuss, in more detail, the modes of retention and selectivity optimization that can be applied to obtain suitable chromatographic resolution. Various strategies were described in the scientific literature [51]. Progress in methodology determines the progress in science with more reliable methods and more reliable results for scientific work. The process of HPLC method development generally uses a series of columns, and selectivity is obtained through modifications of the mobile phase organic modifier, pH, and occasionally ion pairing reagents. Although this can produce robust methods, it can be time-consuming, and the mobile phase may be not compatible with mass spectrometric (MS) detection. For example, the addition of ion pairing reagents and buffer salts are incompatible with liquid chromatography (LC)/MS.

In gradient elution, it is often difficult to predict how resolution and retention will change when we introduce a new gradient program. Computers have been applied in method development for HPLC almost since the inception of the technique. The first example of computer simulation for HPLC was use of a resolution map for isocratic RP-HPLC experiments as described by Laub and Purnell [19,52]. Glajch, Kirkland et al. [19,53] reported the use of mobile phases containing methanol (MeOH), acetonitrile (ACN), tetrahydrofuran (THF), and water in isocratic mode for optimizing solvent type based on the experimental plan. The isocratic approach of Glajch and Kirkland was extended to gradient elution in 1983 [19,54]. Several applications were described, for example,

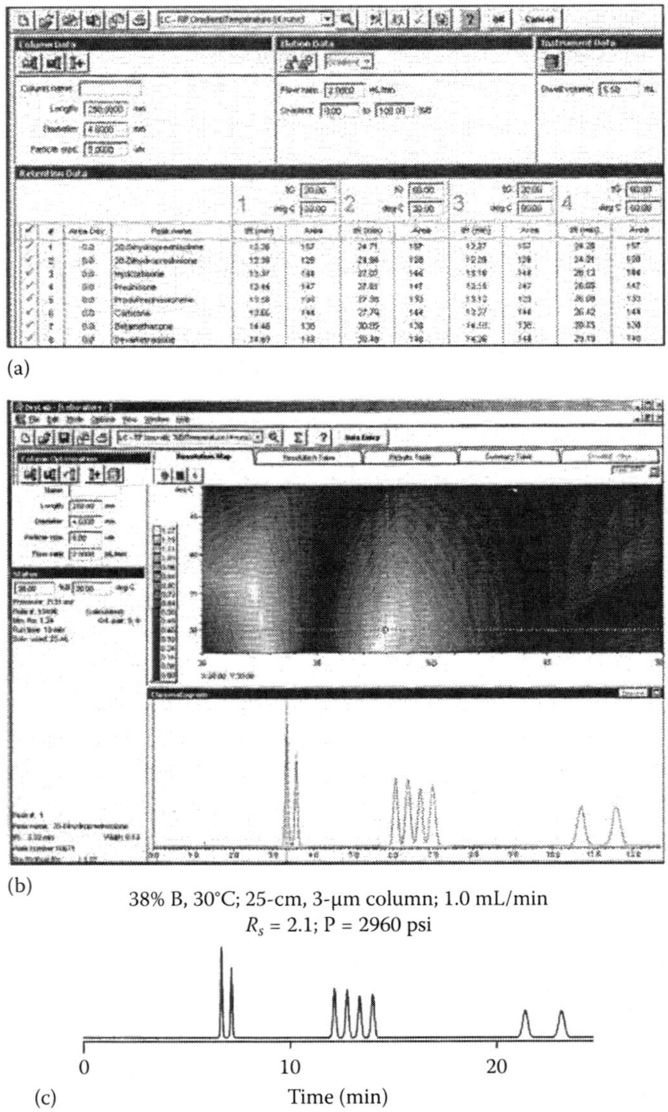

FIGURE 3.15 Examples of data entry (a) (gradient data) and laboratory (b) screens for isocratic computer simulation by means of DryLab for eight-component mixture of corticosteroids (From Dolan, J. W. et al., *J. Chromatogr. A*, 803, 1–31, 1998); conditions: C_{18} column (250 × 4.6-mm, 5-μm particles), mobile phase: acetonitrile-water; 38%B, 30°, 2.0 mL/min; (c) computer simulation for separation of sample with conditions with (b) except for a change in particle size (3 μm) and flow rate (1.0 mL/min). (Snyder, L. R., Kirkland, J. J., Dolan, J. W.: *Introduction to Modern Liquid Chromatography*. 2010. Copyright Wiley-VCH Verlag GmbH & Co. KGaA. Reproduced with permission.)

four-solvent mobile phase mapping by Glajch, Kirkland et al. [53], mapping of pH plus ion-pair concentration by Deming et al. [55], and the application of a simplex algorithm to the combined effects of %B and flow rate by Berridge [56]. In 1990, Snyder and Glajch published a review on computer-assisted method development [57], which appeared in *Journal of Chromatography* as volume no. 485, consisting of 43 papers in a broad scientific context.

Snyder's gradient elution theory was applied in the DryLab® software. The DryLab software is based on some semiempirical relationships. In RP-HPLC systems, retention can be described as Equation 3.4. Therefore, based on Equation 3.21,

$$\log k = \frac{a+b}{T_k} \tag{3.21}$$

where a, b are constants, T_k is temperature (K), and it is possible to predict isocratic retention as a function of ϕ_{mod} and gradient retention as a function of gradient program, column size, and flow rate (on the basis of only two experimental runs) [58].

For isocratic or gradient elution, it is possible to approximate the peak bandwidths by means of the Knox equation when the sample molecular weight, temperature, mobile-phase composition, and "column conditions" are known.

The 1985 DryLab software was introduced and subsequently expanded into the most comprehensive and widely used computer-simulation program [59]. The advanced software for gradient elution was introduced in 1987. The DryLab 2000 software (LC Resources, Walnut Creek, CA) can be used for predictions of separation as a function of pH and either t_G or %B.

Modern DryLab software includes modeling options for RP isocratic %B, gradient LC, the optimum pH, isocratic ternary conditions, ionic strength, additive/buffer concentration, and temperature. The DryLab program can also be applied in gas chromatography. Examples of applications of DryLab software are illustrated in Figures 3.15 and 3.16 [19,60].

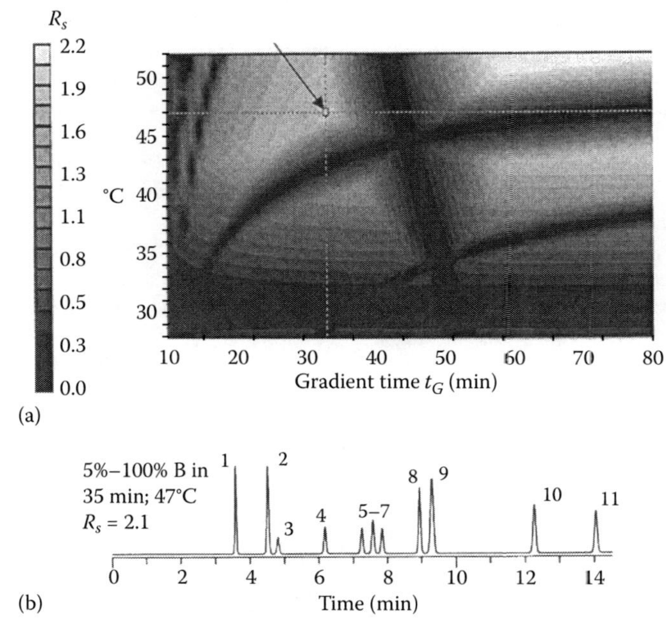

FIGURE 3.16 Computer simulation for optimization of 11-component mixture of gradient time and temperature (a) resolution map; (b) optimized separation. (Snyder, L. R., Kirkland, J. J., Dolan, J. W.: *Introduction to Modern Liquid Chromatography*. 2010. Copyright Wiley-VCH Verlag GmbH & Co. KGaA. Reproduced with permission.)

Another software program for computer-assisted HPLC method development and optimization is ChromSword. The ChromSword software provides an alternative option for computer simulation, with which one or more experimental runs are replaced by predictions of retention, based on molecular structure. The ChromSword software is based on Equation 3.22 [61–64]:

$$\ln k = a(V)^{2/3} + b(\Delta G) + c \tag{3.22}$$

where k is the retention factor; V is the partial molar volume of the modifier in water; ΔG is the energy of interaction with water; and a, b, and c are parameters of the sorbent/eluent system.

The main advantage of the ChromSword software is the reduction of the number of necessary preliminary experiments. For example, in 2-D experiments by using 2-D resolution maps, two chromatographic parameters can be optimized simultaneously, allowing analysis of resolution to determine the limiting factor for peak resolution.

Details about DryLab and ChromSword software and their applications for optimization of HPLC experiments for pesticide analysis can be found in Chapters 7 and 9 (and also other software and computer-assisted methods).

REFERENCES

1. Snyder, L. R., *Principles of Adsorption Chromatography*, Marcel Dekker, New York, 1968.
2. Soczewiński, E., Solvent composition effects in thin-layer chromatography systems of the type silica gel-electron donor solvent, *Anal. Chem.* 41 (1), 179–182, 1969.
3. Soczewiński, E., Quantitative retention-eluent composition relationships in partition and adsorption chromatography, in: *A Century of Separation Science*, Ed., Issaq, H., Marcel Dekker, New York, Chapter 11, pp. 179–195, 2002.
4. Soczewiński, E., Mechanistic molecular model of liquid–solid chromatography retention-eluent composition relationships, *J. Chromatogr. A* 965, 109–116, 2002.
5. Martin, A. J. P., Synge, R. L. M., A new form of chromatogram employing two liquid phases, *Biochem. J.* 35, 1358–1368, 1941.
6. James, A. T., Martin, A. J. P., Gas–liquid partition chromatography: The separation and micro-estimation of volatile fatty acids from formic acid to dodecanoic acid, *Biochem. J.* 50, 679–690, 1952.
7. Scott, R. P., Kucera, P., Solute interactions with the mobile and stationary phases in liquid–solid chromatography, *J. Chromatogr.* 112, 425, 1975.
8. Scott, R. P., The role of molecular interactions in chromatography, *J. Chromatogr.* 122, 35, 1976.
9. Kowalska, T., A thermodynamic interpretation of the behavior of higher fatty alcohols and acids upon the chromatographic paper, *Microchem. J.* 29, 375–390, 1984.
10. Kowalska, T., Bestimmung der Aktivitätskoeffizienten im chromatographischen Modell "binärer Lösungen," *Monatsh. Chem.* 116, 1129–1132, 1985.
11. Kowalska, T., Eine neue Möglichkeit zur Voraussage der Retentionsparameter bei der Adsorptionsdünnschichtchromatographie, *Fat Sci. Technol.* 90, 259–263, 1988.
12. Ościk, J., Chojnacka, G., Investigations on the adsorption process in thin-layer chromatography by using two-component mobile phases, *J. Chromatogr.* 93, 167–176, 1974.
13. Ościk, J., Chojnacka, G., The use of thin-layer chromatography in investigations of the adsorption of some aromatic hydrocarbons, *Chromatographia* 11, 731–735, 1978.
14. Kowalska, T., Podgórny, A., Estimation of the degree of self-association of water using a novel retention model, *Chromatographia* 31, 387–392, 1991.
15. Snyder, L. R., Dolan, J. W., Gant, J. R., Gradient elution in high-performance liquid chromatography. I. Theoretical basis for reversed-phase systems, *J. Chromatogr.* 165, 3–30, 1979.
16. Soczewiński, E., Wachtmeister, C. A., The relation between the composition of certain ternary two-phase solvent systems and R_M values, *J. Chromatogr.* 7, 311–320, 1962.
17. Snyder, L. R., Carr, P. W., Rutan, S. C., Solvatochromically based solvent-selectivity triangle, *J. Chromatogr. A* 656, 537–547, 1993.
18. Tuzimski, T., Use of planar chromatography in pesticide residue analysis in: *Handbook of Pesticides Methods of Pesticide Residue Analysis*, Nollet L. M. L., Rathore, H. S. (Eds.) CRC Press, Chapter 9, pp. 187–264, 2010.

19. Snyder, L. R., Kirkland, J. J., Dolan, J. W., *Introduction to Modern Liquid Chromatography*, A John Wiley & Sons, Inc., Hoboken, NJ, 2010.
20. Schoenmakers, P. J., Billiet, H. A. H., Tijssen, R., De Galan, L., Gradient selection in reversed-phase liquid chromatography, *J. Chromatogr.* 149, 519–537, 1978.
21. Croes, K., Steffens, A., Marchand, D. H., Snyder, L. R., Relevance of π–π and dipole–dipole interactions for retention on cyano and phenyl columns in reversed-phase liquid chromatography, *J. Chromatogr. A* 1098, 123–130, 2005.
22. Braumann, T., Weber, G., Grimme, L. H., Quantitative structure—Activity relationships for herbicides: Reversed-phase liquid chromatographic retention parameter, log k_w, versus liquid–liquid partition coefficient as a model of the hydrophobicity of phenylureas, *s*-triazines and phenoxycarbonic acid derivatives, *J. Chromatogr.* 261, 329–343, 1983.
23. Quarry, M. A., Grob, R. L., Snyder, L. R., Measurement and use of retention data from high-performance gradient elution: Correction for "non-ideal" processes originating within the column, *J. Chromatogr.* 285, 19–61, 1984.
24. Trappe, W., *J. Biochem.* 305, 150–154, 1940.
25. Snyder, L. R., Solvent selectivity in adsorption chromatography on alumina: Non-donor solvents and solutes, *J. Chromatogr.* 63, 15–44, 1971.
26. Gocan, S., Mobile phase in thin-layer chromatography, in *Modern Thin-Layer Chromatography*, Ed. Grinberg, E., Chromatographic Science Series, 52, Marcel Dekker, New York, pp. 427–434, 1990.
27. Stahl, E., Ed., Dünnschicht-Chromatographie, Springer Verlag, Berlin 1st edn., 1962 and 2nd edn., 1967.
28. Abboutt, D., Andrews, R. S., *An Introduction to Chromatography*, Longmans, Green, London, p. 27, 1965.
29. Neher, R., in *Thin-Layer Chromatography*, Ed. Marini-Bettolo B. G., Elsevier, Amsterdam, the Netherlands, 1964.
30. Saunders, D. L., Solvent selection in adsorption liquid chromatography, *Anal. Chem.* 46, 470–473, 1974.
31. Snyder, L. R., *Principles of Adsorption Chromatography*, Marcel Dekker, New York, 1968.
32. Snyder, L. R., Linear elution adsorption chromatography. VII. Gradient elution theory, *J. Chromatogr.* 13, 415–434, 1964.
33. Meyer, V., Palamareva, M. D., New graph of binary mixture solvent strength in adsorption liquid chromatography, *J. Chromatogr.* 641, 391–395, 1993.
34. Lide, D. R., Frederikse, H. P. R., Eds., Fluid properties, in *CRC Handbook of Chemistry and Physics*, CRC Press, Boca Raton, FL, 1995.
35. Héron, S., Tchapla, A., Propriétés et caractérisations des phases stationnaires et phases mobiles de chromatographie liquide à polarité de phases inversée, *Analusis* 21, 327–347, 1993.
36. Rohrschneider, L., Solvent characterization by gas–liquid partition coefficients of selected solutes, *Anal. Chem.* 45, 1241–1247, 1973.
37. McReynolds, W. O., Characterization of some liquid phases, *J. Chromatogr. Sci.* 8, 685–691, 1970.
38. Jandera, P., Churáček, J., Gradient elution in liquid chromatography. I. The influence of the composition of the mobile phase on the capacity ratio (retention volume, bandwidth, and resolution) in isocratic elution—theoretical considerations, *J. Chromatogr.* 91, 207, 1974.
39. Karger, B. L., Gant, R., Hartkopf, A., Weiner, P. H., Hydrophobic effects in reversed-phase liquid chromatography, *J. Chromatogr.* 128, 65–78, 1976.
40. Schoenmakers, P. J., Billiet, H. A. H., De Galan, L., Influence of organic modifiers on the rentention behaviour in reversed-phase liquid chromatography and its consequences for gradient elution, *J. Chromatogr.* 179–195, 519–537, 1979.
41. Poole, C. F., Dias, N. C., Practitioner's guide to method development in thin-layer chromatography, *J. Chromatogr., A* 892, 123–142, 2000.
42. Snyder, L. R., Classification of the solvent properties of common liquids, *J. Chromatogr.* 92, 223–230, 1974.
43. Snyder, L. R., Classification of the solvent properties of common liquids, *J. Chromatogr. Sci.* 16, 223–234, 1978.
44. Nyiredy, Sz., Solid-liquid extraction strategy on the basis of solvent characterization, *Chromatographia* 51, S-288–S-296, 2000.
45. Nyiredy, Sz., Dallenbach-Tölke, K., Sticher, O., The "PRISMA" optimization system in planar chromatography, *J. Planar Chromatogr.—Mod. TLC* 1, 336–342, 1988.
46. Dallenbach-Tölke, K., Nyiredy, Sz., Meier, B., Sticher, O., Optimization of overpressured layer chromatography of polar naturally occurring compounds by the "PRISMA" model, *J. Chromatogr.* 365, 63–72, 1986.

47. Nyiredy, Sz., Dallenbach-Tölke, K., Sticher, O., Correlation and prediction of the k′ values for mobile phase optimization in HPLC, *J. Liquid Chromatogr.* 12, 95–116, 1989.
48. Nyiredy, Sz., Fatér, Z. S., Automatic mobile phase optimization, using the "PRISMA" model, for the TLC separation of apolar compounds, *J. Planar Chromatogr.—Mod. TLC* 8, 341–345, 1995.
49. Siouffi, A. M., Abbou, M., *Optimization of the mobile phase* in: *Planar Chromatography, A Retrospective View for the Third Millennium*, Nyiredy, Sz. (Ed.), Springer Scientific Publisher, Budapest, Chapter 3, pp. 47–67, 2001.
50. Nyiredy, Sz., Meier, B., Erdelmeier, C. A. J., Sticher, O., "PRISMA": A geometrical design for solvent optimization in HPLC, *J. High Resolut. Chromatogr.* 8, 186–188, 1985.
51. Snyder, R. L., Kirkland, J. J., Glajch, J. L., *Practical HPLC Method Development*, Wiley & Sons, New York, 1997.
52. Laub, R. J., Purnell, J. H., Quantitative optimization of system variables for chromatographic separations: The column temperature, *J. Chromatogr.* 161, 49–57, 1978.
53. Glajch, J. L., Kirkland, J. J., Squire, K. M., Minor, J. M., Optimization of solvent strength and selectivity for reversed-phase liquid chromatography using an interactive mixture-design statistical technique, *J. Chromatogr.* 199, 57–79, 1980.
54. Kirkland, J. J., Glajch, J. L., Optimization of mobile phases for multisolvent gradient elution liquid chromatography, *J. Chromatogr.* 255, 27–39, 1983.
55. Sachok, B., Kong, R. C., Deming, S. N., Multifactor optimization of reversed-phase liquid chromatographic separations, *J. Chromatogr.* 199, 317–326, 1980.
56. Berridge, J. C., Unattended optimisation of reversed-phase high-performance liquid chromatographic separations using the modified simplex algorithm, *J. Chromatogr.* 244, 1–14, 1982.
57. Glajch, J. L., Snyder, L. R., *Computer-Assisted Method Development for High-Performance Liquid Chromatography*, Elsevier, Amsterdam, 1990.
58. Snyder, L. R., Dolan, J. W., *High-Performance Gradient Elution: The Practical Application of the Linear-Solvent-Strengh Model*, J. Wiley & Sons, Hoboken, NJ, 2007.
59. Molnar, I., Computer design of separation strategies by reversed-phase liquid chromatography: Development of DryLab software, *J. Chromatogr. A* 965, 175–194, 2002.
60. Dolan, J. W., Snyder, L. R., Djordjevic, M., Hill, D. W., Saunders, D. L., Van Heukelem, L., Waeghe, T. J., Simultaneous variation of temperature and gradient steepness for reversed-phase high-performance liquid chromatography method development: I. Application to 14 different samples using computer simulation, *J. Chromatogr. A* 803, 1–31, 1998.
61. Galushko, S. V., Calculation of retention and selectivity in reversed phase liquid chromatography, *J. Chromatogr. A* 552, 91–102, 1991.
62. Galushko, S. V., The calculation of retention and selectivity in reversed phase liquid chromatography. II. Methanol-water eluents, *Chromatographia* 36, 39–41, 1993.
63. Galushko, S. V., Kamenchuk, A. A., Pit, G. L., Calculation of retention in reversed-phase liquid chromatography: IV. ChromDream software for the selection of initial conditions and for simulating chromatographic behaviour, *J. Chromatogr. A* 660, 47–59, 1994.
64. Galushko, S. V., Kamenchuk, A. A., Pit, G. L., Software for method development in reversed-phase liquid chromatography, *Am. Lab.* 27, 421–432, 1995.

4 Choice of the Mode of Chromatographic Method for Analysis of Pesticides on the Basis of the Properties of Analytes

Piotr Kawczak, Tomasz Bączek, and Roman Kaliszan

CONTENTS

4.1 INTRODUCTION

Pesticides considered to be synthetic or natural substances are used to control pests or unwanted harmful organisms and are very common in the environment. Pesticides are at the same time a direct threat to the health of living humans. According to the Stockholm Convention on Persistent Organic Pollutants, nine of the 12 most dangerous and persistent organic chemicals are pesticides. Due to the diversity of chemical structures (even only between organochlorines, organophosphorus, and carbamate pesticides) continuous monitoring and identification, along with residue analysis, of pesticides is important.

The above processes can be mainly observed by the use and application of modern, automated analytical methods: high-performance liquid chromatography (HPLC) coupled to gas chromatography (GC) or/and mass spectrometry (MS). In turn, chromatographic techniques become the most crucial and the most convenient tools in the determination of pesticides in environmental and food samples.

4.2 MEASUREMENTS OF ACID-BASE CHARACTER BY HPLC METHODS: MODERN HPLC METHODS IN DETERMINATION OF BOTH pK_a AND LIPOPHILICITY

GC, HPLC, supercritical fluid chromatography (SFC), thin-layer chromatography (TLC), and combinations of these methods with MS are the basis of pesticide analysis.

GC can be applied during the analysis of compounds, both polar and nonpolar, and therefore, it seems perfect for simultaneous determination of residues of chlorinated pesticides, organophosphorus carbamates, and pyrethroids. HPLC can be considered as important as GC in pesticide residue analysis. HPLC is suitable for compounds possessing low volatility as well as polar compounds and, finally, for those which are thermally unstable (carbamates, herbicides, neonicotinoides) [1].

Adsorption is probably the most important mode of interaction between soil and pesticides and is responsible for the control of the pesticides' concentration in the soil/liquid phase. Adsorption processes vary from complete reversibility to total irreversibility. The extent of adsorption depends on the properties of soil and the compounds themselves, which include size, configuration, shape, molecular structure, solubility, polarity, polarizability, chemical functions, and charge distribution of interacting species as well as the acid-base nature of the pesticide molecule [2].

Pesticides and their metabolites adsorbed by ionic bonding or cation exchange exist in the cationic form in solution or can be protonated and become cationic. Ionic bonding comprises ionized or easily ionizable carboxylic and phenolic hydroxyl groups of humic substances. Bipyridilium pesticides (e.g., diquat and paraquat) bind to soil humic substances by ion exchange and by their cationic group. They form highly unreactive and stable bonds with the carboxyl groups of the humic substances. The effect of pH on binding has been reported for less basic pesticides [3]. It becomes cationic, at the same time, depending on the pesticides' basicity and the pH of the system as well as governing the degree of ionization of acidic groups of the humic substances. Next, Weber et al. [4], working with the *s*-triazine herbicides, secured maximum adsorption of basic compounds occurring at pH values close to their pK_a value [5].

In parallel, biologically convenient active compound testing methods used for lipophilicity and acidity determination are highly expected in modern pharmaceutical research. Reversed-phase (RP)-HPLC might be particularly useful for the determination of both pK_a and the apparent (pH-dependent) *n*-octanol–water partition coefficient to be applicable in high-throughput analysis of multicomponent mixtures. The pharmacokinetic properties of biologically active substances strongly depend on their lipophilicity. In the case of ionizable agents, the dissociation of the drug in aqueous compartments of a living system separated by lipid membranes is also of great importance. It is significant for the processes determining pharmacokinetic phase action (i.e., absorption, distribution, metabolism, and elimination). This implies a need for reliable procedures for the measurements of lipophilicity parameters and pK_a values that are convenient and fast and which request relatively small amounts of a given, often rare and valuable, substance. The RP-HPLC pH-gradient methods previously elaborated by Kaliszan and coworkers [6–11] are unique in that respect, allowing simultaneous determination of the important biorelevant physicochemical properties of biologically active compounds. There only exist minimum requirements for compounds [12]: availability in minute amounts only without prior isolation of an individual component of complex mixtures (which are presently often provided by combinatorial and other modern synthetic chemistry methods).

Retention of ionizable analytes in RP-HPLC can depend significantly on the pH and organic modifier content of the mobile phase. Various theoretical aspects of the isocratic and gradient RP-HPLC mode have been studied. Theoretical and experimental studies of pH influence on isocratic retention have been demonstrated by numerous researchers [13–16]. The rules were also established to provide a reliable way for optimizing isocratic separations with respect to pH.

Recently, the modern theory of combined pH/organic modifier gradient has been elaborated [10,11]. It is a unified, comprehensive theory of chromatographic retention for pH and/or organic modifier content changes during a chromatographic run. The proposed model of combined pH/ organic modifier gradient has been successfully applied to optimize separations in various chromatographic modes, that is, isocratic, gradient, or when both pH and organic modifier content change in a nonlinear manner were taken advantage of [17].

The pH gradient is a new separation technique useful to analyze the ionogenic analytes in RP-HPLC that include the programmed linear or nonlinear changes of the mobile phase pH with constant organic modifier content during the chromatographic separation. The increase of pH for acids and decrease of pH for bases, both affect the degree of analyte dissociation and lead to the changes in analyte retention. The pH gradient has standard features of the gradient RP-HPLC, such as peak compression, improved peak sensitivity, and minimized peak tailing. On the other hand, the degree of peak compression highly depends on the parameters of the pH gradient, like gradient steepness is more difficult to describe theoretically than in the case of classical organic modifier gradient [18].

The general equation describing the dependence of the analyte retention during isocratic and gradient conditions is as follows:

$$\int_0^{t'_R} \frac{1}{t_0}\frac{dt}{k_i} = 1 \tag{4.1}$$

where k_i is the instantaneous retention factor referring to the isocratic retention factor, k, which would be obtained with the mobile phase composition actually present at a column inlet; t_0 is a hold-up time; and $t'_R = t_R - t_0$ representing an adjusted retention time. In the case of a combined pH/ organic modifier gradient, the instantaneous retention factor is a function of time:

$$k_i(t) = k(\phi(t), \text{pH}(t)) \tag{4.2}$$

where $\phi(t)$ represents the organic modifier, and pH(t) represents pH changes with time at the column inlet.

By combining several further equations and equaling S values (constant; characteristic of each form of the analyte and the chromatographic system involved) one obtains the final model:

$$\int_{t_d}^{t'_R} \frac{10^{S\phi(t)}}{t_0} \frac{1+\sum_{r=1}^{n} 10^{r\text{pH}(t)-\text{p}K_{a,\text{chrom}}(r)}}{k_{w,0}+\sum_{r=1}^{n} k_{w,r}10^{r\text{pH}(t)-\text{p}K_{a,\text{chrom}}(r)}}\text{dt} = 1 \tag{4.3}$$

where $\text{p}K_{a,\text{chrom}}(r) = \sum_{i=1}^{r} \text{p}K_{a,i,\text{chrom}}$ is a sum of first r chromatographic $\text{p}K_{a,i,\text{chrom}}$. The $\text{p}K_{a,i,\text{chrom}}$ should be interpreted as a resultant of $\text{p}K_a$ changes during the organic modifier gradient, and $k_{w,r}$ denotes the retention factor in a neat water eluent.

For a neutral compound, Equation 4.3 reduces to a well-known equation describing retention of analyte in the organic modifier gradient:

$$\int_{t_d}^{t'_R} \frac{10^{S\phi(t)}}{t_0} \frac{1}{k_{w,0}}\text{dt} = 1 \tag{4.4}$$

In the case of monoprotic acid $[(HA)\leftrightarrow(H^+) + (A^-)]$ and base $[(BH^+)\leftrightarrow(H^+) + (B)]$, Equation 4.3 becomes

$$\int_{t_d}^{t'_R} \frac{10^{S\phi(t)}}{t_0} \frac{1+10^{\mathrm{pH}(t)-\mathrm{p}K_{a,1\mathrm{chrom}}} +}{k_{w,0}+k_{w,1}10^{\mathrm{pH}(t)-\mathrm{p}K_{a,1\mathrm{chrom}}}} \mathrm{dt} = 1 \tag{4.5}$$

The differentiation between acidic and basic analyte is done based on the relationship between $k_{w,r}$, that is, $k_{w,0}(HA) > k_{w,1}(A^-)$ and $k_{w,0}(BH^+) < k_{w,1}(B)$.

Similarly, for an analyte with two dissociation steps, such as acidic ($H_2A \leftrightarrow H^+ + HA^-$; $HA^- \leftrightarrow H^+ + A^{2-}$), basic $\left(BH_2^+ \leftrightarrow H^+ + BH^+ \leftrightarrow H^+ + B\right)$, and amphoteric ($HABH^+ \leftrightarrow H^+ + HAB$; $HAB \leftrightarrow H^+ + AB^-$) compounds, Equation 4.3 becomes

$$\int_{t_d}^{t'_R} \frac{10^{S\phi(t)}}{t_0} \frac{1+10^{\mathrm{pH}(t)-\mathrm{p}K_{a,1\mathrm{chrom}}}+10^{2\mathrm{pH}(t)-\mathrm{p}K_{a,1\mathrm{chrom}}-\mathrm{p}K_{a,2\mathrm{chrom}}}}{k_{w,0}+k_{w,1}10^{\mathrm{pH}(t)-\mathrm{p}K_{a,1\mathrm{chrom}}}+k_{w,2}10^{2\mathrm{pH}(t)-\mathrm{p}K_{a,1\mathrm{chrom}}-\mathrm{p}K_{a,2\mathrm{chrom}}}} \mathrm{dt} = 1 \tag{4.6}$$

The differentiation between analyte forms is done based on the relationship between $k_{w,r}$, and for diprotic acid, one has $k_{w,0}(H_2A) > k_{w,1}(HA^-) > k_{w,2}(A^{2-})$; for diprotic base, one has $k_{w,0}\left(BH_2^+\right) < k_{w,1}(BH^+) < k_{w,2}(B)$; and for an amphoteric analyte, one has $k_{w,0}(HABH^+) < k_{w,1}(HAB) > k_{w,2}(AB^-)$ [17–19].

4.3 MEASUREMENTS OF H-BOND ACIDITY AND BASICITY AND THE DETERMINATION OF MOLECULAR DESCRIPTION BY HPLC

A biologically active compound can be characterized by molecular descriptors. These descriptors can be rapidly estimated from its structure by a fragment scheme and used to predict physicochemical and transport properties of drug candidates (e.g., log *P*, solubility, gastrointestinal absorption, permeability, and blood–brain distribution). *n*-Octanol–water partition (log *P*) and distribution (log *D*) coefficients are the most widely used measures of lipophilicity, and several different schemes have been provided to explain log *P* quantitatively in terms of molecular size and polar parameters (e.g., H-bonding terms) [20–22].

Moreover, appropriate solvation equations can be interpreted to provide a qualitative chemical insight into biological partition and transport mechanisms. Quantitative experimental measures of H-bond strengths can be derived from partition measurements in different organic solvents. However, recent drug discovery applications require high-throughput methods and computational approaches to H-bonding that are generally used. Some calculations are simple, and H-bonding contributions are crudely estimated from simple parameters, such as the number of H-bond donors and acceptors. The Abraham's approach [23] based on the solvation equation correlates solute properties, such as solubility, partitioning, cell permeability, blood–brain distribution, and human intestinal absorption, with a set of five molecular descriptors, which are based on the physically meaningful theoretical cavity model of solute–solvent interactions: *E* is excess molar refraction, which models dispersion force interactions arising from the greater polarizability of π- and *n*-electrons (E = 0 for saturated alkanes); *S* is solute polarity/polarizability (due to solute–solvent interactions between bond dipoles and induced dipoles); *A* is solute H-bond acidity; *B* is solute H-bond basicity; and *V* is McGowan characteristic molar volume.

The H-bonding parameters are summation terms, which are relevant to the behavior of solutes in solvents. The acidity, *A*, is related to the strength and number of H-bonds formed by donor groups in the solute when they interact with single pairs of acceptor groups in solvent molecules. On the other hand, the basicity, B, is related to the strength and number of H-bonds formed by the single pairs of acceptor groups in the solute when they interact with donor solvents. The range of solutes for which descriptors are currently available can be considered now relatively large and comprises compounds that range from simple gases (e.g., hydrogen, nitrogen) to complex drugs and pesticides [23].

The solute descriptors can be combined in a linear free-energy relationship (LFER):

$$SP = c + eE + sS + aA + bB + vV \tag{4.7}$$

where the dependent variable, SP, is a solute property in a given system. For example, it might be log P for a set of solutes in a given water–solvent partition system. The coefficients in the equations can be found by the method of multiple linear regression. As the solute descriptors (E, S, A, B, and V) represent the solute influence on various solute–solvent phase interactions, the regression coefficients c, e, s, a, b, and v correspond to the complementary effect of the solvent phases on these interactions. The coefficients can then be regarded as system constants characterizing the stationary phase and contain chemical information on the stationary phase. Abraham's descriptors should provide a consistent and complete set to derive the LFER relationships. In turn, a quantitative structure–retention relationship model is going to explain properties that are not sensitive to molecular shape or conformation (i.e., solution–phase properties that involve rapid molecular motions). A large number of LFER equations exist in view of a wide variety of physical, physiological, and toxicological aspects in which rate of diffusion (log k), equilibrium partitioning (log P, log K_{ow}), or nonspecific interactions (–log toxicity, –log MIC) can play a very significant role [23,24].

Humic substances, with numerous oxygen- and hydroxyl-containing functional groups can form H-bonds with complementary groups on pesticide molecules. In that way, pesticide molecules compete with water for these binding sites. H-bonding is suggested to be playing a vital role in the adsorption of a number of nonionic polar pesticides, including substituted ureas and phenylcarbamates [25]. Acidic and anionic pesticides as well as esters can interact with soil organic matter by H-bonding at pH values below their pK_a in nonionized forms through their –COOH, –COOR groups [26]. Hydrophobic retention is then not necessary to explain an active adsorption mechanism, but it also can be regarded as a partitioning between a solvent and a nonspecific surface [2].

The role of the n-octanol–water partition coefficient (K_{ow}) for organic compounds has become one of the most important in predictive environmental studies. This physicochemical parameter can be used in evaluation models for the prediction of distribution among environmental compartments or in equations for estimation of bioaccumulation in animals and plants and to predict thermodynamic properties (such as heat of formation or critical volume) as well as in predictions of the toxic effects of a substance in quantitative structure–activity relationship (QSAR) studies. QSAR has been a common approach to understanding the link between the behavior of interest, for example, sorption and chemical, structural properties. The QSAR approach has been used to predict degradation of pollutants. The compounds have been classified there according to their degradability. In studies of the environmental fate of organic chemicals, the logarithm of the partition coefficient between n-octanol and water, log P, correlates with water solubility and soil/sediment adsorption coefficients, bioconcentration factors for aquatic organisms [27,28].

The fragmental constant approach proposed by Leo et al. [29] was based on the additive–constitutive nature of log K_{ow}. Suitable software, such as the Clog P, for example, HyperChem (HyperCube, Inc., Gainesville, FL, USA) or ACD/ChemSketch (Advance Chemistry Development, Inc., Toronto, Canada) is available for the calculation. The most common methods were classified as "fragment constant" methods in which a structure was divided into fragments (atoms or larger functional groups). The values for each group were also summed together (sometimes with structural correction factors) to yield the log P estimate. The range of applicability of this method was considered to be wider than other procedures. In fact, it was proved a reliable tool for highly lipophilic classes of compounds, such as organochlorines and pyrethroids. The reliability of this method can be highly improved when reliable experimental log K_{ow} values of a congeneric chemical are known. In this case, the calculation procedure can be simplified, applying the known molecule as a starting point. The method yielded a good agreement with direct measurements when used on neutral species. On the contrary, it tended to be inaccurate for compounds showing several functional groups with interactions not sufficiently defined. Moreover, for relatively complex molecules, also comprising pesticides, the breakdown of the molecule can be produced. In few cases, large or unusual fragments, fragmental values were not available by the method or in the software library [27,28,30].

n-Octanol–water partition coefficients and aqueous solubilities are considered important ones because they can be used to predict soil/sediment adsorption coefficients and bioconcentration factors for nonionic pesticides from aqueous solution [31]. The relationships between water solubility (S_w) and K_{ow} have been also additionally, extensively studied [32–34]. Unfortunately, the sizes of the studied databases were often relatively small, mainly designed for congeneric classes of chemicals. Therefore, special care should be taken during the selection of the most appropriate equation. It should be also noted that a nonnegligible source of error of this method is due to the quality of solubility data. Therefore, solubility figures must be selected very carefully.

The ability of the Weighted Holistic Invariant Molecular (WHIM) and GEometry, Topology, and Atom-Weights AssemblY (GETAWAY) descriptors to explain the effect of molecular structure on the retention of pesticides in RP-HPLC was also investigated [35]. GETAWAY descriptors outperformed WHIM descriptors when RP-HPLC behavior of structurally different solutes was modeled as a function of mobile phase composition. On the other hand, WHIM and GETAWAY descriptors' performance was comparable when retention of structurally related ionizable compounds was described as a function of both eluent composition and acidity. The explanatory power of these kinds of theoretical descriptors was finally just only slightly worse in comparison to quantum-chemical descriptors (however, computation time was enormously shorter).

Quantitative structure–property relationship (QSPR) studies based on QSAR and started from the calculation and selection of descriptors, to find the relationships between retention times of the selected compounds and derivation of mathematical models involving these multivariate data in order to be useful for predictive purposes in a chromatographic system were studied [36–38].

In the literature, there are also reports about the application of QSPR/QSAR in the chromatographic studies of pesticides [39–49]. Vapnik and coworkers [50] applied a computational method: support vector machine (SVM). SVM is a new algorithm developed on the basis of the machine learning community. SVM has been extended to solve regression problems and significantly improved the performance of QSPR studies due to the extraordinary opportunities to interpret the nonlinear relationships between molecular structure and activities. The SVM technique is considered to be a simple, sensitive, and inexpensive method that can predict accurately the chemical property of analytes, such as retention time. The proposed model could identify and provide some insights into calculated descriptors related to retention time. It seems that SVM-based modeling methods could produce more accurate QSPR models in comparison to linear regression methods because they have the ability to handle the possible nonlinear relationships during the training process. Additionally, a nonlinear SVM model based on the same sets of descriptors appeared to show better predictive ability than a heuristic method used in parallel [51,52]. Methods for lipophilicity measurements of different types of pesticides applying gradient and isocratic liquid chromatography (LC) techniques are provided in details in Table 4.1.

TABLE 4.1
Application of Gradient and Isocratic LC Techniques in Lipophilicity Measurements of Different Types of Pesticides

Determined Substances	Chromatographic Mode	References
Acetamides	Gradient RP-8/water-methanol	[54]
	Gradient silica gel impregnated by paraffin oil/different eluents	[55]
	Gradient RP-18/buffered methanol (pH 5.3, 7.3, and 8.3)	[56]
	Gradient RP-18/different eluents	[57]
	Thin layer chromatography (TLC) and micellar liquid chromatography (MLC): RP-18W/water-Brij35-tetrahydrofuran	[58]
Acetanilides	Gradient starch, cellulose/different eluents	[59]
	Isocratic RP-18/water-methanol	[60]
Biphenyls	Isocratic RP-18/water-methanol	[61]
Carbametes	Isocratic RP-18/water-methanol	[62]
Fungicides, growth regulators, herbicides	Isocratic and gradient RP-18/water-methanol	[63]
Fungicides, herbicides, insecticides	Gradient RP-18/water-methanol	[64]
	Isocratic and gradient alumina/*n*-hexane-dioxan	[65]
	Isocratic RP-18/water-methanol	[66]
	Isocratic soil/water-methanol	[67]
	Gradient ODP/buffered methanol	[68]
Herbicides	Micellar electrokinetic chromatography (MEKC) and MLC: RP-18/water-Brij 35	[69]
Herbicides	Isocratic RP-18/water-methanol	[70]
Pesticides	Isocratic RP-18/water-acetonitrile	
Pesticides	Isocratic RP-18/water-methanol	[71]
	Gradient PEE_{sil}/water-methanol	[72]
	Isocratic and gradient RP-18/water-methanol	[73]
	Gradient soil/water methanol	[74]
	Gradient silica, soil, mixed layers/different eluents	[75]
	Biopartitioning micellar chromatography (BMC): RP-18/buff. Brij 35 (pH 7.4)	[76]
	Microemulsion electrokinetic capillary *chromatography* (MEEKC) and vesicle electrokinetic chromatography (VEKC)	[77,78]
	Isocratic and gradient RP-18/water-methanol	[79]
	BMC: RP-18/buff. Brij 35 (pH 7.0)	[80]
	Isocratic ODS, RP-CN, soils/different eluents	[81]
	BMC: RP-18/water-Brij 35	[40]
Phenylureas	Isocratic C-18-PS-ZrO_2/water-methanol	[82]
	Isocratic RP-18/water-methanol	
S-triazines	*N*-octanol coated column/buffer saturated with *n*-octanol,	[83]
	Gradient RP-18e/water-methanol	[84,85]
	Isocratic RP-18/water-methanol	
	Isocratic RP-18/water-acetonitrile	
	Gradient RP-8 and RP-18/different eluents	
Thiazoles	Isocratic RP-18/buffered methanol (pH 4 and 7.4)	[86,87]
	Gradient RP-8 and RP-18W/different eluents	
Thioamides	Isocratic and gradient RP-18/water-methanol	[88]
Thiobenzanilides	Isocratic different systems	[89,90]
	Gradient different systems	[91,92]

(*Continued*)

TABLE 4.1 (CONTINUED)
Application of Gradient and Isocratic LC Techniques in Lipophilicity Measurements of Different Types of Pesticides

Determined Substances	Chromatographic Mode	References
Triazoles	Isocratic RP-18/water-methanol	[93–97]
	Gradient different systems	[98]
	Isocratic RP-18/buffered dioxane (pH 3.5)	

Source: Janicka, M., Chapter 11. Liquid chromatography in studying lipophylicity and bioactivity of pesticides, Stoytcheva, M., ed., *Pesticides—Strategies for Pesticides Analysis*, InTech, 2011.

4.4 INFLUENCE OF THE PROPERTIES OF ANALYTES AND MATRIX ON THE CHOICE OF CHROMATOGRAPHIC METHOD AND SYSTEM

Due to the relatively low concentrations of pesticide residues in food samples, their preparation requires not only the isolation of an analyte from a complex matrix, it also requires the appropriate enrichment before the final determination. Frequently used sample preparation methods include solid–phase extraction (SPE) or liquid–liquid extraction (LLE). The main problem during the analysis of pesticides, regarding their small concentrations, is associated with the diversity of their properties. Pesticides should not be treated as a homogeneous group of pollutants; they differ with many properties.

Determination of pesticide residues is usually performed with the use of chromatographic techniques: GC, LC, or TLC along with selective and specific detectors. In the analysis of insecticides (nitrogen- and organophosphorus), one can perform GC with nitrogen-phosphorus detector (GC-NPD). In the case of organochlorine insecticides, electron capture detectors are used. In addition, it appears useful to couple GC to MS. In the case of positive qualitative and quantitative determination (for example by GC-NPD), it is necessary to confirm it by other independent methods, such as GC using MS as a detector. Confirmation with the use of an alternative method is further necessary due to the insufficient reliability of identification based on retention times only. More polar pesticides (for example, phenoxy acids) can be determined using HPLC along with a diode array detector. Other HPLC detectors useful for the determination of pesticides may be as follows: UV, fluorescence, and amperometric. Definitely, less frequently used methods are SFC, capillary electrophoresis, flow injection analysis, and chemiluminescent methods. Some of the pesticides can also be determined by immunoenzymatic methods [99].

According to the literature, a variety of sample-preparation techniques have also been used for the extraction of transformation products of pesticides from different matrices. Nevertheless, SPE is mainly applied. However, other extraction techniques that can be found in pesticide-related literature are LLE, anion-exchange disk, TLC, pressurized liquid extraction (also known as accelerated solvent extraction), solid–liquid extraction, solid-phase microextraction, matrix solid–phase dispersion, supported liquid membrane, and molecularly imprinted polymers. For techniques in relation to matrices under study, the current literature positions can be classified into three main groups: environmental, biological, and food matrices. Water and soil are the most common matrices studied in environmental monitoring whereas urine is widely analyzed as a biological sample. Finally, in the food samples, a wide variety of crops and matrices have been studied [100–105].

Trace analysis of organic contaminants in food and/or biotic samples typically consists of the following steps: the first is solation of analytes from the sample matrix; the second step is removing of bulk coextracts from crude extract, and the third step is identification and quantification of target analytes. When a not sufficiently specific detector is used, that is, when the combination of retention time along with detection principles does not avoid false positive results, an additional confirmation step should be implemented. Separation techniques, represented nowadays mainly by GC and LC, are mainly applied to accomplish final determination steps. Although the instrumental configuration and setting of operational conditions predetermine greatly performance characteristics of a respective method, in many cases the key role is a choice of the appropriate sample preparation strategy corresponding to the first two steps [106].

In general terms, the wider range of physicochemical properties of target analytes, the more complicated is an efficient removal of co-isolated matrix components from a particular crude extract. Depending on their physicochemical nature (molecular size, polarity, thermal stability, volatility), the pesticide substances may interfere in various stages of chromatographic processes. The character of phenomena responsible for adverse effects depends finally on the quality of different analytical data of GC- and LC-based methods.

In GC, an efficient separation of matrix components avoiding both the identification and quantitation problems can be achieved by modern comprehensive GC × GC Coupling this technique with fast time-of-flight MS is a challenging option for the further development of residue analysis. As regards the injector-related problems, the use of the programmable temperature vaporizer injection technique may efficiently reduce matrix-induced chromatographic enhancement. The novel dirty matrix introduction technique [106] can also achieve a smart solution to problems caused by a nonvolatile matrix.

In LC-MS, which represents the prominent analytical technique in determination of polar and/or thermo-labile residues, great attention has to be paid to the compensation of enhancement or suppression of the pesticide analyte signal by co-eluting the matrix. The use of echo-peak calibration may represent a good compromise in solution of these adverse phenomena. Nowadays, LC-MS instruments employing atmospheric-pressure ionization (API) are probably the most common technique used in pesticide trace analysis. Among API techniques, electrospray ionization (ESI) and external and atmospheric pressure chemical ionization are the most often applied ionization techniques. Analogously to GC, co-eluting matrix components may interfere with the detection process. To achieve high sensitivity and selectivity of target analyte detection, tandem mass spectrometry (MS/MS) employing either tandem-in-time MS/MS (ion trap analyzers) or tandem-in-space MS/MS (e.g., triple stage quadrupoles) is a preferred option by most experts working in the field of pesticide trace analysis. In any case, as a part of the validation procedure, the evaluation of an influence of the sample matrix on the quality of generated data is a crucial issue [106].

LC-MS is a highly selective method taking into account the use of selected ion monitoring and multiple reaction-monitoring modes. Only the signal of interest is registered, leaving out the information on the occurrence of all other compounds. This provides the illusion that other substances co-eluting with the pesticide analyte do not interfere. However, other nonpesticide compounds—although invisible in the LC-MS signal—may and very often do interfere. The suppression or enhancement of ionization by the co-eluting compounds occurs in the ESI source before any MS detection, and it is thus, in principle, impossible to compensate it by MS. A change of ionization efficiency in the presence of other compounds is called the matrix effect. This is defined by the International Union of Pure and Applied Chemistry as "the combined effect of all components of the sample other than the analyte on the measurement of the quantity." One challenge of sample preparation is to overcome these matrix effects. One of the most popular methods worldwide is the quick, easy, cheap, effective, rugged, and safe (QuEChERS) method. Since its introduction, the method was modified and extended several times and is a part of the diverse official method collections in different variants. The compounds that cause the matrix effect may appear as normal chromatographic peaks or as broad bands. The matrix effect and recovery are both dependent on the matrix. For obtaining reliable results during validation, both matrix effect and recovery should be studied for all the analyzed samples [107,108].

The occurrence of matrix-induced effects in GC and their extent can be simultaneously influenced by many factors: (i) pesticide character, that is, higher apparent recoveries together with poorer precision of repeated injections; (ii) matrix type—matrix-induced effects recorded for particular pesticides were clearly proven to depend on matrix type (i.e., type of co-extracts in sample); (iii) analyte or matrix concentration—unacceptable accuracy of measurements was encountered especially at lower concentration levels of analytes and/or at a higher matrix concentration in the sample; (iv) the state-of-the-art of the GC system—relative GC responses (the response of standard in pure solvent) of most pesticides were gradually diminished over time when these analytes were injected in to matrix-containing solutions [109].

Many compounds are not affected by matrix-induced enhancement, as these compounds are either thermally stable or have limited potential for adsorption interactions in hot vaporizing injectors. This can also take place because the matrix is unable to provide a significant protecting effect. Most of these compounds are polar and/or strong hydrogen-bond acids and/or bases exemplified by the presence of phosphate (–PO), hydroxyl, amino, imidazole, benzimidazole, carbamate (–O–CO–NH–), and urea (–NH–CO–NH–) functional groups. Organophosphorous pesticides with a (–PS) group are not as susceptible to matrix-induced enhancement as those with a (–PO) group. Most studies demonstrating a matrix-induced enhancement have employed either vegetable or animal food matrices of several types. However, matrix-induced enhancement is not limited to food analyses only. These are simply the matrix types most commonly studied together with organophosphorous pesticides because of their role as a source of pesticides in the human diet. Therefore, regulatory authorities need to assess potential health effects arising from these contaminants in the food chain. Matrix-induced enhancement can result in the reporting of false results in pesticide residue analysis and in estimation of recoveries of sample preparation procedures when matrix-free standards are used for calibration. The response enhancement effect can vary for different pesticides, matrix types, and concentrations and, of course, for the system parameters related to the type, design, and operating conditions of the injector. A number of methods can be provided during the measurements of control of the enhancement effect. However, for calibration purposes, the most effective solutions are the use of either matrix-matched standards or analyte protectants. In addition, LC-MS/MS becomes more useful among variable approaches used in pesticide residue laboratories as the one having the potential to replace GC for those compounds affected by matrix-induced enhancement [110].

4.5 CONCLUSIONS

Pesticides are among the most dangerous chemical substances in the environment, mainly because of their resistance to degradation as well as long-term disadvantageous effects in living organisms. Their presence in food possesses a particular threat to health after exposure. Therefore, it is necessary to monitor and study pesticide residues in food.

A number of so far developed testing procedures and techniques for pesticide sample preparation allows the determination of residues of the compound of interest in a wide variety of food and agricultural products.

Determination of pesticide residues in food is complicated due to its presence in the matrix sample and low concentrations of pesticides in the matrix. That leads to the requirement for labor- and time-consuming methods. Therefore, research continues to improve already used techniques and to develop new ones, which, in a simple, cheap, effective, and safe for the environment way, would allow the determination, at the same time, of pesticides derived from various chemical groups.

ACKNOWLEDGMENTS

This work was supported by grants from the Polish National Service Centre Nos. N 405 630 038 and 2011/03/B/NZ1/03113.

REFERENCES

1. Fodor-Csorba, K., Chromatographic methods for the determination of pesticides in foods, *J. Chromatogr. A*, 624, 353–367, 1992.

2. Gevao, B., Semple, K. T., Jones, K. C., Bound pesticide residues in soils: A review, *Environ. Pollut.*, 108, 3–14, 2000.
3. Weber, J. B., Weed, S. B., Adsorption and desorption of Diquat, Paraquat, and Prometone by montmorillonitic and kaolinitic clay minerals, *Soil Sci. Soc. Am. Pro.*, 32, 485–487, 1968.
4. Weber, J. B., Weed, S. B., Ward, T. M., Adsorption of *s*-triazines by soil organic matter, *Weed Sci.*, 17, 417–421, 1969.
5. Gevao, B., Semple, K. T., Jones, K. C., Bound pesticide residues in soils: A review, *Environ. Pollut.*, 108, 3–14, 2000.
6. Kaliszan, R., Haber, P., Bączek, T., Siluk, D., Gradient HPLC in the determination of drug lipophilicity and acidity, *Pure Appl. Chem.*, 73, 1465–1475, 2001.
7. Kaliszan, R., Haber, P., Bączek, T., Siluk, D., Valko, K., Lipophilicity and pK_a estimates from gradient high-performance liquid chromatography, *J. Chromatogr. A*, 965, 117–127, 2002.
8. Kaliszan, R., Wiczling, P., Markuszewski, M. J., pH gradient reversed-phase HPLC, *Anal. Chem.*, 76, 749–760, 2004.
9. Wiczling, P., Markuszewski, M. J., Kaliszan, R., Determination of pK_a by pH gradient reversed-phase HPLC, *Anal. Chem.*, 76, 3069–3077, 2004.
10. Wiczling, P., Markuszewski, M. J., Kaliszan, M., Kaliszan, R., pH/organic solvent double-gradient reversed-phase HPLC, *Anal. Chem.*, 77, 449–458, 2005.
11. Wiczling, P., Kawczak, P., Nasal, A., Kaliszan, R., Simultaneous determination of pK_a and lipophilicity by gradient RP HPLC, *Anal. Chem.*, 78, 239–249, 2006.
12. Wiczling, P., Nasal, A., Kubik, Ł., Kaliszan, R., A new pH/organic modifier gradient RP HPLC method for convenient determination of lipophilicity and acidity of drugs as applied to established imidazoline agents, *Eur. J. Pharm. Sci.*, 47, 1–5, 2012.
13. Horváth, C., Melander, W., Molnár, I., Liquid chromatography of ionogenic substances with nonpolar stationary phases, *Anal. Chem.*, 49, 142–156, 1977.
14. Rittich, B., Pirochtova, M. J., Chromatographic behaviour of aromatic acids in reversed-phase high-performance liquid chromatography, *J. Chromatogr. A*, 523, 227–233, 1990.
15. Lopez Marques, R. M., Schoenmakers, P. J., Modelling retention in reversed-phase liquid chromatography as a function of pH and solvent composition, *J. Chromatogr. A*, 592, 157–182, 1992.
16. Jano, I., Hardcastle, J. E., Zhao, K., Vermillion-Salsbury, R., General equation for calculating the dissociation constants of polyprotic acids and bases from measured retention factors in high-performance liquid chromatography, *J. Chromatogr. A*, 762, 63–72, 1997.
17. Wiczling, P., Kaliszan, R., Influence of pH on retention in linear organic modifier gradient RP HPLC, *Anal. Chem.*, 80, 7855–7861, 2008.
18. Wiczling, P., Kaliszan, R., pH gradient as a tool for the separation of ionizable analytes in reversed-phase high-performance chromatography, *Anal. Chem.*, 82, 3692–3698 2010.
19. Wiczling, P., Waszczuk-Jankowska, M., Markuszewski, M. J., Kaliszan, R., The application of gradient reversed-phase high-performance liquid chromatography to the pK_a and log k_w determination of polyprotic analytes, *J. Chromatogr. A*, 1214, 109–114, 2008.
20. El-Tayar, A., Testa, B., Carrupt P.-A., Polar intermolecular interactions encoded in partition coefficients: An indirect estimation of hydrogen-bond parameters of polyfunctional solutes, *J. Phys. Chem.*, 96, 1455–1459, 1992.
21. Abraham, M. H., Scales of hydrogen bonding—Their construct on and application to physicochemical and biochemical processes, *Chem. Soc. Rev.*, 22, 73–83, 1993.
22. Platts, J. A., Abraham, M. H., Butina, D., Hersey, A., Estimation of molecular linear free energy relationship descriptors by a group contribution approach. 2. Prediction of partition coefficients, *J. Chem. Inf. Comput. Sci.*, 40, 71–80, 2000.
23. Abraham, M. H., Ibrahim, A., Zissimos, A. M., Zhao, Y. H., Comer, J., Reynolds, D. P., Application of hydrogen bonding calculations in property based drug design, *Drug Discov. Today*, 7, 1056–1063, 2002.
24. Clarke, E. D., Beyond physical properties—Application of Abraham descriptors and LFER analysis in agrochemical research, *Bioorg. Med. Chem.*, 17, 4153–4159, 2009.
25. Senesi, N., Testini, C., The environmental fate of herbicides: The role of humic substances, *Ecol. Bull.* 35, 477–490, 1983.
26. Khan, S. U., Interactions of humic substances with chlorinated phenoxyacetic and benzoic acids, *Environ. Lett.*, 4, 141–148, 1973.
27. Finizio, A., Vighi, M.; Sandroni, D., Determination of *n*-octanol/water partition coefficient (K_{ow}) of pesticide critical review and comparison of methods, *Chemosphere*, 34, 131–161, 1997.

28. Benfenati, E., Gini, G., Piclin, N., Roncaglioni, A., Varí, M. R., Predicting log *P* of pesticides using different software, *Chemosphere*, 53, 1155–1164, 2003.
29. Leo, A., Hansch, C., Helkins, D., Partition coefficient and their uses, *Chem. Rev.* 71, 525–538, 1971.
30. Worrall, F., Thomsen, M., Quantum vs. topological descriptors in the development of molecular models of groundwater pollution by pesticides, *Chemosphere*, 54, 585–596, 2004.
31. Briggs, G. G., Theoretical and experimental relationships between soil adsorption, octanol–water partition coefficients, water solubilities, bioconcentration factors, and the parachor, *J. Agric. Food Chem.*, 29, 1050–1059, 1981.
32. Nirmalakhandan, N. N., Speece, R. E., Prediction of aqueous solubility of organic chemicals based on molecular structure, *Environ Sci. Technol.*, 22, 328–338, 1988.
33. Patil, G. S., Prediction of aqueous solubility and octanol–water partition coefficient for pesticides based on their molecular structure, *J. Hazard. Mater.*, 36, 35–43, 1994.
34. Kumbar, S. G., Kulkarni, A. R., Dave, A. M., Aminabhavi, T. M., An assessment of solubility profiles of structurally similar hazardous pesticide in water + methanol mixture and co-solvent effect on partition coefficient. *J. Hazard. Mater.*, 89, 233–239, 2002.
35. D'Archivio, A. A., Maggi, M. A., Mazzeo, P., Ruggieri, F., Quantitative structure–retention relationships of pesticides in reversed-phase high-performance liquid chromatography based on WHIM and GETAWAY molecular descriptors, *Anal. Chim. Acta*, 628, 162–172, 2008.
36. Kaliszan, R., *Quantitative Structure-Chromatographic Retention Relationships*, Wiley, New York, 1987.
37. Kaliszan, R., *Structure and Retention in Chromatographic Approach*, Harwood Academic Publishers, Amsterdam, 1997.
38. Kaliszan, R., QSRR: Quantitative structure–(chromatographic) retention relationships, *Chem. Rev.*, 107, 3212–3246, 2007.
39. Gramatica, P., Corradi, M., Consonni, V., Modelling and prediction of soil sorption coeffcients of non-ionic organic pesticides by molecular descriptors, *Chemosphere*, 41, 763–777, 2000.
40. Ma, W., Luan, F., Zhang, H., Zhang, X., Liu, M., Hu, Z., Fan, B., Quantitative structure–property relationships for pesticides in biopartitioning micellar chromatography, *J. Chromatogr. A*, 1113, 140–147, 2006.
41. Toropov, A. A., Benfenati, E., QSAR models for Daphnia toxicity of pesticides based on combinations of topological parameters of molecular structures, *Bioorg. Med. Chem.*, 14, 2779–2788, 2006.
42. D'Archivio, A. A., Ruggieri, F., Mazzeo, P., Tettamanti, E., Modelling of retention of pesticides in reversed-phase high-performance liquid chromatography: Quantitative structure–retention relationships based on solute quantum-chemical descriptors and experimental (solvatochromic and spin-probe) mobile phase descriptors, *Anal. Chim. Acta*, 593, 140–151, 2007.
43. Aschi, M., D'Archivio, A. A., Maggi, M. A., Mazzeo, P., Ruggieri, F., Quantitative structure–retention relationships of pesticides in reversed-phase high-performance liquid chromatography, *Anal. Chim. Acta*, 582, 235–242, 2007.
44. Ghasemi, J., Asadpour, S., Abdolmaleki, A., Prediction of gas chromatography/electron capture detector retention times of chlorinated pesticides, herbicides, and organohalides by multivariate chemometrics methods, *Anal. Chim. Acta*, 588, 200–206, 2007.
45. Hadjmohammadi, M. R., Fatemi, M. H., Kamel, K., Quantitative structure–property relationship study of retention time of some pesticides in gas chromatography, *J. Chromatogr. Sci.*, 45, 400–404, 2007.
46. Hu, R., Yin, C., Wang, Y., Lu, C., Ge, T., QSPR study on GC relative retention time of organic pesticides on different chromatographic columns, *J. Sep. Sci.*, 31, 2434–2443, 2008.
47. Zhao, C., Boriani, E., Chana, A., Roncaglioni, A., Benfenati, E., A new hybrid system of QSAR models for predicting bioconcentration factors (BCF), *Chemosphere*, 73, 1701–1707, 2008.
48. Bhattacharjee, A. K., Gordon, J. A., Marek, E., Campbell, A., Gordon, R. K., 3D-QSAR studies of 2,2-diphenylpropionates to aid discovery of novel potent muscarinic antagonists, *Bioorg. Med. Chem.*, 17, 3999–4012, 2009.
49. Kar, S., Roy, K., QSAR modeling of toxicity of diverse organic chemicals to Daphnia magna using 2D and 3D descriptors, *J. Hazard. Mater.*, 177, 344–351, 2010.
50. Vapnik, V. N., *Statistical Learning Theory*, Wiley, New York, 1998.
51. Dashtbozorgi, Z., Golmohammadi, H., Konoz, E., Support vector regression based QSPR for the prediction of retention time of pesticide residues in gas chromatography–mass spectroscopy, *Microchem. J.*, 106, 51–60, 2013.
52. Li, X., Luan, F., Si, H., Hu, Z., Liu, M., Prediction of retention times for a large set of pesticides or toxicants based on support vector machine and the heuristic method, *Toxicol. Lett.*, 175, 136–144, 2007.
53. Janicka, M., Chapter 11. Liquid chromatography in studying lipophylicity and bioactivity of pesticides, Stoytcheva, M., ed., *Pesticides–Strategies for Pesticides Analysis*, InTech, 2011.

54. Cimpan, G., Irimie, F., Gocan, S., Prediction of the lipophilicity of some N-hydroxyethylamides of aryloxyalkylene and pyridine carboxylic acids by reversed phase thin-layer chromatography, *J. Planar Chromatogr.*, 11, 342–345, 1998.
55. Djaković-Sekulić, T. L., Perišić-Janjić, N. U., Petrović, S. D., Normal- and reversed phase chromatography of para-substituted propanoic acid amides, *J. Planar Chromatogr.*, 15, 274–279, 2002.
56. Janicka, M., Ościk-Mendyk, B., Tarasiuk, B., Planar chromatography in studies of the hydrophobic properties of some new herbicides, *J. Planar Chromatogr.*, 17, 186–191, 2004.
57. Perišić-Janjić, N. U., Vastag, G., Tomić, J., Petrović, S., Effect of the physicochemical properties of *N,N*-disubstituted-2-phenylacetamide derivatives on their retention behavior in RP-TLC, *J. Planar Chromatogr.*, 20, 353–359, 2007.
58. Janicka, M., Pietras-Ożga, D., Chromatographic evaluation of lipophilicity of *N*-phenyl-trichloroacetamide derivatives using micellar TLC and OPLC, *J. Planar Chromatogr.*, 23, 396–399, 2010.
59. Djaković-Sekulić, T. L., Perišić-Janjić, N. U., Petrović, S. D., The retention behavior of some anilides on unconventional TLC supports, *J. Planar Chromatogr.*, 10, 298–304, 1997.
60. Djaković-Sekulić, T. L., Petrović, S. M., Perišić-Janjić, N. U., Petrović, S. D., HPLC behavior and hydrophobic parameters of some anilides, *Chromatographia*, 54, 60–64, 2001.
61. De Kock, A. C., Lord, D. A., A simple procedure for determining octanol–water patition coefficients using reverse phase high performance liquid chromatography (RPHPLC), *Chemosphere*, 16, 133–142, 1987.
62. Yamagami, C., Katashiba, N., Hydrophobicity parameters determined by reversed-phase liquid chromatography. Prediction of log *P* values for phenyl *N*-methyl and phenyl *N,N*-dimethylcarbamates, *Chem. Pharm. Bull.*, 44, 1338–1343, 1996.
63. Zhang, L., Zhang, M., Wang, L. X., Wang, Q. S., Relationship between the lipophilicity and specific hydrophobic surface area of some pesticides by RP-HPLC and HPTLC, *Chromatographia*, 52, 305–308, 2000.
64. Darwish, Y., Cserháti, T., Forgács, E., Relationship between lipophilicity and specific hydrophobic surface area of a non-homologous series of pesticides, *J. Planar. Chromatogr.*, 6, 458–462, 1993.
65. Cserháti, T., Forgács, E., Relationship between the high-performance liquid and thin-layer chromatographic retention of non-homologous series of pesticides on an alumina support, *J. Chromatogr. A*, 668, 495–500, 1994.
66. Cserháti, T., Forgács, E., Relationship between the hydrophobicity and specific hydrophobic surface area of pesticides determined by high-performance liquid chromatography compared with reversed-phase thin-layer chromatography, *J. Chromatogr. A*, 771, 105–109, 1997.
67. Xu, F., Liang, X., Lin, B., Su, F., Schramm, K.-W., Kettrup, A., Soil column chromatography for correlation between capacity factors and soil organic partition coefficients for eight pesticides, *Chemosphere*, 39, 2239–2248, 1999.
68. Donovan, S. F., Pescatore, M. C., Method for measuring the logarithm of the octanol–water partition coefficient by using short octadecyl–poly(vinyl alcohol) high performance liquid chromatography columns, *J. Chromatogr. A*, 952, 47–61, 2002.
69. Martín-Biosca, Y., Escuder-Gilabert, L., Marina, M. L., Sagrado, S., Villanueva-Camañas, R. M., Quantitative retention- and migration-toxicity relationships of phenoxy acid herbicides in micellar liquid chromatography and micellar electrokinetic chromatography, *Anal. Chim. Acta*, 443, 191–203, 2001.
70. Braumann, T., Weber, G., Grimme, H., Quantitative structure–activity relationships for herbicides: Reversed-phase liquid chromatographic retention parameter, log k_w, versus liquid–liquid partition coefficient as a model of the hydrophobicity of phenylureas, *s*-triazines and phenoxycarbonic acid derivatives, *J. Chromatogr. A*, 261, 329–343, 1983.
71. Hsieh, M.-M., Dorsey, J. G., Bioavailability estimation by reversed-phase liquid chromatography: High bonding density C-18 phases for modeling biopartitioning processes, *Anal. Chem.*, 67, 48–57, 1995.
72. Forgács, E., Cserháti, T., Use of principal component analysis for studying the separation of pesticides on polyethylene-coated silica columns, *J. Chromatogr. A*, 797, 33–39, 1998.
73. Verbruggen, E. M. J., Klamer, H. J. C., Villerius, L., Brinkman, U. A. Th., Hermens, J. L. M., Gradient elution reversed-phase high-performance liquid chromatography for fractionation of complex mixtures of organic micropollutants according tohydrophobicity using isocratic retention parameters, *J. Chromatogr. A*, 835, 19–27, 1999.
74. Ravanel, R., Liégeois, M. H., Chevallier, D., Tissut, M., Soil thin-layer chromatography and pesticide mobility through soil microstructures: new technical approach, *J. Chromatogr. A*, 864, 145–154, 1999.
75. Mohammad, A., Khan, I. A., Jabeen, N., Thin-layer chromatographic studies of the mobility of pesticides through soil-containing static flat-beds, *J. Planar Chromatogr.*, 14, 283–290, 2001.

76. Escuder-Gilabert, L., Martín-Biosca, Y., Sagrado, S., Villanueva-Camañas, R. M., Medina-Hernández, M. J., Biopartitioning micellar chromatography to predict ecotoxicity, *Anal. Chim. Acta*, 448, 173–185, 2001.
77. Klotz, W., Schure, M. R., Foley, J. P., Determination of octanol–water partition coefficients of pesticides by microemulsion electrokinetic chromatography, *J. Chromatogr. A*, 930, 145–154, 2001.
78. Klotz, W., Schure, M. R., Foley, J. P., Rapid estimation of octanol–water partition coefficients using synthesized vesicles in electrokinetic chromatography, *J. Chromatogr. A*, 962, 207–219, 2002.
79. Pyka, A., Miszczyk, M., Chromatographic evaluation of the lipophilic properties of selected pesticides, *Chromatographia*, 61, 1, 37–42, 2005.
80. Bermudez-Saldaña, J. M., Escuder-Gilabert, L., Medina-Hernández, M. J., Villanueva-Camañas, R. M., Sagrado, S., Chromatographic evaluation of the toxicity in fish of pesticides, *J. Chromatogr. B*, 814, 115–125, 2005.
81. Bermúdez-Sadaña, J. M., Escuder-Gilabert, L., Medina-Hernández, M. J., Villanueva-Camañas, R. M., Sagrado, S., Chromatographic estimation of the soil sorption coefficients of organic compounds, *Trends Anal. Chem.*, 25, 122–132, 2006.
82. Xia, Y., Guo, Y., Wang, H., Wang, Q., Zuo, Y., Quantitative structure-retention relationships of benzoylphenylureas on polystyrene-octadecene-encapsulated zirconia stationary phase in reversed-phase high performance liquid chromatography, *J. Sep. Sci.*, 28, 73–77, 2005.
83. Kaune, A., Brüggemenn, R., Kettrup, A., High-performance liquid chromatographic measurement of the 1-octanol–water partition coefficient of *s*-triazine herbicides and some of their degradation products, *J. Chromatogr. A*, 805, 119–126, 1998.
84. Janicka, M., Perišić-Janjić, N. U., Różyło, J. K., Thin-layer and over-pressured layer chromatography for evaluation of the hydrophobicity of *s*-triazine derivatives, *J. Planar Chromatogr.*, 17, 468–475, 2004.
85. Janicka, M., Kwietniewski, L., Perišić-Janjić, N. U., Determination of retention factors of *s*-triazines homologous series in water using numerical method basing on Ościk's equation, *Chromatographia*, 63, 87–93, 2006.
86. Matysiak, J., Żabińska, A., Różyło, J. K., Niewiadomy, A., Determination of the lipiophylicity of bioactive 2-phenylbenzothiazoles by RPTLC. *J. Planar Chromatogr.*, 15, 380–383, 2002.
87. Matysiak, J., Niewiadomy, A., Sęczyna, B., Różyło, J. K., Relationships between LC retention, octanol–water partition coefficient, and fungistatic properties of 2-(2,4-dihydroxyphenyl)benzothiazoles, *J. AOAC Int.*, 87, 579–586, 2004.
88. Kostecka, M., Niewiadomy, A., Czeczko, R., Evaluation of N-substituted 2,4-dihydroxyphenylthioamide fungicide lipophilicity using the chromatographic techniques HPLC and HPTLC, *Chromatographia*, 62, 121–126, 2005.
89. Niewiadomy, A., Matysiak, J., Żabińska, A., Różyło, J. K., Sęczyna, B., Jóźwiak, K., Reversed-phase high-performance liquid chromatography in quantitative structure–activity relationship studies of new fungicides, *J. Chromatogr. A*, 828, 431–438, 1998.
90. Józwiak, K., Szumiło, H., Sęczyna, B., Niewiadomy, A., RPHPLC as a tool for determining the congenericity of a set of 2,4-dihydroxythiobenzanilide derivatives, *Chromatographia*, 52, 159–161, 2000.
91. Matysiak, J., Niewiadomy, A., Żabińska, A., Różyło, J. K., Structure and retention of 2,4-dihydroxythiobenzanilides in a reversed-phase system, *J. Chromatogr. A*, 830, 491–496, 1999.
92. Janicka, M., Kwietniewski, L., Matysiak, J., A new method for estimations log k_w values and solute biological activity, *J. Planar Chromatogr.*, 13, 285–289, 2000.
93. Perišić-Janjić, N. U., Ačanski, M. M., Janjić, N. J., Lazarević, M. D., Dimova, V., Study of the lipophilicity of some 1,2,4-triazole derivatives by RPHPLC and TLC, *J. Planar Chromatogr.*, 13, 281–284, 2000.
94. Perišić-Janjić, N. U., Jevrić, L. L., Bončić-Carčić, G., Jovanović, B. Ž., Study of the lipophilicity and retention behavior of some *s*-triazine derivatives on aminoplast and cellulose, *J. Planar Chromatogr.*, 14, 277–282, 2001.
95. Perišić-Janjić, N. U., Jovanović, B. Ž., Reverse-phase thin-layer chromatographic behavior of some *s*-triazine derivatives, *J. Planar Chromatogr.*, 16, 71–75, 2003.
96. Perišić-Janjić, N. U., Jovanović, B. Ž., Janjić, N. J., Rajković, O. S., Antonović, D. G., Study of the retention behavior of newly synthesized *s*-triazine derivatives in RP TLC systems, and the lipophilicity of the compounds, *J. Planar Chromatogr.*, 16, 425–432, 2003.
97. Perišić-Janjić, N. U., Djaković-Sekulić, T. L., Jevrić, L. L., Jovanović, B. Ž., Study of quantitative structure–retention relationships for *s*-triazine derivatives in different RP HPTLC systems, *J. Planar Chromatogr.*, 18, 212–216, 2005.

98. Sztanke, K., Tuzimski, T., Rzymowska, J., Pasternak, K., Kendefer-Szerszeń, M., Synthesis, determination of the lipophilicity, anticancer and antimicrobial properties of some fused 1,2,4-triazole derivatives, *Eur. J. Med. Chem.*, 43, 404–419, 2008.
99. Sadowska-Rociek, A., Cieślik, E., Stosowane techniki i najnowsze trendy w oznaczaniu pozostałości pestycydów w żywności metodą chromatografii gazowej, *Chemia. Dydaktyka. Ekologia. Meteorologia.*, 13, 33–38, 2008.
100. Martínez Vidal, J. L., Plaza-Bolaños, P., Romero-González, R., Garrido Frenich, A., Determination of pesticide transformation products: A review of extraction and detection methods, *J. Chromatogr. A*, 1216, 6767–6788, 2009.
101. Chung, S. W. C., Chen, B. L. S., Determination of organochlorine pesticide residues in fatty foods: A critical review on the analytical methods and their testing capabilities, *J. Chromatogr. A*,1128, 5555–5567, 2011.
102. Koesukwiwat, U., Lehotay, S. J., Leepipatpiboon, N., Fast, low-pressure gas chromatography triple quadrupole tandem mass spectrometry for analysis of 150 pesticide residues in fruits and vegetables, *J. Chromatogr. A*, 1218, 7039–7050, 2011.
103. Radišić, M., Grujić, S., Vasiljević, T., Laušević, M., Determination of selected pesticides in fruit juices by matrix solid-phase dispersion and liquid chromatography–tandem mass spectrometry, *Food Chem.*, 113, 712–719, 2009.
104. Gilbert-López, B., García-Reyes, J. F., Molina-Díaz, A., Sample treatment and determination of pesticide residues in fatty vegetable matrices: A review, *Talanta*, 79, 109–123, 2009.
105. Beyer, A., Biziuk, M., Applications of sample preparation techniques in the analysis of pesticides and PCBs in food, *Food Chem.*, 108, 669–680, 2008.
106. Hajšlová, J., Zrostlíková, J., Matrix effects in (ultra)trace analysis of pesticide residues in food and biotic matrices, *J. Chromatogr. A*, 1000, 181–197, 2003.
107. Kruve, A., Künnapas, A., Herodes, K., Leito, I., Matrix effects in pesticide multi-residue analysis by liquid chromatography–mass spectrometry, *J. Chromatogr. A*, 1187, 58–66, 2008.
108. Kittlaus, S., Schimanke, J., Kempe, G., Speer, K., Assessment of sample cleanup and matrix effects in the pesticide residue analysis of foods using postcolumn infusion in liquid chromatography–tandem mass spectrometry, *J. Chromatogr. A*, 1218, 8399–8410, 2012.
109. Hajšlová, J., Holadová, K., Kocourek, V., Poustka, J., Godula, M., Cuhra, P., Kempný, M., Matrix-induced effects: A critical point in the gas chromatographic analysis of pesticide residues, *J. Chromatogr. A*, 800, 283–295, 1998.
110. Poole C. F., Matrix-induced response enhancement in pesticide residue analysis by gas chromatography, *J. Chromatogr. A*, 1158, 241–250, 2007.

5 Choice of the Mode of Sample Preparation in the Analysis of Pesticides on the Basis of the Properties of the Matrix

Robert E. Smith, Kevin Tran, Chris Sack, and Kristy M. Richards

CONTENTS

5.1 INTRODUCTION

In this chapter, only human-made pesticides will be discussed, even though more than 99.9% of the pesticides that we consume in our diets are natural [1], as are many bacterial toxins and other natural substances that help prime the innate immune system and strengthen our natural defenses against infection and disease [2]. Mechanisms evolved to do this. However, human-made pesticides and other environmental toxins have not been around long enough for protective mechanisms to evolve, and they may contribute to the large increase in autoimmune diseases in recent decades [3–6].

As in any analysis, the choice of the mode of sample preparation depends on the properties of the matrix and the needs of the study. For example, some workers simply dilute the samples or do a standard addition with no other sample preparation when analyzing avocado, black tea, oranges, arugula, and cucumbers [7,8]. When analyzing thousands of foods for hundreds of pesticides, fungicides, herbicides, and industrial chemical samples in the USFDA's Total Diet Study [9], speed, efficiency, ruggedness, minimum solvent use, and maximum throughput are emphasized, so the QuEChERS method is used to clean up most samples [10]. However, when there were only a few precious samples from a remote environment, such as the atmosphere above the Tibetan plateau, the older, more time-consuming Soxhlet extraction of high-volume polyurethane foam (PUF) from air samplers was used, followed by cleanup using an alumina/silica column and anhydrous sodium sulfate [11].

Still, the more the analyst knows about the matrix and the analytes, the better. This might be relatively easy when monitoring groundwater in areas where certain pesticides have been used for decades, or nearly impossible when analyzing environmental sludge and other solid waste from locations with unknown previous exposure to pesticides. For example, groundwater samples in the Platô de Neópolis, State of Sergipe, Brazil, were analyzed for 16 pesticides after extracting them with a solid phase microextraction (SPME) polyacrylate fiber [12]. Methyl parathion, bifenthrin, pyraclostrobin, and azoxystrobin were found [12]. A variety of solid wastes were analyzed for chlorinated pesticides and other environmental toxins after extracting them with an accelerated solvent extractor (ASE) [13]. So the physical properties, compounds present, and particle sizes of samples can differ significantly and affect the choice of sample preparation method needed.

5.2 TYPES OF SAMPLES

Samples can range from fresh air and water to solid waste and sludge. Analytes may need to be extracted and concentrated from the cleanest samples that have little or no interfering compounds in them. On the other hand, analytes may need to be extracted selectively from solid waste and sludge and then diluted before being analyzed.

5.3 SAMPLE PREPARATION FOR SOLID AND SEMISOLID SAMPLES

The first step is to try to homogenize the sample as well as possible by macerating, chopping, mechanical grinding, pulverizing, and/or blending, often in the presence of an extraction solvent, such as acetone, ethyl acetate, or acetonitrile. Fatty foods may need to be frozen by adding dry ice to the grinder. Some seeds, such as those of the Brazilian fruit acai, are so hard that they need to be frozen in liquid nitrogen and ground in a Micro-Mill, then sieved through a fine mesh sieve. Such samples need to be milled before adding the solvent. However, many methods of sample preparation have been designed to use a minimum of the sample. Such methods require extra care when homogenizing the sample and should be repeated as many times as possible to ensure that the sample is as homogeneous as possible. Still, the needs of the study will dictate the decision on how many replicates to analyze.

5.3.1 Extraction of Pesticides from Solid Matrices

Depending on the type of sample and the needs of the study, samples can be extracted with solvent by shaking them in a centrifuge tube in a mechanical shaker for a few minutes at room temperature and pressure. For pesticides that are sprayed on the surface of fruits and vegetables, this may be sufficient. However, for animal fat, solid waste, and sludge, the extraction method may need to be more extensive. Either manual or automated Soxhlet extraction uses a boiling solvent that is continuously refluxed through the sample for as many as 8 hours. Supercritical fluid extraction (SFE) usually uses supercritical CO_2 at high pressure. Pressurized liquid extraction (PLE) uses high temperature and pressure. PLE methods include ASE, pressurized fluid extraction, pressurized hot solvent extraction, subcritical solvent extraction, and hot water extraction. They all use solvents that are heated to the near-supercritical region, in which they show better extraction properties due to the decrease of surface tension and the increase of solubility and diffusion rate into the sample showed by the solvent at high temperatures. Pressure keeps the solvent below its boiling point and, in solid samples, such as cereals, forces its penetration into the pores of the sample [14]. Some of the different extraction methods will be discussed next.

5.3.2 Soxhlet and Soxtec Extraction

Soxhlet extractions are named after the inventor, Franz von Soxhlet [15]. A Soxhlet extractor contains a round-bottom flask or solvent cup, into which the extracting solvent is placed. The sample is wrapped in a filter that is shaped like a thimble or small cup, and it is placed into a condenser that is attached to the flask. Cold water is circulated through the outer part of the condenser. Once heat is applied and the solvent starts to boil, pure solvent enters the thimble containing the sample as it passes through and extracts the analytes. In the continuous mode, the solvent recondenses in the condenser and falls back into the round-bottom flask. This continues for a specified period of time. Alternatively, some Soxhlet extraction systems have a valve between the larger cup that holds the solvent and the condenser that holds the sample. This enables four different modes of Soxhlet extraction: hot, standard, warm, and continuous flow [16]. In the hot extraction mode, as soon as the solvent level in the extraction vessel reaches an optical sensor, a heating source in the upper level is switched on. As soon as the solvent level in the extraction chamber has reached the optical sensor, solvent is permitted to enter by briefly opening the valve. This ensures the solvent level in the extraction chamber remains nearly constant. As a result, the sample remains in boiling solvent throughout the entire extraction period.

In the standard Soxhlet extraction mode, as soon as the solvent level reaches the optical sensor, the valve opens, and the solvent containing the dissolved analytes flows back into the solvent cup. This optimizes the exchange between analytes and solvent. In the warm Soxhlet extraction, the solubilities of the analytes are increased by heating the condensed solvent in the extraction chamber. This reduces the time needed to do the extraction. In the continuous flow method, the solvent is evaporated. It is condensed in the condenser, in which it flows into the sample as in the other three modes. However, the glass valve is opened so that the solvent does not accumulate in the extraction chamber but continuously flows back into the solvent cup. Thus, enrichment of the analytes in the extract is avoided [16].

After the sample is extracted, it is rinsed, and the valve is opened. The sample tube can be lifted up automatically. Rinsing ensures that all sample residues are washed off the outer side of the sample tube and the inner side of the extraction chamber. The final step is drying, in which the valve is closed while heating continues at a low level. The solvent evaporates, is condensed in the condenser, and is collected in the empty extraction chamber. This allows the solvent to be removed almost completely and in the shortest possible time. The concentrated extract can then be analyzed [16].

When automated, it is called a Soxtec extraction. In either case, the solvent is chosen so that it extracts the analytes as selectively as possible while co-extracting as few of the other compounds in the sample matrix as possible. The rate of extraction needs to be carefully controlled when using highly volatile solvents, such as acetone or methanol. If the extraction rate is too high, the solvent can condense faster than the pass-through rate, causing the solvent to accumulate in the sample holder and eventually spill over the side and take dissolved material with it. Moreover, several small but fast extractions will give higher recoveries for all but the most soluble analytes than a single longer extraction.

5.3.3 Supercritical Fluid Extraction

SFE uses a compound, such as CO_2, that is a gas at room temperature and pressure but can be converted to a supercritical fluid at elevated pressure in a sealed container. Once an extraction is completed and the container is opened, the CO_2 evaporates off, avoiding the production of hazardous waste that is produced by solvent extractions, so it is much more environmentally friendly. The critical point for CO_2 is 30.9°C and 73.8 bar. At pressures above 73.8 bar, it becomes a supercritical fluid with much higher diffusivity, enabling it to penetrate samples. It does become much less polar, acting much like cyclohexane. The polarity can be increased by adding small amounts of cosolvents. Also, extractions can be done at lower temperatures in a nonoxidizing environment, which is important for thermally labile or easily oxidizable analytes.

In a SFE, fluid is pumped at a pressure exceeding the critical point into a sealed container containing the sample, and the temperature is increased. After a short time, the container is opened, and the analytes are collected in a small volume of organic solvent or on a solid adsorbent trap. This can be done in a static, dynamic, or recirculating mode. In the static extraction mode, the container is filled with the supercritical fluid, pressurized, and allowed to equilibrate with the sample. In the dynamic mode, the supercritical fluid is passed continuously through the extraction cell. In the recirculating mode, the same fluid is repeatedly pumped through the sample and, after a specified number of cycles, it is pumped out to the collection system [17].

Even though SFE is a quick and efficient way of extracting solid samples, the apparatus is relatively expensive, it can be difficult to optimize, and lipids (such as triglycerides in food) are co-extracted. Still, it has been used to extract pesticides in white and wild rice. An in-line aminopropyl solid phase cartridge was used to clean the sample, removing many compounds in the sample matrices [18].

5.3.4 Pressurized Liquid Extraction

In the different forms of PLE, the sample is mixed with a drying or inert sorbent, such as sodium sulfate, HydroMatrix™, (diatomaceous earth) is packed in a stainless-steel cell and extracted with an appropriate solvent at temperatures up to 200°C and pressures up to 20 MPa [19]. PLE is used in a U.S. Environmental Protection Agency (EPA) method [20]. Usually, it is used in the static, rather than dynamic mode to minimize sample dilution and solvent consumption.

A very popular way of doing PLE is to use an ASE [21]. Extractions are done at elevated temperature and pressure in sealed containers. This greatly increases the solubility of many compounds in the extracting solvent. Often, a sample is mixed with enough HydroMatrix (Sigma Aldrich, St. Louis, MO, USA) to reach the 100-mL mark in the sample cell used in an ASE (ThermoFisher, Sunnyvale, CA, USA). Then, approximately 40 mL of solvent can be added. The temperature and pressure can be increased to desired values over an approximately three-minute time period (static time). Next, the solvent can be flushed out into a collection vessel using compressed gas. This can be repeated as many times as needed (often three times) as long as all the solvent extracts are combined. The ASE can be used to extract pesticides and other environmental toxins from solid waste samples using acetone:hexane (1:1, v/v), tolune, or dichloromethane:acetone (1:1, v/v) [22]. Usually, a 2 × 5 min extraction with acetone:hexane (1:1, v/v) is suitable for extracting chlorinated pesticides from soil. However, for difficult matrices that are contaminated with relatively high concentrations of pesticides, tolune may be better [22]. For fatty foods, an ASE can be used with dichloromethane:hexane (1:1, v/v) for extraction. Fats (triglycerides) can be removed from the extract by passing it through a size exclusion chromatography (SEC) column.

5.3.5 Microwave-Assisted Extraction

Microwave-assisted extraction (MAE), also known as microwave-assisted solvent extraction (MASE), is an environmentally friendly extraction method that uses much less solvent and electricity and generates much less hazardous waste than other methods. It also requires less time and sample. However, as with all methods that require small amounts of sample, MAE and MASE require thorough blending, grinding, and/or mixing of solid samples to ensure that they are as homogeneous as possible. Microwaves are nonionizing electromagnetic radiation with a frequency between 300 and 300,000 MHz that generate heat due to ionic conduction and dipole rotation. Ionic conductance is the electrophoretic migration of ions when electricity is applied, and dipole rotation is when dipoles realign. The resistance to these movements causes friction and heats up the sample. Microwaves heat the extracting solvent directly, compared to conventional heating, which first heats up the flask or cup that contains the sample. This does require the presence of a dielectric compound [23]. The greater the dielectric constant and dissipation factor (tan δ, the ratio of the sample's dielectric loss, ε'' to its dielectric constant) the more thermal energy is released and the more rapid is the heating for a given frequency. That is, the dielectric constant is a measure of a compound's ability to absorb microwave energy, and the loss factor its ability to dissipate the absorbed energy, given by ε'. So the effect of microwave energy depends on the nature of both the solvent and the matrix. Table 5.1 shows the dielectric properties of several solvents used for MAE [24].

More polar compounds have higher dielectric losses, so microwave heating is selective. That is, the solvent should have a high selectivity toward the analytes of interest and exclude unwanted matrix components. Also, analytes can be extracted into a single solvent or mixture of solvents that absorb microwave energy strongly into a solvent mixture of high and low dielectric losses or into a solvent that does not absorb microwaves well.

Ionic conduction produces heat because the medium resists the flowing of ions. The migration of dissolved ions causes collisions between molecules as the directions of ions change as many times as the field changes sign. The dipole rotation is related to the alternative movement of polar molecules, which try to line up with the electric field in the electromagnetic radiation. Multiple collisions from this agitation of molecules will cause energy to be released, producing an increase in temperature [25]. The extraction heating process may occur by three different mechanisms. The sample can be immersed in a single solvent or mixture of solvents that absorbs microwave energy strongly. Alternatively, samples can be extracted in a combined solvent containing solvents with

TABLE 5.1
Dielectric Properties of Several Solvents Used in MAE

Solvent	ε'	ε'	tan δ ($\times 10^4$)
Hexane	1.89	0.00019	0.10
Ethyl acetate	6.02	3.2	5312
Acetone	21.1	11.5	5555
Methanol	23.9	15.3	6400
Ethanol	24.3	6.1	2500
Acetonitrile	37.5	2.3	620
Water	76.7	12.0	1570

Source: Popp, P. et al., *J. Chromatogr. A*, 774, 203–211, 1997.

both high and low dielectric losses mixed in various proportions. Finally, samples that have a high dielectric loss (samples with high water content) can be extracted in a solvent that is transparent to microwaves. For example, when pesticides were extracted from sediments, the moisture content of the sample was an important parameter. The best recoveries of analytes were obtained when the total moisture content was 15%. It was concluded that acceptable recoveries of organochlorine pesticides (OCPs) extracted with hexane could not be obtained when the moisture content was too high or too low [26]. Samples can be extracted in either open or closed vessels using focused MAE or pressurized MAE (PMAE) in a sealed container. For PMAE, there is a magnetron tube, an oven that contains a turntable into which samples are placed, devices that monitor the temperature and pressure, and several electronic components. To perform an extraction, samples are placed into extraction vessels that are lined with Teflon perfluoroalkyl; solvent is added, and the container is closed. The microwave generator is turned on to start a preextraction step, in which the solvent is heated to the preselected temperature. This usually takes about two minutes. Then, the sample and solvent are irradiated with more microwaves in a 10–30 minute static extraction step. After cooling it down, an internal standard can be added, and a cleanup step can be added. One system enables 12 extraction vessels to be extracted simultaneously using microwave power up to 900 W. Several commercially available microwave systems were reviewed previously [27].

Most MAE extractions use mixtures of a solvent with low polarity and water, which is often in biological samples themselves. In general, a higher volume of solvent will increase the recovery of the analytes in conventional extractions. However, this may lead to lower recoveries in MAE, probably due to inadequate mixing of the solvent with the matrix by the microwaves. So the selection of solvent volume depends on the type and the size of the sample, but there must be enough solvent to immerse the entire sample. Also, the water content of the sample is a key factor because of the high dipole moment of the water molecules, which leads to high efficiency in heating the sample. However, it is important to control the water content of the matrix to obtain reproducible results. Other possible components of the matrix (such as iron or Fe^{2+}) can cause arcing due to the absorption of microwave energy. The organic carbon content of the matrix can hinder the extraction due to strong interactions between the analytes and sample matrix that can be difficult to disrupt. The choice of microwave power and the irradiation time depend on the type of sample and solvent used. Usually, the high-power microwaves should reduce the exposure time needed. However, in some cases, a very high-power microwave decreases the extraction efficiency by degrading the sample or rapidly boiling off the solvent in open-vessel systems. Still, extraction times in MAE are much shorter than those of classical extraction techniques. Usually, increasing extraction times above the optimal range does not improve extraction efficiency, and, in some cases, may even decrease the recoveries of thermally labile compounds. Still, in most cases, higher temperatures produce improved extraction efficiency due to the increased diffusivity of the solvent into the matrix and the enhanced desorption of the analytes from the matrix. In closed systems, the pressure is also an important variable, especially for thermolabile analytes, which may be decomposed at high temperatures [23]. OCPs; organophosphorus pesticides (OPPs); dichlorodiphenyltrichloroethane (DDT); a common breakdown product of DDT, *p,p′*-dichlorodiphenyldichloroethylene (*p,p′*-DDE); triazines; imidazolinones; phenyl ureas; sulfonyl ureas; hexaconazole; dimethomorph; dacthal; chlorpyrifos; chlorothalonil; diazinon; permethrin; methoxychlor; and azinphos-methyl have been extracted from a variety of samples using microwave extraction [23].

For example, OCPs were extracted from sediments using isooctane/acetonitrile, 1:1, v/v with higher recoveries that Soxhlet extraction for six hours. Another group used hexane-acetone, 1:1, v/v to extract OCPs from marine sediments and soils. Still another group used toluene and water to extract more DDTs from aged marine sediments. Another group used a mixture of hexane and water at 115°C for 10 minutes, and others used tetrahydrofuran at 100°C. To extract pesticides from fatty tissues, such as seal blubber and pork fat, ethyl acetate:cyclohexane, 1:1, v/v was used. OCPs, OPPs, and five other pesticides were extracted from water using solid phase extraction (SPE) disks, followed by MAE to elute them using acetone at 100°C for five minutes [23].

The triazines, atrazine, desethylatrazine, desisopropylatrazine, and simazine were extracted from agricultural soils along with surface and groundwater using dichloromethane:methanol 9:1, v/v [23]. Others extracted triazines from soils using water, followed by an enzyme-linked immunosorbent assay. Still others used either methanol or acetone-hexane, 1:1, v/v to extract triazines from soil samples, but water was just as efficient. Imidazolinone herbicides were extracted from soils and crops using pure water and an aqueous solution buffered at pH 10 at 125°C for three minutes. Sulfonyl- and phenylurea herbicides were extracted with dichloromethane:methanol, 9:1, v/v. The fungicide dimethomorph was extracted from soil samples using acetonitrile:water, 9:1, v/v at 125°C for three minutes. Dacthal, chlorpyriphos, chlorothalonil, diazinon, permethrin, methoxychlor, and azinphos-methyl were extracted from several crops using a 1:2 mixture of isopropyl alcohol plus light petroleum at 100°C and 10 minutes [23].

Fuentes and coworkers used atmospheric pressure MAE to extract OPPs from olive and avocado oil, followed by SPE or low-temperature precipitation as a cleanup step. A simple glass system equipped with an air-cooled condenser was used as the extraction vessel. The pesticides were partitioned between acetonitrile and hexane. Mao and coworkers extracted OCPs, OPPs, pyrethroid, and carbamate pesticides from the oriental herbal medicine *Radix astragali* using acetonitrile in an MAE, followed by cleanup using a primary and secondary amine (PSA) sorbent plus $MgSO_4$ followed by gas chromatography–mass spectrometry (GC-MS) analysis [28]. Fang and coworkers used both MAE and a mixed mode polymeric cation exchange SPE cartridge to extract pesticides and melamine from infant milk formula, followed by (liquid chromatography (LC)-MS analysis [27]. They used a mixture of methanol and water and adjusted the pH. For pesticides with pK_a values from 2.35 to 5.27, the solution was made alkaline, but the easily hydrolyzable atrazine was extracted at neutral pH and methanol with 2%–22% water. For all but atrazine extractions, the solutions were acidified with hydrogen chloride to protonate the amines and increase their affinity for the SPE sorbent. They obtained higher recoveries of spiked pesticides using this method than a Soxhlet extraction [29]. Satpathy and coworkers used MAE with dispersive SPE (d-SPE) and GC-MS with deconvolution software to determine the concentrations of 72 different pesticides in 35 different fruits and vegetables [30]. Coscallá and coworkers used MAE to extract airborne pesticides from quartz filters using ethyl acetate at 50°C, followed by gel permeation chromatography (GPC) cleanup and GC-MS analysis [31]. Wang and coworkers used an absorbing microwave SPE device packed with activated carbon to extract OPPs in fruits and vegetables with hexane at 60°C for 10 minutes [32]. So, MAE has been shown to be fast, simple, and effective while using less solvent and producing less hazardous waste. It has been said that it is easier to use and less expensive than SFE and PLE [21].

5.4 SAMPLE PREPARATION FOR LIQUID SAMPLES

5.4.1 Single-Drop Microextraction

Single-drop microextraction (SDME) uses a drop of a solvent to extract analytes. The first report on SDME described using a 1.3-μL drop of chloroform containing the colorimetric reagent methylene blue suspended inside a flowing aqueous drop from which the analyte, sodium dodecyl sulfate, was extracted as an ion pair [33]. This is also called direct immersion SDME. It is applicable to liquid samples containing analytes that are nonpolar or only slightly polar. To ensure the stability of the drop, it is important to remove any insoluble or particulate matter from the sample and use an organic solvent that is not miscible with water and has as low a vapor pressure as possible. Toluene and *n*-octane are good solvents for nonpolar compounds. Moreover, ionic liquids may be used because they have low vapor pressure and high viscosity, so they can form larger and more reproducible extraction drops. High-performance liquid chromatography (HPLC) is used to analyze ionic liquid extracts but their nonvolatility makes them unsuitable for GC [34]. So the first applications of SDME enabled analysts to simultaneously extract and analyze samples in a windowless optical detection system [35]. This was followed by methods in which the micro-drop was continuously renewed by continuously pumping a reagent solution. It was held as a sessile at the end of a silica capillary. It served the functions of sample gas extraction, reactor for a chromogenic reaction, and windowless optical cell for fiber-based absorbance detection. Later, SDME became much more useful as GC detection was added. A GC syringe can be used to suspend a 1-μL drop of solvent at the end of the needle. It was retracted back into the syringe and analyzed by injection directly into the GC [36]. When optimized, SDME extracts only the analytes of interest from aqueous solutions.

5.4.2 Hollow-Fiber Membrane Liquid–Phase Microextraction

Hollow fiber (HF) membrane liquid-phase microextraction (LPME) is used to extract analytes from aqueous samples and into a supported liquid membrane that is in the pores in the wall of a small porous HF. The analytes also enter an acceptor phase that is inside the lumen of the HF. When the acceptor phase is organic, it can be a two-phase extraction system that is analyzed by GC. Alternatively, the acceptor phase can be aqueous, making it a three-phase system that can be analyzed by HPLC, LC-MS/MS, or capillary electrophoresis. Three-phase extractions with an extraction and back extraction are also possible [37]. For two- and three-phase extractions, the solvent should be immiscible with water, have low volatility and good extraction efficiency of analytes, and be able to properly immobilize pores in the fiber. Toluene, undecane, 1-octanol, and dihexyl ether have been used. More recently, ionic liquids have been used. When present in the pores of a supported membrane, ionic liquids could not be displaced, and the supported ionic liquid membrane was quite stable when stirred mildly [38]. Also, ionic liquids such as 1-octyl-3-methylimidazolium hexafluorophosphate ($[C_8MIM][PF_6]$) have high affinities for polar compounds, and they can also transport some organic compounds selectively [39].

Due to the partial purification and enrichment of analytes and very low consumption of organic solvent, there is substantial interest in LPME [40]. In some LPME methods, including SDME, a small drop of water-immiscible solvent is used to extract analytes. However, these drops can be dislodged from the needle of the syringe during the extraction. So Pedersen-Bjergaard and Rasmussen developed a HF to hold the organic solvent that accepts the analytes from the donor aqueous solution [41].

OCPs have been extracted using HF-LPME, and the results compared favorably to the USEPA Method 508, in which the analytes in water are extracted with dichloromethane in a separatory funnel. After isolating, drying, and concentrating the extract, the solvent is exchanged into methyl *tert*-butyl ether before analysis by GC-electron capture detection [42,43]. Dynamic HF-LPME has also been used to extract OPPs from lake water [44] and OCPs from soil samples [45].

5.4.3 Solvent Bar Microextraction

Solvent bar microextraction is when an organic solvent is confined to a HF membrane that is sealed at both ends [46,47]. It is placed into the aqueous sample solution to extract analytes. Due to the vigorous tumbling of the solvent bar in the agitated sample, mass transfer between the organic and aqueous phases is facilitated, enabling a more efficient extraction. After extraction, the solvent bar is taken out, and the ends are trimmed off. An aliquot of the organic solvent is withdrawn into a micro-syringe and is injected into a GC or GC-MS. OCPs were extracted from wine by solvent bar microextraction using *n*-tetradecane as the acceptor phase, which provided a 1900- to 7100-fold enhancement of analyte concentrations and detection at fg/mL levels by GC-MS [48]. There is also a three-phase solvent bar microextraction in which a third, aqueous phase is the final acceptor of the analytes [49]. It is confined within the lumen of the HF, and the pores of the fiber are impregnated with organic solvent. By using an aqueous acceptor phase, analysis by HPLC, LC-MS, and/or electrophoresis can be done. Another approach was to use monolithic materials, such as silica, to extract analytes [50]. It enabled a high rate of mass transfer and separation of analytes. Moreover, the high porosity of the pores allows large molecules and organic solvents to penetrate easily. Microwaves can be added to improve extraction efficiency of water-soluble analytes from solid samples. The water then acted as the donor phase or "sample solution" for further extraction into a polypropylene HF containing 1-octanol, toluene, *o*-xylene, or hexane [51]. In another twist, a homemade 12-mL glass extraction tube containing 5 mL of water was used with microwaves to extract OPPs from soil, followed by extraction into toluene and analysis by GC [52].

5.4.4 Liquid–Liquid–Liquid Microextraction

In liquid–liquid–liquid microextraction, or three-phase liquid microextraction, analytes are extracted from an aqueous sample through an organic solvent and then further into an aqueous acceptor solution [53]. This extraction mode is limited to ionizable compounds. For example, uncharged amines can be extracted from water and into toluene by increasing the pH of the aqueous phase, then back extracted into an acceptor aqueous phase that is acidic, so the amines will be protonated. In the first application of liquid–liquid–liquid microextraction, a polypropylene HF was dipped in the organic solvent a few times to immobilize the solvent in the pores, and excess solvent was removed. The acceptor aqueous solution was then added to fill the lumen. The HF was filled with the aqueous sample, and analytes were extracted from it [53]. In another application, phenoxy herbicides were extracted from milk using a polyproplylene HF that was flame-sealed on one end [54].

5.4.5 Solid Phase Extraction

SPE is when a solid phase is used to extract analytes from liquid samples. This can be done with a disposable column, filter, disk, pipette tip, or 96-well plate. SPE can be quite useful in concentrating analytes from relatively clean samples, such as water, by extracting analytes onto an SPE cartridge, then eluting them with a much smaller volume. The columns or cartridges can be filled with PSAs, octadecylsilica, silica, alumina, or any number of other sorbents. Moreover, a porous polystyrene/divinyl benzene (PS/DVB) column can be used in SEC, also known as GPC, to remove fats from fatty foods. For example, one method for determining the concentrations of pesticides in fatty foods used hexane for the extraction, followed by SEC to remove the larger triglycerides from the smaller analytes. In SEC, small compounds can fit into the pores of the PS/DVB and are retained while larger triglycerides don't fit into the pores and elute first.

The solvent used for extraction and the columns used for cleanup depend on the amount of fat in the food sample. That is, multiple residues in foods with <2% fat and >75% water can be extracted using method 302 in the Pesticide Analytical Manual (PAM), and foods with <2% fat and <75% water use PAM method 303. Foods with >2% fat are extracted using PAM method 304 [55]. High-purity solvents must be used to prevent interferences. Method 302 uses acetone, followed by HydroMatrix to remove water and may include a liquid–liquid partition with petroleum ether or methylene chloride, CH_2Cl_2. Cleanup can use a florisil column with one CH_2Cl_2 elution or mized ether eluents for relatively nonpolar analytes or a combination of charcoal, Celite, and magnesium oxide for polar residues. When determining the concentrations of N-methyl carbamates, silanized Celite is used instead of just Celite. Alternatively, a C-18 cartridge can be used. Samples containing polar and nonpolar residues can be cleaned up using a strong anion exchange or PSA cartridge.

The PAM method 303 uses acetonitrile, followed by partition into petroleum ether with high water content to extract analytes from fruits and vegetables with low sugar (<5%) and low fat (<2%). It calls for extraction with acetonitrile and partition into petroleum ether for eggs and extraction with a mixture of acetonitrile and water followed by partition into petroleum ether for dried egg whites, grains, and other foods with <75% moisture. Fruits and other foods with 5%–15% sugar are extracted with acetonitrile and water followed by partition into petroleum ether. Fruits and other foods with >15% sugar are extracted with heated acetonitrile and water followed by partition into petroleum ether. Florisil is used for cleanup. Three diethyl ether or petroleum ether eluents are used for nonpolar residues, and three CH_2Cl_2 eluents are used to recover even more analytes.

The PAM method 304 uses a mixture of petroleum ether and sodium sulfate for animal tissues and fatty fish, and filtering is used for butter and oils. Many cheeses and eggs can form emulsions that are not broken by centrifugation, so 1 mL of water is added for every 2 g of sample before blending. Then, 25–100 g of diced cheese or 25–50 g of eggs (to provide about 3 g fat) is placed in a blender jar containing about 2 g sodium or potassium oxalate and 100 mL ethyl or methyl alcohol and is blended for 2–3 minutes. Evaporated milk is diluted with equal volumes of water. About 100 g of either the hydrated milk or a fluid milk sample is placed into a centrifuge bottle containing 100 mL ethyl or methyl alcohol and about 1 g sodium or potassium oxalate. After centrifugation, the solvent layer is transferred to a separatory funnel containing 500–600 mL water and 30 mL saturated sodium chloride solution. This is re-extracted twice by shaking vigorously with 50-mL portions of ethyl ether:petroleum ether, 1:1, v/v. This is then centrifuged. The solvent layer is transferred to a separatory funnel, and the aqueous phase is discarded. The organic phase is rewashed with two 100-mL portions of water. If emulsions form, about 5 mL of water saturated with NaCl is added. The ether phase is poured through a 25 mm by 50 mm column of sodium sulfate, and the eluent is collected in a Kuderna-Danish tube, so the solvent can be evaporated off. The fat is then removed. The solvent is removed when the total amount of fat is expected to be <3 g. When the total fat is >3 g, the solvent is not evaporated off. Instead, 15 mL of petroleum ether is added followed by 30 mL of acetonitrile that is saturated with petroleum ether. After the layers separate, the acetonitrile phase is drained into a 1-L separator containing 650 mL water, 40 mL saturated sodium chloride solution, and 100 mL petroleum ether. The petroleum ether solution is extracted with three additional portions of acetonitrile saturated with petroleum ether, shaking vigorously for one minute each time, and all extracts are combined in the 1-L separator. This is held horizontally and mixed thoroughly for 30–45 seconds. After the two layers separate, the aqueous phase is drained into a second 1-L separator, to which 100 mL of petroleum ether is added. The aqueous phase is discarded, and the petroleum ether phase is combined with the other petroleum ether in the first 1-L separator. This is washed two more times with 100-mL portions of water. The washings are discarded, and the petroleum ether layer is drained through a 25 mm by 50 mm column containing sodium sulfate and into a Kuderna-Danish tube. The solvent volume is reduced to 5–10 mL for transfer to a Florisil column containing about 0.5 inches (1.27 cm) of sodium sulfate. After passing the sample through it at about 5 mL/min, the container is rinsed with two 5-mL portions of petroleum ether and elute with about 5 mL/min with 200 mL 6% ethyl ether/petroleum ether. The Kuderna-Danish column is changed, and the column is eluted with about 5 mL/min with 200 mL 15% ethyl ether/petroleum ether. The Kuderna-Danish column is changed again, and the column is eluted with about 200 mL 50% ethyl ether/petroleum ether. The volume is reduced, and the concentrated solution is ready for GC or GC-MS analysis.

Other approaches can be used, too. SPE devices can contain weak anion exchange materials, such as aminopropyl (NH_2) or diethylaminopropyl (DEA). Yang and coworkers used a combination of NH_2 and carbon black SPE cartridges to clean up acetonitrile extracts of raspberries, strawberries, grapes, and blueberries before analyzing them for 88 pesticides [54]. They found that pigments were absorbed by the carbon black, and the NH_2 cartridge eliminated sugars and proteins. They also found that extraction with acetone or ethyl acetate provided good recoveries of pesticides that were spiked into samples, but they co-extracted so many matrix components (lipids and pigments) that they harmed column performance after repetitive injections [56]. Others used a highly cross-linked PS/DVB and silica modified with DEA to clean up and preconcentrate pesticides that were extracted from fresh fruits and vegetables using acetone [57]. To extract compounds such as OPPs that are too polar to bind to nonpolar sorbents like C-18, a highly cross-linked and hydroxylated PS/DVB-based SPE cartridge can be used [58]. On the other hand, pesticides and other environmental contaminants of lower polarity in natural waters were extracted using a C-18 cartridge [59]. More recently, a PSA column plus 4 g of anhydrous $MgSO_4$ and 1 g of NaCl has been used to clean up acetonitrile extracts of many different types of foods when analyzing them for hundreds of pesticides in the QuEChERS (Quick, Easy, Cheap, Effective, Rugged, and Safe) method [10]. Moreover, nanomaterials such as single-walled, multiwalled, and oxidized multiwalled carbon nanotubes have been used due to their high surface area, stability, and wide, large-scale availability [60]. Another SPE method uses mixed hemimicelles (hemimicelles and admicelles) that are made by absorbing ionic surfactants onto mineral oxides. Hemimicelles consist of a surfactant monolayer that is adsorbed head down onto the surface of the oppositely charged metal oxide. Admicelles have the bilayer structure of normal micelles with an ionic head group on the outermost surface. They can provide high extraction yields, easy elution of analytes, high breakthrough volumes, and high flow rate for sample loading.

Another new approach is to use molecular imprinting polymer (MIP) that can make polymers with cavities that can extract OPPS. They are synthesized based on noncovalent interactions between target molecules that act as a template with functional monomers. Once the template is removed, selective molecular recognition sites become available for the selective rebinding of target analytes with excellent specificity and stability [61]. A noncovalent MIP was used to extract dimethoate pesticide from tea leaves, which enabled the separation of dimethoate from an impurity [62].

Another approach is to couple a SPME fiber coated with polyacrylate (PA) with headspace GC. This was used to extract 36 common pesticides and break down products from agricultural soils [63]. Another group used a PA-coated SPME fiber to extract lindane, heptachlor, and two heptachlor transformation products in groundwater [64]. Another group optimized and validated a method using a polydimethylsiloxane (PDMS)-coated SPME fiber to extract 46 pesticides with different structures and polarities from water [65]. Newer materials have also been described for coating SPME fibers [66]. Finally, another modification of SPME is when the fiber is attached to a stir bar. This will be discussed further in Section 5.4.7.

5.4.6 Solid Phase Microextraction

SPME onto chemically modified fused silica fibers was invented by Pawliszyn and coworkers [67]. The original paper described using uncoated fused silica and polyimide-coated fused silica. Since then, PDMS, PDMS-DVB, PA, Carboxen-PDMS, Carbowax-DVB, Carbowax templated resin, and DVB-PDMS-Carboxen have become popular [68]. An SPME apparatus resembles a modified syringe, in which a 1- to 2-cm-long retractable SPME fiber is placed. After sampling, the fiber is retracted into a metal needle so it can be transferred to a GC or LC. The absorbed analytes can be thermally desorbed from the SPME fibers and analyzed by GC, GC-MS, HPLC, LC-MS, or capillary electrophoresis. For thermal desorption and GC analysis, a narrow-bore (0.75 mm i.d.) unpacked injection liner is used to ensure a high linear gas flow, reduce desorption time, and prevent peak broadening. Because no solvent is used for sample preparation, injections are carried out in the splitless mode. Also, the GC septa can be damaged with large (24-gauge) SPME guide needles. To avoid septum coring, predrilled high-temperature GC septa or a micro-seal septumless system can be used. For HPLC or LC-MS, a multiport injection valve and special desorption chamber can be used as the mobile phase desorbs the analytes.

SPME is fast and requires no solvent when extracting and preconcentrating pesticides. The sampling fibers can be reused many times. Since the original invention, an in-tube SPME/HPLC system (SPMS) was introduced, in which an open tubular fused-silica capillary column is used as the SPMS device that is suitable for automation [69]. Analytes are extracted into the internally coated stationary phase of a capillary column and then desorbed by the mobile phase or a static desorption solvent. This column is placed between the injection loop and needle of the LC autosampler. An important difference between fiber and in-tube SPME is that analytes are sorbed onto the surface of a SPME fiber, but they are sorbed on the inner surface of the in-tube device. So particulate matter should be filtered out of samples before using the in-tube SPME, but it is not required for the fibers.

Recently, new chemically bonded sorbent materials, such as chemically bonded carbon nanotubes on a modified gold substrate [70], covalently bonded polyaminithiophenol on a gold substrate

[71], and chemically bonded ionic liquid on a fused-silica substrate [72] have been described for SPME. Conducting polymers, such as polypyrrole and its derivatives, have also been described [73,74], but their applicability has been restricted due to their relatively low thermal stability (200°C) [75]. A more recent approach has been to use a nanostructured self-doped polypyrrole that contained covalently bound sulfo groups to extract OCPs into a SPME fiber [76].

5.4.7 Stir Bar Sorption Extraction

Stir bar sorption extraction (SBSE) uses a stir bar that has a polymeric coating to extract analytes. It has also been described as a type of SPME that can be used to extract larger sample volumes [69]. As with SPME, analytes can be desorbed thermally for GC analysis or by a solvent for LC, and PDMS is the most popular sorbent in SBSE. So the basic principles of SPME and SBME are the same, but the stir bar can extract samples with 50–250 times the volume as SPME fibers alone [77]. Stir bars that are 1 or 2 cm long can be coated with a 0.5 or 1 mm layer of sorbent. A magnetic rod is usually encapsulated in a glass jacket on which a PDMS siloxane layer is added. This avoids direct contact between the metal and PDMS, preventing metal-catalyzed degradation of the PDMS coating. However, even thinner layers (30 μm) of PDMS have been made using sol-gel technology, and larger (8 cm long) stir bars have been coated with 250 μL of PDMS. SBSE can also be used for *in situ* derivatization of more polar analytes.

SBSE has been used to extract pesticides from aqueous samples. For example, a multiresidue method for pesticides, PCBs, and PAHs was validated in accordance with ISO/EN 17025 [78]. That is, a 100-mL sample was saturated with NaCl and extracted for 14 hours using a 2-cm stir bar coated with a 0.5-mm-thick film of PDMS. This method was used in an inter-laboratory trial, and the results showed excellent agreement with results obtained by classical methods [79]. SBSE using a PDMS coating is used to extract pesticides from nonfatty foods and beverages (<2% fat), such as wine, honey, fruits, and vegetables. Fruits and vegetables can be homogenized using a water-miscible solvent. An aliquot of the extract can be diluted with water, followed by SBSE. For honey, 1 g was diluted into 10 mL of water before SBSE in one study, and another diluted 2.5 g of honey 1/10, extracted, then desorbed with 1 mL of methanol. SBSE showed higher concentration ability and greater accuracy and sensitivity than SPME [80].

Even though SBSE can be used for many different kinds of pesticides, often it is not possible to find optimum extraction conditions for all analytes. For the least polar pesticides, an organic modifier (acetonitrile or methanol) was added to reduce wall adsorption and matrix effects. However, for relatively polar pesticides, high modifier content can cause unacceptably low recovery of spiked samples. So, Ochiai et al. [81], took two aliquots from the extracts of fruit and vegetables, and the ratio of methanol to water was adjusted to either 50% or 20%. Each aliquot was extracted, and both stir bars were simultaneously desorbed and analyzed by GC-MS. This method was validated for 85 pesticides [81].

5.4.8 Polymer-Coated Hollow Fiber Microextraction

Polymer-coated HF microextraction based on an amphiphilic polyhydroxylated polyparaphenylene that was coated onto the surface of a porous polypropylene HF membrane was introduced by Basheer and coworkers for extracting OCPs from water [82]. The same group subsequently introduced an on-site preparation of polymer-coated multifibers for the extraction of OPPs from sea water [83]. They coated Technora fibers (each strand consisted of 1000 filaments, each with a diameter of about 9.23 μm) with a functional conjugated polymer that they named 2-(9,9-bis(6-bromo-2-ethylhexyl)9-H-fluoren-2-yl)benzene-1,4-diamine. Even though this name does not contain the word "polymer" or the prefix "poly," the authors still called it a polymer. They described the synthesis of 2,5-dibromo-4-nitroaniline and its reduction to 2,5-dibromo-4-aminoaniline. They said that the polymerization was conducted by a Suzuki coupling reaction but did not explicitly state the monomer that was used. They simply said that the reaction used a tetrakis (triphenylphosphine) palladium catalyst in a mixture (3:2, v/v) of aqueous (2 M) potassium carbonate and tetrahydrofuran under a nitrogen atmosphere at 75°C–80°C for 72 hours [83]. So this method may be difficult to reproduce, and there are not near as many publications about coated HF microextraction than there are about other methods of extraction that are described in this chapter.

5.4.9 Matrix Solid Phase Dispersion Extraction

Matrix solid phase dispersion (MSPD) was introduced by Barker and coworkers [82]. It uses mechanical forces from the grinding of samples with irregularly shaped particles (silica- or polymer-based solid supports) with the lipid-solubilizing capacity of a support-bound polymer to produce a mixture from which analytes that are dispersed into the sample matrix components can be selectively isolated [84]. MSPD used for the extraction of pesticides from olives using aminopropyl as the sorbent material was introduced by Ferrer and coworkers [85]. It provided an advantage of not requiring SEC to remove fats, which is often required in older methods of analyzing fatty foods. In MSPD, a fine dispersion of the sample is mixed with a sorbent material such as silica, alumina, or C-18 using a mortar and pestle. After blending, the mixture is packed into a minicolumn, in which the analytes are eluted by a relatively small volume of a suitable eluting solvent. This can be combined with a co-column to enable further removal of unwanted compounds in the sample matrix. The co-column material is packed into the bottom of the same column of the MSPD sorbent. Thus, MSPD enables the development of extraction and cleanup steps. It reduces the amount of solvents used and speeds up the sample treatment process [85]. Torres and coworkers used MSPD to extract insecticides from oranges using ethyl acetate and silica [86]. Stafford and Lin [87] used MSPD to determine oxamyl and methomyl residues in apples, oranges, soybean leaves, insects, and river water.

5.5 SAMPLE PREPARATION FOR GAS SAMPLES

Even though pesticides are not usually thought of as being in the air, an estimated 30%–50% of sprayed pesticides enter the atmosphere [88]. That is, soil particles containing pesticides can enter the troposphere and can remain there for several weeks. Winds can even carry pesticides to remote locations far from the places where they were initially applied. To prepare gas or air samples for analysis, first it is necessary to extract the analytes onto a filter that is placed inside a large volume sampler, such as the XAD or PUF samplers [11]. They can collect suspended airborne particulates as well as trap airborne vapors that contain pesticides at the relatively high flow rates of 13–30 m^3/h [88]. However, they are noisy. So for indoors, a lower volume, quieter air sampler can be used. In windy environments, passive air filters containing semipermeable membrane devices can also be used. In any case, particles can be collected on glass fiber filters, and compounds in the gas phase can be collected on PUF plugs [11]. Moreover, analytes can be deposited on absorbents, such as XAD, Carbopack, Carbotrap, Carboxen, Tenax TA, Chromosorb, or silica gel, that are deposited onto filters or in stainless-steel or glass containers that are placed between PUFs.

Once the analytes are collected onto a filter, they must be extracted into a solvent before being analyzed by LC-MS/MS and/or GC-MS/MS. Unlike water, soil, and food samples that can be thoroughly mixed, it is usually impractical to grind up filters to obtain a nearly homogeneous sample. So, extraction techniques that use small sample sizes may not be practical. To avoid problems due to the inherent inhomogeneity of a filter, often it is more reliable and convenient to extract the entire filter with a suitable solvent. This can be done by Soxhlet or Soxtec extraction (see Section 5.3.2 for details), ASE, ultrasound-assisted extraction or MAE. Another alternative is to avoid using solvents by using thermal desorption prior to GC analysis.

For LC-MS/MS analysis, solvent extraction is needed. To improve sensitivity, the volume of the solvent can be reduced by evaporation or sorption onto solid absorbents.

5.6 DERIVATIZATION OF ANALYTES

Fluorescence detection of pesticides has the potential to be highly selective and sensitive, but most pesticides have no inherent fluorescence. Exceptions to this are the benzimidazoles (such as 2-aminobenzimidazole and carbendazim), which were extracted from soils and detected by fluorescence after separation by HPLC [89]. Analytes can be derivatized in liquid–liquid extraction, SPE, and SPME to transform them into more easily analyzable compounds. It can do this by increasing the volatility for GC or by reducing the polarity of some analytes, thus increasing extraction efficiency. Direct derivatization as well as derivatization on SPME fibers and derivatization in a GC port can be done. For *in situ* derivatization using SPME fibers, a derivatization agent is added to the sample matrix. Derivatization occurs and the SPME fiber extracts the derivatized analytes from the sample solution. On the other hand, SPME fibers can extract pyrethroid insecticides, which can be converted to fluorescent derivatives by exposure to UV light in postcolumn photo-induced fluorimetry [90]. More recently, a sol-gel based SPME fiber that was coated with an amino compound was used to extract pyrethoids from water samples, prior to HPLC separation and photochemically induced fluorescence detection [91].

Even though they are not pesticides, mycotoxins are important analytes that can be found in foods, especially nuts and grains. Often, they are detected by online postcolumn photochemical derivatization and fluorescence detection [92].

One of the most widely used pesticides, glyphosate, can be converted into a fluorescent derivative before injecting it onto an LC column by reacting it with FMOC-Cl or 9-fluorenylmethylchloroformiate. Glyphosate can also be detected after postcolumn derivatization with *o*-phthalaldehyde (OPA) [93]. Fluorescent derivatives of carbamate pesticides often are prepared by postcolumn hydrolysis to form methylamine, which reacts with OPA to form a fluorescent product [94]. Although mercaptoethanol was added to OPA, now the nucleophilic compound called Thiofluor™ is more popular [95]. Methylamine reacts with OPA and Thiofluor™ to form the fluorescent 1-methyl-2-dimethylethylamine thioisoindole derivative [96]. Avermectins can also be detected by fluorescence after being derivatized with a mixture of 1-methylimidazole, trifluoroacetic anhydride, and acetic acid [97].

More recently, nanotechnology has been used. Organophosphorothioate pesticides were detected using CdTe quantum dots that have dithizone attached to them. The natural fluorescence of the quantum dots is quenched when dithizone is bound. However, when there are organophosphorothioate pesticides, such as chlorpyrifos, present in a sample, they can displace the dithizone and cause the dots to fluoresce [98]. Another approach was to use gold nanoparticles coated with rhodamine B to analyze foods for OPPs and carbamates [99]. Detection was based on the ability of these pesticides to bind to the enzyme acetylcholine esterase (AChE) and inhibit it. That is, they prevent the AChE-catalyzed production of thiocholine when acetylthiocholine is added, which turns the coated nanoparticles from red to purple and recovers the natural fluorescence of the rhodamine that was bound to the gold nanoparticles. So, when pesticides such as carbaryl are present, the color of the nanoparticles stays red, and the fluorescence remains quenched [99].

5.7 FACTORS AFFECTING THE ACCURATE QUANTIFICATION OF PESTICIDE RESIDUES

As in any type of analysis, the accuracy is often limited by factors that can be deceptively easy to overlook or ignore. First and foremost, no analysis can be better than the sample. That is, if a sample is not homogeneous or if it is not prepared properly, even the most sophisticated LC-MS/MS analysis will produce inaccurate results. So many methods that became popular because they require very little sample can be limited if extra care is not taken to stir, chop, homogenize, or blend a solid or liquid sample. Outdoor air samples can be even more problematic as the distribution of particulate matter can vary considerably depending on the wind speed, humidity, and temperature of the air. Moreover, it will not be deposited uniformly onto the absorbing filter. So, one should not take a small portion of the filter and assume that it is homogeneous. Instead, the whole filter should be extracted with a suitable solvent.

Another factor that is seldom discussed is the importance of regulatory compliance. Once a method is validated and becomes widely accepted, it can be quite difficult to get a better method approved. For example, when analyzing fatty fruits or vegetables (such as olives) for pesticides that are sprayed on their surfaces, it can be relatively easy to solubilize the analytes by shaking the samples briefly with a suitable solvent. To validate the method, pesticide standards dissolved in the same solvent are added or spiked onto the samples. Because they are also on the surface, they can be solubilized and detected easily. The solvent does not have to penetrate into the interior of the sample to get acceptable recoveries. Moreover, lower amounts of matrix components will be solubilized if the solvent never penetrates the sample. This might give accurate results for olives but could be problematic for olive oil or for meat. That is, livestock can ingest pesticides when they eat. The pesticides do not stay on the skin of the animals; it enters into their cells. Lipophilic analytes can be especially difficult to solubilize if they are inside of fat cells. Still, pesticides that are spiked onto the surface will be solubilized easily. Because fewer matrix components will be co-solubilized, there is much less chance for ion suppression and low recovery when analyzed by LC-MS/MS. Still, an ASE can do the extraction at high temperature and pressure, enabling solvents to penetrate fat cells and extract any pesticides that might be in them. However, more matrix components can be co-extracted, and they can cause ion suppression, especially when standards are spiked at very low levels. So even though PLE with an ASE may find higher levels of pesticides in some foods than other methods that extract at room temperature, it may not be acceptable to some if it does not produce adequate recovery of standards spiked at very low levels. This is especially true if another method has been validated and shown to produce acceptable spike recoveries at the lowest concentration possible. So accuracy may be sacrificed when analyzing some samples that have relatively high amounts of pesticides so that apparently better results are obtained for samples that have the lowest amounts of pesticides.

Another factor that can affect accuracy is the objective of the study. That is, if just a few samples are being analyzed, methods that offer speed and efficiency may not be very useful. However, when analyzing hundreds of samples for hundreds of analytes, speed, efficiency, and minimizing solvent use can become quite important. Still, speed and efficiency can be sacrificed if specifications are written so that the analyst must obtain accuracies or spike recoveries that are too strict. That is, it may be possible to get spike recoveries between 90% and 100% for some relatively clean matrices (like some fruit juices), but very difficult with other matrices, especially soil, sludge, and fatty foods. Regulators must decide how to balance the desire to spend as much time as it takes to quantify the lowest levels of pesticides with the desire to analyze as many samples as possible and emphasize the search for samples that might contain relatively high levels of pesticides or other toxins. For this reason, there are rapid screening methods and detailed, more time-consuming methods for accurate quantitation. Sometimes the need to protect the public from high levels of toxins can be best served by rapid methods that may sacrifice some accuracy, especially at ng/g levels of pesticides.

5.7.1 Factors Affecting the Accurate Quantification of Pesticide Residues in Fatty Matrices

So, one of the biggest factors affecting accurate quantification of pesticides in fatty matrices is the location of the pesticides. In fatty fruits and vegetables, pesticides are sprayed onto the surface and can be easily extracted at room temperature and pressure. On the other hand, when lipophilic pesticides are present in meat, they will not just be on the surface; they will also be inside the fat cells. So they can't be extracted easily with acetonitrile, which is used in the QuEChERs method. However, the recoveries of spiked standards may be quite good, especially if the standards were dissolved in acetonitrile and never reach the interior of the fat of the meat. Even standard addition will not be suitable for determining which method is best for sample preparation. If the standards do not penetrate the fat cells of the meat, they will be easily solubilized. Nearly linear plots of detector response versus the amount of added analyte may be obtained for both PLE and simply shaking the sample with acetonitrile, followed by a QuEChERS cleanup. The detector response may be much higher when extracting with an ASE. That is, fatty meat that has high levels of incurred pesticides and no added standard might get a higher detector response that when extracting with acetonitrile at room temperature and pressure.

Another important factor for sample preparation of fatty foods is that some of them are not easily chopped, ground, or blended at room temperature. So some of them have to be mixed with dry ice when placed in a sample grinder. Once frozen, they can be homogenized more easily.

5.7.2 Factors Affecting the Accurate Quantification of Pesticide Residues in Nonfatty Matrices

Unlike fatty foods, nonfatty foods often present few problems for sample preparation and the extraction of pesticides. However, soil, sludge, and even industrial wastewater can present problems even if they contain little or no fat. They can contain humic substances or industrial chemicals that can be strongly retained on an HPLC or LC-MS column, thus damaging or destroying it after just a few analyses. Moreover, they might contain matrix components that can cause ion suppression when using MS for detection or false positives with UV-Vis detection.

5.8 NEW TRENDS IN SAMPLE PREPARATION METHODS FOR PESTICIDE RESIDUE ANALYSIS: QuEChERS (QUICK, EASY, CHEAP, EFFECTIVE, RUGGED, AND SAFE) TECHNIQUE

The QuEChERS (Quick, Easy, Cheap, Effective, Rugged, and Safe) method was developed for extracting pesticides in fruits and vegetables and was introduced in 2003 [100]. It uses acetonitrile to extract samples containing water by adding salts to make these normally miscible solvents separate into two phases. It has become quite popular due to its speed, low cost, and small use of solvents.

The original version of the QuEChERS method was done at a neutral pH and a small volume (10 mL) of acetonitrile, followed by adding 4 g of anhydrous $MgSO_4$ plus 1 g of NaCl. Residual water and polar matrix compounds were removed using a d-SPE cleanup using a sorbent that contained PSAs [100]. Subsequent workers added octadecylsilica (C-18) [101] to the acetonitrile extract, followed by shaking and centrifugation to remove matrix interferences.

Even though only a few pesticides were separated by GC in the original study, hundreds of pesticides in different types of foods have been quantified since then [102–104]. Still, this method gave lower stability and/or recoveries of some pesticides, depending on pH of the matrix [100]. The original method was modified by using a pH 5 buffer to achieve acceptably high recoveries (>70%) for certain pH-dependent pesticides pymetrozine, imazalil, and thiabendazole in a variety of fruits and vegetables [105]. The method was further modified by using a relatively strong acetate buffer or a weaker citrate buffer [106]. Both versions of these methods went through extensive interlaboratory studies [107] for dozens of pesticides in various fortified and incurred samples in using different types of GC-MS and LC-MS/MS conditions and instruments for separation and detection. Both methods met statistical criteria for acceptability from independent organizations with the acetate-buffering version becoming the AOAC Official Method 2007.01 [108] and the citrate-buffering version being named European Committee for Standardization (CEN) Standard Method EN 15662 [109]. Moreover, an interlaboratory validation was done using LC-MS/MS [10].

The QuEChERS approach is very flexible, so it can be modified to accommodate analytes with different properties and matrix compositions along with different equipment that is available in diverse laboratories. Acceptable recoveries can be achieved for many pesticides in a variety of matrices even if different ratios and types of sample size, solvent, salts, and sorbents are used in the modifications. The ruggedness of the QuEChERS approach has been evaluated in the original [100] and subsequent publications [10,110–114]. Still, when analyzing food for many pesticides with widely varying physicochemical properties, the sample preparation method may require using different sorbents [115–117].

One of the more interesting applications was when the QuEChERS method was used with acetonitrile and a dispersive SPE cleanup to discover that there are relatively high amounts of carbendazim and other pesticides in honeybees [118]. In another application, the QuEChERS method was used to determine the concentrations of the natural pyrethrins cinerin I and II, jasmolin I and II, and pyrethrin I and II as well as two pyrethroid insecticides, cypermethrin and deltamethrin, in fin and nonfin fish products [119]. Others used the QuEChERS method to extract OCPs in soil using acetonitrile, followed by liquid–liquid partition into n-hexane. The hexane extracts were suitable for determination using GC-MS/MS [120]. Still others used a modification designed for fatty foods to prepare olives for multiresidue pesticide analysis. That is, after extracting with acetonitrile, dispersive solid phase extraction cleanup was done using graphitized carbon black, PSA, and C-18 sorbents [121]. They also used an advanced software package for carrying out method development for multiple reaction monitoring (MRM). It overcame several limitations and drawbacks associated with MS/MS methods (time segment boundaries, tedious method development/manual scheduling, and acquisition limitations).

More recently, QuEChERS was used to prepare cereals for the analysis of 219 pesticides by GC-MS [122]. Different buffers and solvents were evaluated for their abilities to improve the recovery of standards that were spiked into these samples that had <75% moisture. It was found that the citrate buffer used in the CEN method EN 15662 [123] was better than the acetate buffer used in the AOAC method [107]. Another group optimized the QuEChERS method so that it could prepare Brazilian apples, strawberries, and tomatoes for GC-ECD analysis of pesticides [124]. They found that the optimum sample weight and solvent volume were 18 g and 10 mL, respectively [124]. Another group validated a slightly modified QuEChERS method to analyze Colombian tomatoes (*Solanum lycopersicum*), tamarillos (*S. betaceum*), and goldenberries (*Physalis peruviana*) [125]. They used the AEN/CTN method [126], which uses 4 g of $MgSO_4$, 1 g of NaCl, 1 g of sodium citrate, and 0.5 g of sodium citrate sesquihydrate [125,126]. The method was validated for 24 pesticides in tomatoes, 33 in tamarillos, and 28 in goldenberries [125]. In still another study, the QuEChERS method was used to prepare peppers for the quantification of dimethomorph [126]. They found that washing and parboiling could reduce its concentration in peppers [127]. Additional recent research has shown that ammonium formate may be better than other salts because it is volatile, and the ammonium can form adducts, making the MS detection easier [128]. They found that using 7.5 g of ammonium formate and 15 mL of 5% (v/v) formic acid in acetonitrile for the extraction of 15 g of sample (5 g for wheat grain) was best [128].

The QuEChERS method has also been used recently to prepare samples for mycotoxins [129], aflatoxins [130,131], and antifungal compounds [132]. For mycotoxins in rice, a modified QuEChERS method, in which acidified aqueous acetonitrile was added, followed by salting-out (liquid partitioning) with $MgSO_4$, NaCl, and citrate buffer salts was used [129]. The extracts of rice samples were cleaned up by d-SPE with $MgSO_4$, PSA, C_{18}, and neutral alumina [129]. They found that the QuEChERS method for cleaning up samples was faster and more efficient than affinity columns [129]. In one of the studies, the partly purified extracts of dairy products were concentrated using dispersive liquid–liquid microextraction [130]. In another, 15 antifungal compounds in lactic acid bacteria were quantified after using a QuEChERS sample cleanup [132].

5.9 CONCLUSIONS

The properties of the sample matrix and the goals of the study or analysis can affect the type of sample preparation that should be used. Many different methods can be used for sample preparation.

This work should not be taken as reflecting FDA policy or regulations.

REFERENCES

1. Ames, B. N., Profet, M., and Gold, L. S., Dietary pesticides (99.99% all natural), *Proc. Natl. Acad. Sci. USA*, 87, 7777–7781, 1990.
2. Leslie, M., Gut microbes keep rare immune cells in line, *Sci.*, 335, 1428, 2012.
3. Nakazawa, D. K., *The Autoimmune Epidemic*, Touchstone, New York, 2008.
4. Smith, R. E., *Medicinal Chemistry—A Fusion of Traditional and Western Medicine*, pp. 509–520, Bentham Science, U.A.E., 2013.
5. Costenbader, K., and Laden, F., What do pesticides, farming, and dose effects have to do with the risk of developing connective tissue disease? *Arthrit. Care Res.*, 63, 175–177, 2011.
6. Corsinia, E., Sokootib, M., Galli, C. L., Morettoc, A., and Colosiob, C., Pesticide induced immunotoxicity in humans: A comprehensive review of the existing evidence, *Toxicol.*, 307, 123–135, 2013.
7. Stahnke, H., Kittlaus, S., Guñther Kempe, G., and Lutz Alder, L., Reduction of matrix effects in liquid chromatography–electrospray ionization–mass spectrometry by dilution of the sample extracts: How much dilution is needed? *Anal. Chem.*, 84, 1474–1482, 2012.
8. Frenich, A. G. et al., Compensation for matrix effects in gas chromatography–tandem mass spectrometry using a single point standard addition, *J. Chromatogr. A*, 1216, 4798–4808, 2009.
9. Bolger, M. et al., US Food and Drug Administration's program for chemical contaminants in food, *ACS Symp. Series*, 1020, 7–24, 2009.
10. Sack, C. et al., Collaborative validation of the QuEChERS procedure for the determination of pesticides in food by LC-MS/MS, *J. Agric. Food Chem.*, 59, 6383, 2013.
11. Gong, P., Wang, X., Sheng, J., and Yao, T., Variations of organochlorine pesticides and polychlorinated biphenyls in atmosphere of the Tibetan Plateau: Role of the monsoon system, *Atmosph. Environ.*, 44, 2518–2523.
12. Filho, A. M., dos Santos, F. N., and Pereira, P. A. P., Development, validation and application of a method based on DI-SPME and GC–MS for determination of pesticides of different chemical groups in surface and groundwater samples, *Microchem. J.*, 96, 139–145, 2010.
13. Poppa, P., Keil, P., Möder, M., Paschke, A., and Thuss, U., Application of accelerated solvent extraction followed by gas chromatography, high-performance liquid chromatography and gas chromatography–mass spectrometry for the determination of polycyclic aromatic hydrocarbons, chlorinated pesticides and polychlorinated dibenzo-*p*-dioxins and dibenzofurans in solid wastes, *J. Chromatogr. A*, 774, 203–211, 1997.
14. Zanella, R., Prestes, O. D., Friggi, C. A., Martins, M. L., and Adaime, M. B., An overview about recent advances in sample preparation techniques for pesticide residues analysis in cereals and feedstuffs, *Pesticides—Recent Trends in Pesticide Residue Assay*, Academypublish.org, 2012.
15. Soxhlet, F., Die gewichtsanalytische Bestimmung des Milchfettes, *Polytechnisches J.*, 232, 461–465, 1879.
16. Büchi Analytical Inc., *Extraction System B-11*, Büchi Analytical Inc., New Castle, DE, 2013 http://www.buchi.com/Extraction-System-B-811-Standa.317.0.html.
17. Zanella, R. et al., Modern sample preparation methods for pesticide multiresidue determination in foods of animal origin by chromatographic–mass spectrometric techniques, in *The Impact of Pesticides*. AcademyPublish.org, 2012.
18. Aguilera, A., Rodríguez, M., Brotons, M., Boulaid, M., and Valverde, A., Evaluation of supercritical fluid extraction/aminopropyl solid-phase "in-line" cleanup for analysis of pesticide residues in rice, *J. Agric. Food Chem.*, 53, 9374–9382, 2005.
19. Ramos, L., Critical overview of selected contemporary sample preparation techniques, *J. Chromatogr. A*, 1221, 84–98, 2012.
20. EPA Method 3545 (July 1995) Pressurized Fluid Extraction, Test Methods for Evaluating Solid Waste, 3rd ed., Update III; EPA SW-846:US GPO, Washington, DC, 1995.
21. Richter, B. E. et al., Accelerated solvent extraction: A technique for sample preparation, *Anal. Chem.*, 68, 1033–1039, 1996.
22. Popp, P., Keil, P., Moder, M., Paschke, A., and Thuss, U., Application of accelerated solvent extraction followed by gas chromatography, high-performance liquid chromatography and gas chromatography–mass spectrometry for the determination of polycyclic aromatic hydrocarbons, chlorinated pesticides and polychlorinated dibenzo-*p*-dioxins and dibenzofurans in solid wastes, *J. Chromatogr. A*, 774, 203–211, 1997.
23. Eskilsson, C. S., and Bjōrklund, E., Analytical-scale microwave-assisted extraction, *J. Chromatog. A*, 902, 227–250, 2000.

24. Dean, J. R., Fitzpatrick, L., and Heslop, C., in: A. J. Handley (Ed.), *Extraction Methods in Organic Analysis*, Sheffield Academic Press, Sheffield, 1999, p. 166.
25. Sanchez-Prado, L., Garcia-Jares, C., and Llompart, M., Microwave-assisted extraction: Application to the determination of emerging pollutants in solid samples, *J. Chromatogr. A*, 1217, 2390–2414, 2010.
26. Molins, C., Hogendoorn, E. A., Heusinkveld, H. A. G., Van Zoonen, P., and Baumann, R. A., Microwave assisted solvent extraction (MASE) of organochlorine pesticides from soil samples, *Int. J. Environ. Anal. Chem.*, 68, 155, 1997.
27. Erickson, B., Standardizing the world with microwaves, *Anal. Chem.*, 70, 467A, 1998.
28. Mao, X., Wana, Y., Yan, A., Shen, M., and Wei, Y., Simultaneous determination of organophosphorus, organochlorine, pyrethriod and carbamate pesticides in *Radix astragali* by microwave-assisted extraction/dispersive-solid phase extraction coupled with GC–MS, *Talanta*, 97, 131–141, 2012.
29. Fang, G., Lau, H. F., Law, W. S., and Li, S. F. Y., Systematic optimisation of coupled microwave-assisted extraction-solid phase extraction for the determination of pesticides in infant milk formula via LC–MS/MS, *Food Chem.*, 134, 2473–2480, 2012.
30. Satpathy, G., Tyagi, Y. K., and Gupta, R. K., A novel optimised and validated method for analysis of multi-residues of pesticides in fruits and vegetables by microwave-assisted extraction (MAE)–dispersive solid-phase extraction (d-SPE)–retention time locked (RTL)–gas chromatography–mass spectrometry with deconvolution reporting software (DRS), *Food Chem.*, 127, 1300–1308, 2011.
31. Coscollà, C., Castillo, M., Pastor, M., and Yusà, V., Determination of 40 currently used pesticides in airborne particulate matter (PM 10) by microwave-assisted extraction and gas chromatography coupled to triple quadrupole mass spectrometry, *Anal. Chim. Acta*, 693, 72–81, 2011.
32. Wang, Z. et al., An absorbing microwave micro-solid-phase extraction device used in non-polar solvent microwave-assisted extraction for the determination of organophosphorus pesticides, *Anal. Chim. Acta*, 760, 60–68, 2013.
33. Liu, H., and Dasgupta, P. K., Analytical chemistry in a drop. Solvent extraction in a microdrop, *Anal. Chem.*, 68, 1817, 1996.
34. Yao, C., Pitner, W. R., and Anderson, J. L., Ionic liquids containing the tris(pentafluoroethyl)trifluorophosphate anion: A new class of highly selective and ultra hydrophobic solvents for the extraction of polycyclic aromatic hydrocarbons using single drop microextraction, *Anal. Chem.*, 81, 5054–5063, 2009.
35. Jain, A., and Verma, K. K., Recent advances in applications of single-drop microextraction: A review, *Anal. Chim. Acta*, 706, 37–65, 2011.
36. Jeannot, M. A., and Cantwell, F. F., Mass transfer characteristics of solvent extraction into a single drop at the tip of a syringe needle, *Anal. Chem.*, 69, 235–239, 1997.
37. Peng, J.-P., Liu, J.-F., Hu, X.-L., and Jiang, G.-B., Direct determination of chlorophenols in environmental water samples by hollow fiber supported ionic liquid membrane extraction coupled with high-performance liquid chromatography, *J. Chromatogr. A*, 1139, 165–170, 2007.
38. Fortunato, R., Afonso, C. A. M., Benavente, J., Rodriguez-Castellon, E., and Creso, J. G., Stability of supported ionic liquid membranes as studied by x-ray photoelectron spectroscopy, *J. Membr. Sci.*, 256, 216–223, 2005.
39. Peng, J.-P., Liu, J.-F., Hu, X.-L., and Jiang, G.-B., Direct determination of chlorophenols in environmental water samples by hollow fiber supported ionic liquid membrane extraction coupled with high-performance liquid chromatography, *J. Chromatogr. A*, 1139, 165–170, 2007.
40. Lee J., Lee, H. K., Rasmussen, K. E., and Pedersen-Bjergaard, S., Environmental and bioanalytical applications of hollow fiber membrane liquid-phase microextraction: A review, *Anal. Chim. Acta*, 624, 253–268, 2008.
41. Pedersen-Bjergaard, S., and Rasmussen, K. E., Liquid–liquid–liquid microextraction for sample preparation of biological fluids prior to capillary electrophoresis, *Anal. Chem.*, 71, 2650–2656, 1999.
42. Basheer, C., Suresh, V., Renu, R., and Lee, H. K., Development and application of polymer-coated hollow fiber membrane microextraction to the determination of organochlorine pesticides in water, *J. Chromatogr. A*, 1033, 213–220, 2004.
43. United States Environmental Protection Agency Method 508: Determination of chlorinated pesticides in water by gas chromatography with an electron capture detector, Cincinnati, OH, 1989.
44. Chen, P. S., and Huang, S. D., Determination of ethoprop, diazinon, disulfoton and fenthion using dynamic hollow fiber-protected liquid-phase microextraction coupled with gas chromatography–mass spectrometry, *Talanta*, 69, 669–675, 2006.
45. Hou, L., and Lee, H. K., Determination of pesticides in soil by liquid-phase microextraction and gas chromatography–mass spectrometry, *J. Chromatogr. A*, 1038, 37–42, 2004.
46. Jiang, X., and Lee, H. K., Solvent bar microextraction, *Anal. Chem.*, 76, 5591–5596, 2004.

47. Xua, L., and Lee, H. K., Solvent-bar microextraction—Using a silica monolith as the extractant phase holder, *J. Chromatogr. A*, 1216, 5483–5488, 2009.
48. Chia, K.-J., and Huang, S.-D., Analysis of organochlorine pesticides in wine by solvent bar microextraction coupled with gas chromatography with tandem mass spectrometry detection, *Rapid Commun. Mass Spectrom.*, 20, 118–124, 2006.
49. Melwanki, M. B., Huang, S.-D., and Fuh, M.-R., Three-phase solvent bar microextraction and determination of trace amounts of clenbuterol in human urine by liquid chromatography and electrospray tandem mass spectrometry, *Talanta*, 72, 373–377, 2007.
50. Xua, L., and Lee, H. K., Solvent-bar microextraction—Using a silica monolith as the extractant phase holder, *J. Chromatogr. A*, 1216, 5483–5488, 2009.
51. Guo, L., and Lee, H. K., Microwave assisted extraction combined with solvent bar microextraction for one-step solvent-minimized extraction, cleanup and preconcentration of polycyclic aromatic hydrocarbons in soil samples, *J. Chromatogr. A*, 1286, 9–15, 2013.
52. Su, Y.-S., Yan, C.-T., Ponnusamy, V. K., and Jen, J.-F., Novel solvent-free microwave-assisted extraction coupled with low-density solvent-based in-tube ultrasound-assisted emulsification microextraction for the fast analysis of organophosphorus pesticides in soils, *J. Sep. Sci.*, 36, 2339–2347, 2013.
53. Sarafraz-Yazdi, A., and Amiri, A., Liquid-phase microextraction, *Trens Anal. Chem.*, 29, 1–14, 2010.
54. A Zhu, L., Ee, K. H., Zhao, L., and Lee, H. K., Analysis of phenoxy herbicides in bovine milk by means of liquid–liquid–liquid microextraction with a hollow-fiber membrane, *J. Chromatogr. A*, 963, 335–343, 2002.
55. Pesticide Analytical Manual, Vol. 1, Chapter 3, Multiclass MRMs.
56. Yang, X. et al., Multiresidue method for determination of 88 pesticides in berry fruits using solid-phase extraction and gas chromatography–mass spectrometry: Determination of 88 pesticides in berries using SPE and GC–MS, *Food Chem.*, 127, 855–865, 2011.
57. Štajnbaher, D., and Zupancic-Kralj, L., Multiresidue method for determination of 90 pesticides in fresh fruits and vegetables using solid-phase extraction and gas chromatography–mass spectrometry, *J. Chromatogr. A*, 1015, 185–198, 2003.
58. Kuivinene, J., and Bengtsson, S., Solid-phase extraction and cleanup of organophosphorus pesticide residues in bovine muscle with gas chromatographic detection, *J. Chromat. Sci.*, 40, 392–396, 2002.
59. Rodriguez-Mozaz, S., de Alda, M. J. L., and Barceló, D., Monitoring of estrogens, pesticides and bisphenol A in natural waters and drinking water treatment plants by solid-phase extraction–liquid chromatography–mass spectrometry, *J. Chromatogr. A*, 1045, 85–92, 2004.
60. Chen, J., Duan, C., and Guan, Y., Sorptive extraction techniques in sample preparation for organophosphorus pesticides in complex matrices, *J. Chromatogr. B*, 878, 1216–1225, 2010.
61. Kugimiya, A., and Takei, H., Selectivity and recovery performance of phosphate-selective molecularly imprinted polymer, *Anal. Chim. Acta*, 606, 252–256, 2008.
62. Lv, Y., Lin, Z., Feng, W., Zhou, X., and Tan, T., Selective recognition and large enrichment of dimethoate from tea leaves by molecularly imprinted polymers, *Biochem. Eng. J.*, 36, 221–229, 2007.
63. Fernandez-Alvarez, M. et al., Simultaneous determination of traces of pyrethroids, organochlorines and other main plant protection agents in agricultural soils by headspace solid-phase microextraction–gas chromatography, *J. Chromatogr. A*, 1188, 154–163, 2008.
64. McManus, S.-L., Coxon, C. E., Richards, K. G., and Danaher, M., Quantitative solid phase microextraction—Gas chromatography mass spectrometry analysis of the pesticides lindane, heptachlor and two heptachlor transformation products in groundwater, *J. Chromatogr. A*, 1284, 1–3, 2013.
65. Beceiro-González, E. et al., Optimisation and validation of a solid-phase microextraction method for simultaneous determination of different types of pesticides in water by gas chromatography–mass spectrometry, *J. Chromatogr. A*, 1141, 165–173, 2007.
66. Xu, J. et al., New materials in solid-phase microextraction, *Trends Anal. Chem.*, 47, 68–83, 2013.
67. Arthur, C. L., and Pawliszyn, J., Solid phase microextraction with thermal desorption using fused silica optical fibers, *Anal. Chem.*, 62, 2145–2146, 1990.
68. Vas, G., and Vékey, K., Solid-phase microextraction: A powerful sample preparation tool prior to mass spectrometric analysis, *J. Mass Spectrom.*, 39, 223–254, 2004.
69. Kataoka, H., Automated sample preparation using in-tube solid-phase microextraction and its application—A review, *Anal. Bioanal. Chem.*, 373, 31–45, 2002.
70. Bagheri, H., Ayazi, Z., and Sistani, H., Chemically bonded carbon nanotubes on modified gold substrate as novel unbreakable solid phase microextraction fiber, *Microchim. Acta*, 174, 295–301, 2011.
71. Amini, R., Rouhollahi, A., Adibi, M., and Mehdinia, A., A novel reusable ionic liquid chemically bonded fused-silica fiber for headspace solid-phase microextraction/gas chromatography flame ionization detection of methyl tert-butyl ether in a gasoline sample, *J. Chromatogr. A*, 1218:130–136, 2011.

72. Mehdinia, A., Mohammadi, A., Davarani, S. H., and Banitaba, M., Application of self-assembled monolayers in the preparation of solid-phase microextraction coatings, *Chromatographia*, 74, 421–427, 2011.
73. Szultka, M., Kegler, R., Fuchs, P., Olszowy, P., Miekisch, W., Schubert, J., Buszewski, B., and Mundkowski, R., Polypyrrole solid phase microextraction: A new approach to rapid sample preparation for the monitoring of antibiotic drugs, *Anal. Chim. Acta*, 667, 77–82, 2010.
74. Olszowy, P., Szultka, M., Nowaczyk, J., and Buszewski, B., A new way of solid-phase microextraction fibers preparation for selected antibiotic drug determination by HPLC–MS, *J. Chromatogr. B*, 879, 2542–2548, 2011.
75. Alizadeh, N., Zarabadipour, H., and Mohammadi, A., Headspace solid-phase microextraction using an electrochemically deposited dodecylsulfate-doped polypyrrole film to determine of phenolic compounds in water, *Anal. Chim. Acta*, 605, 159–165, 2007.
76. Mehdinia, A., Bashour, F., Roohi, F., and Jabbari, A., A strategy to enhance the thermal stability of a nanostructured polypyrrole-based coating for solid phase microextraction, *Microchim. Acta*, 177, 301–308, 2012.
77. David, F., and Sandra, P., Stir bar sorptive extraction for trace analysis, *J. Chromatogr. A*, 1152, 54–69, 2007.
78. Leon, V. M., Alvarez, B., Cobollo, M. A., Munoz, S., and Valor, I., Analysis of 35 priority semivolatile compounds in water by stir bar sorptive extraction–thermal desorption–gas chromatography–mass spectrometry I. Method optimization, *J. Chromatogr. A*, 999, 91–101, 2003.
79. Leon, V. M., Llorca-Porcel, J., Alvarez, B., Cobollo, M. A., Munoz, S., and Valor, I., Analysis of 35 priority semivolatile compounds in water by stir bar sorptive extraction–thermal desorption–gas chromatography–mass spectrometry: Part II: Method validation, *Anal. Chim. Acta*, 558, 261–266, 2006.
80. Blasco, C., Fernández, M., Picó, Y., and Font, G., Comparison of solid-phase microextraction and stir bar sorptive extraction for determining six organophosphorus insecticides in honey by liquid chromatography–mass spectrometry, *J. Chromatogr. A*, 1030, 77–85, 2004.
81. Ochiai, N., Sasamoto, K., Kanda, H., Yamagami, T., David, F., Tienpont, B., and Sandra, P., Optimization of a multi-residue screening method for the determination of 85 pesticides in selected food matrices by stir bar sorptive extraction and thermal desorption GC-MS, *J. Sep. Sci.*, 28, 1083–1092, 2005.
82. Basheer, C., Suresh, V., Renu, R., and Lee, H. K., Development and application of polymer-coated hollow fiber membrane microextraction to the determination of organochlorine pesticides in water, *J. Chromatogr. A*, 1033, 213–220, 2004.
83. Basheer, C., Balaji, G., Chua, S. H., Valiyaveettil, S., and Lee, H. K., Novel on-site sample preparation approach with a portable agitator using functional polymer-coated multi-fibers for the microextraction of organophosphorus pesticides in seawater, *J. Chromatogr. A*, 2011, 654–661, 2011.
84. Barker, S. A., Long, A. R., and Hines II, M. E., Disruption and fractionation of biological materials by matrix solid-phase dispersion, *J. Chromatogr.*, 629, 23–34, 1993.
85. Ferrer, C., Gómez, M. C., García-Reyes, J. F., Ferrer, I., Thurman, E. M., and Fernández-Alba, A. R., Determination of pesticide residues in olives and olive oil by matrix solid-phase dispersion followed by gas chromatography/mass spectrometry and liquid chromatography/tandem mass spectrometry, *J. Chromatogr. A*, 1069, 183–194, 2005.
86. Torres C. M., Pico, Y., Redondo, M. J., and Mafies, J., Matrix solid-phase dispersion extraction procedure for multiresidue pesticide analysis in oranges, *J. Chromatogr. A*, 719, 95–103, 1996.
87. Stafford, S. C., and Lin, W., Determination of oxamyl and methomyl by high-performance liquid chromatography using a single-stage postcolumn derivatization reaction and fluorescence detection, *J. Agr. Food Chem.*, 40, 1026–1029, 1992.
88. Kosikowska, M., and Biziuk, M., Review of the determination of pesticide residues in ambient air, *Trends Anal. Chem.*, 29, 1064–1072, 2010.
89. Asensio-Ramos, M., Hernández-Borges, J., Borges-Miquel, T. M., and Rodríguez-Delgado, M. A., Ionic liquid-dispersive liquid–liquid microextraction for the simultaneous determination of pesticides and metabolites in soils using high-performance liquid chromatography and fluorescence detection, *J. Chromatogr. A*, 1218, 4808–4816, 2011.
90. Vázquez, P. P., Mughari, A. R., and Galera, M. M., Solid-phase microextraction (SPME) for the determination of pyrethroids in cucumber and watermelon using liquid chromatography combined with post-column photochemically induced fluorimetry derivatization and fluorescence detection, *Anal. Chim. Acta*, 607, 74–82, 2008.
91. Bagheri, H., Ghanbamejad, H., and Khalilian, F., Immersed sol-gel based amino-functionalized SPME fiber and HPLC combined with post-column photochemically induced fluorimetry derivatization and fluorescence detection of pyrethroid insecticides from water samples, *J. Sep. Sci.*, 32, 2912–2918, 2009.

92. Kong, W.-J., Liu, S.-Y., Qiu, F., Xiao, X.-H., and Yang, M.-H., Simultaneous multi-mycotoxin determination in nutmeg by ultrasound-assisted solid–liquid extraction and immunoaffinity column clean-up coupled with liquid chromatography and on-line post-column photochemical derivatization-fluorescence detection, *Analyst*, 138, 2729–2739, 2013.
93. Mallat, E., and Barceló, D., Analysis and degradation study of glyphosate and of aminomethylphosphonic acid in natural waters by means of polymeric and ion-exchange solid-phase extraction columns followed by ion chromatography–post-column derivatization with fluorescence detection, *J. Chromatogr. A*, 823, 129–136, 1998.
94. Hill, K. M., Hollowell, R. H., and dal Cortivo, L. A., Determination of N-methylcarbamate pesticides in well water by liquid chromatography with postcolumn fluorescence derivatization, *Anal. Chem.*, 56, 2465–2468.
95. Vassilakis, I., Tsipi, D., and Scoullos, M., Determination of a variety of chemical classes of pesticides in surface and ground waters by off-line solid-phase extraction, gas chromatography with electron-capture and nitrogen–phosphorus detection, and high-performance liquid chromatography with postcolumn derivatization and fluorescence detection, *J. Chromatogr. A*, 823, 49–58, 1998.
96. Pickering Laboratories, Carbamates, http://www.pickeringlabs.com/lib/supported/CarbamateManual .pdf Accessed Sept., 2013.
97. Rübensam, G., Barreto, F., Hoff, R. B., and Pizzolato, T. M., Determination of avermectin and milbemycin residues in bovine muscle by liquid chromatography-tandem mass spectrometry and fluorescence detection using solvent extraction and low temperature cleanup, *Food Contr.*, 29, 55–60, 2013.
98. Zhang, K., Mei, Q., Guan, G., Liu, B., Wang, S., and Zhang, Z., Ligand replacement-induced fluorescence switch of quantum dots for ultrasensitive detection of organophosphorothioate pesticides, *Anal. Chem.*, 82, 9579–9586, 2010.
99. Liu, D., Chen, W., Wei, J., Li, X., Wang, Z., and Jiang, X., A highly sensitive, dual-readout assay based on gold nanoparticles for organophosphorus and carbamate pesticides, *Anal. Chem.*, 84, 4185–4191, 2012.
100. Anastassiades, M., Lehotay, S. J., Stajnbaher, D., and Schenck, F. J., Fast and easy multiresidue method employing acetonitrile extraction/partitioning and "dispersive solid-phase extraction" for the determination of pesticide residues in produce, *J. AOAC Int.*, 86, 412–431, 2003.
101. Lehotay, S., and Mastovska, K., Evaluation of two fast and easy methods for pesticide residue analysis in fatty food matrices, *J. AOAC Intl.*, 88, 630-638, 2005.
102. Wilkowska, A., and Biziuk, M., Determination of pesticide residues in food matrices using the QuEChERS methodology, *Food Chem.*, 125, 803–812, 2011.
103. Prestes, O. D. et al., QuEChERS—Um método moderno de preparo de amostra para determinação multirresíduo de pesticidas em alimentos por métodos cromatográficos acoplados à espectrometria de massas, *Química Nova*, 32, 1620–1634, 2009.
104. Kolberg, D. I., Prestes, O. D., Adaime, M. B., and Zanella, R., Development of a fast multiresidue method for the determination of pesticides in dry samples (wheat grains, flour and bran) using QuEChERS based method and GC–MS, *Food Chem.*, 125, 1436–1442, 2011.
105. Lehotay, S. J., Mastovská, K., and Lightfield, A. R., Use of buffering and other means to improve results of problematic pesticides in a fast and easy method for residue analysis of fruits and vegetables, *J. AOAC Int.*, 88, 615–629, 2005.
106. Anastassiades, M., Scherbaum, E., Tasdelen, B., and Stajnbaher, D., in: Ohkawa, H., Miyagawa, H., Lee, P. W. (eds.), *Crop Protection, Public Health, Environmental Safety*, Wiley-VCH, Weinheim, Germany, 2007, p. 439.
107. Lehotay, S. J. et al., Determination of pesticide residues in foods by acetonitrile extraction and partitioning with magnesium sulfate: Collaborative study, *J. AOAC Intl.*, 90, 485–520, 2007.
108. AOAC International, Official method 2007.01: Pesticide residues in foods by acetonitrile extraction and partitioning with magnesium sulphate, *J. AOAC Intl.*, 2007.
109. European Committee for Standardization, CEN, CEN/TC 275 15662:2008: Foods of plant origin—Determination of pesticide residues using GC-MS and/or LC-MS/MS following acetonitrile extraction/partitioning and clean-up by dispersive SPE–QuEChERS-method. European Union; 2008.
110. Lehotay, S. J. et al., Comparison of QuEChERS sample preparation methods for the analysis of pesticide residues in fruits and vegetables, *J. Chromatogr. A*, 1217, 2548–2560, 2010.
111. Aguilera-Luiz, M. M., Vidal, J. L. M., Romero-González, R., and Frenich, A. G., Multi-residue determination of veterinary drugs in milk by ultra-high-pressure liquid chromatography–tandem mass spectrometry, *J. Chromatogr. A*, 1205, 10–16, 2008.
112. Cunha, S. C., Lehotay, S. J., Mastovska, K., Fernandes, J. O., Beatriz, M., and Oliveira, P. P., Evaluation of the QuECHERS sample preparation approach for the analysis of pesticide residues in olives, *J. Sep. Sci.*, 4, 620–632, 2007.

113. Diez, C., Traag, W. A., Zommer, P., Marinero, P., and Atienza, J., Comparison of an acetonitrile extraction/partitioning and "dispersive solid-phase extraction" method with classical multi-residue methods for the extraction of herbicide residues in barley samples, *J. Chromatogr. A*, 1131, 11–23, 2006.
114. Lehotay, S. J., De Kok, A., Hiemstra, M., and Van Bodegraven, P., Validation of a fast and easy method for the determination of residues from 229 pesticides in fruits and vegetables using gas and liquid chromatography and mass spectrometric detection, *J. AOAC Int.*, 88, 615–629, 2005.
115. Li, J., Dong, F., Xu, J., Liu, X., Li, Y., Shan, W., and Zheng, Y., Enantioselective determination of triazole fungicide simeconazole in vegetables, fruits, and cereals using modified QuEChERS (quick, easy, cheap, effective, rugged and safe) coupled to gas chromatography/tandem mass spectrometry, *Anal. Chim. Acta*, 702, 127–135, 2011.
116. Koesukwiwat, U., Lehotay, S. J., Miao, S., and Leepipatpiboon, N., High throughput analysis of 150 pesticides in fruits and vegetables using QuEChERS and low-pressure gas chromatography-time-of-flight mass spectrometry, *J. Chromatogr. A*, 1217, 6692–6703, 2010.
117. Wong, J. et al., Development and interlaboratory validation of a QuEChERS-based liquid chromatography–tandem mass spectrometry method for multiresidue pesticide analysis, *J. Agr. Food Chem.*, 58, 5897–5903, 2010.
118. Wiest L. et al., Multi-residue analysis of 80 environmental contaminants in honeys, honeybees and pollens by one extraction procedure followed by liquid and gas chromatography coupled with mass spectrometric detection, *J. Chromatogr. A*, 1218, 5743–5756, 2010.
119. Lazartigues, A., Wiest, L., Baudot, R., Thomas, M., Feidt, C., and Cren-Olivé, C., Multiresidue method to quantify pesticides in fish muscle by QuEChERS based extraction and LC-MS/MS, *Anal. Bioanal. Chem.*, 400, 2185–2193, 2011.
120. Rashid, A., Nawaz, S., Barker, H., Ahmad, I., and Ashraf, M., Development of a simple extraction and clean-up procedure for determination of organochlorine pesticides in soil using gas chromatography–tandem mass spectrometry, *J. Chromatogr. A*, 1217, 2933–2939, 2010.
121. Gilbert-López, B. et al., Large-scale pesticide testing in olives by liquid chromatography–electrospray tandem mass spectrometry using two sample preparation methods based on matrix solid-phase dispersion and QuEChERS, *J. Chromatogr. A*, 1217, 6022–6035, 2010.
122. He, Z. et al., Multiresidue analysis of over 200 pesticides in cereals using a QuEChERS and gas chromatography–tandem mass spectrometry-based method, *Food Chem.*, 169, 372-80, 2015.
123. Paya, P. et al., Analysis of pesticide residues using the Quick Easy Cheap Effective Rugged and Safe (QuEChERS) pesticide multiresidue method in combination with gas and liquid chromatography and tandem mass spectrometric detection, *Anal. Bioanal. Chem.*, 389, 1697–1714, 2007.
124. Lorenz, J. G. et al., Multivariate optimization of the QuEChERS-GC-ECD method and pesticide investigation residues in apples, strawberries, and tomatoes produced in Brazilian South, *J. Braz. Chem. Soc.*, 25, 1583–1591, 2014.
125. Restrepo, A. R., Ortiz, A. F. G., Ossa, D. U. H., and Mesa, G. A. P., QuEChERS GC–MS validation and monitoring of pesticide residues in different foods in the tomato classification group, *Food Chem.*, 158, 153–161, 2014.
126. AEN/CTN. Determinación de residuos de plaguicidas utilizando GC–MS y/o LC–MS/MS seguido de extracción/división de acetonitrilo y método de purificación dispersiva SPE-QuEChERS (Vol. UNE-EN 15662). Madrid, CEN, 2008, p. 87.
127. Kim, S.-W. et al., The effect of household processing on the decline pattern of dimethomorph in pepper fruits and leaves, *Food Control*, 50, 118–124, 2015.
128. González-Curbelo, M. A. et al., Use of ammonium formate in QuEChERS for high-throughput analysis of pesticides in food by fast, low-pressure gas chromatography and liquid chromatography tandem mass spectrometry, *J. Chromatogr. A*, 1358, 75–84, 2014.
129. Koesukwiwat, U., Sanguankaew, K., and Leepipatpiboon, N., Evaluation of a modified QuEChERS method for analysis of mycotoxins in rice, *Food Chem.*, 153, 44–51, 2014.
130. Sirhan, A. Y. et al., QuEChERS-HPLC method for aflatoxin detection of domestic and imported food in Jordan, *J. Liq. Chrom. Rel. Tech.*, 37, 321–342, 2014.
131. Karaseva, N. M., Amelin, V. G., and Tret'yakov, A. V., QuEChERS coupled to dispersive liquid–liquid microextraction for the determination of aflatoxins B1 and M1 in dairy foods by HPLC, *J. Anal. Chem.*, 69, 461–466, 2014.
132. Brosnan, B., Coffey, A., Arendt, E. K., and Furey, A., The QuEChERS approach in a novel application for the identification of antifungal compounds produced by lactic acid bacteria cultures, *Talanta*, 129, 364–373, 2014.

6 Selection of the Mode of Stationary Phases and Columns for Analysis of Pesticides

Bogusław Buszewski and Szymon Bocian

CONTENTS

6.1 INTRODUCTION

The following chapter deals with the different modes of liquid chromatography methods and the selection of stationary phases, which may be applied for pesticide separations. In the following, the influence of the geometric parameters, such as particle size and column dimensions, on the separation selectivity, efficiency, and back pressure are all discussed.

Although the majority of liquid chromatographic analyses are carried out in a reversed-phase (RP) system, in liquid chromatography, four types of separation mechanisms may be distinguished: adsorption, partition, ion-replaceable, and sieve mechanism (partition, ion exchange, and size exclusion). The basic property, which is utilized to separate substances in high-performance liquid chromatography (HPLC), is the difference in solute interactions with the stationary phase and the mobile phase. Depending on the type of such interaction, the chromatographic separation can be carried out in different modes of liquid chromatography. The applied mode of the liquid chromatography for pesticide analysis has to be chosen according to the polarity of the given solutes.

6.2 ADSORPTION MODE

Adsorption means interaction of solute molecules with the adsorbent surface and occurs both in packed column (HPLC) and in thin layer (planar) chromatography. A system in which a polar adsorbent (e.g., silica, alumina, Florisil, bonded phases such as amino, diol, CN, enantioselective sorbents) is in contact with nonaqueous less polar eluents is called a normal-phase (NP) system while systems with nonpolar sorbents (e.g., octyl, octadecyl) and polar aqueous eluents are called reversed-phase (RP) systems.

In the literature, a lot of studies may be found in which separation of pesticides is performed in a NP system. The introduction of RP packing means that, every year, more and more pesticide analysis is carried out in RP systems.

Because the adsorption mode works on the basis of the differences in the polarities of the species, RP chromatography, which is based on hydrophobic interactions, offers much better separation selectivity.

6.2.1 SILICA GEL

Porous silica gel is the most important material applied in liquid phase separation sciences, either as an adsorbent or as the support for bonded stationary phases. The most important advantages of silica-based packing materials are their physical stability and a well-defined and controllable pore structure and morphology (totally porous, core shell, or monolithic). These properties assure rapid mass transfer, good loadability, and high reproducibility [1].

Silica gel is a silicon dioxide with the formula SiO_2. The active sites on the silica gel surface that control the separation are the silanols: hydroxyl groups bonded to the silicon atom (≡Si-OH).

Even when the porous silica particles have been exhaustively derivativized, a significant fraction of silanol groups remain unreacted and are accessible for interaction with polar and charged solutes. One can recognize four types of surface silanol groups: isolated, vicinal, germinal, and siloxane. The presence of residual silanols can have a negative influence on the separation of polar analytes, especially basic compounds and biopolymers [2].

Various modification methods with different silanes lead to obtaining stationary phases with diverse functional groups.

Silica gel used in liquid chromatography offers a wide range of pore diameters and specific surface areas. Depending on the synthesis methodology, there is an inverse relationship between the pore size and the surface area. As the pore size increases, the surface area decreases. A silica gel with the most common 100 Å pores usually offers a specific surface area in the range of 300–400 m^2/g. Silica with larger pore size (e.g., 300 Å) will have a smaller surface area (around 100 m^2/g). Such material possesses fewer silanols, so it would be less active. As a result, the retention in an HPLC column or on a TLC plate is shorter. It would also result in a smaller mass capacity. Wider pores offer a better possibility for surface modification. However, the smaller surface area would mean less chemically bonded phase attached to the surface and provide lower retention and less loading capacity on larger-pore-size packing with the same bonded phase attached to the silica gel with narrow pores and higher specific surface area.

Consideration of the adsorbent pore size is important when choosing a suitable HPLC column packing for separation of sample components with different molecular weight. Silica gels with pore size in the range of 60–100 Å are used for small molecules whereas 300 Å silica is applied for protein separation.

The most important disadvantage of silica gel is that it can dissolve under certain conditions, which results in damaging of silica gel packing in a chromatographic column. This process occurs when the column is packed with pure silica gel or a chemically modified silica gel. This is because any bonding chemistry, even when the material is endcapped, leaves a significant number of silanols on the surface (usually more than 40% of the original value). The silica-based columns can be operated in the pH range of 3–9.5. However, even at pH 7, the dissolving can be significant. The higher the ionic strength and the higher the pH, the faster the dissolving of silica materials.

The older silica gels are now referred to as Type A silica gels. With this type of manufacture, the silica contained more heavy metals as potassium ions instead of sodium ions. In the 1990s, manufacturers began to use an alternative way to make silica gel. Silica gel is transformed to a volatile organosilane, such as tetraethyoxysilane. This could be distilled, leaving behind heavy metals. This pure organosilane is then hydrolyzed to create a silica sol, which could be formed into spherical particles. This newer, purer silica is referred to as Type B silica gel. The advantage of Type B silica gel is that it offers more symmetrical peaks, and it is much more stable in higher pH (11–12) ionic mobile phases. At the present time, most manufacturers use only Type B silica gel in the preparation of all their column packings.

6.2.1.1 Irregular

At the beginning of analytical and preparative chromatography, only irregular silica gels were used. Now, irregular silica gels, having typical particle sizes of 40–63 and 60–200 μm, remain interesting for flash chromatography (laboratory and large-scale processes) and also for sample preparation processes, for example, solid phase extraction (SPE) [3].

Irregular silica gel is prepared with a sophisticated milling technology. The shapes of the particles are more homogenous. The shape of the particles is cubes. Corners and edges are rounded. It was simple to produce but needed to be packed under high pressure to obtain a suitable packed bed that would perform well and be reproducible. The lower the average particle size, the higher the column performance and the back pressure. The broader the particle size distribution, the higher is the back pressure. Thus, particle size distribution has to be as narrow as possible. Irregular silica was sized to a narrow range with a mean particle diameter of 10 ± 2 μm or even smaller. Nevertheless, the column packed with spherical silica gel offers better chromatographic properties than the column packed with irregular particles.

6.2.1.2 Spherical

The spherical particles of silica gel were introduced about five years after the introduction of HPLC columns packed with irregular silica. During the manufacturing process, the addition of surfactants and other chemicals allowed obtaining the silica sol of microspheres. The microspheres agglomerate into larger spherical silica particles, which may be screened or air-classified to the desired particle size range.

Over the years, the average particle size of the HPLC columns has decreased from 10 μm, through 5 μm and 3.5 μm, up to even sub-2 μm particle size (presently with average particle diameters of 1.3 μm) in the ultra-HPLC systems.

A batch of silica gel is always constituted of particles with slightly different particle size. As indicated by the manufacturer, there is always a plus or minus range to the average particle size. The reason is that it is neither possible nor economical to produce a packing of a single diameter. However, a narrow range of particle size offers better chromatographic properties (higher efficiency and more symmetrical peaks).

6.2.1.3 Monolithic

Until 2001, HPLC columns have been made of particulate materials, usually silica gels. By their very small particle size when packed tightly into a steel column, a significant resistance to the flow of the mobile phase was created, and a higher back pressure was generated. To overcome these limitations, the monolithic rods of highly porous silica were synthesized by E. Merck in Darmstadt, Germany, and it commercialized the work of Japanese researchers. Based on this new technology, highly porous monolithic rods of silica with a bimodal pore structure (macroporous and mezoporous) can be obtained. Each macropore has an average diameter of 2 µm, and they together form a dense network of pores, which allows a rapid flow of eluent at low pressure, which results in dramatically reducing the back pressure and separation time.

The mesopores form the fine porous structure (13 nm in diameter) of the column interior and create a very large surface area on which adsorption of the target compounds can take place. Owing to the very high porosity of these columns, very high flow rates can be applied with very low pressure. The column of 100 mm × 4.6 mm may be operated with flow rate of 9 ml/min at a pressure of 153 bars (using 20% ACN in water).

Monolithic columns provide efficiency of up to 100,000 plates per meter, which is similar to 3.5-µm packed beds. They are made with Type B silica gel; hence they are stable at pH up to 12. Monolith silica gel may be also chemically modified. Reversed chemically bonded stationary phases are commercially available.

6.2.2 Silica Hybrid

Recent developments in the composite materials during silica sol manufacturing have led to obtaining materials with extended chemical stability to higher pH values [4]. They were obtained by an introduction of organosilanes to form organosilica particles, using, for example, tetraethoxysilane with organosilanes.

These materials introduce a slightly different character for separation purposes than silica and also have an extended life due to being more resistant to dissolving. These materials are referred to as hybrid silica gels. They were commercialized as Xterra and XBridge.

6.2.3 Alumina

Alumina (aluminum oxide Al_2O_3) has been used as a column chromatography packing for many years. As with silica gels, alumina could be made into smaller particle size ranges for use in HPLC. Alumina has similar properties to silica gel as a stationary phase; however, it is not nearly so popular. Aluminum oxide exists in many forms whereas γ-aluminum oxide is the most popular. Different forms of alumina arise during the heat treatment of aluminum hydroxide or aluminum oxy hydroxide. Alumina is offered as either irregular or spherical particles.

Their most important advantage is that it could be used with high pH, up to pH 14, mobile phases without dissolving. Alumina does not provide the required efficiencies (number of theoretical plates) that are realized using silica gel. Three types of alumina are available in the market: acidic, basic, and neutral, depending on chemical treatment. Depending on the surface composition of Al, O, and H atoms, alumina may exhibit different surface characteristics, which provide many sorption possibilities, depending on the mobile phase used.

6.2.4 Other Inorganic Oxides

Apart from silica and alumina, only a few other inorganic oxides, such as zirconia and titania, have been used in HPLC. Of these two materials, only zirconia has been commercialized. Zirconia (ZrO_2) is more resistant to dissolving at higher pH, so it is recommended for extremely alkaline mobile phases.

The last group of packing materials for HPLC are porous graphitic carbons. These materials have been available for a number of years. Their properties allow the use in the RP as well as the NP modes. Graphitic carbons are stable in pH range from 1 to 14 in a wide range of mobile phase compositions.

6.3 REVERSED PHASE MODE

Nowadays, approximately 80% of all chromatographic separations are done in RP systems. The reason is good solubility of many analyzed compounds in aqueous, aqueous–methanol, or acetonitrile solvents. The separations of many classes of compounds have been accomplished on RP bonded silica gels due to better selectivity in the comparison with NP systems.

6.3.1 Silica Gel Bonded with Hydrophobic Phases

Various modification methods with different silanes lead to obtaining stationary phases with diverse functional groups. The use of monofunctional modifiers leads to the formation of monomeric structures of the chemically bonded phase. The use of di- or trifunctional modifiers leads to the formation of cross-linked *polymeric* structures [5].

Alkyl ligands, the most commonly used chemically bonded phases in RP-HPLC, have hydrophobic characteristics. Those ligands are the hydrophobic adsorption centers for organic modifiers and solute molecules. The important parameters of chemically bonded phases, which determine the chromatographic properties, are the structure of the bonded ligands, their length, coverage density (surface concentration), and coverage homogeneity [6,7].

The hydrophobicity of the stationary phase and residual silanol activity are important parameters of silica-based chromatographic adsorbents. Their relative importance may be measured by standard tests as described by Tanaka [8], Engelhardt [9], Walters [10], Galushko [11], and several other researchers [12–15]. Structures of chemically bonded phases used in RP liquid chromatography are illustrated in Figure 6.1.

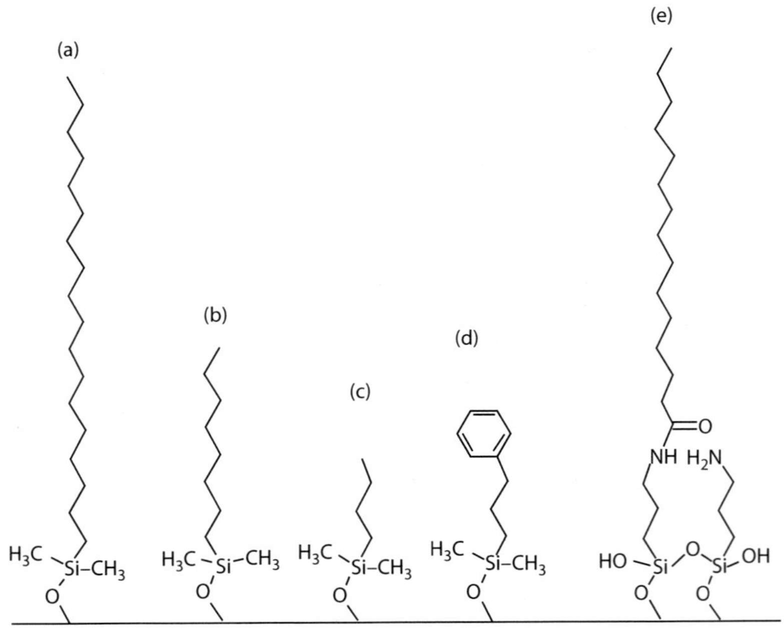

FIGURE 6.1 Structures of chemically bonded phases used in RP liquid chromatography: (a) octadecyl (C18), (b) octyl (C8), (c) butyl (C4), (d) phenyl, and (e) *N*-acylamide.

6.3.1.1 Bonded C2, C4, C8, C18

Silica-based stationary phases used in RP liquid chromatography differ in their chemical properties resulting from chemical modification of the silanol groups by attaching a nonpolar alkyl chain with different lengths, for example, C4, C8, C18, etc.

The stationary phases with the shortest hydrophobic ligands offered in the market are trimethylmethoxysilanol (TMS) packings. Material with C2 bonded ligands is also available. Many manufacturers offer a C4 stationary phase or a modified version—a propyl chain with a polar group (cyano, amino) on the top.

When slightly higher hydrophobicity of the stationary phase is needed, the octyl (C8) bonded phase may be a good choice. Using the C8 bonded phase for many compounds, the organic content in the mobile phase could be reduced for the same retention time as obtained on a C18 column. For many separations, C8 material offers better chromatographic properties than C18 material.

The C8 stationary phase contains significantly less carbon on the silica gel surface; thus, it provides less retention. Many separations can benefit when performed on a C8 bonded phase due to shorter retention times and lower organic content in the mobile phase. Additionally, the equilibration times are faster using a gradient elution [16]. It has to be mentioned that the retention properties of C8 adsorbents strongly depend on their coverage densities.

The most common RP adsorbents are octadecyl (C18) materials. They are also the most popular material for liquid chromatography analysis of pesticides. Every manufacturer of HPLC packings has a C18 adsorbent in the offer. However, the manufacturing process differs from one producer to another. Unfortunately, despite the similar surface chemistry, the chromatographic properties of different C18 packings provide significantly different chromatographic properties. The properties of C18 material depend not only on the coverage density but also on the type of the bonding. Monomeric and "polmeric" material exhibits different properties as a result of a different number of residual silanols on the surface. The hydrolysis of side groups also provides some additional silanols, which participate in chromatographic properties (see Figure 6.2).

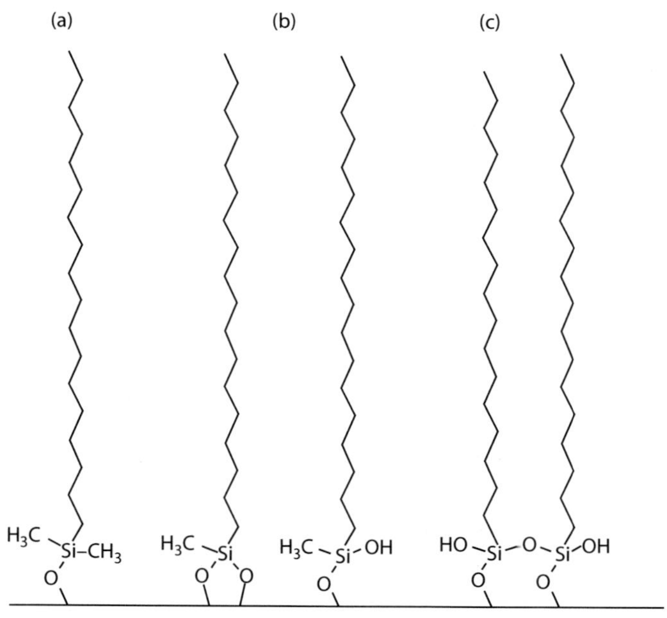

FIGURE 6.2 Different structures of octadecyl-bonded ligands attached to silica gel during C18 stationary phase synthesis: (a) monofunctional silane and (b, c) bifunctional silane.

6.3.1.2 Bonded Phenyl or Diphenyl

Phenyl- and diphenyl-bonded stationary phases have been available for many years. The idea of this type of material is to promote the π–π interaction of the aromatic rings with a solute as an additional force in the RP retention mechanism. The presence of phenyl-bonded ligands changes the solvation process, which offers different selectivity of the chromatographic system. This special selectivity will be observed when analyzed compounds possess aromatic functionalities in their structure, which enable π–π interactions.

Other types of phenyl-bonded stationary phases are fluorinated phenyl phases. This material is especially recommended for solutes that possess some halogens in their structure as many pesticides do, for example.

6.3.1.3 Bonded Longer Chain Length

In the literature, some information can be found about alkyl bonded phases longer than C18. All of them consist of an even number of carbon atoms, for example, 22 or 30. From the practical point of view, only the C30 bonded phase was commercialized. Such hydrophobic materials have limited application, especially for the separation of carotenoids. Generally, these phases exhibit higher retention for polar and nonpolar analytes than most polar-embedded and even high-coverage C18 phases. Due to a higher degree of surface shielding, long-chain bonded phases also offer greater pH stability in comparison with C8 and C18 phases. The long-chain bonded phases are more resistant to phase collapse under high aqueous conditions than the C18 phase. This phenomenon is a result of conformation changes of the bonded phase with temperature, which is a hamper due to the higher melting point of the triacontyl ligand [17].

6.3.1.4 Bonded Mixed Hydrophobic/Hydrophilic Phases

When water-rich mobile phases (>85% water or a buffer) are percolated through a chromatographic column with hydrophobic packings, the performance seems to indicate that the bonded ligands might be collapsing. This problem is more significant when the stationary phase is more hydrophobic. It may be solved by an increase of organic content in the mobile phase, which should improve solvation and bring the bonded ligands back to the original conformation. To avoid this procedure, which reduces the retention and selectivity of the separation, some manufacturers introduce stationary phases incorporated with some polar groups mixed in with the original alkyl ligands, for example, C18 or C8, or they add some polar groups during the endcapping procedure. This produces variation in the bonding. Such materials are specially labeled, which suggests that they can work in water-rich mobile phases, for example, the "aqua" stationary phase.

Another method to increase the polarity of hydrophobic adsorbent is the synthesis of *N*-acylamide stationary phases. These types of adsorbents consist of a hydrophobic alkyl ligand (e.g., C12 or C18) bonded to an aminopropyl silica using an amide bond. Such structure of the stationary phase offers better solvation and thus better stability and efficiency in water-rich mobile phases.

In the group of hydrophobic/hydrophilic phases, the cholesterol-bonded material has to be mentioned [18]. This material offers specific properties, but unfortunately, they are not so popular as other stationary phases. In the literature, one may find information about application of cholesterol-bonded packing for analysis of fungicides [19].

6.3.2 Silica Gel Coated with Hydrophobic Phases

Although the silanization procedure is the common method for silica gel modification to obtain a chemically bonded stationary phase, one manufacturer has been reported to have produced a silica gel with a polymeric coating. Such methodology offers greater resistance to any dissolving effect of the mobile phases in acidic and alkaline solution. Unfortunately, such material possesses one significant disadvantage: slow mass transfer and, as a result, poor efficiency and very slow equilibration from one mobile phase or pH to another one. This disadvantage is observed when the polymeric layer is coated on a silica support or the support is totally polymeric.

The main advantage of this material is that pH stability does not compensate lower efficiency; thus, these materials are rarely used, and only few may be found in the literature.

6.3.3 Alumina Coated with Hydrophobic Phases

Alumina, similar to silica, may also be coated with a hydrophobic phase. Although it is pH stable, it does not contain any stable hydroxyl groups onto which some ligands can be bonded. One way to impart a RP surface to this support is a polybutadiene polymerization on the spherical surface. The greatest disadvantage of this method is slow mass transfer when the coated material is packed into the chromatographic column.

6.3.4 Polymeric Supports

Although the silica gel is the most common stationary phase support, a number of totally porous polymeric supports have been created during the last 50 years. In some cases, the efficiencies of these materials are comparable with silica.

Polymeric supports are usually copolymers of styrene-divinylbenzene, polymerized methylacrylate, acrylamide, and others. Organic polymers, due to their hydrophobicity, may be used for chromatographic separations without further derivatization or can be modified during their manufacture with alkyl chains (e.g., C18, C8, C4), phenyl groups, or some ion exchange groups.

The most important property of polymeric support is the resistance to acidic and alkaline solutions. Unfortunately, they offer slower mass transfer and need longer equilibration. They can also expand or contract with different solvents. This causes high back pressure or a broadened peak when polymers are expanded or contracted, respectively. Chromatographic columns with such packings are available in hardware with adjustable end-fittings. Special end-fittings allow minimizing the effect of expansion or contraction of the column bed, according to the operating instructions given by the manufacturer.

6.4 NORMAL BONDED PHASE MODE

Most of the chromatographic separations seem to be done on silica gel bonded with hydrophobic moiety. However, the separation of polar components is easier in a NP system. Reproducible separations may be done using not only pure silica gel, but also on silica with bonded chains that contains a polar group, such as amide, $-NH_2$, –CN, Diol, and $-NO_2$ [12]. Structures of chemically bonded phases used in NP liquid chromatography are shown in Figure 6.3.

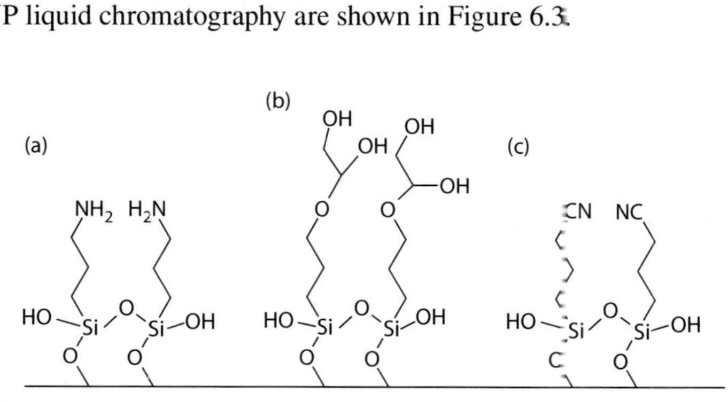

FIGURE 6.3 Structures of chemically bonded phases used in NP liquid chromatography: (a) aminopropyl, (b) DIOL, and (c) cyanoropyl.

6.4.1 Silica Gel Bonded with Hydrophilic Phases

The preparation of the polar chemically bonded stationary phases is similar to that of the synthesis of RPs. Usually, a silane with a single main chain with a polar functional group and one, two, or three active groups can react with silanols. In the case of single bonded silanes, two methyl groups are often used to shield the silica surface. The idea of the synthesis is to obtain a polar material, so these stationary phases are usually not endcapped. In the polar stationary phase, the residual silanols contribute in their own way to the final selectivity exhibited by the packing material.

Although silica gel bonded with hydrophilic phases has been available since the beginning of HPLC packing, its use in NP chromatographic methods accounts for not more than 15%. However, since 1990, these polar stationary phases have been applied in hydrophilic interaction liquid chromatography (HILIC) for separation of polar compounds in acetonitrile–water (buffer) mobile phases. Nowadays, a number of new hydrophilic packing material is introduced to the literature every year.

6.4.1.1 Silica Gel Bonded with Cyano (CN) Groups

The main types of cyano-bonded stationary phases are cyano-propyl chains. According to the length of the ligand, they are very similar to a bonded C4 phase. The presence of a cyano group causes these materials to be slightly more polar than C4 stationary phase. This type of material is produced by few manufacturers, and it is not in wide use. Due to the low polarity of this adsorbent, it may also be used in RP chromatography.

6.4.1.2 Silica Gel Bonded with Amino (NH_2) Groups

Amino-bonded stationary phases are mostly synthesized bonding amino-propyl silane to silica gel. Such material may contain single ligands or the ligands may be cross-linked, forming a "polymeric" layer. The amino-bonded stationary phase may be used as a stationary phase or as an intermediate to bond any number of other chemical groups, for example, to fatty acids or cholesterol. It may also be easily converted to a quaternary form, which possesses anion-exchange properties.

The amino-bonded stationary phase might appear to be a RP material, but due to poor stability in water-rich mobile phase, it is actually a NP adsorbent. Some of the most common groups of compounds that may be analyzed on this stationary phase are mono- and polysaccharides.

6.4.1.3 Silica Gel Bonded with Diol Groups

In NP mode, the diol stationary phase is an alternative to pure silica. The bonded hydroxyl groups provide good selectivity without excessive retention because hydrogen bonding with the diol layer is not as strong as in the case of the silanols. Diol stationary phase usually provides improved reproducibility in comparison with bare silica. Diol stationary phases are also suitable for separations using RP techniques or molecular weight determination of proteins by gel filtration.

Diol-bonded phase is a reactive group, which may form esters. It should be remembered that analysis of carboxylic acids is not recommended on this material.

6.4.1.4 Silica Gel Bonded with Ion Exchanger

Some pesticides exhibit an ionic nature. Due to this, silica gel has been modified to obtain ion-exchange stationary phases. Unfortunately, ion chromatography needs to use buffers as mobile phases. In such conditions, the silica-based ion exchangers (as well as other silica-based stationary phases) can dissolve. Particularly, the silica gel support will dissolve and leave the bonded ion exchangers intact. As a result, broader and more split peaks are formed.

Dissolving of the silica support can be prevented by using a precolumn packed with pure large-particle silica gel. The precolumn should be placed between the pump and the injector. This precolumn presaturates the mobile phase with silicate, which should hamper the silica support in the analytical ion-exchange column against dissolving [20].

6.4.2 Polymers Bonded with Hydrophilic Phases

Polymeric supports are used for the ion-exchange packing synthesis. Such materials are generally synthesized as intermediates during the initial particle-forming process, followed by reactions that convert the intermediate to an ion exchanger.

The synthesis and the separation of small polymer particles are much difficult than the synthesis of silica particles. Thus, the size of polymeric support is rather not less than 5 μm. Another difficulty connected with small polymer particles is the efficiency of the packing procedure into the chromatographic column.

Hydrophilic phases synthesized on the polymeric support may contain a number of different functional groups. These materials may be divided into two groups cationic and anionic. The anion exchangers usually contain quaternary and tertiary amine groups whereas in the structure of cation exchangers the carboxylic and sulphonic groups are presented. The polymer-bonded stationary phases exhibit good pH and thermal stability. However, regarding the structure of pesticides, the ion exchangers are rarely used for ionic pesticide analysis.

6.4.3 Cellulose

Cellulose is, basically, another material that found its application in liquid chromatography. In the case of pesticide analysis, cellulose has often been used in the separation in TLC. Two kinds of cellulose are commonly used in TLC: native cellulose that has between 400 and 500 units per chain and microcrystalline cellulose that is prepared by the partial hydrolysis of regenerated cellulose. The microcrystalline cellulose has shorter chains, usually between 40 and 200 units.

Spherical cellulose may be also used as a bonding support for chiral HPLC columns. The main disadvantage of cellulose is the fact that it does not have the mechanical strength. When packed into a chromatographic column, it may collapse, giving high back pressure or limited flow rates.

6.5 STATIONARY PHASES APPLIED IN NP SYSTEMS AND RP SYSTEMS

Although the hydrophobic stationary phase is used in RP systems and polar materials were dedicated to NP systems, it is possible to use polar stationary phase in combination with RP solvents. Such a system is called hydrophilic interaction liquid chromatography (HILIC). In such a system, the separation of polar compounds on the polar stationary phase may be performed using a mobile phase consisting of a huge amount (70%–95% v/v) of acetonitrile in water [21]. Another chromatographic system in which a polar stationary phase may be applied is an aqueous NP (ANP) system [22].

Despite the HILIC mode, some stationary phases that are dedicated to NP systems may be operated in RP systems, for example, a stationary phase with a bonded amine group. Additionally, when the HILIC method is developed, a number of new stationary phases are synthesized every year. Such stationary phases contain in their structures both hydrophobic alkyl chains and polar groups. This combination of different functionalities allows using these materials in a different mode of HPLC. The presence of polar groups reduces the retention in RP systems but increase the retention in NP, HILIC, and ANP systems.

6.6 CHIRAL-BONDED PHASE MODE

A significant number of pesticides possess an asymmetric center. For analysis of such compounds, when each isomer has to be determined, the chiral stationary phases should be applied. Many different chiral stationary phases are now available on the market for enantiomer separation. The different chiral packings are often referred to as chiral stationary phases (CSPs). In the case of chiral pesticide analysis, the most common are saccharide-base materials, including cellulose-*tris*-(3,5-dimethylphenylcarbamate), cellulose-*tris*-(phenylcarbamate), and amylose-*tris*-(3,5-dimethylphenylcarbamate) [23].

Before starting work with CSPs, it is recommended to contact the manufacturers of a given packing and to consult on the details of how to use them effectively. Usually, special mobile phases or additives are necessary to generate the selectivity desired.

CSPs are the most expensive chromatographic columns on the market. They may cost up to five times more than standard RP columns. The high price of these packings results in more complicated synthesis procedures. Some of the phases that are bonded have first to be chirally separated themselves. This allows the bonding of only the material with the correct configuration that allows the selective separation. All of the CSPs have spatial-design considerations in the bonded chiral selector to enhance three-point interaction. Such an attracted isomer is delayed in moving through the column. This allows the second isomer to move through faster.

The main classes of CSPs are the brush-type ones, cellulose esters and carbamate, cyclodextrin, macrocyclic glycoproteins and protein, ligand exchange, and crown ethers [24].

6.6.1 Silica Gel Bonded with Chiral Phases

The majority of CSPs are synthesized using silica gel as a support. Despite the trend toward particle size reduction, the CSPs are usually made using 5 μm and larger silica particles for analytical work and for chromatographic columns with larger internal diameters (IDs), respectively.

The cyclodextrin stationary phases may be divided into three cyclic forms: alpha (which consists of six units), beta (seven units), and gamma (eight units). Enantiomers fit into the barrel shape of these molecules to different degrees, which allows the separation.

An interesting group of CSPs creates macrocyclic glycoproteins. They are excellent chiral selectors due to their complex structure and chiral centers. Proteins are other compounds that have the ability to promote chiral separations. The most common are three proteins bonded to silica: human serum albumin, cellobiohydrolase, and α-1 acid glycoprotein.

Crown ethers are macrocyclic molecules containing several ether groups. They are oligomers of ethylene glycol. CSPs contain derivativized crown ether: (+) or (–)-18-crown-6 tetracarboxylic acid. Such material is applied for the separation of amino acids and of compounds containing primary amines.

6.6.2 Cellulose-Bonded Chiral Phases

Cellulose exhibits chiral selectivity, which is a result of its ordered structure [25]. This chiral selectivity can be greatly enhanced by bonding it with various functional groups through a carbamate linkage. Similar modification may be carried out using amylose. Such material has to be dissolved and coated onto spherical particles of silica gel.

6.7 ION EXCHANGE MODE

The majority of stationary phases for separation of ionic compounds used today are based on organic polymers and silica gel. In contrast to stationary phases prepared on the silica gel, organic polymers have much higher stability toward extreme pH conditions. The silica-based anion exchangers can be operated only in the pH range 2–9.5, and polymeric ion exchangers are stable in all pH ranges.

Surface silanols may act as a cation exchanger; thus, porous silica can be considered to be a weak cation exchanger [14,26,27]. At a pH of 7.5, almost all surface silanols should be dissociated. At a pH of 2.5, only the isolated silanols are ionized (stronger acidic properties). However, all the dissociated silanols in the pores probably cannot participate in the ion exchange process [14]. Properties of porous silica induce the cation exchange properties of RP adsorbents, which are prepared on the silica gel support.

6.7.1 Silica Gel Bonded with Ion Exchangers

Ion exchange stationary phases may be synthesized onto silica gel. Unfortunately, the silica material has limited pH stability and exhibits a tendency to dissolve under the buffer mobile-phase conditions. The most important advantage of silica-based anion exchangers is significantly higher efficiency in comparison with polymeric supports. The dissolving of the silica may be reduced by the use of a precolumn (between the pump and the injector) filled with silica gel. Detailed explanation is described in Sections 6.12.6 and 6.12.7.

6.7.2 Cellulose Bonded with Ion Exchangers

The first support applied for ion exchanger synthesis was cellulose. Nowadays, the polymers are the most common support. As was mentioned earlier, cellulose has no mechanical stability. When packed in a column, it is not stable under the higher pressure that is used in modern HPLC (IC).

6.7.3 Polymers Bonded with Ion Exchangers

A key requirement for a liquid chromatography packing is a minimum mechanical strength that makes the packing compatible with the typical pressure drops required for using a column with 5–10 μm particles. Four types of polymeric packings have a sufficient strength and are commercially available: styrene/divinylbenzene copolymers [28,29], ethylvinylbenzene/divinylbenzene copolymers [30], polyvinyl, and polymethacrylate resins [31,32]. These resins are the most important of all the organic compounds that were tested for their suitability as substrate materials in the manufacturing process for polymer-based ion exchangers. In contrast to silica-based column packings [33], organic polymers are employed as the predominant support material used in ion chromatography. These materials have a much higher stability in extreme pH conditions. Organic polymers are also stable in the alkaline pH region.

6.7.4 Bonded Phases for Ion Exchange

Ion exchange bonded phases are usually silica-based and are formed by the direct reaction of an appropriate silane with a silica support. The bonded silane must contain a functional group, which can either be converted to an ion exchange group or permitted to attach an ion exchange group to it. The most common ion exchange functional groups are carboxylic acids and amino groups. They are frequently used to provide ionogenic properties to bonded phases.

Ion exchange bonded phases onto silica gel have a very limited pH range over which they can be operated effectively. Thus, ion exchange resins are preferentially chosen as the stationary phase for both column liquid chromatography and TLC.

Bonded phases for ion exchange may be prepared using celluloses as a support. It is done in a similar manner to the way such groups are attached to polystyrene resins.

6.7.4.1 Strong and Weak Anion Exchangers

Anion and cation exchangers are classified as strong or weak. This classification is based on how much the ionization state of the functional groups changes with pH. A strong ion exchanger over a wide pH range has the same charge density on the surface. The charge density of a weak ion exchanger changes with pH. As a result, the selectivity and the capacity of a weak ion exchanger are different at different pH of the mobile phase.

Packings that contain quaternary ammonium groups possess a positive charge in pH range 1–14; thus, they are strong anion exchangers. On the other hand, tertiary, secondary, and primary amino groups can be positively charged below pH ~ 9. They are weak anion exchangers.

6.7.4.2 Strong and Weak Cation Exchangers

Cation exchangers may also be divided into two groups: strongly acidic (typically sulfonic acid groups, e.g., sodium polystyrene) and weakly acidic (mostly carboxylic acid groups). Although the ionization of sulfonic groups is constant over a broad pH range, the cation exchangers containing carboxylic groups may be successfully operated in a pH higher than 5.

6.8 SILICA GEL ENDCAPPING

The presence of silanols participating in the retention mechanism may change the local concentration of the eluent component in the mobile and stationary phases. The presence of residual accessible silanols can have a negative influence on the separation of polar analytes, especially basic compounds and biopolymers. The influence on nonpolar solute retention is rather low [34]. The data on the number of surface silanol groups per surface area unit published in the literature range from 6 to 9 μmol/m^2 [1,35,36]. These differences depend on the means of measurement. On the other hand, the calculated value of surface silanols for ß-cristobalit (111 or 100 plane of the octaeder) is 7.6 μmol/m^2 or 4.55 OH groups per nm^2 [35]. We can assume that the concentration of silanols in fully hydroxylated silica is up to 8 μmol/m^2 [2,37–40]. This value is commonly used as a reference value.

Despite the silanol groups that are present on the unmodified silica gel surface, the new silanol groups may be introduced during the silanization procedure [41]. Bonding of organosilanes leads to siloxane group creation. Additionally, the use of bifunctional or trifunctional silanes provides extra silanol groups due to hydrolyses of a functional group (connected to the silicon atom in the silane), which cannot create bonds with the surface [42].

6.8.1 Classical Endcapping

In reality, from the total amount of residual silanols—which are not bonded with silanes—only about 5% exist as strong polar adsorption centers, and they are accessible for adsorption of water and basic compounds [43]. This is caused by the shielding properties of the methyl or isopropyl group connected to the silicon atom in the silanes [6,44,45]. The most commonly used endcapping reagent is a trimethylsilyl group because it is the smallest in size and can penetrate between the ligands to the silica gel surface. Trimethylsilyl groups deactivate silanol groups when bonding, and they shield other silanols from the mobile phase components.

Some manufacturers claim to perform double or triple endcapping procedures. Due to the fact that this procedure is limited by sterical hindrance, it does not eliminate all of the residual silanols.

6.8.2 Polar Embedded Endcapping

Some manufacturers have modified the endcapping procedure to aid in keeping the hydrophobic stationary phases solvated in a water-rich mobile phase. This was accomplished by adding a polar group to the endcapping reagent. The amount of endcapping reagent that bonds to residual silanols is also governed by the same factors as the classical procedure. The polar embedded groups cause such material, even when they possess octadecyl hydrophobic ligands, not to lose their stability and selectivity in extremely high water content in the mobile phase [46].

6.9 PARTICLE SIZE VARIATIONS

From the beginning of modern HPLC, smaller and smaller particles of silica were made and packed into chromatographic columns. First, for producing cuts of larger particles suitable for column chromatography, mesh screens were applied. Unfortunately, this could not be used for particle size ranges lower than 10 μm. To solve this problem, the air classification was used to sort the first generation of packings in the 15 to 10 μm ranges.

Over the years, columns with smaller and smaller particle sizes were offered on the market. The advantage is higher efficiency in the resulting chromatograms. This allows performance of the separations in a better and faster way. Using smaller particles, there is a smaller loss of efficiency when the flow rate is increased. This allows an increase of the flow rate for a faster separation without loss of resolution. Smaller particles lead to increased efficiency, reduce the time of the analysis, and lower cost per analysis. The disadvantage of small particles is an increase of back pressure.

6.9.1 10, 5, 3, and Sub-2 μm Sorbents

One of the most important characteristics of HPLC packings is a narrow particle size distribution, which allows obtaining better efficiency of packed columns. The particle size also plays an important role. The column efficiency depends not only on a narrow size distribution of particles, but also on the packing techniques to produce well-packed HPLC columns that would be suitable and reproducible for application in HPLC systems.

Since the 1960s, a decrease of synthesized particle size has been observed, which is a result of production and separation methods. During the last 50 years, the particle size decreased from 20 to 3 μm. During the last few years, the production of sub-2-μm particles for UHPLC technology was started. It is interesting to note that the superficially porous packings have been revived again, built on a 1.7-μm solid core with a 0.5-μm-thick shell. Nowadays, such materials are dedicated to ultra-high-performance liquid chromatography (UHPLC) systems. The old version of these materials consisted of a solid core of 30- to 40-μm glass coated with microfine silica, which was RP bonded. These packings offered two advantages. They were easy to pack and exhibited better speed of mass transfer on and off the thin shell surface [16].

6.9.2 Particle Size Effect on Efficiency

The column efficiency is reversibly proportional to the particle size: the smaller the particle used, the better the efficiency. Unfortunately, the decrease of the particle size causes a dramatic increase of back pressure. Thus, smaller particles may be, or even should be, packed into shorter and narrower columns. Smaller particles offer better efficiency in a wider range of the mobile phase linear velocity compared with bigger particles.

By decreasing the particle size, the analysis may be done faster using a shorter and narrower column, which results in cost reduction per analysis. Additionally, the increase of the efficiency offers a lower limit of the detection and lower limit of quantification. A theoretical Van Deemter plot showing the relationship between flow rate and height equivalent to a theoretical plate (HETP) for different particle size is presented in Figure 6.4.

6.9.3 Particle Size Effect on Back Pressure

As was mentioned previously, the price that has to be paid for using columns with smaller particle size is the back pressure that is generated when the mobile phase is pressed through the chromatographic column. Additionally, the column ID is reduced proportionally to the particle size. The application of a larger-diameter column may reduce the back pressure, but it lacks sense from the practical point of view. Thus, the columns packed with particles lower than 3 μm are reserved for UHPLC.

To emphasize the scale of the changes, the same analysis that takes 10 minutes at the flow of 1 ml/min using a 5 μm particle column (4.6 × 150 mm) may be performed in less than 1.5 minutes

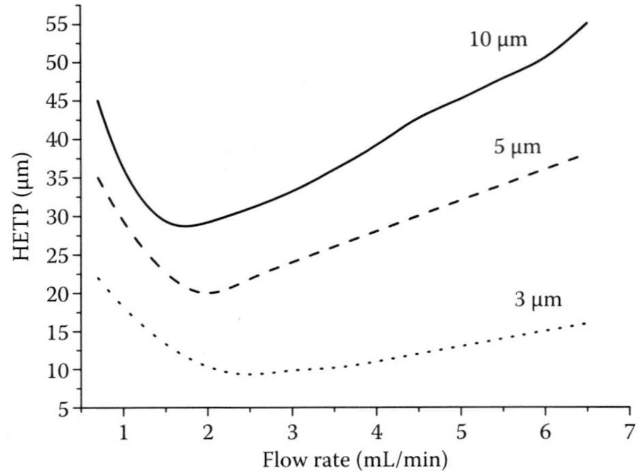

FIGURE 6.4 Theoretical Van Deemter plot showing the relationship between flow rate and HETP for different particle size.

using a flow rate 0.4 ml/min using a 1.7 μm particle column (2 × 100 mm). Unfortunately, the back pressure is increased from 100 to 900 bars.

6.9.3.1 UHPLC Systems

The reduction of the size of particles used for column packing and the increase of the back pressure require improved chromatographs. About five years ago, a new generation of chromatographic pumps, injectors, and mixers was designed, and the technique was called UHPLC. Such equipment could work with sub-2-μm particles. At this time, these systems allow the chromatographer to use pressures of up to 1300 bar. UHPLC offers higher efficiencies, shorter time of analysis, and reduction of solvent consumption.

Using UHPLC, extra care should be taken in sample and mobile phase preparation. Columns with smaller particles and narrow capillaries are less resistant to some impurities from the sample and form the mobile phase. Thus, working with UHPLC, the filtration of all liquids used is necessary.

6.10 COLUMN HARDWARE

When a stationary phase is made, it has to be packed under high pressure into special columns constructed of resistant materials. All of the components of the column hardware (the tubing, the fittings, and the frits) have to be made of materials resistant to the mobile phase components (solvents and buffers) used during chromatographic analyses in HPLC systems. The majority of column hardware and the components are made of stainless steel. Sometimes, a polymer, such as polyether ether ketone (PEEK), can be used, especially for ion exchange columns.

6.10.1 Stainless Steel

Stainless steel is most common for column hardware production. The inside of the column has to be polished to a mirror-like sheen. It allows the minimization of wall effects (less-well-packed areas that can cause tailing and band broadening) [16].

The main differences in column appearance between different manufactures are due to the special design of the end-fittings. End-fittings are produced by each manufacturer according to its own specifications.

The connecting tubing depth (ferrule lock distance) is different from one manufacturer's columns to another's. Thus, the plastic ferrules (usually made of PEEK) can be used on the connecting tubing. PEEK ferrules can then move freely on the tube and allow self-adjustment to the correct depth. This allows a good liquid seal that eliminates any dead space.

The PEEK ferrules can be operated with HPLC systems to pressures up to 250 bars. Sometimes, depending on the combinations used by the manufacturer, they can exhibit higher pressure resistance.

6.10.2 Polymer (PEEK®, Other)

Many parts of the HPLC system and column hardware can be manufactured with polymers, for example, PEEK. This solution is used when sensitivity to any metals is to be avoided. Polymers used need to be stable in aqueous solutions or mobile phases consisting of polar organics and water. Polymer parts have to exhibit suitable pressure resistance. According to column hardware, it makes no difference whether the HPLC packing is into a metal or plastic. PEEK columns are usually used for ion exchange stationary phases and for some HILIC packing. The majority of RP stationary phases are packed into stainless steel columns.

6.11 COLUMN FITTINGS AND CONNECTIONS

The market for column fittings and connections is extremely large. There are a number of options when selecting the hardware in which the packing can be ordered. Each manufacturer has a proprietary combination of fittings and column designs. Additionally, a huge amount of different column fittings is offered by manufacturers that produce the HPLC accessories.

6.11.1 Columns Complete with Fittings

The typical column hardware is a stainless steel pipe that is complete with the end-fittings necessary to connect the column to capillaries in liquid chromatograph. Each manufacturer designs its end-fittings independently.

To make a correct connection without any leak, the connecting tubing is equipped with a ferrule. The ferrule may be stainless steel or polymer. The stainless steel ferrule has to lock onto the connecting tubing and fit snugly to the coned area in the column end-fitting. Unfortunately, a stainless steel ferrule, when it is locked onto the stainless steel connecting tubing, cannot be moved, and it may be usable only for a given type of end-fitting. This problem may be solved by using the polymer ferrule or a ferrule fitting combination that is self-adjusted to any of the end-fittings. This solution is available from most laboratory distributors. However, it has to be mentioned that the ferrule is pressure-resistant. It is extremely important in UHPLC systems in which polymer ferrules cannot be used. For such systems, the stainless steel or combined stainless steel–polymer ferrules are applied.

6.11.2 Cartridge Columns

The end-fittings are the most expensive part of the chromatographic column hardware. To reduce the column costs, manufacturers start to offer cartridge columns. The cartridge columns are chromatographic columns without end-fittings that fit into a holder. This fitting holder can be used over and over again with different cartridge columns. From the chromatographic separation point of view, there is no difference between a standard column and a complete cartridge column with fittings when the stationary phase is the same and purchased from one manufacturer.

6.11.3 Reducing Dead Volume

During chromatographic separation, the dead volume of the system has a significant influence on the separation parameters, for example, efficiency. The void volume of the system is a whole volume occupied by the mobile phase. This happens when any connecting pieces are not well fitted together, and a void volume exists, which acts like a dilution vessel. When the sample runs through this vessel, the volume of liquid in this dead space adds to the sample volume. As a result, broader peaks are observed. This results in lower resolution, possible overlapping of peaks, and poor efficiency.

The most important dead volumes are between the injector and the HPLC column and between the column and the detector. Thus, any connections in these areas must be performed correctly. The guard column, when used, should be as small as possible. The capillaries should be narrow and as short as possible to connect all of the parts. The connection between the column and the detector also has to be free of extra dead volume. It has to be remembered that on the column outlet, the back pressure is atmospheric, so connections need only be finger-tightened to be leak free.

All parts have to be tightened gently but not over-tightened, especially when the connection of any component is done for the first time. Any connection at the high pressure of an HPLC system needs to be made to become "leak free" at the operated pressures.

6.11.4 Avoiding Leaking Connections

Ferrules, which seal connections in liquid chromatography, after few uses, may be destroyed and have be replaced by new ones. Each time the stainless steel ferrule is tightened into a fitting, it distorts more due to the pressure used to make the liquid seal. At all times, all parts have to be tightened gently and not over-tightened. Ferrules need to be tightened enough so that there is no leakage at the operated pressures.

On the market, connecting pieces with gripping areas are available, which allow finger tightening. They often have a larger diameter. It has to be remembered that this solution cannot be used in UHPLC due to high back pressure. For UHPLC systems, the stainless steel and component connecting pieces are recommended, which are usually tightened using a wrench.

6.12 COLUMN SIZES

In the beginning of HPLC when large particle silica was used to pack the analytical column, the most common column size was 4.6 × 250 mm. This was fine for research and development laboratories, which needed the high efficiency generated by the longer column. The length compensated for the large particle size.

Together with the introduction of smaller particles, the column sizes also decreased to 150, 50, 25 mm, and shorter dimensions. Additionally, the ID decreased as well from 4.6 to 3, 2.1, and narrower columns. These changes were done without loss of efficiency.

6.12.1 Column Lengths

Today on the market, various types of chromatographic columns with different combinations of lengths and IDs are available. Using a long column, the time of analysis is always longer and the solvent consumption is bigger. For a 250-mm-long column, separation times averaged 15–20 minutes, and up to two hours for a complex sample. Nowadays, with short columns, most analyses can be done in five minutes, and for a complex sample, 20–30 minutes should be enough. The application of shorter and narrower columns increases the sample through output and lowers the cost per sample analyzed. This makes the chromatographic analysis economical and more environmentally friendly.

6.12.1.1 Analytical: 300–250 mm

The efficiency of the column (number of theoretical plates) may be calculated using a chromatographic column in isocratic conditions (a mobile phase whose composition does not change during analysis). That efficiency depends on the column length (assuming the same packing material). A longer column gives higher efficiency than the shorter one.

The long column (250–300 mm) is usually packed with 5 μm silica particles or bigger ones. Such a column may be used with a common HPLC system. The packing of such a long column using smaller particles will result in a strong increase in the back pressure, which can make it impossible to operate in a standard HPLC system. Theoretically, a long column packed with small particles may be operated using an UHPLC system; however, this is not done in practice. UHPLC systems are designed for fast analyses using short columns packed with very small particles.

6.12.1.2 Fast: 150–25 mm

The long 250-mm column usually offers a higher resolution than is necessary for the separation. Thus, the idea is to pack a shorter column with the same packing material. This offers the same selectivity with a small decrease in the resolution as a result of the reduction in efficiency. This allows the separation of a given mixture in a shorter time. To elute the sample in a shorter time, it is much better to use to a shorter column than to use a longer one at a higher flow rate. This guarantees the reduction of the back pressure, which increases the lifetime of the system.

As a result of further speed optimization, the shorter column (e.g., 75 mm) was packed with smaller particles (e.g., 3 μm). The reduction of the particle size results in an increase of efficiency and, thus, the resolution whereas a shorter column provides a shorter time of analysis. Such analysis is faster and cheaper due to the lower solvent consumption. It might be a perfect choice for the final analytical column.

6.12.1.3 Fastest: <25 mm

These shorter columns are packed with 3-μm or sub-2-μm particles. For UHPLC, a 2.1 × 25 mm column filled with 1.7- or 1.6-μm particles (depending on the manufacturer) would give very fast separations. The smallest silica particles, with a diameter of 1.3 μm, were introduced in the market in 2012. Using a short column, one must remember that the void volume of the system has a big influence on the separation parameters, especially on resolution and efficiency. This influence is higher when the ratio of column void volume to system void volume is lower. The shorter column needs the chromatograph with the smaller void volume. Thus, the 2.1 × 25 mm column will be extremely sensitive to the void volume of the system.

6.12.2 Column IDs

The ID of a HPLC column is an important parameter. Larger columns are usually used in industrial applications. Low-ID columns have improved sensitivity and lower solvent consumption, and they are used for analytical purposes. Larger ID columns, over 10 mm, are used in a preparative and semipreparative scale. Analytical scale columns have an ID in the range of 4.6–2.1 mm. Narrow-bore columns (1–2 mm ID) are used for applications when more sensitivity is desired, for example, in a liquid chromatography–mass spectrometry connection. Capillary columns with IDs lower than 0.3 mm are used almost exclusively with mass spectrometry. They are usually made from fused silica capillaries rather than the stainless steel tubing of larger columns.

For HPLC in a preparative scale, much wider columns are used. The ID of a preparative column is usually higher than 10 mm. It is caused by a need for a higher column loading to separate large quantities of a sample. Of course, their cost can be considerably high.

6.12.2.1 Classical: 4.6, 3.9, 3.0 mm

Typically, conventional analytical columns of 4.0 or 4.6 mm ID are used. These diameters are still the most popularly purchased, generally because these IDs have been written into quality control and quality assurance protocols.

One of the most important advantages of such a column diameter is lower back pressure in comparison with smaller-diameter column hardware. Unfortunately, with the operation of the classical column, the solvent consumption is higher.

6.12.2.2 Modern: 3.0, 2.1, 1.0 mm

By scaling down from an analytical 4.6 mm to a 3.2 mm ID column, it is possible to significantly reduce the flow rate and solvent volume needed to reach the same linear velocity. For example, if a 4.6-mm ID column is operated at a 1.0 ml/min flow rate, the same linear velocity is achieved with a 3.0-mm column using a flow rate of 0.42 ml/min. This results in a reduction of 52% of solvent for one analysis. Using a 2.1-mm ID column, the amount of solvent use could actually be reduced by 80% in comparison with a 4.6-mm ID column.

6.12.2.3 Special: Capillary <0.5 mm

The application of narrower columns for given amounts of the injected sample produces taller peaks. Taller peaks result in lower detection limits. For the same amount of sample injected, the peak height is inversely proportional to the cross-sectional area of the column.

Although small-diameter HPLC columns are available, they are used most often when the amount of sample is very limited and directed to detection with various mass spectrometers or laser-induced fluorescence detectors. The application of such a column is also limited to the special micro-volume HPLC systems. The bandwidths of sample components being separated are very, very small [16].

6.12.3 Flow Rates of Columns with Different IDs

When the diameter of the used column changes, the flow rate applied has to be considered.

When a wider-diameter column is replaced by a narrower one, using the same flow rate as on the initial column, the linear flow rate will increase proportionally to the radius squared of the column. Remembering that the column efficiency changes with the linear flow rate according to the van Deemter plot, this would result in different efficiencies (usually lower). As a result, the separation could look very different.

In this case, it is necessary to reduce the flow rate to the calculated value to make sure that the linear velocities are made equivalent in each column. This allows obtaining the same efficiency and should offer the same chromatographic separation. Also, the analysis time will be the same, assuming the same length of both columns. The opposite situation is when the narrow column is replaced by a wider one. Then, the flow rate should be increased to the proper one.

A detailed calculation of the equivalent linear flow rate for columns with different ID may be simply done using Equation 6.1:

$$F_{C^2} = \left(\frac{d_{C^2}}{d_{C^1}}\right)^2 F_{C^1} \tag{6.1}$$

Example results for columns with different diameters are listed in Table 6.1.

TABLE 6.1
Flow Rates for Equivalent Linear Flow Rate for Columns with Different IDs

Column Diameter (mm)	Volumetric Flow Rate (ml/min)
25	29.5
10	4.7
4.6	1
3	0.42
2.1	0.21

6.12.4 Bandwidths of Columns with Different IDs

As was mentioned in the previous section, the limit of detection decreases with the decrease of the column ID. Using narrower columns, the injected sample bands moved through the column, and due to interaction with the stationary phase (adsorption/desorption), the bands are narrower and higher. This is the main reason to change the column from an analytical to a smaller one.

The ratio of the peak height changing column diameter is related to the radius squared of the different column sizes when the volume of the injected samples is the same. It has to be mentioned that the smaller-ID column contains less packing material. Thus, column overloading might become a problem. In such a case, the sample size, both the mass in the sample and the injected volume, typically have to be reduced.

6.12.4.1 Requirements for Special Detector Settings

New short columns packed with smaller particle sizes provide smaller sample bands. The HPLC detection system has to be able to deal with these narrow peak widths. Standard HPLC systems have been made to use wider (4.6 mm ID) columns. The older detection system was able to be adjusted to a sampling rate of 1–3 Hz (typically a data point every 400 ms) [47,48].

Using the narrow column, the bandwidths of a given compound are narrow. To obtain the correct signal, the detector has to be set up for a high frequency of data acquisition. For the modern column used in the UHPLC system, it is recommended to use a frequency of data acquisition of 20 Hz or higher to obtain enough data points for each peak.

Whenever smaller columns are used, the narrower peaks may be dispersed (broadened) within the detector cell. Thus, for a column with a 2.1 mm ID, the semimicro detection cell is recommended as it possesses a much smaller volume.

For fast HPLC (UHPLC) the sampling rate has to be increased from 3 to 20 Hz (from 400 to 50 ms) to capture a more complete profile of the peak. Another detector setting that has to be taken into account is a time constant. It allows the improvement of background noise by signal averaging, typically set at two seconds. Again, for fast HPLC, this value should be reduced to 0.05 second [16].

6.12.4.2 Requirements for Special HPLC Systems

The void volume of the system (capillaries, mixer, and so on) has an influence on the separation efficiency. Such influence is higher when the void volume of the column is lower. Using the small diameter or short columns, the reduction of their efficiency caused by the void volume of the system is more significant.

Another problem that is connected with decreasing column diameter is that the column capacity decreases. If the sample volume and its concentration are high, then it is possible to overload the column. When the peak is too high, the detector may be overloaded, and the detection signal may be not proportional to the concentration.

6.12.5 Performance of Columns with Different Lengths and IDs

The efficiency of the separation depends on the column length and the particle size. Columns packed with smaller particles offer higher efficiency. The increase of the column length also increases the efficiency of the separation. However, longer columns provide higher retention, and thus, the time of the analysis is longer and solvent consumption is higher. Thus, when possible, most chromatographers try to use a short column packed with small particles to obtain satisfactory efficiency and shorter retention time.

6.12.5.1 Effect of Column Size on Efficiency

As was mentioned previously, the efficiency of the separation is a function of particle size of the packing and the length of the column only. The ID of the column does not influence the column efficiency. However, it has to be mentioned that a given column may be packed perfectly or poorly, so the quality of the packing procedure also has a significant influence on the efficiency. Fortunately, most of the columns offered on the market are tested before the sale. The ID of the chromatographic column may influence the efficiency only when incorrect flow rate is applied. As was discussed in the previous section, the flow rate has to be adjusted properly to a given column ID.

6.12.5.2 Effect of Column Size on Separation

To obtain a good separation, it is necessary to separate all compounds to the baseline. It is possible when the used column at a given mobile phase composition offers proper retention, selectivity, and efficiency. All parameters influence the resolution. Although the retention may be easily controlled by mobile phase composition, the selectivity depends strongly on the stationary phase. The influence of the mobile phase on the selectivity is governed by changes of the retention. The last parameter, the efficiency, depends only on the chromatographic column (its length and particle size). Additionally, the limit of detection also depends on the column efficiency.

6.13 COLUMN CARE FOR LONGER LIFE

A chromatographic column is getting used. Unfortunately, it is quite an expensive part, especially when it contains chiral packing material. Thus, it is justified to use the column for as long as possible. However, the column may only be used when it offers satisfying selectivity and efficiency. It is reasonable to use the column carefully for its longer life.

The most important thing is that all components (solvents, samples) have to be miscible and soluble in one another. The sample has to dissolve completely in the mobile phase and in any mobile phases in an applied gradient elution. It is forbidden to pump a solvent through a column if it is not miscible with the previous solvent. It is especially important when changing a NP system solvent to a RP system solvent and vice versa.

6.13.1 Standards and Pure Components

It is reasonable to use the purest standards, additives, and solvents for the mobile phase. Although many catalogs indicate the purity of a standard, this has to be tested to see that it does not contain any impurities that would compromise the analysis. Sometimes it might be necessary to clean up the standard. This may be done in column chromatography ("flash" chromatography), TLC, or preparative HPLC.

Most of the standards are available from more than one manufacturer, and at any time it is possible to check their specifications to choose the highest purity.

6.13.2 Environmental Samples

In chromatography of environmental samples, a good sample cleanup is essential for the long life of the chromatographic system, especially the column. Good sample cleanup limits the problem with chromatographic separations, such as problems with reproducibility, loss of efficiency, and high back pressure after few chromatographic runs. All of these effects indicate the collection of impurities on the column inlet. It has to be mentioned that the column packed with smaller particles, for example, sub-2 μm, is more sensitive to impurities than the 5-μm-particles column.

There are many sample preparation methods that can be used, depending on the source of the sample, the type, and the cleanup necessary to get reproducible results. Usually, sample preparation is the most time-consuming step of the analysis even if it is automated. It has to be emphasized that when working with environmental samples, the time spent on sample preparation will pay off in the future. This may eliminate many problems that can arise during HPLC analysis. It has to be remembered that pollution prevention is more easy than removing impurities from your system [16].

6.13.3 Mobile Phase Components

The mobile phase and any components have to be as clean as possible to avoid contributing to the collection of impurities in the analytical column. The amount of solvent pumped through the column is significant. A hundred analyses, each 15 minutes long using a flow rate of 1 ml/min gives 1.5 L of solvent. It has to be considered as a potential source of impurities if the purity of the solvent used is poor. It is extremely important if the mobile phase, besides the solvent, also contains some dissolved additives, such as salts, buffers, or ion-pair reagents. In this case, such components have to be of high purity (for HPLC). When some additives are dissolved in the mobile phase solvent, the filtration of the mobile phase is recommended.

6.13.3.1 Organic Solvents

A large number of manufacturers offer HPLC-grade solvents. The general specifications of solvents for HPLC are that they have to be filtered to eliminate particulate matter using 0.2-μm membrane filters.

All impurities that exhibit UV adsorption have to be removed because HPLC solvents have to exhibit transparency in the UV region. It is especially important to offer a low cutoff. Acetonitrile, for example, can be used from 200 nm to longer wavelengths.

Some solvents, for example, chloroform, are not stable, and they can break down with time. Thus, some preservatives are added. The preservatives must not interfere with the separated compounds or change the selectivity of the separation.

6.13.3.2 Water

Nowadays, approximately 80% of all chromatographic separations are done in RP systems. In such systems, the mobile phase contains water and some polar organic modifier. Thus, HPLC-grade water is necessary to carry out the mobile phases. HPLC-grade water can be purchased from the manufacturers who produce HPLC solvents. However, most laboratories use water-purification systems that produce high-purity water with high resistance (low conductivity). There exist a number of different water purification systems on the market. Most of them contain cartridges with ion exchangers and carbon filters. All systems have monitors showing the resistance of the purified water.

6.13.3.3 Buffer Salts

The pH of the mobile phase is an important parameter that usually has to be controlled. Thus, most mobile phases have a buffer additive to obtain fixed and stable pH. Controlling the pH of the mobile phase provides better separation, better peak shape, and higher reproducibility, especially if some basis or acidic compounds are separated. The pH adjustment affects especially the components that can ionize. Ionized compounds do not interact with the stationary phase in RP-HPLC; thus, they are usually eluted near the void volume. Adjusting the pH to the value that hampers the ionization causes the nonionized form to exhibit higher retention.

To prepare a mobile phase with adjusted pH, a HPLC-grade buffer has to be used. Such high-purity buffers are available from many manufacturers. They are made to have low particulates and low UV absorption, similar to the HPLC-grade solvents. It has to be mentioned that even when using HPLC-grade buffers, after dissolving in the mobile phase, they should be filtered with a 0.2-μm membrane filter before use in the HPLC system.

6.13.3.4 Acids and Bases

In some RP methods, the ionization of sample components may be suppressed by the addition of an acid or base. To suppress the ionization, a relatively small amount, usually 0.1%, is needed.

Weak organic bases, diethylamine or triethylamine, are used to prevent any attack on the silica support. Weak acids, such as acetic, trifluoroacetic, formic, and phosphoric acids, have been used without deleterious effects on the long-term stability of the HPLC column. Acids and bases mentioned here are available from manufacturers as high purity for HPLC [16].

6.13.4 Filtration of Solvent Mixtures

As was mentioned previously, even if using only HPLC-grade buffers for the mobile phase preparation, filtration before use is always recommended. The porosity of the filter should be chosen according to the column used. The smaller the particle size in the HPLC column, the smaller the filter pores should be. The most commonly used membrane filter is 0.5 μm for 3–5 μm packed columns. When using sub-2-μm particles, it is recommended to use 0.2-μm membrane filters [16].

A wide range of membrane filter materials is available, and at any time, the compatibility of the membrane filter with the mobile phase has to be considered. Using an incompatible filter, the dissolving of the membrane may occur, and the component of the membrane will create an impurity in the mobile phase.

6.13.5 Apparent pH of Solvent Mixtures

The term pH is defined in water condition. In the mixture of water with organic solvent, the term "apparent pH" (pH*) is used [49]. When the protons (H^+ ions) are solvated with the combination of water and organic solvent instead of pure water, the pH changes.

In most combinations of buffer and organic solvents, the apparent pH increases. The increase corresponds to the buffer concentration and the volume fractions of organic solvent and water. The apparent pH is usually form 0.2 to 0.5 higher than when measured for aqueous buffer solution.

Due to the difference between pH and pH* values, it is important to correctly state how a buffered mobile phase was prepared. When the mobile phase is stated as a methanol–0.1 M KH_2PO_4, pH 4.2, adjusted using 1 M H_3PO_4 (80:20), the apparent pH of the mobile phase was 4.8 [16]. In the correct way, first it is necessary to adjust the pH of the buffer and next mix it with organic solvent in a given ratio. Next, the apparent pH can be measured using a glass pH electrode. It gives a close approximation of the apparent pH.

6.13.6 Use of Guard Columns

When some extracts or other natural samples are analyzed using HPLC, even if a sample preparation technique is used, some impurities may be present in the sample. Such impurities will collect on the inlet of the analytical column. Some of them can be washed off the column with another mobile phase or wash solvent. Others may be irreversibly bound, which affects column deterioration and decreases the selectivity and efficiency. It may also cause an increase of back pressure.

At any time, especially when natural samples are analyzed, it is a good idea to incorporate a guard column ahead of the analytical column. It keeps the analytical column from collecting the impurities. The guard columns (cartridges) are installed in front of an analytical column. The column is usually very short, around 20 times shorter than the analytical column. The guard column might be packed into its own hardware or might be a part of the cartridge system. The guard column is used to increase the back pressure, which is a result of a significant amount of impurities accumulated on the inlet. In this case, the guard column should be replaced by a new one.

Additionally, a guard column provides saturation of the mobile phase with silica by bleeding silica into the mobile phase instead of from the analytical column, which takes place when the buffered mobile phases are used. This can be achieved without a decrease of resolution and performance of analytical columns. In common opinion, a guard column can increase the life of chromatographic columns by a factor of four [50]. In order to obtain maximum protection, it is recommended to use a guard column that contains the same packing material as the analytical column.

6.13.7 Use of Precolumns

As was stated previously, the major drawback of silica gel is limited pH stability. Silica gel may dissolve if the ionic strength of a buffered mobile phase is high and if the pH is higher than 11.

To prevent this phenomenon, a precolumn filled with large-particle silica gel (40–63 μm) may be placed before the analytical column. As the mobile phase passes through the precolumn, the mobile phase will slowly dissolve the silica; thus, when the mobile phase enters the analytical column, it is already saturated with silicate and cannot dissolve more from the analytical column [20]. When a buffered mobile phase is used, the silica dissolving and presence of silicates in the mobile phase cannot be avoided. Thus, it is better to bleed the precolumn than the analytical column.

It has to be mentioned that, in the literature, both terms *guard column* and *precolumn* are used interchangeably, despite the fact that the origins of both are different.

6.13.8 Column Cleaning and Regeneration

During chromatographic analyses, some sample components or mobile phase components may be strongly retained in the column, usually in the inlet part. Such a situation results in decrease of the resolution and performance of chromatographic separation, and it may provide broadened or split peaks.

In such a case, the first option is to use the analytical column in reversed flow. This places the clean bottom at the top to become the inlet. As a result, samples are introduced into a relatively clean packed bed. It should be mentioned that the inlet bed determines the placement of the sample bands that then are carried through the column for the separation [16]. Additionally, this usually allows cleaning the column from impurities adsorbed on a frit and in the first part of the packed bed.

When using a preparative column with an ID higher than 8 mm, the reversing of the column will not work because these columns have different inlet and outlet fittings.

Each column has to be equilibrated before use. Time of equilibration has to be long enough, and it may differ from one column to another. Some new columns take longer to equilibrate than others. Generally, RP columns equilibrate in 15 to 30 minutes (depending on the flow rate). In buffered conditions, time needed for equilibration is usually longer. When the mobile phase contains a buffer or an ion-pairing reagent (for example using an ion-exchange column), the best option is to equilibrate it overnight. The equilibration is complete when the baseline is stable and the retention times of a given compound do not change in the following injections.

Another possibility is to wash the column of any impurities that might have collected at the inlet. Most of the impurities are collected on the guard column, if it is used. If yes, it should be replaced. When the problem is solved, a new guard column should be used. If its replacement did not improve the separation problems, it seems that impurities were carried into the analytical column, where they were collected.

In this case, removing the analytical column from the LC system and washing it in reverse flow is a must.

The column washing requires a minimum of 20 column volumes of each solvent in a series. A column volume may be estimated as approximately half the volume of the empty column (calculated as the volume of a cylinder).

Cleaning solvents should be chosen depending on the support and bonding chemistry. The recommended series of cleaning solvents summarized in Table 6.2 [16]. More detailed description may be found in the work of F. Rabel [20]. At any time, the cleaning procedure is delivered with a given column in operation guide.

6.13.9 Column Storage

When a chromatographic column is not being used, it should be saturated with organic solvent or with a mixture of a high concentration of organic solvent in water. These conditions should be applied when the column is replaced from the equipment or even if it is connected to the LC system but is not in use. For storage, the chromatographic column should not have any acids, bases, or salts.

TABLE 6.2
Cleaning Solvent Series for Different Chromatographic Systems

NP System	RP System	Ion Exchange System
Heptane	Water methanol	Water
Chloroform	Chloroform	0.1 M Buffer
Ethyl acetate	Methanol	Water
Acetone	Water	0.01 M Sulfuric acid
Ethanol	0.01 Sulfuric acid	Water
Water	Water	Acetone
		Water
		0.01 M Disodium EDTA

Source: Rabel, F., Stationary phases and columns in analysis of primary and secondary metabolites. In: M. Waksmundzka-Hajnos and J. Sherma, editors. *High Performance Liquid Chromatography in Phytochemical Analysis.* CRC Press, Boca Raton, FL, 2011.

The column-washing procedure is usually provided by the manufacturer. For example, if a mobile phase contains some buffer in an acetonitrile–water mixture (75:25 v/v), before shutting down the system, the column should be washed with about 10 column volumes of a mixture of acetonitrile–water (75:25 v/v). Such a wash should remove any salts from the entire system to prevent attack on the stationary phase and prevent any corrosion of the stainless steel components of the LC system. The washing procedure prevents any salt crystallization around the pump and injection components, which eliminates scratching of the liquid seals and prevents leaks from occurring.

However, if possible, the washing procedure should be prolonged. Some researchers suggest the best procedure to clean up the column after acids and salts is washing it with pure water overnight and then short washing with an organic water mixture.

A chromatographic column may also contain some impurities of strongly adsorbed substances from a sample. Thus, when storing the column, it is recommended that the column should be washed to remove as many impurities as possible. The chromatographic column may be stored in the solvent in which it was supplied from the manufacturer; however, it is more convenient to store it in the solvent system that it usually uses (methanol or acetonitrile).

REFERENCES

1. Unger, K. K., *Porous Silica*. Elsevier, Amsterdam, 1979.
2. McCalley, D. V., Selection of suitable stationary phases and optimum conditions for their application in the separation of basic compounds by reversed-phase HPLC, *J. Sep. Sci.*, 26, 187–200, 2003.
3. Telepchak, M. J., August, T. F., and Chaney, G., *Forensic and Clinical Applications of Solid Phase Extraction: Forensic Science and Medicine.* Humana Press Inc., Totowa, NJ, 2004.
4. Neue, U. D., Serowik, E., Iraneta, P., Alden, B. A., and Walter, T. H., Universal procedure for the assessment of the reproducibility and the classification of silica-based reversed-phase packings: I. Assessment of the reproducibility of reversed-phase packings, *J. Chromatogr. A*, 849, 87–100, 1999.
5. Unger, K. K., *Packings and Stationary Phases in Chromatographic Techniques*. Marcel Dekker, Inc., New York and Basel, 1990.
6. Nawrocki, J., The silanol group and its role in liquid chromatography, *J. Chromatogr. A*, 779, 29–71, 1997.
7. Nawrocki, J., and Buszewski, B., Influence of silica surface chemistry and structure on the properties, structure and coverage of alkyl-bonded phases for high-performance liquid chromatography, *J. Chromatogr.*, 449, 1–25, 1988.
8. Kimata, K., Iwaguchi, K., Onishi, S., Jinno, K., Eksteen, R., Hosoya, K., Araki, M., and Tanaka, N., Chromatographic characterization of silica C18 packing materials. Correlation between a preparation method and retention behavior of stationary phase, *J. Chromatogr. Sci.*, 27, 721–728, 1989.

9. Engelhardt, H., Low, H., and Gotzinger, W., Chromatographic characterization of silica-based reversed phases, *J. Chromatogr. A*, 544, 371–379, 1991.
10. Walters, M. J., Classification of octadecyl-bonded liquid chromatography columns, *J. Assoc. Off. Anal. Chem.*, 70, 465–469, 1987.
11. Galushko, S. V., The calculation of retention and selectivity in reversed-phase liquid chromatography II. Methanol–water eluents, *Chromatographia*, 36, 39–42, 1993.
12. Neue, U. D., *HPLC Columns. Theory, technology and practice*. Wiley-VCH, New York, 1997.
13. Claessens, H. A., van Straten, M. A., Cramers, C. A., Jezierska, M., and Buszewski, B., Comparative study of test methods for reversed-phase columns for high-performance liquid chromatography, *J. Chromatogr. A*, 826, 135–156, 1998.
14. Engelhardt, H., Blay, C., and Saar, J., Reversed phase chromatography—The mystery of surface silanols, *Chromatographia*, 62, S19–S29, 2005.
15. Sadek, P. C., and Carr, P. W., A simple test for characterizing silanophilic interactions of reversed-phase columns based on the chromatographic behavior of cyclic tetraaza compounds, *J. Chromatogr. Sci.*, 21, 314–320, 1983.
16. Rabel, F., Stationary phases and columns in analysis of primary and secondary metabolites. *In* M. Waksmundzka-Hajnos and J. Sherma, editors. *High Performance Liquid Chromatography in Phytochemical Analysis*. CRC Press, Boca Raton, FL, 2011.
17. Majors, R. E., and Przybyciel, M., Columns for reversed-phase LC separation in highly aqueous mobile phases, *LCGC North America*, 20, 584–593, 2002.
18. Bocian, S., Matyska, M. T., Pesek, J. J., and Buszewski, B., Study of the retention and selectivity of cholesterol bonded phases with different linkage spacers, *J. Chromatogr. A*, 1217, 6891–6897, 2010.
19. Noga, S., Michel, M., and Buszewski, B., Effect of functionalized stationary phases on the mechanism of retention of fungicides in RP-LC elution, *Chromatographia*, 73, 857–864, 2011.
20. Rabel, F., Use and maintenance of microparticle high performance liquid chromatography columns, *J. Chromatogr. Sci.*, 18, 394–408, 1980.
21. Alpert, A. J., Hydrophilic-interaction chromatography for the separation of peptides, nucleic acids, and other polar compounds, *J. Chromatogr.*, 449, 177–196, 1990.
22. Pesek, J. J., and Matyska, M., How to retain polar and nonpolar compounds on the same HPLC stationary phase with an isocratic mobile phase, *LC GC North America*, 24, 296–303, 2006.
23. Zhou, Z. Q., Wang, P., Jiang, S. R., Wang, M., and Yang, L., Preparation of polysaccharide-based chiral stationary phases and the direct separation of six chiral pesticides and related intermediates, *J. Liq. Chromatogr. R. T.*, 26, 2873–2880, 2003.
24. Aboul-Enein, H. Y., and Ali, I., *Chiral Separations by Liquid Chromatography: Theory and Applications*, Chromatographic Science, CRC Press, Taylor & Francis Group, Boca Raton, FL, 2003.
25. Yuan, L.-M., Xu, Z.-G., Ai, P., Chang, Y.-X., and Azam, A. K. M. F., Effect of chiral additive on formation of cellulose triacetate chiral stationary phase in HPLC, *Anal. Chim. Acta*, 555, 152–155, 2005.
26. Ahrland, S., Grenthe, I., and Noren, B., The ion-exchange properties of silica gel, *Acta Chem. Scand.*, 14, 1077–1090, 1960.
27. Iler, R. K., Ion exchange properties of a crystalline hydrated silica, *J. Colloid Sci.*, 19 (7), 648–657, 1964.
28. Weiss, J., and Jensen, D., Modern stationary phases for ion chromatography, *Anal. Bioanal. Chem.*, 375, 81–98, 2003.
29. Gawdzik, B., Matynia, T., and Osypiuk, J., Porous copolymer of the methacrylic ester of dihydroxydiphenylmethane diglycidyl ether and divinylbenzene as an HPLC packing, *Chromatographia*, 47, 509–514, 1998.
30. Weiss, J., Reinhard, S., Pohl, C., Saini, C., and Narayaran, L., Stationary phase for the determination of fluoride and other inorganic anions, *J. Chromatogr. A*, 706, 81–92, 1995.
31. Nair, L. M., Saari-Nordhaus, R., and Montgomery, R. M., Applications of a new methacrylate-based anion stationary phase for the separation of inorganic anions, *J. Chromatogr. A*, 789, 127–134, 1997.
32. Weiss, J., *Handbook of Ion Chromatography*, 3rd ed. VCH, Weincheim, 2004.
33. Matsushita, S., Tada, Y., Baba, N., and Hosako, K., High performance ion chromatography of anions, *J. Chromatogr.*, 259, 459–464, 1983.
34. McCormick, R. M., and Karger, B. L., Distribution phenomena of mobile-phase components and determination of dead volume in reversed-phase liquid chromatography, *Anal. Chem.*, 52, 2249–2257, 1980.
35. Iler, R. K., *The Chemistry of Silica*. Wiley, New York, 1979.
36. Fóti, G., and Kovats, E. S., Chromatographic study of the silanol population at the surface of derivatized silica, *Langmuir*, 5, 232–239, 1989.
37. Gilpin, R. K., Jaroniec, M., and Lin, S., Studies of the surface composition of phenyl and cyanopropyl bonded phases under reversed-phase liquid chromatographic conditions using alkanoate and perfluoroalkanoate esters, *Anal. Chem.*, 63, 2849–2852, 1991.

38. Nahum, A., and Horvath, C., Surface silanols in silica-bonded hydrocarbonaceous stationary phases: I. Dual retention mechanism in reversed-phase chromatography, *J. Chromatogr.*, 203, 53–63, 1981.
39. Bij, K. E., Horvath, C., Melander, W. R., and Nahum, A., Surface silanols in silica-bonded hydrocarbonaceous stationary phases: II. Irregular retention behavior and effect of silanol masking, *J. Chromatogr.*, 203, 65–84, 1981.
40. Scholten, A. B., Claessens, H. A., de Haan, J. W., and Cramers, C. A., Chromatographic activity of residual silanols of alkylsilane derivatized silica surfaces, *J. Chromatogr. A*, 759, 37–46, 1997.
41. Scully, N. M., Healy, L. O., O'Mahony, T., Glennon, J. D., Dietrich, B., and Albert, K., Effect of silane reagent functionality for fluorinated alkyl and phenyl silica bonded stationary phases prepared in supercritical carbon dioxide, *J. Chromatogr. A*, 1191, 99–107, 2008.
42. Pursch, M., Sander, L. C., and Albert, K., Chain order and mobility of high-density C18 phases by solid-state NMR spectroscopy and liquid chromatography, *Anal. Chem.*, 68, 4107–4113, 1996.
43. Buszewski, B., Bocian, S., and Felinger, A., Excess isotherms as a new way for characterization of the columns for reversed-phase liquid chromatography, *J. Chromatogr. A*, 1191, 72–77, 2008.
44. Kohler, J., and Kirkland, J. J., Improved silica-based column packings for high-performance liquid chromatography, *J. Chromatogr.*, 385, 125–150, 1987.
45. Maus, M., and Engelhardt, H., Thermal treatment of silica and its influence on chromatographic selectivity, *J. Chromatogr.*, 371, 235–242, 1986.
46. Wilson, N. S., Gilroy, J., Dolan, J. W., and Snyder, L. R., Column selectivity in reversed-phase liquid chromatography: VI. Columns with embedded or end-capping polar groups, *J. Chromatogr. A*, 1026, 91–100, 2004.
47. Rabel, F., Instrumentation for small bore liquid chromatography, *J. Chromatogr. Sci.*, 23, 247–252, 1985.
48. Kucera, P., editor. 1984. *Microcolumn High Performance Liquid Chromatography, J. Chromatogr. Library*, Vol. 28. Elsevier, New York.
49. Rosés, M., Determination of the pH of binary mobile phases for reversed phase liquid chromatography, *J. Chromatogr. A*, 1037, 283–298, 2004.
50. MicroSolv Technology. *Specifications: Guard Columns*.

7 Selection of the Type of Mobile Phases for Analysis of Nonionic Analytes

Reversed- and Normal-Phase HPLC

Tomasz Tuzimski

CONTENTS

7.1 INTRODUCTION

Neutral substances are analytes that, regardless of mobile phase pH, remain in a nonionized form. In other words, by neutral solutes/analytes, we mean those that contain no molecules that carry a positive or negative charge—usually as the result of ionization of an acid or a base. Therefore, their chromatographic behavior is similar to those of nonionizable compounds.

Selection of the type of mobile phase for the analysis of nonionic (neutral) analytes depends on the molecular structure of the analyte. The first choice for the separation of nonionic (as well as ionic) analytes is usually separation in reversed-phase (RP) chromatography. Less frequently, normal-phase (NP) systems are applied in the analysis of this type of compound. In this chapter, the separation of neutral substances in both types of chromatographic systems is discussed.

7.2 NONIONIC PESTICIDES IN ENVIRONMENTAL SAMPLES

Most pesticides are compounds that behave as nonionic or neutral analytes. The majority are nonpolar or weakly polar substances or weak acids or bases. Pesticide molecules often do not contain polar groups, or polar groups are substituted by nonpolar fragments, for example, alkyl chains. In addition, the halogen atoms in the molecules of pesticides influence the polarity of analytes.

7.3 METHOD DEVELOPMENT FOR NONIONIC ANALYTES

The main challenge in the separation of complex samples by liquid chromatography is selection of the experimental conditions, in both NP and RP systems that can provide suitable band spacing and optimal resolution.

Experiments can be carried out under the following conditions and are usually optimized by

- Variation of the mobile phase (solvents; pH, especially for ionic analytes)
- Choice of different columns (stationary phases)
- Variation of temperature

7.4 REVERSED-PHASE HIGH-PERFORMANCE LIQUID CHROMATOGRAPHY

Separation in RP systems by high-performance liquid chromatography (HPLC) is usually more convenient, robust, and versatile. In RP-HPLC, the mobile phase is, in most cases, a mixture of water and organic solvents, such as methanol and acetonitrile (used most frequently), dioxane and tetrahydrofuran (less useful), and acetone and isoporopanol (used very rarely).

7.4.1 Isocratic

Isocratic separation works well for many samples, and it represents the simplest and most convenient form of liquid chromatography.

Sample retention k in isocratic elution is usually controlled by varying the mobile phase composition. The usual separation goal is to obtain average values of k in the range $1 \leq k \leq 10$ for all peaks during the experiment. This ideal range of values of k for all solutes/analytes is the main aim of chromatographic separation because it corresponds to narrower, taller peaks for improved detection as well as short run times. Some details are also described in Chapters 3 and 9.

For most samples to obtain a "perfect" chromatogram (with an optimal range of k for analytes, acceptable selectivity [values of α], and resolution) the experiment can be realized in the following steps [1]:

- Mobile phase strength (%B) is varied until the right retention range is achieved ($1 \leq k \leq 10$); the first conditions that should be explored for the improvement of retention and selectivity are changes in %B (usually ± 10%) and temperature (e.g., 20°C–50°C)
- If some peaks are still overlapped and poorly separated, other conditions can be varied to improve selectivity (see also Chapters 3 and 9)
- Column conditions are varied (e.g., column length, particle size) and/or flow rate of mobile phase

7.4.2 Programming Analysis

Peak capacity is of much greater importance for separations of multicomponent samples. It is seldom possible to separate such samples with an acceptable resolution of all peaks, so peak capacity becomes a better measure of overall separation than the values of R_s. Therefore, separations of samples containing a very large number of components are usually carried out by gradient elution, for which the concept of peak capacity is more relevant (see also Chapters 3 and 9). The peak capacity of separation should not be confused with the number of solutes/analytes separated at $R_s = 1$ because it is rarely possible to achieve a regular spacing of peaks as in Figure 7.1a [1].

The required peak capacity (PC_{req}) for the separation of all peaks of a n-component sample is shown in Figure 7.1b [1,2]. For isocratic separation, peak capacity is given as Equation 7.1 [1]:

$$PC = 1 + \left(\frac{N^{0.5}}{4}\right)\ln\left(\frac{t_{R,z}}{t_0}\right) = 1 + 0.575N^{0.5}\log\left(\frac{t_{R,z}}{t_0}\right) \tag{7.1}$$

where $t_{R,z}$ refers to the retention time of the last peak in the chromatogram.

We can obtain more separated analytes during a single experiment in the case of a more efficient system—with narrower peaks of separated analytes (especially ionic compounds) and with optimal peak asymmetry factor (As) or/and peak tailing factor (TF) as illustrated in Figure 7.2.

Programmed analysis can be realized by programming of solvent composition (gradient elution) as illustrated in Figure 7.3 [1,3].

Sometimes, we can have problems with obtaining "optimal" separations of components (Figure 7.4) [1]:

Gradient elution can be carried out with different gradient shapes (by gradient shape, we mean the way in which mobile phase composition changes with the time during a gradient run) as illustrated in Figure 7.5 [4]:

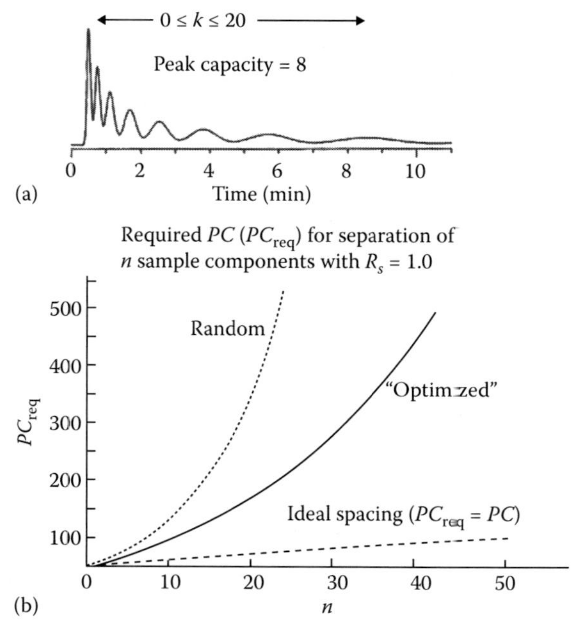

FIGURE 7.1 Peak capacity. (a) Example of peak capacity (PC) for a separation where $PC = 8$, $N = 100$, and $R_s = 1$ for every peak; (b) peak capacity required for the separation of a sample that contains n components [2]; "ideal spacing" is from Equation 7.1. (Snyder, L. R., Kirkland, J. J., Dolan, J. W.: *Introduction to Modern Liquid Chromatography*. 2010. Copyright Wiley-VCH Verlag GmbH & Co. KGaA. Reproduced with permission.)

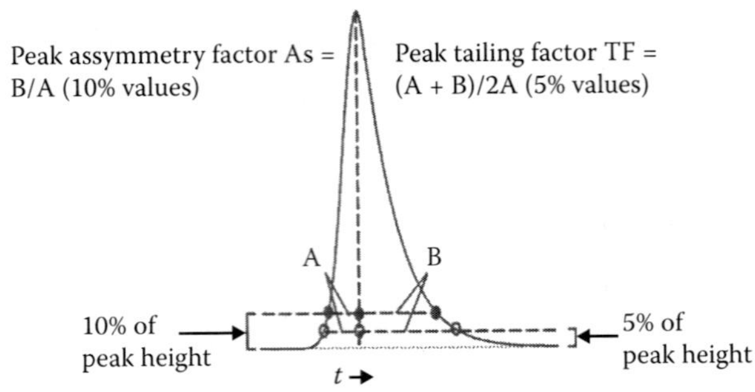

FIGURE 7.2 Measurement and effects of peak tailing. Definitions of peak asymmetry and tailing factors (As and TF). (Snyder, L. R., Kirkland, J. J., Dolan, J. W.: *Introduction to Modern Liquid Chromatography*. 2010. Copyright Wiley-VCH Verlag GmbH & Co. KGaA. Reproduced with permission.)

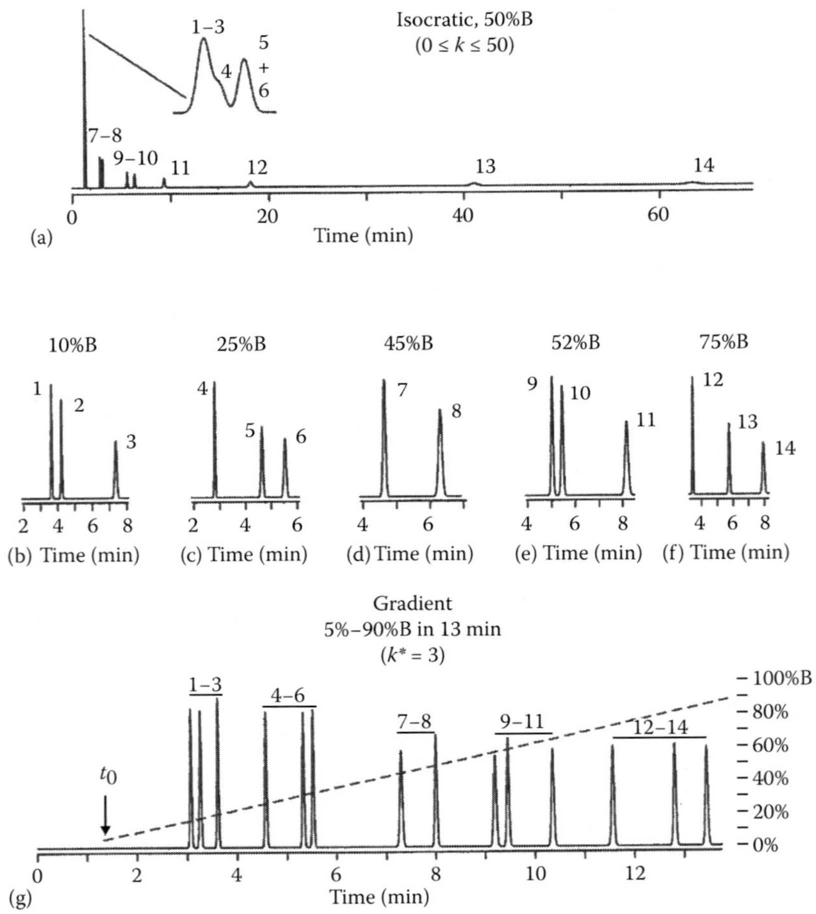

FIGURE 7.3 Example of the general elution problem. Sample: Fourteen toxicology standards. Conditions: 250 × 4.6 mm (5 μm) C18 column; mobile phase is ACN (B) and pH-2.5 phosphate buffer (A); 65°C; 2 ml/min. (a) Isocratic separation with 50%B; (b–f), isocratic separation of indicated compounds (peaks) with 10%B, 25%B, 45%B, 52%B, and 75%B, respectively ($k \approx 3$); (g) gradient elution as indicated. (Chromatograms are computer simulations based on the experimental data of Zhu, P. L. et al., *J. Chromatogr. A*, 756, 21–39, 1996. Snyder, L. R., Kirkland, J. J., Dolan, J. W.: *Introduction to Modern Liquid Chromatography*. 2010. Copyright Wiley-VCH Verlag GmbH & Co. KGaA. Reproduced with permission.)

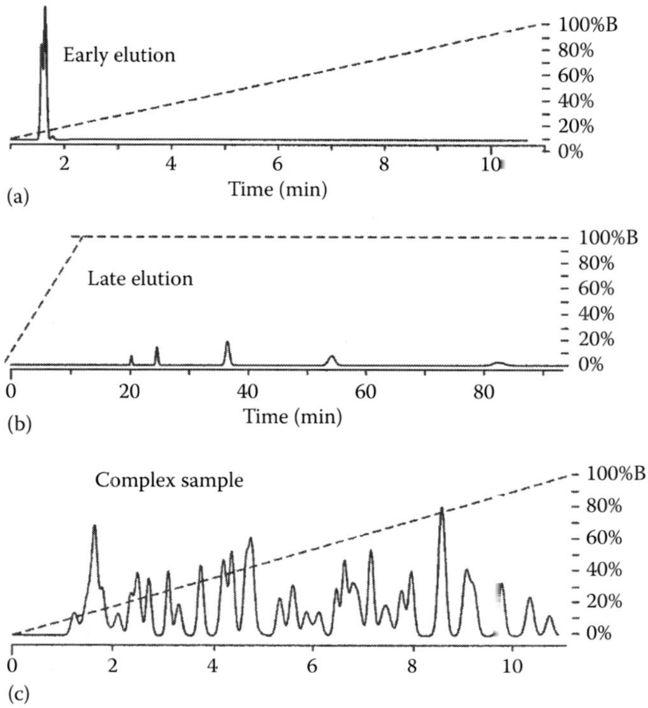

FIGURE 7.4 Potential problems in gradient elution. (a) Nonretentive sample; (b) excessively retentive sample; (c) sample contains too many components. Gradient indicated by (---). (Snyder, L. R., Kirkland, J. J., Dolan, J. W.: *Introduction to Modern Liquid Chromatography*. 2010. Copyright Wiley-VCH Verlag GmbH & Co. KGaA. Reproduced with permission.)

Most gradient separations use [1,4]

- Linear gradient (a, h, g), which is strongly recommended during the initial stages of method development
- Curved gradients (b, c)
- Segmented gradients (d)
- Gradient delay or "isocratic hold" (e)
- Step gradient (f)

The gradient can be "programmed" by a gradient program; we refer to the description of how mobile phase composition changes with time. For example, see Ref. [4],

- A linear gradient presents the simplest example, for example, a gradient from 10%B to 80%B in 20 minutes (Figure 7.5g), which can also be described as 10%B to 80%B in 0–20 minutes (10%B at time 0 to 80%B at 20 minutes).
- Segmented programs described by values of %B and time for each linear segment in the gradient, for example, a gradient 5%B–25%B–40%B–100%B at 0–5–15–20 minutes (Figure 7.5h).

7.4.3 Retention: Eluent Composition Relationships

A single solvent rarely provides suitable separation selectivity and retention in chromatographic systems. A typical mobile phase is selected by adjusting an appropriate qualitative and quantitative composition of a two-component (binary), ternary complex, or more component mixture. The

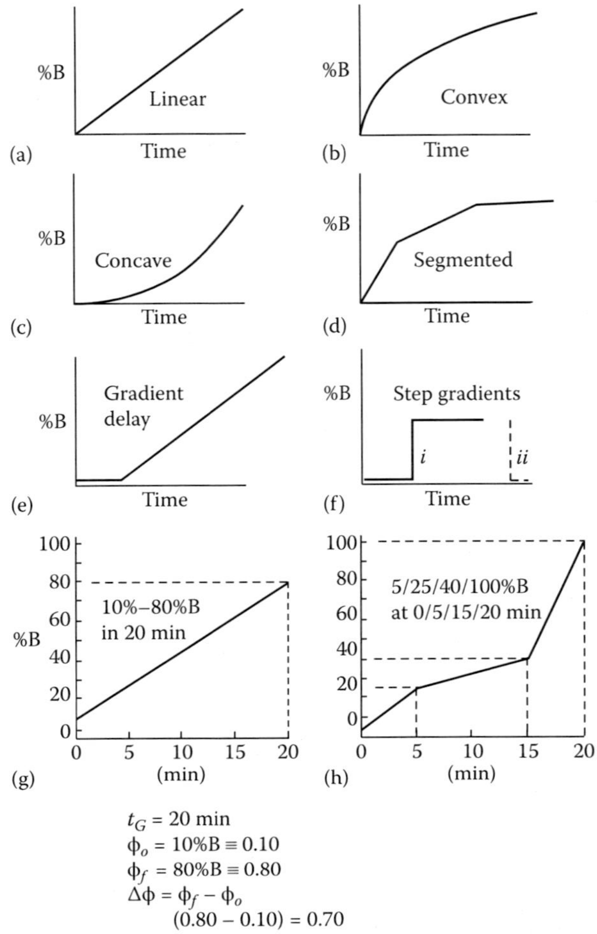

FIGURE 7.5 Illustration of different gradient shapes (plots of %B vs. time). (a) Linear. Curved gradients: (b) convex and (c) concave. (d) Segmented, (e) gradient delay, or "isocratic hold," (f) step gradient, (g) linear gradient from 10% to 80% B in 20 min (10%–80% B in 0–20 min), (h) segmented program 5%-25%-40%-100% B at 0-5-15-20 min. (Snyder, L. R., Kirkland, J. J., Dolan, J. W.: *Introduction to Modern Liquid Chromatography*. 2010. Copyright Wiley-VCH Verlag GmbH & Co. KGaA. Reproduced with permission.)

dependence of retention on the composition of the mobile phase can be predicted using a few popular approaches reported in the literature and used in laboratory practice.

As noted in Chapter 3, the retention in RP-HPLC varies with mobile phase %B as described by Equation 3.4, in which k_w refers to the extrapolated value of k for 0%B (water as mobile phase), S is a constant for a given solute when only %B is varied, and ϕ is the volume fraction of organic solvent B (polar modifier) in the mobile phase ($\phi \equiv 0.01\%B$). For the obtained optimal chromatogram, the separation can be regulated by selected value of %B and, simultaneously, values of k for the sample that are within a desired range (e.g., $1 \leq k \leq 10$).

Changes in percentage of modifier in the mobile phase often lead to changes in relative retention with maximum resolution occurring for an intermediate value of %B. To take maximum advantage of solvent-strength selectivity, the allowable retention range can be expanded from $1 \leq k \leq 10$ to $0.5 \leq k \leq 20$ [1].

7.4.4 Selectivity

The mobile and stationary phase compositions in HPLC are more likely to affect a particular separation than any other factors, for example, flow rate. The most frequently used columns in RP-HPLC experiments are the C18 or C8 bonded phases. Mobile phase optimization procedures normally are based on the solvent selectivity triangle first described by Snyder [5] and Rohrschneider [6] in which organic modifiers are classified according to their ability to interact with the solute as a proton donor (acidic), a proton acceptor (basic), or dipole. The triangle was named the Snyder-Rohrschneider solvent selectivity triangle (SST). Solvents that are located far away from each other in this triangle have the largest differences in these properties, and hence, greater differences in selectivity will be found using them as the mobile phase.

Some empirical scales of solvent polarity based on kinetic or spectroscopic measurements have been described [7] to present their ability for molecular interactions.

There are several solvatochromic classifications of solvents, which are based on spectroscopic measurements of their different solvatochromic parameters [7–12]. The ET(30) scale [7] is based on the charge–transfer interaction of the 2,6-diphenyl-4-(2,4,6-triphenyl-N-pyridino) phenolate molecule (known as Dimroth and Reichardt's betaine scale). The Z scale [8,9] is based on the charge–transfer interaction of the N-ethyl-4-methocycarbonyl pyridinium iodine molecule (developed by Kosower and Mohammad). The scale based on Kamlet–Taft solvatochromic parameters has gained growing popularity in the literature and laboratory practice [10–13]. The following parameters can be distinguished in this scale: dipolarity/polarizability (π^*) and hydrogen-bond acidity (α) and basicity (β) (see Table 7.1) [1,5–29].

The solvatochromic parameters are average values for a number of selected solutes and somewhat independent of solute identity. Some representative values for solvatochromic parameters of common solvents used in TLC are summarized in Table 7.2 [1,17,30,31].

These parameters were normalized in a similar way as x_d, x_e, x_n parameters of Snyder. The values of α, β, and π^* for each solvent were summed up and divided by the resulting sum. Then fractional parameters were obtained (fractional interaction coefficients): α/Σ (acidity), β/Σ (basicity), and π^*/Σ (dipolarity). These values were plotted on a triangle diagram similar to as in the Snyder-Rohrschneider SST (see also Chapter 3).

More comprehensive representation of parameters characterizing solvent properties can be expressed based on Abraham's model in which the following Equation 7.2 is used [11–14]:

$$\log K_L = c + l \log L^{16} + rR_2 + s\pi_2^H + a\sum \alpha_2^H + b\sum \beta_2^H \tag{7.2}$$

where $\log K_L$ is the gas–liquid distribution constant, $\log L^{16}$ is the distribution constant for the solute between a gas and n-hexadecane at 298 K, R_2 is the excess molar refraction (in $cm^3/10$), π_2^H is the ability of the solute to stabilize a neighboring dipole by virtue of its capacity for orientation and induction interactions, $\sum \alpha_2^H$ is the effective hydrogen-bond acidity of the solute, and $\sum \beta_2^H$ is the hydrogen-bond basicity of the solute.

All these parameters, with the exception of $\log K_L$, are the solute descriptors. As can be seen, the parameters s, a, and b represent polar interactions of the solvent molecule with a solute as dipole–dipole, hydrogen-bond basicity, and hydrogen-bond acidity, respectively; the parameter r represents the ability of the solvent molecule to interact with n- or π-electrons of the solute molecule. In addition to previous classifications of solvents, this model takes into account molecular interactions with cavity formation in the solvent for solute molecule and dispersion interactions between solvent and solute. These effects are presented by constants c and l. The values of the discussed parameters are given in Table 7.2 [1,17,30,31]. The chromatographer can compare these data and others in this table that can be helpful for optimization of retention and separation selectivity.

7.4.5 Optimizing

Reliable results of this analysis can be obtained with HPLC mode when the resolution, R_S, of sample components is satisfactory, at least greater than 1.5 (see Chapter 3).

TABLE 7.1
Parameters Applied for Characterization of Solvents for Liquid Chromatography

	Snyder's Classification Based on Selectivity Triangle ($S_{S\ RP}$ is an Empirical Solvent Strength Parameter Used in RP System)						Kamlet-Taft and Coworkers' Classification					Abraham's Model Classification: System Constant for Distribution between Gas Phase and Solvent					
	Selectivity Group	Solvent Strength		Solvent Selectivity			Solvatochromic Parameters					(Abraham's Model)					
Solvent		(P')	$(S_{S\ RP})$	x_e	x_d	x_n	$E_{t(30)}$	E_T^N	π_1^*	α_1	β_1	r	s	a	b	l	c
n-Butyl ether	I	2.1		0.44	0.18	0.38	33.0	0.071	0.27	0	0.46						
Diisopropyl ether		2.4		0.48	0.14	0.38	34.1	0.105	0.27	0	0.48						
Methyl tert.-butyl ether		2.7					34.7	0.124									
Diethyl ether		2.8		0.53	0.13	0.34	34.6	0.117	0.27	0	0.47						
n-Butanol	II	3.9		0.59	0.19	0.25	49.7	0.586	0.47	0.79	0.88						
2-Propanol		3.9		0.55	0.19	0.27	48.4	0.546	0.48	0.76	0.95						
1-Propanol		4.0		0.54	0.19	0.27	50.7	0.617	0.52	0.78							
Ethanol		4.3	3.6	0.52	0.19	0.29	51.9	0.654	0.54	0.83	0.77	−0.21	0.79	3.63	1.31	0.85	0.01
Methanol		5.1	3.0	0.48	0.22	0.31	55.4	0.762	0.60	0.93	0.62	−0.22	1.17	3.70	1.43	0.77	0

Tetrahydrofuran	III	4.0	4.4	0.38	0.20	0.42	37.4	0.207	0.58	0	0.55						
Pyridine		5.3		0.41	0.22	0.36	40.5	0.302	0.87	0	0.64					
Methoxyethanol		5.5		0.38	0.24	0.38										
Dimethylformamide		6.4		0.39	0.21	0.40	43.2	0.386	0.88	0	0.69					
Acetic acid	IV	6.0		0.39	0.31	0.30	51.7	0.648	0.64	1.12						
Formamide		9.6		0.38	0.33	0.30	55.8	0.775	0.97	0.71						
Dichloromethane	V	4.3		0.27	0.33	0.40	40.7	0.309	0.82	0.30	0					
1,1-Dichloroethane		4.5		0.30	0.21	0.49	41.3	0.327	0.81	0	0					
Ethyl acetate	VI	4.4		0.34	0.23	0.43	38.1	0.228	0.55	0	0.45					
Methyl ethyl ketone		4.7		0.35	0.22	0.43			0.67	0.06	0.48					
Dioxane		4.8	3.5	0.36	0.24	0.40	36	0.164	0.55	0	0.37					
Acetone		5.1	3.4	0.35	0.23	0.42	42.2	0.355	0.71	0.08	0.48					
Acetonitrile		5.8	3.1	0.31	0.27	0.42	45.6	0.460	0.75	0.19	0.31	−0.22	2.19	2.38	0.41	0.73	0
Toluene	VII	2.4		0.25	0.28	0.47	33.9	0.099	0.54	0	0.11	−0.22	0.94	0.47	0.10	1.01	0.12
Benzene		2.7		0.23	0.32	0.45	34.3	0.111	0.59	0	0.10	−0.31	1.05	0.47	0.17	1.02	0.11
Nitrobenzene		4.4		0.26	0.30	0.44	41.2	0.324	1.01	0	0.39					
Nitromethane		6.0		0.28	0.31	0.40	46.3	0.481								
Chloroform	VIII	4.3		0.31	0.35	0.34	39.1	0.259	0.58	0.44	0	−0.60	1.26	1.37	0.98	0.98	0.17
Dodecafluoroheptanol		8.8		0.33	0.40	0.27										
Water		10.2	0	0.37	0.37	0.25	63.01	1.000	1.09	1.17	0.18	0.82	2.74	3.90	4.80	−2.13	−1.27

Note: See Refs. [1,5–29].

TABLE 7.2
Solvent Selectivity Characteristics

	Normalized Selectivity[a]				
Solvent	**H-B Acidity α/Σ**	**H-B Basicity β/Σ**	**Dipolarity π^*/Σ**	**P'^{b}**	**ε^{c}**
Acetic acid	0.54	0.15	0.31	6.0	6.2
Acetonitrile	0.15	0.25	0.60	5.8	37.5
Alkanes	0.00	0.00	0.00	0.1	1.9
Chloroform	0.43	0.00	0.57	4.1	4.8
Dimethylsulfoxide	0.00	0.43	0.57	7.2	4.7
Ethanol	0.39	0.36	0.25	4.3	24.6
Ethylacetate	0.00	0.45	0.55	4.4	6.0
Ethylene chloride	0.00	0.00	1.00	3.5	10.4
Methanol	0.43	0.29	0.28	5.1	32.7
Methylene chloride	0.27	0.00	0.73	3.1	8.9
Methyl-*t*-butyl ether	0.00	≈0.6	≈0.4	≈2.4	≈4
Nitrometane	0.17	0.19	0.64	6.0	35.9
Propanol (*n*- or *iso*)	0.36	0.40	0.24	3.9	6.0
Tetrahydrofuran	0.00	0.49	0.51	4.0	7.6
Triethylamine	0.00	0.84	0.16	1.9	2.4
Water	0.43	0.18	0.45	10.2	80

[a] Values from [30], where Σ refers to the sum of values of α, β, and π^* for each solvent.
[b] Polarity index; values from Snyder, L. R., *J. Chromatogr. Sci.*, 16, 223–234, 1978.
[c] Dielectric constant; values from Riddick, J. A., Bunger, W. B., *Organic Solvents*, Wiley-Interscience, New York, 1970.

Although improvements in the resolution may be explored by changing the solvent-type selectivity, the new mobile phase must have a similar solvent strength in order to maintain comparable (optimal) values of *k* and run time. Considering miscibility, ease of use, availability and reasonable boiling point as well, the best organic solvent choice for RP-HPLC experiments are acetonitrile, methanol, and tetrahydrofuran. The new mobile phase must have a similar solvent strength in order to maintain comparable values of *k* and time of single experiment. As illustrated in Figure 7.6, when changing solvent type, we can estimate the necessary change in %B values for the new B solvent. In Figure 7.6 similar values of %B values for different solvents, acetonitrile (ACN), methanol (MeOH), and tetrahydrofuran (THF), fall on vertical lines [1,32].

As illustrated in Figure 7.7, by changing the modifier, we can get a different selectivity of separated components of a sample (mixture) [1,15].

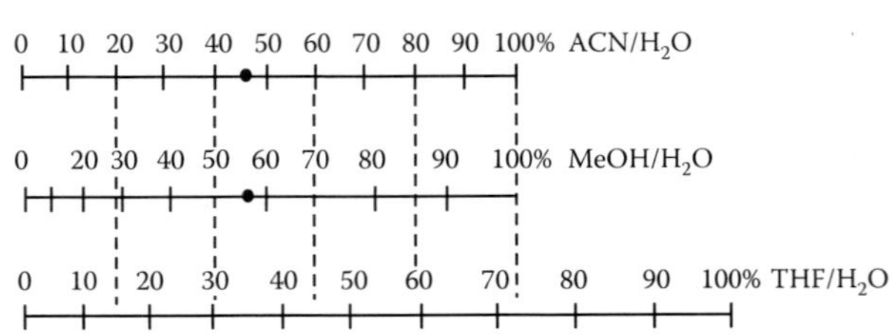

FIGURE 7.6 Solvent-strength nomograph for RP-HPLC. (Adapted from Schoenmakers, P. J. et al., *J. Chromatogr.*, 205, 13–30, 1981.) Two mobile phases of equal strength (46% ACN and 57% MeOH) marked by •, as an example. (Snyder, L. R., Kirkland, J. J., Dolan, J. W.: *Introduction to Modern Liquid Chromatography*. 2010. Copyright Wiley-VCH Verlag GmbH & Co. KGaA. Reproduced with permission.)

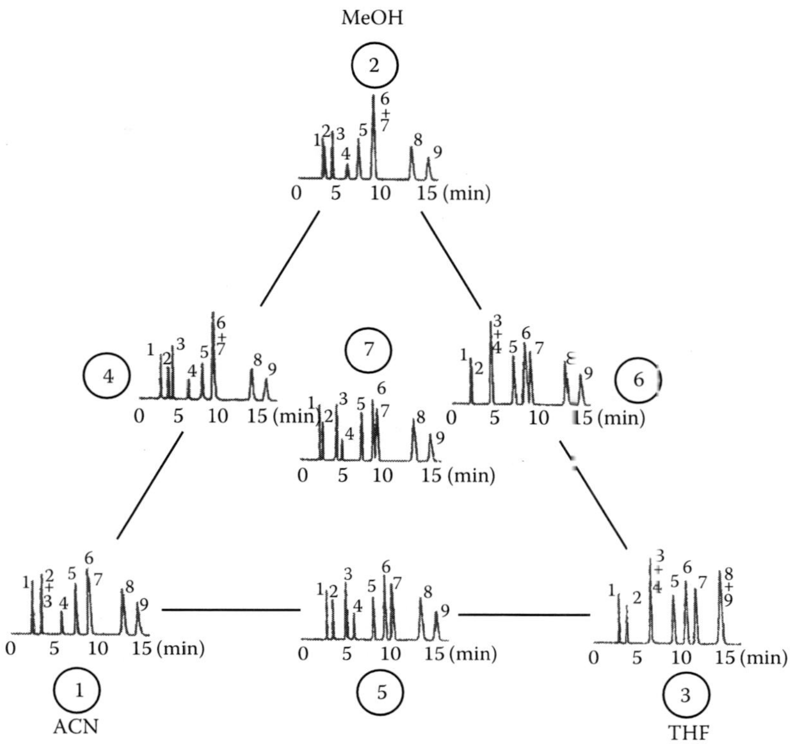

FIGURE 7.7 Use of seven solvent-type selectivity experiments for the separation of a mixture of nine substituted naphtalenes. Sample substituents are 1, 1-$NHCOCH_3$; 2, 2-SO_2CH_3; 3, 2-OH; 4, 1-$COCH_3$; 5, 1-NO_2; 6, 2-OCH_3; 7, -H (naphtalene); 8, 1-SCH_3; 9, 1-Cl. Conditions: 150 × 4.6 mm C8 column; 40°C; 2.0 ml/min. Mobile phases (circled): 1, exchange: 1, ACN; 2, MeOH; 2 exchange: 1, ACN; 2, MeOH; 3, 39% tetrahydrofuran/water, 4, 1:1 mixture of 1 and 2; 5, 1:1 mixture of 2 and 3; 6, 1:1 mixture of 1 and 3; 7, 1:1:1 mixture of 1, 2, and 3. (Recreated from data of Glajch, J. L. et al., *J. Chromatogr.*, 199, 57–79, 1980. Snyder, L. R., Kirkland, J. J., Dolan, J. W.: *Introduction to Modern Liquid Chromatography*. 2010. Copyright Wiley-VCH Verlag GmbH & Co. KGaA. Reproduced with permission.)

The effect of a change in temperature on selectivity is usually minor for RP-HPLC separation of neutral (nonionic) analytes (samples). However, sometimes by changing the values of both temperature and percentage of modifier in the mobile phase, we can obtain better results (Figure 7.8).

By combination of experimental conditions (%B, type of solvent, and temperature), we can obtain optimum selectivity and maximum resolution. The use of simultaneous variation of percentage of modifier in mobile phase (%B) and temperature (T) has some advantages [1]:

- Only initial experiments are required once a value of %B for a reasonable range in k has been established (as illustrated in Figure 7.8a, experiments 1 and 2 are carried out first, that is, a change in %B only, and both $0.5 \leq k \leq 20$ and resolution is adequate; if acceptable separation cannot be attained in this way; experiments 3 and 4 are carried out next, repeating experiments 1 and 2 at higher temperature T).
- With online mixing of the A and B solvents, all four experiments can be carried out automatically without operator intervention.
- There are no experimental problems associated with other means of optimizing selectivity.
- The procedure is "friendly for the researcher" by achieving the desired selectivity and resolution of sample.
- Peak matching tends to be easier than for other experimental designs.

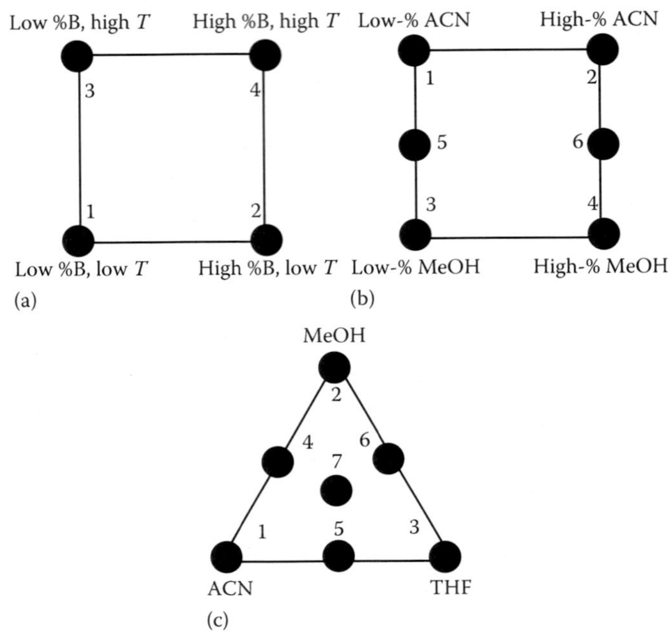

For each of the above runs in (a)–(c), %-water is varied to maintain $0.5 \le k \le 20$.

FIGURE 7.8 Experimental designs for the simultaneous optimization of various separation conditions for optimum selectivity. (a) Solvent strength (%B) and temperature (T); (b) solvent strength and solvent type (MeOH and ACN); (c) solvent type (MeOH, ACN, and THF). (Snyder, L. R., Kirkland, J. J., Dolan, J. W.: *Introduction to Modern Liquid Chromatography*. 2010. Copyright Wiley-VCH Verlag GmbH & Co. KGaA. Reproduced with permission.)

7.4.6 Applications

Gradient elution can be used for separation of pesticides and their quantitative analysis in environmental samples. An example is illustrated in Figure 7.9 [33–42]:

Several papers based on Snyder's selectivity triangle describe the use of mixture designs to optimize the mobile phases [43,44]. The optimal stationary phase combination can be predicted using serially connected columns (cyano, phenyl, and octadecyl) and the principle of the "PRISMA" model. Its application for mobile phase optimization for separation of 12 pesticides was described

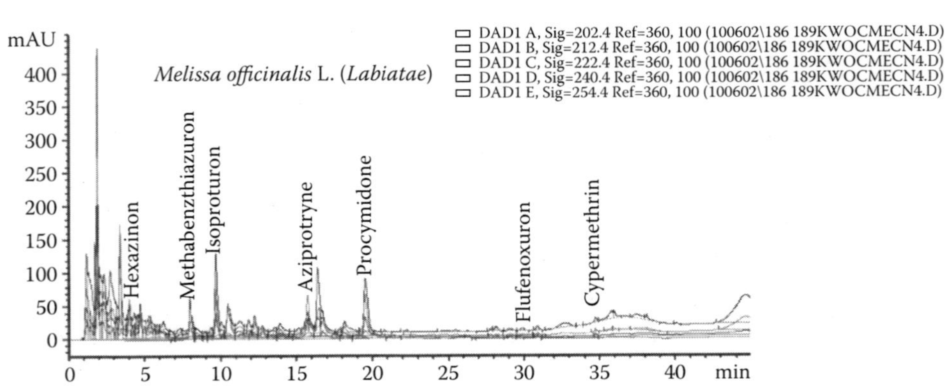

FIGURE 7.9 Chromatogram obtained by HPLC-DAD after SPE from *M. officinalis* L. (*Labiatae*). (From Tuzimski, T., *J. Sep. Sci.* 34, 27–36, 2011. With permission.)

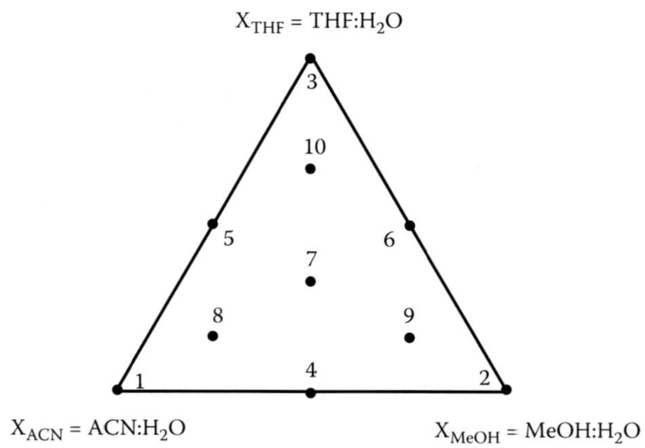

FIGURE 7.10 Simplex-centroid design with axial check indicating the mobile phase composition used for model building and validation (for the separation of components of a mixture of pesticides): X_{ACN} ACN:H_2O 30:70 (v/v), X_{MeOH} MeOH:H_2O 45:55 (v/v), X_{THF} THF:H_2O 30:70 (v/v). (From Breitkreitz, M. C. et al., *J. Chromatogr. A* 1216, 1439–1449, 2009. With permission.)

by Nyiredy et al. [43]. A statistical approach for the simultaneous optimization of the mobile and stationary phases used in RP-HPLC experiments was reported [44]. The optimization of mobile phase selectivity was carried out simultaneously for stationary and mobile phases using statistical design techniques that allowed the construction of models describing the influence of the variables as well their interactions on the compounds (pesticides) [44]. Mixture designs using aqueous mixtures of three organic modifiers—acetonitrile, methanol, and tetrahydrofuran—were reported. Organic modifiers were selected simultaneously with column type optimization (C8 and C18) to obtain the best separation of an 11-component mixture of pesticides [44] by quadratic or special cubic mixture models. The authors [44] described the models, which were evaluated by ANOVA in terms of regression significance. The models were determined assuming completely random execution of experiments and using a split-plot approach that takes into account the randomization restrictions actually employed to facilitate laboratory work. Objective functions as well as elementary criteria, such as retention time, resolution, and relative retention factors (the relative retention factor is defined as the ratio of retention time of the more retained and the less retained compounds) were tested as potential response values [44]. An example of simplex-centroid design with axial checkpoints, indicating the mobile phase compositions used for model building and validation for separation of mixture of pesticides, is illustrated in Figure 7.10 [44] and Tables 7.3 and 7.4.

Instead of using an objective response function, combined models were built for elementary chromatographic criteria (retention factors, resolution, and relative retention) of each solute or pair of solutes and, after their validation, the global separation was accomplished by means of Derringer's desirability functions. For pesticides, a 15:15:70 (v/v/v) ACN:THF:H_2O mixture with the C8 column provide excellent resolution of all peaks (Figure 7.11) [44].

7.5 NORMAL-PHASE HPLC

NP-HPLC has several practical advantages [45]:

- Because of lower viscosity, pressure drop across the column is lower than with aqueous-organic mobile phases used in RP-HPLC.
- Columns are usually more stable in organic than in aqueous-organic solvents.

TABLE 7.3
Experimentally Observed and Predicted Values for the Combined Models of Peak Pair Relative Retention Factors for the Verification Points of the Mobile Phase Mixture Design for the C8 Column

	Point 8		Point 9		Point 10	
Compound Pair	**Relative Retention Factor (Predicted)**	**Relative Retention Factor (Observed)**	**Relative Retention Factor (Predicted)**	**Relative Retention Factor (Observed)**	**Relative Retention Factor (Predicted)**	**Relative Retention Factor (Observed)**
Ametryn/cyanazine	1.16	1.11	1.52	1.47	1.26	1.32
Ametryn/simazine	1.26	1.26	1.49	1.45	1.48	1.50
Simazine/cyanazine	1.13	1.14	1.08	1.02	1.11	1.12
Simazine/carbaryl	1.89	1.87	1.51	1.53	1.40	1.39
Ametryn/carbaryl	1.42	1.48	1.11	1.06	1.18	1.00
Cyanazine/carbaryl	1.64	1.63	1.55	1.55	1.29	1.24

Source: Breitkreitz, M. C. et al., *J. Chromatogr. A* 1216, 1439–1449, 2009.

TABLE 7.4
Experimentally Observed and Predicted Values for the Combined Models of Peak Pair Relative Retention Factors for the Verification Points of the Mobile Phase Mixture Design for the C18 Column

	Point 8		Point 9		Point 10	
Compound Pair	**Relative Retention Factor (Predicted)**	**Relative Retention Factor (Observed)**	**Relative Retention Factor (Predicted)**	**Relative Retention Factor (Observed)**	**Relative Retention Factor (Predicted)**	**Relative Retention Factor (Observed)**
Ametryn/cyanazine	1.26	1.29	1.77	1.75	1.41	1.43
Ametryn/simazine	1.36	1.38	1.60	1.60	1.53	1.52
Simazine/cyanazine	1.05	1.06	1.05	1.09	1.05	1.09
Simazine/carbaryl	1.87	1.81	1.50	1.52	1.32	1.32
Ametryn/carbaryl	1.28	1.29	1.09	1.03	1.19	1.18
Cyanazine/carbaryl	1.70	1.70	1.64	1.67	1.29	1.22

Source: Breitkreitz, M. C. et al., *J. Chromatogr. A* 1216, 1439–1449, 2009.

- Columns packed with unmodified inorganic adsorbent are not subject to "bleeding," that is, to gradual loss of the stationary phase, which decreases slowly the retention during the lifetime of a chemically bonded column.
- Some samples are more soluble or less prone to decompose in organic mobile phases.

However, RP-HPLC generally offers better selectivity for the separation of molecules with different sizes of their hydrocarbon part. Separation, identification, and qualitative analysis of pesticides in NP systems by HPLC is very rarely used, especially in analysis of natural samples, due to the properties of the analytes themselves as well as the nature of the matrix. Chromatography on polar adsorbents suffers from a specific inconvenience [45]:

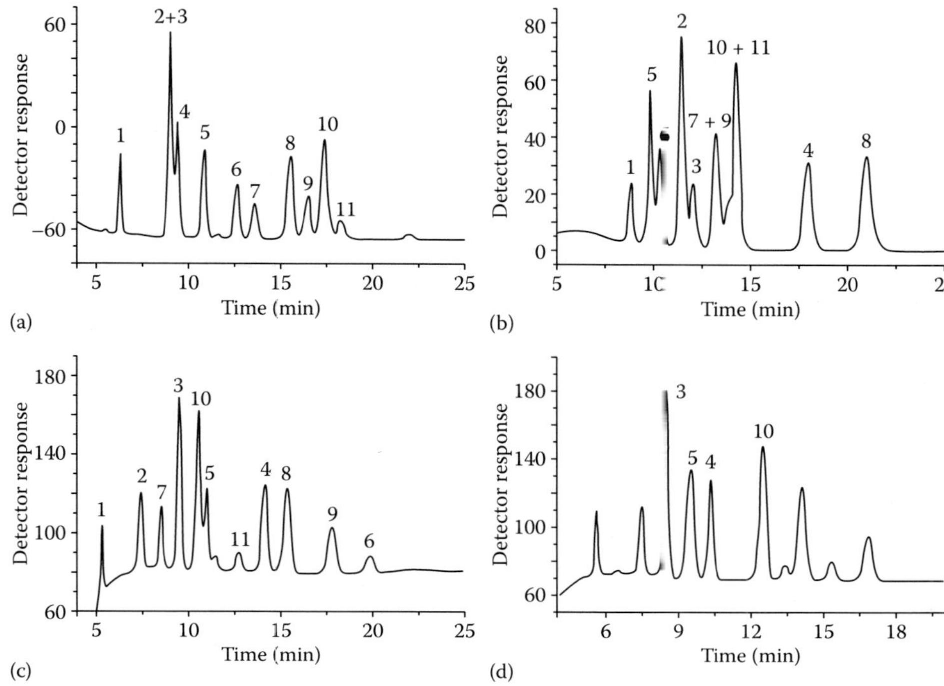

FIGURE 7.11 Chromatograms obtained in the separation of the 11 pesticides [imazetapir (1), imazaquim (2), simazine (3), ametryn (4), cyanazine (5), thiophanate (6), metsulfuron (7), atrazine (8), bentazone (9), carbaryl (10), and carboxin (11)] using the C8 column and mobile phase compositions (a) X_{ACN} ACN:H_2O 30:70 (v/v), (b) X_{MeOH} MeOH:H_2O 45:55 (v/v), (c) X_{THF} THF:H_2O 30:70 (v/v), (d) ACN:THF:H_2O 15:15:70 (v/v/v). The compounds of interest are: 3, 4, 5, and 10 (210 nm, $T = 40°C$, flow rate = 1 ml/min). (From Breitkreitz, M. C. et al., *J. Chromatogr. A* 1216, 1439–1449, 2009. With permission.)

- That is, preferential adsorption of more polar solvent, especially water, which is often connected with long equilibration times if the separation conditions are changed.
- To get reproducible results, it is necessary to keep up constant adsorbent activity, which can be accomplished using mobile phases prepared from "isohydric" organic solvents with equilibrium water concentrations.
- The reproducibility in NP-HPLC can be significantly improved by using dehydrated solvents kept dry over activated molecular sieves and filtered just before use to improve the reproducibility and by accurate temperature control during separation.

The polarity and the elution strength, that is, the ability to enhance the elution, generally increases in the following order of the most common solvents used in NP-HPLC [45]: hexane (heptane) ≈ octane < methyl chloride < methyl-*t*-butyl ether < ethyl acetate < dioxane < acetonitrile ≈ tetrahydrofuran < 1- or 2-propanol < methanol.

Large changes in selectivity in NP-HPLC separations can be achieved by selecting the solvent with the appropriate type of polar interactions. The specificity of NP-HPLC is the result of the direct formation of a strong molecular interaction between sample molecules and the adsorbent. An example of these interactions on silica stationary phase is illustrated in Figure 7.12.

The interactions of the mobile phase, containing more polar solvent (tetrahydrofuran) and more polar solute (phenol), with surface silanols of the stationary phase will be stronger with localized adsorption (Figure 7.12c and d) in contrast to the weaker and less specific interactions between the mobile phase containing less polar solvent (dichloromethane) and less polar solute (chlorobenzene)

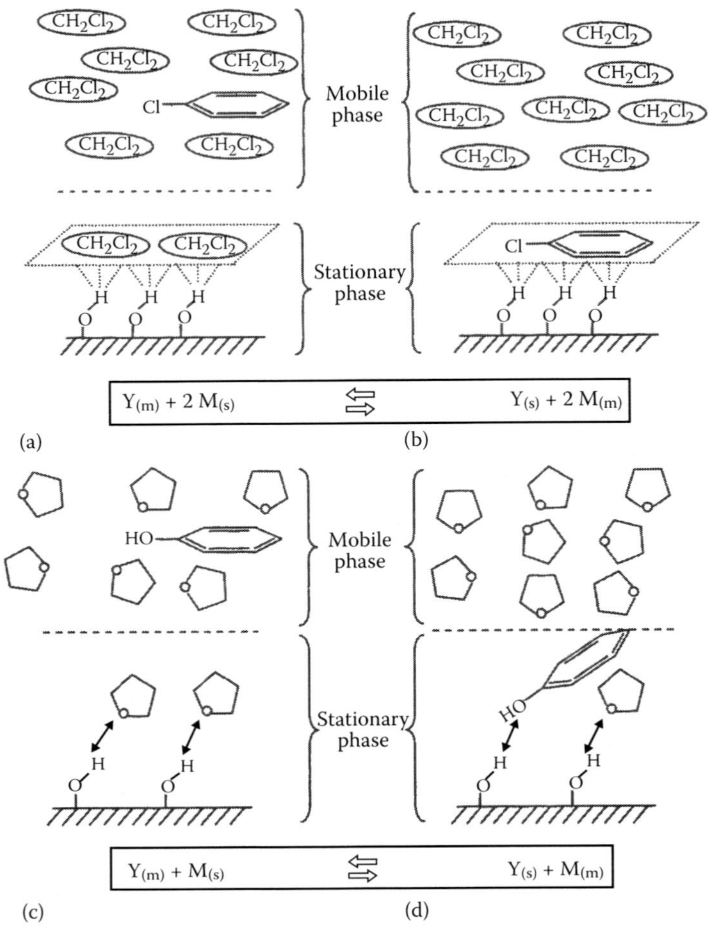

FIGURE 7.12 Hypothetical examples of solute retention on silica for chlorobenzene (a, b nonlocalized) and phenol (c, d localized). Mobile phase in (a, b) is a less-polar solvent (CH_2Cl_2); mobile phase in (c, d) is a more-polar solvent (tetrahydrofuran, THF). (Snyder, L. R., Kirkland, J. J., Dolan, J. W.: *Introduction to Modern Liquid Chromatography*. 2010. Copyright Wiley-VCH Verlag GmbH & Co. KGaA. Reproduced with permission.)

as illustrated in Figure 7.12a and b [1]. Details about retention, selectivity, and optimization of NP-HPLC systems can be found in Chapter 3.

Today, NP systems are useful mainly for analytical separations by thin-layer chromatography. Retention, selectivity, and optimization in these types of systems are described in other chapters [46,47].

7.5.1 Isocratic

NP systems in HPLC are used very rarely, especially with isocratic elution. NP adsorption chromatography is the oldest liquid chromatographic mode, using either an inorganic adsorbent (silica or, less often, alumina) or a moderately polar bonded phase (cyanopropyl, diol, or aminopropyl) chemically bonded on a silica gel support.

A perhydro-26-membered hexaazamacrocycle-based silica stationary phase for HPLC was prepared using 3-glycidoxypropyltrimethoxysilane as a coupling reagent by He et al. [48]. The chromatographic performance and retention mechanism of the new phase were evaluated in RP and NP modes, using different solute probes, including aromatic compounds, organophosphorus pesticides,

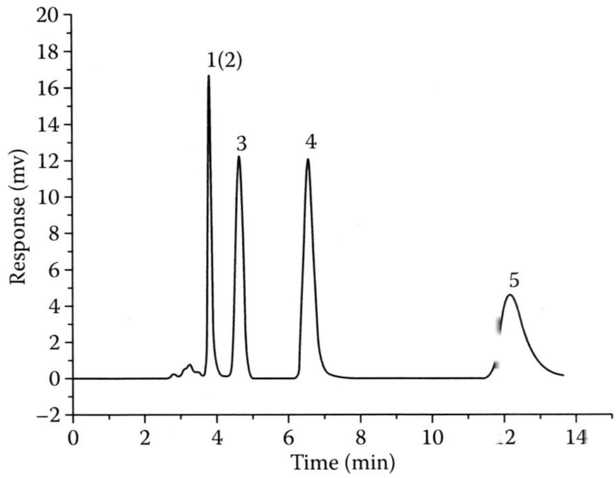

FIGURE 7.13 Chromatograms obtained in the separation of four carbamate pesticides using the mobile phase containing 60/40 (v/v) hexane/isopropyl alcohol on a perhydro-26-membered hexaazamacrocycle-based silica stationary phase (From He, L. et al., *J. Chromatogr. A.* 1217 (38), 5971–5977, 2010. With permission.)

carbamate pesticides, and phenols. Multiple interactions increased the retention of carbamate pesticides and selectivity of the new stationary phase, thus separation of carbamate pesticides on the new adsorbent under the isocratic NP condition was expected. Figure 7.13 represents the separation of four carbamate pesticides using the mobile phase containing 60/40 (v/v) hexane/isopropyl alcohol. The new phase could provide various interaction sites for different solutes, such as hydrophobic, hydrogen bonding, π–π, dipole–dipole interactions, and acid–base equilibrium. The presence of phenyl rings, secondary amino groups, and alkyl linkers in the resulting material made it suitable for the separation of the above-mentioned analytes by multimode retention mechanisms [48].

7.5.2 Retention

Retention in HPLC with NP and RP systems is different for analytes containing different substituents in their molecules. All polar adsorbents investigated showed, for nonaqueous eluents, only occasionally larger differences in selectivity. Greater differences could be expected for RP systems with aqueous eluents. A good illustration of individual effects of differences in molecular structures in NP and RP systems is presented in Table 7.5 from Snyder, Kirkland, and Glajch's monograph *Practical HPLC Method Development* [49], which lists selectivity coefficients of benzene derivatives (C_6H_5X) relative to benzene (C_6H_5H) for octadecyl silica (RP) and plain silica (NP). For instance, alkyl substituents cause marked changes in retention in RP systems and insignificant changes in NP systems [50,51].

7.5.3 Selectivity

Comparison of selectivity of NP and RP systems applied on a perhydro-26-membered hexaaza-macrocycle-based silica stationary phase for HPLC for selected pesticides is tabularized in Table 7.6 [48].

7.5.4 Optimizing

A single solvent only rarely provides suitable separation selectivity and retention in NP systems, which should be adjusted by selecting an appropriate composition of a two- or a multicomponent mobile phase. The dependence of retention on the composition of the mobile phase can be described

TABLE 7.5
Retention as a Function of Sample Molecular Structure for Substituted Benzenes: Effect on *k* of Different Substituent Groups

	Relative Value of k^a			
	RPC		NPC	
Group	30% ACN	60% ACN	Hexane	CH_2Cl_2
Phenyl	12.3	3.2	13	1.5
–Br	2.8	1.7	0.7	0.6
$–CH_3$	2.5	1.5	1.2	0.9
$–CH_2–$	2.2	1.5	1.0	0.7
–Cl	2.3	1.5	0.7	0.4
–F	1.3	1.0	0.8	0.7
$–OCH_3$	1.1	1.0	24	3.5
–H[b]	(1.0)	(1.0)	(1.0)	(1.0)
$–CO_2CH_3$	0.9	0.8	390	13
–CN	0.5	0.6	310	10
–CHO	0.4	0.6	410	13
–OH	0.2	0.3	1400	60
$–NH_2$	0.2	0.4	6700	180
$–CONH_2$	0.1	0.2	90,000	1800
$–SO_2NH_2$	0.1	0.2	–	–

Source: Snyder R. L., Kirkland, J. J., Glajch, J. L.: *Practical HPLC Method Development*. Appendix. Table III.1. 1997. Copyright Wiley-VCH Verlag GmbH & Co. KGaA. Reproduced with permission.

[a] $k_{C_6H_5X}/k_{C_6H_6} = \alpha$.
[b] Reference solute – benzene.

using theoretical models of adsorption. With some simplification, both the Snyder and Soczewiński models lead to an identical equation describing the retention (retention factor, *k*) as a function of the concentration of the stronger (more polar) solvent, ϕ_{mod}, in binary mobile phases comprised of two solvents of different polarities (Equation 7.3) [52]:

$$\log k = \log k_0 - m \log \phi_{mod} \tag{7.3}$$

where k_0 and *m* are experimental constants (k_0 being the retention factor in pure strong solvent).

Equation 7.3 also can be derived on the basis of the molecular statistical–mechanical theory of adsorption chromatography [53]. Equation 7.3 applies to systems in which the solute retention is very high in the pure nonpolar solvent (diluent). If this is not the case, another retention Equation 7.4 was derived from the original Snyder model [54,55]:

$$k = (a + b\phi)^{-m} \tag{7.4}$$

where *a*, *b*, and *m* are experimental constants.

The constants *a*, *b*, and *m* depend on the solute and on the chromatographic system (Equation 7.5):

$$a = 1/(k_a)^m \tag{7.5}$$

where k_a is the retention factor in pure nonpolar solvent.

TABLE 7.6
Comparison of k, α, and R_s of Analytes between RP- and NP-HPLC

	RP-HPLC[a]			NP-HPLC[b]		
	k	α	R_s	k	α	R_s
Methomyl	0.36	–	–	3.05	2.58	6.55
Carbofuran	0.65	1.81	2.50	0.55	2.04	2.79
Isoprocarb	0.95	1.46	2.19	0.27	–	–
Fenobucarb	1.20	1.26	1.54	0.27	1.00	0
Carbaryl	2.61	2.18	5.99	1.18	2.15	4.30
	RP-HPLC[c]			**NP-HPLC[d]**		
	k	α	R_s	k	α	R_s
Phorate	1.40	–	–	0.08	–	–
Parathion	1.80	1.29	1.74	0.36	1.50	1.62
Phoxim	2.35	1.31	1.97	0.24	3.00	2.34
Chlorpyrifos	3.52	1.50	3.29	0.08	–	–

Source: He, L. et al., *J. Chromatogr. A.* 1217 (38), 5971–5977, 2010.

[a] Methanol-water (40/60, v/v), L'GlySil column, flow rate, 1 ml min^{-1}, UV at 254 nm.
[b] Hexane/isopropyl alcohol (40/60, v/v), L'GlySil column, flow rate, 1 ml min^{-1}, UV at 254 nm.
[c] Methanol-water (45/55, v/v), L'GlySil column, flow rate, 1 ml min^{-1}, UV at 254 nm.
[d] Hexane/isopropyl alcohol (70/30, v/v), L'GlySil column, flow rate, 1 ml min^{-1}, UV at 254 nm.

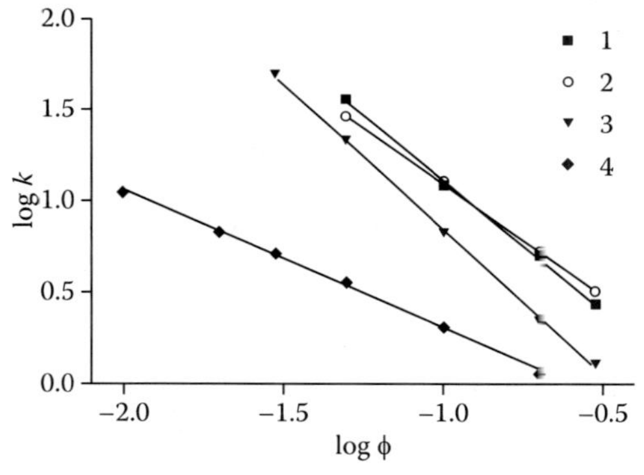

FIGURE 7.14 Dependence of retention factors, k, of phenylurea herbicides (1-metoxuron, 2-deschlorometoxuron, 3-desphenuron, and 4-linuron) on the concentration, ϕ (% vol. × 10^{-2}), of 2-propanol in n-heptane on silica gel (Separon SGX, column; 150 × 3.3 mm ID; 7 μm) at 40°C. Points: experimental data; lines: best-fit plots of two-parameter from Equation 7.3. (Reprinted from *Handbook of Analytical Separations*, Valko, K. (Ed.), *Separation Methods in Drug Synthesis and Purification, Volume 1*, Smith, R. G. (Series Ed.), Copyright 2000, with permission from Elsevier.)

The suitability of Equations 7.3 and 7.4 to describe experimental NP chromatographic systems data is illustrated in Figure 7.14 [45].

7.5.5 Applications

NP systems in HPLC are used very rarely. Examples with separation of a mixture of pesticides by isocratic and gradient elution on a silica stationary phase is illustrated in Figures 7.15 and 7.16, respectively [45,56].

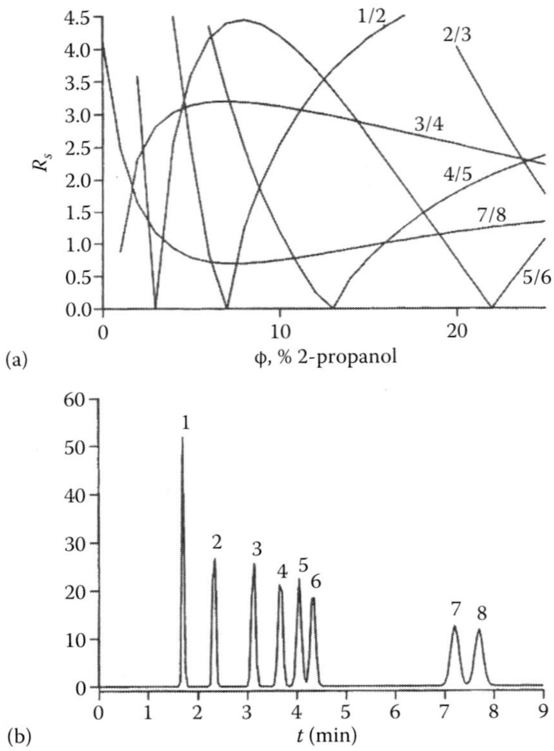

FIGURE 7.15 (a) The window diagram (the dependence of the resolution on the concentration of 2-propanol in *n*-heptane as the mobile phase) for a mixture of eight phenylurea herbicides (neburon (1), chlorbromuron [2], 3-chloro-4-methylphenylurea [3], desphenuron [4], isoproturon [5], diuron [6], metoxuron [7], and deschlorometoxuron [8]) on the concentration of 2-propanol in *n*-heptane on silica gel (Separon SGX, column (150 × 3.3 mm ID; 7.5 μm). (b) The separation with optimized concentration 19% 2-propanol in the mobile phase for maximum resolution (column plate number $N = 5000$, $T = 40°C$, flow rate = 1 ml/min). (Reprinted from *Handbook of Analytical Separations*, Valko, K. (Ed.), *Separation Methods in Drug Synthesis and Purification, Volume 1*, Smith, R.G. (Series Ed.), Copyright 2000, with permission from Elsevier.)

Figure 7.15 shows an example of a "window diagram" used for the optimization of a binary mobile phase for NP-HPLC separation of eight phenylurea herbicides.

On the basis the window diagram, we can obtain optimal values of R_s (e.g., $R_s > 1.5$). In this "optimal region of the window diagram" we can obtain the optimal operation parameter, which is selected to yield the maximum resolution of the "critical" pair of adjacent peaks most difficult to separate or the desired resolution for all adjacent peaks in the chromatogram obtained in the shortest run time [45].

An example of the window diagram for optimization of NP gradient elution chromatography on a silica stationary phase is shown in Figure 7.16a. The gradient separations with the two optimized initial concentrations of 2-propanol are shown in Figure 7.16b and c.

7.6 "PSEUDO" RP-HPLC

Strongly retained, very hydrophobic samples may require a water-free mobile phase (nonaqueous RP chromatography) called also "pseudo" RP-HPLC. Separation by nonaqueous RP chromatography is reserved for very hydrophobic samples that are retained strongly and not eluted by 100% acetonitrile as a mobile phase. Very hydrophobic samples are often insoluble in aqueous solutions and in the separation process are applied mobile phases, which consist of a more polar solvent (ACN, MeOH) and a less polar organic solvent (THF, methylene chloride, methyl-*t*-butyl ether, or other

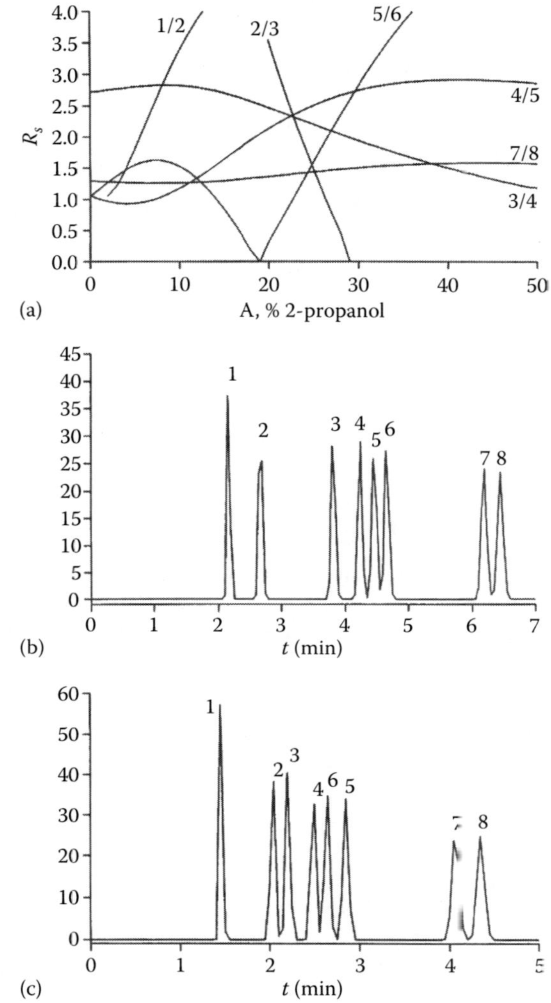

FIGURE 7.16 The resolution window diagram for the gradient elution separation of a mixture of eight phenylurea herbicides (neburon [1], chlorbromuron [2], 3-chloro-4-methylphenylurea [3], desphenuron [4], isoproturon [5], diuron [6], metoxuron [7], and deschlorometoxuron [8]) in dependence on the initial concentration of 2-propanol in *n*-heptane at the start of the gradient: (a) with optimum gradient volume V_G = 10 ml (column plate number N = 5000); (b, c) the separation of a mixture of eight phenylurea herbicides with optimized gradient elution conditions with a flow rate = 1 ml/min: maximum resolution (a); gradients from 12% to 38.6% 2-propanol in *n*-heptane in seven minutes (b); gradients from 25% to 37.5% 2-propanol in *n*-heptane in five minutes (c). (Reprinted from *Handbook of Analytical Separations*, Valko, K. (Ed.), *Separation Methods in Drug Synthesis and Purification, Volume 1*, Smith, R.G. (Series Ed.), Copyright 2000, with permission from Elsevier.)

less polar organic solvents). Pseudo RP-HPLC experiments for separation and quantitative analysis are not practically useful. (Details about HILIC systems can be found in Chapter 8.)

REFERENCES

1. Snyder, L. R., Kirkland, J. J., Dolan, J. W., *Introduction to Modern Liquid Chromatography*, John Wiley & Sons, Inc., Hoboken, New Jersey, 2010.
2. Dolan, J. W., Snyder, L. R., Djordjevic, N. M., Hill, D. W., Van Heukelem, L., Waeghe, T. J., Reversed-phase liquid chromatographic separation of complex samples by optimizing temperature and gradient time. II. Two-run assay procedures, *J. Chromatogr. A.*, 857 (1–2), 21–39, 1999.

3. Zhu, P. L., Snyder, L. R., Dolan, J. W., Djordjevic, N. M., Hill, D. W., Sander, L. C., Waeghe, T. J., Combined use of temperature and solvent strength in reversed-phase gradient elution. I. Predicting separation as a function of temperature and gradient conditions, *J. Chromatogr. A.*, 756, 21–39, 1996.
4. Snyder, L. R., Dolan, J. W., *High-Performance Gradient Elution. The Practical Application of the Linear-Solvent-Strength Model*, John Wiley & Sons, Hoboken, NJ, 2007.
5. Snyder, L. R., Classification of the solvent properties of common liquids, *J. Chromatogr.*, 92, 223–230, 1974.
6. Rohrschneider, L., Solvent characterization by gas–liquid partition coefficients of selected solutes, *Anal. Chem.*, 45, 1241–1247, 1973.
7. Johnson, B. P., Khaledi, M. G., Dorsey, J. G., Solvatochromic solvent polarity measurements and retention in reversed-phase liquid chromatography, *Anal. Chem.*, 58, 2354–2365, 1986.
8. Kosower, E. M., The effect of solvent on spectra. I. A new empirical measure of solvent polarity: Z-values, *J. Am. Chem. Soc.*, 80, 3253–3260, 1958.
9. Kosower, E. M., Mohammad, M., The solvent effect on a electron-transfer reaction of pyridinyl radicals, *J. Am. Chem. Soc.*, 90, 3271–3272, 1968.
10. Kamlet, M. J., Abboud, J. L. M., Abraham, M. H., Taft, R. W., Linear solvation energy relationships. 23. A comprehensive collection of the solvatochromic parameters, π^*, α, and β, and some methods for simplifying in generalized solvatochromic equation, *J. Org. Chem.*, 48, 2877–2887, 1983.
11. Kamlet, M. J., Abboud, J. L. M., Taft, R. W., The solvatochromic comparison method. 6. The π^* scale of solvent polarities, *J. Am. Chem. Soc.*, 99, 6027–6038, 1977.
12. Laurence, C., Nicolet, P., Dalati, M. T., Abboud, J. L. M., Notario, R., The empirical treatment of solvent–solute interactions: 15 years of π^*, *J. Phys. Chem.*, 98, 5807–5816, 1994.
13. Taft, R. W., Kamlet, M. J., The solvatochromic comparison method. 2. The α-scale of solvent hydrogen bond donor (HBD) acidities, *J. Am. Chem. Soc.*, 98, 2886–2894, 1976.
14. Abraham, M. H., Poole, C. F., Poole S. K., Classification of stationary phases and other material by gas chromatography, *J. Chromatogr. A*, 842, 79–114, 1999.
15. Glajch, J. L., Kirkland, J. J., Squire, K. M., Minor, J. M., Optimization of solvent strength and selectivity for reversed-phase liquid chromatography using an interactive mixture-design statistical technique, *J. Chromatogr.*, 199, 57–79, 1980.
16. Poole, C. F., Dias, N. C., Practitioner's guide to method development in thin-layer chromatography, *J. Chromatogr. A*, 892, 123–142, 2000.
17. Snyder, L. R., Classification of the solvent properties of common liquids, *J. Chromatogr. Sci.*, 16, 223–234, 1978.
18. Snyder, L. R., Glajch, J. L., Kirkland, J. J., Theoretical basis for systematic optimization of the mobile phase selectivity in liquid–solid chromatography. Solvent–solute localization effects, *J. Chromatogr.*, 218, 299–326, 1981.
19. Kowalska, T., Klama, B., On a deficiency of the concept of the eluotropic series, *J. Planar Chromatogr.*, 10, 353–357, 1997.
20. Abraham, M. H., Whiting, G. S., Shuely, W. J., Doherty, R. M., The solubility of gases and vapours in ethanol: The connection between gaseous solubility and water–solvent partition, *Can. J. Chem.*, 76, 703–709, 1998.
21. Abraham, M. H., Whiting, G. S., Carr, P. W., Quyang, H., Hydrogen bonding. Part 45. The solubility of gases and vapours in methanol at 298 K: An LFER analysis, *J. Chem. Soc., Perkin Trans.*, 2, 1385–1390, 1998.
22. Abraham, M. H., Platts, J. A., Hersey, A., Leo, A. J., Taft, R. W., Correlation and estimation of gas–chloroform and water–chloroform partition coefficients by a linear free energy relationship method, *J. Pharm. Sci.*, 88, 670–679, 1999.
23. Dallenbach-Tölke, K., Nyiredy, Sz., Meier, B., Sticher, O., Optimization of overpressured layer chromatography of polar naturally occurring compounds by the "PRISMA" model, *J. Chromatogr.*, 365, 63–72, 1986.
24. Nyiredy, Sz., Dallenbach-Tölke, K., Sticher, O., The "PRISMA" optimization system in planar chromatography, *J. Planar Chromatogr.—Mod. TLC*, 1, 336–342, 1988.
25. Nyiredy, Sz., Dallenbach-Tölke, K., Sticher, O., Correlation and prediction of the k′ values for mobile phase optimization in HPLC, *J. Liquid Chromatogr.*, 12, 95–116, 1989.
26. Nyiredy, Sz., Fatér, Z. S., Automatic mobile phase optimization, using the "PRISMA" model, for the TLC separation of apolar compounds, *J. Planar Chromatogr.—Mod. TLC*, 8, 341–345, 1995.
27. Nyiredy, Sz., Solid–liquid extraction strategy on the basis of solvent characterization, *Chromatographia*, 51, S-288–S-296, 2000.

28. Nyiredy, Sz., Separation strategies of plant constituents–current status, *J. Chromatogr. B*, 812, 35–51, 2004.
29. Reich, E., George, T., Method development in HPTLC, *J. Planar Chromatogr.—Mod. TLC*, 10, 273–280, 1997.
30. Snyder, L. R., Carr, P. W., Rutan, S. C., Solvatochromically based solvent-selectivity triangle, *J. Chromatogr.*, 656, 537–547, 1993.
31. Riddick, J. A., Bunger, W. B., *Organic Solvents*, Wiley-Interscience, New York, 1970.
32. Schoenmakers, P. J., Billiet, H. A. H., de Galan, L., Use of gradient elution for rapid selection of isocratic conditions in reversed-phase high performance liquid chromatography, *J. Chromatogr.*, 205, 13–30, 1981.
33. Tuzimski, T., Determination of pesticides in water samples from the Wieprz-Krzna Canal in the Łęczyńsko-Włodawskie Lake District of Southeastern Poland by thin-layer chromatography with diode array scanning and high performance column liquid chromatography with diode array detection, *J. AOAC Int.*, 91 (5), 1203–1209, 2008.
34. Tuzimski, T., Application of SPE–HPLC–DAD and SPE–TLC–DAD to the determination of pesticides in real water samples, *J. Sep. Sci.*, 31 (20), 3537–3542, 2008.
35. Tuzimski, T., Application of SPE–HPLC–DAD and SPE–HPTLC–DAD to the analysis of pesticides in lake water, *J. Planar Chromatogr.—Mod. TLC*, 22 (4), 235–240, 2009.
36. Tuzimski, T., Sobczyński, J., Application of HPLC–DAD and TLC–DAD after SPE to the quantitative analysis of pesticides in water samples, *J. Liq. Chromatogr. Relat. Technol.*, 32 (9), 1241–1258, 2009.
37. Tuzimski, T., Application of HPLC and TLC with diode array detection after SPE to the determination of pesticides in water samples from the Zemborzycki Reservoir in (Lublin, southeastern Poland), *J. AOAC Int.*, 93 (6), 1748–1756, 2010.
38. Tuzimski, T., New procedure for determination of analytes in complex mixtures by multidimensional planar chromatography in combination with diode array scanning densitometry (MDPC–DAD) and high performance liquid chromatography coupled with DAD detector (HPLC–DAD), *J. Planar Chromatogr.—Mod. TLC*, 23 (3), 184–189, 2010.
39. Tuzimski, T., Determination of clofentezine in medical herbs extracts by chromatographic methods combined with diode array scanning densitometry, *J. Sep. Sci.*, 33, 1954–1958, 2010.
40. Tuzimski, T., Determination of analytes in medical herbs extracts by SPE coupled with two-dimensional planar chromatography in combination with diode array scanning densitometry and HPLC-diode array detector, *J. Sep. Sci.*, 34, 27–36, 2011.
41. Tuzimski, T., Determination of pesticides in wines samples by HPLC-DAD and HPTLC-DAD, *J. Liq. Chromatogr. and Relat. Technol.*, 35, 1415–1428, 2012.
42. Tuzimski, T., Rejczak, T., Determination of pesticides in sunflower seeds by high-performance liquid chromatography coupled with a diode array detector, *J. AOAC Int.*, 97 (4), 1012–1020, 2014.
43. Nyiredy, Sz., Szücs, Z., Szepesy, L., Stationary-phase optimized selectivity LC (SOS-LC): Separation examples and practical aspects, *Chromatographia* 63 (13), S3–S9, 2006.
44. Breitkreitz, M. C., Jardim, I. C. S. F., Bruns, R. E., Combined column–mobile phase–mixture statistical design optimization of high-performance liquid chromatographic analysis of multicomponent systems, *J. Chromatogr. A*, 1216, 1439–1449, 2009.
45. Valko, K. (Ed.), *Handbook of Analytical Separations, Separation Methods in Drug Synthesis and Purification, Volume 1*, Smith, R.G. (Series Ed.), Elsevier, Amsterdam, The Netherlands, 2000.
46. Dzido, T. H., Tuzimski, T., Chambers, Sample Application, and Chromatogram Development. In: *Thin-Layer Chromatography in Phytochemistry*, Waksmundzka-Hajnos, M., Sherma, J., Kowalska, T. (eds.), Boca Raton 2008, CRC Press Taylor & Francis Group, Chapter 7, pp. 119–174.
47. Tuzimski, T., Use of planar chromatography in pesticide residue analysis. In: *Handbook of pesticides: Methods of pesticide residues analysis*, Nollet, L. M. L., Rathore, H. S. (eds.), Boca Raton, 2010, CRC Press Taylor & Francis Group, Chapter 9, pp. 187–264.
48. He, L., Zhang, J., Sun, Y., Liu, J., Jiang, X., Qu, L., A multiple-function stationary phase based on perhydro-26-membered hexaazamacrocycle for high-performance liquid chromatography, *J. Chromatogr. A.*, 1217 (38), 5971–5977, 2010.
49. Snyder R. L., Kirkland, J. J., Glajch, J. L., *Practical HPLC Method Development*, Wiley & Sons, New York, 1997, Appendix. Table III.1.
50. Smith, R. M., Functional group contributions to the retention of analytes in reversed-phase high-performance liquid chromatography, *J. Chromatogr. A*, 656, 381–415, 1993.
51. Snyder, L. R., *Principles of Adsorption Chromatography*, Marcel Dekker, New York, 1968, p. 264.

52. Jandera, P., Churáček, J., Gradient elution in liquid chromatography I. The influence of the composition of the mobile phase on the capacity ratio (retention volume, bandwidth, and resolution) in isocratic elution—Theoretical considerations, *J. Chromatogr.*, 91, 207–221, 1974.
53. Martire, D. E., Boehm, R. E., Molecular theory of liquid adsorption chromatography, *J. Liquid Chromatogr.*, 3, 753–774, 1980.
54. Jandera, P., Janderova, M., Churáček, J., Gradient elution in liquid chromatography: VIII. Selection of the optimal composition of the mobile phase in liquid chromatography under isocratic conditions, *J. Chromatogr.*, 148, 79–97, 1978.
55. Jandera, P., Churáček, J., Liquid chromatography with programmed composition of the mobile phase, *Adv. Chromatogr.*, 19, 125–260, 1981.
56. Glajch, J. L, Kirkland, J. J., Snyder, L. R., Practical optimization of solvent selectivity in liquid–solid chromatography using a mixture-design statistical technique, *J. Chromatogr.*, 238, 269–280, 1982.

8 Selection of the Mobile Phases for Analysis of Ionic Analytes

Reversed-Phase, Ion-Pair, Ion-Exchange, Ion-Exclusion HPLC

Monika Waksmundzka-Hajnos and Anna Petruczynik

CONTENTS

8.1 INTRODUCTION

Analysis of pesticides and their metabolites, especially in different matrices, continues to be an active research area closely related to food safety and environmental issues. Despite the fact that the use of pesticides in agricultural practice has resulted in a substantial increase in crop yield, pesticides and their metabolites can affect human health, pollute the natural environment, and disturb the equilibrium of the ecosystem. Their toxic effects are compound-specific and include several mechanisms of action. Pesticides comprise a large number of substances that belong to completely different chemical groups. The nature of the samples necessitates the use of preconcentration, cleanup, and separation techniques. Therefore, various extraction methods are suited for matrix separation and analyte preconcentration, followed by cleanup and chromatographic procedures with various detection techniques and their combinations for the actual determination.

In pesticide testing, the development of multiresidue methods, which allows proper control of a large number of pesticides in a unique analysis, is basically the main applied strategy. However, the different physicochemical properties (the pesticides belong to different groups of chemical substances having a broad range of polarity and acidic characteristics) cause difficulties in the development of methodologies that cover all the analytes under study. Sometimes, different sample treatment methodologies are necessary. The selection of sample preparation and analysis methodology is highly dependent on both analyte and sample nature.

8.2 ACIDIC AND BASIC ANALYTES

Most pesticides in their structure have one or more acidic or basic groups. In this chapter, we present an overview of the separation of selected ionic pesticides from different chemical groups by high-performance liquid chromatography (HPLC), thus indicating the range of useful methods available for ionic pesticide analysis.

The separation of ionic analytes tends to be more complicated and difficult to understand. Also, the analysis of these compounds is often associated with problems not encountered with neutral compounds.

HPLC methods for the determination of pesticides can often employ reversed-phase (RP) chromatography with C18 or C8 columns and aqueous mobile phases, followed by, for example, UV absorption, UV diode array, mass spectrometric, or fluorescence detection. The chromatographic analysis of ionizable compounds by RP liquid chromatography (LC) is difficult because these compounds exist in solution as neutral and ionic forms, which interact differently with the stationary phase (ionic, H-bond, and hydrophobic interactions). For this reason, their separation and analysis is difficult. The occurrence of ionizable compounds in both forms leads to tailing of peaks, low system efficiency, and poor reproducibility of retention data. In RP chromatography, compound retention increases for more hydrophobic forms. When ionic compounds undergo ionization, they become more hydrophilic, and their retention in RP systems is reduced. In systems at low pH, retention for an acid is increased and for a base it is decreased. When pH is varied over a sufficiently wide range, sample ionization and retention exhibit a characteristic S-shape plot for acids and a reversed S-shape plot for bases. At the midpoint of this log k versus pH curve, the pH is equal to the pK_a value of the compound.

The capacity factor of an ionizable compound is a function of pH and the volume fraction of the organic modifier in the mobile phase, and additionally, the retention of such a compound depends on processes such as ion pairing with other ions, a solvophobic effect, etc.

Especially poor results are obtained for basic compounds because their cationic forms interact with free silanol groups on the silica layer. It has been observed that retention occurs by an ion-exchange process that involves protonated solutes and ionized silanols.

Protonated basic compounds can interact with residual silanol groups of the stationary phase, as shown in Equation 8.1:

$$XH^+ + SiO^-Na^+ \leftrightarrow Na^+ + SiO^-XH^+ \tag{8.1}$$

Obtained asymmetric peaks can be explained in terms of kinetic phenomena, that is, when the kinetics of mass transfer of one type of column site is slower than the other [1]. For basic solutes, the kinetics of the ion-exchange interaction with silanol groups may be slower than those with the alkyl ligands, giving rise to peak tailing [2].

It is generally desirable to minimize these silanol interactions by the appropriate choice of experimental conditions. Silanol interactions can be reduced by the selection of a stationary phase that is designed for basic samples with a reduced number of very acidic silanols that favor the retention process. There are several methods to reduce ionization of compounds and silanol effect:

- Use of a low pH mobile phase because, at these pH values, the ionization of silanol groups is suppressed [3].
- Use of a high pH mobile phase in which ionization of basic compounds is suppressed [4].
- Addition of an ion-pairing reagent to the mobile phase causing the formation of neutral associates [5].
- Addition of a silanol blocker to the eluent, that is, pK_a silanol blocker > pK_a basic analyte [6].
- Selecting a type of stationary phase [7].

An effective alternative to RP-LC for the separation of samples containing very polar compounds is Hydrophilic Interaction Chromatography (HILIC). Good results for many authors were also obtained using ion-exchange chromatography.

8.3 RP SEPARATION OF IONIC ANALYTES

Most HPLC procedures for separating ionic pesticides use an alkyl silica-bonded stationary phase (most often C18). In a RP-LC system, neutral compounds are much more retained than ionic compounds.

The RP-LC analysis of ionic compounds can be problematic. Both the properties of the eluent and the stationary phase can influence the chromatographic performance. Therefore, selection of suitable experimental conditions for the analysis of ionic substances can be difficult.

RP-LC based on a silica support with organic solvents and buffers or acid and base additives in a mobile phase have been widely and successfully used to retain and separate hydrophobic and moderate hydrophilic compounds. Advantages of RP systems are, for example, short equilibrium times, the possibility of using water-rich eluents and samples, and the possibility of performing gradient analysis.

Acidic compounds in their ionized forms are poorly retained in RP due to electrostatic repulsion between anions and the deprotonated residual silanol groups present on the silica-based hydrophobic sorbent surface. In the current practice, this problem is overcome via two approaches: the separation of acidic compounds in low pH mobile phases and/or at low organic solvent content and, rarely, the addition of ion-pair reagents.

Basic compounds in ionized form interact with free silanol groups by an ion-exchange process. These interactions are reduced by the use of a mobile phase with different additions (buffers at acidic or basic pH, ion-pair reagents, silanol blockers) or the selection of stationary phases.

Numerous studies have been devoted to further understanding retention behavior of ionizable compounds in the presence of different additions to the mobile phase. Many ionic compounds with bulk nonpolar parts of molecules can be separated in RP systems using mobile phases containing ionic additives in pure water or in mixed aqueous-organic mobile phases with low concentrations of an organic modifier. Strong organic acids or bases are often not sufficiently retained in RP-HPLC with simple buffer or salt additives. A consistent adsorption mechanism of acido-basic compounds on silica C18 stationary phases has not yet been established, and many controversies regarding the adsorption of ionizable compounds are still lingering.

The retention parameters and separation selectivity of ionic compounds can be controlled by the change of eluent composition: modifier kind and concentration, addition of buffers at different pH, ion-pair kind and concentration, or, rarely, silanol blocker kind and concentration. Additionally, the retention and selectivity can be controlled by the change of the type of stationary phase. Recent major progress in column technology has resulted in highly efficient columns that provide excellent separations of compounds. During the last decades, many chromatographers have put lots of effort into improving the chemical stability of silica-based stationary phases. Polymer-coated, horizontally polymerized, bidentate, and polar-embedded stationary phases are typical examples. In recent years, a unique class of organic/inorganic hybrid silica materials has been developed.

Some ionic pesticides were separated in RP systems with eluents containing only an organic modifier and water or buffer at neutral pH. Magner et al. determined polar pesticides in water samples using mixture of acetonitrile (MeCN) and water on a C18 analytical column [8]. Tuzimski analyzed pesticides of different properties (neutral and acidic or basic) in eluent systems containing only an organic modifier and water in water samples and medicinal herb extracts [9,10]. A mobile phase containing phosphate buffer at pH 7 was used for separation of some pesticides in water samples [11]. A mixture of methanol (MeOH), acetonitrile (MeCN), and phosphate buffer at pH 7 was applied for analysis of chloroacetanilide herbicide metabolites in water by HPLC-diode array detection and HPLC-mass spectrometry (MS) methods [12]. Good precision and accuracy were obtained for both methods in analysis of surface water and groundwater.

8.3.1 pH, Buffer Type, and Concentration

Changing the pH values of the mobile phase strongly influences the retention of ionic compounds. Figure 8.1 illustrates the dependencies between the pH of the eluent and the retention time of selected pesticides [13].

For an ionizable analyte, that is, a compound with acid–base properties, the ratio between the concentrations of neutral and ionic forms, and therefore chromatographic retention, depends on the pH of the mobile phase and on the pK_a of the compound. The observed retention factor (k) is an average of the retention factors of the acid and basic forms of the compound (k_{HA} and k_A, respectively), that is, the retention factors that are obtained when the analyte is completely in its acidic or basic form [14]. The expression derived from this assumption is Equation 8.2:

$$k = \frac{k_{HA} + k_A 10^{pH - pK_a}}{1 + 10^{pH - pK_a}} \quad (8.2)$$

where k_{HA} is the retention factor of the acid form of an ionizable compound, k_A is the retention factor of the basic form of an ionizable compound, and K_a is the thermodynamic acidity constant.

The retention time for ionizable compounds is described by Equation 8.3:

$$t_r = \frac{t_{r(A)} + t_r 10^{pK_a - pH}}{1 + 10^{pK_a - pH}} \quad (8.3)$$

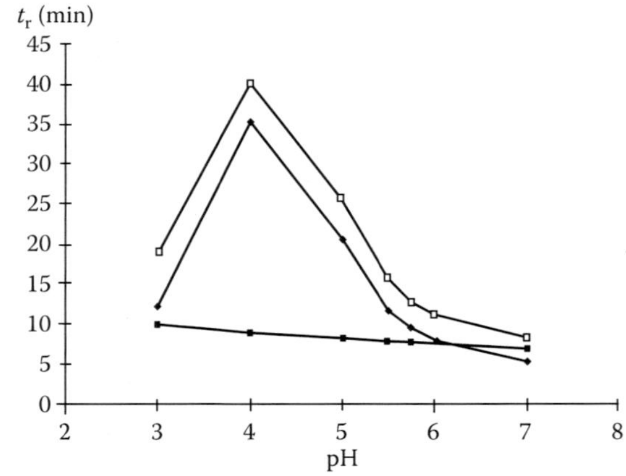

FIGURE 8.1 Dependencies of t_r versus buffer pH. Chromatographic condition: Adsorbosphere NH_2 250 × 4.6 mm ID 5 μm column from Alltech (Carnforth, UK); mobile phase: 0.05 M phosphate buffer acetonitrile (35:65; v/v). Pesticides: ■—Aminomethylphosphonic acid, □—glyphosate, ♦—glufosinate. (From Sancho, J. V. et al., *J. Chromatogr. A*, 737, 75–83, 1996. With permission.)

From Equations 8.2 and 8.3, it can be seen that the retention of a weak acid depends on three constant parameters: the dissociation constant of the acid, the retention time of the acid, and the retention time of the conjugate base.

Investigations showed that the nature and concentration of the buffer can have a significant influence on retention, peak shape, and chromatographic system efficiency for ionic compounds. Careful pH control and measurement of the mobile phase is essential for a reproducible and successful chromatographic separation and analysis of ionizable compounds. The measurement of pH in chromatographic mobile phases has been a constant subject of discussion for many years. Three different methods are common in pH measurement of eluents: measurement of pH in the aqueous buffer before addition of the organic modifier, measurement of pH in the mobile phase prepared by mixing the aqueous buffer and organic modifier after pH calibration with standard solutions prepared in the same mobile phase solvent, and measurement of pH in the mobile phase prepared by mixing the aqueous buffer and organic modifier after pH calibration with aqueous standard solutions [15]. The three measurement methods have their advantages and disadvantages.

In selecting a buffer, several considerations should be kept in mind: solubility, stability, buffer capacity, UV absorbance, suitability for LC-MS, and interaction with the sample and/or column.

Buffer ions can interact with residual silanols and with ionic analytes; it is obvious that the choice of the buffer can significantly influence results.

Higher buffer concentrations provide increased buffer capacity but are hardly soluble in the mobile phase, especially with higher concentrations of organic modifiers. The influence of the phosphate buffer concentration on retention time is presented in Figure 8.2 [13]. Usually buffer concentrations of about 25 mM are optimal.

The pH variation when adding an organic modifier to the aqueous buffer depends on the particular buffered system, on its concentration, and on the fraction of organic solvent in the mixture. Buffered solutions prepared from anionic and neutral (uncharged) acids (e.g., HAc/Ac$^-$, $H_2PO_4^-/HPO_4^{2-}$ buffers) increase their pH value when acetonitrile or methanol is added whereas buffers from cationic acids (e.g., NH_4^+/NH_3 buffers) show the reverse trend [15].

The amount and nature of the modifier also influences the pK of ionic analytes. As a general rule for both buffering species and acid–base analytes, the pK_a values of neutral (HA ↔ H$^+$ + A$^-$) and anionic (HA$^-$ ↔ H$^+$ + A^{2-}) acids increase when increasing the organic solvent content in the hydro-organic mixture whereas protonated bases (cationic acids) (BH$^+$ ↔ H$^+$ + B) slightly decrease

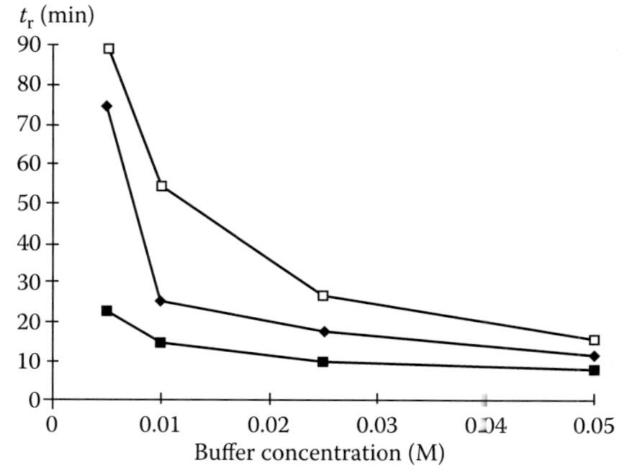

FIGURE 8.2 Dependencies of t_r versus buffer concentration. Chromatographic condition: Adsorbosphere NH_2 250 × 4.6 mm ID 5 μm column from Alltech (Carnforth, UK); mobile phase: phosphate buffer at pH 5.5 acetonitrile (35:65; v/v). Pesticides: ■—Aminomethylphosphonic acid, □—glyphosate, ♦—glufosinate. (From Sancho, J. V. et al., *J. Chromatogr. A*, 737, 75–83, 1996. With permission.)

their pK_a values from 0% to 90% of methanol or 60% of acetonitrile and then increase the pK_a up to values higher than the aqueous pK_a [16].

8.3.1.1 Application of Eluent at Acidic pH

Chromatographic systems containing eluent at acidic pH (addition of different acids or buffers at acidic pH) were often used to analyze pesticides. Such systems were applied to the analysis of both acidic and basic compounds. For acidic pesticides in the mobile phase at low pH, the ionization of compounds was suppressed. For basic pesticides, use of the addition of an acidic buffer or acids to the mobile phase suppressed ionization of the free silanol groups, which limited ion exchange between free silanols and analyte cations. The use of low pH mobile phases (pH 3.0 or less), especially with pure silica Type B phases, can successfully suppress the ionization of silanol groups and their detrimental interactions with protonated bases.

In an acidic mobile phase, more symmetrical peaks and an increase of system efficiency were obtained for both groups of substances.

Two basic pesticides—epoxiconazole and novaluron—in soil samples were analyzed on a C18 column in an eluent system containing the addition of citric acid [17]. The addition of acetic acid to a mobile phase containing MeOH and water was used for the simultaneous determination of pesticide residues in vegetables [18]. Weak bases, sulcotrione and mesotrione, and their degradation products were successfully analyzed in an eluent system containing the addition of trifluoroacetic acid [19]. The addition of the same acid was used to analyze aryloxyphenoxypropionic acid herbicides from drinking, spring, and groundwater [20].

The addition of formic acid to a mobile phase was often applied. Ethylenethiourea, the primary degradation product of ethylenebisdithiocarbamates, which include some of the most widely used fungicides in agriculture, were analyzed in human urine in an eluent system containing the addition of formic acid [21]. Addition of formic acid to a mobile phase was also used for analysis of neonicotinoid insecticides in grains [22] and some triazine pesticides in drinking and surface water [23]. Acidic organophosphorus pesticides diazinon and malathion were quantified in an eluent system containing 0.1% formic acid [24]. A mobile phase system containing the addition of formic acid was applied to the determination of diazinon and its degradation products [25]. For determination of prohexadione in cabbage and apple, an eluent containing a mixture of MeCN, water, and 0.2% formic acid was applied [26].

The acidic herbicides (quinclorac, bentazone, 2,4-D) were determined in agriculture water in chromatographic systems containing the addition of phosphoric acid [27]. The addition of phosphoric acid to the mobile phase was also used for analysis of triazine pesticide residues in water [28].

Addition of phosphoric or formic acids to eluent systems was applied to analysis of acidic herbicide (2, 4-D; dicamba; 4-CPA) residues in food crops [29].

A mixture of basic, acidic, and neutral pesticides in cucumber samples was analyzed in a mixture of MeCN, water, and 0.2% formic acid by the ultra-HPLC method [30]. In a similar system, 32 pesticides of different properties in fruits and vegetables were analyzed [31]. The addition of trifluoroacetic acid to a mobile phase was used to determine a mixture of acidic and neutral pesticides in lettuce [32]. The mobile phase containing formic acid was applied to analysis of pesticide residues of different properties in fruits and vegetables [33]. Sulfonylurea herbicides, acidic or basic substances, were successfully determined on a C18 column in eluent containing a mixture of MeCN, water, and acetic acid [34]. A mixture of 16 pesticides of different properties was separated on a C18 column with eluent containing MeCN, water, and phosphoric acid [35].

Addition of buffers is also often applied. The addition of formate buffer at pH 2.5 to a mobile phase was applied to the analysis of chlorophenoxy acid herbicides [36]. Determination of some hydrophilic herbicides was carried out in an eluent system containing MeCN and formate buffer at pH 3.7 [37]. Acidic pesticides in river water samples were also determined in eluent containing formate buffer at pH 3 [38]. In similar chromatographic systems, acidic pesticides and their transformation products were analyzed in rice samples [39].

Addition of phosphate buffer at pH 2.5 was used for analysis of different pesticides (including triazine derivatives) in soil [40]. The chromatograms obtained for soil extract spiked by a mixture of pesticides are presented in Figure 8.3. The addition of phosphate buffer at pH 3 in an aqueous-organic mobile phase was applied for analysis of different polar pesticides [41]. Good separation of sulfonylurea herbicides—weak acids—was achieved using a mobile phase containing MeOH and phosphate buffer at pH 5.91 [42]. Phosphate buffer at pH 2.3 was added to a mobile phase for determination of glyphosate and its primary metabolite, aminomethylphosphonic acid, from juices [43]. Benzimidazolic fungicides in water samples were separated using a mobile phase with the addition of phosphate buffer at pH 4 [44]. A mixture of MeCN and phosphate buffer at pH 2.95 was used for determination of acidic pesticides: 2,4-D and paraquat [45].

Kim et al. applied the addition of an acetate buffer to a mobile phase for analysis of a sulfonylurea herbicide and its metabolites [46]. The addition of an acetate buffer to a mobile phase was also used for analysis of weak bases, triazine derivative pesticides [47]. Acidic pesticides glyphosate and glufosinate were analyzed on a C18 column with a mixture of MeCN and acetate buffer at pH 4.8 [48].

Diaz et al. determined nitrophenol pesticides on a C18 column with a mobile phase containing a mixture of MeOH, MeCN, THF, and a buffer at pH 2.7 prepared from acetic acid and sodium perchlorate [49].

Mayer-Helm determined 52 pesticides of different properties using eluent containing MeOH and formate buffer [50]. Pesticides having different properties—weak acids or bases—were successfully

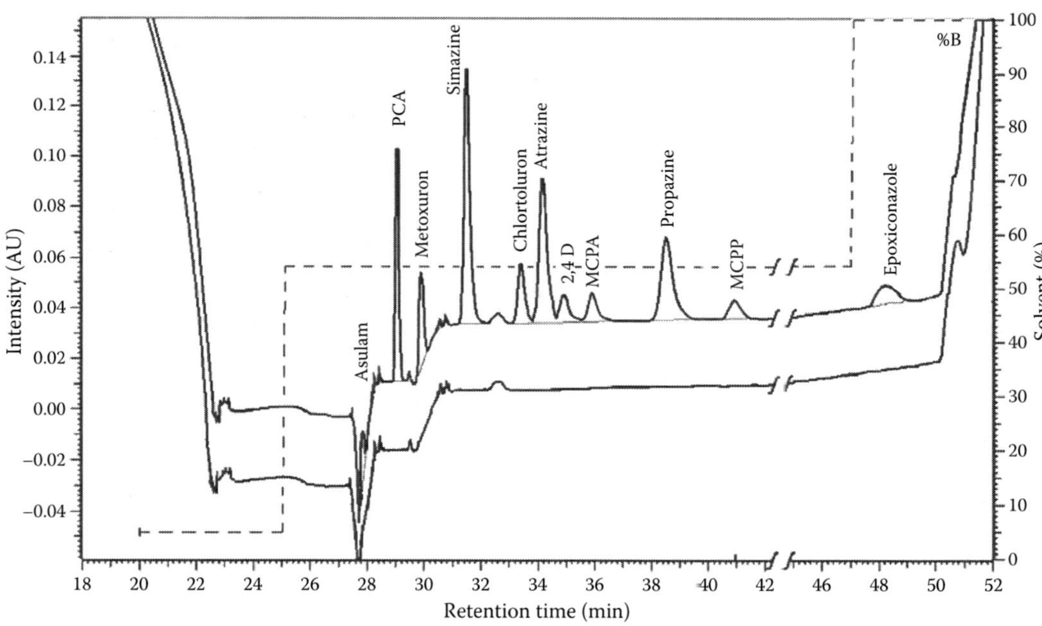

FIGURE 8.3 Chromatogram of soil extract spiked by pesticides with concentrations 5 μg/L. Chromatographic condition: Purospher RP 18e (125 × 4 mm) analytical column with a Purospher RP 18e (464 mm) precolumn; mobile phase: 0.05 M phosphate buffer acetonitrile (35:65; v/v). Mobile phase components were A (100% aqueous phosphoric acid/sodium phosphate buffer pH 2.50 [20 mM]) and B (100% methanol). Mobile phases were mixed according to the same gradient program: 0.0 to 20 min sample delivery by D line, alternatively for evaluation of instrumental blank isocratic 5% B in A, 20.1 to 25.0 min isocratic 5% B in A, at 25.1 min step gradient to 53% B in A, then hold isocratic 53% B in A to 50.0 min, at 50.1 min step gradient to 100% B in A, then hold up to 65 min isocratic 100% B in A, from 65.1 to 68.0 min back to 5% B in A, at 68.1 to 72.0 min isocratic 5% B for column re-equilibration and between runs 1 min re-equilibration. (From Hutta, M. et al., *J. Sep. Sci.* 32, 2034–2042, 2009. With permission.)

determined in infant milk formula in a chromatographic system containing a mixture of MeCN and formate buffer as the eluent [51].

8.3.1.2 Application of Eluent at Basic pH

Silanol interaction between free silanol groups on a stationary phase surface and cations of basic analytes can be reduced by using a high-pH mobile phase when free silanols are in ionized form, but ionization of basic pesticides is suppressed. The separation of basic compounds with mobile phases at high pH, in many cases, results in extended retention, excellent peak shapes, and good chromatographic efficiency.

RP separations are usually performed using silica-based stationary phases, which are stabile in the pH range 2–8. The use of mobile phases at more basic pH, allowing effective suppression of analyte dissociation, is possible only by using specially prepared, resistant to high pH, stationary phases.

Melo et al. analyzed basic pesticides in tomatoes on a C18 column with eluent containing MeCN and an aqueous solution of ammonia [52]. A similar chromatographic system was used for separation of pesticides in grapes [53]. Figure 8.4 shows a chromatogram of the spiked grape extract.

Thiophanate-methyl, carbendazim, and 2-aminobenzimidazole were determined in spiked natural water samples in an eluent system containing the addition of ammonia [54].

Some basic pesticides in soil samples were simultaneously analyzed in a system containing a C18 stationary phase and a mixture of MeCN and phosphate buffer at pH 8.7 as the eluent [55].

Nurmi and Pellinen have determined basic pesticides in wastewater by UHPLC using a C18 column and a mobile phase containing MeOH and ammonium buffer at pH 9.5 [56].

8.3.1.3 Application of Eluent with Addition of Salts

In order to ensure appropriate ionic strength and pH of eluents for analysis of ionic compounds, the addition of different salts may be used. The retention of these species depends mostly on such processes as ion pairing with other ions, solvophobic effects, and co-ion exclusion on residual silanol groups. Interaction between cations of basic solutes and free silanols is based on an ion-exchange mechanism. For this reason, the ionic strength critically impacts and influences the adsorption behavior of analyte cations on free silanol groups.

Ascenzo et al. have analyzed acidic arylphenoxypropionic herbicides in aqueous environmental samples using the addition of ammonium acetate to a mobile phase containing MeCN and water

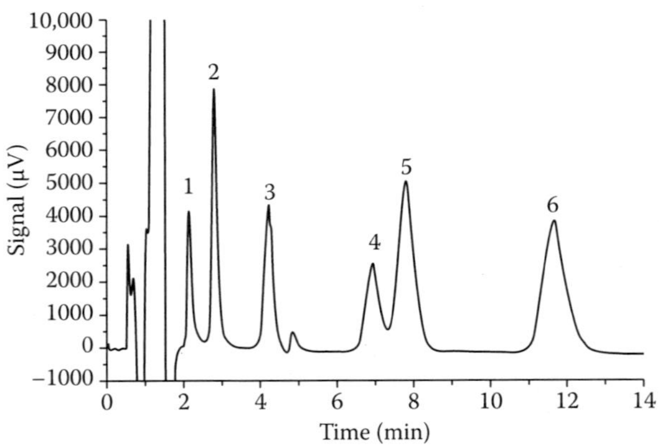

FIGURE 8.4 Chromatogram of the extract from grapes spiked by pesticides with concentrations of 1000 µg/kg. Chromatographic condition: Purospher RP 18 (125 mm × 4 mm) analytical column with precolumn (3mm × 3mm); mobile phase, CH_3CN: 0.01% aqueous NH_4OH pH 8.4 (35:65, v/v); flow rate, 0.7 ml min−1; detection, 235 nm. Pesticides: (1) benomyl, (2) tebuthiuron, (3) simazine, (4) atrazine, (5), diuron, and (6) ametryn. (From Melo, L. F. C. et al., *J. Chromatogr. A*, 1032, 51–58, 2004. With permission.)

[20]. A mixture of MeCN, water, and ammonium acetate as the mobile phase was used for determination of glyphosate and its major degradation product aminomethylphosphonic acid in atmosphere [57] and water samples [58]. A similar chromatographic system was applied to determination of some fungicides in wine samples [59]. Glyphosate and aminomethylphosphonic acid were also analyzed in rat plasma in an eluent containing the addition of ammonium formate [60].

Benoit and Preston described a method of determination of atrazine in soil [61]. The HPLC analysis was performed on a C18 column in an eluent system containing MeOH, water, and ammonium acetate.

Addition of ammonium acetate to a mobile phase was also applied for separation of neutral and weak acidic pesticides in tobacco [62]. Pesticides of different properties were analyzed in food samples in an eluent system containing MeOH, water, and ammonium formate [63]. A similar chromatographic system was used for screening analysis of 300 pesticides of different properties in water samples [64].

In a similar chromatographic system, determination of some acidic pesticides (e.g., imazalil, carbendazin) was performed in fruit samples [65] and of weak basic pesticide—amitraz and its degradation products—in fruits [66].

8.3.2 Solvents as Eluent Modifiers

Solvent type is expected to affect selectivity for ionic samples in the same way as for neutral samples. A change in solvent is a potentially useful variable for optimizing separation. The solvent strength increases as solvent polarity decreases. Relative solvent strengths are as follows: water < methanol < acetonitrile < ethanol < tetrahydrofuran < propanol. Acetonitrile is the best initial choice of organic modifier of a mobile phase in a RP system because of its UV transparency at low wavelengths and its low viscosity. However, for analysis of ionic compounds, methanol may be preferred due to the greater solubility of most buffers in methanol–water mixtures compared to mobile phases containing MeCN or THF. The choice of organic modifier concentration should ensure the optimal k values and separation selectivity for analytes.

For RP-LC analysis of ionic pesticides, methanol or acetonitrile were often applied as organic modifiers in mobile phases containing different additions [67–69]

Some authors compared determination and resolution of pesticides in eluent systems containing different organic modifiers. For analysis of pesticides of different properties (ionic and neutral) by LC-MS, acetonitrile and methanol as organic modifiers in mobile phases containing phosphate buffer, ammonium acetate, or formate were applied [70]. Optimum detectability was obtained with an aqueous 10 mM ammonium formate–methanol as the eluent.

Thirty-seven polar pesticides, mainly triazines, phenylurea herbicides, and phenoxy acids, were determined by LC–MS–MS with methanol and acetonitrile as the organic modifiers [71]. For most investigated pesticides, detection limits were the same, irrespective of the modifier. However, for the phenylurea herbicides propachlor, carbetamide, triadimefon, triadimenol, triethylcitrate, benzothiazole, and metazachlor, the results were much poorer in the presence of acetonitrile.

A mixture of acidic and basic pesticides was separated on a C8 column with a mobile phase containing MeCN, THF, and water [72].

8.3.3 Application of Eluents Containing Silanol Blockers

Toribio et al. used the addition of triethylamine to a mobile phase for enantiomeric separation of some triazole pesticides [73] and obtained an improvement of peak shape for some analytes.

Triethylamine was added to a mobile phase containing acetonitrile and water for analysis of hydramethylnon and avermectin [74].

A mobile phase containing MeOH, water, potassium bromide, and triethylamine and adjusted to pH 3 with 1 M phosphoric acid (to protect the column) was used for the determination of paraquat and diquat in olive oil samples [75]. The similar eluent was applied to analysis of paraquat in mouse tissues [76]. Addition of triethylamine to a mobile phase was applied to multiresidue determination of fluoroquinolones, organophosphorus, and *N*-methyl carbamates simultaneously in porcine tissue [77].

8.4 ION-PAIR CHROMATOGRAPHY

Ion-pair chromatography is a more general and applicable approach that allows the separation of mixtures of very polar and ionic molecules. This is accomplished by adding an ionic reagent to the mobile phase to pair with sample ions of opposite charge. Separation is based on differences in retention of the various ion pairs on the stationary phase. The mobile phase contains an organic modifier, appropriate buffer, and an ion-pairing reagent. Ion-pairing reagents consist of large ionic molecules having a charge opposite to the analyte of interest as well as a substantial hydrophobic region that allows interaction with the stationary phase, plus associated counterions. The retention and selectivity in an ion-pair RP system can be controlled by changing the type and concentration of the ion-pair reagent, mobile phase pH, and type and concentration of the organic modifier. Further variations in retention and selectivity may be obtained by a change of the stationary phase type. Mobile phase pH should be selected to obtain maximal ionization of the solute and ion-pairing reagent molecules to form an ion pair. Changing the type of ion-pair reagents often causes variations in analyte retention and separation selectivity. In a limited range of ion-pair reagent concentrations, a linear relationship between log k and the log of its concentration is obtained. After the surface is saturated by ion-pair reagent ions, a further increase in concentration does not lead to significant changes in retention.

There are several theories explaining the mechanism of adsorption in ion-pair chromatography. In the first theory, ions of the ion-pair reagent react with the ionized solute, forming neutral ion pairs in the mobile phase before its adsorption on the stationary phase. In the second theory, ions of the ion-pair reagent are adsorbed on the hydrophobic stationary phase creating an active ion-exchange surface. Bidlingmeyer et al. proposed an ion-interaction model: a model of a double electric layer formed on the sorbent surface. The retention of the analyte is caused by the charge of the double electric layer formed by the ions of the ion-pairing reagent. Another theory proposed the electrostatic model, which assumes the ion-pairing reagent is fully ionized in the applied pH range and influences, first of all, the retention of the solute's ionized form.

8.4.1 pH and Ion Pairing

Mobile phase pH should be selected to obtain maximal ionization of solute and ion-pairing molecules to possibly form an ion pair. Then pH is pK_a ± 2, the solute molecules exist in ionic and neutral forms, and the adsorption of both forms and ion pairing occurs. Important considerations in selecting a buffer to control mobile phase pH are its buffering capacity, its solubility in useful mobile phases, its ionic strength, and its suitability for use in detection.

8.4.2 Ion Pair Type and Concentration

Retention of ionic compounds can be adjusted by changing the chemical nature and concentration of the ion-pairing reagent and by altering the proportion of the organic modifier in the mobile phase.

In the case of acidic compounds, cation ion-pairing reagents have been used, such as alkylammonium compounds, organic amines, and other basic compounds. In the case of basic compounds, anionic ion-pairing reagents are used, such as sulfonic acids, alkyl sulfonates, or other acids. More hydrophobic ion-pairing reagents saturate the column at a lower mobile phase reagent concentration. A higher concentration of a less hydrophobic reagent in the mobile phase is needed to obtain comparable analyte retention. The use of an ion-pairing reagent containing an alkyl chain elongated by a one-methylene group leads to an increase in the log k value by about 0.2 unit [78]. Introduction of a methyl group to a primary, secondary, or tertiary amine increases the log k value by about 0.4 unit.

The concentration of an ion-pairing reagent in the mobile phase is typically between 0.001 M and 0.05 M. Initially, with the increasing concentration of the ion-pairing reagent, the increasing of ionic solute retention is observed. Improvement of peak shape with an increase of ion-pairing reagent concentration is also observed (Figure 8.5) [79].

In a limited range of concentrations, a linear relationship between log k and the log of the ion-pairing reagent concentration is obtained [80]. After the surface of the stationary phase is saturated by the ion-pairing reagent, a further increase in the concentration does not lead to significant changes in retention (Figure 8.6) [81].

Ion-pair RP chromatography was applied to the determination of many ionic pesticides. Glyphosate, a strong acidic pesticide, was quantified in apples on a C18 column in an eluent system containing MeCN, trifluoroacetic acid (pH 2.5), and the addition of cetyltrimethylammonium bromide as an ion-pairing reagent [82]. Retention of glyphosate derivative was increased with an increase in the cetyltrimethylammonium bromide concentration, and the separation efficiency or peak shape was improved. The appropriate addition of the ion-pairing reagent was 10 mM; higher concentrations of the reagent caused a delay of migration time with peak broadening.

Marr and King have used the addition of heptafluorobutyric acid to the mobile phase for determination of two herbicides, paraquat and diquat, in water [83]. Application of the ion-pairing reagent caused increase retention and improvement of peak shape via suppression of interactions of the ionic compounds with the active silanol sites on the HPLC column. A similar chromatographic system with the addition of heptafluorobutyric acid was used for determination of the same pesticides in olive oil [84].

Castro et al. have performed separation of quaternary ammonium pesticides on a C8 column using a mixture of MeOH, water, and heptafluorobutyric acid [85] The authors additionally studied the effect of column temperature. When the temperature increased, the retention time for some investigated pesticides decreased significantly whereas retention of some others was not affected. Heptafluorobutyric acid was also added to the mobile phase for determination of quaternary ammonium herbicides in soil samples [86].

Wang et al. described simultaneous quantitation of highly polar, water-soluble, and less-volatile herbicides, including glyphosate, glufosinate, paraquat, and diquat, in serum using ion-pair chromatography [87].

The addition of tridecafluoroheptanoic acid as an ion-pairing reagent to a mobile phase containing MeOH and water was used for determination of cyromazine and its metabolite, melanine, in chard samples [88].

Quaternary ammonium pesticides were analyzed on a C8 column in an eluent system containing MeCN, water, and heptafluorobutyric acid [89].

Qian et al. described a chromatographic method for glufosinate analysis in maize samples [90]. The authors discussed the influence of different ion-pairing reagents on the separation. The optimal separation was performed in an eluent system containing MeCN, water, phosphoric acid, and dodecyl trimethyl ammonium bromide as an ion-pairing reagent.

Basic pesticides and their transformation products in water samples were determined on a C18 column with a mixture of MeCN, water, and the addition of heptafluorobutyric acid as an ion-pairing reagent [91].

Ion-pair chromatography was used also for analysis of sulcotrione in soil [92]. The mobile phase was a mixture of MeOH, water, and tetra-n-butylammoniumchloride.

Ion-pairing reagents are most often added to the mobile phase. However, some ion-pairing reagents are not very suitable for MS detection because of their relatively low volatility and the high matrix effect generated when they are introduced continuously by the mobile phase into the atmospheric pressure ionization source. This effect can be minimized by adding the ion-pairing reagent

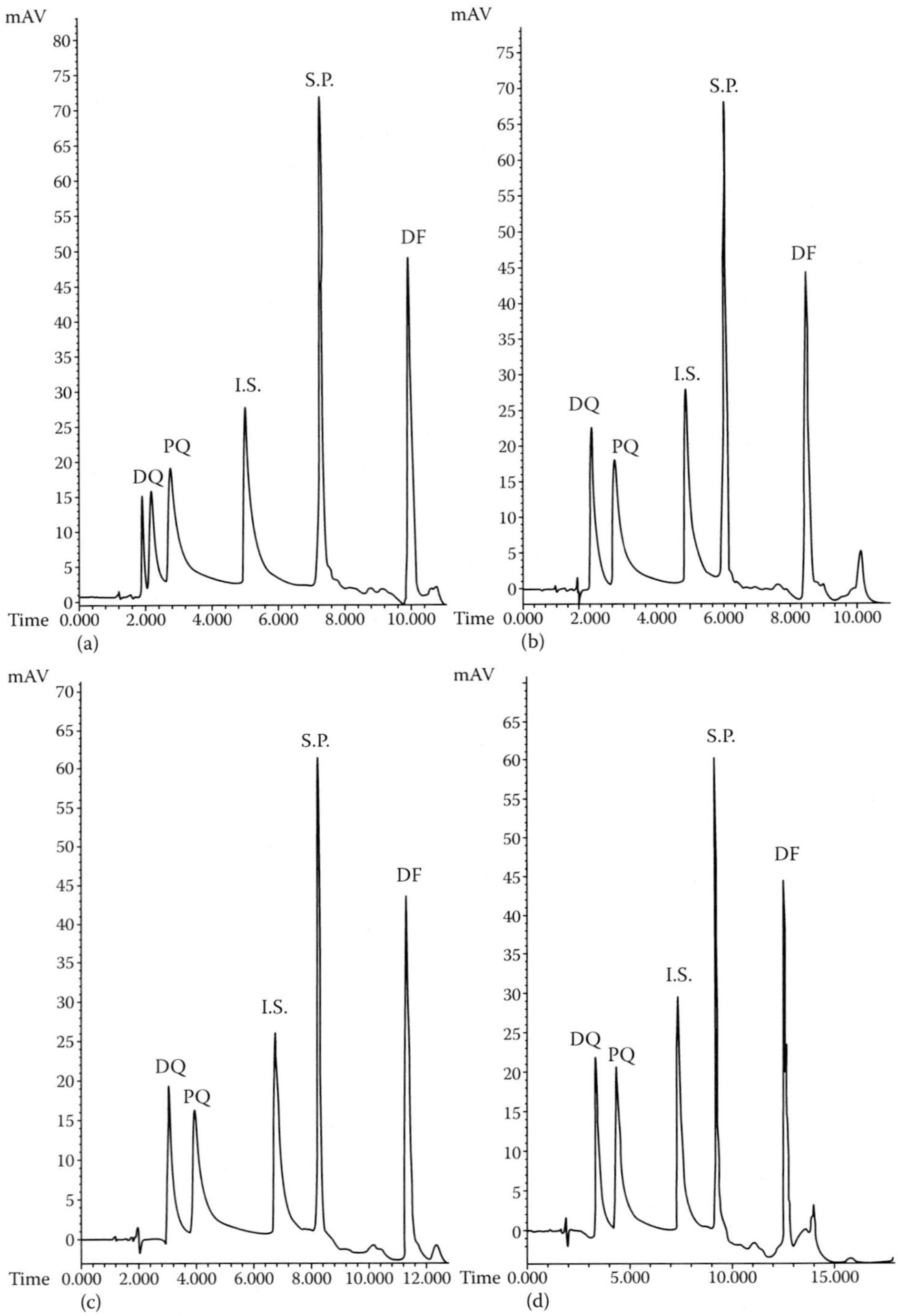

FIGURE 8.5 Chromatogram of standard solution of pesticides (4 mg/L). Chromatographic condition: Kromasil C8 (200 × 21 mm, 5 μm column; Tracer Analitica, Spain); mobile phase: pentafluoropropionic acid at different concentrations, formate buffer at pH 3.3. Acetonitrile linear gradient from 2% to 8.6% in 5 min and an increase to 40% at 5.01 min. (a) 10 mM, (b) 15 mM, (c) 20 mM, (d) 25 mM. I.S.: Internal standard, S.P.: system peak. Pesticides: diquat (DQ), paraquat (PQ), difenzoquat (DF). (From Castro, R. et al., *J. Chromatogr. A*, 830, 145–154, 1999. With permission.)

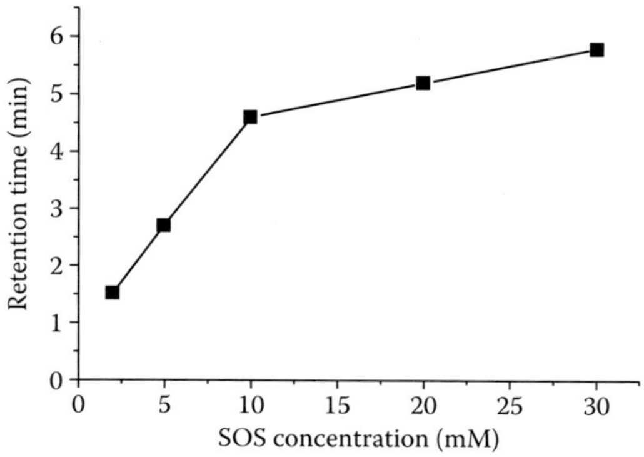

FIGURE 8.6 Dependencies of t_r versus sodium octane sulfonate concentration. Chromatographic condition: ODS C-18 column (25 cm × 4.6 mm ID 5 μm VARIAN (Palo Alto, CA); mobile phase: 40% methanol and 10 mM SOS in 0.05 M orthophosphoric acid, pH 2.8 adjusted with TEA. Pesticide: paraquat. (From Brunetto, M. R. et al., *Talanta* 59, 913–921, 2003. With permission.)

only in the sample extract, avoiding its presence in the mobile phase. The method was successfully applied by Marin et al. for determination of ethephon residues in vegetables [93]. The authors have used the addition of tetrabutylammonium acetate as an ion-pairing reagent.

8.5 SELECTING STATIONARY PHASES

Optimization of the stationary phase for analysis of basic pesticides is mainly achieved by minimizing the interaction between the analyte and residual silanols. Optimization of the stationary phase for analysis of basic compounds is also mainly achieved by minimizing the interaction between the analyte and residual silanols. Silica supports are still superior to other supports in terms of efficiency, performance, and rigidity. However, protonated basic compounds can interact with residual silanol groups on the stationary phases by an ion-exchange mechanism. Thus, in addition to the typical RP retention mechanism, the ion-exchange mechanism also occurs, which often results in asymmetry of peaks, worse separation selectivity, low performance of chromatographic systems, and irreproducible retention.

Besides the commonly used n-alkyl-type RP columns based on the immobilization of n-alkyl-type ligands onto a silica support, alternative hybrid RP-type phases provide additional interaction sites and properties due to embedded functional groups. Alternatively, the introduction of hydrophobic π–π active aromatic moieties to the common n-alkyl chain RP-sites generates a concerted π–π RP retention mechanism, which, as a consequence of the new functionality, diversifies the common RP interaction properties.

The separation of herbicides metamitron, metribuzin, isometiozin, and nitralin in soil samples was performed on a CN column in an eluent system containing MeCN and phosphate buffer at pH 2.5 [94].

Acidic compounds glufosinat, glyphosate, and its metabolite aminomethylphosphonic acid were determined in a system containing an NH_2 stationary phase and a mixture of MeCN and phosphate buffer at pH 5.5 [95].

8.6 ION-EXCHANGE CHROMATOGRAPHY

In some cases, in a RP-LC system, simultaneous and selective separation of a complex mixture of ionizable or ionic compounds is difficult. Therefore, an alternative approach to modulate the retention and separation of very polar, ionizable compounds, an ion exchange chromatography (IEC), is required. In IEC, the separation mode is based on the exchange of ionic analytes with the counterions of the ionic groups attached to the solid support. In IEC, ion-exchange resins were used as the stationary phases. The resin can be functionalized with anions, such as strong acidic sulfonic acid and weak acidic carboxylic acid, to make a cation-exchange column or be functionalized with cations, such as a strong basic tertiary amine group or a weak basic primary amine group, to make an anion-exchange column. The basis of the ion-exchange process is the reversible binding of charged molecules to an oppositely charged insoluble matrix. The ion-exchange process is dependent upon the nature of the functional group anchored to the matrix of the stationary phase. The pK_a of these groups is used to define the strength of the exchanger and is dependent upon the ionized state of the group and its ability to effect a separation. A different retention mechanism in IEC often leads to a quite different separation selectivity of ionic compounds. IEC demands the application of anion-exchange columns for the separation of acidic compounds—stationary phases with attached quaternary ammonium groups—and cation exchange columns for the separation of basic compounds—stationary phases with attached sulfonic or carboxylic groups. The mobile phase in IEC is usually composed of an aqueous solution of acids, bases, or salts and an organic modifier.

The retention of ionic compounds can be controlled by changing the pH, kind, and concentration of the counterion and sometimes by the kind and concentration of the organic modifier.

Separated ions are retained by electrostatic forces on the surface of the stationary phase, where they· compete with ions of the mobile phase bearing a charge of the same sign for the active (ion-exchange) sites.

8.6.1 pH and Buffer Type

The choice of the type of the stationary phase and manipulation of the buffer type and pH can be made to selectively bind ionized molecules. Elution of bound solutes from the stationary phase may be achieved by either adjustment of ionic strength or changes in pH. The eluent pH can have effects on the form of the solute and the eluent component ions. In IEC, only the ionized form of a compound is retained on the stationary phase. The increase of the mobile phase pH leads to stronger ionization and retention of the acidic compounds, and the decrease in pH causes stronger ionization and retention of basic compounds.

The prime component of the buffer added to the eluent is the counterion, which has the role of eluting sample components from the column in a reasonable time. In cation-exchange chromatography, the presence of positively charged ions from a buffer or slats adding to the mobile phase decreases analyte retention. Solute ions on the cation-exchange stationary phase are retained in the following order:

$$Ba^{2+} < Ca^{2+} < Cu^{2+} < Zn^{2+} < Mg^{2+} < K^{+} < NH_4^{+} < Na^{+} < H^{+} < Li^{+}$$

For anions in anion-exchange chromatography, the retention order is as follows:

$$\text{Citrate} < SO_4^{2-} < NO_3^{-} < Br^{-} < SCN^{-} < Cl^{-} < CH_3COO^{-} < OH^{-} < F^{-}$$

8.6.2 Organic Solvents

The eluent used in IEC generally consists of an aqueous solution of a suitable salt or mixture of salts. Sometimes addition of an organic modifier to an aqueous mobile phase is used. In this case, the eluent contains an ionic component for control of the ion-exchange interactions and an organic solvent to control the RP interactions. Application of a mobile phase containing an organic solvent usually results in great differences in retention and separation selectivity of analyzed compounds.

Addition of organic solvents to the eluent is often used to minimize hydrophobic adsorption on the stationary phase and improve solubility of analytes.

Careri et al. have determined chlormequat residues in tomato products on a cation exchange column with a mixture of MeCN and an aqueous solution of ammonium acetate [96].

Cyromazine and its metabolic melaminein were determined in egg and milk samples on a strong cation exchange (SCX) column with eluent containing MeOH and KH_2PO_4 at pH 3 [97].

Separation of glyphosate and its main metabolite aminomethylphosphonic acid in water was carried out on a SCX column, the mobile phase containing an aqueous solution of phosphate buffer at pH 2 [98]. The same pesticide and its metabolites were also determined in water samples on a PCX column in an aqueous solution of KH_2PO_4 as eluent [99].

On a SCX column, separation of chlormequat and mepiquat in tomato, pear, and wheat flour samples was performed [100]. The mobile phase contained a mixture of MeOH, water, and ammonium formate. Chlormequat was also determined in an ion-exchange system containing SCX stationary phase and a mixture of MeOH, water, and the addition of ammonium ethanoate as the eluent [101].

Glyphosate, glufosinate, fosamine, and ethephon—organophosphorus herbicides—at nanogram levels in environmental water samples were determined on an anion-exchange column with an aqueous solution of citric acid as the eluent (Figure 8.7) [102]. The authors optimized various parameters affecting the separation and detection, for example, concentration of the buffer and sample injection volume. The determination of glyphosate and its degradation product, aminomethylphosphonic acid, was performed on a column packed with quaternary ammonium anion-exchange resin [103]. The mobile phase was an aqueous solution of KOH. The chromatographic behavior of glyphosate and aminomethylphosphonic acid was studied separately for the two compounds because they had very different retention times. Each compound was determined with different mobile phase concentrations, and the results were evaluated using the mean of peak retention time, peak area, peak width, asymmetry factor, retention factor *k*, and number of theoretical plates. Sequentially, a multianion standard solution was analyzed with each mobile phase concentration in order to study the possible chromatographic interferences from naturally occurring anions. The multianion solution contained F^- at 0.5 mg/L, Cl^- at 10 mg/L, and NO^{2-}, NO^{3-}, SO_4^{2-}, and HPO_4^{2-} at 1 mg/L, each. After the isocratic elution optimization for the two compounds, a two-step gradient elution program was optimized for the simultaneous determination of aminomethylphosphonic acid and glyphosate.

The determination of glyphosate was also performed on a polymer anion-exchange column with eluent containing citric acid in water [104].

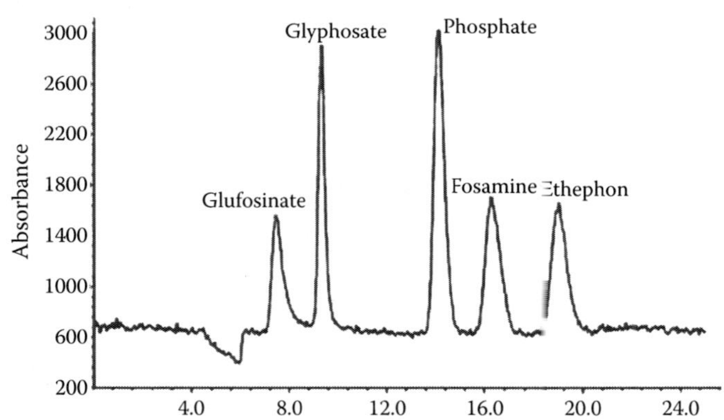

FIGURE 8.7 Chromatogram of standard solution of pesticides (50 μg/L). Chromatographic condition: Dionex IonPac AS16 (4.0 mm × 250 mm) with guard column AG 16; mobile phase: 30 mM citric acid. (From Guo, Z.-X. et al., *Rapid Commun. Mass Spectrom.* 21, 1606–1612, 2007. With permission.)

An automated method based on the online coupling of anion-exchange solid-phase extraction and cation-exchange liquid chromatography followed by postcolumn derivatization and fluorescence detection has been developed for the determination of glyphosate and its primary degradation product, aminomethyl phosphonic acid, in water [105]. The eluent was a 5 mM KH_2 PO_4 solution in water, and the pH was adjusted to 1.9 with orthophosphoric acid.

Ionic pesticides in a complex organophosphate matrix were analyzed on an anion-exchange column with aqueous sodium hydroxide as the eluent [106]. The described chromatographic method was applied for the separation and structure elucidation of anionic compounds in a complex matrix.

The determination of chlormequat residues in food crops was performed on a cation exchange column in an eleunt system containing MeCN, water, and H_2SO_4 [107].

Hau et al. analyzed chlormequat in food samples on a SCX column using a mixture of MeOH, water, and ammonium acetate as the eluent [108].

8.7 HILIC

Highly hydrophilic compounds are usually retained weakly in RP systems, which makes their separation difficult. Moreover, these compounds are often retained too strongly in adsorption, normal phase (NP) chromatography. An alternative for these methods may be the application of HILIC.

HILIC is a LC technique that uses polar stationary phases—silica or a polar-bonded phase—in conjunction with a mobile phase containing an appreciable quantity of water combined with a higher proportion of a less polar solvent (often acetonitrile). The stationary phases used in HILIC are, by definition, polar and include silica gel, sorbents with various polar groups chemically bonded to silica (diol, amide, aminopropyl, zwitterionic phases), and polymers bearing polar functional groups.

HILIC is a technique in which the analytes interact with a hydrophilic stationary phase and are eluted with a relatively hydrophobic eluent in which water is the stronger eluting member.

The retention mechanism in HILIC is complex—consisting of partitioning between a layer of water held on the surface and the bulk mobile phase, specific adsorption on polar functional groups, ionic retention on ionized groups or on ionized silanols of the silica matrix, and sometimes even RP retention on the hydrophobic portions of bonded ligands—and is still debated.

Pesticides are rarely analyzed by HILIC. Lindh et al. analyzed chlormequat in human urine on a HILIC column using a mixture of MeCN, water, and acetate buffer at pH 3.75 [109].

Organophosphorus insecticides in human urine were separated on a HILIC column with isocratic elution by using 93% acetonitrile and 7% 100 mM ammonium acetate in water [110].

Esparza et al. determined two quaternary ammonium growth regulators (chlormequat and mepiquat) in food samples using a HILIC column and a mixture of MeCN and ammonium formate buffer as the eluent [111]. The authors optimized the ionic strength of the mobile phase, and different aqueous buffer solutions were prepared at concentrations ranging from 10 to 100 mM. They concluded that an increased buffer concentration produced more symmetric and narrower chromatographic peaks due to the higher ionic strength.

Polar organophosphorus pesticides acephate, methamidophos, monocrotophos, omethoate, oxydemeton-methyl, and vamidothion in water samples were analyzed on a HILIC silica column with a mixture of MeCN, isopropanol, and ammonium formate buffer as the mobile phase [112].

The determination of dithiocarbamate fungicide residues in fruits and vegetables was developed on a ZIC-pHILIC column with eluent containing MeCN and aqueous ammonia [113].

8.8 ION-EXCLUSION CHROMATOGRAPHY

The ion-exclusion chromatography mechanism is based on the separation of partially ionized species on strong anion-exchange or SCX stationary phases with Donnan exclusion of the analytes from the charged stationary phase [114].

The retention of an analyte is influenced by a large number of parameters. These include the degree of ionization of the analyte, the molecular size and structure of the analyte, the eluent concentration and its pH value, the presence of organic solvents in the eluent, the ionic strength of the eluent, the temperature of the column, the material comprising the ion-exchanger used and its hydrophobicity, the type of ion-exchange functional group on the stationary phase, the degree of cross-linking of the polymer used in the stationary phase, the ion-exchange capacity, and the ionic form of the resin [115].

The analytical column used in ion-exclusion chromatography separations of anionic analytes is usually packed with fully sulfonated, polystyrene-divinylbenzene copolymer. In the case in which cationic analytes are to be separated, the resin is usually fully functionalized with quaternary ammonium groups. Commonly used mobile phases contain water, sodium hydroxide aqueous solution, a mixture of glycerol and water, and eluents containing methanol, xylitol, ethylene glycol, glucose, fructose, sucrose, sorbitol, n-butanol, sulfuric acid, benzoic acid, acetonitrile, and sugar alcohols.

Ion-exclusion chromatography was applied to the determination of sulfur dioxide used as a fungicide in grapes [116]. Separation was performed on an anion exclusion column with eluent containing an aqueous solution of sulfuric acid.

REFERENCES

1. Fornstedt, T., Zhong, G., Guichon, G., Peak tailing and mass transfer kinetics in linear chromatography, *J. Chromatogr. A*, 741, 1–12, 1996.
2. Vervoort, R. J. M., Debets, A. J. J., Claessen, H. A., Cramers, C. A., Jong, J., Optimisation and characterisation of silica-based reversed-phase liquid chromatographic systems for the analysis of basic pharmaceuticals, *J. Chromatogr. A*, 897, 1–22, 2000.
3. McCalley, D. V., Comparative evaluation of bonded-silica reversed-phase columns for high-performance liquid chromatography using strongly basic compounds and alternative organic modifiers buffered at acid pH, *J. Chromatogr. A*, 769, 169–178, 1997.
4. Sellergren, B., Zander, A., Renner, T., Swietlow, A., Rapid method for analysis of nicotine and nicotine-related substances in chewing gum formulations, *J. Chromatogr. A*, 829, 143–152, 1998.
5. LoBrutto, R., Jones, A., Kazakievich, Y. V., Effect of counter-anion concentration on retention in high-performance liquid chromatography of protonated basic analytes, *J. Chromatogr. A*, 913, 189–196, 2001.
6. McCalley, D. V., Comparison of the performance of conventional C18 phases with others of alternative functionality for the analysis of basic compounds by reversed phase high-performance liquid chromatography, *J. Chromatogr. A*, 844, 23–38, 1999.
7. Kirkland, J. J., Development of some stationary phases for reversed-phase HPLC, *J. Chromatogr. A*, 1060, 9–21, 2004.
8. Magner, J. A., Alsberg, T. E., Broman, D., Evaluation of poly(ethylene-co-vinyl acetate-co-carbon monoxide) and polydimethoxylsiloxane for equilibrium sampling of polar organic contaminants in water, *Environ. Toxicol. Chem.*, 28, 1874–1880, 2009.
9. Tuzimski, T., Application of SPE-HPLC-DAD and SPE-TLC-DAD to the determination of pesticides in real water samples, *J. Sep. Sci.*, 31, 3537–3542, 2008.
10. Tuzimski, T., Determination of analytes in medical herbs extracts by SPE coupled with two-dimensional planar chromatography in combination with diode array scanning densitometry and HPLC-diode array detector, *J. Sep. Sci.*, 34, 27–36, 2011.
11. Guenu, S., Hennion, M.-C., Evaluation of new polymeric sorbents with high specific surface areas using an on-line solid-phase extraction-liquid chromatographic system for the trace-level determination of polar pesticides, *J. Chromatogr. A*, 737, 15–24, 1996.
12. Hostetler, K. A., Thurman, E. M., Determination of chloroacetanilide herbicide metabolites in water using high-performance liquid chromatography-diode array detection and high-performance liquid chromatography mass spectrometry, *Sci. Total Environ.*, 248, 147–155, 2000.
13. Sancho, J. V., Hernandez, S. F., Lopez, F. J., Hogendoorn, E. A., Dijkman, E., van Zoonen, P., Rapid determination of glufosinate, glyphosate and aminomethylphosphonic acid in environmental water samples using precolumn fluorogenic labeling and coupled-column liquid chromatography, *J. Chromatogr. A*, 737, 75–83, 1996.

14. Subirats, X., Bosch, E., Roses, M., Retention of ionisable compounds on high-performance liquid chromatography XVI. Estimation of retention with acetonitrile/water mobile phases from aqueous buffer pH and analyte pK_a, *J. Chromatogr. A*, 1121, 170–177, 2006.
15. Rosés, M., Determination of the pH of binary mobile phases for reversed-phase liquid chromatography, *J. Chromatogr. A*, 1037, 283–298, 2004.
16. Rosés, M., Bosch, E., Influence of mobile phase acid–base equilibria on the chromatographic behaviour of protolytic compounds, *J. Chromatogr. A*, 982, 1–30, 2002.
17. Rybar, I., Góra, R., Hutta, M., Method of fast trace microanalysis of the chiral pesticides epoxiconazole and novaluron in soil samples using off-line flow-through extraction and on-column direct large volume injection in reversed phase high performance liquid chromatography, *J. Sep. Sci.*, 30, 3164–3173, 2007.
18. Romero-Gonzalez, R., Pastor-Montoro, E., Martinez-Vidal, J. L., Garrido-Frenich, A., Application of hollow fiber supported liquid membrane extraction to the simultaneous determination of pesticide residues in vegetables by liquid chromatography/mass spectrometry, *Rapid Commun. Mass Spectrom.*, 20, 2701–2708, 2006.
19. Chaabane, H., Vulliet, E., Calvayrac, C., Coste, C.-M., Cooper, J.-F., Behaviour of sulcotrione and mesotrione in two soils, *Pest. Manag. Sci.*, 64, 86–93, 2008.
20. D'Ascenzo, G., Gentili, A., Marchese, S., Perret, D., Determination of arylphenoxypropionic herbicides in water by liquid chromatography–electrospray mass spectrometry, *J. Chromatogr. A*, 813, 285–297, 1998.
21. Jones, K., Patel, K., Cocker, J., Bevan, R., Levy, L., Determination of ethylenethiourea in urine by liquid chromatography–atmospheric pressure chemical ionisation–mass spectrometry for monitoring background levels in the general population, *J. Chromatogr. B*, 878, 2563–2566, 2010.
22. Wang, P., Yang, X., Wang, J., Cui, J., Dong, A. J., Zhao, H. T., Zhang, L. W., Wang, Z. Y., Xu, R. B., Li, W. J., Zhang, Y. C., Zhang, H., Jing, J., Multi-residue method for determination of seven neonicotinoid insecticides in grains using dispersive solid-phase extraction and dispersive liquid–liquid microextraction by high performance liquid chromatography, *Food Chem.*, 134, 1691–1698, 2012.
23. Garcia-Ac, A., Segura, P. A., Viglino, L., Fürtös, A., Gagnon, C., Prévost, M., Sauvé, S., On-line solid-phase extraction of large-volume injections coupled to liquid chromatography-tandem mass spectrometry for the quantitation and confirmation of 14 selected trace organic contaminants in drinking and surface water, *J. Chromatogr. A*, 1216, 8518–8527, 2009.
24. Lazarević-Pašti, T., Colović, M., Savić, J., Momić, T., Vasic, V., Oxidation of diazinon and malathion by myeloperoxidase, *Pest. Biochem. Physiol.*, 100, 140–144, 2011.
25. Ibáñez, M., Sancho, J. V., Pozo, Ó. J., Hernández, F., Use of liquid chromatography quadrupole time-of-flight mass spectrometry in the elucidation of transformation products and metabolites of pesticides. Diazinon as a case study, *Anal. Bioanal. Chem.*, 384, 448–457, 2006.
26. Choi, J.-H., Yoon, H.-J., Do, J.-A., Park, J. H., Kim, Choi, D., An analytical method for prohexadione in Chinese cabbage and apple, *Biomed. Chromatogr.*, 25, 493–497, 2011.
27. Zanella, R., Primel, E. G., Goncalves, F. F., Kurz, M. H. S., Mistura, C. M., Development and validation of a high-performance liquid chromatographic procedure for the determination of herbicide residues in surface and agriculture waters, *J. Sep. Sci.*, 26, 935–938, 2003.
28. Pinto, G. M. F., Jardim, I. C. S. F., Use of solid-phase extraction and high-performance liquid chromatography for the determination of triazine residues in water: Validation of the method, *J. Chromatogr. A*, 869, 463–469, 2000.
29. Shin, E.-H., Choi, J.-H., El-Aty, A. M. A., Khay, S., Kim, S.-J., Im, M. H., Kwon, C.-H., Shim, J.-H., Simultaneous determination of three acidic herbicide residues in food crops using HPLC and confirmation via LC-MS/MS, *Biomed. Chromatogr.*, 25, 124–135, 2011.
30. Wang, J., Du, Z., Yu, W., Qu, S., Detection of seven pesticides in cucumbers using hollow fibre-based liquid-phase microextraction and ultra-high pressure liquid chromatography coupled to tandem mass spectrometry, *J. Chromatogr. A*, 1247, 10–17, 2012.
31. Lehotay, S. J., Son, K. A., Kwon, H., Koesukwiwat, U., Fu, W., Mastovska, K., Hoh, E., Leepipatpiboon, N., Comparison of QuEChERS sample preparation methods for the analysis of pesticide residues in fruits and vegetables, *J. Chromatogr. A*, 1217, 2548–2560, 2010.
32. Melo, A., Aguiar, A., Mansilha, C., Pinho, O., Ferreira, I. M. P. L. V. O., Optimisation of a solid-phase microextraction/HPLC/diode array method for multiple pesticide screening in lettuce, *Food Chem.*, 130, 1090–1097, 2012.
33. Malato, O., Lozano, A., Mezcua, M., Agüera, A., Fernandez-Alba, A. R., Benefits and pitfalls of the application of screening methods for the analysis of pesticide residues in fruits and vegetables, *J. Chromatogr. A*, 1218, 7615–7626, 2011.

34. Perreau, F., Bados, P., Kerhoas, L., Nélieu, S., Einhorn, J., Trace analysis of sulfonylurea herbicides and their metabolites in water using a combination of off-line or on-line solid-phase extraction and liquid chromatography–tandem mass spectrometry, *Anal. Bioanal. Chem.*, 388, 1265–1273, 2007.
35. D'Archivio, A. A., Fanelli, M., Mazzeo, P., Ruggieri, F., Comparison of different sorbents for multiresidue solid-phase extraction of 16 pesticides from ground water coupled with high-performance liquid chromatography, *Talanta*, 71, 25–30, 2007.
36. Moral, A., Caballo, C., Sicilia, M. D., Rubio, S., Highly efficient microextraction of chlorophenoxy acid herbicides in natural waters using a decanoic acid-based nanostructured solvent prior to their quantitation by liquid chromatography–mass spectrometry, *Anal. Chim. Acta*, 709, 59–65, 2012.
37. Tran, A. T. K., Hyne, R. V., Doble, P., Calibration of a passive sampling device for time-integrated sampling of hydrophilic herbicides in aquatic environments, *Environ. Toxicol. Chem.*, 26, 435–443, 2007.
38. Hogenboom, A. C., Hofman, M. P., Jolly, D. A., Niessen, W. M. A. Brinkman, U. A. Th., On-line dual-precolumn-based trace enrichment for the determination of polar and acidic microcontaminants in river water by liquid chromatography with diode-array UV and tandem mass spectrometric detection, *J. Chromatogr. A*, 885, 377–388, 2000.
39. Santos, T. C. R., Rocha, J. C., Barcelo, D., Determination of rice herbicides, their transformation products and clofibric acid using on-line solid-phase extraction followed by liquid chromatography with diode array and atmospheric pressure chemical ionization mass spectrometric detection, *J. Chromatogr. A*, 879, 3–12, 2000.
40. Hutta, M., Chalanyova, M., Halko, R., Góra, R., Dokupilova, S., Rybar, I., Reversed phase liquid chromatography trace analysis of pesticides in soil by on-column sample pumping large volume injection and UV detection, *J. Sep. Sci.*, 32, 2034–2042, 2009.
41. Pichon, V., Charpak, M., Hennion, M.-C., Multiresidue analysis of pesticides using new laminar extraction disks and liquid chromatography and application to the French priority list, *J. Chromatogr. A*, 795, 83–92, 1998.
42. Chao, J.-b., Liu, J.-f., Wen, M.-j., Liu, J.-m., Cai, Y-q., Jiang, G.-b., Determination of sulfonylurea herbicides by continuous-flow liquid membrane extraction on-line coupled with high-performance liquid chromatography, *J. Chromatogr. A*, 955, 183–189, 2002.
43. Khrolenko, M. V., Wieczorek, P. P., Determination of glyphosate and its metabolite aminomethylphosphonic acid in fruit juices using supported-liquid membrane preconcentration method with high-performance liquid chromatography and UV detection after derivatization with *p*-toluenesulphonyl chloride, *J. Chromatogr. A*, 1093, 111–117, 2005.
44. Moral, A., Sicilia, M. D., Rubio, S., Perez-Bendito, D., Sodium dodecyl sulphate-coated alumina for the extraction/preconcentration of benzimidazolic fungicides from natural waters prior to their quantification by liquid chromatography/fluorimetry, *Anal. Chim. Acta*, 569, 132–138, 2006.
45. Sannino, F., Iorio, M., De Martino, A., Pucci, M., Brown, C. D., Capasso, R., Remediation of waters contaminated with ionic herbicides by sorption on polymerin, *Water Research*, 42, 643–652, 2008.
46. Kim, J., Liu, K.-H., Kang, S.-H., Koo, S.-J., Kim, J.-H., Degradation of the sulfonylurea herbicide LGC-42153 in flooded soil, *Pest. Manag. Sci.*, 59, 1037–1042, 2003.
47. Huang, S.-D., Huang, H.-I., Sung, Y.-H., Analysis of triazine in water samples by solid-phase microextraction coupled with high-performance liquid chromatography, *Talanta*, 64, 887–893, 2004.
48. Ibanez, M., Pozo, O. J., Sancho, J. V., Lopez, F. J., Hernandez, F., Residue determination of glyphosate, glufosinate and aminomethylphosphonic acid in water and soil samples by liquid chromatography coupled to electrospray tandem mass spectrometry, *J. Chromatogr. A*, 1081, 145–155, 2005.
49. Diaz, T. G., Guiberteau, A., Ortiz, J. M., Lopez, M. D., Salinas, E., Use of neural networks and diode-array detection to develop an isocratic HPLC method for the analysis of nitrophenol pesticides and related compounds, *Chromatographia*, 53, 40–46, 2001.
50. Mayer-Helm, B., Method development for the determination of 52 pesticides in tobacco by liquid chromatography–tandem mass spectrometry, *J. Chromatogr. A*, 1216, 8953–8959, 2009.
51. Fang, G., Lau, H. F., Law, W. S., Li, S. F. Y., Systematic optimisation of coupled microwave-assisted extraction-solid phase extraction for the determination of pesticides in infant milk formula via LC–MS/MS, *Food Chem.*, 134, 2473–2480, 2012.
52. Melo, L. F. C., Collins, C. H., Jardim, I. C. S. F., High-performance liquid chromatographic determination of pesticides in tomatoes using laboratory-made NH_2 and C18 solid-phase extraction materials, *J. Chromatogr. A*, 1073, 75–81, 2005.
53. Melo, L. F. C., Collins, C. H., Jardim, I. C. S. F., New materials for solid-phase extraction and multiclass high-performance liquid chromatographic analysis of pesticides in grapes, *J. Chromatogr. A*, 1032, 51–58, 2004.

54. Sandahl, M., Mathiasson, L., Jonsson, J. A., Determination of thiophanate-methyl and its metabolites at trace level in spiked natural water using the supported liquid membrane extraction and the microporous membrane liquid–liquid extraction techniques combined on-line with high-performance liquid chromatography, *J. Chromatogr. A*, 893, 123–131, 2000.
55. Asensio-Ramos, M., Hernández-Borges, J., Borges-Miquel, T. M., Rodríguez-Delgado, M. Á., Ionic liquid-dispersive liquid–liquid microextraction for the simultaneous determination of pesticides and metabolites in soils using high-performance liquid chromatography and fluorescence detection, *J. Chromatogr. A*, 1218, 4808–4816, 2011.
56. Nurmi, J., Pellinen, J., Multiresidue method for the analysis of emerging contaminants in wastewater by ultra performance liquid chromatography–time-of-flight mass spectrometry, *J. Chromatogr. A*, 1218, 6712–6719, 2011.
57. Chang, F.-C., Simcik, M. F., Capel, P. D., Occurrence and fate of the herbicide glyphosate and its degradate aminomethylphosphonic acid in the atmosphere, *Environ. Toxicol. Chem.*, 30, 548–555, 2011.
58. Vreeken, R. J., Speksnijder, P., Bobeldijk-Pastorova, I., Noij, Th. H. M., Selective analysis of the herbicides glyphosate and aminomethylphosphonic acid in water by on-line solid-phase extraction–high-performance liquid chromatography–electrospray ionization mass spectrometry, *J. Chromatogr. A*, 794, 187–199, 1998.
59. Fontana, A. R., Rodríguez, I., Ramil, M., Altamirano, J. C., Cela, R., Solid-phase extraction followed by liquid chromatography quadrupole time-of-flight tandem mass spectrometry for the selective determination of fungicides in wine samples, *J. Chromatogr. A*, 1218, 2165–2175, 2011.
60. Bernal, J., Bernal, J. L., Martin, M. T., Nozal, M. J., Anadón, A., Martínez-Larranaga, M. R., Martínez, M. A., Development and validation of a liquid chromatography–fluorescence–mass spectrometry method to measure glyphosate and aminomethylphosphonic acid in rat plasma, *J. Chromatogr. B*, 878, 3290–3296, 2010.
61. Benoit, P., Preston, C. M., Transformation and binding of ^{13}C and ^{14}C-labelled atrazine in relation to straw decomposition in soil, *Eur. J. Soil Sci.*, 51, 43–54, 2000.
62. Mayer-Helm, B., Hofbauer, L., Mu ller, J., Development of a multi-residue method for the determination of 18 carbamates in tobacco by high performance liquid chromatography/positive electrospray ionisation tandem mass spectrometry, *Rapid Commun. Mass Spectrom.*, 20, 529–536, 2006.
63. Lacina, O., Urbanova, J., Poustka, J., Hajslova, J., Identification/quantification of multiple pesticide residues in food plants by ultra-high-performance liquid chromatography-time-of-flight mass spectrometry, *J. Chromatogr. A*, 1217, 648–659, 2010.
64. Greulich, K., Alder, L., Fast multiresidue screening of 300 pesticides in water for human consumption by LC-MS/MS, *Anal. Bioanal. Chem.*, 391, 183–197, 2008.
65. Pico, Y., la Farre, M., Soler, C., Barcelo, D., Identification of unknown pesticides in fruits using ultra-performance liquid chromatography–quadrupole time-of-flight mass spectrometry Imazalil as a case study of quantification, *J. Chromatogr. A*, 1176, 123–134, 2007.
66. Picó, Y., la Farré, M., Tokman, N., Barceló, D., Rapid and sensitive ultra-high-pressure liquid chromatography–quadrupole time-of-flight mass spectrometry for the quantification of amitraz and identification of its degradation products in fruits, *J. Chromatogr. A*, 1203, 36–46, 2008.
67. Chen, L., Yin, L., Song, F., Liu, Z., Zheng, Z., Xing, J., Liu, S., Determination of pesticide residues in ginseng by dispersive liquid–liquid microextraction and ultra high performance liquid chromatography–tandem mass spectrometry, *J. Chromatogr. B*, 917–918, 71–77, 2013.
68. Arienzo, M., Cataldo, D., Ferrara, L., Pesticide residues in fresh-cut vegetables from integrated pest management by ultra performance liquid chromatography coupled to tandem mass spectrometry, *Food Control*, 31, 108e115, 2013.
69. Gilbert-López, B., García-Reyes, J. F., Fernández-Alba, A. R., Molina-Díaz, A., Evaluation of two sample treatment methodologies for large-scale pesticide residue analysis in olive oil by fast liquid chromatography–electrospray mass spectrometry, *J. Chromatogr. A*, 1217, 3736–3747, 2010.
70. Hogenboom, A. C., Hofman, M. P., Kok, S. J., Niessen, W. M. A., Brinkman, U. A. Th., Determination of pesticides in vegetables using large-volume injection column liquid chromatography–electrospray tandem mass Spectrometry, *J. Chromatogr. A*, 892, 379–390, 2000.
71. Geerdink, R. B., Kooistra-Sijpersma, A., Tiesnitscha, J., Kienhuis, P. G. M., Brinkman, U. A. Th., Determination of polar pesticides with atmospheric pressure chemical ionisation mass spectrometry–mass spectrometry using methanol and/or acetonitrile for solid-phase desorption and gradient liquid chromatography, *J. Chromatogr. A*, 863, 147–155, 1999.
72. Breitkreitz, M. C., Jardim, I. C. S. F., Bruns, R. E., Combined column–mobile phase mixture statistical design optimization of high-performance liquid chromatographic analysis of multicomponent systems, *J. Chromatogr. A*, 1216, 1439–1449, 2009.

73. Toribio, L., del Nozal, M. J., Bernal, J. L., Jimenez, J. J., Alonso, C., Chiral separation of some triazole pesticides, *J. Chromatogr. A*, 1046, 249–253, 2004.
74. Reyzer, M. L., Brodbelt, J. S., Analysis of fire ant pesticides in water by solid-phase microextraction and gas chromatography/mass spectrometry or high-performance liquid chromatography/mass spectrometry, *Anal. Chim. Acta*, 436, 11–20, 2001.
75. Zougagh, M., Bouabdallah, M., Salghi, R., Hormatallah, A., Rios, A., Supercritical fluid extraction as an on-line clean-up technique for rapid amperometric screening and alternative liquid chromatography for confirmation of paraquat and diquat in olive oil samples, *J. Chromatogr. A*, 1204, 56–61, 2008.
76. Srivastava, G., Dixit, A., Yadav, S., Patel, D. K., Prakash, O., Singh M. P., Resveratrol potentiates cytochrome P450 2 d22-mediated neuroprotection in maneb- and paraquat-induced parkinsonism in the mouse, *Free Radic. Biol. Med.*, 52, 1294–1306, 2012.
77. Wang, M. S., Mu, H., Bai, Y., Zhang, Y., Liu, H., Multiresidue determination of fluoroquinolones, organophosphorus and *N*-methyl carbamates simultaneously in porcine tissue using MSPD and HPLC–DAD, *J. Chromatogr. B*, 877, 2961–2966, 2009.
78. Crommen, J., Reversed phase ion-pair high-performance liquid chromatography of drugs and related compounds using underivatized silica as the stationary phase, *J. Chromatogr.*, 186, 705–724, 1979.
79. Castro, R., Moyano, E., Galceran, M. T., Ion-pair liquid chromatography–atmospheric pressure ionization mass spectrometry for the determination of quaternary ammonium herbicides, *J. Chromatogr. A*, 830, 145–154, 1999.
80. Low, K. G. C., Bartha, A., Billiet, H. A. H., de Galan, L., Systematic procedure for the determination of the nature of the solutes prior to the selection of the mobile phase parameters for optimization of reversed-phase ion-pair chromatographic separations, *J. Chromatogr.*, 478, 21–38, 1989.
81. Brunetto, M. R., Morales, A. R., Gallignani, M., Burguera, J. L., Burguera, M., Determination of paraquat in human blood plasma using reversed-phase ion-pair high-performance liquid chromatography with direct sample injection, *Talanta*, 59, 913–921, 2003.
82. Qian, K., Tang, T., Shi, T., Li, P., Li, J., Cao, Y., Solid-phase extraction and residue determination of glyphosate in apple by ion-pairing reverse-phase liquid chromatography with pre-column derivatization, *J. Sep. Sci.*, 32, 2394–2400, 2009.
83. Marr, K. J. C., King, J. B., A simple high performance liquid chromatography/ionspray tandem mass spectrometry method for the direct determination of paraquat and diquat in water, *Rapid Commun. Mass Spectrom.*, 11, 479–483 1997.
84. Aramendıa, M. A., Borau, V., Lafont, F., Marinas, A., Marinas, J. M., Moreno, J. M., Porras, J. M., Urbano, F. J., Determination of diquat and paraquat in olive oil by ion-pair liquid chromatography–electrospray ionization mass spectrometry (MRM), *Food Chem.*, 97, 181–188, 2006.
85. Castro, R., Moyano, E., Galceran, M. T., Determination of quaternary ammonium pesticides by liquid chromatography–electrospray tandem mass spectrometry, *J. Chromatogr. A*, 914, 111–121, 2001.
86. Pateiro-Mourea, M., Martinez-Carballo, E., Arias-Estevez, M., Simal-Gandara, J., Determination of quaternary ammonium herbicides in soils: Comparison of digestion, shaking and microwave-assisted extractions, *J. Chromatogr. A*, 1196–1197, 110–116, 2008.
87. Wang, K.-C., Chen, S.-M., Hsu, J.-F., Cheng, S.-G., Lee, C.-K., Simultaneous detection and quantitation of highly water-soluble herbicides in serum using ion-pair liquid chromatography–tandem mass spectrometry, *J. Chromatogr. B*, 876, 211–218, 2008.
88. Sancho, J. V., Ibanez, M., Grimalt, S., Pozo, O. J., Hernandez, F., Residue determination of cyromazine and its metabolite melamine in chard samples by ion-pair liquid chromatography coupled to electrospray tandem mass spectrometry, *Anal. Chim. Acta*, 530, 237–243, 2005.
89. Vega, B., Lopez, F. J. S., Frenich, A. G., Application of internal quality control to the analysis of quaternary ammonium compounds in surface and groundwater from Andalusia (Spain) by liquid chromatography with mass spectrometry, *J. Chromatogr. A*, 1050, 179–184, 2004.
90. Qian, K., He, S., Tang, T., Shi, T., Li, J., Cao, Y., A rapid liquid chromatography method for determination of glufosinate residue in maize after derivatisation, *Food Chem.*, 127, 722–726, 2011.
91. Hernandez, F., Ibanez, M., Pozo, O. J., Sancho, J. V., Investigating the presence of pesticide transformation products in water by using liquid chromatography–mass spectrometry with different mass analyzers, *J. Mass Spectrom.*, 43, 173–184, 2008.
92. Mamy, L., Barriuso, E., Desorption and time-dependent sorption of herbicides in soils, *Eur. J. Soil Sci.*, 58, 174–187, 2007.
93. Marin, J. M., Pozo, O. J., Beltran, J., Hernandez, F., An ion-pairing liquid chromatography/tandem mass spectrometric method for the determination of ethephon residues in vegetables, *Rapid Commun. Mass Spectrom.*, 20, 419–426, 2006.

94. de Erenchun, R., Goicolea, M. A., de Balugera, Z. G., Portela, M. J., Barrio, R. J., Determination of herbicides by reductive amperometric detection in liquid chromatography, *J. Chromatogr. A*, 763, 227–235, 1997.
95. Sancho, J. V., Hernandez, S. F., Lopez, F. J., Hogendoorn, E. A., Dijkman, E., van Zoonen, P., Rapid determination of glufosinate, glyphosate and aminomethylphosphonic acid in environmental water samples using precolumn fluorogenic labeling and coupled-column liquid chromatography, *J. Chromatogr. A*, 737, 75–83, 1996.
96. Careri, M., Elviri, L., Mangia, A., Zagnoni, I., Rapid method for determination of chlormequat residues in tomato products by ion-exchange liquid chromatography/electrospray tandem mass spectrometry, *Rapid Commun. Mass Spectrom.* 16, 1821–1826, 2002.
97. Wang, X., Fang, Q., Liu, S., Chen, L., The application of pseudo template molecularly imprinted polymer to the solid-phase extraction of cyromazine and its metabolic melamine from egg and milk, *J. Sep. Sci.*, 35, 1432–1438, 2012.
98. Piriyapittaya, M., Jayanta, S., Mitra, S., Leepipatpiboon, N., Micro-scale membrane extraction of glyphosate and aminomethylphosphonic acid in water followed by high-performance liquid chromatography and post-column derivatization with fluorescence detector, *J. Chromatogr. A*, 1189, 483–492, 2008.
99. Mallat, E., Barcelo, D., Analysis and degradation study of glyphosate and of aminomethylphosphonic acid in natural waters by means of polymeric and ion-exchange solid-phase extraction columns followed by ion chromatography–post-column derivatization with fluorescence detection, *J. Chromatogr. A*, 823, 129–136, 1998.
100. Riediker, S., Obrist, H., Varga, N., Stadler, R. H., Determination of chlormequat and mepiquat in pear, tomato, and wheat flour using on-line solid-phase extraction (Prospekt) coupled with liquid chromatography–electrospray ionization tandem mass spectrometry, *J. Chromatogr. A*, 966, 15–23, 2002.
101. Evans, C. S., Startin, J. R., Goodall, D. M., Keely, B. J., Improved sensitivity in detection of chlormequat by liquid chromatography–mass spectrometry, *J. Chromatogr. A*, 897, 399–404, 2000.
102. Guo, Z.-X., Cai, Q., Yang, Z., Ion chromatography/inductively coupled plasma mass spectrometry for simultaneous determination of glyphosate, glufosinate, fosamine and ethephon at nanogram levels in water, *Rapid Commun. Mass Spectrom.*, 21, 1606–1612, 2007.
103. Dimitrakopoulos, I. K., Thomaidis, N. S., Megoulas, N. C., Koupparis, M. A., Effect of suppressor current intensity on the determination of glyphosate and aminomethylphosphonic acid by suppressed conductivity ion chromatography, *J. Chromatogr. A*, 1217, 3619–3627, 2010.
104. Guo, Z.-X., Cai, Q., Yang, Z., Determination of glyphosate and phosphate in water by ion chromatography—inductively coupled plasma mass spectrometry detection, *J. Chromatogr. A*, 1100, 160–167, 2005.
105. Patsias, J., Papadopoulou, A., Papadopoulou-Mourkidou, E., Automated trace level determination of glyphosate and aminomethyl phosphonic acid in water by on-line anion-exchange solid-phase extraction followed by cation-exchange liquid chromatography and post-column derivatization, *J. Chromatogr. A*, 932, 83–90, 2001.
106. Mohsin, S. B., Use of ion chromatography–electrospray mass spectrometry for the determination of ionic compounds in agricultural chemicals, *J. Chromatogr. A*, 884, 23–30, 2000.
107. Peeters, M.-C., Defloor, I., Coosemans, J., Delcour, J. A., Ooms, L., Deliever, R., De Vos, D., Simple ion chromatographic method for the determination of chlormequat residues in pears, *J. Chromatogr. A*, 920, 255–259, 2001.
108. Hau, J., Riediker, S., Varga, N., Stadler, R. H., Determination of the plant growth regulator chlormequat in food by liquid chromatography–electrospray ionisation tandem mass spectrometry, *J. Chromatogr. A*, 878, 77–86, 2000.
109. Lindh, C. H., Littorin, M., Johannesson, G., Jönsson, B. A. G., Analysis of chlormequat in human urine as a biomarker of exposure using liquid chromatography triple quadrupole mass spectrometry, *J. Chromatogr. B*, 879, 1551–1556, 2011.
110. Odetokun, M. S., Montesano, M. A., Weerasekera, G., Whitehead Jr., R. D., Needham, L. L., Barr, D. B., Quantification of dialkylphosphate metabolites of organophosphorus insecticides in human urine using 96-well plate sample preparation and high-performance liquid chromatography–electrospray ionization-tandem mass spectrometry, *J. Chromatogr. B*, 878, 2567–2574, 2010.
111. Esparza, X., Moyano, E., Galceran, M. T., Analysis of chlormequat and mepiquat by hydrophilic interaction chromatography coupled to tandem mass spectrometry in food samples, *J. Chromatogr. A*, 1216 4402–4406, 2009.
112. Hayama, T., Yoshida, H., Todoroki, K., Nohta, H., Yamaguchi, M., Determination of polar organophosphorus pesticides in water samples by hydrophilic interaction liquid chromatography with tandem mass spectrometry, *Rapid Commun. Mass Spectrom.*, 22, 2203–2210, 2008.

113. Crnogorac, G., Schwack, W., Determination of dithiocarbamate fungicide residues by liquid chromatography/mass spectrometry and stable isotope dilution assay, *Rapid Commun. Mass Spectrom.*, 21, 4009–4016, 2007.
114. Fischer, K., Kotalik, J., Kettrup, A., Chromatographic properties of the ion-exclusion column IonPac ICE-AS6 and application in environmental analysis part I: Chromatographic properties, *J. Chromatogr. Sci.*, 37, 477–485, 1999.
115. Novic, M., Haddad, P. R., Analyte-stationary phase interactions in ion-exclusion chromatography, *J. Chromatogr. A*, 1118, 19–28, 2006.
116. Kim, H.-J., Conca, K. R., Richardson, M. J., Determination of sulfur dioxide in grapes: Comparison of the Monier-Williams method and two ion exclusion chromatographic methods, *J. Assoc. Off. Anal. Chem.*, 73, 983–989, 1990.

9 Optimization of Normal-Phase and Reversed-Phase Systems for Analysis of Pesticides

Choice of the Mode of Elution—Isocratic and Gradient Elution

Pavel Jandera

CONTENTS

9.1 INTRODUCTION

Successful sample analysis depends on the availability of the instrumentation in the laboratory and on the selection of a suitable chromatographic method. First, an appropriate technique (separation mode) should be selected, depending on the properties of the analyte(s) to be determined; then the development and optimization of the separation method follows. This topic was treated in depth by Snyder et al. [1]. In this chapter, various aspects of the individual liquid chromatographic (LC) modes are examined and approaches for adjusting the experimental variables and optimization of separation conditions are discussed with special attention given to computer-assisted methods.

9.2 ISOCRATIC SEPARATIONS

The first step in the high-performance (HP) LC method of development consists of selecting an adequate separation mode. Many neutral compounds can be separated either by reversed-phase (RP) or by normal-phase (NP) chromatography. The RP system is usually the best first choice because it is likely to result in a satisfactory separation of a great variety of nonpolar, polar, and even ionic compounds. In RP chromatography, the stationary phase is less polar than the mobile phase; sample retention increases as the polarity of the mobile phase increases, and polar analytes are less retained than the nonpolar ones. The polarity effects of the sample and of the stationary and mobile phases are opposite in the RP and in the NP LC systems (Figure 9.1).

Lipophilic samples can often be separated either by nonaqueous (NA) RP chromatography or by organic NP chromatography. Polar compounds are usually too weakly retained in RP systems and often too strongly in traditional NP systems with organic mobile phases (adsorption chromatography). However, adequate retention and resolution can often be achieved on polar columns in aqueous-organic mobile phases—aqueous normal phase (ANP)—or hydrophilic interaction (HILIC) LC systems.

Weak acids or bases can be analyzed by RP or ANP LC with a buffered mobile phase and strong acids or strong bases by ion-pair chromatography (IPC) or ion-exchange chromatography (IEC).

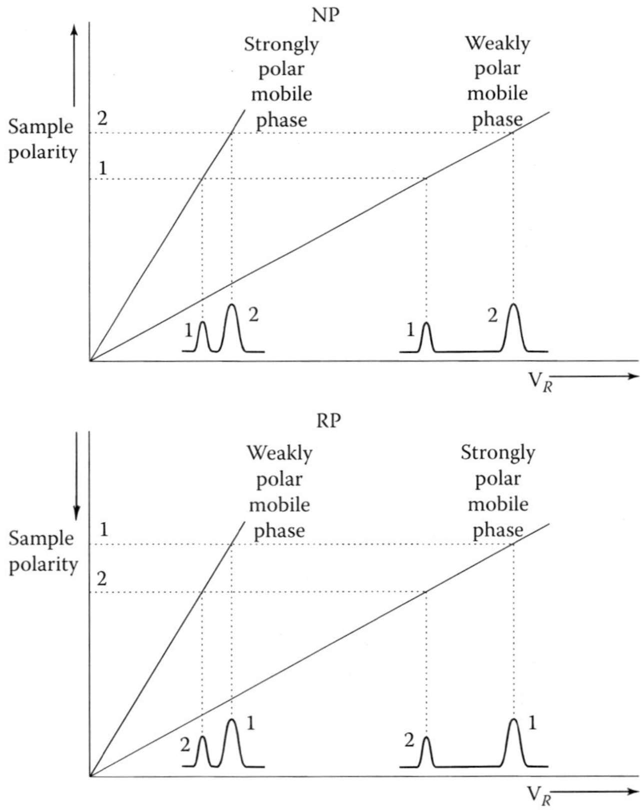

FIGURE 9.1 Schematic diagram of the effects of sample and mobile phase polarities on the retention in NP and RP LC. V_R: retention volume.

Special chiral columns or chiral selector additives to the mobile phase can be used for separation of optical isomers.

Macromolecules are usually separated and characterized by size-exclusion chromatography on columns packed with inert materials (gels) characterized by controlled pore distribution on the basis of different accessibility of the pores for molecules of different sizes, larger molecules eluting first. For some lower polymers with molecular masses in the range 10^3–10^4 Da, "interactive," that is, RP or NP HPLC modes provide better selectivity of separation than size-exclusion chromatography. Many ionizable biopolymers, such as peptides, proteins, oligonucleotides, and nucleic acids, can be separated by IEC or by RP chromatography on wide-pore packing materials with mobile phases containing trifluoroacetic acid or triethylammonium acetate as an ion-pairing reagent.

9.2.1 RPLC

The stationary phase in RP chromatography (RPC)—usually an alkyl immobilized on an inorganic support—is less polar than the aqueous-organic mobile phase. Nonpolar samples are more strongly retained than the polar ones, and the retention increases with increasing polarity of the mobile phase so that very lipophilic samples may require nonaqueous mobile phases.

The most frequently used stationary phases for RP separations are nonpolar or moderately polar stationary phases covalently bonded on silica gel support using reactions of the surface silanol (Si-OH) groups with organosilanes (halogeno- or alkoxy-) to obtain stationary phases with Si-O-Si-R bonds (R is most often C8 or C18 alkyls). The retention generally increases with increasing content of carbon atoms in the chemically bonded phase and with increasing length of the bonded alkyl chains but only up to a certain "critical" length of the bonded alkyls [2].

Rather bulky silanization reagents can chemically modify no more than 50% of the original silanol groups. The residual silanol groups may interact with polar, especially basic, solutes, often causing strong and irreversible retention and poor separation with tailing or distorted peaks, especially for basic compounds. Some residual silanol groups can be removed by a subsequent "endcapping" reaction with small-molecule trimethylchlorosilane or hexamethyldisilazane reagents. Other approaches rely on using diisopropyl or diisobutyl chlorosilane reagents in a single silanization step to provide steric shielding of the residual silanols or on bidentate attachment of C18 or C8 alkyls to the silica gel surface via two reactive groups separated by $-CH_2-CH_2-$ or $-CH_2-CH_2-CH_2-$ bridges. "Bridged" bidentate bonded phases efficiently shield the nonreacted silanol groups and protect the surface of the silica gel support from direct contact with the mobile phase, improving thus the column stability over a broad pH range.

In highly aqueous mobile phases, which are often necessary for RPLC separations of polar compounds, nonpolar bonded alkyls are poorly solvated and may "collapse" and stick together, changing significantly the properties of the bonded stationary phases. Incorporating amide or carbamate groups in the bonded ligands between the alkyl chain and the surface of the silica gel support improves the retention behavior in highly aqueous mobile phases.

Materials with inorganic or porous hydrophobic or (less frequently) hydrophilic organic polymer matrices and graphitized carbon are stable over a broad pH range from 0 to 12–14; hence they are useful for separations of basic compounds. Hybrid particles prepared using both inorganic (silica) and organic (organosiloxanes) components, such as 1,2-bis(siloxy)ethane, incorporated into their skeleton, chemically modified with desired functionality and endcapped in separated steps, possess ethylene bridge moieties shielding residual silanol groups and show improved pH and mechanical stability.

Stationary phases with chemically bonded branched hydrocarbons, perfluoroalkanes, polyethylene glycol, cholesterol, alkylaryl or other groups show different separation selectivities, which can be useful for specific separations. For example, chemically bonded phenyl groups show preferential retention of aromatic compounds and increased shape selectivity for planar and rigid rod-like molecules.

The mobile phase in RPC contains water (a weak eluent) and one or more polar organic solvents (strong eluents), most frequently acetonitrile, methanol, or tetrahydrofuran. By the choice of the organic solvent, selective polar interactions (dipole–dipole, proton-donor, or proton-acceptor) with analytes can be either enhanced or suppressed and the selectivity of separation adjusted. Binary aqueous-organic mobile phases are usually well suited for separations of a variety of samples. Very lipophilic samples may be too strongly retained even in mobile phases with low concentrations of water and may require NARP chromatography with binary organic mobile phases. Ternary or, less often, quaternary mobile phases may offer improved selectivity for some difficult separations.

The retention times, t_R, are controlled by the concentration of the organic solvent in binary aqueous-organic mobile phases (methanol–water or acetonitrile–water). The retention decreases as the concentration of less polar organic solvent increases. A simple semilogarithmic Equation 9.1 is widely used to describe the effect of the volume fraction of methanol or acetonitrile, ϕ, on the retention factors, $k = t_R/t_0 - 1$ [1,3,4]:

$$\log k = a - m\phi \tag{9.1}$$

where t_0 is the column dead (hold-up) time. The constants m and a in Equation 9.1 increase as the polarity of the solute decreases or as its size increases; m increases with decreasing polarity of the organic solvent. According to Equation 9.1, the parameter a should mean the logarithm of k in pure water as the mobile phase. However, deviations from Equation 9.1 are often observed in mobile phases with low concentrations of water. For a more accurate description of retention in highly aqueous mobile phases, the second-order Equation 9.2 was suggested [5]:

$$\log k = a - m\phi + d\phi^2 \tag{9.2}$$

9.2.2 NPLC

In NP chromatography, the stationary phase is more polar than the mobile phase. The retention increases as the polarity of the mobile phase decreases, and polar analytes are more strongly retained than nonpolar ones—the opposite of RPC (Figure 9.1). The column packing is either an inorganic adsorbent (silica gel or, less often, a metal oxide) or a moderately polar bonded phase, such as cyanopropyl $-(CH_2)_3-CN$, diol $-(CH_2)_3-O-CH_2-CHOH-CH_2-OH$, or aminopropyl $-(CH_2)_3-NH_2$), chemically bonded on silica gel or another support. The mobile phases usually are binary mixtures of a nonpolar (e.g., hexane) and one or more strongly or moderately polar solvents. The retention decreases with increasing concentration of the nonpolar solvent (diluent). NP behavior sometimes can also be observed in NARP LC, probably due to the activity of polar residual silanol groups.

Separation selectivity can be adjusted by changing either the mobile or the stationary phases in NPLC. Proton donor–acceptor interactions cause strong retention of basic compounds on silica gel in nonaqueous mobile phases whereas acidic compounds show increased affinities to aminopropyl columns. The elution strength is proportional to the polarity of the mobile phase. Great changes in selectivity of NPLC separations can be achieved by selecting solvents with the appropriate type of selective polar interactions. Lesser changes in separation selectivity may be observed even when changing only the concentration ratio of the two organic solvents in a binary organic mobile phase. The retention in NPLC decreases with increasing concentration of the more polar solvent, ϕ. With some simplification, the effect of the volume fraction, ϕ, of the more polar solvent on the sample retention factor, k, can be described by Equation 9.3 [6]:

$$\log k = \log k_0 - m \log \phi \tag{9.3}$$

The constants k_0 and m depend on the nature of the solute and on the chromatographic system but are independent of ϕ. According to the competitive adsorption model, ϕ is the molar fraction of the more polar solvent, but—to first approximation—it can be expressed in terms of volume fractions.

k_0 is the retention factor in pure polar solvent. The parameter m theoretically corresponds to the number of molecules of the strong solvent necessary to displace one adsorbed sample molecule. Equation 9.3 generally can satisfactorily describe the mobile phase effects on the retention of compounds strongly retained in a pure less polar organic solvent, such as hexane. However, sometimes a more polar solvent should be used as the mobile phase component, which provides the elution strength that cannot be neglected. In such a case, the effects of a binary organic mobile phase on the retention are often more accurately described by the three-parameter Equation 9.4 [7]:

$$k = (a + b\phi)^{-m} \tag{9.4}$$

RPC offers better selectivity than NPLC for the separation of molecules differing in the hydrophobic parts of the molecules, but there are some practical reasons for selecting NP chromatography methods in specific cases: (i) A lower organic mobile phase viscosity offers a lower pressure drop across the column than in aqueous-organic mobile phases used in RPLC at a comparable flow-rate; (ii) HPLC columns are usually more stable and have longer lifetimes in organic solvents than in aqueous-organic mobile phases; (iii) many samples are more soluble or less prone to decompose in organic than in aqueous mobile phases and do not cause injection problems in NPLC, which are occasionally observed in RPLC; (iv) unlike RPLC, NP chromatography enables direct injection of samples extracted into a nonpolar solvent; and (v) NPLC is usually better suited for separation of isomers than RPLC.

9.2.3 HILIC

Strongly polar samples are usually too weakly retained in RPLC to allow their separation, whereas they are often retained too strongly in nonaqueous mobile phases used in conventional adsorption NPLC, and/or are not sufficiently soluble in the nonaqueous mobile phases. This problem can be solved by using polar stationary phases and aqueous-organic mobile phases rich in organic solvents (usually acetonitrile) in the separation mode called HILIC, which can be characterized as NPLC with conventional RP mobile phases [8]. HILIC is becoming increasingly popular in the analysis of carbohydrates, amino acids, peptides, polar drugs, toxins, natural compounds in plant extracts, and polar compounds in environmental, food, and pharmaceutical samples.

Various polar columns are suitable for HILIC applications, such as silica gel or chemically bonded amino-, amido-, cyano-, carbamate-, diol-, polyol-, poly(2-sulphoethyl aspartamide), zwitterionic sulfobetaine, etc., ligands. Silica gel often shows higher selectivity differences for small polar compounds, such as carboxylic acids, nucleosides, and nucleotides, with respect to other polar-bonded stationary phases commonly used in HILIC (amide, amino, aspartamide, or sulfobetaine) [9].

In contrast to RPLC, water is the strong eluting solvent in HILIC mobile phases, in which it is usually contained in concentrations of 1%–30%. Hence, it is preferentially adsorbed on a polar adsorbent, forming a more or less thick diffuse layer. Liquid–liquid partition between the bulk mobile phase and the adsorbed water-rich layer is assumed to participate in the retention process in HILIC besides the adsorption onto the polar adsorbent. Additional effects of ion exchange may contribute to the retention of ionizable compounds (especially on bonded amino, zwitterionic, or weak ion-exchange columns). However, the adsorption and partition retention mechanisms cannot be easily distinguished, and their relative importance may change as the water concentration in the mobile phase gradually increases.

In aqueous NP (HILIC) chromatography, the retention decreases with the increasing concentration of water. The effect of the volume fraction of water, ϕ, on the retention dependence often can be described either by Equation 9.1 or by Equation 9.3 in the HILIC range to first approximation. However, many polar columns show a dual retention mechanism, besides HILIC in highly organic mobile phases, also RP in mobile phases is richer in water, in which the retention increases at higher water concentrations. Consequently, U-shaped plots of the retention times versus the concentration

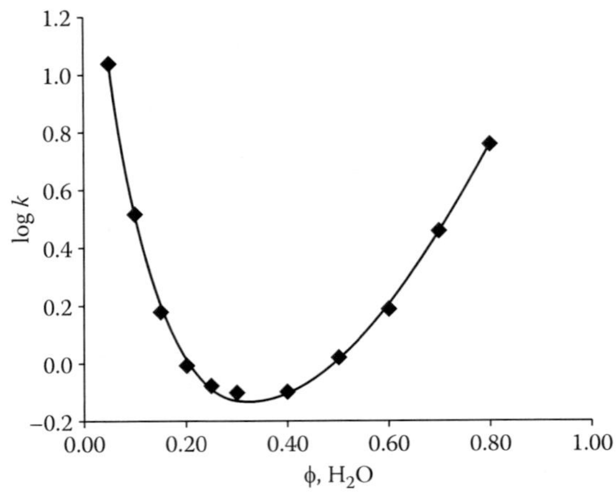

FIGURE 9.2 Dual HILIC-RP mode retention mechanism of protocatechuic acid in aqueous acetonitrile on a monolithic zwitterionic poly(methacrylate) monolithic column (180 × 0.32 mm inner diameter). k: retention factor; ϕ, H_2O: volume fraction of 0.01 mol/L ammonium acetate in acetonitrile. Points: experimental data; full line: predicted from Equation 9.5.

of the organic solvent are often observed on HILIC columns over the full composition range of the mobile phase with a minimum retention corresponding to the transition from the RP to the NP mechanism (see the example in Figure 9.2). The "U-turn" mobile phase composition shifts to higher concentrations of water in the mobile phase when less polar stationary phases are used, allowing HILIC separations in mobile phases with higher water concentrations. Stationary phases with various functionalities show significant differences in retention and separation selectivity. The effects of water on the retention over the full composition range of binary mobile phases can be described—to first approximation—by Equation 9.5, obtained as the combination of Equations 9.1 and 9.3 for the RP and HILIC range, respectively [9]:

$$\log k = a + m_1\phi_w - m_2 \log \phi_w \tag{9.5}$$

where ϕ_w is the volume fraction of water in the mobile phase.

Figure 9.2 illustrates the validity of Equation 9.5 for protocatechuic acid on a zwitterionic poly(methacrylate) monolithic column in aqueous acetonitrile. The suitability of a HILIC system for particular samples can be evaluated using a gradient of increasing concentration of water in acetonitrile (from 2% or 10% to 50% in 10 minutes). Increasing concentration of the organic solvent or of a salt in the mobile phase enhances the retention of polar compounds except for the bonded amino phase, in which the retention of the acid compounds decreases with increasing concentration of salts due to the ion-exchange effects. HILIC mobile phases should be buffered, as pH affects the ionization and hence the retention of weakly acidic or basic polar compounds.

9.2.4 Separation of Ionic Compounds

Ionized compounds are usually much less retained than noncharged compounds in RPLC, and their separation is usually possible only with ionic additives to the mobile phase. In buffered aqueous or aqueous-organic mobile phases, the ionization of weak acids (at pH <7) or bases (at pH >7) can be more or less suppressed to improve the separation and peak symmetry. By adjusting the pH in the range of ±1.5 units around the pK_a, differences in the ionization of the individual sample components often can be utilized to control the separation selectivity, except for strong acids or bases, which are completely ionized and weakly retained over a broad pH range, and their chromatographic behavior is usually little affected by adding a buffer to the mobile phase. Basic compounds can interact with residual silanols in alkyl silica bonded phases, which may cause their irreversible adsorption or late elution as strongly tailing peaks [1].

Strong acids or bases can be separated by RP IPC with ion-pairing reagent mobile phase additives, which contain a strongly acidic or strongly basic group and a bulky hydrocarbon part. Basic substances can usually be separated using C6–C8 alkanesulphonates and acidic substances using tetralkylammonium salts. Ion pair additives significantly increase the retention and improve the peak symmetry through formation of neutral ionic associates with increased affinity to a nonpolar stationary phase. The retention in IPC can be controlled by the type and concentration of the ion-pairing reagent or of the organic solvent in the mobile phase. Increasing the number and size of alkyls in the reagent molecules enhances the retention in the reagent concentration range in between 10^{-4} and 10^{-2} mol/L.

Ion exchange chromatography (IEC) was a traditional technique for LC of ionic compounds. Nowadays, it is used mainly for separations of small inorganic ions or of ionic biopolymers, such as oligonucleotides, nucleic acids, peptides, and proteins, rather than in the analysis of small organic ions, for which RPC or IPC usually offer higher efficiency and better resolution. IEC columns are packed with fine particles of ion exchangers, which contain charged ion-exchange groups covalently attached to a solid matrix, either an organic cross-linked styrene-divinylbenzene or ethyleneglycol-methacrylate co-polymer, or silica gel support, to which a functional group is chemically bonded via a spacer–propyl or phenylpropyl moiety. Strong cation exchangers contain $-SO_3^-$ sulphonate groups and strong anion exchangers $-N(CH_3)_3^+$ quaternary ammonium groups, completely ionized over a broad pH range (pH = 2–12). Weak cation exchangers contain carboxylic or phosphonic acid groups, which are ionized only in alkaline solutions, whereas tertiary or secondary amino groups (e.g., diethyl aminoethyl), of weak anion exchangers are ionized only in acidic mobile phases. Ion-exchange separations require aqueous or aqueous-organic mobile phases, which must contain counterions (10^{-2}–10^{-1} mol/L salts, buffers, ionized acids, or bases), competing with the sample ions for the ion exchange groups. The retention in IEC can be controlled by adjusting the ionic strength of the mobile phase. It decreases with increasing concentration of counterions in the mobile phase and with decreasing ion-exchange capacity of the column (1 to 5 meq/g with organic polymer ion exchangers and 0.3–1 meq/g with silica-based ion exchangers) [1]

To predict the effects of ionic strength on the retention, the stoichiometric model of ion exchange yields the equation formally very similar to the NPLC equation, Equation 9.3, where ϕ stands for the molar concentration of the counterions in the mobile phase, and the parameter m is the stoichiometric coefficient of ion exchange, which is proportional to the charge of the sample ions [1,6].

Weak acids are usually separated by anion-exchange chromatography at pH >7, and weak bases by cation-exchange chromatography at pH <6, and their retention increases with increasing ionization. Varying the pH of the mobile phase can adjust the separation selectivity, and the retention is controlled by the ionic strength.

9.2.5 Selection and Optimization of Isocratic Separation Conditions

Once a suitable HPLC separation mode has been selected, the experimental conditions can be adjusted using either an empirical or systematic method development approach. The separation of two sample compounds with retention times t_{R1} and t_{R2}, respectively, is conveniently characterized in terms of resolution, R_S (Equation 9.6):

$$R_S = \frac{2(t_{R1} - t_{R2})}{w_1 + w_2} = \frac{\sqrt{N}}{4}(r_{1,2} - 1)\frac{k}{k+1} = \text{Efficiency} \times \text{Selectivity} \times \text{Capacity} \quad (9.6)$$

Here, w_1 and w_2 are the bandwidths of the two compounds at the baseline, N is the column efficiency expressed as the number of theoretical plates, $r_{1,2} = k_2/k_1$ is the separation factor, which characterizes the selectivity of separation, and k is the average retention factor of the two sample compounds 1 and 2 as a measure of the capacity contribution to the resolution. The resolution depends on many experimental conditions, which can be adjusted either simultaneously or in subsequent steps [1].

For accurate quantitative analysis, the resolution usually should not be less than 1.5. If the separation is not satisfactory, it can be improved using several approaches:

- Poorly resolved peaks appearing close to the column hold-up volume indicate that the retention is too low and the contribution of the capacity term should be increased, best by decreasing the elution strength of the mobile phase. On the other hand, the elution strength should be increased if the separation takes too long a time.
- If the retention times are adequate and partial separation of the bands is apparent but the bands are relatively broad, the resolution usually can be improved by increasing the column efficiency, that is, the plate number N, by using a longer column or, preferably, a column packed with finer particles.
- If the bands are narrow but not well separated from each other, the separation selectivity, $r_{1,2}$, should be improved. This may be achieved (i) by changing the components of a binary mobile phase, (ii) by using ternary or more complex mobile phases or mobile phase additives inducing specific interactions with sample components, or (iii) by using a HPLC column with another stationary phase, which is usually the most reliable approach.
- If the resolution of early eluted bands is unsatisfactory and/or the separation time of the last eluting compounds is too long, the sample separation is usually improved by temperature or solvent gradients [1].

9.2.5.1 Control of the Separation Efficiency

The efficiency contribution to the resolution, that is, the column plate number, N, is directly proportional to the column length, L, and increases with decreasing particle size of the column packing with increasing column length and, to a lesser extent, with decreasing flow rate of the mobile phase. The dispersion of a solute band as it migrates along the column is characterized by the height equivalent to the theoretical plate, $H = L/N$, and depends on experimental conditions, such as the velocity of the mobile phase, u, described to first approximation by the van Deemter equation, Equation 9.7:

$$H = \mathrm{A} + \mathrm{B}/u + \mathrm{C}u = \lambda d_p + 2\gamma(D_M/u) + \mathrm{c}(d_p^2/D_M)u \tag{9.7}$$

A, B, and C are the additive contributions to band broadening (Figure 9.3).

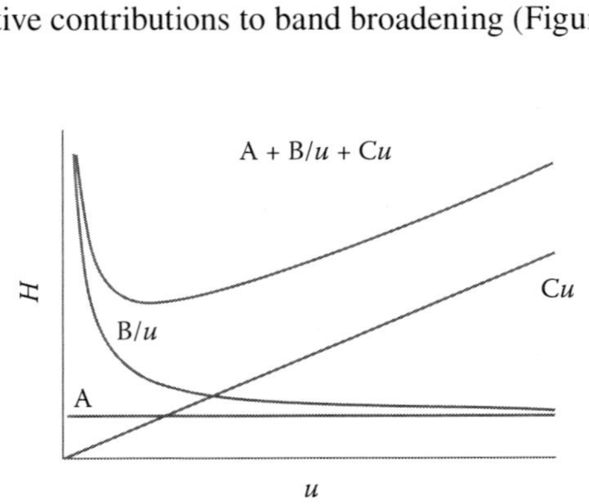

FIGURE 9.3 Contributions of the eddy diffusion (A), molecular diffusion (B), and mass transfer resistance (C) to the (H). u: mobile phase velocity along the column.

The velocity-independent term A characterizes the contribution of eddy (radial) diffusion to band broadening and is a function of the size and the distribution of interparticle channels and of possible nonuniformities in the packed bed (coefficient λ); it is directly proportional to the mean diameter of the column packing particles, d_p. The term B describes the effect of the molecular (longitudinal) diffusion in the axial direction and is directly proportional to the solute diffusion coefficient in the mobile phase, D_M. The "obstruction factor" γ takes into account the hindrance to the rate of diffusion by the particle skeleton.

The third term, C, is a measure of the resistance to mass transfer between the stationary and the mobile phase. To first approximation, it is inversely proportional to the diffusion coefficient, D_M, and directly proportional to the second power of the distance a solute molecule should travel to get from the mobile phase to the interaction site in the particle. For a totally porous particle, this distance is proportional to the mean particle diameter, d_p. More correctly, average pore depth should be used instead, but this quantity is often difficult to determine [10].

The minimum on the *H*–*u* plot corresponds to the best separation efficiency, but the separations are usually performed at a higher than optimum flow rate to decrease the run time. Hence, the right-hand part of the *H*–*u* plot should be as flat as possible to allow fast separations. This can be best achieved on columns packed with very small fully porous particles or core-shell particles [10] as shows the experimental *H*–*u* plots of toluene in 70% methanol on a 2.7-μm porous-shell solid core C18 column (Figure 9.4b); the use of monolithic columns (Chromolith C18) at the same conditions is less efficient (Figure 9.4a).

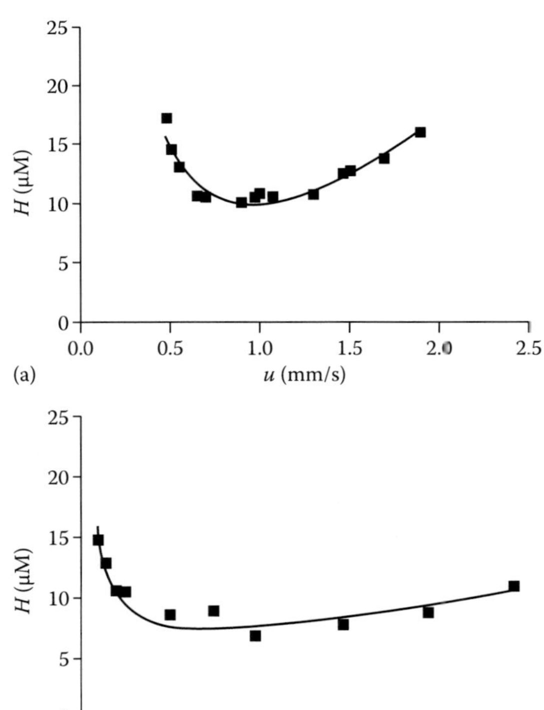

FIGURE 9.4 Experimental van Deemter plots of toluene in 70% methanol on a monolithic C18 column (a) and on a 2.7 μm porous-shell solid core C18 column (b). *H*: height equivalent to a theoretical plate; *u*: mobile phase velocity along the column.

9.2.5.2 Control of the Retention and Separation Selectivity: The Column

The speed versus efficiency performance of various HPLC columns can be optimized using kinetic plots proposed by Poppe [11]. There, the minimum time necessary to produce the desired number of theoretical plates, N_{req}, can be determined from the cross-section of the diagonal lines showing the column hold-up time, t_0, with the plots of log (H/u), characterizing the speed of separation, that is, the time necessary to achieve one theoretical plate for a nonretained compound.

Figure 9.5 shows the plots calculated for columns packed with fully porous particles of different mean particle diameter, d_p, assuming the maximum allowed instrumental pressure, ΔP_{max}, (usually 40 MPa in conventional LC) and the optimum flow velocity, u, corresponding to the minimum height equivalent to the theoretical plate, H. A dashed envelope line, drawn at the minimum N_{req} of the individual plots, divides the plane into two regions. The part on the right side below the dividing line corresponds to the "forbidden" area, in which the desired N_{req} cannot be accomplished in a given column hold-up time, t_0, under the limiting experimental conditions. The graph shows that very high numbers of theoretical plates can be achieved in columns packed with relatively large-diameter particles (10 μm or more), however, using a very long column at the cost of very long separation times. On the other hand, fast separation times <1 min with the efficiency of several thousands of theoretical plates require short columns packed with particles of diameters less than 2 μm, for which the operation pressure of 40 MPa is too low.

The kinetic plots can be moved deeper to the "forbidden" area at a higher operation pressure. This can be accomplished using the ultra-HPLC (UHPLC) technique, for which appropriate instrumentation is now commercially available (with pressure limits up to 150 MPa) and columns packed with particles <2 μm in diameter. A similar effect can be accomplished using conventional equipment with monolithic columns (rods made of a continuous separation media), which exhibit low flow resistance due to large flow pores, so that pressure limitations are less stringent. Alternatively, columns with a thin (sub 1 μm) porous shell on fused-core particles of small size (e.g., 2.7 μm) allow faster separations at the same pressure drop in comparison to the columns of the same length packed with the particles of the same size but with a totally porous structure. The thin shell provides a short diffusion path inside the particles, reducing thus the axial dispersion of solutes and peak broadening [1,10].

Column permeability also improves at elevated temperatures, owing to reduced viscosity of mobile phases.

9.2.5.3 Control of the Retention and Separation Selectivity: The Mobile Phase

For a successful HPLC separation, the appropriate selection of the mobile phase is equally important as the correct choice of the separation column. Once a suitable column is determined for the

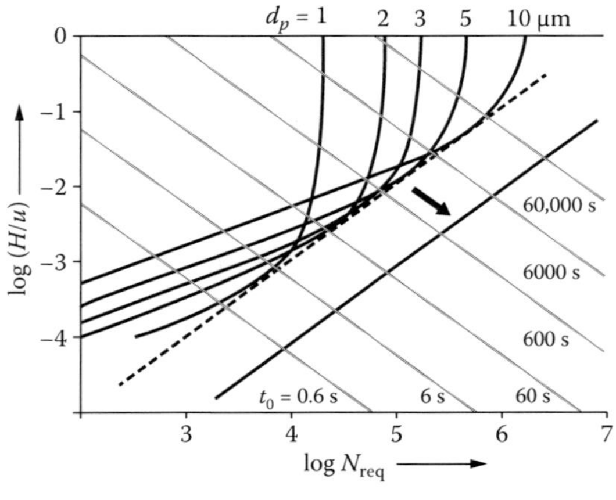

FIGURE 9.5 Kinetic plots for columns packed with particles of mean diameter, d_p. The upper limits of the "forbidden area" at ΔP_{max} 40 MPa (dashed line) and at 150 MPa (full line).

separation of a particular sample, the next step in the HPLC method development is the selection of the mobile phase. Rational systematic optimization may provide a faster and more efficient way to this end than the empirical trial-and-error approach. In NPLC, the elution strength increases whereas in RPLC it decreases with increasing solvent polarity.

Single-component mobile phases do not allow fine adjusting of the elution strength as there is only a limited selection of solvents compatible with UV and other common detection techniques, so that mixed mobile phases composed of solvents with different elution strengths should be used. An increase in the concentration of the stronger eluting component in a binary mobile phase enhances the elution strength and decreases the retention factors of sample solutes. The concentration of the stronger elution component, necessary to achieve the desired retention factor, k, can be predicted or at least estimated theoretically by calculation using, for example, Equation 9.1 or 9.2 in RP systems [1,2] and Equation 9.3 or 9.4 in NP systems [6,7]. In this way, the effects on the retention and separation selectivity are predicted over a broad range of the mobile phase composition, and the optimum mobile phase is determined, using a computer-assisted approach, as discussed in Section 9.3.

Even though adjusting the concentration ratio of the solvents in a binary mobile phase affects mainly the elution times, some changes in the separation selectivity may be observed, too, so that it is only rarely possible to change the selectivity and the retention fully independently of each other when optimizing an HPLC separation. The retention of weak acids and of weak bases in RPLC increases when the pH of the mobile phase is adjusted to suppress their ionization. As retention of weak acids or bases depends on their ionization, a change in pH may be an efficient tool for adjusting their separation selectivity [1].

Ternary and more complex mobile phases contain at least two strong solvents with different predominant selective polar contributions (dipole–dipole, proton-donor, and proton-acceptor) in a weak solvent (diluter)—water in RPLC and hexane or heptane in NPLC. Fine selectivity tuning is often possible by adjusting the concentration ratios of the strong solvents in three- or four-component mobile phases whereas the elution strength is controlled primarily by the concentration of the diluting solvent.

The solvent selectivity triangle (Figure 9.6) is a useful tool for adjusting the equilibrium between the selective dipole–dipole, proton-donor, and proton-acceptor contributions to the polarity of binary, ternary, or quaternary mobile phases. In RPC, the apices of the triangle correspond to pure organic solvents (or to their isoeluotropic mixtures with water). Acetonitrile has predominating dipole–dipole properties, tetrahydrofuran proton-acceptor, and methanol both proton-donor and proton-acceptor properties [12].

The sides of the triangle represent ternary solvent mobile phases, in which the distances from the apices correspond to the concentration ratios of two strong solvents, representing the proportions

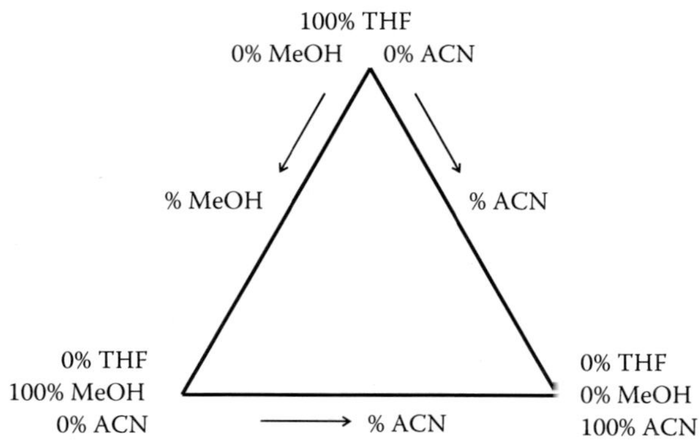

FIGURE 9.6 The solvent selectivity triangle in RPLC. MeOH: methanol; ACN: acetonitrile; THF: tetrahydrofuran.

of the individual selective contributions to the polarity. The coordinates of the points inside the triangle correspond to the proportions of the three strong eluting solvents, controlling the separation selectivity. The selectivity triangle can be applied also in NP chromatography, in which the apices represent a nonlocalizing solvent (dichloromethane), a basic localizing solvent (methyl-*t*-butyl ether), and a nonbasic localizing solvent (acetonitrile or ethyl acetate); *n*-hexane or *n*-heptane being used as diluting solvents to adjust the elution strength of the mixed nonaqueous mobile phases. More details on the computer-assisted mobile phase optimization are given in Section 9.4.

9.3 PROGRAMMED ELUTION TECHNIQUES

9.3.1 General Elution Problem

Practical samples often contain components widely differing in retention, for which HPLC at constant elution conditions (in the isocratic mode) may not yield sufficient separation. If the working conditions are adjusted for adequate retention of strongly retained solutes, some weakly retained components may elute too early as poorly, if at all, separated bands (see example in Figure 9.7b). On the other hand, if the conditions are adjusted for satisfactory separation of weakly retained compounds, the elution of strongly retained sample components may take a very long time; their bandwidths are excessively large, and the detector signal may even drop below the detection limits (Figure 9.7a) [1].

The "general elution problem" can be solved using two or more separated analyses of the sample with differing chromatographic conditions or by preseparating the sample into several fractions, each of which is separated in independent subsequent runs, which can be performed off-line or online, combining different columns and separation principles in multidimensional chromatography (see Section 9.3.3).

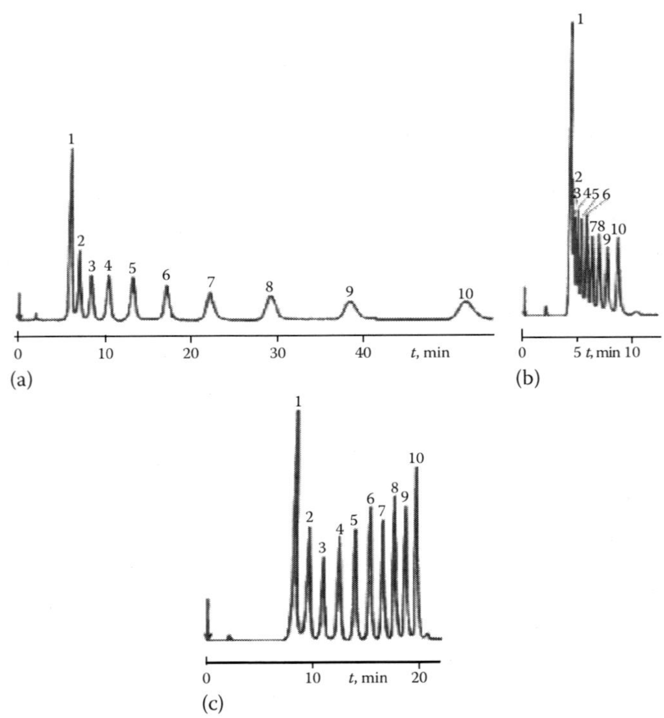

FIGURE 9.7 RP separation mobile phase: (a) 80% methanol in water, (b) 95% methanol in water, (c) linear gradient, 70%–100% methanol in 20 minutes, 1 ml/min. Numbers of the peaks agree with the numbers of carbon atoms in the alkyls.

On a single column, adequate retention of all sample components in one run can be adjusted by changing (programming) separation conditions—mobile phase composition or temperature—during the analysis (Figure 9.7c). Solvent gradients are generally much more efficient to decrease the retention than programmed temperature—the retention factor, k, of low-molecular-weight samples in RPLC decrease by a factor of two to three with a 10% increase in the concentration of organic solvent in an aqueous-organic mobile phase whereas an increase of temperature by 10°C usually leads to a decrease in k of nonionic compounds by 10%–20% [13].

9.3.2 Gradient Elution

The most widely used programming technique in HPLC is gradient elution, in which the elution strength of the mobile phase increases during the chromatographic run. In this way, a broader interval of retention is covered during a single-run gradient elution LC, and sample peaks are generally more regularly spaced than under isocratic conditions. Various sample compounds elute at similar local mobile phase composition at the time the sample zones leave the column. The zones of later eluting compounds are subject to less broadening than in isocratic elution because the migration velocities of the bands along the column accelerate during gradient elution so that all sample compounds eventually are eluted with very similar retention factors, k_e (Figure 9.7c) [14,15].

The main benefits of gradient elution are the following:

- Improved resolution of samples within the whole range of retention times.
- Increased number of peaks resolved within a fixed time of separation (the peak capacity).
- Improved separation of synthetic polymers and biopolymers, whose retention changes markedly even for very small changes in the composition of mobile phases.
- Suppressed peak tailing, especially for basic compounds.
- Initial "scouting" gradient experiments accelerate the development of separation methods.
- Interfering compounds are more easily removed using gradient elution.
- Fast generic gradients can be applied to a large number of samples to provide important information for fast sample screening or for generating impurity profiles.

On the other hand, some extra time is necessary for column re-equilibration after the end of the gradient. Some detection methods (e.g., refractive index or most electrochemical types) and column/mobile phase combinations cannot be used in gradient elution. Besides the UV detector, the only (almost) universal detector that can be used for gradient elution is the evaporative light-scattering detector, which is, however, less sensitive, and its use is restricted to volatile mobile phases and nonvolatile analytes. Gradient elution is usually well compatible with LC/MS techniques.

The gradient LC should accurately mix two or more (up to four) components of the mobile phase according to a preset time program. To this aim, either low-pressure or high-pressure gradient instrumental systems can be used [14].

Solvent gradients are by far the most frequently used gradient elution modes. In classical aqueous-organic RP-HPLC, the concentration(s) of one or more organic solvent(s) in water increase(s). In organic NPLC gradient elution, the concentration(s) of a polar organic solvent(s) in a less polar one increase(s). Because of the strong preferential adsorption of polar solvents in NPLC, the real gradient profile may deviate from the preset one, and column re-equilibration times after the end of the gradient are often long. It should be noted that even traces of water in the mobile phase may significantly decrease the adsorbent activity and the retention. To suppress these effects, gradients should be started at 3% or more of the polar solvent rather than at a zero concentration if possible. Reproducible NP gradient operation requires strict control of temperature, and it is recommended to use dehydrated solvents kept dry over activated molecular sieves and filtered before use [16]. In aqueous-organic NP (HILIC) gradient chromatography, the concentration of water (or of aqueous buffer) as the stronger eluent increases in a polar organic solvent (acetonitrile).

Salt (ionic strength) gradients, with increasing concentration of competing ions, are generally used in ion-exchange gradient LC, for example, for separation of complex peptides, proteins, and other biopolymer samples as a complementary technique to RP solvent gradients. The gradients usually start at a low salt (chloride, sulphate, etc.) concentration and typically run from 0.005 M to 0.5 M. A buffer is used to control the pH; acetonitrile and methanol may be added to improve the resolution and urea to improve the solubility of proteins that are difficult to dissolve. Ion exchangers with less hydrophobic matrices can be used to prevent protein denaturation in aqueous mobile phases.

RP hydrophobic interaction chromatography (HIC) employs gradients of decreasing ionic strength to diminish the retention of biopolymers due to the salting-out effect.

Ion-pairing reagent concentration gradients in the RP mode are not frequently used.

pH gradients are used in IEC, mainly for separations of proteins, peptides, and other ionic biopolymers. Protein molecules carry multiple negative charges and are strongly retained on an anion exchanger at pH higher than *pI*, but as soon as the pH drops below the *pI* during gradient elution, the initially strongly retained protein gets positively charged and is released from the ion exchanger rapidly. Hence, biopolymers elute roughly in the order of their iso-electric points as sharp band zones, comparing favorably to ion-strength gradients.

In a binary gradient, the concentration of the strong eluting component (solvent B) increases, and the concentration of the weak solvent A decreases according to a preset program, either continuously or comprised of a few consequent isocratic or short linear gradient steps. In a linear gradient, the volume fraction ϕ of the stronger eluent (solvent B) changes proportionally to the time t elapsed since the start of the gradient, usually identical with the time of the sample injection [14,15] (Equation 9.8):

$$\phi = \mathrm{A} + \mathrm{B}t \tag{9.8}$$

Here, A is ϕ at the start of the gradient, and B is the slope (steepness, ramp) of the change in ϕ (Equation 9.9):

$$B = \frac{\phi_G - A}{t_G} \tag{9.9}$$

ϕ_G is ϕ at the end of the gradient, in the time $t = t_G$ from the start of the gradient.

The retention data in gradient elution cannot be directly predicted from equations applying for isocratic RPLC, NPLC, or IEC modes (Equations 9.1 through 9.5) as the retention factors, k, are not constant but decrease in the course of gradient elution. However, taking into account a continuous decrease in k, equations enabling calculation of the retention times (or volumes) in various modes of gradient LC combined with various gradient programs were derived [14]. For linear solvent gradients in RPC, to which Equation 9.1 applies, the retention volumes, V_R, can be calculated using Equation 9.10 using the parameters a and m determined in isocratic or gradient scouting experiments [14,15]:

$$V_R = \frac{F_m}{mB}\log\left\{2.31\,m\frac{B}{F_m}\left[V_m 10^{(a-mA)} - V_D\right]+1\right\} + V_m + V_D \tag{9.10}$$

where A is the volume fraction of the organic solvent at the start of the gradient, B is the steepness of the gradient (Equation 9.9), V_m is the volume of the mobile phase in the column and V_D is the instrumental dwell volume between the injector and the gradient mixer filled with the starting mobile phase at the time of injection. The effects of V_D are especially significant with short and narrow columns and fast gradients. The gradient profile affects the retention in a similar way as the concentration of the strong solvent in a binary mobile phase under isocratic conditions: The retention decreases with steeper gradients (shorter gradient times) and gradients starting at higher initial concentrations of the solvent B (gradient concentration range) if the other conditions are kept constant (see Figure 9.8).

Gradient bandwidths, w_g, are usually considerably narrower than the bandwidths of late-eluting peaks under isocratic conditions and can be—to first approximation—estimated as the bandwidths under isocratic conditions at the elution time of the peak maximum with the instantaneous retention factor, k_e, assuming an approximately equal column plate number, N, from Equation 9.11 [17]:

$$w_g = \frac{4V_m}{\sqrt{N}}(1+k_e) = \frac{4V_m}{\sqrt{N}}\left[1+\frac{1}{2.31\,m\dfrac{B}{F_m}V_m + 10^{(mA-a)}}\right] \tag{9.11}$$

(w_g are in volume units). As the retention factors change during gradient elution, the average plate number, N, to be used in Equation 9.11 is determined under isocratic conditions. At increasing gradient steepness, B, and/or the starting concentration of the organic solvent, A, the gradient bandwidths

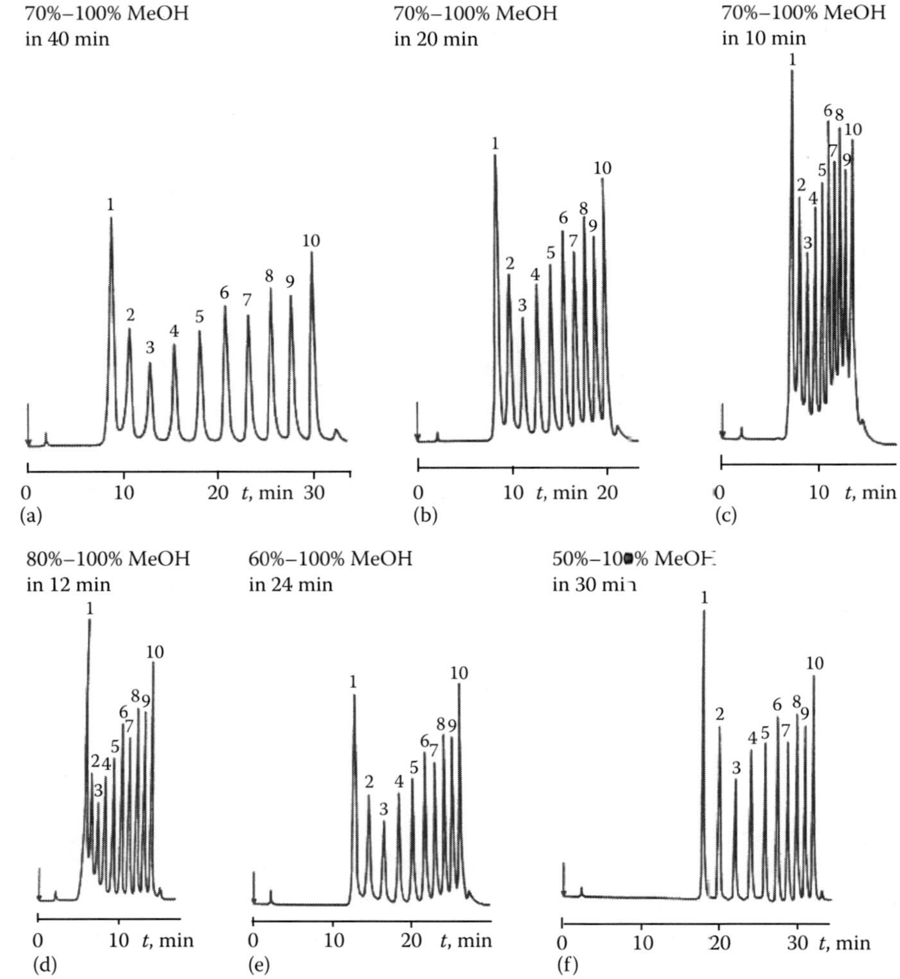

FIGURE 9.8 Gradient separation of 10 homologues on a C18 column. The effects of the (a–c) gradient time (10–40 min) at a constant gradient concentration range (70%–100% methanol) and (d–f) gradient range (20%–50% methanol) at a constant gradient range, 1.67% MeOH/min.

decrease. The experimental bandwidths are often slightly narrower than the w_g, calculated from Equation 9.11, due to the "additional gradient bandwidth suppression" caused by faster migration of the rear edge of the sample zone along the column in a stronger eluent (higher percentage of organic solvent) in comparison to the front edge; however, this effect can be neglected with very narrow peaks.

Introducing Equation 9.10 for retention volumes and Equation 9.11 for bandwidths into Equation 9.6, the effects of the gradient parameters A and B on the resolution, R_S, can be determined and the optimum gradient profile calculated by using a window-diagram approach or the DryLab G software (for more details, see Section 9.4).

If the separation with binary gradients is unsatisfactory, ternary gradients can sometimes improve the selectivity by changing, simultaneously, the concentrations of two solvents with high elution strengths (strong solvents) in a weak solvent. For example, the early-eluting compounds may show poor resolution with the gradients of methanol but are better separated with gradients of acetonitrile in water whereas the separation selectivity for the late-eluting compounds is better with a gradient of methanol than with gradients of acetonitrile in water or vice-versa. In such a case, a ternary gradient with an increasing concentration of methanol and simultaneously decreasing concentration of acetonitrile may improve the overall resolution [18].

The suitability of the LC separation system for resolving complex samples can be characterized by theoretical peak capacity, n_c, which determines the maximum number of peaks that can be accommodated side by side in the chromatogram at a desired degree of resolution (e.g., $R_S = 1.5$). The peak capacity strongly depends on the elution mode. Under isocratic conditions, the baseline bandwidths of the sample solutes, w_i, increase proportionally to increasing retention times at a constant column efficiency for all sample compounds (with approximately equal numbers of theoretical plates, N).

Assuming sufficient and approximately constant column efficiency (the number of theoretical plates, $N > 1000$), Giddings [19] derived a simplified equation, Equation 9.12, that can be used to estimate the theoretical isocratic peak capacity, $n(i)$:

$$n(i) = \frac{t_{R,z} - t_{R,1}}{\sum_{i=1}^{n-1} w_i} + 1 = \frac{\sqrt{N}}{4} \ln\left(\frac{t_{R,z}}{t_{R,1}}\right) + 1 = \frac{\sqrt{N}}{4} \ln\left(\frac{k_z + 1}{k_1 + 1}\right) + 1 \tag{9.12}$$

where k_1 and k_z are the retention factors of the first (1) and of the last (z) eluting compounds, respectively ($k = (t_R - t_m)/t_m$).

With the average bandwidth, w_g, calculated from Equation 9.11, the theoretical peak capacity under RP gradient conditions, $n(g)$, can be calculated from Equation 9.13 [20]:

$$n(g) \cong \frac{\sqrt{N}}{4}\left(\frac{t_{R,z}}{t_{R,1}} - 1\right) + 1 \cong \frac{\sqrt{N}}{4} \frac{1}{t_m} \frac{\Delta t_R}{1 + k_e} + 1 = 1 + \sqrt{\frac{N}{4}} \frac{t_G}{t_m} \frac{1}{1 + \left[10^{(mA-a)} + 2.31\, m \frac{\Delta\phi}{t_G} t_m\right]^{-1}} \tag{9.13}$$

where t_G is the gradient time range, which is considered equal to the elution interval, $t_G = \Delta t_R = t_{R,z} - t_{R,1}$, between the elution times of the first peak, 1, and of the last one, z, assuming that the whole chromatogram is regularly covered by sample peaks, stacked side by side. And t_m is the column hold-up time, often considered equal to the elution time of the first solute. Gradient elution covers a broader range of retention and provides more narrow peaks and almost constant bandwidths, especially for strongly retained compounds [14,15]; consequently, higher peak capacity can be achieved in gradient-elution mode than in isocratic chromatography. In practice, the "theoretical" peak capacity must be several times larger than the number of sample components for a reasonable probability that any peak in the chromatogram represents a single substance.

9.3.3 Column Switching

9.3.3.1 Column Serial Coupling

To improve the separation of complex samples, two or more columns coupled via one or more high-pressure switching valves can be used in combined chromatographic systems with different selectivities. Serially coupled columns can be used for online sample cleanup and/or enrichment on a first short trapping column, directly coupled to the second, analytical column. This arrangement is most frequently used in the analysis of environmental samples with aqueous matrices or in the automated determination of drugs in biological fluids by RPLC [21]. Because water has very low elution strength in RPLC, large sample volumes can be introduced onto a hydrophobic enrichment (trapping) column, usually packed with alkyl silica, while the analytical column is bypassed. All nonpolar and weakly polar sample compounds are strongly retained in a thin upper layer of the stationary phase in the trapping column whereas the aqueous matrix and strongly polar ballast compounds elute from the column and are directed to waste (on-column sample focusing). After washing, the valve is switched into the second position and the less polar trapped compounds are flushed from the trapping column onto the analytical column with the mobile phase used for the separation (Figure 9.9, full line—enrichment step, dashed line—separation step).

To avoid the precipitation of proteins in untreated plasma or serum samples directly injected onto RP columns, a short precolumn packed with restricted access sorbents can be used. The column packing particles contain two bonded layers. The inner pore surface is covered with a nonpolar retentive bonded phase. The external surface bonded layer does not allow penetration of proteins, which elute within the hold-up volume of the precolumn and do not interfere with the separation of target compounds (pharmaceuticals, metabolites) [22].

Another purpose of using coupled columns is improving separations of complex samples and increasing the number of detected sample components by subsequent separation in two or more chromatographic systems with different separation selectivities.

9.3.3.2 Multidimensional HPLC

Usually, samples containing up to a few tens of compounds can be separated on a single column in a unidimensional HPLC system. Combinations of two or more different separation systems can significantly increase the number of resolved compounds in a complex sample [23]. Moreover, group separations of various structurally related classes of compounds can be distinguished and located in different areas of the multidimensional retention space.

Two-dimensional separations can be performed either off-line or in real time using online coupled chromatographic systems, according to the heart-cutting, stop-and-go, or comprehensive scheme. In the off-line 2-D systems, which are the most simple and easy to operate, the fractions collected from the first dimension are stored and usually preconcentrated prior to subsequent introduction onto the

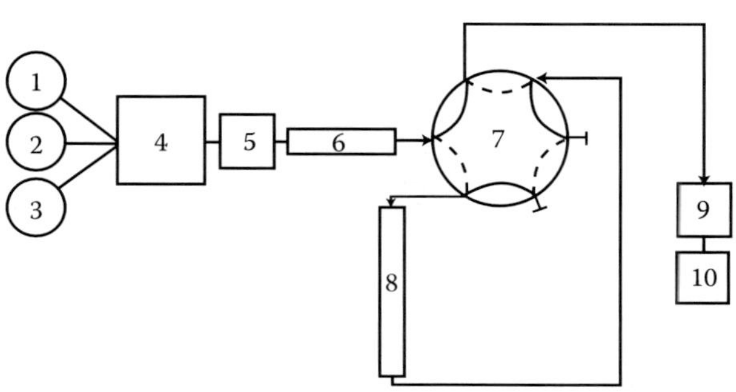

FIGURE 9.9 Online sample cleanup and enrichment setup. 1, 2, 3: mobile phase and washing liquid reservoirs; 4: pump; 5: sample injector; 6: enrichment column; 7: switching valve; 8: analytical column; 9: UV detector; 10: waste reservoir.

second column. However, off-line procedures are usually time-consuming, difficult to automate, and may be subject to sample loss or contamination.

The "heart-cutting" setup is used for online sample transfer between two serially coupled columns, in which only one or a few selected fractions are collected from the first column effluent and directly reinjected into the second dimension separation system for analysis while the remaining effluent is bypassed to the waste. In the "stop-and-go" approach, a selected fraction of the column effluent is transferred to the second dimension; the elution from the first-dimension column is stopped for the time necessary to analyze the fraction on the second-dimension column. After switching an interface valve, the elution in the first dimension is resumed again to continue in the separation of the sample temporarily "parked" in the first-dimension column. The approach is repeated for as many fractions as necessary [24].

In "comprehensive" 2-D HPLC, the whole sample (or its aliquot part) is subject first to the separation in the first dimension and then in the second dimension [24]. The instrumental setup employs two columns coupled via an interface, usually an eight-port or a ten-port switching valve or two six-port switching valves with two identical-volume sampling loops or trapping columns. The subsequent low-volume fractions are analyzed in the real time in multiple repeated alternating cycles, in which the two loops are regularly switched between the collecting and the elution positions (Figure 9.10). While one loop (or trapping column) is collecting a fraction from the first dimension, the previous fraction contained in the other loop is released and analyzed in the second dimension. After the end of the separation of the transferred fraction in the second dimension, the interface modulator valve is switched into the other position, and the function of the loops is interchanged.

Fairchild et al. [25] compared the relative merits of the off-line, stop-and-go, and comprehensive approaches in terms of peak capacity and overall ("aggregate") time necessary to accomplish a 2-D separation. The online approach is best if the desired peak capacity is to be achieved in a short separation time, such as $n_c = 300–500$ in 30 minutes with the second gradient times of 1–2 minutes or even less. When larger peak capacities are needed, long analysis times should be accepted using an off-line approach with longer second-dimension columns or slower second-dimension gradients. The stop-and-go scheme represents a compromise between the off-line and real-time online approaches; peak capacities up to 2000–3000 can be achieved in 2–4 hours.

In 2-D chromatography, the peak capacity significantly increases and theoretically approaches the product of the capacities in the individual dimensions, n_{c1}, and n_{c2} [26] (Equation 9.14):

$$n_{c,2D} = n_{c1} \times n_{c2} \tag{9.14}$$

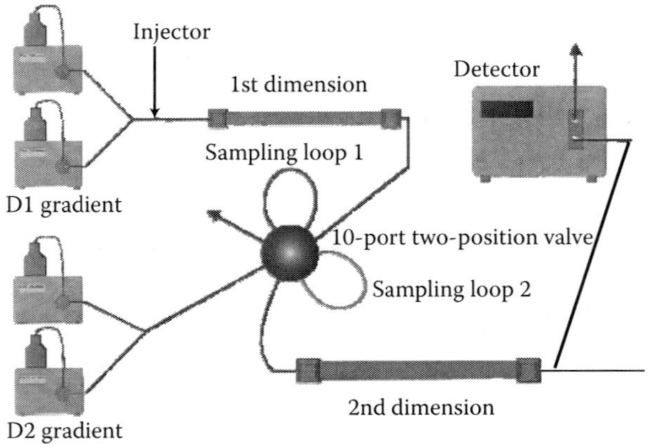

FIGURE 9.10 Scheme of a comprehensive LC × LC setup with a 10-port switching valve interface and two sampling loops.

Unfortunately, in the real online coupled 2-D systems, the compounds separated in the first dimension are remixed in a collected fraction, which diminishes the theoretical 2-D peak capacity predicted from Equation 9.1. According to the Murphy-Schure-Foley rule, at least three to four fractions should be collected per one first-dimension peak width for the second-dimension separation if the resolution obtained in the first dimension is not to be significantly impaired by sample remixing due to "undersampling" [27].

The practical 2-D LC × LC peak capacity decreases with increasing similarity of the separation properties of the columns used in the first and in the second dimension [28]. "Orthogonal" 2-D systems with widely different retention mechanisms should be used to provide statistically independent retention times of sample compounds and best coverage of the available 2-D retention space [23].

Orthogonality, which characterizes the differences between the first- and second-dimension systems, increases in proportion to the dissimilarity of separation selectivities. The orthogonality coefficient, s^2, characterizing the differences in the separation selectivities, is complementary to the correlation coefficient, r^2, between the retention times in two separation systems [29] (Equation 9.15):

$$s^2 = 1 - r^2 \tag{9.15}$$

where $s^2 = 1$ for fully orthogonal systems, whereas $s^2 = 0$ for equivalent separation systems. Unfortunately, even careful selection of separation systems often cannot completely avoid some selectivity correlations for practical samples.

Various LC modes can be combined in 2-D LC × LC systems to suit specific separation problems. Combinations of two RP systems with different separation selectivities (RP × RP) are most frequently used in real-time 2-D online gradient setups. As a rule, nonspecific interactions based on the differences in lipophilicity (in the size of the nonpolar part of the sample molecules) are utilized for the separation on a bonded alkyl silica (or other nonpolar) column in one dimension whereas the other dimension usually employs a column with a more polar stationary phase.

Online combination of strong cation-exchange or anion-exchange chromatography, using a salt concentration gradient in the first dimension, coupled with the RPLC and solvent gradient in the second dimension, and mass spectrometry (MS) or MS/MS identification, are most common in the 2-D separations of peptides today.

The second-dimension analysis in comprehensive LC × LC should be accomplished in a short time equal to the period of the fraction collection, including the time necessary for the fraction transfer, usually one or two minutes or even less. The time constraint seriously limits the number of compounds that can be separated during one second-dimension analysis, that is, the peak capacity in the second dimension is much lower than the peak capacity in the first dimension in the comprehensive LC × LC setup [30,31]. Gradient elution should be used wherever possible in 2-D LC × LC separations as it offers significant increase in peak capacity (Equation 9.11) and enables separation of samples with widely differing retention [32,33].

Mobile phase compatibility may present a serious problem when transferring online fractions between the RP systems with aqueous-organic mobile phases and NP systems using purely organic mobile phases immiscible with water and vice versa. Poor miscibility of mobile phases and deactivation of polar adsorbents, such as bare silica gel, or nonmatching elution strengths in the two modes may seriously impair or completely destroy the separation. Therefore, RP systems are, as a rule, connected off-line to NP or HILIC systems [34]. Online coupling of NP and RP systems is possible using an interface, in which a volatile organic mobile phase from the fractions collected in the first dimension is evaporated under vacuum before the transfer into the second, RP dimension, however, often at a cost of incomplete recovery of sample compounds with low boiling points. A possible alternative solution of the system compatibility problem is in transferring very small fraction volumes online from the first, NPLC or HILIC dimension, to the RPLC second dimension [34,35].

Figure 9.11 shows an example of 2-D separation of phenolic compounds and flavonoids using simultaneous gradient elution with decreasing concentration of acetonitrile in aqueous buffer in

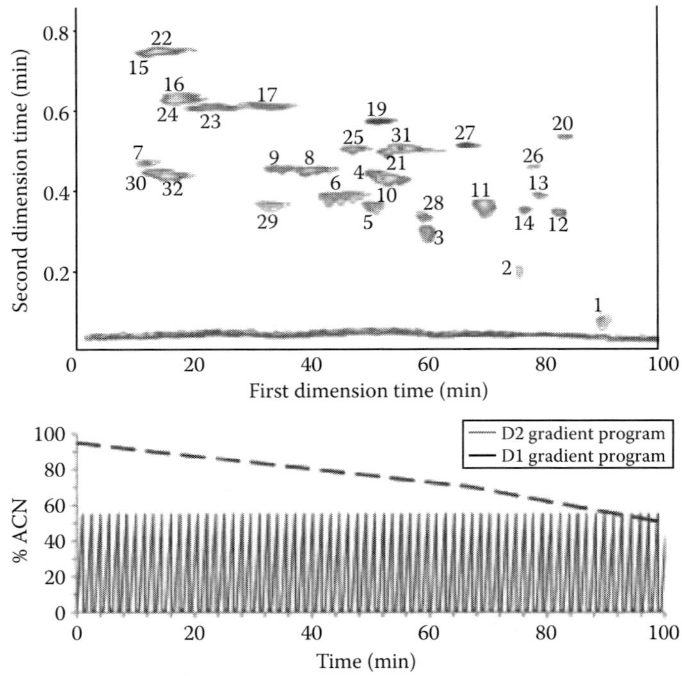

FIGURE 9.11 2-D comprehensive LC × LC separation of 32 phenolic acids and flavones on a zwitterionic monolithic column (210 × 0.53 mm inner diameter) in the first dimension and a Kinetex XB-C18 core-shell column (30 × 3 mm inner diameter) in the second dimension, using simultaneous gradient elution with decreasing, D1 (HILIC) and increasing, D2 (RP) concentration of acetonitrile 0.01 mol/L CH_3COONH_4 (pH = 3.1). The profiles of acetonitrile gradients are shown in the bottom figure (dashed lines: D1; full lines: D2). Interface: 10-port switching valve with two 10-μL sampling loops, switching time 1.5 min (A) or 1 min (B, C).

the first, HILIC dimension (with a zwitterionic monolithic microcolumn) and with increasing concentration of acetonitrile in aqueous buffer in the second, RP dimension (with a core-shell short column) [35].

The objective of the optimization of 2-D separations may be either obtaining the desired resolution (peak capacity) within as short a time as possible or accomplishing the highest possible peak capacity in a fixed analysis time. To this end, suitable separation systems with a high degree of orthogonality should be selected, such as combining the separation principles based on the differences in lipophilicity on a nonpolar column in one dimension and the differences in selective interactions on a more polar column in the other one.

For comprehensive online 2-D separations, matching column dimensions and flow rates should be adjusted. In the first dimension, a long narrow-bore column and a low flow rate provide high peak capacity while keeping the volume of the fractions transferred to the second dimension low enough to preserve the first-dimension resolution. Fast separations in the second dimension can be achieved using short and efficient monolithic columns, columns packed with sub 2 μm porous or core-shell particles at a high flow rate [32,33].

9.3.4 High Temperature Operation and Temperature Gradients

The regulation of temperature is convenient and simple and can be used for optimizing the resolution, for example, using the window diagram strategy, such as the optimization based on adjusting the composition of the mobile phase, but is usually less effective for improving HPLC separations. There are some advantages of high-temperature LC in comparison to ambient operation:

A lower solvent viscosity and increasing diffusion coefficients allow using smaller particle size and/or higher flow rates of the mobile phase at the same operation pressure to provide higher efficiency and/or speed of the analysis. At high temperatures, the speed of UHPLC with sub 2 μm particles is further promoted. Last but not least, at elevated temperatures, the polarity of water decreases so that pure water often can be used as the mobile phase, providing retention comparable to aqueous-organic solvent mixtures containing methanol or acetonitrile.

On the other hand, the useful temperature range is limited for thermally labile compounds and for many HPLC stationary phases chemically bonded on silica gel supports, which may be subject to enhanced hydrolysis and dissolution of the silica support at temperatures higher than 60°C–70°C. More stable columns based on organic polymers, carbon, titanium dioxide, or zirconium dioxide supports with carbon deposited on the surface or modified with C18, polybutadiene, polystyrene, and other ligands show excellent long-term chemical and thermal stability and withstand extended exposure to column temperatures as high as 200°C [36], unfortunately often at a cost of inferior efficiency in comparison to bonded silica gel columns, tailing peaks, or irreversible retention of some compounds. Great effort is focused on the development of new, more stable, silica-based columns, showing improved resistance against hydrolysis and dissolution, such as polydentate-bonded phases with a bridged structure or stationary phases with hybrid organic–silica or hydrosilated silica supports, which will certainly promote extended use of high-temperature HPLC in the future.

The retention in HPLC usually decreases with increasing temperature; however, a large rise in temperature during the run would be required to reduce significantly the retention of strongly retained compounds. The effects of column temperature on the sample retention factor, *k*, can be described by Equation 9.16 [37]:

$$\ln k = -\frac{\Delta H}{RT} + \frac{\Delta S}{R} + \ln \Phi \tag{9.16}$$

where R is the universal gas constant, T is the thermodynamic temperature in Kelvins, and ΔH_0 and ΔS_0 are the molar enthalpy and entropy of a solute transfer from the mobile phase to the stationary phase, respectively. Φ is the phase ratio in the column (the ratio of the volumes of the stationary and the mobile phases). The retention factor, *k*, of small molecules usually decreases by approximately 1% to 2% when the column temperature increases by 1°C, but the temperature effects are much more significant for macromolecular compounds, with which temperature control is an important tool for optimization of separation.

Equation 9.16 applies reasonably well if the retention mechanism does not change over the experimental temperature range (see example in Figure 9.12).

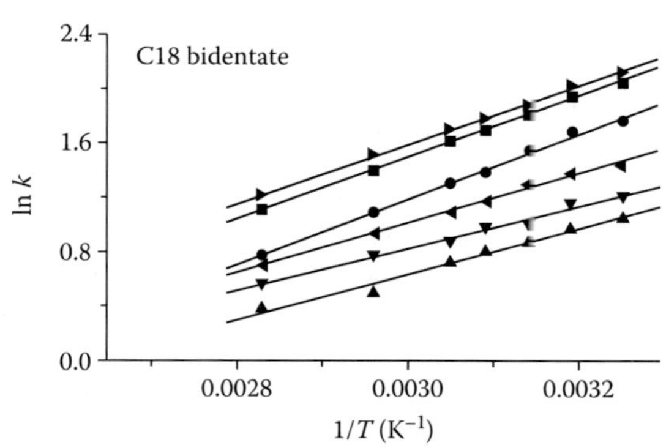

FIGURE 9.12 Temperature effects on the retention factor, *k*, of phenolic acids on a C18 bidentate column. Mobile phase: 10 mM NH_4AC in 85:15% water:acetonitrile (pH 3.26); T: thermodynamic temperature in K.

However, if the sample molecular conformation, the solvation, or the conformation of the stationary phase changes at increasing temperature, nonlinear ln k versus $1/T$ plots may be observed as is often the case with biopolymers (proteins). A change in temperature may cause a change in separation selectivity, $r_{1,2}$, when sample molecules have different functional groups, relative size, or shape.

Temperature programming does not necessitate expensive gradient pumps, but it is rarely used in HPLC in contrast to gas chromatography. One reason is that slow radial heat transfer in columns with an inner diameter larger than 1 mm and slow heat dissipation across the column walls may give rise to undesired radial temperature gradients, resulting in a lower than expected efficiency (plate number) at a high temperature unless the mobile phase is preheated. The heat dissipation is much improved in capillary LC columns, in which temperature programming is more robust than with conventional columns. Consequently, temperature gradients are a promising alternative to solvent gradients in microcolumn and capillary LC, especially for separations of synthetic polymers [38].

The effects of the solvent strength and temperature are often complementary so that simultaneous control of temperature and mobile phase in RPLC is an efficient tool in the optimization of HPLC separations. The approach can be based on the simple Equation 9.17 [39]:

$$\ln k = a_1 + \frac{a_2}{T} + \phi\left(m_1 - \frac{m_2}{T}\right) \tag{9.17}$$

where ϕ is the volume fraction of the organic solvent in the aqueous-organic mobile phase; T is the thermodynamic temperature; and a_1, a_2, and m_1 and m_2 are parameters depending on the solute and the stationary phase. To acquire these parameters, at least four initial experiments should be run at two different temperatures and mobile phase compositions to allow designing the resolution map as a function of both the temperature and the mobile phase composition. In a similar way, simultaneous optimization of temperature and gradient elution can be performed [40].

9.4 COMPUTER-ASSISTED METHOD DEVELOPMENT

Computers significantly facilitate and speed up systematic optimization of binary or more complex mobile phases, pH, or the gradient elution program. Temperature, flow-rate of the mobile phase, the column type, and dimensions can also be subject to computer optimization procedures. Different strategies can be used to optimize either a single operation parameter or more parameters at a time. Multiparameter optimization strategies take into account possible synergistic effects of various operation conditions. The simultaneous multiparameter optimization approach was introduced originally by Glajch et al. [41] for the optimization of the composition of ternary or quaternary mobile phases, especially in RP HPLC, but it also can be used for simultaneous optimization of the composition of the mobile phase and temperature or of the concentration of organic solvent in water and pH. Excellent discussion of this topic can be found elsewhere [42]. A nice collection of the early contributions to the theory and practice of HPLC modeling was compiled by Glajch and Snyder [43].

Single-parameter optimization strategies predict the resolution as a function of the optimized parameter (such as the concentration of the strong solvent in a binary mobile phase, pH, temperature, etc.), using empirical or simple model-based calculations to determine the effect of the optimized parameter on the desired values of relative retention or resolution of the sample compounds. Some initial experimental data necessary for the prediction of retention or resolution of a sample should be determined in several "scouting experiments." By increasing the amount of input data, the accuracy of the prediction improves; on the other hand, the time necessary for the data acquisition increases.

In the "window diagrams" approach, the plots are constructed showing the effects of the optimized operation parameter on the suitable optimization criterion, such as resolution, or the separation factors for all relevant sample components over the experimental range of the optimized parameter. These plots limit the "window" area, showing the allowed range of the adjusted parameter (resolution or separation selectivity) that can be achieved for all adjacent bands in the chromatogram. In the window diagram, the optimum value of the operation parameter can be found, either for the maximum value of the optimization criterion or for the shortest separation time.

The window diagram strategy can be essentially applied for a single-parameter optimization of any experimental condition, most often the composition of binary mobile phases, the gradient time, or the gradient range; it can be used also in a complementary way for the optimization of 2-D separations [44]. An example of a window diagram in Figure 9.13 illustrates the approach for the optimization of the starting concentration of the strong solvent (methanol) in the gradient separation in the RPLC of phenylurea pesticides [45].

Some operation parameters may show synergistic effects on the separation. In this case, simultaneous optimization of two or more parameters at a time can provide better results than their subsequent optimization. Simplex methods can be used for sequential multiparameter optimization in HPLC.

Several parameters, such as the composition of multicomponent mobile phases, can be optimized simultaneously using the overlapping resolution mapping (ORM) strategy. Here, the values of the resolution determined on the basis of several scouting experiments are plotted in a plane defined by

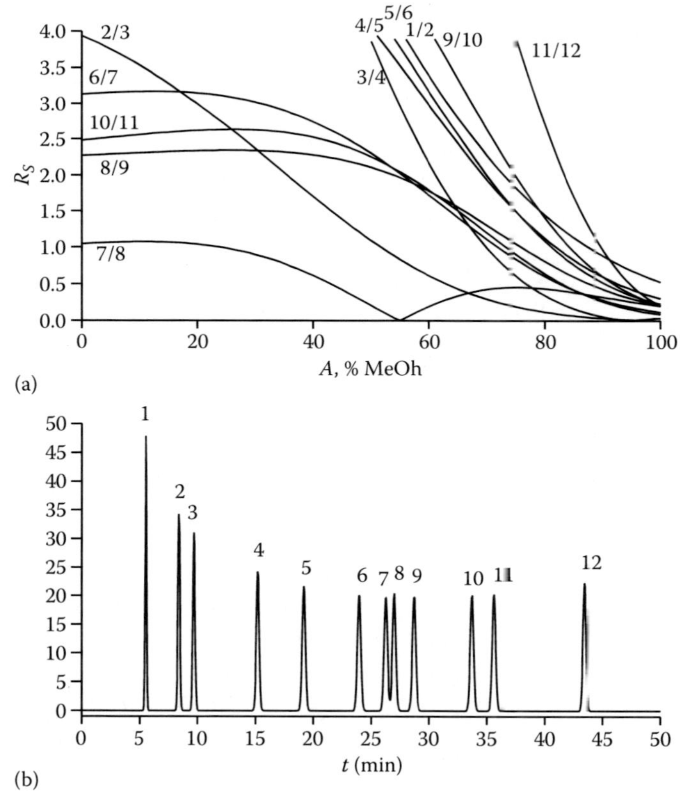

FIGURE 9.13 (a) The window diagram employed for the optimization of gradient separation of phenylurea herbicides on a Separon SGX C18 column (150 × 3.3 mm inner diameter). The volume fraction at the start of the gradient, *A*, was optimized for the optimum constant gradient time (73 mL at 1 mL/min). (b) The separation with the optimized gradient yielding the resolution, $R_S > 1$. Compounds: 1: hydroxymetoxuron; 2: desphenuron; 3: phenuron; 4: metoxuron; 5: monuron; 6: monolinuron; 7: chlorotoluron; 8: metobromuron; 9: diuron; 10: linuron; 11: chlorobromuron; 12: neburon.

the optimized operation parameters for all pairs of compounds with adjacent peaks. The "allowed" parameter range, yielding higher resolution for all sample compounds than the preset threshold,for example, $R_S \geq 1.5$, corresponds to the area outside the overlapping resolution plots. In this range, the optimum operation parameter is determined, yielding either the highest resolution or the desired resolution in the shortest time of separation. The ORM approach has been originally developed for optimization of separation selectivity in mixed three- or four-component mobile phases (methanol, acetonitrile, and tetrahydrofuran in water or in a buffer) in a solvent selectivity triangle (Figure 9.6). Seven or 10 initial experiments with solvent mixtures of approximately equal elution strength selected at regular intervals from the selectivity triangle area are run to "scout" the effect of the mobile phase composition on the resolution. Based on the retention data from the initial experiments, graphs are constructed by interpolation, showing the plots of the desired resolution of all relevant sample component combinations with adjacent peaks in the selectivity triangle space, mapping all combinations of the concentrations of the three organic solvents. From the "resolution map," the composition of the multicomponent mobile phase that provides maximum resolution can be selected. Instead of the concentrations of one or two solvents, any combination of two operation parameters, such as pH of the mobile phase or temperature, can be optimized.

9.4.1 DryLab Software

Probably the best-known commercial mobile phase optimization program based on the resolution map approach is the Dry-Lab software, which can be used to adjust the composition of a binary mobile phase and includes options for optimization of ternary mobile phases, pH, temperature, or of the gradient time and range. The approach was introduced by L. Snyder and coworkers in 1988 for RP separations. The DryLab I software allowed modeling of retention times, band spreading and resolution for optimization of the volume fraction of the strong eluent B (organic solvent) in aqueous-organic mobile phases under isocratic conditions (single-parameter optimization) [46]. The DryLab G software followed, enabling optimization of the concentration ramp of the organic solvent in gradient elution [47]. The next improvement was the DryLab Imp software, enabling modeling the effect of pH, temperature, ionic strength, ion-pairing mobile phase additives, or isocratic ternary mobile phase composition. The tG-T-model allowed 2-D modeling of gradient time (t_G) and temperature (T). Various factors, such as column length and inner diameter, particle size, flow rate, dwell volume, and %B at the start and at the end of the gradient could be included in the modeling [48]. The strong feature of the DryLab is the computer visualization. Both the peak retention times and areas are freshly calculated for all sample compounds at any actual combination of the operation parameters, and both the resolution map and simulated chromatogram are shown on the computer screen, and the user can follow peak movement caused by changing one operation parameter at a time (%B, gradient time, pH) while keeping the other factors constant. The experimental conditions can be modified until the desired sample resolution is obtained. A table with the calculated optimized operation parameters, the expected chromatographic data, and the final simulated optimum chromatogram can be saved and printed.

In some instrumental systems, the DryLab software can be merged with the chromatographic station software, controlling the actual setting of operation parameters to enable automated method optimization. For this purpose, the DryLab model was extended into three dimensions (the Cube model) enabling simultaneous optimization of three operation parameters, for example, temperature, gradient time, and pH, according to an experimental design created on the basis of several initial isocratic or gradient runs (four or more) [49]. The approach needs peak tracking (usually based on the comparison of peak areas after each change of experimental conditions) to guarantee reliable peak assignment to sample components and to avoid misinterpretation of the results of optimization. If correctly used, this strategy may significantly shorten the time necessary for method development.

9.4.2 ChromSword Software

Structure-based commercial optimization software incorporates some features of the "expert system" as the retention is predicted on the basis of the additive contributions of the individual structural elements, and consequently, optimum composition of the mobile phase is suggested for mobile phase optimization in RPC on the basis of the molecular structures of sample components (which should be known) and the earlier investigated behavior of model compounds on various HPLC columns. This approach was pioneered by Galushko, who named the software first ChromDream; later, it was renamed ChromSword and was intended for the development of HPLC methods in drug design and analysis [50]. The AutoChromSword software enables collecting the data in a series of experiments run overnight. However, such predictions are only approximate and do not take into account stereochemical and intramolecular interaction effects, and predicted separation conditions can be used rather as the recommendation for the initial experimental run in the subsequent optimization procedure.

9.4.3 Other Software and Computer-Assisted Methods of Optimization

The retention and the selectivity of separation in RP and NP systems depend primarily on the chemistry of the stationary phase and on the mobile phase, which control the polarity of the separation system. There is no generally accepted definition of polarity, but it is agreed that it includes various selective contributions of dipole–dipole, proton-donor, proton-acceptor, π–π electron, or electrostatic interactions. For rational selection of suitable column chemistry, a linear free energy relationship (LFER) or another quantitative structure-retention relationship (QSRR) approach can be used [51,52].

LFER models are widely used to characterize chemical and biochemical processes and were successfully applied in LC to describe QSRRs and to characterize structural contributions to the retention and selectivity in LC, using multiple linear correlation, such as Equation 9.18 [53]:

$$\log k = (\log k)_0 + m_1 V_X / 100 + s_1 \pi_2^* + a_1 \alpha_2 + b_1 \beta_2 \tag{9.18}$$

The molecular structural descriptors in Equation 9.18 characterize various structural effects influencing the retention: molar volume of a solvated solute, V_X, polarity, π_2^*, hydrogen-bonding basicity, β_2, and hydrogen-bonding acidity, α_2, and Equation 9.18 allows the predictions of retention of different solutes on the same column and in the same mobile phase from the molecular descriptors, which can be found in published databases [53]. The parameters of the LFER equation (Equation 9.6), m_1, s_1, a_1, and b_1, obtained using multivariate simultaneous least-square regressions of the experimental retention data for selected standard compounds are characteristic for specific combinations of the mobile and the stationary phases and therefore can be used for HPLC method development. The stationary phases providing large differences in the LFER parameters for the individual sample components are likely to provide high specific polar selectivity for separations of specific sample types.

To better distinguish the contributions of polar interactions to retention, the LFER model was transformed into the hydrophobic subtraction model (HSM) for RPLC, in which the hydrophobic contribution to the retention is compensated for by relating the solute retention to a standard nonpolar reference compound. This approach was applied to characterize more than 300 stationary phases for RPLC, including silica gel supports with bonded alkyl-, cyanopropyl-, phenylalkyl-, and fluoro-substituted stationary phases and columns with embedded or endcapping polar groups. The LFER and HSM QSRR-based models can be used to characterize and compare the suitability of columns not only for RP, but also for NP and HILIC systems [52].

Recently, the stationary phase optimized selectivity in liquid chromatography (POPLC) strategy was introduced for selecting the suitable column for specific RP separations [54]. The principle is basically similar to the "PRISMA" approach, suggested earlier by Nyiredy et al. [55] for the selection of optimum composition of ternary mobile phases. A sample is injected under the same isocratic conditions on five bonded-phase columns with different chemistries: (i) An endcapped C18 column with predominant hydrophobic interactions; (ii) a C18 column with an embedded amide group, providing additional preferential interactions with carbonyl, carboxylic acid, and amino compounds; (iii) a phenyl column with preferential adsorption of aromatic compounds due to π–π interactions; (iv) a nitrile column with enhanced dipole–dipole interactions; and (v) a C30 column with enhanced retention of planar compounds with respect to nonplanar ones. Based on the experimental retention data, software calculations are performed to design a tailor-made POPLC column combining various lengths of up to 25 serially connected segments of the five basic stationary phases, 10 mm each, to provide optimum resolution of a specific sample. The approach with serially connected stationary phases is easily performed in a common HPLC laboratory because the different stationary phase segments can be easily dismantled and recombined for other separations.

REFERENCES

1. Snyder, L. R., Kirkland, J. J. Glajch, J. L., *Practical HPLC Method Development*, 2nd Ed. New York: Wiley-Interscience, 1997.
2. Berendsen, G. E., De Galan, L., Role of the chain length of chemically bonded phases and the retention mechanism in reversed-phase liquid chromatography. *J. Chromatogr.*, 196, 21–37, 1980.
3. Snyder, L. R., Dolan, J. W., Gant, J. R., Gradient elution in high-performance liquid chromatography: I. Theoretical basis for reversed-phase systems. *J. Chromatogr.*, 165, 3–30, 1979.
4. Jandera, P., Churáček, J., Svoboda, L., Gradient elution in liquid chromatography X. Retention characteristics in reversed-phase gradient elution chromatography. *J. Chromatogr.*, 174, 35–50, 1979.
5. Schoenmakers, P. J., Billiet, H. A. H., De Galan, L. Description of solute retention over the full range of mobile phase compositions in reversed-phase liquid chromatography. *J. Chromatogr.*, 282, 107–121, 1983.
6. Jandera, P., Churáček, J., Gradient elution in liquid chromatography I. The influence of the composition of the mobile phase on the capacity ratio (retention volume, bandwidth, resolution) in isocratic elution—Theoretical considerations. *J. Chromatogr.*, 91, 207–221, 1974.
7. Jandera, P., Janderová, M., Churáček, J., Gradient elution in liquid chromatography VIII. Selection of the optimum composition of the mobile phase in liquid chromatography under isocratic conditions. *J. Chromatogr.*, 148, 79–97, 1978.
8. Alpert, A. J., Hydrophilic-interaction chromatography for the separation of peptides, nucleic acids and other polar compounds. *J. Chromatogr.*, 499, 177–196, 1990.
9. Jandera, P., Stationary and mobile phases in hydrophilic interaction chromatography. *Anal. Chim. Acta*, 692, 1–25, 2011.
10. Neue, U. D. *HPLC Columns: Theory, Technology and Practice.* New York: Wiley-VCH, 1997.
11. Poppe, H., Some reflections on speed and efficiency of modern chromatographic methods. *J. Chromatogr. A*, 778, 3–21, 1997.
12. Glajch, J. L, Kirkland, J. J., Squire, K. M., Minor, J. M., Optimization of solvent strength and selectivity for reversed-phase liquid chromatography using an interactive mixture-design statistical technique. *J. Chromatogr.*, 199, 57–79, 1980.
13. Jandera, P., Blomberg, L., Lundanes, E., Controlling the retention in capillary LC with solvents, temperature, and electric fields. *J. Sep. Sci.*, 27, 1402–1418. 2004.
14. Jandera, P., Churáček, J. *Gradient Elution in Column Liquid Chromatography*. Amsterdam: Elsevier, 1985.
15. Snyder, L. R., Dolan, J. W., *High-Performance Gradient Elution. The Practical Application of the Linear-Solvent-Strength Model*. Hoboken, NJ: Wiley-Interscience, 2007.
16. Jandera, P., Gradient elution in normal-phase HPLC systems. *J. Chromatogr. A*, 965, 239–261, 2002.
17. Jandera, P., Churáček, J., Gradient elution in liquid chromatography II. Retention characteristics (retention volume, bandwidth, resolution, plate number) in solvent-programmed chromatography—Theoretical considerations. *J. Chromatogr.*, 91, 223–235, 1974.

18. Jandera, P., Churáček, J., Colin, H., Gradient elution in liquid chromatography XIV. Theory of ternary gradients in reversed-phase liquid chromatography. *J. Chromatogr.*, 214, 35–46, 1981.
19. Giddings, J. C., Maximum number of components resolvable by gel filtration and other elution chromatographic methods. *Anal. Chem.*, 39, 1027–1028, 1967.
20. Neue, U. D., Carmody, J. L., Cheng, Y. F., Lu, Z., Phoebe, C. H., Wheat, T. E., Design of rapid gradient methods for the analysis of combinatorial chemistry libraries and the preparation of pure compounds. *Adv. Chromatogr.*, 41, 93–136, 2001.
21. Vanvliet, H. P. M., Bootsman, T. C., Frei, R. W., Brinkman, U. A. T. On-line trace enrichment in high-performance liquid chromatography using a pre-column. *J. Chromatogr.*, 185, 483–495, 1979.
22. Haginaka, J., Drug determination in serum by liquid chromatography with restricted access stationary phases. *TRAC-Trends Anal. Chem.*, 10, 17–22, 1991.
23. Blahova, E., Jandera, P., Cacciola, F., Mondello, L., Two-dimensional and serial column reversed-phase separation of phenolic antioxidants on octadecyl-, polyethyleneglycol- and pentafluotophenylpropyl silica columns. *J. Sep. Sci.*, 29, 555–566, 2006.
24. Schoenmakers, P. J., Marriott, P. J., Beens, J. Comprehensive two-dimensional gas chromatography: A powerful and versatile analytical tool. *LCGC Eur.*, 16, 335–339, 2003.
25. Fairchild, J. N., Horvath, K., Guiochon, G., Theoretical advantages and drawbacks of on-line, multidimensional liquid chromatography using multiple columns operated in parallel. *J. Chromatogr. A*, 1216, 6210–6217, 2009.
26. Davis, J. M., Giddings, J. C., Statistical theory of component overlap in multicomponent chromatograms. *Anal. Chem.*, 55, 418–424, 1983.
27. Murphy, R. E., Schure, M. R., Foley, J. P., Effect of sampling rate on resolution in comprehensive two-dimensional liquid chromatography. *Anal. Chem.*, 70, 1585–1594, 1998.
28. Davis, J. M., Stoll, D. R., Carr, P. W., Dependence of effective peak capacity in comprehensive two-dimensional separations on the distribution of peak capacity between the two dimensions. *Anal. Chem.*, 80, 8122–8134, 2008.
29. Neue, U. D., O'Gara, J. E., Mendez, A., Selectivity in reversed-phase separations: Influence of the stationary phase. *J. Chromatogr. A*, 1127, 161–174, 2006.
30. Guiochon, G., Marchetti, N., Mriziq, K., Shalliker, R. A., Implementations of two-dimensional liquid chromatography. *J. Chromatogr. A*, 1189, 109–168, 2008.
31. Jandera, P., Column selectivity for two-dimensional liquid chromatography. *J. Sep. Sci.*, 29, 1763–1783, 2006.
32. Stoll, D. R., Carr, P. W., Fast, comprehensive two-dimensional HPLC separation of tryptic peptides based on high-temperature HPLC. *J. Am. Chem. Soc.*, 127, 5034–5035, 2005.
33. Jandera, P., Programmed elution in comprehensive two-dimensional liquid chromatography—A review. *J. Chromatogr. A*, 1255, 112–129, 2012.
34. Dugo, P., Favoino, O., Lupino, R., Dugo, G., Mondello, L., Comprehensive two-dimensional normal-phase (adsorption)–reversed-phase liquid chromatography. *Anal. Chem.*, 76, 2525–2530, 2004.
35. Jandera, P., Hájek, T., Staňková, M., Vyňuchalová, K., Česla, P., Optimization of comprehensive two-dimensional gradient chromatography coupling in-line hydrophilic interaction and reversed-phase liquid chromatography. *J. Chromatogr. A*, 1268, 91–101, 2012.
36. Greibrokk, T., Andersen, T., High-temperature liquid chromatography. *J. Chromatogr. A*, 1000, 743–755, 2003.
37. Melander, W. R., Horvath, C., Reversed-phase chromatography In: *High-Performance Liquid Chromatography*, vol. 2, ed. C. Horvath. 113–319. New York: Academic Press, 1980.
38. Chen, M. H., Horváth, C., Temperature programming and gradient elution in reversed-phase chromatography with packed capillary columns. *J. Chromatogr. A*, 788, 51–61, 1997.
39. Jandera, P., Krupczyńska, K., Vyňuchalová, K., Buszewski, B., Combined effects of mobile phase composition and temperature on the retention of homologous and polar compounds on polydentate C_8 column. *J. Chromatogr. A*, 1217, 6052–6060, 2010.
40. Zhu, P. L., Dolan, J. W., Snyder, L. R., Hill, D. W., Van Heukelem, L., Waeghe, T. J. Combined use of temperature and solvent strength in reversed-phase gradient elution. 3. Selectivity for ionizable samples as a function of sample type and pH. *J. Chromatogr. A*, 756, 51–62, 1996.
41. Glajch, J. L., Kirkland, J. J., Snyder, L. R. Practical optimization of solvent selectivity in liquid-solid chromatography using a mixture-design statistical technique. *J. Chromatogr.*, 238, 269–280, 1982.
42. Schoenmakers, P. J., *Optimisation of Chromatographic Selectivity*. Amsterdam: Elsevier, 1986.
43. Glajch, J. L., Snyder, L. R. eds. *Computer Assisted Development for High Performance Liquid Chromatography*. Amsterdam: Elsevier, 1990.

44. Jandera, P., Česla, P., Hájek, T., Vohralík, G., Vyňuchalová, K., Fischer, J. Optimization of separation in two-dimensional HPLC by adjusting phase system selectivity and using programmed elution techniques. *J. Chromatogr. A*, 1189, 207–220, 2008.
45. Jandera, P., Predictive calculation methods for optimization of gradient elution using binary and ternary solvent gradients. *J. Chromatogr.*, 485, 113–141, 1989.
46. Snyder, L. R., Dolan, J. W., Lommen, D. C. Drylab computer simulation for high-performance liquid chromatographic method development: I. Isocratic elution. *J. Chromatogr.*, 485, 65–89, 1989.
47. Dolan, J. W., Lommen, D. C., Snyder, L. R. Drylab® computer simulation for high-performance liquid chromatographic method development: II. Gradient elution. *J. Chromatogr.*, 485, 91–112, 1989.
48. Molnár, I., Computerized design of separation strategies by reversed-phase liquid chromatography: Development of DryLab software. *J. Chromatogr. A*, 965, 175–194, 2002.
49. Euerby, M. R., Schad, G., Rieger, H.-J., Molnár, I., 3-dimensional retention modelling of gradient time, ternary solvent strength and temperature of the reversed-phase gradient liquid chromatography of a complex mixture of 22 basic and neutral analytes using DryLab® 2010. *Chromatography Today*, 3, 13, 2010.
50. Galushko, S. V., Kamenchuk, A. A., Pit, G. L. Calculation of retention in reversed-phase liquid chromatography: IV. ChromDream software for the selection of initial conditions and for simulating chromatographic behaviour. *J. Chromatogr. A*, 660, 47–59, 1994.
51. Kaliszan, R., *Quantitative Structure—Chromatographic Retention Relationship*. New York: Wiley, 1987.
52. Snyder, L. R., Dolan, J. W., Carr, P. W. The hydrophobic-subtraction model of reversed-phase column selectivity. *J. Chromatogr. A*, 1060, 77–116, 2004.
53. Abraham, M. H., McGowan, J. C., The use of characteristic volumes to measure cavity terms in reversed phase liquid chromatography. *Chromatographia*, 23, 243–246, 1987.
54. Nyiredy, S., Szucs, Z., Szepesy, L., Stationary phase optimized selectivity liquid chromatography: Basic possibilities of serially connected columns using the "PRISMA" principle. *J. Chromatogr. A*, 1157, 122–130, 2007.
55. Nyiredy, S., Meier, B., Erdelmeier, C. A. J., Sticher, O., PRISMA—A geometrical design for solvent optimization in HPLC. *J. High Resolut. Chromatogr.*, 8, 186–188, 1985.

Section II

Kinetic Study of Pesticides

10 Kinetics Study of Pesticides in the Environment

Application of HPLC to Kinetic Effects of Pesticide Analysis

Łukasz Cieśla

CONTENTS

10.1 KINETICS ASPECTS OF TRANSFORMATION OF ANALYTES

Pesticides are biologically active substances playing an invaluable role in crop management or controlling the population of insects, which are vectors of some diseases, for example, malaria. Unfortunately, apart from their beneficial effects, for example, the increase in food production or disease prevention, pesticides have been blamed for several unfavorable effects exerted on humans and the environment. Some pesticides have been found to be responsible for mutations and carcinogenesis [1]. Developed countries put a ban on using several groups of pesticides, organochlorine pesticides for instance, and set limits for the use of these substances in, for example, the agricultural sector [2]. However, an excessive use of pesticides in developing countries still poses a threat on a global scale as pesticides may undergo long-distance transport from their origins [3]. That is the reason for finding pesticides even in remote areas. Upon entering the environment, pesticides may be transported by flows, colloids, or nanopraticles; those that are volatile may travel long distances to condensate over areas with lower temperatures far from the pollution source [3]. Apart from transport, pesticides may be sorbed onto the surface of soil particles or undergo transformation. Therefore, there is a need to study pesticide transport and fate in the environment. A lot of studies focused only on parent pesticide compounds without taking into account the degradation products formed in different environment compartments, such as soil or groundwater. Studying only the level of parent compounds in the environment may even lead to a false conviction that a given pesticide is safe due to its quick elimination from the environment. However, degradation products may have greater toxicity. It is important to know the products of pesticide transformations as these may have a great impact on the environment and may be even more toxic than the parent compounds. This can be achieved, at least partially, by researching the transformation kinetics of pesticides.

10.1.1 Factors Affecting Reaction Rates: General Considerations on Transformation Kinetics of Pesticides

Before going into the details, some general aspects concerning factors affecting reaction rates will be discussed. The following are the factors influencing reaction rates: nature of the reactants, the ability of the reactants to come into contact with each other, concentration, temperature, and the presence of catalysts.

It is rather obvious that, under given conditions, a reaction rate is determined by the nature of the reactants [4]. Other factors may be controlled in a laboratory; therefore, one can have an influence on reaction rate, namely concentration of the reactants, temperature, and the use of a catalyst. Concentration and temperature have major impacts on reaction rate, which is explained by two models: collision theory and transition state theory.

Pesticides undergo different pathways once entering the environment, namely transformation/degradation, sorption-desorption, volatilization, uptake by plants, runoff to surface waters, and transport to groundwater as enumerated by Gao et al. [3]. Among them, transformation/degradation plays the most important role in the elimination of pesticides from the environment. Biodegradation is the predominant transformation process performed by microorganisms; however, there are abiotic degradation processes as well. This chapter focuses largely on abiotic transformation of pesticides.

As a part of abiotic degradation, pesticides may undergo the following reactions: hydrolysis, oxidation, photolysis, reduction, conjugation, and rearrangement [5]. Among the aforementioned, photolysis and hydrolysis are usually seen as the most important abiotic degradation pathways of pesticides. It also should be stressed that abiotic processes predominate only under certain conditions in environments poor in bacterial flora, such as deep soils [3]. Oxidation may only be observed in the abundance of oxygen; therefore, it can be neglected in the deeper soil layers.

Knowing the kinetics of these processes may be crucial for understanding the impact of a given pesticide and its degradation products on the environment. It is important to know the degradation products before a molecule is completely mineralized. For example, hydrolysis of pesticides is supposed to follow first-order kinetics as, in the majority of cases, pesticide degradation is described by a first-order model. However, even under controlled conditions, pesticide degradation kinetics may exhibit deviation from first-order kinetics.

All the aforementioned reactions (hydrolysis, oxidation, photolysis) and biological degradation processes (biobeds, activated sludge) are also used for removing pesticides and their degradation products from the environment [6].

10.1.1.1 Chemical Nature of Reactants

Under any given set of conditions for a reaction, the rate depends on the nature of the reactants [4]. This also means that any given reaction has a different rate under different conditions. Another factor that influences the collision frequency is the physical state of the reactants. When all the reactants are in the same phase, they can easily come into contact with each other. In case the reactants are in different phases, contact occurs only at the interface between the phases [4]. That can be a common phenomenon for pesticides adsorbed onto the surface of soil molecules, coming into contact with reactant molecules present in the liquid phase. The increased surface area of a solid provides the exposure of more reactant molecules [7].

Different physicochemical properties of pesticide molecules may influence their degradation rates, namely solubility in water, lipophilicity, acid-base character, or soil sorption coefficient.

The influence of the chemical nature of pesticides on their degradation kinetics can be easily noticed while comparing, for example, hydrolysis of different pesticide classes. Organophosphorus pesticides are more likely to undergo alkaline hydrolysis, and phosphorodithioates are prone to acidic hydrolysis [3].

Physicochemical properties of pesticides strongly determine their degradation pathways. As it was well presented in the chapter by Tuzimski, pesticides with low water solubility and high lipophilicity are characterized by higher half-life values in the soil [8]. Organochlorine pesticides, such as DDT, endosulfan, or lindane, are known to possess molecules resistant to chemical degradation in soil [5]. Therefore, these compounds are characterized with high persistence in the environment. These compounds are more prone to photodegradation. Organophosphate and carbamate pesticides are less persistent pesticides; however, danger posed by their presence in the environment should not be diminished.

Chemical nature strongly influences microbial degradation of pesticides. Water-soluble and more polar compounds are degraded more quickly when compared with lipophilic xenobiotics. Positively charged pesticide molecules are more strongly attracted by soil particles; therefore, their rate of biodegradation is lower in comparison with anionic pesticides [5]. Compounds with aliphatic moieties are more easily metabolized by microorganisms [5].

10.1.1.2 Ability of Reactants to Come into Contact with Each Other

It is rather common knowledge that molecules must collide to react. The more molecule collisions, the more often they react. Reaction rate is proportional to the number of collisions, which, in turn, depends on the concentration of reactants. However, the total number of collisions is much greater than the number of effective ones. One of the factors ascertaining effective collision is the appropriate molecular orientation as given in the Arrhenius Equation 10.1

$$k = Ae^{-E_a/RT} \quad (10.1)$$

where k is the rate constant, e is the base of natural logarithms, T is the absolute temperature, R is the universal gas constant, A is the frequency factor, and E_a is the activation energy.

The frequency factor is obtained by multiplying the collision frequency (Z) and an orientation probability factor (p) (Equation 10.2):

$$A = pZ \quad (10.2)$$

The orientation probability factor may be defined as the ratio of effectively oriented collisions to all possible collisions [4]. The more complex the molecule, the lower the p values. The orientation probability factor can range from approximately 10^{-9} for biochemical reactions to nearly 1 for single atoms.

Reaction mechanisms are usually more complex than can be seen in the balanced overall equation. Most reactions occur through a sequence of individual steps, which are called elementary reactions. In order to react, the molecules should collide with enough energy and have effective orientations; uni- and bimolecular reactions are the most common. It means that elementary reactions with only one or two molecules are involved in the steps that make up a reaction mechanism. The probability that three colliding molecules have enough energy and proper orientation is very low; only a few termolecular elementary steps are known.

10.1.1.3 Concentrations of Reactants

One of the major factors influencing reaction rate is reactant concentration. The frequency of collisions depends on the number of molecules: the greater the population of reactants, the likelihood they will collide increases.

Reaction rate is defined as the rate of increase in the molar concentration of the products of a reaction per unit of time or the rate of decrease in the molar concentrations of the reactants [unit: mol/(L sec)] [7].

For a general reaction occurring at a fixed temperature, the rate law is of the form (Equation 10.3):

$$\text{rate} = k[\text{molarity of reactants}]^m \tag{10.3}$$

where k is a rate constant (changing only with temperature), and m is the order of the reaction.

A first-order model has been widely used to describe pesticide degradation kinetics (Equation 10.4):

$$\frac{\mathrm{d}C}{\mathrm{d}t} = -kC \tag{10.4}$$

where C is pesticide concentration, t is time, and k is the degradation rate constant (hydrolysis, oxidation, or biodegradation rate constant). According to this model, the pesticide degradation rate is directly proportional to the concentration of a given pesticide. The degradation process is often characterized by its half-life ($t_{1/2}$), which is characteristic of the reaction at a given temperature. For a first-order reaction, the half-life is independent of the initial concentration and can be derived from the following (Equation 10.5):

$$t_{1/2} = \frac{0.693}{k} \tag{10.5}$$

where k is the reaction rate. The reaction half-life of the majority of pesticides has been found to be within the following range: 0.02 to 4 years [5]. Studying transformation kinetics of pesticides is of practical importance as, nowadays, modern agriculture focuses on using xenobiotics of decreased environmental persistence. In order to know the environmental fate of a given pesticide, detailed studies regarding its transport, sorption, and transformation are needed.

The first-order model has also been widely applied as a description of the biodegradation processes of pesticides. However, this is only a simplification of Michaelis-Menten kinetics, which is used to describe enzyme-catalyzed reactions (biodegradation performed by bacteria) (Equation 10.6):

$$\frac{\mathrm{d}C}{\mathrm{d}t} = \frac{-(V\max C)}{(Km + C)} \tag{10.6}$$

where C is pesticide concentration in solution, $V\max$ is maximum reaction rate, Km is the Michaelis constant, and t is time.

In case of low pesticide concentration and assuming that the population of microbial organisms participating in the biodegradation is not changing, the data may be fitted by the first-order kinetics. A modified Monod equation is used to describe pesticide degradation, taking into account bacterial growth (Equation 10.7):

$$\frac{\mathrm{d}C}{\mathrm{d}t} = \frac{-(\mu\max BC)}{[Y(Ks + C)]} \tag{10.7}$$

where $\mu\max$ is the maximum specific growth rate, B is the density of active microbial cells, Y is the yield coefficient or the amount of biomass produced out of a unit amount of substrate consumed, and Ks is the Monod constant at which the growth rate is half of the maximum rate [9].

Biodegradation studies are usually performed under constant conditions, although in the real field conditions, all the degradation processes are influenced by changing conditions.

According to Osman et al., the kinetics of pesticide degradation in soil is usually biphasic with a rapid degradation rate in the first phase followed by a very slow second phase [2].

Results obtained both in the field and on a laboratory scale confirm pesticide degradation is a first-order reaction with respect to pesticides [10].

10.1.1.4 Temperature Effects

Temperature has a major impact on reaction rates. It affects the reaction rate by increasing the frequency and the energy of collisions. An increase of 10 K doubles or triples the rate [4]. The temperature dependency of the reaction rate is given by the Arrhenius Equation 10.1.

However, not all the molecule collisions are effective. The colliding molecules have to reach a certain minimum energy ascertaining their collision is effective. This energy is called activation energy (E_a). A temperature rise increases the frequency and energy of the molecule collisions, having much lower impact on the frequency. With increasing collision energy, the temperature rise enlarges the fraction of collisions with energy exceeding E_a. The activation energy is used to reach the transition state as explained by the transition state theory [4].

10.1.1.5 Availability of Rate-Accelerating Agents (Catalysts)

Apart from increasing the temperature, a reaction can be accelerated by using a catalyst. A catalyst lowers the activation energy, causing more molecules to undergo effective collisions. Only a small, nonstoichiometric amount of the catalyst is used to speed up the reaction [4]. The activation energy is lowered as the catalysts provide a different mechanism for the reaction.

As mentioned previously, pesticides may undergo the following abiotic degradation processes in the environment: hydrolysis, oxidation, and photolysis. All of them may be affected by many factors with some of them acting as reaction catalysts. For example, pesticide oxidation may be influenced by the following: the presence and concentration of metal ions, soil pH, the level of humic acids, etc. [3]. Several research results indicate that pesticides may chelate metal ions as they contain chemical moieties, which may act as Lewis bases (electron donors). Chelation of metal ions ends up with degradation of these xenobiotics [11,12]. Phosphates, in turn, were found to catalyze hydrolysis of pesticides in aqueous solutions.

Photodegradation of pesticides is normally restricted to the upper layers of soil. However, pesticides may also utilize photo energy in an indirect way, receiving it from other compounds, for example, humic acids or metal oxides [3].

For more specific data concerning rates and mechanisms of chemical reactions, please refer to physical chemistry handbooks [4].

10.1.2 Kinetics Aspects of Transformation of Main Chemical Groups of Pesticides

Currently used pesticides belong to one of the following major groups: organophosphates, carbamates, organochlorines, pyrethroids, triazines, substituted ureas, thiocyanates, phenols, or formamides [8]. The trend nowadays is to use polar pesticides, which more easily undergo degradation and, therefore, are characterized with a low environmental persistence. Apart from biodegradation, which is the most common way of pesticide transformation, the following are the abiotic processes pesticides may undergo upon entering the environment: hydrolysis, photolysis, oxidation, reduction, conjugation, or rearrangement [5]. Studying pesticide transformation pathways is of crucial importance as, apart from the parent molecules, the transformation products may have even greater impact on the environment. These compounds may not only pose a threat to human health, but may also be responsible for disturbing the fragile equilibria in different ecosystems. Depending on their structures as well as environmental factors (e.g., soil pH, the amount of organic matter, presence of humic acids or metal oxides, etc.), pesticides may undergo many chemical reactions. Therefore, it is reasonable to study degradation processes of every newly introduced xenobiotic. Despite a great diversity of pesticide degradation reactions, they can be ascribed to one of the following groups, according to Bansal [5]:

a. Hydrolysis observed for compounds possessing ester, epoxide, and amide functional groups or halogens, for example,

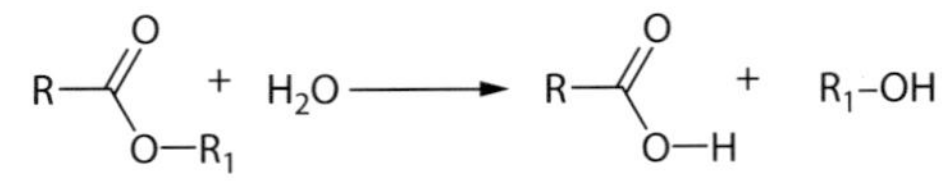

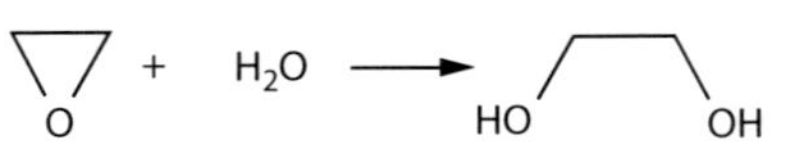

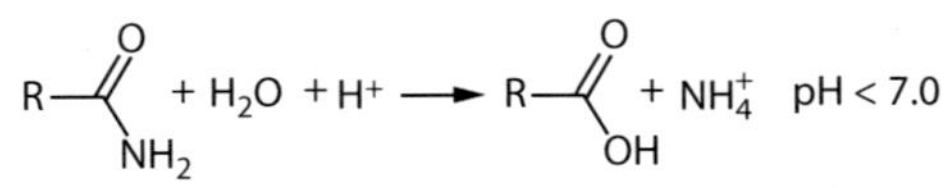

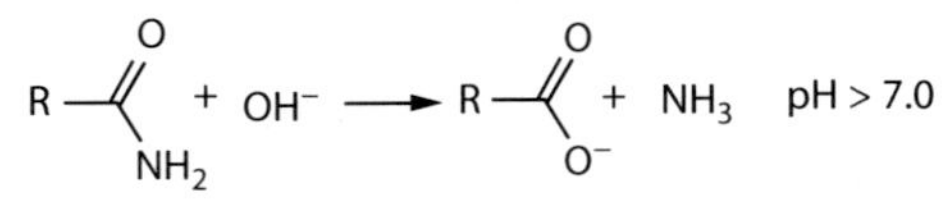

Organophosphate, carbamate, and pyrethroid pesticides are prone to hydrolysis and are characterized with rather low environmental persistence [13].

b. Photodegradation is characteristic of compounds able to absorb light energy, such as those possessing chromophores or forming metal complexes (direct photolysis). However, pesticides may also undergo indirect photolysis in the presence of photosensitizing compounds, such as humic acids, chlorophyll, or xanthone [5]. However, the role of humic acids in the photodegradation of pesticides is still not fully understood as there are also reports indicating that these compounds may slow down photochemical reactions [13]. The presence of constituents able to form free radicals may also initiate indirect photodegradation processes. The examples of such compounds are nitrates and nitrites or iron ions [13]. Photodegradation reactions run with the formation of free radicals.

Hydrolysis and photolysis are considered the major types of degradation reactions that pesticides undergo in the environment [3]. Sandin-España and Sevilla-Moránesses stress that these two chemical processes are mainly responsible for degradation of pesticides in aquatic environments [13].

c. Oxidation may involve oxygen or other oxidizing agents. One of the most common oxidizing agents is chlorine, present, for example, in chlorinated water; enzymatically catalyzed oxidation of pesticides is one of the biodegradation processes performed by microorganisms. The following are exemplary oxidation reactions [5]:
   - Hydroxylation:

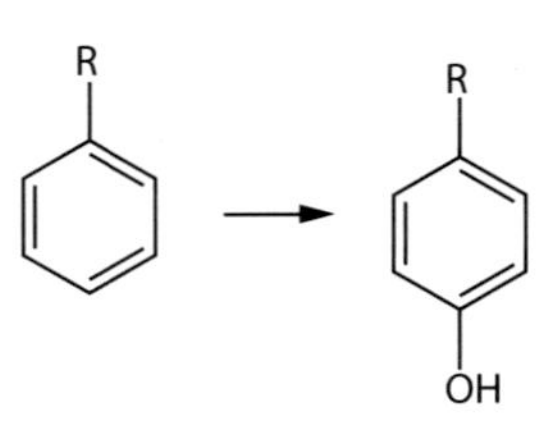

For example,

– Aromatic hydroxylation of carbaryl (carbamate insecticide) [14]:

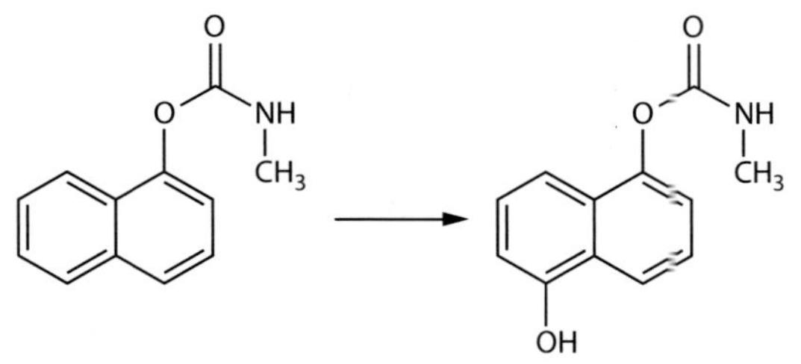

– Aliphatic hydroxylation of DDT [14]:

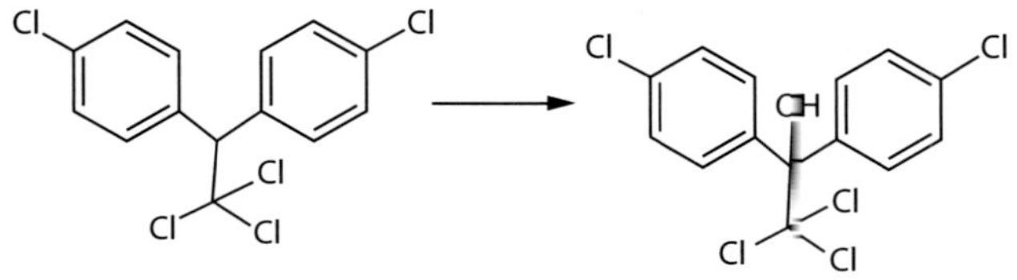

- Epoxidation:

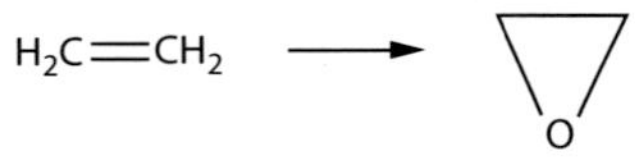

For example, epoxidation of Aldrin (organochlorine insecticide) [14]:

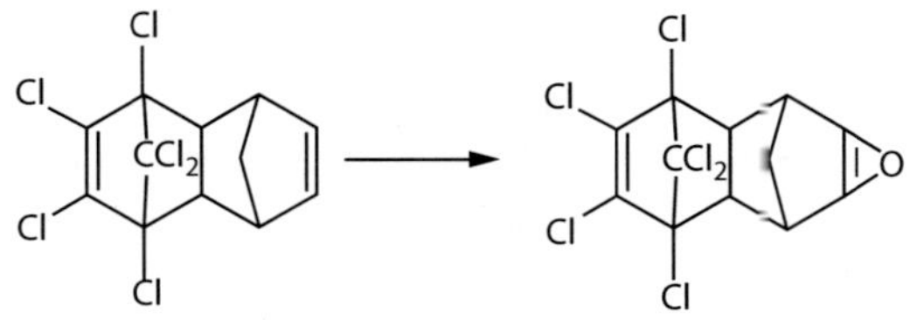

- O-dealkylation:

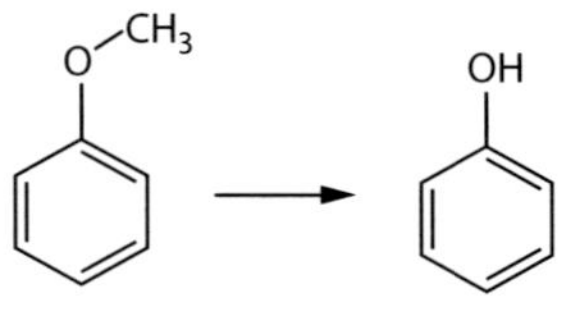

For example, O-dealkylation of methoxychlor [14]:

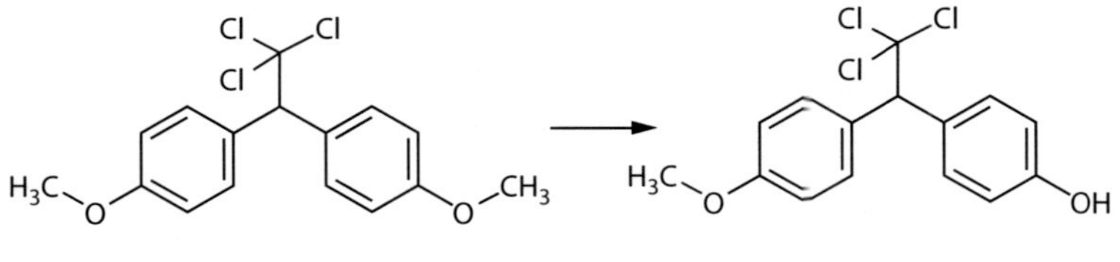

- N-dealkylation:

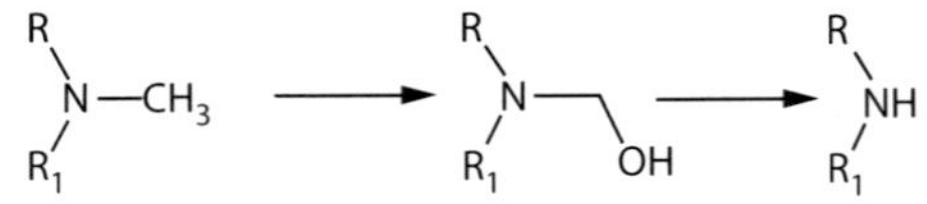

For example, N-dealkylation of propoxur (carbamate insecticide) [14]:

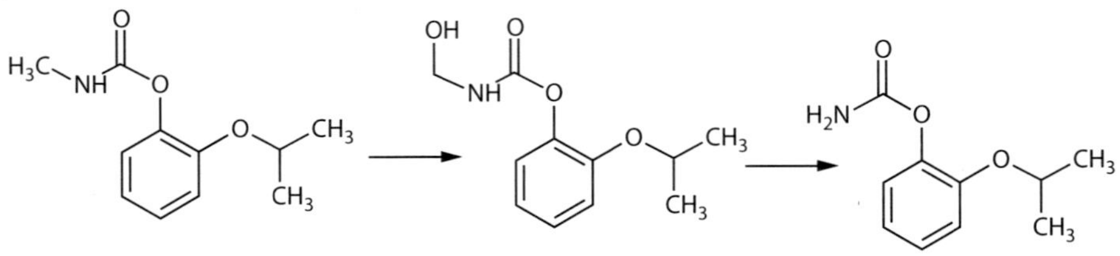

Apart from the main oxidation reactions, pesticides may also undergo S-dealkylation, oxidative desulfuration, amine oxidation, and sulfoxidation [5].

As already mentioned in this chapter, hydrolysis and photodegradation are considered to be the main degradation reactions of pesticides. The following are transformations that may be observed for some pesticides: reduction, conjugation, and rearrangement. For example, reduction reactions may be observed in anaerobic environments, leading to the formation of less polar compounds when compared to parent molecules. Bansal gave a more detailed overview of the degradation reactions of pesticides in the environment [5].

Microbiological degradation of pesticides plays the most important role in their conversion into mineral compounds. Microorganisms (bacteria, fungi, protozoa, and algae) are responsible for complete mineralization of pesticides into carbon dioxide, ammonia, water, mineral salts, etc. [13]. Transformation of pesticides into the final end products may follow two general tracks. The first one occurs when a pesticide is used as the source of energy; therefore, it is transformed by bacterial/fungal primary metabolism. The other one is when pesticide molecules are transformed by the secondary metabolism pathways [5]. Taking into account oxygen abundance, biodegradation processes may be classified as aerobic or anaerobic. The following are the reaction types pesticides may undergo in bacterial or fungal cells: oxidation, reduction, hydrolysis, dealkylation, etc. The principles of these reactions are the same as those described earlier for abiotic degradation. The difference lies in the involvement of enzymes, natural catalysts, making some of the pesticide degradation reactions characteristic for one species only. Usually biodegradation products are more polar and water soluble when compared to the parent compounds as microorganisms transform xenobiotics into forms that can be easily transported and eliminated [13]. The following are the microorganism species most commonly involved in the biodegradation of pesticides: bacteria *Arthrobacter*, *Bacillus*, *Corynebacterium*, *Flavobacterium*, *Pseudomonas*, etc., and fungi *Penicillium*, *Aspergillus*, *Fusarium*, and *Trichoderma*.

10.1.3 Determination of Pesticide Adsorption Capacity

Determination of pesticide adsorption capacity is important for two main reasons. The first relates to pesticide adsorption onto soil particles, which influences pesticide fate in the environment. The other one is connected to the possibility of removing hazardous compounds from, for example, polluted waters [1]. Pesticide molecules strongly bound to the soil particles will not undergo biodegradation [5]. Knowing pesticide adsorption capacity is important to estimate its persistence in the environment. Pesticides characterized with low sorption capacity cause the compounds to be more labile and can, for example, more easily contaminate groundwater. Adsorption of organic molecules onto soil particles can be either of a physical (physisorption) or chemical nature (chemisorption) [3]. The adsorption of pesticide molecules onto soil is believed to be a result of Coulombic attraction between positively charged pesticide molecules and the negatively charged soil particles [3]. Adsorption of charged molecules depends on pH as well as ionic strength. The increase of pH results in increasing the adsorption of positively charged pesticide molecules. Studying temperature dependences may distinguish physisorption from chemisorption [15]. Apart from the chemical properties of soil particles and pesticide molecules, the following are factors influencing xenobiotic sorption: distribution and size of soil particles as well as the amount of organic matter [13].

The adsorption of a pesticide molecule onto soil particles, in some ways, resembles the processes inside a chromatographic column. Therefore, high-performance liquid chromatography (HPLC) may be used to study adsorption kinetics of these xenobiotics. Sorption-desorption processes depend on the chemical properties of pesticide molecules and, as previously mentioned, the characteristics of soil particles. By analogy, with liquid chromatography (LC), these processes seem to resemble adsorption kinetics in reversed-phase, ion-exchange, or ion-exclusion chromatography. For example, negatively charged soil particles may be compared to cation-exchange columns [16]. Lipophilic compounds, in turn, are strongly retained by soils rich in clay and organic matter.

The adsorption of pesticide molecules onto soil particles is believed to be a spontaneous equilibrium process described by the distribution coefficient, K_d, also called the soil sorption coefficient, which is defined as the ratio of the sorbed phase concentration to the solution phase concentration at Equation 10.8:

$$K_d = \frac{Ca}{Cd} \tag{10.8}$$

where K_d is the distribution coefficient, Ca is the amount of pesticide adsorbed (M/M), and Cd is the concentration of pesticide dissolved (M/V) [5].

The higher the K_d value, the more sorbed is a molecule [13]. The pesticide distribution coefficient was found to be related to the levels of clay and organic matter in the soil and is related to the pesticide soil organic partition coefficient (K_{oc}) [5]. The soil sorption coefficient can be calculated with the use of a simple Equation 10.9:

$$K_d = K_{oc}\, OC/100 \tag{10.9}$$

where K_{oc} is the soil organic partition coefficient, and OC is organic carbon content (%) [3].

The sorption capacity also strongly depends on the pesticide chemical structure. The following factors should be taken into consideration when studying adsorption of pesticides: soil pH, octanol-water partition coefficient of a given pesticide, surface area, solubility and polarity of pesticide molecules, size of the soil particles, etc. [3].

A simple distribution coefficient (K_d) is, however, criticized for not being consistent with basic chemistry [17]. Gamble et al. underlined that the use of a distribution coefficient is incorrect for two main reasons. The first one relates to the use of ambiguous sorption data without making a distinction between labile sorption and total sorption [17]. The other reason is that it is difficult to make the assumption that equilibria develop under field conditions. Due to frequent dynamic conditions, kinetics description produces much more reliable data [17].

Very lipophilic compounds, such as organochlorines, are strongly retained by soils rich in clays and organic matter and, therefore, are characterized by long environmental persistence. The adsorption of pesticides with pronounced basic or acidic properties depends on the soil pH, electric charge of the soil molecules, and ionic strength of water leaching through the soil profile.

It is difficult to study pesticide adsorption onto soil particles as natural sorbents are usually mixtures of varying particle sizes or different distribution of organic matter, etc. [15]. These are particles with different physical and chemical properties, which makes developing an adsorption model nearly impossible. Several models have been proposed to study adsorption kinetics based on bulk soil properties [15]. These models usually assume uniform distribution of soil particles as well as organic matter [15]. Li et al. underlined the need to classify soils by particle size fractions for better understanding of soil sorption properties [15]. The following regularity is observed for varying sorbent particle sizes: the finer the sorbent particle size, the higher the sorption capacities [15]. However, the increase in sorption capacity between the different fractions is not large.

Generally, two different approaches have been presented to study the sorption mechanism in environmental and pure system crystals as reported by Gamble et al. [17]. These are HPLC as well as scanning tunneling and fluorescence microscopy, which both may be considered complementary methods.

An online microfiltration-HPLC technique was used for studying the kinetics and adsorption mechanism of atrazine (herbicide) on mineral soil particles [15]. The injection of soil slurries into an HPLC and separate solution-phase analysis permits the resolution of the total amount of pesticide in a soil into its dissolved, labile sorbed, and bound residue components [17,18].

10.1.3.1 Sorbent Characterization: Effect of Sorbent Particle Size

As previously mentioned, soils are mixtures with different particle sizes and varying content (different organic and mineralogical compositions). Many models account for sorption on the bulk soil body [15], making an assumption that there is even distribution of particles across the soil profile. However, in many cases, these assumptions do not hold. Li et al. indicated there is a need to classify soil particles by fractions to better simulate the transport of pollutants [15].

Li et al. showed there are only slight changes in sorption of atrazine (triazine herbicide) by different soil size fractions [15].

Adsorption–desorption processes are also utilized in the enrichment and purification of the analyzed samples. Solid phase extraction (SPE) is the most frequently applied method for the analysis of pesticide residues. There are different materials available as fillings of SPE cartridges, however C18 (alkyl-bonded silica) phases are the most popular [13]. These sorbents are usually used in pesticide residue analysis. The pesticide parent compounds, being more lipophilic, when compared to the transformation products, are strongly retained on such adsorbents. The interaction of pesticide molecules and adsorbent molecules is due to Van der Waals forces. However, due to low selectivity of such adsorbents, other compounds are also enriched in the sorbent and interfere with the compounds of interest [13]. The other group of adsorbents frequently used in the analysis of pesticides is polymers. Polymer adsorbents are characterized with high chemical stability and high loading capacities. Adsorption of the analytes is due to π–π interactions. What is more, the surface of the polymers can be additionally modified, resulting in better selectivity toward a given group of compounds. The use of molecularly imprinted polymer materials is another step ahead in increasing the selectivity of pesticide residue analysis. For example, Chen et al. reported on the synthesis of atrazine molecular imprinted polymers and their use as selective sorbents for SPE [19]. When compared with C18 adsorbent, higher recoveries of atrazine were obtained. Wang et al. used molecularly imprinted polymers with trichlorfon and monocrotophos as the template molecules for simultaneous multiresidue determination of organophosphate pesticides [20] and still others. The increase in the use of molecularly imprinted polymers as sorbents is observed and may be regarded as one of the trends in sample preparation in pesticide residue analysis.

From among the adsorbents commonly applied for the removal of hazardous substances, activated carbon is one of the most extensively applied [1]. Brasquet and Le Cloirec reported that, in the case of aromatic organic compounds, kinetic coefficients obtained for activated carbon fiber are greater when compared to granular activated carbon [21]. It was found that adsorption of pesticides, similarly to other aromatic compounds, followed the pseudo-first order kinetics [1].

In the case of activated carbon, it is assumed that the main force responsible for the adsorption of pesticides is the dispersion force between pesticide π electrons and the surface of the adsorbent [1]. Therefore, pesticides possessing aromatic rings are believed to be better adsorbed when compared to those lacking the aromatic moiety. The pesticide adsorption also increases upon increasing the compound's hydrophobicity, such as in the case of compounds possessing branched alkyl substituents [1].

Ayranaci and Hoda concluded that for the following pesticides: dinoseb, aldicarb, ametryn, and diuron, first-order (Equation 10.1) and second-order (Equation 10.2) models are equally applicable for their adsorption kinetics on activated carbon-cloth (Equations 10.10 and 10.11) [1].

$$\ln c - \ln c_0 = -k_1 t \tag{10.10}$$

$$\frac{1}{c} - \frac{1}{c_0} = k_2 t \tag{10.11}$$

where c_0 is the initial concentration of adsorbate, c is the concentration of adsorbate at any time, t is time, and k_1 and k_2 are rate constants for pseudo-first order (1) and pseudo-second order models (2), respectively [1].

The adsorption isotherms fitted equally well for dinoseb, aldicarb, ametryn, and diuron to the Langmuir and Freundlich models [1].

Activated carbon adsorption is one of the physical processes used for removing pesticide contamination from agricultural wastewaters [10].

More polar pesticide degradation compounds may be enriched with the use of other adsorbents. Silica-bonded phases may be utilized in the analysis of more polar pesticide degradation products, such as the following sorbents: CN-silica, DIOL-silica, or NH_2-silica (amino). CN-silica can be used both in normal and reversed-phase systems for extraction of polar and midpolar compounds. Amino sorbents, in turn, may act as polar or ion-exchange sorbents. Nowadays, spherical fillings of SPE cartridges are also available, which enable obtaining more reproducible results.

10.1.3.2 Adsorption Kinetics

Determination of the adsorption isotherm is a popular experimental way for determining pesticide adsorption potential by soil particles [3]. That is defined as the amount of adsorbate on the adsorbent as a function of its pressure or concentration at a constant temperature. According to Gao et al., there are three mathematical models most frequently used to describe the adsorption of pesticides on soil particles [3]:

Langmuir model (Equation 10.12):

$$q_e = \frac{K_1 Q C_e}{1 + K C_e} \tag{10.12}$$

Freundlich model (Equation 10.13):

$$q_e = K_f C_n^e \tag{10.13}$$

Langmuir-Freundlich model (Equation 10.14):

$$q_e = \frac{K_1 Q C_e^n}{1 + K C_e^n} \tag{10.14}$$

where K_1 and K_f are Langmuir bonding terms related to interaction energies and the Freundlich affinity coefficient, respectively; Q is the Langmuir maximum capacity; C_e is the equilibrium concentration of the sorbate in solution; and n is the Freundlich linearity constant [3].

According to Gamble et al., adsorption of an organic chemical onto the particles of immersed soil is governed by second-order kinetics [8], which was experimentally confirmed [22]. Gamble et al. determined the stoichiometry of labile pesticide sorption for selected herbicides on several soils. The full list of the experiments was given in one of the published reports [34]. As has been previously stressed, the existence of equilibria in soils is of low probability; therefore, empirical models using the following equilibrium parameters: K_d or K_{oc} are not suitable for studying sorption processes of organic molecules on soil particles [22]. According to Gamble et al., the use of the aforementioned empirical parameters is based on ambiguous combinations of labile and nonlabile sorption [22]. There is no relationship between these parameters and the number of labile sorption sites. Gamble concludes that, without clear distinction between labile and nonlabile sorption, the law of mass action cannot be adapted to natural soils [22]. The proposed sorption mechanism is presented in Figure 10.1.

As in the case of other equilibria, pesticide sorption–desorption processes are temperature-dependent. With the rise of temperature, sorption usually decreases [5]. Temperature shows significant influence on labile sorption and intraparticle diffusion, having only a small effect on the labile equilibrium constant [15]. It should be stressed that temperature differently affects pesticide sorption onto dry and immersed soils [22]. A temperature increase results in desorption of water particles from hydrated surfaces, allowing for increased sorption of hydrophobic organic molecules [22]. This competition between water particles and organic molecules is responsible for the existence of labile sorption, which was compared by Gamble to ion-exchange chemistry [22].

In many cases, empirical equations and parameters are still often used to describe sorption of organic molecules on soil particles. There are postulates to replace empirical parameters, for example, distribution coefficients with kinetics models proving the presence of labile sorption, which are not taken into consideration in the case of empirical parameters [22].

Pesticide adsorption onto soil particles influences other processes that these xenobiotics may undergo in the environment, namely volatilization, uptake by plants, or leaching into deeper soil layers. As underlined by Arias-Estévez et al., adsorption of pesticides onto sorbent particles can be divided into two steps: an initial fast step followed by a slower second step, ending with a developed equilibrium [23].

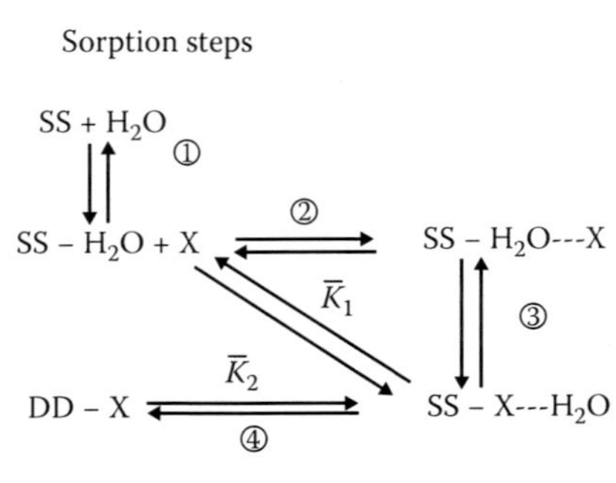

FIGURE 10.1 Proposed sorption mechanism; SS: surface sorption sites; X: an organic chemical molecule; DD: intraparticle sites reached by diffusion. (Reproduced from Gamble, D. S. et al., *Environ. Sci. Technol.*, 34, 120–124, 2000.)

10.2 HPLC ANALYSIS OF KINETICS PRODUCTS OF PESTICIDES

There is a large number of papers dealing with HPLC analysis of pesticides and their degradation products in samples of different origins. Therefore, the aim of this section is to present only some general information regarding HPLC analysis of these xenobiotics. Taking into account the number of papers published on HPLC analysis of pesticides, the discussion will be based on the most recently published data. Analytical approaches optimized for a given compound or group of compounds can be easily looked up in search engines. Problems can be encountered when analyzing newly introduced compounds as there will be no analytical methods available for the analysis of a parent compound, and what is more challenging, there will be no data on the analysis of transformation by-products. In such cases, a new method should be optimized based on the data available for compounds with similar structural, chemical, and physical properties.

Due to the fact that pesticides and their degradation products are present in the environment in quantities that would be difficult to quantify and in complex matrices, sample enrichment and purification steps are usually needed. These steps are performed with the SPE procedure. Apart from SPE, the following techniques are also used as an alternative to liquid–liquid extraction: solid-phase microextraction, matrix solid-phase dispersion, or stir-bar sorbtive extraction [13]. More recently, the quick, easy, cheap, rugged, effective, and safe sample preparation approach has gained much attention in the analysis of pesticide residues [24]. Data on the use of some of the aforementioned techniques are described in previous chapters focusing on the pesticide adsorption processes.

After SPE, the samples are analyzed by means of different chromatographic techniques. Gas chromatography with mass spectrometry (MS) detection is one of the most commonly utilized methods [13]. However, the analysis of pesticide transformation products may be more problematic when compared with the analysis of parent compounds. As was previously mentioned, transformation by-products are usually more polar than parent molecules (polar moieties are introduced or freed during degradation processes of either a biotic or abiotic nature). Also, taking into account the fact that degradation products are usually less volatile and, at the same time, more thermolabile in comparison to parent compounds, the use of LC seems to be a reasonable choice for the analysis of pesticide degradation compounds [13].

As for LC, degradation by-products are usually detected and identified with the use of HPLC coupled with UV/diode array detectors and/or MS. The former detectors may be used when analyzing pesticides and degradation products possessing chromophores. The use of derivatization techniques to introduce a chromophore enabling detection of a given compound may be one of the solutions as recently presented for the determination of glyphosate and its main metabolite aminomethylphosphonic acid residues in soil samples [25]. The two aforementioned xenobiotics were precolumn derivativized with 1,2-naphthoquinone-4-sulfonate. In degradation studies, it is of crucial importance to identify all the by-products formed during the degradation steps. For newly analyzed pesticides, a degradation pathway is usually proposed [6]. It is essential for tracking pesticide fate in the environment. One of the problems associated with chromatographic analysis of degradation products of pesticides is the lack of commercially available standards. The identification of by-products is performed primarily from MS/MS data [6]. The selectivity and sensitivity of the LC-MS technique enables detection and quantification of xenobiotics present in the environment even in low levels. The following are the mass analyzers most frequently used in LC-MS: single quadrupole, triple quadrupole, ion trap, and time-of-flight. A triple-quadrupole mass analyzer is frequently applied for the analysis of pesticide residues in the environment. The use of quadrupole time-of-flight analyzers gives an opportunity to study transformation pathways of pesticides as the use of this mass analyzer allows for assignment of empirical formulas for the analyzed degradation products [13]. As for the column used in the analysis of pesticide residues, in the majority of cases C18 columns are utilized with the use of aqueous-methanol or aqueous-acetonitrile mixtures as mobile phases; however, the use of C8 adsorbents also has been reported (reversed-phase systems) [26].

Formic acid is the most frequently used as a mobile phase additive, and other additives were also utilized, such as acetic acid [25] and ammonium acetate [27,28]. The use of other columns is less frequent but also possible as exemplified by several recent examples: chiral analytical column amylose tris(3,5-dimethylphenylcarbamate) used for the analysis of degradation products of dufulin with mobile phase; *n*-hexane/ethanol (90:10 by volume; normal phase system) [29]; Chiralpak IC column (250 × 4.6 mm inner diameter, cellulose tris-(3,5-dichlorophenylcarbamate) immobilized on silica, mobile phase *n*-hexane/isopropanol (85/15, v/v) [30].

As mentioned previously, the use of molecularly imprinted polymers is one of the trends in sample preparation in the analysis of pesticide residues. The other trend is the use of ultra-HPLC analysis as the number of papers reporting on the application of this technique is increasing. Most recently, the following, exemplary pesticides were analyzed by means of UHPLC in environmental samples: carbamate pesticides [31], fipronil (phenylpyrazole insecticide) and its metabolites [32], dimethomorph and tebuconazole [28], 253 multiclass pesticides [33], malathion [34], and others. When analyzing new compounds without available standards, the probable chemical formulas should be confirmed with the use of other techniques: UV, IR, and NMR spectroscopy.

10.3 CONCLUSIONS

Studying the fate of different pesticides in environmental compartments has now become an important part of experiments that need to be performed for newly introduced xenobiotics. It is now well understood that pesticides may exert harmful effects even after degradation as some of the by-products may be more toxic than parent pesticides. HPLC is one of the indispensable tools that help us understand some of the processes pesticides undergo in the environment. It is important for the examination of pesticide interaction with soil particles (adsorption–desorption processes) and for identification and quantification of pesticides and their degradation products. A lot of different models have been elaborated to follow pesticide fate in the environment. However, the complexity of these processes as well as the introduction of new classes of pesticides will cause HPLC to play an important role in future experiments. Very sensitive and reliable methods will be sought to screen traces of pesticides and their degradation products in the environment. HPLC coupled with tandem MS already has been proven to be ideal for these purposes. Microfiltration HPLC together with microscopy have both been used to set the opportunities to study sorption mechanisms in environmental and crystal systems. Both methods may be used for advancing the research in two directions; however, both focus on adsorption of organic molecules on soil particles.

REFERENCES

1. Ayranaci, E., Hoda, N., Adsorption kinetics and isotherms of pesticides onto activated carbon-cloth, *Chemosphere*, 60, 1600–1607, 2005.
2. Osman, K. A., Ibrahim, G. H., Askar, A. I., Rahman, A., Alkhail, A. A., Biodegradation kinetics of dicofol by selected microorganisms. *Pesticide Biochem. Physiol.*, 91, 180–185, 2008.
3. Gao, J., Wang, Y., Gao, B., Wu L., Chen, H., Environmental fate and transport of pesticides. In: *Pesticides: Evaluation of Environmental Pollution.* Nollet, L. M. L., Rathore, H. S. (Eds.), CRC Press Taylor & Francis Group, Boca Raton, 2012, pp. 29–46.
4. Mortimer, M., Taylor, P. (Eds.), *Chemical Kinetics and Mechanism*, Royal Society of Chemistry, Thomas Graham House, Science Park, Cambridge, UK, 2002.
5. Bansal, O. P., Degradation of pesticides. In: *Pesticides: Evaluation of Environmental Pollution.* Nollet, L. M. L., Rathore, H. S. (Eds.), CRC Press Taylor & Francis Group, Boca Raton, 2012.
6. Ikehata, K., Gamal El-Din, M., Aqueous pesticide degradation by hydrogen peroxide/ultraviolet irradiation and Fenton-type advanced oxidation processes: A review. *J. Environ. Eng. Sci.*, 5, 81–135, 2006.
7. Connors, K. A. (Ed.), *Chemical Kinetics. The Study of Reaction Rates in Solution.* VCH Publishers Inc., New York, 1990.
8. Tuzimski, T., Pesticide residues in the environment. In: *Pesticides: Evaluation of Environmental Pollution.* Nollet, L. M. L., Rathore, H. S. (Eds.), CRC Press Taylor & Francis Group, Boca Raton, 2012.

9. Richter, O., Diekkrűger, B., Nörtersheuser, P. (Eds.), *Environmental Fate Modeling of Pesticides. From the Laboratory to Field Scale*, VCH Verlagsgesselschaft, Weinheim, Germany, 1996.
10. Bourgin, M., Violleau, F., Debrauwer, L., Albet, J., Kinetic aspects and identification of by-products during the ozonation of bitertanol in agricultural wastewaters. *Chemosphere*, 90, 1387–1395, 2013.
11. Nowack, B., Stone, A., Degradation of nitrilotris(methylenephophonic acid) and related (amino)phosphate chelating agents in the presence of manganese and molecular oxygen. *Environ. Sci. Technol.*, 34, 4759–4765, 2000.
12. Nowack, B., Stone, A., Homogenous and heterogenous oxidation of nitrilotrismethylenephosphonic acid (NTMP) in the presence of manganese (II, III) and molecular oxygen. *J. Phys. Chem. B*, 14, 6227–6233, 2002.
13. Sandin-España, P., Sevilla-Morán, B., Pesticide degradation in water. In: *Pesticides: Evaluation of Environmental Pollution.* Nollet, L. M. L., Rathore, H. S. (Eds.), CRC Press Taylor & Francis Group, Boca Raton, 2012.
14. Singh, D. K., *Metabolism or Degradation of Pesticides: Phase I and Phase II Reactions in Pesticides Chemistry and Toxicology*, pp. 62–95, Bentham Science, e-book. doi: 10.2174/97816080513731120101, eISBN: 978-1-60805-137-3, 2012.
15. Li, J., Langford, C. H., Gamble, D. S., Atrazine sorption by a mineral soil: Effects of soil size fractions and temperature. *J. Agric. Food Chem.*, 44, 3680–3684, 1996.
16. Waksmundzka-Hajnos, M., Cieśla, Ł., Separation of ionic analytes: Reversed phase, ion-pair, ion-exchange, and ion-exclusion chromatography. In: *High Performance Liquid Chromatography in Phytochemical Analysis.* Wkamsundzka-Hajnos, M., Sherma, J., (Eds.), CRC Press, Boca Raton, 2011.
17. Gamble, D. S., Barrie Webster, G. R., Lamoureux, M., Quantitative prediction of atrazine sorption in a Manitoba soil using conventional chemical kinetics instead of empirical parameters. *J. Phys. Chem. C*, 114, 20055–20061, 2010.
18. Gamble, D. S., Bruccoleri, A. G., Lindsay, E., Langford, C. H., Chlorothalonil in a quartz sand soil: Speciation and kinetics. *Environ. Sci. Technol.*, 34, 120–124, 2000.
19. Chen, J., Bai, L.-Y., Liu, K.-F., Liu, R.-Q., Zhang, Y.-P., Atrazine molecular imprinted polymers: Comparative analysis by far-infrared and ultraviolet induced polymerization. *Int. J. Mol. Sci.*, 15, 574–587, 2014.
20. Wang, X., Tang, Q., Wang, Q., Qiao, X., Xu, Z., Study of a molecularly imprinted solid-phase extraction coupled with high-performance liquid chromatography for simultaneous determination of trace trichlorfon and monocrotophos residues in vegetables. *J. Sci. Food Agric.*, 94, 1409–1415, 2014.
21. Brasquet, C., Le Cloirec, P., Adsorption onto activated carbon fibers: Application to water and air treatments. *Carbon*, 35, 1307–1313, 1997.
22. Gamble, D. S., Herbicide sorption by immersed soils: Stoichiometry and the law of mass action in support of predictive kinetics. *Environ. Sci. Technol.*, 43, 1930–1934, 2009.
23. Arias-Estévez, M., López-Periago, E., Martínez-Carballo, E., Simal-Gándara, J., Mejuto, J. C., García-Río, L., The mobility and degradation of pesticides in soils and the pollution of groundwater sources. *Agric. Ecosts. Environ.*, 123, 247–260, 2008.
24. Anastassiades, M., Lehotay, S. J., Štajnbaher, D., Schenk, F. J., Fast and easy multiresidue method employing acetonitrile extraction/partitioning and "dispersive solid-phase extraction" for the determination of pesticide residues in produce. *J. AOAC Int.*, 86, 412–431, 2003.
25. Islas, G., Rodriguez, J. A., Mendoza-Huizar, L. H., Pérez-Moreno, F., Carrillo, E. G., Determination of glyphosate and aminomethylphosphonic acid in soils by HPLC with pre-column derivatization using 1,2-naphthoquinone-4-sulfonate. *J. Liq. Chromatogr. Relat. Technol.*, 37, 1298–1309, 2014.
26. Xie, W., Shi, Y., Hou, J., Huang, C., Zhao, D., Pan, L., Dong, S. Simultaneous determination of ethephon, thidiazuron, diuron residues in cotton by using high performance liquid chromatography-tandem mass spectrometry. *Chin. J. Chromatogr. (SePu)*, 32, 179–183, 2014.
27. Chen, L., Chen, J. F., Guo, Y., Li, J., Yang, Y., Xu, L., Fu, F., Study on the simultaneous determination of seven benzoylurea pesticides in Oolong tea and their leaching characteristics during infusing process by HPLC-MS/MS. *Food Chem.*,143, 405–410, 2014.
28. Lv, D. Z., Zhu, Z., Yuan, H. Q., Luo, J. H., Development and validation of a liquid chromatography tandem mass spectrometry (LC-MS/MS) method for the determination of dimethomorph and tebuconazole residue in soil. *Appl. Mech. Mat.*, 522–524, 132–135, 2014.
29. Zhang, K.-K., Hu, D.-Y., Zhu, H.-J., Yang, J.-C., Song, B.-A., Enantioselective degradation of dufulin in four types of soil. *J. Agric. Food Chem.*, 62, 1771–1776, 2014.
30. Qi, Y., Liu, D., Sun, M., Di, S., Wang, P., Zhou, Z., The chiral separation and enantioselective degradation of the chiral herbicide napropamide. *Chirality*, 26, 108–113, 2014.

31. Shi, Z., Hu, J., Li, Q., Zhang, S., Liang, Y., Zhang, H., Graphene based solid phase extraction combined with ultra high performance liquid chromatography–tandem mass spectrometry for carbamate pesticides analysis in environmental water samples. *J. Chromatogr. A*, 1355, 219–227, 2014.
32. Cheng, Y., Dong, F., Liu, X., Xu, J., Meng, W., Liu, N., Chen, Z., Tao, Y., Zheng, Y., Simultaneous determination of fipronil and its major metabolites in corn and soil by ultra-performance liquid chromatography–tandem mass spectrometry. *Anal. Meth.*, 6, 1788–1795, 2014.
33. Kalogridi, E.-C., Christophoridis, C., Bizani, E., Drimaropoulou, G., Fytianos, K., Part I: Temporal and spatial distribution of multiclass pesticide residues in lake waters of Northern Greece: Application of an optimized SPE-UPLC-MS/MS pretreatment and analytical method. *Environ. Sci. Poll. Res.*, 21, 7239–7251, 2014.
34. Naushad, M., Alothman, Z. A., Khan, M. R., Removal of malathion from aqueous solution using De-Acidite FF-I Presin and determination by UPLC-MS/MS: Equilibrium, kinetics and thermodynamics studies. *Talanta*, 115, 15–23, 2013.

11 Phototransformation of Pesticides in the Environment

Davide Vione, Marco Minella, and Claudio Minero

CONTENTS

This chapter presents the main photoinduced transformation processes involving pesticide molecules in sunlit surface waters, on soil, and in the atmosphere (taking into account both the gas and the condensed phases). Moreover, a recently developed model approach to predict pollutant phototransformation in surface waters is presented and described together with an example referred to the herbicide MCPA (4-chloro-2-methylphenoxyacetic acid) and to the production of the toxic transformation intermediate 4-chloro-2-methylphenol.

11.1 PHOTOTRANSFORMATION OF PESTICIDES IN SURFACE WATERS

11.1.1 Photochemical Processes in Surface Waters

The photochemical processes that involve organic contaminants (including pesticides) in surface waters can be divided into direct photolysis and indirect or sensitized photochemistry. The direct photolysis in environmental waters takes place when a compound absorbs sunlight and when sunlight absorption causes transformation. Therefore, only sunlight-absorbing compounds can undergo direct photolysis in the environment [1].

Figure 11.1 gives some insight into the processes that follow radiation absorption by a molecule [2]. Assume that the substrate is initially in the ground vibrational level of the ground electronic state, S_0. For most organic molecules, S_0 is a singlet state. Absorption of a photon can promote the molecule from S_0 to a vibrationally excited state of an electronically excited singlet state. For simplicity, suppose that the electronically excited state is the first singlet state, S_1. After radiation absorption, the molecule reaches a vibrationally excited state of S_1. In some cases, the excess vibrational energy can be high enough to cause bond breaking that triggers the transformation of the molecule. Otherwise, vibrational energy can be lost by relaxation (e.g., by collisions with the solvent) to reach the ground vibrational state of S_1. If the vibrational relaxation is complete, that is, if the molecule gets back to the ground vibrational state of S_0, one speaks of internal conversion. As an alternative, the molecule in the singlet state S_1 can undergo reactions that are, however, little likely due to the short lifetimes of the excited singlet states (usually in the subnanosecond level). Other possible pathways are emission of fluorescence radiation (at higher wavelength, that is, lower energy, than the absorbed one) or the intersystem crossing (ISC) to a triplet state (in Figure 11.1, this is the first triplet state, T_1). The ISC is enabled by the fact that the T_1 energy is lower than that of S_1, but a vibrationally excited T_1 state can have the same or very similar energy as ground-state S_1. The following step is relaxation to the ground vibrational state of T_1, which can continue down to S_0 (internal conversion). As an alternative, the state T_1 can undergo chemical reactions, such as rearrangements; reactions with other dissolved molecules, such as O_2; or reactions with the solvent. There is higher probability that T_1 reacts compared to S_1 because the triplet states are longer-lived than singlet ones (lifetimes are usually in the nanosecond–microsecond level). Some dissolved molecules (for instance humic and fulvic acids) have triplet states with oxidizing capability and can

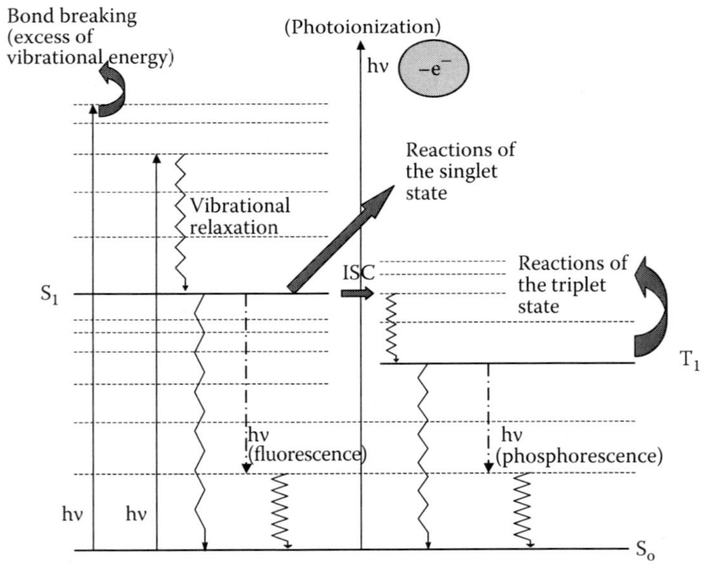

FIGURE 11.1 Schematic of the processes that may follow radiation absorption by a water-dissolved organic compound. Solid horizontal lines represent ground vibrational states of electronic levels, excited vibrational states being dashed. Solid and straight vertical arrows represent radiation absorption processes ($h\nu$ = photon); zigzag arrows are vibrational relaxation processes, and dash-dotted arrows represent light emission (fluorescence or phosphorescence). ISC = intersystem crossing.

induce the degradation of other compounds, including organic pollutants. Molecules with such triplet states are called photosensitizers, and their reactions will be dealt with more extensively when describing the indirect photolysis processes. A final option, which is mainly observed in solid systems or in deep-frozen solutions, is the emission of phosphorescence radiation.

In some cases, the energy of the absorbed radiation is so high that an electron is ejected out of the molecule. The corresponding phenomenon is called photoionization, and it is more common with low-wavelength radiation (e.g., UVC), but some molecules undergo photoionization even with UVB or UVA radiation. The ionization process is usually a first step that is followed by further reactions of the radical cation thus formed, for example, the reaction with the solvent.

Therefore, a sunlight-absorbing molecule can undergo direct photolysis by bond breaking (excess of vibrational energy); reactions involving S_1 or, most notably, T_1; and photoionization. Such reactions can involve the molecule alone (intramolecular processes) or other compounds, including the solvent (intermolecular processes) [2].

In the case of indirect or sensitized photolysis, sunlight is absorbed by photoactive compounds called photosensitizers. The latter produce photoactive transients upon radiation absorption, which can induce the degradation of dissolved compounds, including organic pollutants. The main photosensitizers in surface waters are chromophoric dissolved organic matter (CDOM); nitrate; nitrite; and, probably to a somewhat lesser extent, Fe species and hydrogen peroxide. Reactive transients that are photogenerated and can be involved in substrate degradation are the hydroxyl (•OH) and carbonate radicals ($CO_3^{-\bullet}$), singlet oxygen ($^1O_2$), and the triplet states of CDOM ($^3CDOM^*$) [3,4]. The formation and reactivity of the various transients will now be discussed.

The •OH radical is definitely the most reactive transient that occurs in surface waters. It is produced by photolysis of nitrate and nitrite [5] and by irradiation of CDOM. In the latter case, there is still debate in the literature as to the possible pathway of •OH generation: it could be oxidation of water or of OH^- by $^3CDOM^*$ (Reactions 11.4 and 11.5) [6] or a photo-Fenton process involving complexes between Fe(III) and organic ligands within CDOM (Reactions 11.6–11.9) [7].

$$NO_3^- + h\nu + H^+ \rightarrow {}^\bullet NO_2 + {}^\bullet OH \tag{11.1}$$

$$NO_2^- + h\nu + H^+ \rightarrow {}^\bullet NO + {}^\bullet OH \tag{11.2}$$

$$CDOM + h\nu \rightarrow CDOM^* - (ISC) \rightarrow {}^3CDOM^* \tag{11.3}$$

$$^3CDOM^* + H_2O \rightarrow CDOM\text{-}H^\bullet + {}^\bullet OH \tag{11.4}$$

$$^3CDOM^* + OH^- \rightarrow CDOM^{-\bullet} + {}^\bullet OH \tag{11.5}$$

$$Fe^{III} - L + h\nu \rightarrow Fe^{2+} + L^{+\bullet} \tag{11.6}$$

$$Fe^{2+} + O_2 \rightarrow Fe^{3+} + O_2^{-\bullet} \tag{11.7}$$

$$2\,O_2^{-\bullet} + 2\,H^+ \rightarrow H_2O_2 \tag{11.8}$$

$$Fe^{2+} + H_2O_2 \rightarrow Fe^{3+} + {}^\bullet OH + OH^- \tag{11.9}$$

The radical •OH undergoes very fast reactions with many water-dissolved compounds, including xenobiotics and natural organic molecules (natural dissolved organic matter, hereafter DOM) [8].

For this reason, the steady state [•OH] in surface waters is very low (at or below 10^{-16} M), and this limits the importance of •OH reactions in pollutant transformation. The main •OH scavengers in surface waters are DOM, HCO_3^-, and CO_3^{2-} with nitrite also playing some (limited) role [9,10].

$$^\bullet OH + DOM \rightarrow Products \tag{11.10}$$

$$^\bullet OH + HCO_3^- \rightarrow H_2O + CO_3^{-\bullet} \tag{11.11}$$

$$^\bullet OH + CO_3^{2-} \rightarrow OH^- + CO_3^{-\bullet} \tag{11.12}$$

$$^\bullet OH + NO_2^- \rightarrow OH^- + {}^\bullet NO_2 \tag{11.13}$$

Reactions 11.11 and 11.12 of •OH with bicarbonate and carbonate yield the radical $CO_3^{-\bullet}$, which is also a reactive transient but a less powerful oxidant than •OH. The radical $CO_3^{-\bullet}$ can also be produced upon oxidation of carbonate by $^3CDOM^*$ [4]:

$$^3CDOM^* + CO_3^{2-} \rightarrow CDOM^{-\bullet} + CO_3^{-\bullet} \tag{11.14}$$

Scavenging of $CO_3^{-\bullet}$ mainly takes place upon reaction with DOM. Note that Reactions 11.11 and 11.12 are usually the main sources of $CO_3^{-\bullet}$ in surface waters with Reaction 11.14 usually playing a secondary role. This means that the formation rate of $CO_3^{-\bullet}$ is lower compared to •OH because the rate of •OH formation equals that of its scavenging (steady-state condition), and the main •OH scavenging process is Reaction 11.10 with DOM. Indeed, Reactions 11.11 and 11.12, yielding $CO_3^{-\bullet}$ from •OH, are secondary processes of •OH consumption in most natural waters. However, because the reaction rate constant between DOM and $CO_3^{-\bullet}$ is two to three orders of magnitude lower than the rate constant of •OH with DOM, the steady-state [$CO_3^{-\bullet}$] in surface waters is often considerably higher than the steady-state [•OH]. The higher [$CO_3^{-\bullet}$] compared to [•OH] is usually compensated for by the lower reactivity of $CO_3^{-\bullet}$ toward most pollutants. Nevertheless, easily oxidized compounds, such as anilines and sulfur-containing molecules, can undergo significant degradation by $CO_3^{-\bullet}$ [4,11].

The excited triplet states of CDOM, $^3CDOM^*$, are formed upon radiation absorption by CDOM followed by intersystem crossing. They are quite powerful oxidizing species and play, for instance, a major role in the photoinduced degradation of phenylurea herbicides and sulfonamide antibiotics [12]. It has recently been found that phenolic antioxidants present in DOM may inhibit degradation reactions induced by $^3CDOM^*$. Although it is highly unlikely that DOM significantly scavenges $^3CDOM^*$, these antioxidants could back-reduce, giving back the starting compounds, the pollutant molecules that have initially undergone oxidation upon reaction with $^3CDOM^*$ [13].

Reaction between $^3CDOM^*$ and O_2 produces singlet oxygen ($^1O_2$), which is also a reactive species. Reaction of $^1O_2$ with dissolved compounds (including organic pollutants) is in competition with the deactivation of $^1O_2$ upon collision with the solvent [3].

$$^3CDOM^* + O_2 \rightarrow CDOM + {}^1O_2 \tag{11.15}$$

$$^1O_2 \rightarrow O_2 \tag{11.16}$$

11.1.2 Phototransformation Processes of Some Pesticide Classes

Pesticides constitute an extremely varied class of environmental contaminants, and their fate, including photochemical transformation, has therefore been the object of a huge number of studies. Here, no attempt will be made to tackle the almost impossible task of providing a comprehensive review of the photochemical transformation processes of all known pesticides. On the contrary, examples of phototransformation reactions of some pesticide classes will be provided.

11.1.2.1 Phenoxyacetic Acid Herbicides

Chlorinated phenoxyacetic acid derivatives are extensively used for the protection of cereal crops against broad-leaf weeds. Examples are MCPA (4-chloro-2-methylphenoxyacetic acid), mecoprop [(RS)-2-(4-chloro-2-methylphenoxy)propanoic acid], 2,4-D (2,4-dichlorophenoxyacetic acid), and dichlorprop [(R)-2-(2,4-dichlorophenoxy)propanoic acid]. Moreover, similar compounds bearing a trazole substituent on the alkyl chain are used as fungicides, including triadimefon [(RS)-1-(4-chlorophenoxy)-3,3-dimethyl-1-(1H-1,2,4-triazol-1-yl)butan-2-one] and triadimenol [(1RS,2RS;1RS,2SR)-1-(4-chlorophenoxy)-3,3-dimethyl-1-(1H-1,2,4-triazol-1-yl)butan-2-ol]. The compounds used as herbicides are all characterized by the presence of a –COOH group, which is deprotonated under the pH conditions of surface waters. This is interesting because the products of UV photolysis of the –COOH and –COO$^-$ forms are quite different. The carboxylate forms can undergo photohydrolysis by replacement of the –Cl atom on the aromatic ring by a –OH group, and carboxylic acids mainly undergo a radical rearrangement via a solvent-cage process. This means that irradiation causes the break of a chemical bond that splits the molecule into two radicals, initially surrounded by the cage of water molecules. In such an environment, the radical–radical reaction is highly favored, and it can produce either the starting compounds or rearrangement products [14]. An example of the described reactions is provided for MCPA in Figure 11.2.

An important finding is that irradiation of phenoxyacetic acid herbicides under sunlight also causes the loss of the acid chain to give the corresponding chlorophenol compounds 2,4-chlorophenol from 2,4-D and dichlorprop and 4-chloro-2-methylphenol from MCPA and mecoprop. In a similar way, 4-chlorophenol (as well as 1,2,4-triazole) has been identified upon photolysis of mecoprop and triadimefon [14]. The 4-chloro-2-methylphenol accounts for the increase of toxicity of irradiated MCPA mixtures with irradiation time [14].

Interestingly, among the transformation intermediates of phenoxyacid herbicides, the cited chlorophenols have been found at the highest concentrations in the environment (but they are also impurities of the pesticide formulation, which adds to their environmental occurrence). In nitrate and nitrite-rich waters, such as in flooded paddy fields, herbicide-derived chlorophenols can undergo efficient nitration reactions to produce toxic and potentially mutagenic nitroderivatives. Such compounds have actually been detected in waters of the Rhône delta (Southern France). The nitration

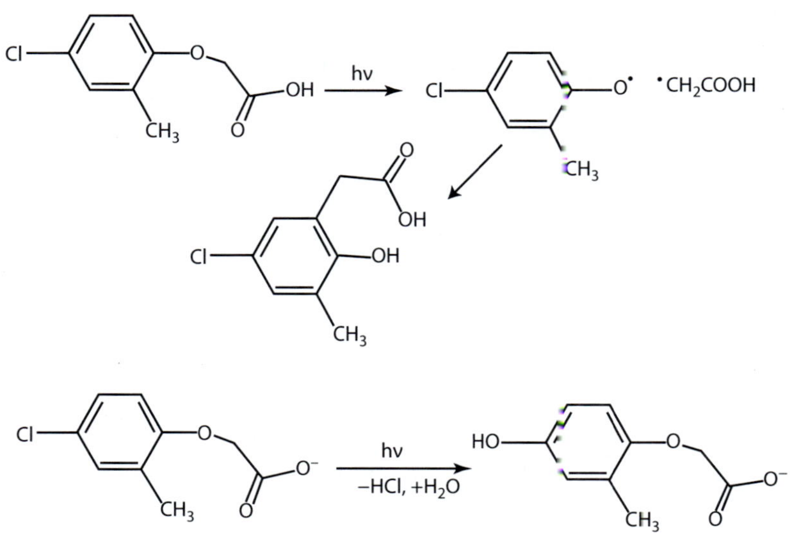

FIGURE 11.2 Direct photolysis processes of the protonated and deprotonated forms of MCPA.

process is most likely induced by $^{\bullet}NO_2$, produced by irradiation of nitrate and nitrite (Reactions 11.1, 11.13) [15–17].

11.1.2.2 Phenylurea Herbicides

This class of herbicides includes compounds of rather widespread use, such as monuron, metobromuron, diuron, and linuron. Their direct photolysis has been shown to proceed mainly by photohydrolysis, that is, the replacement of a halogen atom by a –OH group. Where applicable, the loss of a $-OCH_3$ group from the lateral alkyl chain (demethoxylation) may also take place [18]. Phenylurea herbicides have been shown to undergo transformation in surface waters mainly upon reaction with $^3CDOM^*$ [12], but the detected intermediates are not much different from those of direct photolysis [12,18]. This similarity between intermediates of different photochemical pathways is not uncommon, and it characterizes other compounds as well [19]. Phenylureas are aniline analogs, and the electron couple on the N atom linked to the aromatic ring would increase the ring electron density. Therefore, these compounds are activated to electrophilic and similar processes, and nitration of phenylureas has, for instance, been observed under photochemical conditions in the presence of nitrate and nitrite as $^{\bullet}NO_2$ sources [20].

11.1.2.3 Halogenated Phenol Derivatives

Pesticides belonging to this miscellaneous class would mainly undergo transformation via photohydrolysis. This behavior has been observed, for instance, with bromoxynil (3,5-dibromo-4-hydroxybenzonitrile) as well as its chlorinated and iodinated congeners [21], with dichlorophen [2,2′-methylenebis(4-chlorophenol)] [22], and partially with dicamba (3,6-dichloro-2-methoxybenzoic acid) [23]. In the latter case, the presence of a carboxylic group in *ortho* position to a methoxy one enables a cyclization process that takes place along with photohydrolysis (Figure 11.3).

Photohydrolysis has been detected (but only as a secondary process) in the case of acifluorfen [5-(2-chloro-α,α,α-trifluoro-p-tolyl)-2-nitrobenzoic acid], mainly because the nitrobenzoic acid ring is more reactive than the phenolic one. As a consequence, decarboxylation as well as breaking of the ether bond between the two rings have been observed as the main transformation pathways. Acifluorfen is probably unreactive with $^3CDOM^*$, but it has been found to undergo efficient degradation by $^{\bullet}OH$ [24].

11.1.2.4 Atrazine and Other Triazines

In a study of the direct photolysis of the herbicide atrazine and of its reaction with $^{\bullet}OH$, transformation intermediates by both processes have been identified (they are listed in Table 11.1). The main intermediates of direct photolysis arise from photohydrolysis (replacement of the chlorine atom on the triazine ring with a –OH group) and from oxidation of the lateral alkyl chains. In contrast, no photohydrolysis was observed with $^{\bullet}OH$, and, in addition to oxidation, complete cleavage of the lateral chains was detected. Interestingly, the compounds deriving from lateral-chain oxidation have been observed in both direct photolysis and the $^{\bullet}OH$ reaction [25], which is not uncommon as far as photochemical transformation pathways (direct or indirect) are concerned [19].

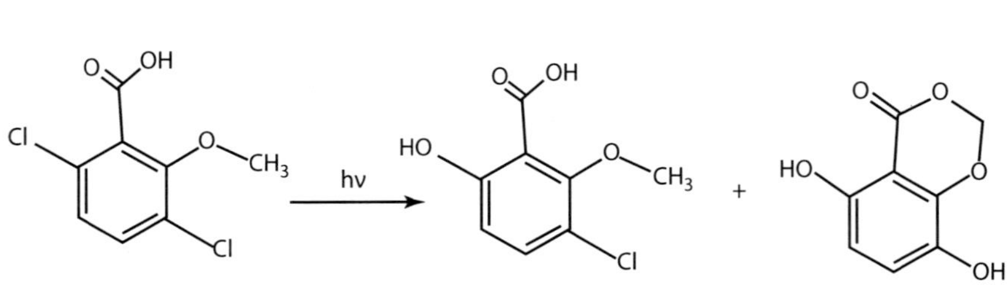

FIGURE 11.3 Direct photolysis of dicamba in aqueous solution.

TABLE 11.1
Main Identified Transformation Intermediates of Atrazine upon Direct Photolysis and Reaction with •OH

Formula	Name	Photolysis	•OH
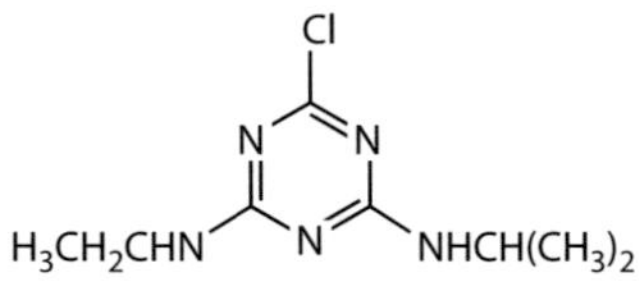	Atrazine (2-chloro-4-ethylamino-6-isopropylamino-s-triazine)	n/a	n/a
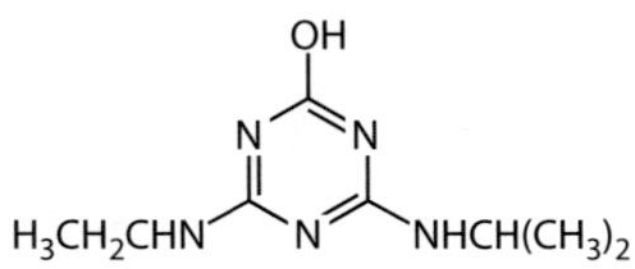	Hydroxyatrazine (4-ethylamino-2-hydroxy-6-isopropylamino-s-triazine)	✓	
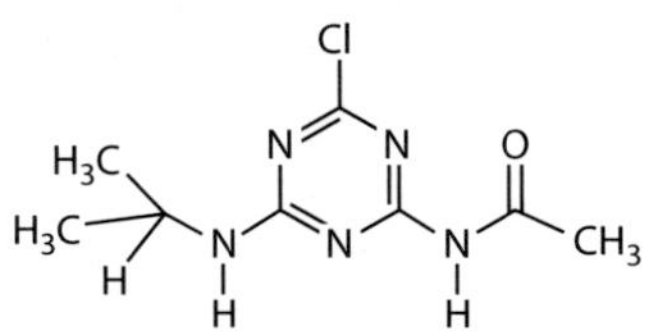	4-Acetamido-2-chloro-6-isopropylamino-s-triazine	✓	✓
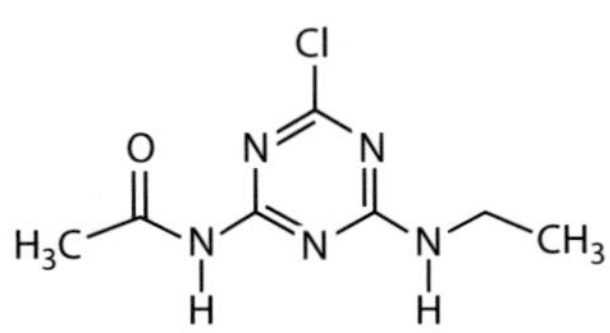	4-Acetamido-2-chloro-6-ethylamino-s-triazine	✓	✓
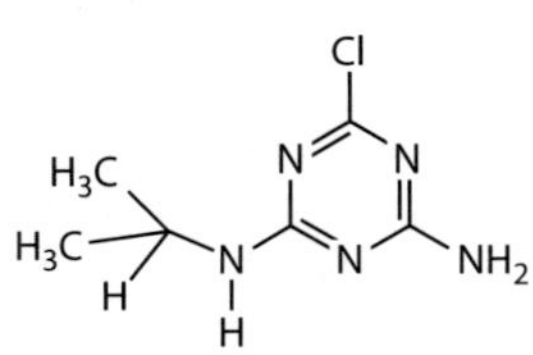	6-Amino-2-chloro-4-isopropylamino-s-triazine		✓
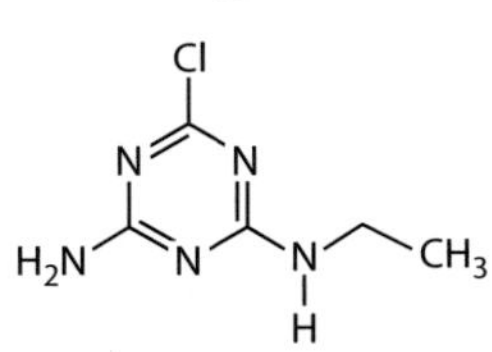	6-Amino-2-chloro-4-ethylamino-s-triazine		✓

Note: n/a: not applicable. ✓: the compound was observed under the reported conditions.

In the case of Irgarol 1051 (2-methylthio-4-tert-butylamino-6-cyclopropylamino-s-triazine), used in antifouling paints, the products of direct and CDOM-sensitized transformation were found to practically coincide. Transformation pathways included modification or cleavage of the lateral chains (dealkylation), oxidation of the methylthio group, or its replacement with –OH [26].

11.1.2.5 Propiconazole

The fungicide propiconazole (1-[2-(2,4-dichlorophenyl)-4-propyl-1,3-dioxolan-2-ylmethyl]-1H-1,2,4-triazole) has been found to undergo a cyclization process upon photolysis, which gives a condensed three-ring structure upon elimination of HCl (Figure 11.4). Moreover, oxidation products have been detected upon irradiation in natural waters [27].

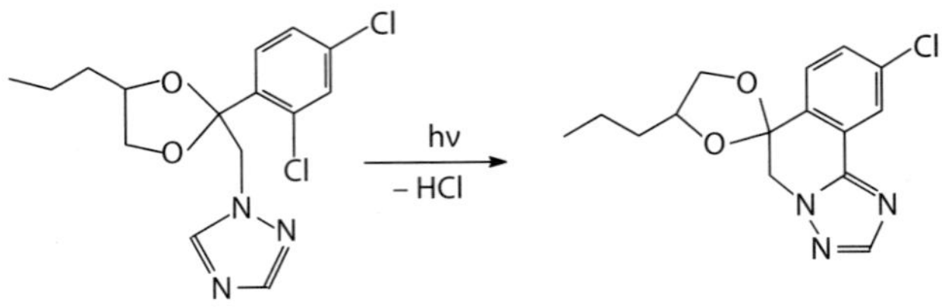

FIGURE 11.4 Photocyclization of propiconazole.

11.1.2.6 Sulfur-Containing Compounds

The fungicide carboxin (5,6-dihydro-2-methyl-1,4-oxathi-ine-3-carboxanilide) undergoes photolysis by oxidation of the sulfur atom to a sulfoxide group. The reaction also proceeds by release of oxanilic acid. Oxidation of the sulfur atom can also take place upon reaction with singlet oxygen and other photoinduced oxidants, and it could play an important role in carboxin photolysis. For instance, the related fungicide oxycarboxin (5,6-dihydro-2-methyl-1,4-oxathi-ine-3-carboxanilide-4,4-dioxide) bears a sulfone group that cannot be further oxidized, which may account for its much slower photodegradation compared to carboxin [28].

The herbicide florasulam (N-(2,6-difluorophenyl)-5-methoxy-8-fluoro(1,2,4)-triazolo-[1,5-c]-pyrimidine-2-sulphonamide) undergoes direct photolysis by release of the difluorophenyl group and by production of a sulfonic acid derivative. However, indirect phototransformation of florasulam in natural waters is considerably faster than direct photolysis. The sensitized process proceeds by difluorophenyl release to form a sulfonamide and/or by replacement with –OH of the methoxy group on the pyrimidine ring. Moreover, the pyrimidine moiety can be disrupted by leaving a carboxylic group linked to the triazole ring [29].

11.1.2.7 Carbamate Insecticides

Carbofuran (2,3-dihydro-2,2-dimethyl-7-benzofuranyl-N-methylcarbamate) undergoes photoinduced cleavage of the carbamic moiety to give a phenolic derivative, followed by photohydrolysis of the furan ring to produce a catechol derivative [3-(2-hydroxy-sec-butyl)catechol] [30]. The photodegradation of carbofuran is inhibited by DOM, partially upon competition with CDOM for sunlight irradiance (which inhibits direct photolysis) and partly upon carbofuran-DOM interaction that inhibits phototransformation. The latter process might involve an enhancement of the thermal deactivation of carbofuran excited states, which would inhibit further chemical reactions [30].

The transformation of carbaryl is enhanced in natural waters compared to ultrapure ones, indicating that indirect photochemistry may be important in addition to direct photolysis. Reaction with $^{\bullet}OH$ and possibly with $^{3}CDOM^{*}$ and/or $^{1}O_2$ are reasonable candidate processes for indirect photochemistry [31].

11.2 PHOTOTRANSFORMATION OF PESTICIDES IN SOIL AND ATMOSPHERE

11.2.1 Photochemical Reactions in Soil

Photodegradation of pesticides in soil is obviously limited by sunlight penetration, which is certainly more difficult below ground than below water. Anyway, photochemical processes would take place in the topmost soil layer, which is often the first portion coming into contact with pesticides. Similarly to surface waters, photoreactions can be divided into direct and indirect photolysis. Direct photolysis on solid surfaces may be different than in solution because of the absence of the cage of water molecules and, sometimes, for the directional effect of surfaces. The water-cage effect usually inhibits direct photolysis because in the solution bulk the photofragments are initially surrounded by water molecules that make photofragment recombination easier (Figure 11.5) [32]. The photofragments often recombine to yield the parent compound although

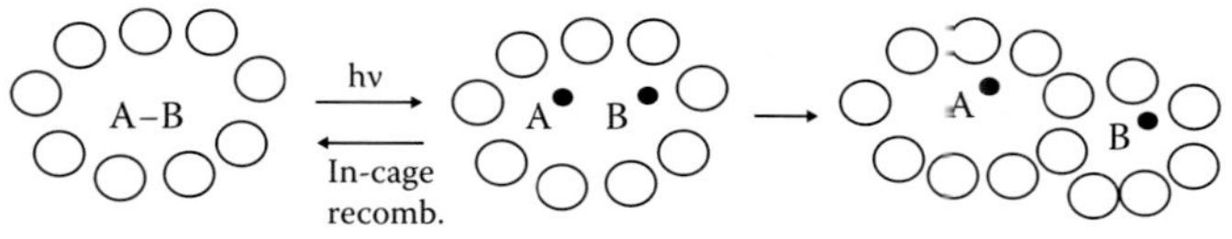

FIGURE 11.5 Schematic diagram of the solvent-cage effect for photochemical reactions in aqueous solution. Open circles represent solvent molecules.

photoisomerization is also possible (see, for instance, Figure 11.2). Being a transformation process, cage photoisomerization does not decrease the photolysis quantum yield. Generally speaking, it could be assumed that the photolysis quantum yields would often be higher in soil than in water, which is counterbalanced by a lower availability of sunlight in soil. Moreover, processes such as photoisomerization and photohydrolysis are more likely to take place in aqueous solution than on the soil surface.

Perhaps the most important difference between water and soil is related to sensitized phototransformation. In the case of soil surfaces, a significant fraction of indirect photochemistry would be triggered by photoactive minerals (most notably the semiconductor oxides) that do not play an important role in surface waters [33]. Examples include TiO_2, ZnO, and Fe(III) (hydr)oxides. The photochemistry of the latter is perhaps more complex because it does not follow a pure semiconductor mechanism (Reactions 11.21 and 11.22) [34].

A mineral having semiconducting properties can be photoactive if its band-gap energy is comparable to the energy of sunlight photons. If this is the case, radiation absorption promotes an electron from the valence to the conduction band, leaving a hole (electron vacancy) in the valence band. The electron and hole can recombine, producing heat, or they can migrate to the semiconductor surface where trapping by surface and subsurface species is possible. Recombination between surface-trapped electrons and holes is still possible, but it is considerably slower than in the semiconductor bulk and enables chemical reactivity to take place. The conduction-band electron (e^-_{CB}) is a reductant and can, for instance, transform molecular oxygen into $O_2^{-\bullet}$. The valence-band hole (h^+_{VB}) is an oxidant and can oxidize compounds that are adsorbed on the semiconductor surface, including pesticide molecules (Figure 11.6).

In the case of the very well-known semiconducting oxide TiO_2, the holes of the valence band can be trapped by surface $\equiv Ti^{4+}–OH^-$ groups (producing $\equiv Ti^{4+}–^\bullet OH$, also called surface-adsorbed $^\bullet OH$) or by subsurface $\equiv Ti^{4+}–O^{2-}–Ti^{4+}\equiv$ species (yielding $\equiv Ti^{4+}–O^{\bullet -}–Ti^{4+}\equiv$, subsurface holes) [35].

The species $\equiv Ti^{4+}–^\bullet OH$ has qualitatively similar (but quantitatively lower) reactivity as free $^\bullet OH$, and $\equiv Ti^{4+}–O^{\bullet -}–Ti^{4+}\equiv$ is mostly involved in electron-transfer processes with adsorbed substrates. Conduction-band e^- can be trapped by Ti^{4+} ions to give Ti^{3+}, also named surface-adsorbed electron.

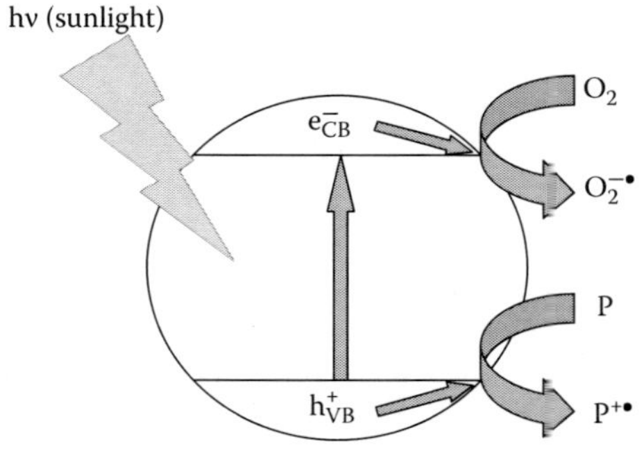

FIGURE 11.6 Processes following radiation absorption by a generic semiconductor oxide. P means pollutant (e.g., pesticide).

Very interestingly, the reductive pathways triggered by e^-_{CB} can produce oxidizing species with the following reaction sequences [36] ($O_2^{-\bullet}$ is produced upon O_2 reduction by e^-_{CB}):

$$O_2^{-\bullet} + H^+ \rightleftarrows HO_2^\bullet \tag{11.17}$$

$$O_2^{-\bullet} + HO_2^\bullet + H^+ \rightarrow H_2O_2 + O_2 \tag{11.18}$$

$$H_2O_2 + h\nu \rightarrow 2\ {}^\bullet OH \tag{11.19}$$

$$H_2O_2 + e^-_{CB} \rightarrow {}^\bullet OH + OH^- \tag{11.20}$$

Reactions 11.17 through 11.20 yield free $^\bullet OH$ in solution rather than surface-adsorbed species. Therefore, rather surprisingly, bulk $^\bullet OH$ in photocatalysis is produced by the reductant e^-_{CB} rather than by the oxidant h^+_{VB} [37].

The semiconductor oxide ZnO has similar behavior as TiO_2 although it has been subjected to many fewer studies. Both ZnO and TiO_2 are characterized by the absorption of sunlight only below 400 nm, thus only environmental UV radiation is available for the described reactions to take place [35].

Fe(III) (hydr)oxides absorb a considerably larger fraction of sunlight (typically, radiation absorption takes place below 550 nm), but this does not imply higher photoactivity compared with ZnO and TiO_2. In fact, despite easier production of e^-_{CB} and h^+_{VB} in Fe(III) compounds, their recombination is much faster and considerably hampers photoactivity. Therefore, semiconductor-like photoactivity of Fe(III) (hydr)oxides is quite limited [34]. However, these compounds can also undergo photolysis of the surface $\equiv Fe^{3+}–OH^-$ groups upon absorption of UV radiation, yielding $^\bullet OH$ and Fe^{2+} that can further react with H_2O_2 to give additional $^\bullet OH$ (Fenton reaction) [38].

$$= Fe^{3+} - OH^- + h\nu \rightarrow Fe^{2+} + {}^\bullet OH \tag{11.21}$$

$$Fe^{2+} + H_2O_2 \rightarrow Fe^{3+} + {}^\bullet OH + OH^- \tag{11.22}$$

The humic fraction of soil could potentially be able to induce pesticide photodegradation via triplet-state reactivity [39]. Indeed, soil-derived humic and fulvic acids are the most photoactive CDOM components in surface waters, and they are definitely more important in soil. Although the lower amount of water available might be unfavorable to such processes, for example, by limiting solute transport [40], this is a potentially important process that has received comparatively little attention by now.

Another important issue is that pesticides can be photodegraded on the leaf surface, which is often the site of their first application. Reactions in the waxy leaf environment might be somewhat different than in water and, as far as direct photolysis is concerned, similar considerations may apply as already seen for topsoil. However, pesticide molecules located deep in the leaf wax could even experience a more important solvent-cage effect than in water because of the much higher solvent viscosity. Photodegradation on the leaf surface has, for instance, been described for sulcotrione [41], mesotrione [42], bentazon, clopyralid, triclopyr [43], nicosulfuron [44], chlorothalonil [45], and cycloxydim [46].

11.2.2 Photochemical Reactions in Atmosphere

Photochemical processes in the atmosphere follow the usual classification of direct and indirect photolysis. They can take place in the gas phase, on the particle surface, and in suspended water droplets, depending on the volatility and the water solubility of the relevant compounds. Compared to the bulk phase of surface waters, direct photolysis processes would be favored in the gas phase, on the particle surface, and at the air–water interface of droplets (but not in the droplet

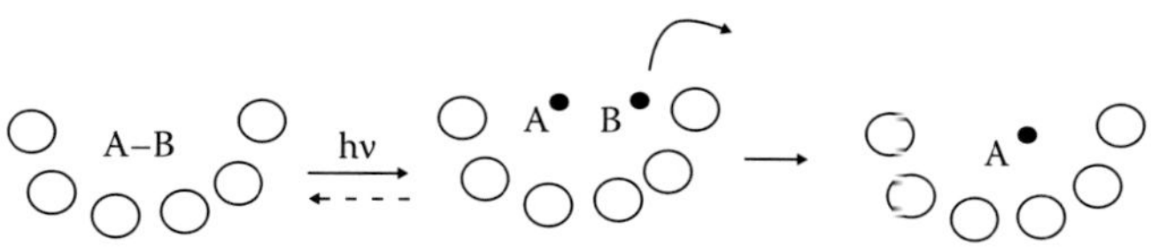

FIGURE 11.7 Direct photolysis process at the air–water interface (to be compared with Figure 11.5).

bulk) because of the absence of the solvent cage (see Figure 11.7 for the case of the air–water interface of droplets) [32].

Enhanced interface photolysis would be operational in surface waters as well, but in that case, the interface has a negligible weight compared to small droplets because of the unfavorable surface-to-volume ratio of large volumes. In the case of particles, direct photolysis can be inhibited by radiation screening effects. Indeed, black carbonaceous particles have been shown to protect adsorbed compounds against direct photolysis, mostly because of sunlight absorption [47].

As far as indirect photochemistry is concerned, the radical •OH will certainly play a more important role in the atmospheric gas phase than in surface waters. The main reason for this is the efficient scavenging of •OH by DOM in aqueous environments, which has no parallel in the atmospheric gas phase. In the latter case, •OH reactivity is mainly a daylight one, and it is triggered by the photolysis of several photoactive compounds: nitrous acid (HONO) in the early morning, formaldehyde later on, and finally, ozone at midday/afternoon [48]. The relevant processes are reported below.

$$HONO + h\nu \rightarrow {}^{\bullet}OH + {}^{\bullet}NO \quad (11.23)$$

$$HCHO + h\nu \rightarrow H^{\bullet} + CHO^{\bullet} \quad (11.24)$$

$$H^{\bullet} + O_2 \rightarrow HO_2^{\bullet} \quad (11.25)$$

$$HO_2^{\bullet} + {}^{\bullet}NO \rightarrow {}^{\bullet}OH + {}^{\bullet}NO_2 \quad (11.26)$$

$$O_3 + h\nu \rightarrow O_2 + O^* \quad (11.27)$$

$$O^* + H_2O \rightarrow 2\ {}^{\bullet}OH \quad (11.28)$$

The radical •OH is mainly involved in electron-transfer processes (which are unlikely in the gas phase, however), hydrogen atom abstraction, and addition to double bonds and aromatic rings [8]. The hydroxyl radical would essentially induce degradation processes during the day. Indeed, the combination of very high reactivity and of extremely low nighttime production ensures that •OH is almost absent from the atmosphere at night. Under such circumstances, gas-phase atmospheric reactivity is dominated by the nitrate radical (${}^{\bullet}NO_3$), which is efficiently photolyzed during the day but can survive in the absence of sunlight. The radical ${}^{\bullet}NO_3$ is produced by reaction between ${}^{\bullet}NO_2$ and ozone [48]:

$${}^{\bullet}NO_2 + O_3 \rightarrow {}^{\bullet}NO_3 + O_2 \quad (11.29)$$

Although less reactive than •OH, ${}^{\bullet}NO_3$ can reach higher concentration values in the atmosphere and can be important in the transformation of reactive pollutants and, most notably, of aromatic compounds. The radical ${}^{\bullet}NO_3$ can also abstract H atoms from aliphatics to produce HNO_3, but this process is considerably less efficient compared to •OH reactions [48].

A further reactant in the atmospheric gas phase is O_3, but it is only important in the transformation of compounds having double C=C bonds. Ozone reactivity with, for example, aromatics or other organic compounds is very low to nil, in particular if compared with •OH and ${}^{\bullet}NO_3$ [49].

In the case of particles, indirect photochemistry processes can be induced by irradiation of semiconductor oxides (see Section 11.2.1) and of nitrate salts (most notably $NaNO_3$ and NH_4NO_3). In the latter case, production of •OH is expected to take place in a similar way as in solution (see Reaction 11.1), and it is enhanced in the presence of water vapor, which would most likely act as a H^+ source [50]. In airborne particulate matter, transformation processes could also be induced by triplet sensitizers [51], such as quinones and aromatic carbonyls, and by aromatic nitroderivatives (e.g., 1-nitronaphthalene) [52]. All these compounds are well known to efficiently produce triplet states under irradiation.

The photochemistry of atmospheric water droplets has many similarities to but also important differences from surface-water photochemistry. First of all, due to the much higher surface-to-volume ratio, interface processes are definitely more important in droplets [32]. Moreover, CDOM that is found in the atmospheric aqueous phase is considerably less reactive than surface-water CDOM. In other words, there is evidence that atmospheric humic-like substances may be considerably less reactive than surface-water humic and fulvic acids [53]. Furthermore, due to the more acidic pH of atmospheric versus surface waters, processes involving Fe species (e.g., photolysis of $FeOH^{2+}$ and the Fenton reaction) would be more important in the atmospheric compartment (with minor exceptions, such as acidic mine-drainage water) [54].

$$FeOH^{2+} + h\nu \rightarrow Fe^{2+} + {}^\bullet OH \tag{11.30}$$

In the case of surface waters, the very low concentration of hydrogen peroxide makes H_2O_2 a minor to negligible •OH source under most circumstances. The situation is completely different in the atmospheric aqueous phase, in which mass transfer from the gas phase makes H_2O_2 photolysis an important •OH source [55,56]. However, differently from surface waters and again due to the much larger surface-to-volume ratio of atmospheric ones, •OH transfer from the gas phase to the aqueous solution is usually the most important source of hydroxyl radicals in atmospheric hydrometeors [57].

11.3 MODELING PESTICIDE PHOTOTRANSFORMATION IN SURFACE WATERS

The model presented here describes the transformation kinetics of a substrate, a generic pollutant P, as a function of water chemistry and substrate reactivity via the main photochemical reaction pathways that are operational in surface waters (direct photolysis and reaction with •OH, $CO_3^{-\bullet}$, $^1O_2$, and $^3CDOM^*$). It also calculates the steady-state concentrations of photogenerated transients in a cylindrical volume of 1 cm^2 surface area and depth *d*. The model may use actual data of the water absorption spectrum or, in their absence, it can approximate the spectrum from the dissolved organic carbon (DOC) values. The DOC, in units of mg C L^{-1}, is a measure of DOM. The different aspects of the model are now described in greater detail.

11.3.1 Surface-Water Absorption Spectrum

It is possible to find a reasonable correlation between the absorption spectrum of surface waters and their content of DOM, expressed as DOC. The following equation holds for the water spectrum, referred to an optical path length of 1 cm [58]:

$$A_1(\lambda) = (0.45 \pm 0.04)\ \text{DOC}\ e^{-(0.015 \pm 0.002)\lambda} \tag{11.31}$$

As an obvious alternative, $A_1(\lambda)$ can be spectrophotometrically determined on a real water sample.

11.3.2 Reaction with •OH [58]

In natural surface waters under sunlight illumination, the main •OH sources are (in order of average importance) CDOM, nitrite, and nitrate. All these species produce •OH upon absorption of sunlight.

The calculation of the photon fluxes absorbed by CDOM, nitrate and nitrite requires taking into account the mutual competition for sunlight irradiance. Actually, CDOM is the main radiation absorber in the 300–500 nm region where nitrite and nitrate also absorb radiation. At a given wavelength λ, the ratio of the photon flux densities absorbed by two different species is equal to the ratio of the respective absorbances. The same is also true for the ratio of the photon flux density absorbed by species to the total photon flux density absorbed by the solution, $p_a^{tot}(\lambda)$ [2]. Accordingly, the following equations hold for the different •OH sources (note that $A_1(\lambda)$ is the specific absorbance of the surface water layer over a 1-cm optical path length in units of cm^{-1}, d is the water column depth in m, $A_{tot}(\lambda)$ the total absorbance of the water column, and $p°(\lambda)$ the spectrum of sunlight, also called the incident photon flux density):

$$A_{tot}(\lambda) = 100 A_1(\lambda)\, d \tag{11.32}$$

$$A_{NO_3^-}(\lambda) = 100\, \varepsilon_{NO_3^-}(\lambda)\, d\, [NO_3^-] \tag{11.33}$$

$$A_{NO_2^-}(\lambda) = 100\, \varepsilon_{NO_2^-}(\lambda)\, d\, [NO_2^-] \tag{11.34}$$

$$A_{CDOM}(\lambda) = A_{tot}(\lambda) - A_{NO_3^-}(\lambda) - A_{NO_2^-} \approx A_{tot}(\lambda) \tag{11.35}$$

$$p_a^{tot}(\lambda) = p°(\lambda)\,(1 - 10^{-A_{tot}(\lambda)}) \tag{11.36}$$

$$p_a^{CDOM}(\lambda) = p_a^{tot}(\lambda)\, A_{CDOM}(\lambda)\, [A_{tot}(\lambda)]^{-1} \approx p_a^{tot}(\lambda) \tag{11.37}$$

$$p_a^{NO_2^-}(\lambda) = p_a^{tot}(\lambda)\, A_{NO_2^-}(\lambda)\, [A_{tot}(\lambda)]^{-1} \tag{11.38}$$

$$p_a^{NO_3^-}(\lambda) = p_a^{tot}(\lambda)\, A_{NO_3^-}(\lambda)\, [A_{tot}(\lambda)]^{-1} \tag{11.39}$$

An important issue is that $p°(\lambda)$ is usually reported in units of Einstein cm^{-2} s^{-1} nm^{-1} (see, for instance, Figure 11.8) [59]; thus, the absorbed photon flux densities are expressed in the same units. To express the formation rates of •OH in M s^{-1}, the absorbed photon fluxes P_a^i should be expressed in Einstein L^{-1} s^{-1}. Integration of $p_a^i(\lambda)$ over wavelength would give units of Einstein cm^{-2} s^{-1} that represent the moles of photons absorbed per unit surface area and unit time.

Assuming a cylindrical volume of unit surface area (1 cm^2) and depth d (expressed in m), the absorbed photon fluxes in Einstein L^{-1} s^{-1} units would be expressed as follows (note that 1 L = 10^3 cm^3, and 1 m = 10^2 cm):

$$P_a^{CDOM} = 10\, d^{-1} \int_\lambda p_a^{CDOM}(\lambda)\, d\lambda \tag{11.40}$$

$$P_a^{NO_2^-} = 10\, d^{-1} \int_\lambda p_a^{NO_2^-}(\lambda)\, d\lambda \tag{11.41}$$

$$P_a^{NO_3^-} = 10\, d^{-1} \int_\lambda p_a^{NO_3^-}(\lambda)\, d\lambda \tag{11.42}$$

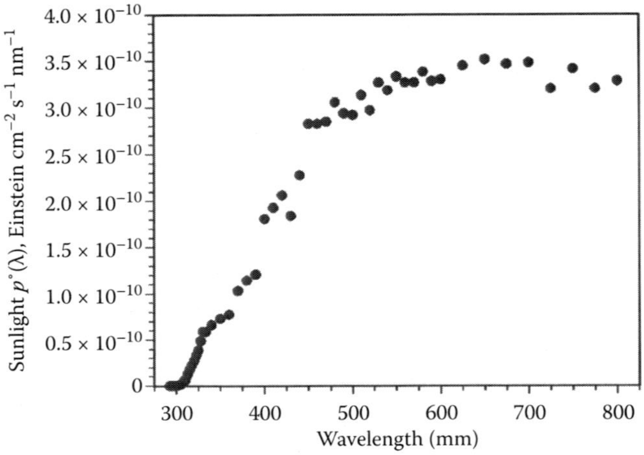

FIGURE 11.8 Sunlight spectral photon flux density at the water surface per unit area. The corresponding UV irradiance is 22 W m^{-2}. (From Frank, R., and Klöpffer, W., *Chemosphere*, 17, 985–994, 1988.)

Various studies have yielded useful correlation between the formation rate of •OH by the photoactive species and the respective absorbed photon fluxes of sunlight. In particular, it has been found that [58,60]

$$R_{\bullet OH}^{CDOM} = (3.0 \pm 0.4)\, 10^{-5}\, P_a^{CDOM} \tag{11.43}$$

$$R_{\bullet OH}^{NO_2^-} = \int_\lambda \Phi_{\bullet OH}^{NO_2^-}(\lambda)\, p_a^{NO_2^-}(\lambda)\, d\lambda \tag{11.44}$$

$$R_{\bullet OH}^{NO_3^-} = (4.3 \pm 0.2)\, 10^{-2}\, \frac{[IC] + 0.0075}{2.25[IC] + 0.0075}\, P_a^{NO_3^-} \tag{11.45}$$

where $[IC] = [H_2CO_3] + [HCO_3^-] + [CO_3^{2-}]$ is the total amount of inorganic carbon. The wavelength-dependent data of $\Phi_{\bullet OH}^{NO_2^-}(\lambda)$ are reported in Table 11.2 [5].

At the present state of knowledge, it is reasonable to hypothesize that CDOM, nitrite, and nitrate generate •OH independently with no mutual interactions. Therefore, the total formation rate of $^\bullet OH(R_{\bullet OH}^{tot})$ is the sum of the contributions of the three species:

$$R_{\bullet OH}^{tot} = R_{\bullet OH}^{CDOM} + R_{\bullet OH}^{NO_2^-} + R_{\bullet OH}^{NO_3^-} \tag{11.46}$$

Accordingly, having as input data *d*, $A_1(\lambda)$, $[NO_3^-]$, $[NO_2^-]$, and $p°(\lambda)$ (the latter referred to a 22 W m^{-2} sunlight UV irradiance, see Figure 11.8), it is possible to model the expected $R_{\bullet OH}^{tot}$ of the sample. The photogenerated •OH radicals could react either with the pollutant P or with the natural scavengers present in surface water (mainly organic matter, bicarbonate, carbonate, and nitrite). The natural scavengers have the following •OH scavenging rate constant [58]:

$$\sum_i k_{Si}[S_i] = 5 \times 10^4\ DOC + 8.5 \times 10^6 [HCO_3^-] + 3.9 \times 10^8 [CO_3^{-2}] + 1.0 \times 10^{10} [NO_2^-]$$

TABLE 11.2
Values of the Quantum Yield of •OH Photoproduction by Nitrite for Different Wavelengths of Environmental Significance

λ, nm	$\Phi_{^\bullet OH}^{NO_2^-}(\lambda)$	λ, nm	$\Phi_{^\bullet OH}^{NO_2^-}(\lambda)$	λ, nm	$\Phi_{^\bullet OH}^{NO_2^-}(\lambda)$
292.5	0.0680	315.0	0.061	350	0.025
295.0	0.0680	317.5	0.058	360	0.025
297.5	0.0680	320.0	0.054	370	0.025
300.0	0.0678	322.5	0.051	380	0.025
302.5	0.0674	325.0	0.047	390	0.025
305.0	0.0668	327.5	0.043	400	0.025
307.5	0.066	330.0	0.038	410	0.025
310.0	0.065	333.3	0.031	420	0.025
312.5	0.063	340.0	0.026	430	0.025

(the scavenging rate constant has units of s^{-1}; DOC is expressed in mg C L^{-1}, and the other concentration values are in molarity). Accordingly, the reaction rate between P and •OH can be expressed as follows:

$$R_P^{^\bullet OH} = R_{^\bullet OH}^{tot} \frac{k_{P,^\bullet OH}[P]}{k_{P,^\bullet OH}[P] + \sum_i k_{Si}[S_i]} \tag{11.47}$$

where $k_{P,^\bullet OH}$ is the second-order reaction rate constant between P and •OH, and [P] is a molar concentration. Note that in the vast majority of environmental cases it would be $k_{P,^\bullet OH}[P] \ll \sum_i k_{Si}[S_i]$; thus, the $k_{P,^\bullet OH}[P]$ term can be neglected at the denominator of Equation 11.47. The pseudo-first order degradation rate constant of P is $k_P = R_{^\bullet OH}^P [P]^{-1}$, and the half-life time is $t_P = \ln 2\, k_P^{-1}$. The time t_P is expressed in seconds of continuous irradiation under sunlight at 22 W m^{-2} UV irradiance (see Figure 11.8 for the sunlight spectrum). It has been shown that the sunlight energy reaching the ground on a summer sunny day (SSD), such as July 15, at 45°N latitude corresponds to 10 h = 3.6×10^4 s of continuous irradiation at 22 W m^{-2} UV irradiance [61]. Accordingly, the half-life time of P, because of the reaction with •OH, would be expressed as follows in SSD units:

$$\tau_{P,^\bullet OH}^{SSD} = \frac{\ln 2 \sum_i k_{Si}[S_i]}{3.6 \times 10^4 R_{^\bullet OH}^{tot} k_{P,^\bullet OH}} = 1.9 \times 10^{-5} \frac{\sum_i k_{Si}[S_i]}{R_{^\bullet OH}^{tot} k_{P,^\bullet OH}} \tag{11.48}$$

Note that $1.9 \times 10^{-5} = \ln 2(3.6 \times 10^4)^{-1}$. The steady-state [•OH] under 22 W m^{-2} UV irradiance would be:

$$[^\bullet OH] = \frac{R_{^\bullet OH}^{tot}}{\sum_i k_{Si}[S_i]} \tag{11.49}$$

11.3.3 Direct Photolysis

The calculation of the photon flux absorbed by P requires taking into account the mutual competition for sunlight irradiance between P and the other water components (mostly CDOM, which is the main sunlight absorber in the spectral region of interest, around 300–500 nm) [62,63].

Under the Lambert-Beer approximation, at a given wavelength λ, the ratio of the photon flux densities absorbed by two different species is equal to the ratio of the respective absorbances [2]. Accordingly, the photon flux absorbed by P in a water column of depth d (expressed in m) can be obtained as follows (note that $A_1(\lambda)$ is the specific absorbance of the surface water sample over a 1-cm optical path length; $A_{tot}(\lambda)$ the total absorbance of the water column; $p°(\lambda)$ the spectrum of sunlight, referred to a UV irradiance of 22 W m^{-2} as per Figure 11.8; $\varepsilon_P(\lambda)$ the molar absorption coefficient of P, in units of M^{-1} cm^{-1}; and $p_a^P(\lambda)$ its absorbed spectral photon flux density—it is also $p_a^P(\lambda) \ll p_a^{tot}(\lambda)$ and $A_P(\lambda) \ll A_{tot}(\lambda)$ in the very vast majority of the environmental cases):

$$A_{tot}(\lambda) = 100\, A_1(\lambda)\, d \tag{11.50}$$

$$A_P(\lambda) = 100\, \varepsilon_P(\lambda)\, d\, [P] \tag{11.51}$$

$$p_a^{tot}(\lambda) = p°(\lambda)(1 - 10^{-A_{tot}(\lambda)}) \tag{11.52}$$

$$p_a^P(\lambda) = p_a^{tot}(\lambda)\, A_P(\lambda)\, [A_{tot}(\lambda)]^{-1} \tag{11.53}$$

To express the rate of P photolysis in M s^{-1}, the absorbed photon flux P_a^P should be expressed in Einstein L^{-1} s^{-1}. Integration of $p_a^P(\lambda)$ over wavelength gives units of Einstein cm^{-2} s^{-1} that represent the moles of photons absorbed per unit surface area and unit time. Assuming a cylindrical volume of unit surface area (1 cm^2) and depth d (expressed in m), the absorbed photon flux in Einstein L^{-1} s^{-1} units would be expressed as follows (note that 1 L = 10^3 cm^3 and 1 m = 10^2 cm):

$$P_a^P = 10\, d^{-1} \int_\lambda p_a^P(\lambda)\, d\lambda \tag{11.54}$$

The rate of photolysis of P, expressed in M s^{-1}, is (note that 1 L = 10^3 cm^3, and 1 m = 10^2 cm):

$$\text{Rate}_P = 10\, d^{-1} \int_\lambda \Phi_P(\lambda)\, p_a^P(\lambda)\, d\lambda \tag{11.55}$$

where $\Phi_P(\lambda)$ is the photolysis quantum yield of P in the relevant wavelength interval, and d is expressed in cm. If only a single average value for Φ_P is known, it can be brought out of the integral as a constant. The pseudo-first order degradation rate constant of P is $k_P = \text{Rate}_P\, [P]^{-1}$, which corresponds to a half-life time of $t_P = \ln 2(k_P)^{-1}$. The time t_P is expressed in seconds of continuous irradiation under sunlight at 22 W m^{-2} UV irradiance. The sunlight energy reaching the ground in a SSD, such as July 15, at 45°N latitude corresponds to 10 h = 3.6 × 10^4 s continuous irradiation at 22 W m^{-2} UV irradiance [61]. Accordingly, the half-life time expressed in SSD units would be given by (note that $V = 0.1d$):

$$\tau_P^{SSD} = (3.6\times10^4)^{-1}\ln 2(k_P)^{-1} = 1.9\times10^{-5}[P](Rate_P)^{-1}$$
$$= 1.9\times10^{-5}[P]V\left(\int_\lambda \Phi_P(\lambda)p_a^P(\lambda)\,d\lambda\right)^{-1}$$
$$= 1.9\times10^{-5}[P]V\left(\int_\lambda \Phi_P(\lambda)\,p_a^{tot}(\lambda)\,A_P(\lambda)[A_{tot}(\lambda)]^{-1}\,d\lambda\right)^{-1}$$
$$= \frac{1.9\times10^{-5}V[P]}{\int_\lambda \Phi_P(\lambda)p^\circ(\lambda)(1-10^{-100A_1(\lambda)d})\frac{\varepsilon_P(\lambda)}{A_1(\lambda)}d\lambda} \tag{11.56}$$

Note that $1.9\times10^{-5} = (\ln 2)(3.6\times10^4)^{-1}$.

11.3.4 Reaction with $CO_3^{-\bullet}$

The radical $CO_3^{-\bullet}$ can be produced upon oxidation of carbonate and bicarbonate by •OH, upon carbonate oxidation by $^3CDOM^*$, and possibly also from irradiated Fe(III) oxide colloids and carbonate [64]. However, as far as the latter process is concerned, there is still insufficient knowledge about the Fe speciation in surface waters to enable a proper modeling. The main sink of the carbonate radical in surface waters is the reaction with DOM, which is considerably slower than that between DOM and •OH.

$$^\bullet OH + CO_3^{2-} \rightarrow OH^- + CO_3^{-\bullet} \quad [k_{27} = 3.9\times10^8\ M^{-1}\ s^{-1}] \tag{11.57}$$

$$^\bullet OH + HCO_3^- \rightarrow H_2O + CO_3^{-\bullet} \quad [k_{28} = 8.5\times10^6\ M^{-1}\ s^{-1}] \tag{11.58}$$

$$^3CDOM^* + CO_3^{2-} \rightarrow CDOM^{-\bullet} + CO_3^{-\bullet} \quad [k_{29} \approx 1\times10^5\ M^{-1}\ s^{-1}] \tag{11.59}$$

$$DOM + CO_3^{-\bullet} \rightarrow DOM^{+\bullet} + CO_3^{2-} \quad [k_{30} \approx 10^2 (mg\ C)^{-1}\ s^{-1}] \tag{11.60}$$

The formation rate of $CO_3^{-\bullet}$ in Reactions 11.57 and 11.58 is given by the formation rate of •OH times the fraction of •OH that reacts with carbonate and bicarbonate, as follows:

$$R_{CO_3^{-\bullet}}^{\bullet OH} = R_{\bullet OH}^{tot}\frac{8.5\times10^6[HCO_3^-]+3.9\times10^8[CO_3^{2-}]}{5\times10^4\,DOC+1.0\times10^{10}[NO_2^-]+8.5\times10^6[HCO_3^-]+3.9\times10^8[CO_3^{2-}]} \tag{11.61}$$

The formation of $CO_3^{-\bullet}$ in Reaction 11.59 is given by [64]

$$R_{CO_3^{-\bullet}}^{CDOM} = 6.5\times10^{-3}[CO_3^{2-}]P_a^{CDOM} \tag{11.62}$$

The total formation rate of $CO_3^{-\bullet}$ is $R_{CO_3^{-\bullet}}^{tot} = R_{CO_3^{-\bullet}}^{\bullet OH} + R_{CO_3^{-\bullet}}^{CDOM}$. The transformation rate of P by $CO_3^{-\bullet}$ is given by the fraction of $CO_3^{-\bullet}$ that reacts with P in competition with Reaction 11.30 between $CO_3^{-\bullet}$ and DOM [64]:

$$R_{P,CO_3^{-\bullet}} = \frac{R_{CO_3^{-\bullet}}^{tot}\, k_{P,CO_3^{-\bullet}}[P]}{k_{30}\ DOC + k_{P,CO_3^{-\bullet}}[P]} \tag{11.63}$$

where $k_{P,CO_3^{-\bullet}}$ is the second-order reaction rate constant between P and $CO_3^{-\bullet}$. In the very vast majority of the environmental cases, it is $k_{P,CO_3^{-\bullet}}[P] \ll k_{30}$ DOC.

In a pseudo-first order approximation, the rate constant of P transformation is $k_P = R_{P,CO_3^{-\bullet}}[P]^{-1}$, and the half-life time is $t_P = \ln 2\, k_P^{-1}$. Considering the usual conversion (≈10 h) between a constant 22 W m^{-2} sunlight UV irradiance and a SSD unit, the following expression for $\tau_{NCP,CO_3^{-\bullet}}^{SSD}$ is obtained:

$$\tau_{P,CO_3^{-\bullet}}^{SSD} = 1.9\times10^{-5}\left(\frac{k_{30}\ DOC}{R_{CO_3^{-\bullet}}^{tot}\, k_{P,CO_3^{-\bullet}}}\right) \tag{11.64}$$

Note that $1.9 \times 10^{-5} = \ln 2(3.6 \times 10^{4})^{-1}$. The steady-state $[CO_3^{-\bullet}]$ under 22 W m^{-2} UV irradiance would be

$$[CO_3^{-\bullet}] = \frac{R_{CO_3^{-\bullet}}^{tot}}{k_{30}\ DOC} \tag{11.65}$$

11.3.5 Reaction with $^1O_2$

The formation of singlet oxygen in surface waters arises from energy transfer between ground-state molecular oxygen and the excited triplet states of CDOM (^{3}CDOM*). Accordingly, irradiated CDOM is practically the only source of $^1O_2$ in aquatic systems. In contrast, the main $^1O_2$ sink is the energy loss to ground-state O_2 by collision with water molecules with a pseudo-first order rate constant $k_{^1O_2} = 2.5\times10^5\ s^{-1}$. Dissolved species, including DOM, that is certainly able to react with $^1O_2$ would play a minor role as sinks of $^1O_2$ in aquatic systems. The main processes involving $^1O_2$ and P in surface waters would be the following [65]:

$$^3CDOM^* + O_2 \rightarrow CDOM + {}^1O_2 \tag{11.66}$$

$$^1O_2 + H_2O \rightarrow O_2 + H_2O + heat \tag{11.67}$$

$$^1O_2 + P \rightarrow Products \tag{11.68}$$

In the Rhône delta waters, it has been found that the formation rate of $^1O_2$ by CDOM is $R_{^1O_2}^{CDOM} = 1.25\times10^{-3}\ P_a^{CDOM}$ [66]. Considering the competition between the deactivation of $^1O_2$ by collision with the solvent (Reaction 11.67) and Reaction 11.68 with P, one gets the following expression for the degradation rate of P by $^1O_2$ (note that $k_{P,^1O_2}[P] \ll k_{^1O_2}$):

$$R_P^{^1O_2} = R_{^1O_2}^{CDOM} \frac{k_{P,^1O_2}[P]}{k_{^1O_2}} \tag{11.69}$$

In a pseudo-first order approximation, the rate constant of P transformation is $k_P = R_P^{^1O_2}[P]^{-1}$, and the half-life time is $t_P = \ln 2\, k_P^{-1}$. Considering the usual conversion (≈10 h) between a constant 22 W m^{-2} sunlight UV irradiance and a SSD unit, the following expression for $\tau_{P,^1O_2}^{SSD}$ is obtained (remembering that $R_{^1O_2}^{CDOM} = 1.25\times10^{-3}P_a^{CDOM}$ and that $P_a^{CDOM} = 10^3 d^{-1}\int_\lambda p_a^{CDOM}(\lambda)\,d\lambda$):

$$\tau_{P,^1O_2}^{SSD} = \frac{4.81}{R_{^1O_2}^{CDOM} k_{P,^1O_2}} = \frac{3.85d}{k_{P,^1O_2}\int_\lambda p_a^{CDOM}(\lambda)\,d\lambda} \tag{11.70}$$

Note that $3.85 = (\ln 2)k_{^1O_2}(1.25\times10^{-3}\times3.60\times10^4\times10^3)^{-1}$. The steady-state [$^1O_2$] under 22 W m^{-2} UV irradiance would be

$$[^1O_2] = \frac{R_{^1O_2}^{CDOM}}{k_{^1O_2}} \tag{11.71}$$

11.3.6 Reaction with ^{3}CDOM*

The formation of excited triplet states of CDOM (^{3}CDOM*) in surface waters is a direct consequence of radiation absorption by CDOM [66]. In aerated solution, ^{3}CDOM* could undergo thermal deactivation or reaction with O_2, and a pseudo-first order quenching rate constant $k_{^3CDOM*} \approx 5\times10^5\ s^{-1}$ has been observed. The quenching of ^{3}CDOM* would be in competition with the reaction between ^{3}CDOM* and P [65]:

$$CDOM + h\nu \rightarrow {}^3CDOM^* \tag{11.72}$$

$$^3CDOM^* - (O_2) \rightarrow \text{Deactivation and } {}^1O_2 \text{ production} \tag{11.73}$$

$$^3CDOM^* + P \rightarrow \text{Products} \tag{11.74}$$

In the Rhône delta waters, it has been found that the formation rate of ^{3}CDOM* is $R_{^3CDOM*} = 1.28\times10^{-3}P_a^{CDOM}$ [66]. Considering the competition between Reaction 11.74 with P and other processes (Reaction 11.73), the following expression for the degradation rate of P by ^{3}CDOM* is obtained (note that $k_{P,^3CDOM*}[P] \ll k_{^3CDOM*}$, where $k_{P,^3CDOM*}$ is the second-order reaction rate constant between P and ^{3}CDOM*):

$$R_P^{^3CDOM*} = R_{^3CDOM*}\frac{k_{P,^3CDOM*}[P]}{k_{^3CDOM*}} \tag{11.75}$$

In a pseudo-first order approximation, the rate constant of P transformation is $k_P = R_P^{^3CDOM*}[P]^{-1}$, and the half-life time is $t_P = \ln 2\, k_P^{-1}$. Considering the usual conversion (≈10 h) between a constant 22 W m^{-2} sunlight UV irradiance and a SSD unit, one gets the following expression for $\tau_{P,^3CDOM*}^{SSD}$ (remembering that $P_a^{CDOM} = 10^3 d^{-1}\int_\lambda p_a^{CDOM}(\lambda)\,d\lambda$):

$$\tau_{P,^3CDOM*}^{SSD} = \frac{7.52d}{k_{P,^3CDOM*}\int_\lambda p_a^{CDOM}(\lambda)\,d\lambda} \tag{11.76}$$

Note that 7.52 = (ln 2) $k_{^3\text{CDOM*}}$ $(1.28 \times 10^{-3} \times 3.60 \times 10^4 \times 10^3)^{-1}$. The steady-state [^{3}CDOM*] under 22 W m^{-2} UV irradiance would be:

$$[^3\text{CDOM*}] = \frac{R_{^3\text{CDOM*}}}{k_{^3\text{CDOM*}}} \tag{11.77}$$

11.3.7 Formation of Intermediates

In the photochemical process *ph* (direct photolysis or reaction with •OH, $^1O_2$, $CO_3^{-\bullet}$, ^{3}CDOM*), the pollutant P could produce the intermediate I with yield y_I^{ph}, experimentally determined as the ratio between the initial formation rate of I and the initial transformation rate of P [19]. The pseudo-first order rate constant of I formation in the process *ph* is $(k_I^{ph})' = y_I^{ph} k_P^{ph}$, where k_P^{ph} is the (model-derived) first-order transformation rate constant of P in the process *ph*. The production of I from P often takes place via more than one process. Therefore, the overall rate constant of I formation is

$$(k_I)' = \sum_{ph} (k_I^{ph})' = \sum_{ph} (y_I^{ph} k_I^{ph}) \tag{11.78}$$

One can also obtain the overall yield of I formation from P (y_I) as [19]

$$y_i = (k_I)'(k_P)^{-1} = \frac{\sum_{ph} (y_I^{ph} k_I^{ph})}{\sum_{ph} k_P^{ph}} \tag{11.79}$$

11.3.8 Meaning of Water Depth in Model

An important issue is that the model was not designed to make depth profiles of the transformation kinetics or of the concentration of reactive transients. Therefore, when setting depth as a variable, one actually compares different water bodies, each with its own depth value. This means that, for example, 1 m depth, the model returns the average [•OH] (or the steady-state concentration of other species) in the first 1 m of the water column. It should be underlined that it is the average concentration in the first 1 m of the column and *not* the point concentration at 1 m. One can also obtain the transformation kinetics of dissolved species in the hypothesis of thorough mixing in the water column because the model applies to well-mixed shallow waters or to the top mixing layer of stratified water bodies. A key issue is that, if one wants to determine the photochemical reaction kinetics due to, for example, the reaction with •OH in the first 1 m of the water column, the needed value is the average [•OH] value (as determined by the model) and not the point [•OH] at 1 m.

11.3.9 Main Approximations of Model

Surface waters represent an extremely complex and varied series of environments, and the present attempt to describe their photochemical behavior had to include a number of assumptions and approximations. The main ones are listed below.

- The model considers well-mixed water. Therefore, it applies to shallow water environments and to the well-mixed epilimnion of stratified ones.
- The Lambert-Beer approximation does not take radiation scattering into account. Therefore, the model applies to clear waters rather than to highly turbid ones.

- The data on which the modeling of the surface-water absorption spectrum is based (Equation 11.31) were obtained for lake water in NW Italy. There is evidence that applicability is much wider, but more accurate results for a particular environment can be obtained if the actual water spectrum is available.
- The quantum yields for the formation of •OH by CDOM are average values for NW Italian lakes. The corresponding data of $^1O_2$ and ^{3}CDOM* have been obtained in the Rhône delta (S. France), and the value of $CO_3^{-•}$ formation from ^{3}CDOM* is from Lake Greifensee (Switzerland). In different environments, different values may be found. The best scenario is obviously attained when one has actual data measured in the water environment under study.
- The scavenging rate constants of •OH and $CO_3^{-•}$ by DOM are average values from the literature. The same consideration as above also applies here.

11.3.10 Model Application to Herbicide MCPA

MCPA (4-chloro-2-methylphenoxyacetic acid) is a phenoxyacetic acid herbicide that undergoes photochemical transformation in the environment by direct photolysis and reaction with •OH [67,68]. The main phototransformation intermediate is the toxic 4-chloro-2-methylphenol (CMP), which is formed from the parent compound by the two photochemical processes with different yields (0.3 for the direct photolysis and 0.5 for •OH). MCPA has a second-order reaction rate constant with •OH of $6.6 \times 10^9\ M^{-1}\ s^{-1}$ [69], and the photolysis quantum yield depends on the DOC content of the solution. The main reason for this is that MCPA photolysis proceeds through reactions of its triplet state, which can be reduced to the radical anion (and the radical anion recycled back to initial MCPA by O_2) in the presence of dissolved organic compounds (S). The whole reaction set is as follows [68]:

$$\text{MCPA} + h\nu \rightarrow {}^1\text{MCPA}^* - (\text{ISC}) \rightarrow {}^3\text{MCPA}^* \tag{11.80}$$

$$^3\text{MCPA}^* + \text{S} \rightarrow \text{MCPA}^{-•} + \text{S}^{+•} \tag{11.81}$$

$$\text{MCPA}^{-•} + O_2 \rightarrow \text{MCPA} + O_2^{-•} \tag{11.82}$$

Therefore, the photolysis quantum yield of MCPA decreases with increasing DOC. An experimental assessment of the trend of Φ_{MCPA} versus DOC gave the following results [68]:

$$\Phi_{MCPA} = \frac{(2.3 \pm 0.7) \times 10^{-5} + (4.3 \pm 0.1) \times 10^{-6}\,\text{DOC}}{(4.1 \pm 1.3) \times 10^{-5} + 1.4 \times 10^{-5}\,\text{DOC}} \tag{11.83}$$

With the above values for the •OH reaction rate constant and the photolysis quantum yield, it is possible to model the half-life time of MCPA and the yield of CMP and MCPA (ηCMP) as a function of the chemical composition and depth of surface waters. Figure 11.9 reports the half-life time of MCPA (units of SSD) as a function of depth and DOC (Figure 11.9a) and as a function of nitrate and DOC (Figure 11.9b). Figure 11.10 reports the yield η_{CMP} under the same conditions as for the previous figure.

Figure 11.9a shows that the modeled half-life time of MCPA (in the order of days to some months) increases with increasing DOC and depth. The increase with DOC is accounted for by the fact that CDOM and DOM are more concentrated at elevated DOC. CDOM inhibits MCPA direct photolysis by competing for sunlight irradiance, and DOM decreases the photolysis quantum yield (Equation 11.83) and scavenges •OH. The increase of the half-life time with depth is due to the fact that the bottom layers of a deeper water body are less illuminated by sunlight, which does not favor the light-induced processes.

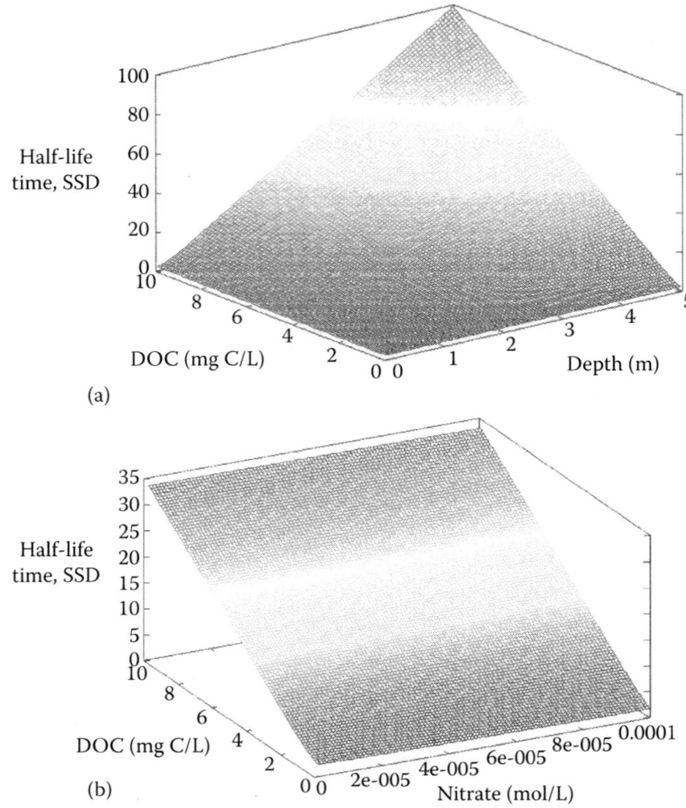

FIGURE 11.9 Modeled half-life time (SSD) of MCPA as a function of (a) DOC and depth (other water parameters: 1 μM nitrate, 10 nM nitrite, 2 mM bicarbonate, 10 μM carbonate), and (b) DOC and nitrate (other water parameters: 2 m depth, 10 nM nitrite, 2 mM bicarbonate, 10 μM carbonate).

Figure 11.9b shows that, in addition to increasing with increasing DOC, the half-life time decreases with increasing nitrate that is a •OH source and enhances MCPA transformation.

Figure 11.10a shows that the yield of CMP from MCPA increases with depth and DOC, which both favor •OH reactions over the direct photolysis (remember that the •OH yield, 0.5, is higher than the yield by direct photolysis, 0.3). The reason is that MCPA mainly absorbs UVB radiation, which has poor penetration inside the water body. In contrast, the •OH sources CDOM and nitrite also absorb significantly in the UVA region (and CDOM absorbs in the visible as well).

Figure 11.10b shows that the yield increases with increasing nitrate as •OH source (which obviously enhances degradation by •OH) and that it has an interesting trend with DOC. At low nitrate, the yield increases with increasing DOC because, under such conditions, CDOM is the main •OH source, and •OH formation by CDOM offsets •OH scavenging by DOM. At high nitrate, the trend of η_{CMP} versus DOC has a minimum because the initial increase of DOC has the main effect of scavenging the •OH radicals produced by nitrate, thereby inhibiting the •OH + MCPA reaction more than the direct photolysis. At high DOC, CDOM becomes the main •OH source, and an increase of DOC inhibits MCPA direct photolysis (through competition between CDOM and MCPA for irradiance and because DOM decreases the photolysis quantum yield) to a higher extent than it inhibits •OH reactions.

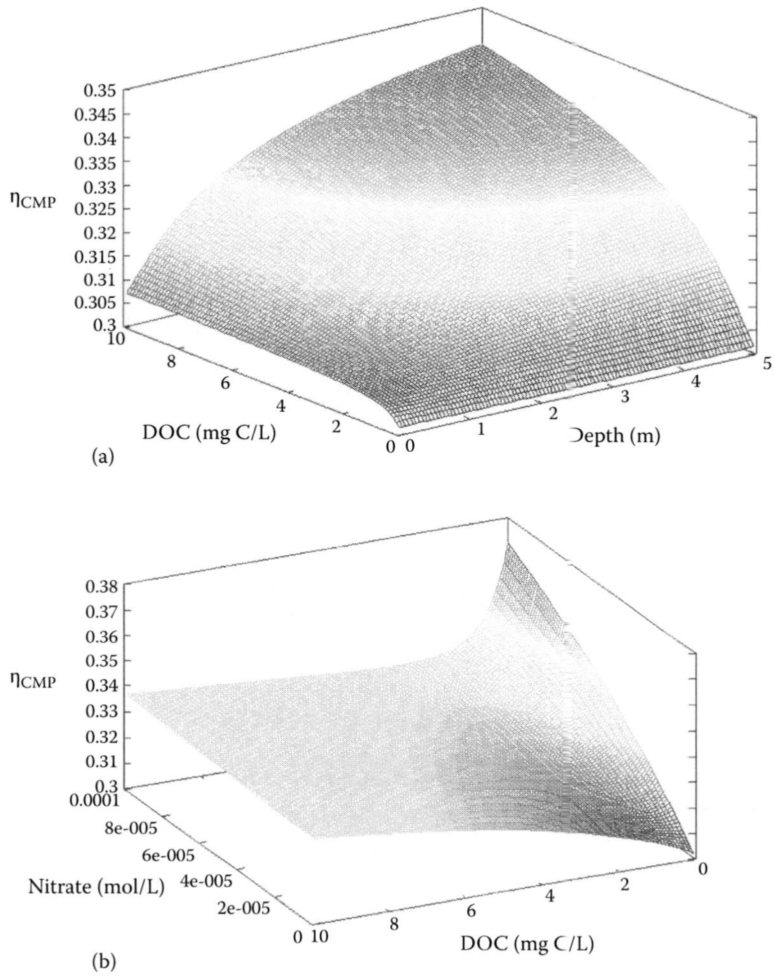

FIGURE 11.10 Modeled CMP yield from MCPA as a function of (a) DOC and depth (other water parameters: 1 μM nitrate, 10 nM nitrite, 2 mM bicarbonate, 10 μM carbonate) and (b) DOC and nitrate (2 m depth, 10 nM nitrite, 2 mM bicarbonate, 10 μM carbonate).

REFERENCES

1. Wayne, R. P., Basic concepts of photochemical transformations. In: Boule, P., Bahnemann, D. W., Robertson, P. K. J. (Eds.), *The Handbook of Environmental Chemistry Vol. 2.M—Environmental Photochemistry Part II*, Springer, Berlin, 2005, pp. 1–47.
2. Braslavsky, S. E., Glossary of terms used in photochemistry, 3rd edition, *Pure Appl. Chem.*, 79, 293–465, 2007.
3. Boreen, A. L., Arnold, W. A., and McNeill, K., Photodegradation of pharmaceuticals in the aquatic environment: A review, *Aquat. Sci.*, 65, 320–341, 2003.
4. Canonica, S., Kohn, T., Mac, M., Real, F. J., Wirz, J., and Von Gunten, U., Photosensitizer method to determine rate constants for the reaction of carbonate radical with organic compounds, *Environ. Sci. Technol.*, 39, 9182–9188, 2005.
5. Mack, J., and Bolton, J. R., Photochemistry of nitrite and nitrate in aqueous solution, *J. Photochem. Photobiol. A: Chem.*, 128, 1–13, 1999.

6. Sur, B., Rolle, M., Minero, C., Maurino, V., Vione, D., Brigante, M., and Mailhot, G., Formation of hydroxyl radicals by irradiated 1-nitronaphthalene (1NN): Oxidation of hydroxyl ions and water by the 1NN triplet state, *Photochem. Photobiol. Sci.*, 10, 1817–1824, 2011.
7. Vermilyea, A. W., and Voelker, B. M., Photo-Fenton reaction at near neutral pH, *Environ. Sci. Technol.*, 43, 6927–6933, 2009.
8. Buxton, G. V., Greenstock, C. L., Helman, W. P., and Ross, A. B., Critical review of rate constants for reactions of hydrated electrons, hydrogen atoms and hydroxyl radicals ($^\bullet OH/^\bullet O^-$) in aqueous solution, *J. Phys. Chem. Ref. Data*, 17, 1027–1284, 1988.
9. Brezonik, P. L., and Fulkerson-Brekken, J., Nitrate-induced photolysis in natural waters: Controls on concentrations of hydroxyl radical photo-intermediates by natural scavenging agents, *Environ. Sci. Technol.*, 32, 3004–3010, 1998.
10. Vione, D., Falletti, G., Maurino, V., Minero, C., Pelizzetti, E., Malandrino, M., Ajassa, R., Olariu, R. I., and Arsene, C., Sources and sinks of hydroxyl radicals upon irradiation of natural water samples, *Environ. Sci. Technol.*, 40, 3775–3781, 2006.
11. Bouillon, R. C., and Miller, W. L., Photodegradation of dimethyl sulfide (DMS) in natural waters: Laboratory assessment of the nitrate-photolysis-induced DMS oxidation, *Environ. Sci. Technol.*, 39, 9471–9477, 2005.
12. Canonica, S., Hellrung, B., Muller, P., and Wirz, J., Aqueous oxidation of phenylurea herbicides by triplet aromatic ketones, *Environ. Sci. Technol.*, 40, 6636–6641, 2006.
13. Wenk, J., and Canonica, S., Phenolic antioxidants inhibit the triplet-induced transformation of anilines and sulfonamide antibiotics in aqueous solution, *Environ. Sci. Technol.*, 46, 5455–5462, 2012.
14. Grabner, G., and Richard, C., Mechanisms of direct photolysis of biocides based on halogenated phenols and anilines. In: Boule, P., Bahnemann, D. W., Robertson, P. K. J. (Eds.), *The Handbook of Environmental Chemistry Vol. 2.M—Environmental Photochemistry Part II*, Springer, Berlin, 2005, pp. 161–192.
15. Chiron, S., Minero, C., and Vione, D., Occurrence of 2,4-dichlorophenol and of 2,4-dichloro-6-nitrophenol in the Rhône River Delta (Southern France), *Environ. Sci. Technol.*, 41, 3127–3133, 2007.
16. Chiron, S., Comoretto, L., Rinaldi, E., Maurino, V., Minero, C., and Vione, D., Pesticide by-products in the Rhône delta (Southern France). The case of 4-chloro-2-methylphenol and of its nitroderivative, *Chemosphere*, 74, 599–604, 2009.
17. Maddigapu, P. R., Vione, D., Ravizzoli, B., Minero, C., Maurino, V., Comoretto, L., and Chiron, S., Laboratory and field evidence of the photonitration of 4-chlorophenol to 2-nitro-4-chlorophenol and of the associated bicarbonate effect, *Environ. Sci. Pollut. Res.*, 17, 1063–1069, 2010.
18. Richard, C., and Canonica, S., Aquatic phototransformation of organic contaminants induced by coloured dissolved natural organic matter. In: Boule, P., Bahnemann, D. W., Robertson, P. K. J. (Eds.), *The Handbook of Environmental Chemistry Vol. 2.M—Environmental Photochemistry Part II*, Springer, Berlin, 2005, pp. 299–323.
19. De Laurentiis, E., Chiron, S., Kouras-Hadef, S., Richard, C., Minella, M., Maurino, V., Minero, C., and Vione, D., Photochemical fate of carbamazepine in surface freshwaters: Laboratory measures and modelling, *Environ. Sci. Technol.*, 46, 8164–8173, 2012.
20. Nelieu, S., Perreau, F., Bonnemoy, F., Ollitrault, M., Azam, D., Lagadic, L., Bohatier, J., Einhorn, J., Sunlight nitrate-induced photodegradation of chlorotoluron: Evidence of the process in aquatic mesocosms, *Environ. Sci. Technol.*, 43, 3148–3154, 2009.
21. Machado, F., Collin, L., and Boule, P., Photolysis of bromoxynil (3,5-dibromo-4-hydroxybenzonitrile) in aqueous solution, *Pest. Sci.*, 45, 107–110, 1995.
22. Mansfield, E., and Richard, C., Phototransformation of dichlorophen in aqueous phase, *Pest. Sci.*, 48, 73–76, 1996.
23. Aguer, J. P., Blachère, F., Boule, P., Garaudee, S., and Guillard, C., Photolysis of dicamba (3,6-dichloro-2-methoxybenzoic acid) in aqueous solution and dispersed on solid supports, *Int. J. Photoenergy* 2, 81–86, 2000.
24. Vialaton, D., Baglio, D., Paya-Perez, A., and Richard, C., Photochemical transformation of acifluorfen under laboratory and natural conditions, *Pest. Manag. Sci.*, 57, 372–379, 2001.
25. Torrents, A., Anderson, B. G., Bilboulian, S., Johnson, W. E., and Hapeman, C. J., Atrazine photolysis: Mechanistic investigations of direct and nitrate-mediated hydroxy radical processes and the influence of dissolved organic carbon from the Chesapeake Bay, *Environ. Sci. Technol.*, 31, 1476–1482, 1997.
26. Sakkas, V. A., Lambropoulou, D. A., and Albanis, T. A., Photochemical degradation study of Irgarol 1051 in natural waters: Influence of humic and fulvic substances on the reaction, *J. Photochem. Photobiol. A: Chem.*, 147, 135–141, 2002.

27. Vialaton, D., Pilichowski, J. F., Baglio, D., Paya-Perez, A., Larsen, B., and Richard, C., Phototransformation of propiconazole in aqueous media, *J. Agric. Food Chem.*, 11, 5377–5382, 2001.
28. Hustert, K., Moza, P. N., and Kettrup, A., Photochemical degradation of carboxin and oxycarboxin in the presence of humic substances and soil, *Chemosphere*, 38, 3423–3429, 1999.
29. Krieger, M. S., Yoder, R. N., and Gibson, R., Photolytic degradation of florasulam on soil and in water, *J. Agric. Food Chem.*, 48, 3710–3717, 2000.
30. Bachman, J., and Patterson, H. H., Photodecomposition of the carbamate pesticide carbofuran: Kinetics and the influence of dissolved organic matter, *Environ. Sci. Technol.*, 33, 874–881, 1999.
31. Miller, P. L., and Chin, Y. P., Photoinduced degradation of carbaryl in a wetland surface water, *J. Agric. Food Chem.*, 23, 6758–6765, 2002.
32. Nissenson, P., Dabdub, D., Das, R., Maurino, V., Minero, C., and Vione, D., Evidence of the water-cage effect on the photolysis of NO_3^- and $FeOH^{2+}$. Implications of this effect and of H_2O_2 surface accumulation on photochemistry at the air–water interface of atmospheric droplets, *Atmos. Environ.*, 44, 4859–4866, 2010.
33. Menager, M., and Sarakha, M., Simulated solar light phototransformation of organophosphorus azinphos methyl at the surface of clays and goethite, *Environ. Sci. Technol.*, 47, 765–772, 2013.
34. Faust, B. C., Hoffmann, M. R., and Bahnemann, D. W., Photocatalytic oxidation of sulfur dioxide in aqueous suspensions of α-Fe_2O_3, *J. Phys. Chem.*, 93, 6371–6381, 1989.
35. Serpone, N., and Pelizzetti, E., *Photocatalysis: Fundamentals and Applications*, Wiley, NY, 1989.
36. Hoffmann, M. R., Martin, S. T., Choi, W. Y., and Bahnemann, D. W., Environmental applications of semiconductor photocatalysis, *Chem. Rev.*, 95, 69–96, 1995.
37. Maurino, V., Minero, C., Mariella, G., and Pelizzetti, E., Sustained production of H_2O_2 on irradiated TiO_2-fluoride systems, *Chem. Commun.*, 20, 2627–2629, 2005.
38. Vione, D., Maurino, V., Minero, C., Borghesi, D., Lucchiari, M., and Pelizzetti, E., New processes in the environmental chemistry of nitrite. 2. The role of hydrogen peroxide, *Environ. Sci. Technol.*, 37, 4635–4641, 2003.
39. Gieguzynska, E., Amine-Khodja, A., Trubetskoj, O. A., Trubetskaya, O. E., Guyot, G., Ter Halle, A., Golebiowska, D., and Richard, C., Compositional differences between soil humic acids extracted by various methods as evidenced by photosensitizing and electrophoretic properties, *Chemosphere*, 75, 1082–1088, 2009.
40. Wang, J. X., Chen, S., Quan, X., and Zhao, Y., Investigation of pentachlorophenol vertical transportation in soil column during its phototransformation on the soil surface, *Wat. Air Soil Pollut.*, 189, 103–112, 2008.
41. Ter Halle, A., Drncova, D., and Richard, C., Phototransformation of the herbicide sulcotrione on maize cuticular wax, *Environ. Sci. Technol.*, 40, 2989–2995, 2006.
42. Lavieille, D., Ter Halle, A., and Richard, C., Understanding mesotrione photochemistry when applied on leaves, *Environ. Chem.*, 5, 420–425, 2008.
43. Eyheraguibel, B., Ter Halle, A., and Richard, C., Photodegradation of bentazon, clopyralid, and triclopyr on model leaves: Importance of a systematic evaluation of pesticide photostability on crops, *J. Agric. Food Chem.*, 57, 1960–1966, 2009.
44. Ter Halle, A., Lavieille, D., and Richard, C., The effect of mixing two herbicides mesotrione and nicosulfuron on their photochemical reactivity on cuticular wax film, *Chemosphere*, 79, 482–487, 2010.
45. Monadjemi, S., El Roz, M., Richard, C., and Ter Halle, A., Photoreduction of chlorothalonil fungicide on plant leaf models, *Environ. Sci. Technol.*, 45, 9582–9589, 2011.
46. Monadjemi, S, Ter Halle, A., and Richard, C., Reactivity of cycloxydim toward singlet oxygen in solution and on wax film, *Chemosphere*, 89, 269–273, 2012.
47. Vione, D., Maurino, V., Minero, C., Pelizzetti, E., Harrison, M. A. J., Olariu, R. I., and Arsene, C., Photochemical reactions in the tropospheric aqueous phase and on particulate matter, *Chem. Soc. Rev.*, 35, 441–453, 2006.
48. Hoffmann, M. R., Homogeneous and heterogeneous photochemistry in the atmosphere. In: Boule, P., Bahnemann, D. W., Robertson, P. K. J. (Eds.), *The Handbook of Environmental Chemistry Vol. 2.M—Environmental Photochemistry Part II*, Springer, Berlin, 2005, pp. 77–118.
49. Wallington, T. J., and Nielsen, O. J., Atrmospheric photooxidation of gas phase air pollutants. In: Boule, P., Bahnemann, D. W., Robertson, P. K. J. (Eds.), *The Handbook of Environmental Chemistry Vol. 2.M—Environmental Photochemistry Part II*, Springer, Berlin, 2005, pp. 119–160.
50. Borghesi, D., Vione, D., Maurino, V., and Minero, C., Transformations of benzene photoinduced by nitrate salts and iron oxide, *J. Atmos. Chem.*, 52, 259–281, 2005.

51. Maurino, V., Bedini, A., Borghesi, D., Vione, D., and Minero, C., Phenol transformation photosensitised by quinoid compounds, *Phys. Chem. Chem. Phys.*, 13, 11213–11221, 2011.
52. Brigante, M., Charbouillot, T., Vione, D., and Mailhot, G., Photochemistry of 1-nitronaphthalene: A potential source of singlet oxygen and radical species in atmospheric waters, *J. Phys. Chem. A*, 114, 2830–2836, 2010.
53. Albinet, A., Minero, C., and Vione, D., Photochemical generation of reactive species upon irradiation of rainwater: Negligible photoactivity of dissolved organic matter, *Sci. Total Environ.*, 408, 3367–3373, 2010.
54. Parazols, M., Marinoni, A., Amato, P., Abida, O., Laj, P., and Mailhot, G., Speciation and role of iron in cloud droplets at the puy de Dome station, *J. Atmos. Chem.*, 54, 267–281, 2006.
55. Marinoni, A., Parazols, M., Brigante, M., Deguillaume, L., Amato, P., Delort, A. M., Laj, P., and Mailhot, G., Hydrogen peroxide in natural cloud water: Sources and photoreactivity, *Atmos. Res.*, 101, 256–263, 2011.
56. Vione, D., Maurino, V., Minero, C., Pelizzetti, E., The atmospheric chemistry of hydrogen peroxide: A review, *Ann. Chim. (Rome)*, 93, 477–488, 2003.
57. Warneck, P., The relative importance of various pathways for the oxidation of sulphur dioxide and nitrogen dioxide in sunlit continental fair weather clouds, *Phys. Chem. Chem. Phys.*, 1, 5471–5483, 1999.
58. Vione, D., Das, R., Rubertelli, F., Maurino, V., Minero, C., Barbati, S., and Chiron, S., Modelling the occurrence and reactivity of hydroxyl radicals in surface waters: Implications for the fate of selected pesticides, *Intern. J. Environ. Anal. Chem.*, 90, 258–273, 2010.
59. Frank, R., and Klöpffer, W., Spectral solar photo irradiance in Central Europe and the adjacent north Sea, *Chemosphere*, 17, 985–994, 1988.
60. Vione, D., Khanra, S., Cucu Man, S., Maddigapu, P. R., Das, R., Arsene, C., Olariu, R. I., Maurino, V., and Minero, C., Inhibition vs. enhancement of the nitrate-induced phototransformation of organic substrates by the $^{\cdot}OH$ scavengers bicarbonate and carbonate, *Wat. Res.* 43, 4718–4728, 2009.
61. Minero, C., Chiron, S., Falletti, G., Maurino, V., Pelizzetti, E., Ajassa, R., Carlotti, M. E., and Vione, D., Photochemical processes involving nitrite in surface water samples, *Aquat. Sci.* 69, 71–85, 2007.
62. Vione, D., Feitosa-Felizzola, J., Minero, C., and Chiron, S., Phototransformation of selected human-used macrolides in surface water: Kinetics, model predictions and degradation pathways, *Wat. Res.* 43, 1959–1967, 2009.
63. Vione, D., Minella, M., Minero, C., Maurino, V., Picco, P., Marchetto, A., and Tartari, G., Photodegradation of nitrite in lake waters: Role of dissolved organic matter, *Environ. Chem.*, 6, 407–415, 2009.
64. Vione, D., Maurino, V., Minero, C., Carlotti, M. E., Chiron, S., and Barbati, S., Modelling the occurrence and reactivity of the carbonate radical in surface freshwater, *C. R. Chimie*, 12, 865–871, 2009.
65. Vione, D., Das, R., Rubertelli, F., Maurino, V., and Minero, C., Modeling of indirect phototransformation processes in surface waters. In: Pignataro, B. (Ed.), *Ideas in Chemistry and Molecular Sciences: Advances in Synthetic Chemistry*, Wiley-VCH, Weinheim, Germany, 2010, pp. 203–234.
66. Al-Housari, F., Vione, D., Chiron, S., and Barbati, S., Reactive photoinduced species in estuarine waters. Characterization of hydroxyl radical, singlet oxygen and dissolved organic matter triplet state in natural oxidation processes, *Photochem. Photobiol. Sci*, 9, 78–86, 2010.
67. Zertal, A., Sehili, T., and Boule, P., Photochemical behaviour of 4-chloro-2-methylphenoxyacetic acid: Influence of pH and irradiation wavelength, *J. Photochem. Photobiol. A: Chem.*, 146, 37–48, 2001.
68. Vione, D., Khanra, S., Das, R., Minero, C., Maurino, V., Brigante, M., and Mailhot, G., Effect of dissolved organic compounds on the photodegradation of the herbicide MCPA in aqueous solution, *Wat. Res.*, 44, 6053–6062, 2010.
69. Benitez, F. J., Acero, J. L., Real, F. J., Roman, S., Oxidation of MCPA and 2,4-D by UV radiation, ozone, and the combinations UV/H_2O_2 and O_3/H_2O_2, *J. Environ. Sci. Health B*, 39, 393–409, 2004.

Section III

Applications of HPLC and UPLC to Separation and Analysis of Pesticides from Various Classes

12 Sample Preparation for Determination of Pesticides by High-Performance Liquid Chromatography and Liquid Chromatography–Tandem Mass Spectrometry

Robert E. Smith, Kevin Tran, Chris Sack, and Kristy M. Richards

CONTENTS

12.1 PESTICIDES IN ENVIRONMENTAL SAMPLES: METHODS, PROBLEMS, AND NEW TRENDS

One of the most important applications of pesticide analysis is the analysis of environmental samples. Manufacturers of pesticides must analyze wastewaters. Communities analyze air, water, and soil samples, and they find them—even in remote locations far from where they were originally used. International organizations have analyzed honeybees and their pollen to show that neonicotinoid and phenylpyrazole pesticides contribute to colony collapse disorder (CCD) [1]. The medical community and the general public are quite interested in finding the concentrations of pesticides in people, especially babies, children, and pregnant women. Even the popular press and mass media are quite interested although they can be easily confused. For example, it has been reported that children with attention deficit hyperactivity disorder (ADHD) have higher levels of pesticide metabolites in their urine and blood than children who do not [2]. Much of the mass media reported that this showed that pesticides can cause ADHD even though a principal rule of logic is that correlation does not imply causality. Just because two things are correlated, it tells us nothing about which one is the cause and which is the effect or if they both have the same cause. For example, it is just as likely that ADHD indirectly causes elevated levels of pesticides. Concerned, loving parents of children who have behavioral disorders (such as ADHD and autism spectral disorders) may be less likely to take their children to restaurants where they have to sit still and consume foods with high caloric, saturated fat, and sugar contents (but all low in pesticides) than "normal" children who consume the typical American diet. Instead, such parents might be more likely to feed their children healthy fruits and vegetables that do have some human-made pesticides in them. Some parents may even realize that 99.9% of the pesticides that we consume in our diets are "natural" and not made by humans [3]. Still, pesticides may help cause not just ADHD, but also autism spectral disorders [4] and autoimmune diseases [5]. It should be noted that modern medicine is becoming a fusion of traditional and Western medicine. That is, complex problems (such as autoimmune diseases) have complex causes. Seldom is there a single cause or a single cure for many diseases [5]. So there is a trend to increase the analysis of environmental samples and inform the public about them. There is also a trend to analyze samples from remote regions, to show that pollution can spread across the globe. As for sample preparation, the QuEChERS method (quick, easy, cheap, effective, rugged, and safe) has become quite popular for many food samples [6]. However, solid phase extraction (SPE) is the most commonly used method for preparing relatively clean water samples for analysis. Still, many others have started using direct injection with no sample preparation or enrichment of analytes because modern tandem mass spectrometers can provide excellent sensitivity (0.1 μg/L) in the multiple-reaction monitoring mode [7]. Another approach is to use standard addition, but it is often considered to be labor-intensive because several aliquots of each sample must have standards added to them. So one group has developed an automated standard addition method for the determination of 29 polar pesticide metabolites in wastewater and groundwater [8]. To help automate analysis, 96-well plates made from polytetrafluoroethylene coated with solid phase microextraction (SPME) fibers were used to analyze cucumbers for pesticides [9]. In the next sections, sample preparation for determining pesticides in water, air, sludge, and soil will be discussed.

12.2 SAMPLE PREPARATION METHODS FOR PESTICIDES IN WATER

Many different methods are used to prepare aqueous samples for the determination of pesticides. Many types of SPE cartridges are available, including the most popular one: octadecylsilica, also known as ODS, and C18 [10]. However, the recovery of many analytes can be better when polymeric SPE cartridges are used. A copolymer of divinylbenzene and N-vinylpyrrolidone is quite popular. Divinylbenzene is hydrophobic, and N-vinylpyrrolidone is hydrophilic. Pesticides and their degradation products in the Danube River were analyzed by liquid chromatography-tandem mass spectrometry (LC-MS/MS) after cleanup with an Oasis hydrophilic-lipophilic balanced (HLB) SPE cartridge that was packed with the HLB sorbent, made of copoly(divinylbenze/N-vinylpyrrolidone) [11]. It is wettable (by water) and useful for extracting a wide range of acidic, basic, and neutral compounds. They were able to load 400 mL of river water onto a cartridge, elute the analytes with 8 mL of methanol (CH_3OH), and reduce the volume to 0.50 mL before doing the LC-MS/MS analysis [11]. Others used the HLB SPE cartridge to extract pesticides in surface waters in the Jucar, Ebro, Llobregat, and Guadalquiver rivers in Spain [12]. They loaded 200 mL of each sample onto the cartridge that was preconditioned with 5 mL of 1:1 CH_2Cl_2/CH_3OH (v/v) followed by 10 mL of deionized water. Next, the cartridges were dried under vacuum for 10 min to remove residual water, and analytes were eluted with 10 mL of 1:1 CH_2Cl_2/CH_3OH (v/v). Extracts were evaporated to dryness and redissolved in 1 mL of CH_3OH before doing an LC-MS/MS analysis [12].

Others used stir bar sorptive extraction to extract pesticides and their metabolites from a small river that was contaminated [13]. The stir bars were coated with polydimethylsiloxane (PDMS). Analytes were desorbed best by sonicating the stir bars with 1:1 CH_3CN/CH_3OH (v/v). First, 10% NaCl was added to 20 mL of prefiltered water. Then, the extraction was done for 3 h at 800 rpm. Desorption was by sonicating with 200 mL of 1:1 CH_3CN/CH_3OH (v/v) for 15 min at room temperature. The method was validated. Excellent linearity (r^2 = 0.998) and low limits of detection of 0.02–1 μg/L were obtained [13].

A new format of fast three-phase microextraction was used for the first time on chlorophenols in environmental water samples. It combined low-density solvent-based dispersive liquid–liquid microextraction (DLLME) and single-drop microextraction (SDME) [14]. It used a 2 min DLLME preextraction and a 10 min SDME back-extraction. A portion of the low-density solvent (toluene) was injected into the aqueous sample (donor phase) with CH_3OH as the dispersing solvent. The analytes were preextracted into the organic phase within 2 min. A thin layer of the organic phase formed on the top of the aqueous phase after 2 min of centrifugation. Then, a drop of acceptor solution was introduced into the upper layer and used for back-extraction [14].

Oxidized single-walled carbon nanohorns (o-SWNHs) that were immobilized on the pores of a hollow fiber (HF) were introduced recently for the direct immersion SPME of triazines from tap, bottled, and river water samples [15]. Nanoparticles were oxidized by microwave irradiation to obtain a dispersion in CH_3OH. Then, a porous HF was immersed into the methanolic dispersion of the o-SWNHs and ultrasonicated, thus immobilizing the o-SWNHs in the pores of the HF. A stainless-steel wire was placed inside the fiber so the o-SWNHs-HF could be immersed vertically into the sample. The triazines were eluted from the o-SWNHs using 150 μL of CH_3OH [15].

12.3 SAMPLE PREPARATION METHODS FOR PESTICIDES IN ATMOSPHERE

The extensive use of pesticides has caused their spread throughout the atmosphere and into remote regions from the Arctic through the Himalayas and to the Antarctic [16–21]. Atmospheric and ambient air can be sampled using a dynamic high-volume air sampler with flow rates of 13–30 m^3/h using pumps and flow meters or collected passively by passive air samplers [22]. If only particulate matter is to be collected, samples are often fractionated based on particle size because this can affect the acidity or basicity of the particles. Fractionation minimizes interactions between different types of particles of different pH. Air samplers will have a sampling module that will contain a filter, such as glass and/or quartz fiber [22], and one or more absorbents, such as graphitized carbon black, a polyurethane foam, PDMS, anion-exchange resins, and polymer glass coated samplers [21]. For example, a semicrystalline poly DPPO, or poly(2,6-diphenyl-p-phenylene oxide), has been used to sample air in the Canadian prairie [23]. Regardless of the apparatus that is used, it must be prepared properly by being cleaned—often by Soxhlet extraction with a suitable solvent or mixture of solvents—followed by drying and storing the filters in polyethylene bags. The resin was stored in polyethylene bottles. The filters can be heated to eliminate volatile organic compounds. After cooling, they can be weighed and stored in bags made from polyethylene or aluminum foil until used. After being used, the collected samples can be put into clean polyethylene bags, and the resins go into bottles or glass jars with Teflon caps and are stored in the dark at a low temperature, usually −18°C [22].

The next step is to remove the absorbed pesticides so they can be injected onto an LC-MS or gas chromatograph (GC)-MS. This can be done by Soxhlet extraction, accelerated solvent extraction (ASE), agitation-assisted liquid extraction, ultrasound/sonication, or by microwave-assisted extraction. These are discussed in more detail in Chapter 5. For GC-MS analysis, analytes can be thermally desorbed under a stream of unreactive gas or by using evacuated canisters. For LC-MS analysis, the solubilized extracts can be preconcentrated by evaporating the solvent under vacuum or a stream of gas. If needed, they can be cleaned up by filtering the extract, adsorbing the analytes on a solid sorbent, or by gel permeation chromatography.

12.4 SAMPLE PREPARATION METHODS FOR PESTICIDES IN SLUDGE

Sewage sludge is a major by-product of wastewater treatment. It is rich in organic and inorganic nutrients, so it is often applied to agricultural land as fertilizer [24]. However, sludge may contain pathogens, heavy metals, and organic pollutants, so its use is regulated [25]. Organochlorine pesticides are lipophilic, so they tend to be adsorbed onto the surface of organic matter that is present in sludge. Usually, sludge samples are centrifuged and freeze-dried (lyophilized) to remove as much moisture as possible. Analytes can be extracted from the filtered and dried samples by supercritical fluid extraction (SFE), ultrasound, Soxhlet, Soxtec, matrix solid phase dispersion (MSPD), or pressurized liquid extraction (PLE, often using an ASE). Lipids (mostly triglycerides) and other compounds with a sufficiently high molecular weight are often separated by size exclusion chromatography (SEC), also known as gel permeation chromatography (GPC). A GPC or SEC column is packed with a porous poly(styrene-*co*-divinylbenzene) resin. Smaller molecules fit into the pores and are retained while larger molecules elute first. That is, triglycerides and other lipophilic compounds would be strongly retained to C18 columns used in LC-MS and are often impossible to remove without GPC. Next, silica, florisil, and/or strong anion-exchange SPE cartridges or disposable columns can be used to remove other potential matrix interferences that could damage column performance. Also, a sorbent that contains primary and secondary amines (PSA) has been used to clean up acetonitrile (CH_3CN) extracts of lyophilized sludge [26], similar to their use in the QuEChERS method that is more often used for preparing foods for pesticide analysis [6].

Wastewaters can also be analyzed for pesticides. A polar monolithic coating was used for the stir bar sorptive extraction of pesticides and other contaminants in wastewater [27]. The coating was a poly(poly(ethylene glycol) methacrylate-*co*-pentaerythritol triacrylate). It was able to extract and desorb most of the analytes more effectively and quickly than recently commercialized polar stir bars due to its polar behavior and suitable mechanical and physical properties [27].

12.5 SAMPLE PREPARATION METHODS FOR PESTICIDES IN SOIL AND RIVER SEDIMENTS

When analyzing soil samples, it is important to know their properties. Soil consists of about 45%–50% minerals, 20%–25% water, 25%–30% air, and different amounts of organic substances (humic and nonhumic) [28]. The specific surface area can range from 10–40 m^2/g for sandy loam to 150–250 m^2/g for soil containing clay. Moreover, soils contain high molecular weight fats and waxes, elemental sulfur, and many other substances with molecular weights similar to those of pesticides and herbicides. To extract acidic herbicides, such as (4-chloro-2-methylphenoxy) acetic acid or MCPA and its main metabolite in soil, 4-chloro-2-methylphenol, extraction with aqueous KOH was used, followed by cleanup using extraction disks packed with a C18-modified silica gel [29]. Others extracted phenoxy acid herbicides of soils using an aqueous NaOH solution and a coupled SPE-high performance (HP) LC system [30]. Others compared a Soxhlet extraction with microwave-assisted extraction (MAE) and found that MAE using CH_3OH-H_2O (4:1) with 2% triethylamine provided good recovery of analytes and used less solvent [31]. Still others used MAE and CH_2Cl_2:CH_3OH-trifluoroacetic acid (90:10:0.1, v/v/v), followed by online cleanup using a restricted-access material that combined reverse-phase separation of analytes with a relatively low molecular weight and size exclusion of larger compounds to extract 10 acidic pesticides from soil samples [32]. Others used an automated online Oasis HLB SPE cartridge to extract pesticides from oil and sediment samples [33].

PLE with an ASE can be used to extract pesticides from soil and river sediments after filtering and air drying them [34]. Polar and acidic pesticides have been extracted from soil containing 4.4% organic material using a mixture of acetone and 30% pentafluorobenzyl bromide (PFBBr) at 100°C, and pyrimidine pesticides were extracted using 1:1 CH_3CN/CH_2Cl_2 (v/v) also at 100°C and 10 MPascal pressure. Aged soils with 7% organic material were extracted with 100:1:0.5 toluene/acetic anhydride/pyridine (v/v/v). Pesticides were extracted from sea sand using 4:1 CH_3OH/H_2O (v/v) at 150°C, followed by cleanup with a C18 SPE cartridge. Chloroacetanilides and s-triazines were extracted from air-dried agricultural soil using 100% CH_3OH at 125°C [34]. PLE was also used to extract pesticides from sediments near nesting sites for smallmouth bass in the Potomac River [35].

A modified QuEChERs method was used by another group to extract pesticides from Czech and Moravian river sediments [36]. Stir bar sorptive extraction was used to extract 15 pesticides or selected metabolites from different families (herbicides, insecticides, fungicides) in surface water samples [37]. Extraction time, stirring speed, aqueous medium ionic strength and polarity, along with back desorption solvent and time were optimized.

12.6 SAMPLE PREPARATION METHODS FOR FRUITS, VEGETABLES, AND MEDICINAL PLANTS

Even though fruits and vegetables have been analyzed for decades for pesticides and inorganic arsenic (As) by the U.S. Food and Drug Administration (FDA)'s Total Diet Research Program [6], recently there has been much interest in organometallic arsenic compounds that are either pesticides that were once used to control boll weevils in areas where cotton was grown or their breakdown products, especially inorganic arsenic [38]. Inorganic arsenic has been classified by the International Agency for Research on Cancer as a group 1 carcinogen [39]. The most common forms or species of As in terrestrial habitats are arsenite $\left(AsO_3^{-3}\right)$ and arsenate $\left(AsO_4^{3-}\right)$, monomethylarsonic acid, and dimethylarsinic acid [40]. The FDA and U.S. Department of Agriculture (USDA) have been analyzing rice, apple juice, and many other foods for arsenic since 1991 in the Total Diet Study and National Residue Program, but recent concerns have inspired an increase in the number of samples analyzed [41–43]. There has been heightened public concern about As in rice, fruit juices, and chicken [43]. Lead arsenate was used as a pesticide but was banned in the United States in 1988 although dimethylarsinate, also known as cacodylic acid, is still in use. Moreover, roxarsone and nitarsone are approved for treating coccidiosis (a common parasitic disease in poultry) and to facilitate faster growth in poultry. However, low levels of arsenic are ubiquitous in the environment, so the U.S. Environmental Protection Agency set a limit of 10 μg/g (ppm) in drinking water and the U.S. standards for total arsenic in chicken muscle and liver are 0.5 and 2.0 μg/g. So the major exposure to As in the United States is through food. Physicians have been advised to tell their patients who obtain their drinking water from wells to have it tested for As, and they should diversify their diets. This may be difficult to follow as testing can be expensive and difficult to obtain in many places. Also, rice and rice-based products are important for people with celiac disease and for infants who are lactose intolerant. Rice cereals are also among the first solid foods introduced to infants although many pediatricians now recommend introducing other foods instead [43].

Rice from Southeast Asia can be contaminated by As that is present in the tube wells used to cultivate rice [44]. Arsenic is sequestered in iron plaque on root surfaces in plants and is regulated by phosphorus status. Still there is much variation in the As content of different varieties of rice, which offers hope for breeding genotypes of rice that have lower concentrations of As in them [44]. Moreover, As contamination of rice and rice products (that are considered by some to be functional or super foods) has become a global issue [44]. To extract arsenic from rice, 2 M trifluoroacetic acid was used, followed by ion chromatography (a form of HPLC) for separation of the different species of arsenic (organic and inorganic) and inductively coupled plasma–mass spectrometry (ICP-MS) for detection and quantitation [45]. Others used HNO_3 to digest food samples and solubilize the As, but the concentration used depended on whether total As was being determined (concentrated HNO_3) or As speciation was being done (1% HNO_3) [40,46]. That is, samples of raw rice were washed with ultrapure water. Then, they were dried at 70°C until a constant weight was reached. All husks were removed and the grains powdered. For total As, 0.1–0.2 g of samples were weighed into quartz glass digestion tubes. Then, 2.5 mL of concentrated HNO_3 was added and the mixture allowed to stand overnight at room temperature, followed by MAE at 120°C until the extracts were clear. Finally, the total volume was increased to 10 mL by carefully adding ultrapure water. For As speciation, about 0.20 g of milled samples was weighed into 50 mL polypropylene digestion tubes, and 10 mL of 1% HNO_3 was added. The mixture was allowed to stand overnight. Then, they were heated in a microwave-accelerated reaction system, first to 55°C and then to 75°C, with holding times of 10 min. Finally, the digest was taken up to 95°C and maintained at this temperature for 30 min before cooling. After reaching room temperature, samples were centrifuged. The supernatant was collected and passed through a 0.45 μm nylon filter. To minimize any species transformation, samples were analyzed within a few hours of filtration and kept in the dark and on ice until being analyzed [46].

On the other hand, the QuEChERs method is widely used to extract organic pesticides from fruits, vegetables, and medicinal plants [6,47,48]. It was shown to be superior to the Luke method (AOAC 985.22) and MSPD when analyzing fruits and vegetables for 14 pesticide residues [49]. Carbamates, organophosphorus (OP), nitrogen–sulfur–oxygen, and a few organochlorine pesticides were extracted from milk, salmon, fish, shrimp, almond nuts, olive oil, and avocado using the QuEChERs procedure [50]. However, a fatty layer tends to form between the upper CH_3CN and lower aqueous phases. Many lipophilic pesticides tend to stay in the fatty intermediate layer and are not recovered. To correct for this, a modified QuEChERs method was used to extract even more pesticides from avocados [51]. They used a modified version of AOAC Official Method 2009.01 (also called the "buffered QuEChERS" method) that uses acidified CH_3CN and sodium acetate to improve recovery for base-sensitive pesticides, such as chlorthalonil and tolyfluanid. They increased the ratio of solvent to sample, thus improving the extraction efficiency of the analytes. So about 3 g of each avocado sample were placed in a centrifuge tube. Then, 5 mL of water and 25 mL of 1% acetic acid in CH_3CN were added and the mixture shaken for 10 min at 1000 strokes/min in a Geno grinder. Next, 1.5 g of sodium acetate and 6 g of $MgSO_4$ were added to the tube, followed by shaking another 10 min and centrifugation. Finally, 1 mL of the top layer containing CH_3CN was transferred to an autosampler vial and 1 μL of it was injected on an LC-MS [51].

An automated dispersive SPE (dSPE) cleanup procedure was used recently with the QuEChERs method to extract pesticides from dried botanical dietary supplements [52]. The automated dSPE cleanup used a mixture of C18 and PSA sorbents with anhydrous $MgSO_4$ and online LC-MS/MS analysis [52].

Also, HF microporous membrane liquid–liquid extraction (HF-MMLLE) has been used to extract pesticides from orange juice [53]. HF-MMLLE was used for the LLME with a polypropylene porous membrane as a solid support for the solvent. The solvent forms a renewable liquid membrane on the membrane walls. Analytes were desorbed into 50 μL of 1:1 CH_3OH/acetone (v/v) for 2 min in an ultrasonic bath [53].

More recently, carbon nanotubes (CNTs) in HF-SPME were used to extract carbamate pesticides from apples [54]. The CNTs were dispersed in water by adding a surfactant. They were held in the pores of the polypropylene HF by capillary forces and sonication. The SPME device, which was wetted with 1-octanol, was placed in stirred apple samples to extract the carbamates. After extraction, analytes were desorbed into 25 μL of CH_3OH in an end-sealed pipette tip and analyzed using HPLC diode array detection [54].

Fatty fruits and/or vegetables, such as açaí, olives, and avocados, can be more difficult to clean up due to the presence of triglycerides, which would be strongly retained on a C18 column and damage it. Triglycerides are esters of glycerol and fatty acids. The fatty acids are converted to fatty acyls when they react with glycerol. So, fats in foods and oils that have not turned rancid do not have any free fatty acids in them even though much of the literature on food chemistry uses the terms "fatty acids" and "free fatty acids" as if they were synonymous with fatty acyls or triglycerides [5]. So when a group recently reported synthesizing a new amine modified graphene as a dispersive SPE material to clean up or remove fatty acids in oils, they were really removing triglycerides that contain fatty acyls [55]. Still, when this material was used in combination with a modified QuEChERs method, it was able to clean up four oil crops in a validated method [55].

12.7 SAMPLE PREPARATION METHODS FOR PESTICIDES IN MILK, MEAT, FISH, AND ANIMAL FEED

Several methods can be used to prepare milk, meat, fish, and animal feed for pesticide analysis. Several of these [solid–liquid extraction (SLE) and LLE, QuEChERs, MSPD, MAE, SFE, PLE, SPE, SPME, dSPE, and GPC] were discussed in a recent review [56]. For the past two decades, SLE and LLE have been the most widely used. The original QuEChERs method was developed for extracting analytes from fruits and vegetables [57]. It uses CH_3CN to extract pesticides. In most aqueous solutions, CH_3CN mixes with water at all proportions. However, by adding salts (NaCl and $MgSO_4$) to increase the ionic strength, CH_3CN and many analytes can be salted out to form a separate (upper) phase that is cleaned up by the addition of a PSA sorbent. The original method made no adjustments to the pH of the sample and was intended for GC analysis. This method was modified by adjusting the pH to 5 with a mixture of acetic acid and sodium acetate and adding an LC-MS/MS analysis, which improved recoveries of acid or base-sensitive pesticides, such as folpet, dichlofluanid, chlorothalonil, and pymetrozine [58]. Another approach that emerged used citrate instead of acetate to control the pH [59]. Acetate is now used in AOAC Official Method 2007.01 [60], and citrate is used in the European Committee for Standardization (CEN) Standard Method EN 15662 [61].

A modified QuEChERs method was developed and validated for 38 compounds in milk using the same CH_3CN, NaCl, and $MgSO_4$ specified in the earlier QuEChERs methods [62]. The upper phase containing CH_3CN was concentrated into dimethyl sulfoxide (DMSO) to keep all the analytes in solution, prior to UHPLC-MS/MS analysis with rapid polarity switching so that both positively and negatively charged ions could be detected [62]. The QuEChERs method was also used to extract pyrethroid insecticides from 5 g portions of fish products using 5 mL of 1% glacial acetic acid in CH_3CN and sonication (ultrasound-assisted extraction) [56].

Still, other methods can be used. MAE and SPE were used in combination to extract pesticides from infant milk formula [63]. All of the pesticides were optimally extracted at 102°C for 20 min with 4.56 mL of 0.1% of water in CH_3OH at pH 12 in a sealed microwave vessel. The MAE/SPE method used less solvent and provided higher recoveries than Soxhlet/SPE [63].

12.8 SAMPLE PREPARATION METHODS FOR BIOLOGICAL MATERIALS

Some of the most important applications of pesticide analysis are the analysis of biological materials, such as human bodily fluids and tissues, as well as honeybees. That is, bodily fluids and tissues may need to be analyzed in toxicology studies whereas honeybees need to be analyzed to study the honeybee CCD. For example, dialkylphosphate metabolites of OP insecticides were quantified in human urine using a 96-well plate SPE using weak anion-exchange cartridges [64]. The extraction procedure was automated using a 96-well plate extraction unit containing a weak anion-exchange SPE cartridge plate. There were stations in the unit for pipette tips along with reservoirs containing 1% formic acid in CH_3OH, 1% formic acid in water, and 20% triethylamine (TEA) in CH_3CN and a vacuum box. The 96-well cartridge plate was preconditioned with 0.9 mL of 1% formic acid in CH_3OH, followed by 0.9 mL of 1% formic acid in water. Then the vacuum was applied, aspirating the samples from the storage wells and dispensing them onto the 96-well cartridge plate. Then, the 96-well cartridge plate was washed with 0.9 mL of water, followed by 0.9 mL of CH_3OH to remove potentially interfering compounds in the matrix. The analytes were eluted with 1.35 mL of 20% TEA in CH_3CN into a square 96-well plate. The extracts were concentrated to dryness, redissolved in 50 μL of CH_3CN and transferred to autoinjection vials for LC-MS/MS analysis [64].

A modified QuEChERs method that specified using 2% TEA in CH_3CN was used to extract neonicotinoid pesticides and their metabolites from honeybees and their products [65]. Another modified QuEChERs method included adding some hexane when analyzing pollen and the bees themselves to remove triglycerides, which interfere with the LC separation of analytes [66]. That is, about 5 g of ground honeybees or 2 g of pollen were mixed with 10 mL of CH_3CN, H_2O (3 mL for honeybees and 8 mL for pollen), 3 mL of hexane, and citrate QueChERs salts. Upon centrifugation, three phases formed. The middle (CH_3CN) phase was placed into a centrifuge tube containing PSA and C18 sorbents. This was thoroughly mixed, centrifuged, and analyzed by LC-MS/MS and GC–time-of-flight MS [66].

Others modified the QuEChERs method by using 3 g of sample [67]. They melted comb wax and reduced the particle size of beebread with a high-speed disperser before adding 27 mL of extraction solution (44% deionized water, 55% CH_3CN, and 1% glacial acetic acid), internal standard, 6 g of anhydrous $MgSO_4$, and 1.5 g of anhydrous sodium acetate. After reducing the volume of each extract by evaporation, 1 mL of it was added to a small centrifuge tube containing 0.05 g PSA, 0.05 g C18, and 0.15 g $MgSO_4$. After centrifuging, the supernatant was analyzed by LC-MS/MS [67].

Honeybee colonies have been disappearing in a CCD in the United States, Brazil, Canada, and parts of Europe. Honeybees (*Apis mellifera*) don't just make honey, they pollinate more than 90 flowering crops, including apples, nuts, avocados, soybeans, asparagus, broccoli, celery, squash, cucumbers, citrus fruits, peaches, kiwi, cherries, blueberries, cranberries, strawberries, cantaloupe, and other melons. Even cattle, which feed on alfalfa, depend on bees. So this could be the biggest threat to our food supply. Honeybees are raised by beekeepers around the world and were introduced to North America by English settlers in the seventeenth century. CCD may be caused by a virus [68], pesticides [69], a decrease in biodiversity in flowering plants [70], and a parasitic fly [71]. However, complex problems, such as the honeybee CCD, often have multiple, interacting causes [5,72]. Still, the commonly used pesticide, thiamethoxam (a neonicotinoid systemic pesticide) was shown to cause high mortality due to homing failure at levels that could put a colony at risk of collapse [73]. Moreover, neonicotinoid insecticides also reduce the growth of colonies of the native bumblebee (*Bombus terrestris*) and the production of queens [74]. Others showed that the insecticides imidacloprid and λ-cyhalothrin caused fewer adult worker bumblebees to emerge from pupae in their colonies and higher mortality of the adult workers in the nest [75].

Pesticides and their metabolites (especially As) can be toxic to people, too. However, it would be unethical to measure the toxicokinetics of pesticides (or any toxins) in humans, so such studies are often done on rodents. For example, the toxicokinetics of the pyrethroid insecticide permethrin were studied in rats after administering oral and intravenous doses and analyzing blood, brain, and liver [76]. Permethrin was extracted using *n*-pentane, which was then evaporated off. The residue was redissolved in CH_3OH before doing the HPLC analyses. Similarly, others analyzed rat tissues and blood plasma for metabolites of pyrethroid insecticides by simply adding 50 μL of sample to 350 μL of CH_3CN, followed by shaking and centrifuging [77]. No further cleanup was required because toxicokinetics are studied using relatively high doses, so there were higher concentrations of the analyte than what would be present as a residue on food. That is, the U.S. National Institute of Environmental Health Sciences measures the toxicities of many pesticides and other chemicals as part of the Environmental Toxicology Program. It is concerned with the possible toxicities that might occur in a small minority of the human population. It would be impractical to test millions of rodents, so instead, very large doses of test chemicals are given in hopes that it will shed light on the potentially very small fraction of the human population that might be highly susceptible to their toxicities [5].

Unfortunately, some people can be exposed to very high levels of pesticides, especially in suicides. In those cases, analyte concentrations will also be relatively high. So blood serum has been analyzed for pesticides after SPE cleanup using a matrix cation exchange cartridge that was preconditioned with 1 mL CH_3OH, followed by 1 mL of 0.1 M phosphate buffer, pH 7.0 [78]. Then, 1 mL of blood serum was loaded on the cartridge, which was washed with 1 mL of water. The analytes were eluted in two steps, starting with 1 mL CH_3OH, followed by a second washing with 1 mL of 0.1 M HCl and a second elution with 1 mL CH_3OH followed by 1 mL of CH_3OH containing 5% NH_4OH [78]. Others simply added CH_3CN (which denatures and removes proteins) to blood serum, filtered it, and injected the filtrate on an LC-MS/MS [79].

Whole blood has been analyzed for pesticides after SPE cleanup using a mixed mode anion exchange cartridge called WAX [80]. That is, 1 mL of whole blood was pipetted into a 10 mL screw-top vial, to which were added, sequentially, 100 μL of a 20 mg/L solution of albendazole internal standard in a mixture of 7:3 10 mM pH 3.0 ammonium formate buffer/CH_3CN (v/v) and 2 mL of CH_3CN. The mixture was mixed thoroughly and then centrifuged. The supernatant was then collected in another vial and volume reduced to <1 mL. Then, this was diluted with 1 mL of 0.5 M pH 7 phosphate buffer, mixed, and loaded on an Oasis WAX cartridge previously conditioned with 2 mL of CH_3OH and 2 mL of H_2O. The WAX cartridge was washed with 1 mL of deionized water three times. After drying, a first elution was carried out with 3 mL of CH_3OH followed by another elution with 3 mL of a 2% ammonia solution in 1:4 CH_3OH/CH_3CN (v/v). The two eluates were pooled and evaporated to dryness. The residue was dissolved in 80 μL of 7:3 aqueous 10 mM pH 3 ammonium formate buffer/CH_3CN (v/v). Two μL of this solution was injected into the LC-MS [80].

Others used the QuEChERs method to extract the OP insecticide disulfoton from human whole blood and urine [81]. That is, 0.5 mL of whole blood or urine was diluted threefold with distilled water. The diluted sample was placed into a 4.5-mL plastic tube with 0.5 g of a prepacked extraction packet containing 6 g of $MgSO_4$ and 1.5 g of sodium acetate, two ceramic beads, and 1 mL of CH_3CN containing an internal standard (100 ng/mL D_{10}-disulfoton). The mixture was vigorously shaken for 30 s by hand and centrifuged at 3000 × g for 5 min. The supernatant (600 μL) was transferred to a 2 mL centrifuge tube containing a SPE sorbent containing 25 mg of PSA, 25 mg of end-capped C18, and 150 mg of $MgSO_4$ for sample cleanup. The contents of the tube were mixed for 10 s and centrifuged at 3000 × g for 1 min. The upper layer was transferred into a clean vial, and 10 μL was injected into the LC-MS/MS [81].

Another group recently reported extracting nonpolar and polar pesticides simultaneously from blood serum using a simple and fast monolithic spin column containing mixed-mode C-C18 cartridges [82]. The mixed-mode spin column was preconditioned with 0.4 mL of CH_3CN at 705 × g for 1 min, followed by 0.4 mL of water at 705 × g for 1 min. The sample was then applied and centrifuged for 1 min at 705 × g. The analytes were eluted with 0.1 mL of CH_3CN for 1 min, and the resulting supernatants were injected into the LC-MS/MS system [82].

Others have determined pesticides and their human metabolites in urine. This has been reviewed by Hernández and others [83]. Usually, LLE or SPE cleanup is required [83]. One group took 2 mL of urine and added isotopically labeled internal standards to obtain a final concentration of about 12.5 ng/mL [84]. To hydrolyze possible glucuronide or sulfate-conjugated metabolites, 800 units of the enzyme β-glucuronidase was added. This enzyme catalyzes the hydrolytic cleavage of glucuronide and sulfate that are covalently attached to pesticides (and other compounds). The samples were incubated for 17 h at 37°C and then extracted using an Oasis HLB SPE cartridge that was preconditioned with 1 mL of CH_3OH, followed by 1 mL of 1% acetic acid. To eliminate interfering components, the cartridge was washed with 1 mL of a 5% CH_3OH in 1% acetic acid solution. The cartridge was dried for about 30 s using a vacuum. Then, 1.5 mL of CH_3OH was pushed through the cartridge and the effluent collected. Next, 2 mL of CH_3CN was added to the methanol fraction, and the combined extract was concentrated to dryness and reconstituted in 50 μL of CH_3CN before doing LC-MS/MS analysis [84].

It is also possible to analyze saliva to look for exposure and metabolism of As-based pesticides. To prepare the samples for analysis, one group mixed 0.300 mL of saliva with 1 mL of concentrated HNO_3 then heated it for 12 h at 60°C [85]. The acid in the sample was evaporated off by heating at 80°C until about 100 μL of the solution remained. This was diluted to 1.5 mL with deionized water and analyzed for total arsenic by ICP-MS. For arsenic speciation, 0.500 mL of saliva was mixed with 1 mL of deionized water, followed by sonication. This was filtered and injected on a C18 column, which was connected to an ICP-MS [85].

Many people are also interested in estimating human exposure to pesticides. One group measured 12 biomarkers of pyrethroid insecticides and phenoxyacetic acid herbicides [86]. To hydrolyze any conjugates of Phase III metabolism [5] enough β-glucuronidase (in pH 5 acetate buffer) was added to urine samples to produce a minimum of 1000 units of enzyme activity per sample. Then,

1 mL of each sample was placed in individual wells in a 96-well plate. Then, 50 μL of ^{13}C-labeled internal standard and 750 μL of buffered β-glucuronidase solution were added. This mixture was incubated at 37°C for at least 6 h, but typically overnight. An automated SPE was done using a liquid handling station. So, a HLB 96-well plate was washed and preconditioned with 500 μL acetone and 500 μL of 1% (v/v) glacial acetic acid in deionized water. Then, the sample mixture was pipetted to the SPE plate. The pipette tips were exchanged, and the SPE plate was washed with a solution of 25% (v/v) CH_3OH in 1% (v/v) acetic acid. The SPE bed was dried. A 96-well collection plate was placed below the SPE plate, and the extraction was done by passing 750 μL acetone through the SPE plate in two separate 325 μL aliquots. The extracts were concentrated to dryness. Before reconstitution, 10 μL of an injection standard (3-chloro-2-phenoxybenzoic acid) was added to each extract. On the liquid handling station, 110 μL of reconstitution solution of 25% (v/v) CH_3OH in water was added to the extracts to give a final extraction volume of 120 μL. Finally, the extracts were transferred to autosampler vials for analysis by LC-MS/MS [86]. In another study, LC-MS was used for untargeted profiling of metabolites of pesticides in urine samples from women in early pregnancy as part of an exposomics tool [87]. For sample preparation, urine samples were simply diluted with an equal volume of H_2O–CH_3OH–CH_3CH_2COOH (95:5:0.1, v/v) and injected on the LC-MS [87].

Still, many people realize that 99.99% of all the dietary pesticides are natural [3], so they may question the need to analyze samples for pesticides or look for human exposure. At least one group is more interested in looking for biomarkers that can indicate exposure to pesticides. Still, many OP pesticides can inhibit the enzyme acetylcholine esterase (AchE), not just in insects but also in humans. Moreover, many toxins (such as nerve gas) act by inhibiting AchE, which catalyzes the hydrolysis of the neurotransmitter acetylcholine (Ach). That is, AchE can excite neurons, but it must be rapidly hydrolyzed to prevent sustained overexcitement, causing tremors and seizures [5]. One group developed a CNT-based electrochemical sensor to assay salivary AChE, a biomarker for exposure to pesticides [88]. They diluted saliva samples tenfold into a phosphate-buffered saline solution containing acetylthiocholine, and a model OP pesticide, paraoxon, was added (at 0.7 and 7 nM concentrations) and incubated for up to 120 min [88].

Many other groups are interested in toxicoproteomics, or the identification of as many proteins as possible after being exposed to a toxin [5]. To help provide data on the possible link between pesticide exposure and cancer, one group reviewed the toxicoproteomics of pesticides [89]. This may help to find important biomarkers that may show the possible cancer-causing effects of pesticides.

However, it should be noted that urine (unlike blood) can be relatively dilute or concentrated, depending on the amount of water and other substances that are consumed. To control for this, the concentration of creatinine is often measured at the same time because its concentration in urine is a direct measure of how much urine has been diluted in people with normal kidney function. On the other hand, the concentrations of creatinine and many other compounds, metals, and ions are carefully regulated in normal homeostasis [5]. However, creatinine levels in the blood can be elevated in people with chronic kidney disease.

Scientists are also discovering the importance of commensurate bacteria in human health [5]. Recently, it was discovered that a species of probiotic bacteria that is in the human gut (*Lactobacillus casei*) can produce metabolites of the commonly used fungicide fenhexamid [90]. They analyzed cell culture samples after cleaning them up by SPE with C18 cartridges after adding 1 mL of CH_3CN to 5 mL of sample. Metabolites were identified using a metabolomics software package that extracted molecular features of chlorine-containing compounds from LC-MS data using an untargeted compound search algorithm [90].

12.9 CONCLUSIONS

Environmental samples, water, the atmosphere, fruits, vegetables, medicinal plants, food, animal feed, soil, blood, urine, and saliva often require sample preparation by a variety of methods that are discussed in Chapter 5. However, some may question the need for being concerned about human-made pesticides when 99.99% of the pesticides that we consume in our diets are natural [3]. However, there is much evidence to support the idea that human-made pesticides can help cause many different diseases, such as Parkinson's, Alzheimer's, and amyotrophic lateral sclerosis as well as birth defects and reproductive disorders [91].

REFERENCES

1. Johnson, R. M., Ellis, M. D., Mullen, C. A., and Frazier, M., Pesticides and honey bee toxicity—USA, *Apidol.* 41, 312–331, 2010.
2. Bouchard, M. F., Bellinger, D. C., Wright, R. O., and Weisskopf, M. G., Attention-deficit/hyperactivity disorder and urinary metabolites of organophosphate pesticides, *Pediatr.* 125, e1270–e1277, 2010.
3. Ames, B. N., Profet, M., and Gold, L. S., Dietary pesticides (99.99% all natural), *Proc. Natl. Acad. Sci. USA*, 87, 7777–7781, 1990.
4. Zafeiriou, D. I., Ververi, A., and Vargiami, E., Childhood autism and associated comorbidities, *Brain Develop.*, 29, 257–272, 2007.
5. Smith, R. E., *Medicinal Chemistry—A Fusion of Traditional and Western Medicine*, pp. 9–17, 20–22, 53, 73–75, 125–127, 509–520, Bentham Science, U.A.E., 2013.
6. Sack, C. et al., Collaborative validation of the QuEChERS procedure for the determination of pesticides in food by LC-MS/MS, *J. Agric. Food Chem.*, 59, 6383, 2011.
7. Reemstma, T., Alder, L., and Banasiak, U., A multimethod for the determination of 150 pesticide metabolites in surface water and groundwater using direct injection liquid chromatography–mass spectrometry, *J. Chromatogr. A*, 1271, 95–104, 2013.
8. Kowal, S., Balsaa, P., Werres, F., and Schmidt, T. C., Fully automated standard addition method for the quantification of 29 polar pesticide metabolites in different water bodies using LC-MS/MS, *Anal. Bioanal. Chem.*, 405, 6337–6351, 2013.
9. Bagheria, H., Ali Es'haghia, A., Ali Es-haghib, A., and Mesbahia, N., A high-throughput approach for the determination of pesticide residues in cucumber samples using solid-phase microextraction on 96-well plate, *Anal. Chim. Acata*, 740, 36–42, 2012.
10. Primel, E. G., Caldas, S. S., and Escarrone, A. L. V., Multi-residue analytical methods for the determination of pesticides and PPCPs in water by LC-MS/MS: A review, *Cent. Eur. J. Chem.*, 10(3), 876–899, 2012.
11. Loos, R., Locoro, G., and Contini, S., Occurrence of polar organic contaminants in the dissolved water phase of the Danube River and its major tributaries using SPE-LC-MS^2 analysis, *Water Res.*, 44, 2325–2335, 2010.
12. Masiá, A., Ibañez, M., Blasco, C., Sancho, J. V., Picó, Y., and Hernández, F., Combined use of liquid chromatography triple quadrupole mass spectrometry and liquid chromatography quadrupole time-of-flight mass spectrometry in systematic screening of pesticides and other contaminants in water samples, *Anal. Chim. Acta*, 761, 117–127, 2013.
13. Margoum, C., Guillemain, C., Yang, X., and Coquery, M., Stir bar sorptive extraction coupled to liquid chromatography-tandem mass spectrometry for the determination of pesticides in water samples: Method validation and measurement uncertainty, *Talanta*, 116, 1–7, 2013.
14. Li, X., Xue, A., Chen, C., and Li, S., Low-density solvent-based dispersive liquid–liquid microextraction combined with single-drop microextraction for the fast determination of chlorophenols in environmental water samples by high performance liquid chromatography-ultraviolet detection, *J. Chromatogr. A*, 1280, 9–15, 2013.
15. Jiménez-Soto, J. M., Cárdenas, S., and Valcárcel, M., Oxidized single-walled carbon nanohorns (o-SWNHs) as sorbent for porous hollow fiber direct immersion solid-phase microextraction for the determination of triazines in waters, *Anal. Bioanal. Chem.*, 405, 2561–2669, 2013.
16. Galbán-Malagón, C., Cabrerizo, A., Caballero, G., and Dachs, J., Atmospheric occurrence and deposition of hexachlorobenzene and hexachlorocyclohexanes in the Southern Ocean and Antarctic Peninsula, *Atmosph. Env.*, 80, 41–49, 2013.
17. Zhang, L., Dickhut, R., DeMaster, D., Pohl, K., and Lohmann, R. Organochlorine pollutants in western Antarctic peninsula sediments and benthic deposit feeders, *Env. Sci. Technol.*, 41, 3884–3890, 2013.
18. Xiao, H. et al., Field evaluation of a flow-through sampler for measuring pesticides and brominated flame retardants in the Arctic atmosphere, *Env. Sci. Technol.*, 46, 7669–7676, 2012.
19. Becker, S., Halsall, C. J., Tych, W., Kallenborn, R., Schlabach, M., and Manø, S., Changing sources and environmental factors reduce the rates of decline of organochlorine pesticides in the Arctic atmosphere, *Atmos. Chem. Phys.*, 12, 4033–4044, 2012.

20. Sheng, J. et al., Monsoon-driven transport of organochlorine pesticides and polychlorinated biphenyls to the Tibetan plateau: Three year atmospheric monitoring study, *Env. Sci. Technol.*, 47, 3199–3208, 2013.
21. Coscollà, C., Hart, E., Pastor, A., and Yusà, V., LC-MS characterization of contemporary pesticides in PM10 of Valencia Region, Spain, *Atmosph. Env.*, 77, 394–403, 2013.
22. Kosikowska, M., and Biziuk, M., Review of the determination of pesticide residues in ambient air, *Trends Anal. Chem.*, 29, 1064–1072, 2010.
23. Raina, R., and Smith, E., Determination of azole fungicides in atmospheric samples collected in the Canadian prairies by LC/MS/MS, *J. AOAC Intl.*, 95, 1350–1356, 2012.
24. Tadeo, J. L., Sánchez-Brunete, C., Albero, B., and Garcia-Valcárcel, A. I., Determination of pesticide residues in sewage sludge: A review, *J. AOAC Intl.*, 93, 1692–1702.
25. European Commission, http://ec.europa.eu/environment/waste/sludge/pdf/sludge_en.pdf.
26. García-Valcárcel, A. I., and Tadeo, J. L., A combination of ultrasonic assisted extraction with LC–MS/MS for the determination of organophosphorus pesticides in sludge, *Anal. Chim. Acta*, 641, 117–123, 2009.
27. Gilart, N., Cormack, P. A. G., Marcéa, R. M., Borrulla, F., and Fontanals, N., Preparation of a polar monolithic coating for stir bar sorptive extraction of emerging contaminants from wastewaters, *J. Chromatogr. A*, 1295, 42–47, 2013.
28. Macutkiewicz, E., Rompa, M., and Zygmunt, B., Sample preparation and chromatographic analysis of acidic herbicides in soils and sediments, *Crit. Rev. Anal. Chem.*, 33, 1–17, 2003.
29. Pozo, O., Pitarch, E., Sancho, J. V., and Hernandez, F. F., Determination of the herbicide 4-chloro-2-methylphenoxyacetic acid and its main metabolite, 4-chloro-2-methylphenol in water and soil by liquid chromatography–electrospray tandem mass spectrometry, *J. Chromatogr. A*, 923, 75–85, 2001.
30. Meier, M., Hamann, R., and Kettrup, A., Determination of phenoxy acid herbicides by high-performance liquid chromatography and on-line enrichment, *Fresen. Z. Anal. Chem.*, 334, 235–237, 1989.
31. Alonso, M. C. et al., Determination of priority phenolic compounds in soil samples by various extraction methods followed by liquid chromatography–atmospheric pressure chemical ionisation mass spectrometry, *J. Chrom. A*, 823, 231–239, 1998.
32. Hogendoorn, E. A. et al., Microwave assisted solvent extraction and coupled-column reversed-phase liquid chromatography with UV detection: Use of an analytical restricted-access-medium column for the efficient multi-residue analysis of acidic pesticides in soils, *J. Chrom. A*, 938, 23–33, 2001.
33. Togola, A., Bara, N., and Coureau, C., Advantages of online SPE coupled with UPLC/MS/MS for determining the fate of pesticides and pharmaceutical compounds, *Anal. Bioanal. Chem.*, 1–11, 2013.
34. Ramos, L., Kristenson, E. M., and Brinkman, U. A. T., Current use of pressurised liquid extraction and subcritical water extraction in environmental analysis, *J. Chromatogr. A.*, 975, 3–29, 2002.
35. Kolpin, D. W. et al., Chemical contaminants in water and sediment near fish nesting sites in the Potomac River basin: Determining potential exposures to smallmouth bass (*Micropterus dolomieu*), *Sci. Total Env.*, 443, 700–716, 2013.
36. Kvicalova, W. et al., Application of different extraction methods for the determination of selected pesticide residues in sediments, *Bull. Environ. Contam. Toxicol.*, 89, 21–26, 2012.
37. Margoum, C., Guillemain, C., Yang, X., and Coquery, M., Stir bar sorptive extraction coupled to liquid chromatography–tandem mass spectrometry for the determination of pesticides in water samples: Method validation and measurement uncertainty, *Talanta*, 116, 1–7, 2013.
38. Potera, C., U.S. rice serves up arsenic, *Env. Health Persp.*, 115, A296, 2007. Many of these regions in the USA are now used to grow rice, which has been found to contain some of them.
39. IARC, (International Agency for Cancer Research), Some drinking-water disinfectants and contaminants, including arsenic, 84. IARC, Geneva, Switzerland, 2004.
40. Sommella, A., Deacon, C., Norton, G., Pigna, M., Violante, A., and Meharg, A. A., Total arsenic, inorganic arsenic, and other elements concentrations in Italian rice grain varies with origin and type, *Env. Pollut.*, 181, 38–43, 2013.
41. U.S. FDA, Arsenic in rice and rice products, http://www.fda.gov/Food/FoodborneIllnessContaminants/Metals/ucm319870.htm.
42. Questions & Answers: Arsenic in Rice and Rice Products, http://www.fda.gov/Food/FoodborneIllnessContaminants/Metals/ucm319948.htm.
43. Navas-Acien, A., and Nachman, K., Public health responses to arsenic in rice and other foods, *JAMA Int. Med.*, 173, 1395–1396, 2013.
44. Meharg, A. A., Arsenic in rice—Understanding a new disaster for South-East Asia, *Trends Plant Sci.*, 9, 415–417, 2004.

45. Heitkemper, D. T., Vela, N. P., Stewart, K. R., and Wetphal, C. S., Determination of total and speciated arsenic in rice by ion chromatography and inductively coupled plasma mass spectrometry, *J. Anal. Spctr.*, 16, 299–306, 2001.
46. Zhu, Y.-G. et al., High percentage inorganic arsenic content of mining impacted and nonimpacted Chinese rice, *Envrion. Sci. Technol.*, 42, 5008–5013, 2008.
47. Lehotay, S. J. et al., Comparison of QuEChERS sample preparation methods for the analysis of pesticide residues in fruits and vegetables, *J. Chromatogr. A*, 1217, 2548–2560, 2010.
48. Tran, K., Eide, D., Nickols, S. M., Cromer, M. R., and Smith, R. E., Finding of pesticides in fashionable fruit juices by LC-MS/MS and GC-MS/MS, *Food Chem.*, 134, 2398–2405, 2012.
49. Kruve, A., Künnapas, A., Herodes, K., and Leito, I., Matrix effects in pesticide multi-residue analysis by liquid chromatography–mass spectrometry, *J. Chromatogr. A*, 1187, 58–66, 2008.
50. Lin, Y., Laboratory Information Bulletin 4472, Analysis of twenty-six pesticide residues in fatty food products using a modified QuEChERS method and LC-MS/MS, U.S. Food and Drug Administration, Division of Field Science: Rockville, MD, 2011.
51. Chamkasem, N., Ollis, L. W., Harmon, T., Sookwang Lee, S., and Mercer, G., Analysis of 136 pesticides in avocado using a modified QuEChERS method with LC-MS/MS and GC-MS/MS, *J. Agr. Food Chem.*, 61, 2315–2329, 2013.
52. Yang Chen, Y. et al., Multiresidue pesticide analysis of dried botanical dietary supplements using an automated dispersive SPE cleanup for QuEChERS and high-performance liquid chromatography–tandem mass spectrometry, *J. Agr. Food Chem.*, 60, 9991–9999, 2012.
53. Bedendo, G. C., Jardim, I. C. S. F., and Carasek, E., Multiresidue determination of pesticides in industrial and fresh orange juice by hollow fiber microporous membrane liquid–liquid extraction and detection by liquid chromatography–electrospray-tandem mass spectrometry, *Talanta*, 88, 573–580, 2012.
54. Song, X.-Y., Shi, Y.-P., and Chen, J., Carbon nanotubes-reinforced hollow fibre solid-phase microextraction coupled with high performance liquid chromatography for the determination of carbamate pesticides in apples, *Food Chem.*, 139, 246–252, 2013.
55. Guan, W. et al., Amine modified graphene as reversed-dispersive solid phase extraction materials combined with liquid chromatography–tandem mass spectrometry for pesticide multi-residue analysis in oil crops, *J. Chrom. A*, 1286, 1–8, 2013.
56. Zanella, R., Prestes, O. D., Friggi, C. A., Martins, M. L., and Adaime, M. B., Modern sample preparation methods for pesticide multiresidue determination in foods of animal origin by chromatographic–mass spectrometric techniques, in *The Impact of Pesticides*, pp. 355–379, AcademyPublish.org, New York, 2012.
57. Anastassiades, M., Lehotay, S. J., Stajnbaher, D., and Schenck, F. J., Fast and easy multiresidue method employing acetonitrile extraction/partitioning and "dispersive solid-phase extraction" for the determination of pesticide residues in produce, *J. AOAC Int.*, 86, 412–431, 2003.
58. Lehotay, S. J., Mastovská, K., and Lightfield, A. R., Use of buffering and other means to improve results of problematic pesticides in a fast and easy method for residue analysis of fruits and vegetables, *J. AOAC Intl.*, 88, 615–629, 2005.
59. Anastassiades, M., Scherbaum, E., Tasdelen, B., and Stajnbaher, D., in Ohkawa, H., Miyagawa, H., Lee, P. W. (Eds.), *Crop Protection, Public Health, Environmental Safety*, p. 439, Wiley-VCH, Weinheim, Germany, 2007.
60. AOAC International, *Official Method 2007.01: Pesticide residues in foods by acetonitrile extraction and partitioning with magnesium sulphate*, AOAC International, 2007
61. European Committee for Standardization, *CEN, CEN/TC 275 15662:2008: Foods of plant origin—Determination of pesticide residues using GC-MS and/or LC-MS/MS following acetonitrile extraction/partitioning and cleanup by dispersive SPE-QuEChERS-method*, European Union, 2008.
62. Díez, C., Traag, W. A., Zommer, P., Marinero, P., and Atienza, J., Comparison of an acetonitrile extraction/partitioning and "dispersive solid-phase extraction" method with classical multi-residue methods for the extraction of herbicide residues in barley samples, *J. Chromatogr. A*, 1131, 11–23, 2006.
63. Fang, G., Lau, H. F., Law, W. S., Li, S. F. Y., Systematic optimisation of coupled microwave-assisted extraction-solid phase extraction for the determination of pesticides in infant milk formula via LC–MS/MS, *Food Chem.*, 134, 2473–2480, 2012.
64. Odetokun, M. S. et al., Quantification of dialkylphosphate metabolites of organophosphorus insecticides in human urine using 96-well plate sample preparation and high-performance liquid chromatography–electrospray ionization-tandem mass spectrometry, *J. Chromatogr. A*, 878, 2567–2574, 2010.
65. Kamel, A., Refined methodology for the determination of neonicotinoid pesticides and their metabolites in honey bees and bee products by liquid chromatography–tandem mass spectrometry (LC-MS/MS), *J. Agric. Food Chem.*, 58, 5926–5931, 2010.

66. Wiest, L. et al., Multi-residue analysis of 80 environmental contaminants in honeys, honeybees and pollens by one extraction procedure followed by liquid and gas chromatography coupled with mass spectrometric detection, *J. Chromatogr. A*, 1218, 5743–5756, 2010.
67. Mullin, C. A. et al., High levels of miticides and agrochemicals in North American apiaries: Implications for honey bee health, *PLoS One*, 5, e9754, 2010.
68. Cox-Foster, D. L. et al., A metagenomic analysis of microbes in honeybee colony collapse disorder, *Sci.*, 318, 283–287, 2008.
69. Johnson, R. M., Ellis, M. D., Mullin, C. A., and Frazier, M., Pesticides and honey bee toxicity—USA, *Apidol.*, 41, 312–331, 2010.
70. Levy, S., The pollinator crisis: What is best for bees? *Nature*, 479, 164–165, 2011.
71. Core, A. et al., A new threat to honey bees, the parasitic phorid fly *Apocephalus Borealis*, *PLoS ONE*, 12, e29639, 2012.
72. CCD Steering Committee, *Colony Collapse Disorder Progress Report*, USDA, Washington, D.C. 2012.
73. Henry, M. et al., A common pesticide decreases foraging success and survival in honey bees, *Sci.*, 336, 348–350, 2012.
74. Whitehorn, P. R., O'Conner, S., Wackers, F. L., and Goulson, G., Neonicotinoid pesticide reduces bumble bee colony growth and queen production, *Sci.*, 336, 351–352, 2012.
75. Osborne, J. L., Bumblebees and pesticides, *Nature*, 491, 43–45, 2012.
76. Anadón, A., Martinez-Larrañaga, M. R., Diaz, M. J., and Bringas, P. Toxicokinetics of permethrin in rats, *Toxicol. Appl. Pharmacol.*, 110, 1–8, 1991.
77. Ueyama, J. et al., Toxicokinetics of pyrethroid metabolites in male and female rats, *Env. Toxicol. Phamacol.*, 30, 88–91, 2010.
78. Lacassie, E. et al., Sensitive and specific multiresidue methods for the determination of pesticides in various classes in clinical and forensic toxicology, *Forens. Sci. Intl.*, 121, 116–125, 2001.
79. Inoue, S. et al., Rapid simultaneous determination for organophosphorus pesticides in human serum by LC–MS, *J. Pharm. Biomed. Anal.*, 44, 258–264, 2008.
80. Dulaurent, S., Moesch, C., Marquet, P., Gaulier, J.-M., and Lachâtre, G., Screening of pesticides in blood with liquid chromatography–linear ion trap mass spectrometry, *Anal. Bioanal. Chem.*, 396, 2235–2249, 2010.
81. Usui, K. et al., Rapid determination of disulfoton and its oxidative metabolites in human whole blood and urine using QuEChERS extraction and liquid chromatography–tandem mass spectrometry, *Legal Med.*, 14, 309–316, 2012.
82. Saito, T. et al., Rapid determination of polar and non-polar pesticides in human serum, using mixed-mode C-C_{18} monolithic spin column extraction and LC–MS/MS, *Chromatographia*, 76, 781–789, 2013.
83. Hernández, F., Sancho, J. V., and Pozo, O. J., Critical review of the application of liquid chromatography/mass spectrometry to the determination of pesticide residues in biological samples, *Anal. Bioanal. Chem.*, 382, 934–946, 2005.
84. Olsson, A. O. et al., A liquid chromatography–tandem mass spectrometry multiresidue method for quantification of specific metabolites of organophosphorus pesticides, synthetic pyrethroids, selected herbicides, and DEET in human urine, *Anal. Chem.*, 76, 2453–2461, 2004.
85. Yuan, C. et al., Arsenic speciation analysis in human saliva, *Clin. Chem.*, 54, 163–171, 2008.
86. Jamin, E. L. et al., Untargeted profiling of pesticide metabolites by LC–HRMS: An exposomics tool for human exposure evaluation, *Anal. Bioanal. Chem.*, 1–13, 2013.
87. Davis, M. D. et al., Semi-automated solid phase extraction method for the mass spectrometric quantification of 12 specific metabolites of organophosphorus pesticides, synthetic pyrethroids, and select herbicides in human urine, *J. Chromatogr. B*, 929, 18–26, 2013.
88. Yuan, C. et al., Arsenic speciation analysis in human saliva, *Clin. Chem.*, 54, 163–171, 2008.
89. George, J., and Shukla, Y., Pesticides and cancer: Insights into toxicoproteomic-based findings, *J. Proteom.*, 74, 2713–2722.
90. Lénárt, J., Bujna, E., Kovács, B., Békefi, E., Száraz, L., and Dernovics, M., Metabolomic approach assisted high resolution LC–ESI-MS based identification of a xenobiotic derivative of fenhexamid produced by *Lactobacillus casei*, *J. Agr. Food Chem.*, 61, 8969–8975, 2013.
91. Mostafalou, S., and Abdollahi, M., Pesticides and human chronic diseases: Evidences, mechanisms, and perspectives, *Toxicol. Appl. Pharmacol.*, 268, 157–177.

13 Quantitative Analysis and Method Validation

Renato Zanella, Osmar Damian Prestes, Manoel Leonardo Martins, and Martha Bohrer Adaime

CONTENTS

13.1 INTRODUCTION

Quantitative analysis, the main application of high-performance liquid chromatography (HPLC), determines how much of each substance is present in a sample. Quantitative analysis makes certain demands on the HPLC apparatus, particularly the injection system, in which a representative sample must be transferred to the column, and the detector, which must have a known response defined by a response index. To obtain a value proportional to the mass of solute present, a response factor must be used for each substance determined, and these response factors are obtained by a calibration. Almost all peak area measurements are more precise than peak height measurements and are used more often for quantitative assessment. In general, quantitative measurements are performed using external standard calibration, but matrix-matched calibration is frequently used to determine pesticide residues by HPLC, especially in food samples.

Analytical methods need to be validated or revalidated before their routine application or whenever the conditions or method parameters change from the original scope. Method validation is the process involving laboratory experiments through which performance characteristics and limitations of a method are established and influences that may change these characteristics are identified and measured. The validation should follow a plan that includes the scope, required performance characteristics, and acceptance limits of the analytical method. Validation parameters usually examined are linearity, analytical range, selectivity/specificity, limit of detection, limit of quantification, accuracy, precision, and ruggedness/robustness. A validation report describing all experimental conditions and the complete statistical analysis should be generated. Validation results help to achieve high-quality data and can be used to judge the reliability, consistency, and accuracy of analytical data. Method validation has received considerable attention from the analytical community and from regulatory agencies. The ISO/IEC 17025 accreditation standard [1], for example, now places more emphasis on method validation. Considering the importance of this topic, this chapter describes the most relevant aspects related to quantitative analysis and validation of analytical methods for determining pesticide residues by HPLC.

13.2 ANALYTICAL QUANTIFICATION IN HPLC

As with other analytical techniques, quantification in HPLC involves the conversion of an analytical response from a detector into a quantity and then into the concentration of the analytes in a sample. This task can be improved by a chromatographic separation, in which the chemical species of interest is separated from other substances (interferences) present in the injected solution, but it can also be negatively or positively influenced by peak asymmetry and partial co-elution, peak superposition, baseline noise and shifts, detector acquisition rate, and matrix effects.

The mathematical and statistical methods for calculation of analyte concentration should be well fitted for the purpose in each case. The choice of calibration technique must allow a high degree of confidence. For either calibration approach, different calibration calculation techniques may be used. The most frequently used are linear calibration through origin or response factor method (RF), linear least squares regression, weighted least squares regression, quadratic and polynomial regression (weighted or not), and multivariate calibration. Application details for each calibration calculation technique will be presented together with the applicable calibration techniques.

13.2.1 External Calibration

External calibration, or external standard calibration, is the most common approach to calibration, in both HPLC and other analytical techniques and consists of a simple comparison of detector responses from the target compounds in the samples with responses of these compounds present in calibration standards [2].

Peak areas (or peak heights) from compounds in the samples are compared to peak areas (or peak heights) of the standards. To calculate results with this calibration technique, analysis of the standard mixture needs to be performed under the same conditions used for the samples. One way to establish this comparison is using the calibration factor (CF), as shown in Equation 13.1:

$$\mathrm{CF} = \frac{\text{peak area (or height) of the compound in the standard}}{\text{mass of the compound injected}} \tag{13.1}$$

The final result can be calculated from the relationship between the compound area in the samples and the CF of each compound, as expressed by Equation 13.2:

$$X_s = \frac{A_s}{\overline{\mathrm{CF}}} \tag{13.2}$$

where X_s is the calculated mass of the analyte in the sample aliquot, A_s is the peak area (or height) of the analyte in the sample, and $\overline{\mathrm{CF}}$ = the average calibration factor.

The average value of CF can be obtained through several successive injections into the chromatographic system of an amount of the analyte of interest to produce a signal very close to the signal of the analyte in the sample (single point calibration); the internal standard may be injected before, after, or interchangeably with injections of the sample extracts.

An improved performance can be obtained by injecting quantities of the standard compound to produce, for example, signals equivalent to 80%, 100%, and 120% of the compound signal in the sample or in a narrow range just above and below the analyte signal in the sample. This technique requires calibration of the chromatographic system stability with a relative standard deviation (RSD) of less than 20%, calculated from the number (n) of injections. Another requirement is the establishment of little or no matrix interference. In practical terms, this means that a calibration line drawn straight through the areas of the analyte in the standards at different concentrations should pass exactly through the origin (0,0).

Therefore, the most widespread calibration technique in HPLC methods is the linear least squares regression or, as it is more commonly called, the "calibration curve." This method consists of a mathematical model that describes the relationship between expected and measured values via minimization of the sum of the squared residuals (deviations between observed and expected values). In order to perform this type of calibration, at least five concentration levels of each compound should be injected into the HPLC system, usually from the near-to-minimum amount detectable to a high concentration in which the detector response lacks linearity. This range is called the linear range or work range and must be statistically tested for residuals as described in more detail in Section 13.3.5.

Figure 13.1 shows a schematic approach to construct a calibration curve based on the peak areas obtained from analytical solutions in five concentration levels. The chromatograms obtained by liquid chromatography–tandem mass spectrometry (LC-MS/MS) represent the signal of one analyte from each analytical solution and from a sample with unknown concentration. This ideal case represents a perfect fit of regression with a determination coefficient (r^2) equal to 1 and passing through the origin (b = 0). The peak area of a sample extract (A_s) is converted into a concentration (C_s) of 0.5 mg L^{-1} by a simple reading of the correspondent data in the graph or using regression calculations.

Least squares regression calibrations are typically derived from a minimum of five standards of varying concentration and are applicable to data sets in which the measurement uncertainty is relatively constant across the calibration. A linear least squares regression attempts to construct a linear equation as follows Equation 13.3:

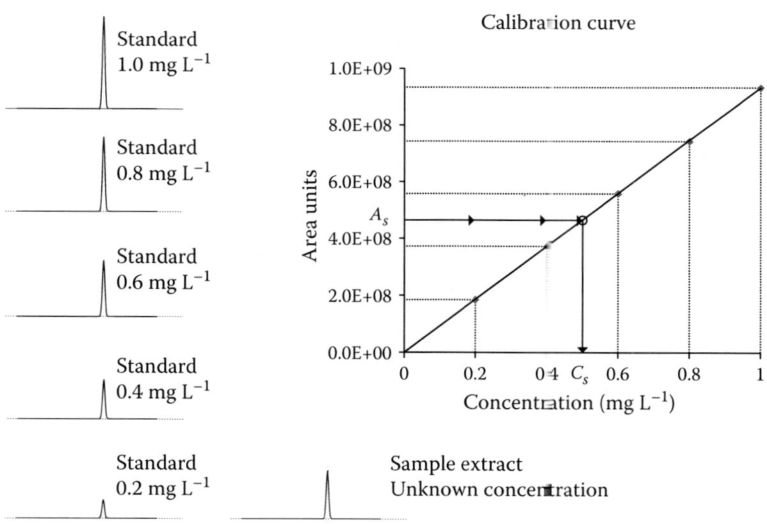

FIGURE 13.1 A linear calibration curve constructed from peak areas from injection of analytical solutions with five concentration levels and a chromatogram of a sample extract. A_s: analyte peak area in the sample; C_s: analyte concentration in the sample.

$$y = ax + b \tag{13.3}$$

where a is the regression coefficient or the slope of the line, b is the y-intercept or linear coefficient, y is the response for the calibration standard, and x is the amount (mass or concentration) of analyte in the calibration standard.

The "a" and "b" factors can be obtained by the minimum least squares method applying Equations 13.4 and 13.5:

$$b = \left(\frac{\sum_y}{n}\right) - a\left(\frac{\sum_x}{n}\right) \tag{13.4}$$

$$a = \frac{\sum_{xy} - \left(\sum_x \times \frac{\sum_y}{n}\right)}{\sum_{xx} - \left(\frac{\sum_x}{n}\right)^2} \tag{13.5}$$

where y is the response and x is the amount or concentration.

$$\sum_y = y_1 + y_2 + \ldots + y_n$$

$$\sum_{yy} = y_1^2 + y_2^2 + \ldots + y_n^2$$

$$\sum_x = x_1 + x_2 + \ldots + x_n$$

$$\sum_{xx} = x_1^2 + x_2^2 + \ldots + x_n^2$$

$$\sum_{xy} = x_1y_1 + x_2y_2 + \ldots + x_ny_n$$

The first evaluation of the linear regression model is through the correlation coefficient (r) obtained from Equation 13.6. The r value measures or describes the relationship between the y-responses and x-values.

$$r = \frac{n\sum_{xy} - \left(\sum_x \times \sum_y\right)}{\sqrt{\left[n\sum_{xx} - \left(\sum_x\right)^2\right] \times \left[n\sum_{yy} - \left(\sum_y\right)^2\right]}} \tag{13.6}$$

The square of r (r^2), also called the determination coefficient, gives a percentage of the correlation between y- and x-values in the calibration curve. In analytical chemistry, the minimum acceptable r^2 is 0.99, but this parameter isn't enough for evaluation of the calibration curve and ultimately for evaluation of the linearity method. Although the fitness evaluation of a linear calibration curve is usually more related to instrument response to a standard, in the method validation, the parameter linearity invokes the evaluation of the whole analytical procedure. Mathematically, both can be checked the same way, which will be discussed in Section 13.3.5.

If the linear model does not describe the data well, other calibration models can be used, such as weighted least squares regression, quadratic and polynomial regression, weighted or not, and multivariate calibration.

13.2.2 Matrix-Matched Calibration

The combination of LC and MS into a LC-MS system has revolutionized the approach to quantitative determination of pesticide residues in complex samples such as food and environmental matrices [3]. Many works have reported that the presence of a co-extracted matrix can severely affect the quantification procedures based on electrospray ionization (ESI) and atmospheric pressure chemical ionization LC-MS methods. This phenomenon is called the matrix effect (ME), and it is considered to be either an unexpected suppression or an enhancement of the analyte response induced by the co-eluting matrix. It can heavily affect the reproducibility, linearity, and accuracy of the method leading to erroneous quantification [4].

Co-extracts are present throughout analysis, frequently causing a poor signal or poor accuracy in the results. Kebarle et al. [5] discussed this phenomenon in detail, suggesting that organic compounds present in the sample exceeding around 10^{-5} mol L^{-1} may compete with the analyte when accessing the droplet surface in gas phase emission. However, because not only the quantity of the organic compounds, but also their quality have a strong influence on the signal, no correlation can be considered from the total dissolved organic carbon content with regard to signal suppression or enhancement. MEs have been observed in other detection systems, such as in fluorescence and electron-capture techniques in which detector quenching by matrix components may lead to MEs. Flame dampening caused by lipids and other plasma constituents observed in flame ionization detectors (nitrogen/phosphorus detectors or flame photometric detectors) and often used in gas chromatographic methods can also cause matrix-related irreproducibility. The difference in response between the solvent sample and the postextraction spiked sample is called the "*absolute matrix effect*," and the difference in response between various lots of postextraction spiked samples is called the "*relative matrix effect*."

Different methods can be used to detect the presence of a ME, and different ways are required for absolute MEs and for relative MEs. The most widely used method to evaluate an absolute ME is to compare the signal response of an analyte in a standard solution and in a postextraction spiked sample. The differences in signal indicate suppression or enhancement. The *ME* value can be estimated using Equation 13.7:

$$ME(\%) = \frac{B}{A} \times 100 \tag{13.7}$$

where A is the peak area of the standard solution, and B is the peak area of a sample extract spiked at the same concentration of the standard.

Figure 13.2 shows a very clear example of experimental evidence that a ME can severely compromise quantitative data generated by LC-MS/MS if this effect will not be compensated.

The extension of intensity suppression/enhancement is typically around 0%–30%, but in some cases, it can be around 100% or higher when using the most common extraction solvents. As a consequence, the response of an analyte in pure solvent can differ significantly from that in the matrix sample. The differences in response can be seen in Figure 13.2, in which rice extracts did not show

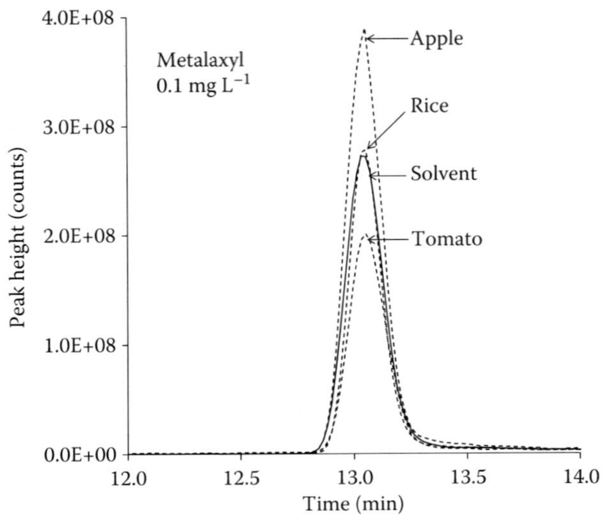

FIGURE 13.2 ME in the LC-MS/MS (ESI) analysis of metalaxyl at 0.1 mg L^{-1} in different matrices. The compound was isolated by QuEChERS acetate method.

signal differences. Although the most usual case is a decrease in response of the analyte due to signal suppression (e.g., tomato extract), enhancement may also occur (e.g., apple extract).

Various operational improvements have been suggested to minimize the interferences of co-extractive matrix compounds. Several modifications in sample preparation can reduce the presence of interfering components in the final extract. However, the results may be compromised with very complex matrices, in which a variety of co-extractives with different chemical properties are present. Another way is to modify the chromatographic conditions to shift the retention time of the target analytes farther from the area of the chromatogram affected by the ME.

The potential for MEs to occur should be assessed during method validation. They are notoriously variable in occurrence and intensity, but some techniques are particularly prone to them. If the techniques used are not inherently free from such effects, calibration should be matrix matched routinely to correct for the systematic error introduced by the ME. Extracts of a blank matrix, preferably of the same type as the sample, may be used for calibration purposes. The comparison between matrix-matched and solvent calibration curves also permits the visualization of the ME on the signal response. Figure 13.3 shows the differences between signal responses in different matrix-matched calibration curves obtained from QuEChERS extracts.

The ME value can also be estimated using Equation 13.8:

$$\text{ME}(\%) = \frac{\text{slope } B}{\text{slope } A} \times 100 \tag{13.8}$$

where slope A is the slope of the solvent calibration curve, and slope B is the slope of the matrix-matched calibration curve at the same concentrations.

During recent years, different strategies have been suggested to compensate for interferences of co-eluting matrix compounds. The standard addition method is probably the most effective procedure to minimize the adverse impact of the matrix on signal response. The labeled internal standards can also achieve adequate quantitative results. Calibration with matrix matching continues to be the most widely used method. Unfortunately, the method is time-consuming and appropriate blanks (i.e., material free of residues of the target analyte) may be unavailable. However, it is

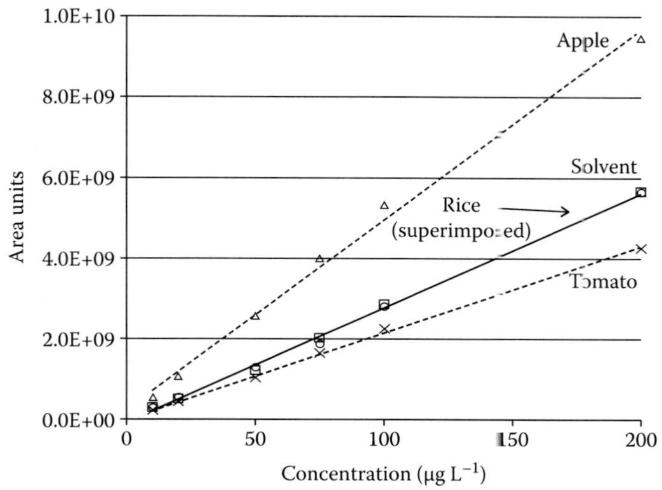

FIGURE 13.3 Calibration curves in solvent and in apple, rice, and tomato MEs (matrix-matched calibration).

important to point out that there is no universal strategy, and in many cases, several approaches must be combined to achieve adequate quantitative results.

13.2.3 Standard Additions

With the standard addition technique, each sample must be run a second time added to a known amount of analyte, and the result is calculated from the relationship between the analyte area in the sample and in the sample plus analyte. This technique can correct variations from the analytical instrument if added just before injection in the sample extract or from the whole analytical process if the standard is added before sample preparation.

In a routine analysis, this technique is more laborious than the external or internal calibration approaches but can be very useful for samples that should occasionally be analyzed, avoiding the construction of an entire calibration curve.

The analyte added to the real test sample permits an automatic compensation of unknown MEs. Another advantage of the standard addition technique is the calculation of results from two consecutive chromatographic runs, requiring less stability (short time) from the HPLC system, instead of the long time necessary between two successive external calibrations.

The drawbacks and conditions for the standard addition technique are the following:

- The analyte must be available in sufficient quantity and purity to be added to the test samples.
- Because the result is calculated from the ratio of two equally imprecise responses, the random error increases by a factor of $\sqrt{2} = 1.41$. This effect can be reduced by using more than one added standard.
- The response must be linear and free from contributions from other components, such as overlapping peaks, with which the bias associated with the perpendicular technique is not compensated in the standard addition technique.

Figure 13.4 shows a hypothetical result from three injections of the sample, sample plus analyte (0.5 mg L^{-1}), and sample plus double concentration of analyte (1.0 mg L^{-1}). The result can be obtained visually by plotting one regression line from the *y*-axis (peak area) through interception

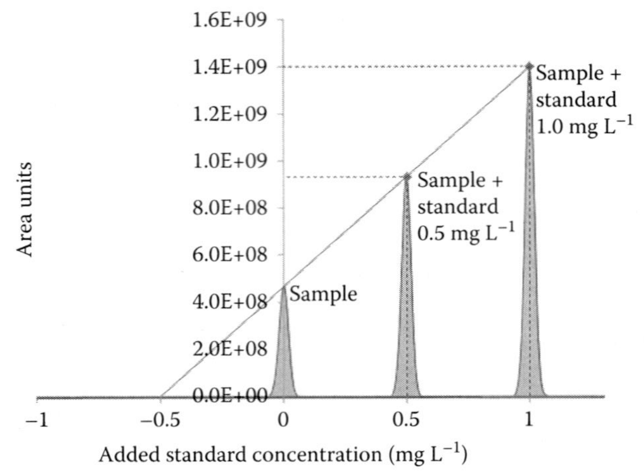

FIGURE 13.4 Representation of standard addition technique from one sample added with two different concentrations of the analyte.

of the *x*-axis (concentration), where the concentration is the absolute value read, which, in this case, corresponds to 0.5 mg L^{-1}.

The result also can be calculated taking into account only one sample addition, using Equation 13.9:

$$C_{\text{sample}} = C_{\text{added}} \times \frac{S_{\text{sample}}}{S_{\text{sample+added}} - S_{\text{sample}}} \tag{13.9}$$

where *C* is the concentration, and *S* is the signal, peak area, or peak height.

13.2.4 Internal Standard Calibration

The concept of an internal standard calibration is quite simple: you simply add a known amount of an internal standard to every sample, both calibrators and unknowns, and instead of basing the calibration on the absolute response of the analyte, the calibration uses the ratio of response between the analyte and the internal standard. To be a good internal standard, some characteristics are needed:

a. Available in sufficient purity and quantity
b. Not present in the test samples
c. Different from but closely similar to the analyte
d. Offering a peak in the chromatogram that is easily quantified (preferably free from overlap)

An internal standard provides a practical means of correction for instrument variations and drift. The advantages of internal standard calibration include the fact that it can be used to compensate the routine variation in the response of the chromatographic system as well as variations in the exact volume of the sample or sample extract introduced into the chromatographic system. Examples of internal standards include brominated or fluorinated compounds and stable isotopically labeled analogs of target compounds, such as a compound containing one or more deuterium atoms instead of a hydrogen atom or a ^{13}C atom instead of one or more ^{12}C atoms.

The isotopically labeled compounds are most often employed in methods that use MS detection systems because the detector can differentiate between the target compound and the internal standard based on the added mass of the internal standard even when these compounds elute from the chromatographic system at the same retention time.

The main disadvantage is that internal standards must be compounds that are not found in the samples to be analyzed, and they must produce an unambiguous response on the chromatographic detector system. Using a classical detector in HPLC, the peak of the internal standard must be resolved from analytes and from interferences. The use of a mass spectrometric detector makes internal standard calibration practical because the masses of the internal standard can be resolved from those of the analytes. Both situations are shown in Figure 13.5.

The corrected amount (or concentration) of one analyte is obtained from the RF calculated for each target compound relative to the internal standard as follows (Equation 13.10):

$$\mathrm{RF} = \frac{A_s \times C_{is}}{A_{is} \times C_s} \tag{13.10}$$

where A_s is the peak area (or height) of the analyte, A_{is} is the peak area (or height) of the internal standard, C_s is the mass of the analyte in the sample aliquot injected into the instrument, and C_{is} is the mass of the internal standard in the sample aliquot injected into the instrument.

The concentration of each analyte in the sample can be obtained from Equation 13.11:

$$X_s = \frac{A_s}{\overline{\mathrm{RF}}} \times \frac{C_{is}}{A_{is}} \tag{13.11}$$

where X_s is the calculated mass of the analyte in the sample aliquot, A_s is the peak area (or height) of the analyte in the sample, A_{is} is the peak area (or height) of the internal standard in the sample, C_{is} is the mass of the internal standard in the sample, and $\overline{\mathrm{RF}}$ is the average response factor.

The concentration can be directly calculated from Equation 13.12:

$$C_k = C_{is}\left[\frac{A_k}{A_{is}}\right] \times \left[\frac{S_{is}}{S_k}\right] \tag{13.12}$$

where C_k is the concentration of the analyte in test sample k, C_{is} is the concentration of the internal standard in the test sample, A_k is the peak area of the analyte in test sample k, A_{is} is the peak area of the internal standard for test sample k, S_{is} is the sensitivity of the response for the internal standard, and S_k is the sensitivity of the response for the analyte.

The idea is that the internal standard responds to variations in the chromatographic conditions in exactly the same way as the analyte so that its response varies to the same degree. The desired effect is that the ratio of their responses is much less susceptible to erratic variations and hence more precise. This can be seen in Figure 13.6, in which the determination coefficient (r^2) increases from

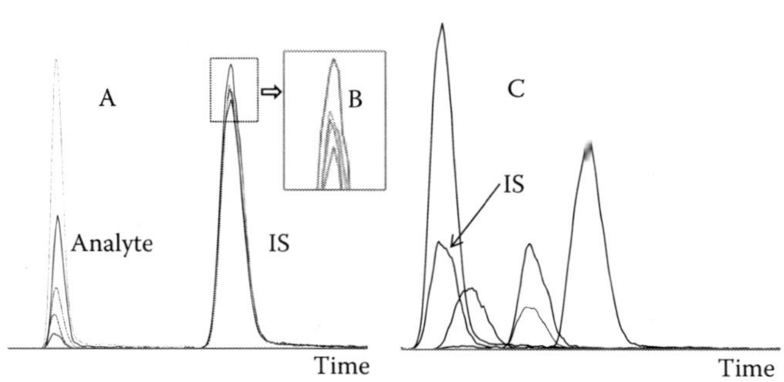

FIGURE 13.5 HPLC chromatograms for (A) an analyte in five concentration levels and for internal standard with constant concentration; (B) zoom window for details of chromatogram A showing the small variations in internal standard peak and (C) co-eluted and near eluted analytes resolved by mass spectrometry.

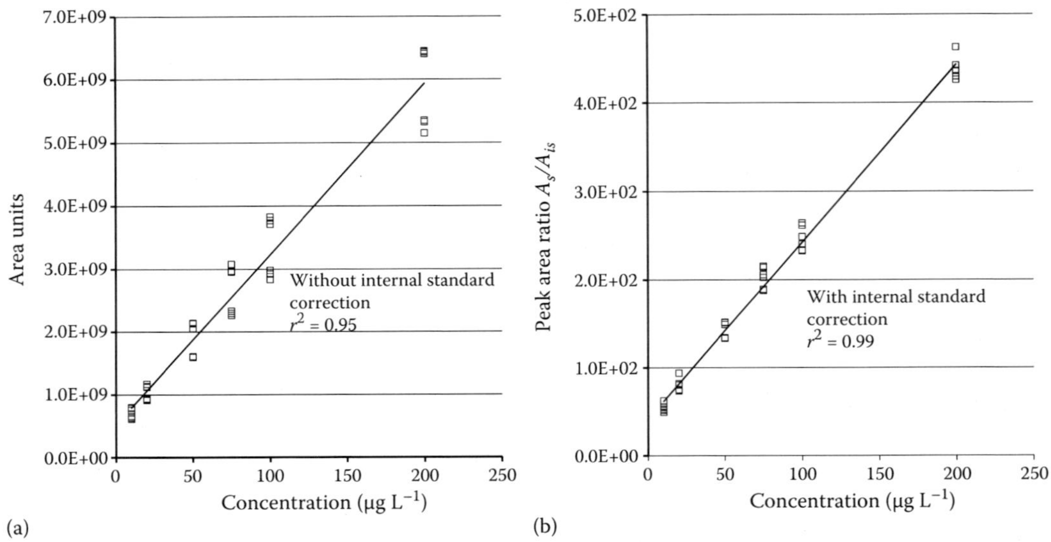

FIGURE 13.6 Calibration curves of fipronil determined by LC-MS/MS with six concentration levels (10, 20, 50, 75, 100, and 200 μg L^{-1}) and six injections each. (a) Without internal standard correction and (b) with the use of internal standard.

0.95 for the calibration curve without internal standard correction to 0.99 for the calibration curve with the use of an internal standard.

In addition to normalizing the response of the target compound to the response of the internal standard in that sample or extract for that injection, the retention times of the target compound and the internal standard may be used to calculate the relative retention time (RRT) of the target compound. The RRT is expressed as a unitless quantity (Equation 13.13):

$$\text{RRT} = \frac{\text{Retention time of the analyte}}{\text{Retention time of the internal standard}} \tag{13.13}$$

In the preparation of analytical solutions for internal standard calibration, some details are important:

- The concentration of the internal standard is the same in all calibration standards whereas the concentrations of the target analytes will vary in the linear range.
- The internal standard solution will contain one or more internal standards, and the concentration of the individual internal standards may differ within the spiking solution. The concentration of internal standards can be different.
- The mass of each internal standard added to each sample extract immediately before analysis must be the same as the mass of the internal standard in each calibration standard.
- The volume of the solution added to the sample extracts should be such that minimal dilution of the extract occurs (e.g., 10 μL of solution added to a 1 mL final extract results in only a negligible 0.1% change in the final extract volume, which can be ignored in the calculations).
- An ideal internal standard concentration would yield a response factor of 1 for each analyte. However, this is not practical when dealing with more than a few target analytes. Therefore, as a general rule, the internal standard should produce an instrument response that is no more than 100 times that produced by the least responsive target analyte at the same concentration. This should result in a minimum response factor of approximately 0.01 for the least responsive target compound.

13.2.5 Internal Standard Normalization

The calibration technique by internal standard normalization is useful to correct for sensitivity changes from one analyte to another in the application of internal standard calibration. This technique provides a normalized analytical signal for all analytes in relation to a reference compound added to the sample (internal standard). This technique consists of a combination of double external–internal calibrations in order to benefit from the advantages of both while minimizing their drawbacks.

Four steps are required to perform the calibration by internal standard normalization:

a. Multiple external standard measurement to determine calibration factors of all analytes (CF) and internal standards (CF_{is}) (Equations 13.14 and 13.15):

$$CF = \frac{X_{a,std}}{Y_{a,std}} \tag{13.14}$$

$$CF_{is} = \frac{X_{is,std}}{Y_{is,std}} \tag{13.15}$$

where $X_{a,std}$ is the concentration (or amount) of analyte in the standard, $Y_{a,std}$ is the area (or height) from the analyte in the standard, $X_{is,std}$ is the concentration (or amount) of internal standard, and $Y_{is,std}$ is the area (or height) from the internal standard.

b. Determination of the normalization factor (NF) of the analytes in relation to the internal standard (Equation 13.16):

$$NF = \frac{CF}{CF_{is}} \tag{13.16}$$

c. Determination of the CF of the internal standard added to the sample (Equation 13.17):

$$CF = NF\ CF_{is} \tag{13.17}$$

d. Quantification of analytes by NF (Equation 13.18):

$$X_{a,spl} = CF\ Y_{a,spl} \tag{13.18}$$

where $X_{a,spl}$ is the concentration (or amount) of analyte in the sample, and $Y_{a,spl}$ is the area (or height) from analyte in the sample.

Very few publications report the use of calibration by internal normalization in analytical quantification, especially in HPLC, but one significant application example of this calibration technique was published by Lehotay et al. [7], who compared this technique with other calculation techniques for the determination of pesticide residues in foods by LC-MS/MS.

13.3 METHOD VALIDATION

An analytical method is the series of procedures carried out from receipt of a sample to the production of the final result. Validation is the process of verifying that a method is fit for the intended purpose. The method may be developed in-house, taken from the literature, or obtained from a third party. The method may then be adapted or modified to match the requirements and capabilities of the laboratory and/or the purpose for which the method will be used [8].

Typically, validation follows the development of a method, and it is assumed that requirements such as calibration, system suitability, analyte stability, etc., have been established satisfactorily. When validating and using a method of analysis, measurements must be made within the calibrated range of the detection system used. In general, validation will precede practical application of the method, but subsequent performance verification is an important continuing aspect of the process. Requirements for performance verification data are a subset of the requirements for method validation [8].

Guidelines have been published for validation of analytical methods for various purposes. The principles described in this section are considered practical and suitable for validation of pesticide residue analytical methods. The analyst should decide on the degree of validation required to demonstrate that the method is fit for the intended purpose and should produce the necessary validation data accordingly. For instance, requirements for testing for compliance with maximum residue limits (MRLs) and providing data for intake estimation may be quite different [9].

Valid analytical data are essential for satisfactory monitoring and control of pesticide residues [10]. The analyst must generate information to show that a method intended for these purposes is capable of providing adequate specificity, accuracy, and precision at relevant analyte concentrations and in appropriate matrices [8]. This information is known collectively as validation and provides the basic evidence to support the validity of the results subsequently generated using the method. In practice, validation of a method cannot encompass the whole range of analytical variables encountered in its use [10]. Accordingly, performance validation data, often referred to as internal/analytical quality control data, will be required to provide evidence of the ongoing performance of the method and analyst. Performance validation also provides a continuing check on the effects of minor modifications and on method transfer between analysts [9].

Concepts of method validation continue to evolve and are currently under consideration for food analysis in the European Community and in the Codex Alimentarius Commission. Comprehensive overviews of validation requirements have been published, identifying many parameters by which method performance may be judged [8,9]. Other authors have contributed with specific statistical and computational techniques to assist with the process of method validation. However, although the parameters to be assessed are clearly defined, few criteria (e.g., specified limits for accuracy or precision) are provided to define the acceptability of a method. In part, this may be because acceptability is determined by the purpose served by the method, and thus a broad overview of validation cannot address the differing requirements of each specific area of analysis [11]. This chapter specifically addresses validation for pesticide residue analysis, but these parameters can be used for other trace analyses as well.

Validation of methods by interlaboratory study has become impractical in most cases. Even when it is practical, it is usually impossible to validate all combinations of analyte, analyte concentration, and sample matrix to which the method may be applied. Published methods may be supported by validation data, but the information is usually limited in scope, and in most cases, further in-house validation data will have to be generated. Method validation, whether in-house or interlaboratory, has rarely incorporated rigorous investigation of sample processing, extraction efficiency, or specificity. These omissions are serious as the procedures may have a considerable influence on the validity of the results obtained. Users of methods should have ready access to validation data to ensure they do not unknowingly exceed the boundaries of validation. Knowingly exceeding the boundaries of validation may be undesirable but, if unavoidable, the analyst must provide this information with the results [10].

In a perfect world, all analytical methods would provide quantitative results of negligible uncertainty and at low cost. In reality, cost and practicality mean that some degree of compromise must be accepted. Methods are often loosely termed "quantitative," "semiquantitative," or "qualitative," depending on the accuracy and precision achievable. The looseness of these terms is useful, and methods should be allocated in an appropriate status on the basis of the method validation criteria satisfied. Somewhat similar distinctions can be made based on the limits of quantification and detection, estimated during method validation, but these limits may give a false impression of what is achievable routinely as they may vary considerably when the method is in use.

Some analytical procedures are more difficult and expensive to characterize than others, but all validation is costly. Within a single laboratory, it is likely that, over a period of time, certain reagents, equipment, and so on will be changed from those used at method validation. It is impracticable and probably unnecessary to revalidate a method to account for all minor changes or minor extensions made. These minor changes should be checked through performance validation whereas revalidation of methods should be limited to major changes or extensions. This flexible approach places considerable responsibility on analysts and laboratory managers in classifying changes as major or minor [10].

Method validation information may be needed to support accreditation or publication of the method or to defend results generated from its use, but there is little international agreement on the exact requirements. There is increasing emphasis on international standards as a means of removing trade barriers and thus increasing reliance on analytical data to determine compliance with the standards. Against this backdrop, the validation procedures should indicate the minimum requirements for method validation in terms of the parameters studied and criteria for acceptability, aiming for requirements that are simple, rational, and affordable to support the validity and claimed status of the method.

13.3.1 Limits of Detection and Quantification

When analysis is carried out at low analyte concentrations, as in pesticide residue analysis, it is important to determine the lowest level of analyte that can be confidently detected by the method in question. For validation, it is usually sufficient to indicate the level at which detection becomes problematic. The limit of detection (LOD) is the level at which a measured value is larger than the uncertainty associated with it. It is the lowest concentration of analyte in a sample that can be detected but not necessarily quantified. In chromatography, LOD is considered to be the analyte concentration or mass that produces a peak with a height at least three times that of the baseline noise level, also expressed as a signal-to-noise (S/N) ratio of three [12].

Limit of quantification (LOQ) is the lowest concentration or mass of the analyte that has been validated with acceptable accuracy by applying the complete analytical method. Figure 13.7 shows the example for LOD and LOQ determination based on S/N ratio for the pesticide azoxystrobin when it was analyzed by LC-MS/MS. Another definition for LOQ is the lowest concentration of an analyte that can be measured with a stated degree of confidence. This has been defined as the level measured in one field blank plus 10 standard deviations of this measure; however, it is recommended

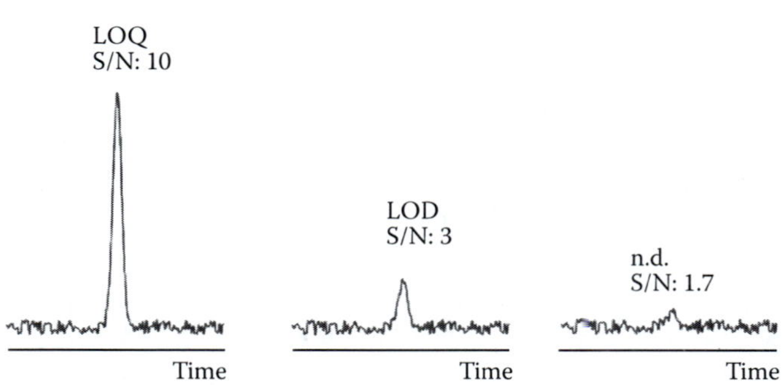

FIGURE 13.7 LOQ and LOD determination based on S/N ratio for azoxystrobin pesticide analyzed by LC-MS/MS. n.d.: not detected.

that this value be established in the laboratory by repeated analysis of an appropriate spiked blank sample. In collaborative studies, the LOQ of the method should be considered the lowest level successfully analyzed in the study. The LOQ is described as the minimum injected amount that affords precision; in other words, an acceptable level of repeatability and trueness, that is, a peak height 10 to 20 times that of the baseline noise. The LOQ can then be expressed as the lowest validated concentration in terms of accuracy and precision. The EURACHEM approach is to inject six samples of decreasing concentrations of analyte. The calculated RSD is plotted against the concentration, and the amount that corresponds to a predetermined RSD is defined as the LOQ [13].

As an initial criterion to estimate the LOQ, a minimum S/N ratio of 10 is adopted. For MS/MS detection, the S/N ratio of the less-intense transition, corresponding in all cases to the qualifier ion, is considered. The LOQ is established as the concentration for which the S/N requirement is fulfilled [14]. Therefore, this level can be accurately quantified and identified. This LOQ is not strictly the method LOQ but the minimum concentration that has been demonstrated to be accurately quantified with the method. In several cases, a common LOQ for all analyzed pesticides can be established.

Tsochatzis et al. [15] developed an HPLC–diode array detector method for pesticide residue analysis in rice paddy water and estimated the LOD as the analyte concentration resulting in a S/N ratio of three and verified by analysis (six independent replicates) of the pesticide mixture fortified at the lower level, calculated as three times the standard deviation (SD) of these analysis. The LOQ was defined as the analyte concentration resulting in a S/N of 10 and verified by the aforementioned procedure applied for LOD. The LOD varied from 0.1 $\mu g\ L^{-1}$ for propanil and tricyclazole to 0.8 $\mu g\ L^{-1}$ for molinate. On the other hand, LOQ values varied from 0.3 $\mu g\ L^{-1}$ for tricyclazole to 2.0 $\mu g\ L^{-1}$ for molinate and profoxydim. The upper linear ranges were 10.0 $\mu g\ L^{-1}$ for azoxystrobin and propanil, 20.0 $\mu g\ L^{-1}$ for penoxsulam and tricyclazole, and 40.0 $\mu g\ L^{-1}$ for 3,4-dichloroaniline, molinate, profoxydim, cyhalofop-butyl, and deltamethrin.

Carneiro et al. [16] defined LOD as the lowest concentration of analyte that could be differentiated from the matrix signal with S/N greater than six. The LOQ was based on the trueness and precision data obtained via recovery determinations and was defined as the lowest validated spike level meeting the requirements of recovery within the range of 70% to 120% and RSD $\leq$ 20%. For all 128 pesticides evaluated in this work, the method LOD and LOQ ranged between 5.0 and 10 $\mu g\ kg^{-1}$, respectively, except for fenamiphos and mevinphos (LOD = 7.5 $\mu g\ kg^{-1}$ and LOQ = 25 $\mu g\ kg^{-1}$).

13.3.2 Accuracy/Trueness

Accuracy expresses the closeness of an analytical result to a true value and is often described using the components of trueness and precision [17]. In the last few years, different methods have been suggested to evaluate method accuracy. The main ways normally used to determine the accuracy are the following [18]:

- Determination of analyte concentration in a particular reference material and comparing the result with the certified value
- Percentage of analyte recovery in blank matrix samples spiked with known analytical concentrations
- Comparison between results from the method under validation with a reference method
- Determination of analytical concentration in the sample by means of the standard addition technique

For some methods, the true value cannot be determined exactly, and it may be possible to use an accepted reference value to determine this value if suitable reference materials are available or if the reference value can be determined by comparison with another method. Another way is to analyze a sample with known concentrations for assessing accuracy, such as a certified reference material (CRM) and comparing the measured value with the true value as supplied for the CRM.

During pesticide residue method validation, the most used approach for assessing accuracy is the percentage recovery of known amounts of analyte spiked into a sample matrix by either

- The assay method that involves spiking analyte in blank matrices. Spiked samples are generally prepared at three levels in the range of 50% to 150% of the target concentration. The blank matrix has to mimic representative samples in all respects whenever possible. The analyte determination should be done using the same quantification procedure as will be used in the final method.
- The standard addition method if a blank sample cannot be prepared without the analyte being present. Mean recovery should be appropriate to the concentration tested, encompassing a range 20% below the lowest expected concentration and 20% above the highest expected concentration.

Recoveries should be appropriate to the concentration tested and depend largely on the sample matrix, sample-processing procedure, and analyte concentration. Several documents, such as the EC document SANCO/3030/99, present tables with expected recoveries for analyte concentrations in samples [19].

When accuracy is evaluated in a set of test results, it involves a combination of random error, estimated as precision, and a common systematic error, named trueness or bias (ISO 5725-1) [20]. This effectively means that spiked recovery experiments should be undertaken to check the accuracy of the method. A minimum of five replicates is required to check the precision at the reporting limit, to check the sensitivity of the method, and at least one other higher level, for example, the MRL. The (method) LOQ is defined as the lowest validated spike level meeting the method performance acceptability criteria, which for pesticide residue analysis, are mean recoveries for each representative commodity in the range of 70% to 120% with RSD ≤ 20% [9].

Other approaches to demonstrate that the analytical method complies with performance criteria may be used, provided that they achieve the same level and quality of information. When the residue definition incorporates two or more analytes, if possible, the method should be validated for all analytes included in the residue definition. The measure of trueness is normally expressed as "bias," the closeness of agreement between the average value obtained from a series of test results, expressed as mean recovery, and the accepted reference or true value (ISO 5725-1) [9].

Accuracy can be defined as a measure of how closely a determined value or mean value approximates the true value of the analyte. This is best supported through analysis of standard reference materials; however, the availability of such materials, especially for pesticides in foods, is usually extremely limited. Normally, the recovery of added analytes to blank samples over an appropriate range of concentrations is taken as an indication of accuracy. For pesticide residue compliance work, the concentration range chosen should certainly bracket the MRL. It should also be recognized that analytes added to field samples may behave differently (typically showing higher recovery) from field-incurred residues. For analysis at the μg kg^{-1} level, recoveries between 70% and 120% are generally considered acceptable [9].

Carneiro et al. [16] evaluated the trueness during pesticide residue analysis in bananas using ultra-HPLC-MS/MS by recovery experiments of blank samples spiked at four levels (10.0; 25.0; 50.0 and 100.0 μg kg^{-1}) with six replicates per level due to the lack of CRMs. Then, the trueness, in terms of recovery, and the precision (repeatability), in terms of RSD%, were estimated. Almost all results showed recoveries in the range considered acceptable (70% to 120%) except methamidophos at the 10 μg kg^{-1} level. However, the recovery (67.5%) is very close to the acceptable range and other parameters, such as intermediate precision (6.5%) and measurement uncertainty (35.7%), are satisfactory for this analyte. Furthermore, to ensure that the method is truly reproducible, the methamidophos was monitored during routine assays. In this same level, fenamiphos and mevinphos did not show acceptable parameters for the recovery assays. The majority of results showed RSD lower than 10% for all levels of fortification. Only three analytes presented RSD between 15% and 20% in the LOQ level.

13.3.3 Precision

In contrast with accuracy, which is the closeness of results to a measured or true value, precision is the closeness of results of multiple analyses to each other. The ISO International Vocabulary of Basic and General Terms in Metrology (ISO-VIM) defined accuracy as the closeness of agreement between quantity values obtained by replicate measurements of a quantity under specified conditions [21]. Precision is the closeness of agreement between independent analytical results obtained by applying the experimental procedure under stipulated conditions. The smaller the random component of the experimental errors that can affect the results, the more precise the method. A measure of precision (or imprecision) is the RSD. For SANCO 12495/2011, the measure of precision usually is expressed in terms of imprecision and computed as the SD of the test result. It may also be defined as the value below which the absolute difference between two single test results on identical material, obtained under the above conditions, may be expected to lie with a specified probability (e.g., 95%) [9].

The determination of precision is one of the principal parameters in the process of achieving repeatability and reproducibility during method validation. The precision indicates the random error or the degree of dispersion of a set of individual measurements by means of the SD, the variance, or the coefficient of variation. For validation of chromatographic methods, it is recommended measuring a minimum of three concentration levels (low, medium, and high) prepared in triplicate and covering the whole analytical range under study (three levels × three replicates per level = nine determinations) [18].

Repeatability aims to ensure that contributing factors to the variability of results, such as the operator, equipment, calibration, and environmental considerations, remain constant and have little or no contribution to the final results. Repeatability, obtained with a minimum of five replicate determinations, with mean, RSD%, and number of determinations reported, consists of two factors [12]:

1. Intralaboratory assay: repeated analysis of an independently prepared sample on the same day by the same operator in the same laboratory
2. Intermediate precision: repeated analysis of an independently prepared sample by different operators on different days in the same laboratory

13.3.4 Ruggedness/Robustness

The ruggedness is defined as the constancy of the results when external method factors, such as the analyst, instruments, laboratories, reagents, or days, are varied. Ruggedness is a measure of reproducibility of test results under normal, expected operational conditions from laboratory to laboratory and from analyst to analyst. Ruggedness cannot be erroneously used as a synonym of robustness [18]. The measure of an analytical method is how well it stands up to less-than-perfect implementation. At any given point in a method, there may be a step in which an error may have a large effect on the performance of the method. These steps should be identified and their potential influence on the method evaluated through deliberate variations to the method and subsequent analysis of the results. This identifies the parts of the method most susceptible to significant variation. It also identifies the steps that may be taken to further improve the method. This should be evaluated during method development [12].

Robustness is the constancy of the results when internal characteristics (no external factors as in ruggedness) such as flow rate, column temperature, injection volume, mobile phase composition, or any other variable inherent to the method of analysis are varied deliberately [18].

13.3.5 Linearity

Linearity is defined as the interval between the upper and lower levels of concentration between which the detector response is proportional to the analyte quantity and in which it is possible to determine the concentration with precision and accuracy. The linear working range is evaluated by the ratio between the analyte concentration and its analytical response. Its determination is performed using analytical solutions in solvent or fortified sample blanks.

Figure 13.8 shows an analytical calibration curve constructed from peak area responses of the pesticide mephospholan obtained by LC-MS/MS on which it is possible to see the good correlation of the peak areas with the analyte concentration.

The straight line passing through the shortest distance between points of each of the seven replicates injected for each concentration level has a relationship expressed by a correlation coefficient of 0.9930. This line is defined by the equation $y = ax + b$, whose values are determined by the equations presented in Section 13.2.1 for the linear calibration model.

In Figure 13.8, it can also be seen that the variability between the peak areas obtained for each of the replicates increases with the concentration of the analyte. Because this may be due to a visual scale of the graph and because applying the method of least squares implies that the residuals have the same variance, a condition called homoscedasticity it is necessary to statistically test whether the data meet this condition. This can be done with the Cochran test, using Equation 13.19:

$$C_{calc} = \frac{s_{max}^2}{\sum_{i=1}^{n} s_1^2} \qquad (13.19)$$

where C_{calc} is the calculated value for C of the Cochran test, and s^2 is the variance.

The value of C_{calc} is compared with the value of C obtained from the table of critical values for the Cochran test for variance outliers [22]. If $C_{calc} < C_{tab}$, it is possible to conclude that the variance is equivalent and the data is homoscedastic; in other words, the accuracy of measurements is independent of concentration.

A calibration model for data with good correlation ($r^2 > 0.99$) and satisfying the condition of homoscedasticity should be able to predict the concentration values from the analytical response, which can be viewed from a graphic residual prediction drawn from the comparison between the percentage values of analytical response and the predicted values, which is shown in Figure 13.9.

SANCO [9] recommends that if individual residuals deviate more than ±20% from the calibration curve, an alternative calibration function should be used. For concentrations near or above the MRL, the deviation should be within ±10%.

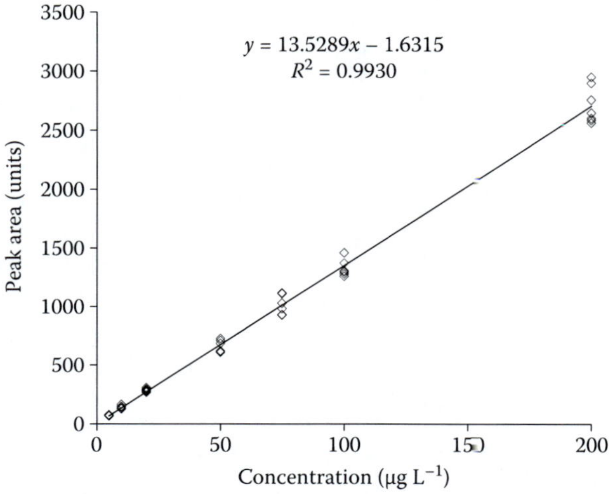

FIGURE 13.8 Calibration curve for the pesticide mephospholan prepared in the matrix extract of wheat samples. Each concentration level was injected seven times in LC-MS/MS.

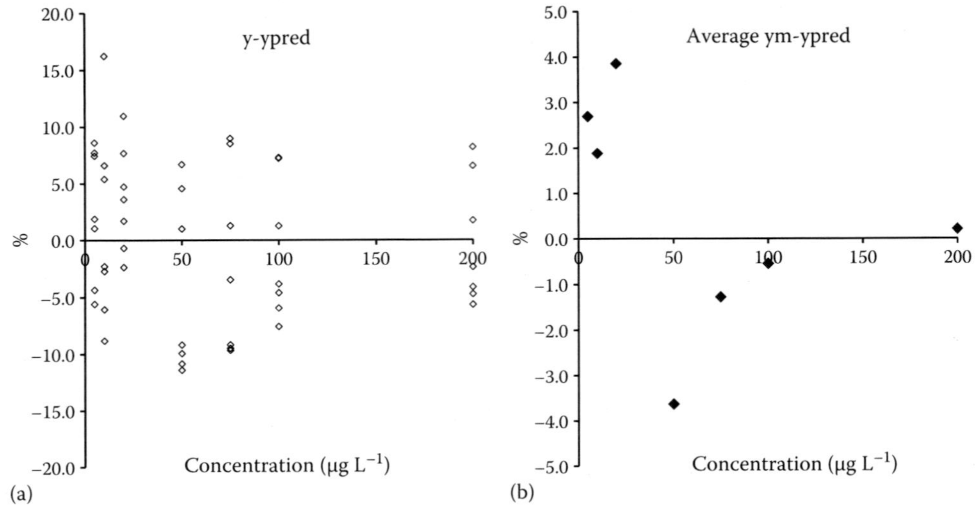

FIGURE 13.9 Plot of the residuals of a linear calibration obtained from (a) individual and (b) mean values for the pesticide mephospholan in blank wheat matrix extract prepared using QuEChERS method. Each extract was injected seven times in LC-MS/MS.

A more robust way to check whether the residual prediction is distributed normally in order to check the linearity of the method and variance decomposition is to use the variance test, also called ANOVA. Some factors affecting the linearity of a method are an interfering matrix, overload of the chromatographic column, detector saturation, dirty detector, noise, and contamination of the material used in the analysis or carryover. Factors affecting the accuracy of measurements can be measurement error of the injected volume, volumetric error in the preparation of replicates, or detector noise. When the range of linearity of one validated method is not appropriate for the intended use and the problems listed above are corrected, the analytic method can be re-optimized, and a division of the linear curve in two smaller segments, use of quadratic models, polynomials, or multivariate calibration can be adopted.

13.3.6 Specificity

Specificity can be defined as the ability of the detector, supported by the selectivity of the extraction, cleanup, derivatization, or separation, if necessary, to provide signals that effectively identify the analyte [9].

The validation parameter selectivity refers to the extent to which a method can determine a particular analyte in a complex mixture without interference from other components in the mixture. This definition is often wrongly used as equivalent to specificity, which is considered to be the ultimate in selectivity; it means that no interferences are supposed to occur. Araujo [18] made an excellent review about these validation parameters. Selectivity can be classified as low, high, bad, partial, good, etc., in order to choose the appropriate category for a particular purpose. The term "specificity" refers always to 100% selectivity or, conversely, 0% interferences. The most used methods to measure selectivity are the following:

- Comparison between chromatograms of blank samples with and without the analytes
- Comparison of the signal response obtained after injection of analytical solutions with and without all the possible interferents
- Analysis of CRMs

13.4 QUALITY CONTROL

There are two main key issues in quality control (QC) to be applied during validation of an analytical method and also during its use. One is related to quality assurance, that is, all the technical requirements to be met to ensure the quality of obtained analytical data, which include the environmental conditions of the sites at which the assays are conducted; the training, skills, and experience of the analysts; the maintenance and calibration of the equipment; the QC procedures; the use of validated methods; the traceability of measurements; the determination of measurement uncertainty; the verification procedures; the procedures for presentation of results; the registration of deviations; the registration of preventive and corrective actions; the use of reagents; and the specification and verification of supplies, reagents, standards, and reference materials before use. These items should be documented within the quality management system of the laboratory [23].

The other issue is defined as QC and refers to procedures, activities, and operational techniques to ensure the results in terms of accuracy and precision, including the following:

- Proficiency tests
- Analysis of CRMs
- Interlaboratory studies
- Use of fortified blank samples (spiked samples)
- Analysis of blind samples
- Use of QC samples and control charts
- Analysis of blanks
- Analysis in duplicate

Participation in proficiency tests and interlaboratory studies is related with global QC and full validation of method, but these activities occur very often in routine analysis [24]. Therefore, an important part of QC is to monitor day-to-day or batch-to-batch the analytical performance in terms of accuracy and precision for the routine use of CRMs and QC samples. Fortified samples are recommended, and the results can be adequately monitored using control charts.

Another important aspect of routine QC is cross-contamination during sample storage, handling, extraction, and in the equipment. In routine analysis, this can be checked by sample blanks and blind blank samples. The process of quality assurance should demonstrate that the method and the analytical instrument provide accurate and precise results or whether deterioration occurs. For this, the quality procedure should include tests that provide information on the characteristic performance of the method. Parameters examined are carryover, column statement, accuracy, and precision [25].

For HPLC determinations, the stability of the chromatographic system and the response of the detector must be frequently checked, either for quantitative measures with standard and QC samples or by evaluation of chromatogram profiles to confirm if the peaks look normal (Gaussian), the response obtained is comparable to the response from previous calibrations, the nontarget peaks are present in calibration analyses, and the HPLC guard column needs replacement [2]. The level and the type of QC should be appropriate to each type and quantity of measurements in the laboratory routine in order to ensure the validity of the performed analysis.

13.5 CONCLUSIONS

The determination of pesticide residues by HPLC requires the establishment of a calibration curve for each analyte, in which the detector response can be correlated with the concentration. For quantitative measurements, the more complex the matrix, the greater the need for attention to the selective detector response because interferences present in a sample matrix can negatively or positively influence detector response. In this sense, MS-MS combines selectivity with sensitivity and is the most used detector in pesticide residue analysis by HPLC. If the matrix interferes with the intensity of the signal, then a procedure that can compensate for this needs to be implemented. Matrix-matched calibration is frequently used in HPLC routine analysis of pesticide residues in complex matrices, such as food, to avoid inaccurate results.

To achieve high-quality data, analytical methods need to be validated or revalidated before their routine application or whenever the conditions or method parameters change from the original scope. In this procedure, the limitations are established, and influences that may change the method characteristics are identified and measured. Method validation should follow a plan that includes the scope, required performance characteristics and acceptance limits. Parameters usually examined are linearity, analytical range, selectivity/specificity, limits of detection and quantification, accuracy, precision, and ruggedness/robustness. Validation results can be used to judge the reliability, consistency, and accuracy of analytical data and has received considerable attention from the analytical community and from regulatory agencies. There is increasing emphasis on harmonized standards to enhance reliance on analytical data and determine compliance with international standards.

A well-defined and documented validation process provides regulatory agencies with evidence that the system and test method are suitable for their intended use. It also ensures that the guidelines established meet method validation requirements and specifications. Method validation data may be needed to support accreditation or publication of methods or to defend results.

REFERENCES

1. ISO/IEC 17025, General Requirements for the Competence of Testing and Calibration Laboratories, 2005.
2. US Environmental Protection Agency (EPA), Method 8000C Revision 3, Washington, DC, 2003.
3. Cappiello, A., Famiglini, G., Palma, P., Pierini, E., Termopoli, V., Trufelli, H., Overcoming matrix effects in liquid chromatography-mass spectrometry, *Anal. Chem.*, 80, 9343–9348, 2008.
4. Srneraglia, J., Baldrey, S. I., Watson, D., Matrix Effects and Selectivity Issues in LC-MS-MS, *Chromatographia*, 55, 95–99, 2002.
5. Kebarle, P., Tang, L., LC/MS/MS—A Powerful Analytical Technique, *Anal. Chem.*, 65, 972A–986A, 1993.
6. Fernandez-Alba, A., Chromatographic-Mass Spectrometric Food Analysis for Trace Determination of Pesticide Residues, Volume 43 (Comprehensive Analytical Chemistry), Elsevier Science, Amsterdam, 2005.
7. Lehotay, S. J., Mastovska, K., Lightfield, A. R., Gattes, R. A., Multi-Analyst, Multi-Matrix Performance of the QuEChERS Approach for Pesticide Residues in Foods and Feeds Using HPLC/MS/MS Analysis with Different Calibration Techniques, *J. AOAC Intern.*, 93, 355–367, 2010.
8. Codex Alimentarius Commission. Guidelines on Good Laboratory Practice in Residue Analysis: CAC/GL 40-1993, Rev. 1-2003. Rome: FAO/WHO, 2003.
9. DG-SANCO, Comission of the European Communities (2013). Document no. SANCO/12571/2013. Guidance document on analytical quality control and validation procedures for pesticide residues analysis in food and feed. 2013. 42 p.
10. Hill, A. R. C., Reynolds, S. L., Guidelines for In-House Validation of Analytical Methods for Pesticide Residues in Food and Animal Feeds, *Analyst*, 124, 953–958, 1999.
11. European Comission. Guidelines for Performance Criteria and Validation Procedures of Analytical Methods Used in Controls of Food Contact Materials, 2009.
12. Health and Safety Executive. Guidelines for Validation of Analytical Methods for Non-Agricultural Pesticide Active Ingredients and Products, 2002.
13. Eurachem. The Fitness for Purpose of Analytical Methods, a Laboratory Guide to Method Validation and Related Topics, 1998.
14. European Comission. Regulation (EC) no. 396/2005 of the European Parliament and of the Council of 23 February, 2005 on Maximum Residue Levels of Pesticides in or on Food and Feed of Plant and Animal Origin and Amending Council Directive 91/414/EEC, 2005.
15. Tsochatzis, E. D., Tzimou-Tsitouridou, R., Menkissoglu-Spiroudi, U., Karpouzas, D. G., Papageorgiou, M., Development and Validation of an HPLC-DAD Method for the Simultaneous Determination of Most Common Rice Pesticides in Paddy Water Systems, *Intern. J. Environ. Anal. Chem.*, 92, 548–560, 2012.

16. Carneiro, R. P., Oliveira, F. A. S., Madureira, F. D., Silva, G., Souza, W. R., Lopes, R. P., Development and Method Validation for Determination of 128 Pesticides in Bananas by Modified QuEChERS and UHPLC–MS/MS Analysis, *Food Control*, 33, 413–423, 2013.
17. Analytical Methods Committee. Terminology—the Key to Understanding Analytical Science. Part 1: Accuracy, Precision and Uncertainty, Royal Society of Chemistry, 2003.
18. Araujo, P., Key Aspects of Analytical Method Validation and Linearity Evaluation, *J. Chromatogr. B*, 877, 2224–2234, 2009.
19. European Comission. Technical Material and Preparations: Guidance for Generating and Reporting Methods of Analysis in Support of Pre- and Post-Registration Data Requirements for Annex II (Part A, Section 4) and Annex III (Part A, Section 5) of Directive 91/414, 2000.
20. International Organization for Standardization. ISO 5725-1:1994. Accuracy (Trueness and Precision) of Measurement Methods and Results. Part 1: General Principles and Definitions, 1994.
21. International Organization for Standardization. International Vocabulary of Metrology, 2012.
22. Kanji, G. K., 100 Statistical Tests. London: SAGE Publication Ltd., 1993.
23. Eurachem/CITAC, Guide to Quality in Analytical Chemistry—An Aid to Accreditation, 2002.
24. Bockstaele, E. V., Loose, M. D., Taverniers, I., Trends in Quality in the Analytical Laboratory. II. Analytical Method Validation and Quality Assurance, *TrAC Trends Anal. Chem.*, 23, 535–552, 2004.
25. Masson, P., Quality Control Techniques for Routine Analysis with Liquid Chromatography in Laboratories, *J. Chromatogr. A*, 1158, 168–173, 2007.

14 Analysis of Pesticides by HPLC-UV, HPLC-DAD (HPLC-PDA), and Other Detection Methods

Tomasz Tuzimski

CONTENTS

14.1 PROPERTIES OF ANALYTES DETERMINED BY HIGH-PERFORMANCE LIQUID CHROMATOGRAPHY COUPLED WITH SPECTROPHOTOMETRIC DETECTION

Monitoring of environmental samples is usually restricted to known or suspected compounds or compound classes ("target analysis") for which specific analytical methods have been developed and optimized. Identification of unknown compounds ("nontarget analysis") appears to be much more difficult. In this case, sample extraction, isolation, and separation must take into account the very different (and a prori unknown) physical and chemical properties of the individual organic compounds, and the instrumental method must have high separation efficiency and provide optimum structural information.

For analytes of high to medium volatility that are thermally stable, gas chromatography coupled to mass spectrometry (GC-MS) or tandem mass spectrometry (GC-MS/MS) is the method of choice as a technique that combines high separation efficiency with the structural specificity of MS. However, polar, nonvolatile, or thermally labile analytes in natural samples are more difficult to identify [1,2]. For GC analysis (including GC-MS or GC-MS/MS), these compounds can be derivatized to enhance their volatility [1,2]. As an example, carbamate pesticides are thermally labile, and they need to be derivatized to form more thermally stable derivatives before GC analysis to avoid their breakdown to amines and phenols in the injection port [3,4]. Commonly used derivatization reactions are silylation, acetylation, and alkylation [4]. On-column derivatization is a very convenient procedure because this method is simple and involves a fast, one-step operation [3,4].

High-performance liquid chromatography (HPLC) is a method that appears to be well suited for separation, identification, and quantitative analysis of substances, especially of polar or thermally labile compounds.

Properties of analytes (e.g., the ability of the analyte to absorb or to fluoresce) that will be determined by the HPLC system can influence the choice of suitable detection techniques.

A universal detector measures a property that is common to all substances. There are four general techniques that are used for HPLC detection [5]:

1. Bulk property (detectors) or differential measurement: The detector measures a change in this property as a differential measurement between the mobile phase containing the sample/analyte and the one without the substance (solvent vs. solvent + solute). The examples of these types of detectors are refractive index detectors, having the advantage of detecting all compounds. On the other hand, this feature can be a disadvantage as all the components of the sample that are eluted from the column will generate a detector signal. Then, additional chromatographic selectivity may be needed to compensate for the lack of detection selectivity.
2. Sample-specific (detectors): These detectors respond to unique property of the solute or at least are not common to all the analytes.
   Examples of these detectors are as follows:
   - Spectrophotomeric detectors: These rely on the ability of the analyte to absorb in the UV/VIS range [UV detectors, UV/VIS detectors, diode array detectors (DAD), photodiode detectors (PDA), which function by relying on the interaction of electromagnetic radiation with matter in the range from near UV (190 nm) to near infrared (1100 nm)].
   - Fluorescence detectors: These detectors rely on the ability of the analyte to produce fluorescence.
   - Conductivity detectors: These detectors measures conductivity of electricity.
   - Conductivity detector measures electronic resistance and measured value is directly proportional to the concentration of ions present in the solution. (Thus it is generally used for ion chromatography).
   - Electrochemical detectors: These react under specific conditions.
3. Mobile-phase modification detectors: These detectors change the mobile phase to produce a change in the properties of the analyte, including the following:
   - Specific liquid-phase chemical reaction with the analyte (e.g., reaction detectors)
   - Gas-phase reaction (e.g., corona discharge, mass spectrometric detectors)
   - Creation of analyte particles suspended in gas phase (e.g., evaporating, light-scattering detectors)
4. Hyphenated techniques in which the HPLC system is coupled with an independent analytical instrument to provide detection, for example, with MS (HPLC-MS); other techniques are LC-IR, LC-FTIR, and LC-NMR

Most popular in this respect is the ability of the analyte, which possesses chromophore groups, to absorb. Almost all aromatic compounds with one or more double bonds can be detected at wavelengths of <215 nm, and the preponderance of aliphatic compounds presents significant absorbance at ≤ 205 nm.

14.2 BASIC DETECTOR REQUIREMENTS, DETECTOR CRITERIA, AND PROPERTIES OF DETECTORS

HPLC coupled with different detectors has become an essential tool in the modern laboratory for the analysis of pesticides in a variety of samples.

The ideal HPLC detector should be characterized as [5]

- Having high sensitivity and predictable response
- Responding to all solutes or having a predictable specificity
- Being unaffected by changes in temperature and carrier flow
- Responding independently from the mobile phase
- Not contributing to extra-column peak broadening
- Being reliable and convenient to use
- Having a response that increases linearly with the amount of solute
- Being nondestructive for the solute
- Provide qualitative information on the detected peak

Of course, in practice, there seem to be no ideal detectors that can be applied to identification and quantitative analysis for any type of analytes.

Detectors can be classified as

- Universal detectors, which detect all sample components
- Selective detectors, which detect a particular group (class) of analytes
- Specific detectors, which detect only one compound

Another classification of detectors is connected with the proportionality of the signal to the quantity of the component in a measuring cell of the detector. In such a case, detectors are classified as

- A concentration detector if the signal is proportional to the concentration of the solute
- A mass detector if the signal is proportional to the mass stream

However, a practical consequence of such a division is both the influence of the eluent flow velocity on the surface and the range of concentration profiles registered by the detector [6].

The ability of the detector to provide precise and accurate quantitative data is a function of the signal size generated by the analyte, background noise, and, to a certain extent, the baseline drift.

14.3 DETECTION METHODS IN QUALITATIVE AND QUANTITATIVE ANALYSIS

Although the signal-to-noise ratio (S/N) is a measure of the inherent quality of the detector signal, the minimum detectable mass or concentration is frequently the limiting factor in the usefulness of a detector for a particular application. For quantitative analysis in HPLC systems, the detector response must be related to the amount of analyte present. If the analyte response (y) is plotted against the analyte concentration (x), the simplest, most convenient, and most reliable relationship is $y = mx$, where the slope (m) is a constant defined as the *sensitivity*. In proper use, *sensitivity* is the slope of a calibration plot, that is, the change in signal per unit change in concentration (or mass) that can be measured. The above relationship between analyte response and analyte amount is termed *linear* [5].

For the best use over a wide range of sample (analyte) concentrations, a wide *linear dynamic range* (the concentration range over which the detector output is proportional to analyte concentration, e.g., 10^5 for UV detection) is desired so that both major and trace components can be determined in a single analysis over a wide concentration range. For example, with a stability-indicating method, peaks ≥ 0.1% of the response of the active ingredient (=100%) must be reported, which would require a linear range of at least $100/0.1 = 10^3$ [5].

The *limit of detection* (*LOD*) is the smallest signal that can be discerned from the noise. Often a *S/N* of three is equated to the *LOD*. The *limit of quantification* (*quantitation*) (*LOQ*) is the smallest signal that can be measured with the required precision for the method. The *LOQ* (sometimes called the *lower limit of quantification*) is often defined as *S/N* ≥ 10. However, a value of *S/N* ≥ 50 may be chosen for high-precision methods. Both values, *LOD* and *LOQ*, are directly related to the concentration (or mass) of solute (analyte) in the detector cell [5].

14.4 DESCRIPTION OF SELECTED METHODS AND THEIR APPLICATION IN LIQUID CHROMATOGRAPHY OF MAIN CHEMICAL GROUPS OF PESTICIDES

14.4.1 HPLC-UV

The most popular detectors, widely used in modern HPLC with high sensitivity for many solutes (analytes), that must absorb in the ultraviolet (UV) or visible (VIS) region are photometers based on UV and VIS light absorption. A fundamental dependence that measurements using such types of detectors are based on is the Beer-Lambert law, which combines absorbance with the concentration, path length, and molar absorptivity. Sample concentration in the flow cell is related to the fraction of light transmitted through the cell and expressed as Equation 14.1 [5]:

$$\log\frac{(I_0)}{(I)} = \varepsilon bc \tag{14.1}$$

where I_0 is the incident light intensity, I is the intensity of the transmitted light, ε is the molar absorptivity (or molar extinction coefficient) of the sample, b is the cell path-length (cm), and c is the sample concentration (moles/L).

Light-absorption HPLC detectors are customarily designed to provide an output in absorbance that is linearly proportional to sample concentration in the flow cell (Equation 14.2) [5]:

$$A = \log\frac{(I_0)}{(I)} = \varepsilon bc \tag{14.2}$$

where A is the absorbance.

Readers can find other details in the fundamental book [5].

UV detectors have the following advantages [5]:

- Capability of producing very high sensitivity (for the samples that absorb in the UV)
- Good linear range ($>10^5$)
- Ability of being made up with small cell volume to minimize extra-column band broadening
- Relatively insensitive to mobile-phase flow and temperature changes
- High reliability
- Being easy to operate
- Nondestructivity of the sample
- Widely varying response to different solutes
- Compatibility with gradient elution
- Free selection of detection wavelength
- International troubleshooting and calibration servicing

Applied in the HPLC systems, their single solvents or mixtures—used as mobile phases—also affect detection. Mobile phase solvents are of primary concern as their properties must often fall within narrow limits of acceptable performance. The solvents used as the mobile phase will preferable have an absorbance of $A < 0.2$ AU at the wavelength used for the detection of the analytes in the sample. On one hand, a lower absorbance may mean improved assay precision and better results with gradient elution, but on the other hand, higher absorbances may be acceptable for some isocratic separations. Table 14.1 summarizes values of the solvent absorbance at different wavelengths

TABLE 14.1
Absorbance as a Function of Wavelength of Various Solvents and Buffers Used for RP-LC

	Absorbance of Indicated Wavelength (nm)								
	200	**205**	**210**	**215**	**220**	**230**	**240**	**250**	**260**
				Solvents					
Acetonitrile	0.06	0.02	0.02	0.01	0.00	0.00	0.00	0.00	0.00
Methanol	1.0+	1.0	0.53	0.35	0.23	0.10	0.04	0.02	0.01
Methanol (degassed)	1.0+	0.76	0.35	0.21	0.15	0.06	0.02	0.00	0.00
Tetrahydrofuran	1.0+	1.0+	1.0+	0.85	0.70	0.49	0.30	0.17	0.09
Isopropanol	1.0+	0.98	0.46	0.29	0.21	0.11	0.05	0.03	0.02
				Buffers					
				Acetate					
Acetic acid, 1%	1.0+	1.0+	1.0+	1.0+	1.0+	0.87	0.14	0.01	0.00
Ammonium salt 10 nM	1.0+	0.94	0.53	0.29	0.15	0.02	0.00	0.00	0.00
				Carbonate					
$(NH_4)HCO_3$, 10 mM	0.41	0.10	0.01	0.00	0.00	0.00	0.00	0.00	0.00
				Formate					
Sodium salt, 10 mM	1.00	0.73	0.53	0.33	0.20	0.03	0.01	0.01	0.01
				Phosphate					
H_3PO_4, 10 mM	0.00	0.00	0.00	0.00	0.00	0.00	0.00	0.00	0.00
KH_2PO_4, 10 mM	0.03	0.00	0.00	0.00	0.00	0.00	0.00	0.00	0.00
K_2HPO_4, 10 mM	0.53	0.16	0.05	0.01	0.00	0.00	0.00	0.00	0.00
$(NH_4)_2HPO_4$, 10 mM	0.37	0.13	0.03	0.00	0.00	0.00	0.00	0.00	0.00
Sodium salt, pH 6.8, 10 mM	0.20	0.08	0.02	0.01	0.00	0.00	0.00	0.00	0.00
				Trifluoroacetic Acid					
0.1% in water	1.0+	0.78	0.54	0.34	0.20	0.06	0.02	0.00	0.00
0.1% in ACN	0.29	0.33	0.37	0.38	0.37	0.25	0.12	0.04	0.01

Source: Snyder, L. R., Kirkland, J. J., Dolan, J. W. *Introduction to Modern Liquid Chromatography*. 2010. Copyright Wiley-VCH Verlag GmbH & Co. KGaA. Reproduced with permission; Li, B. J., *LCGC* 10, 856, 1992; *High-Purity Solvent Guide*, Burdick & Jackson Laboratories, Muskegon, MI, 1980.

(200–260 nm) for solvents that are used in reversed-phase (RP) chromatography, which has dominated HPLC applications [5,7,8].

UV absorbance, as a function of wavelength of various solvents and buffers applied in RP-HPLC systems, is summarized in Table 14.2 [5,8–10].

Practical details can be found in Appendix I [5].

UV detectors come in three configurations:

- Fixed-wavelength detectors: relying on distinct wavelengths of light generated from the lamp (they are not widely used today)
- Variable-wavelength and DADs: selecting one or more wavelengths generated from a broad-spectrum lamp

HPLC coupled with an ultraviolet detector (HPLC-UV), especially in the case of the fixed-wavelength technique, is now rarely applied in pesticide residue analysis [11–23].

TABLE 14.2
Miscellaneous Solvent Properties

Solvent	UV Cutoff (nm)[a]	Refractive Index RI[b]	Viscosity (cP)[c]	Boiling Point (°C)[d]	ε (Silica)[e]
Acetone	330	1.359	0.36	56	0.53
Acetonitrile	190	1.344	0.38	82	0.52
1-Butanol	215	1.399	2.98	118	0.40
1-Chlorobutane	220	1.402	0.45	78	0.20
Chloroform	245	1.446	0.57	61	0.26
Cyclohexane	200	1.424	1.00	81	0.00
Dimethyl formamide	268	1.430	0.92	153	–
Dimethylsulfoxide	268	1.478	2.24	189	0.50
1,4-Dioxane	215	1.422	1.37	101	0.51
Ethyl acetate	256	1.372	0.45	77	0.48
Heptane	200	1.388	0.40	98	0.00
Hexane	195	1.375	0.31	69	0.00
Isooctane	215	1.391	0.50	99	0.00
Methanol	205	1.328	0.55	65	0.70
Methyl-*t*-buthyl ether	210	1.369	0.27	55	0.48
Methylethyl ketone	329	1.379	0.43	80	0.40
Methylene chloride	233	1.424	0.44	40	0.30
i-Propanol	205	1.377	2.40	82	0.60
n-Propanol	210	1.386	2.30	97	0.60
Tetrahydrofuran	212	1.407	0.55	66	0.53
Toluene	284	1.497	0.59	111	0.22
Water	190	1.333	1.00	100	

Source: Snyder, L. R., Kirkland, J. J., Dolan, J. W. *Introduction to Modern Liquid Chromatography*. 2010. Copyright Wiley-VCH Verlag GmbH & Co. KGaA. Reproduced with permission; *High-Purity Solvent Guide*, Burdick & Jackson Laboratories, Muskegon, MI, 1980; Riddick, J. A., Bunger, W. B.: *Organic Solvents*. 1970. Copyright Wiley-VCH Verlag GmbH & Co. KGaA. Reproduced with permission; Snyder, L. R. in *High-Performance Liquid Chromatography: Advances and Perspectives*, Vol. 3, Horväth, C., (ed.), Academic Press, New York, 1983, p. 157.

[a] Wavelength at which solvent absorbs 1.0 AU in a 10-mm cell. UV cutoff: for UV detection; useful solvents depend on wavelength required for sample detection.

[b] Refractive index: for RI detection; low values generally preferred.

[c] Viscosity: determines column pressure drop; low values of viscosity desirable.

[d] Boiling point. Affects pump performance and safety; higher boiling solvents preferred.

[e] Solvent strength parameter. Polarity: determines solvent strength for optimal k values ($1 \leq k \leq 10$).

14.4.2 HPLC-DAD (HPLC-PDA)

The most widely used detectors in HPLC today are the variable-wavelength UV and DADs or PDAs. A broad-spectrum UV lamp (typically a deuterium one) is directed through a slit onto a diffraction grating. The grating spreads the light out into its component wavelengths, and the grating is then rotated to direct a single wavelength (or narrow range of wavelengths) of light through the slit and detector cell and onto a photodetector [5]. For detection in the visible region, a tungsten lamp is used instead of a deuterium one.

The signals from the individual photodiodes (512 or 1024 diodes are common) are processed to generate a spectrum of the analyte. As the spectra are generated at the same time (vs. single-wavelength, monitoring with the variable-wavelength detector), the DAD can contribute to peak identification.

The DAD (PDA) type of detector has the following advantages:

- Can be operated to collect data at one or more wavelengths across a chromatogram
- Can be operated to collect spectra on one or more analytes in the run
- Possibility of peak-purity determination

Examples of application of HPLC coupled with DADs or PDAs to analysis of pesticides are summarized in Table 14.3 [24–45] and presented in Figure 14.1 [42].

Identification of the components of complex samples by comparison of their retention factors alone is decisively not sufficient owing to similar retention of various accompanying compounds in the sample. Therefore, it is necessary to confirm their presence in the sample by comparison of the spectra of the analytes with those in spectra libraries and determination of the peaks purity. DADs (PDAs) allow obtaining spectra of the analyte and purity of the analyte. Diode-array UV detectors allow simultaneous collection of chromatograms at the different wavelengths during a single run [$A = f(t, \lambda)$]. Therefore, DAD provides more information on sample composition than is provided by a single wavelength run. The UV spectrum of each separated peak is also obtained as an important tool for selecting an optimum wavelength to verify peak purity and peak identity. DADs can also be used to examine the chromatograms at different wavelengths, which enables group classification. In the environmental analysis, for example, pesticide residue analysis, the analytes can be identified on the basis of their retention times and by comparison between the UV spectrum of the reference compound in the library and the UV spectrum of the detected peak in the sample.

Some examples are illustrated in Figures 14.2 through 14.4 [43].

HPLC-DAD analysis of pesticides in wine samples was performed on a C18 column with the gradient elution program providing adequate separation for the 10 following pesticides: atrazine, buturon, terbuthylazine, prometryn, bitertanol, procymidone, hexaflumuron, lufenuron, flufenoxuron, and a-cypermethrin. The retention data for the solutes on C18 phases (Eclipse XDB-C18) with acetonitrile-water as the mobile phase is illustrated in Figure 14.2 [43]. The chromatogram (Figure 14.2) shows separation of 10 pesticides at concentrations of 1 μg mL^{-1} [43]. Figure 14.3 shows a chromatogram with fortified samples of wine at concentrations of pesticides equal to 200 ng mL^{-1} of wine. Almost all pesticides (without atrazine) are separated from the components of matrices. The methanol eluates after solid phase extraction (SPE) from wine samples were analyzed by HPLC-DAD (Figure 14.4) [43].

Pesticide identification was accomplished on the basis of the retention times of the analytes and by comparison between the UV spectra of the reference compounds in the library and the UV spectra of the detected peaks of the sample (Figure 14.5) [43]. A match equal to or higher than 990 is fixed to confirm identification between both spectra for the pesticides determined. If the peaks of an analyte are pure, then the surface area under the compared spectra of the standard and the analyte is light gray. If the peak of the analyte is contaminated, the surface area would be dark gray. Because these peaks are pure, the calculated surface areas of the compared peaks are light gray as illustrated in Figure 14.6 [43].

In HPLC-DAD experiments for quantitative analysis of pesticides in wine samples, the best fit for the calibration lines was found when the calibration data were analyzed using linear regression. The calibration plots were linear between 0.05 μg mL^{-1} and 9.6 μg mL^{-1} for analyzed pesticides, the correlation coefficients, *r*, were ≥0.9991 for almost all pesticides and 0.9976 for atrazine determined by HPLC-DAD. The LODs equal 0.03 μg mL^{-1} and 0.27 μg mL^{-1}, and the LOQs were 0.10 μg mL^{-1} and 0.81 μg mL^{-1} for terbuthylazine (the lowest values) and atrazine (the highest values), respectively. The quantities of the prometrin determined were in the ranges of 1.5–2 ng mL^{-1} of wine [43].

The use of a DAD detector is especially useful in analysis of samples with complicated matrices by obtaining UV spectra and evaluation of purity of peaks on the chromatograms. The matrix contains 1024 photodiodes, which corresponds to a nominal difference of 0.9 nm in the UV range. In the visual and near infrared range, the difference is somewhat greater. To correct this optical

TABLE 14.3
Examples of Application of HPLC-DAD in Pesticide Residue Analysis

Pesticides	Stationary Phase	Mobile Phase (v/v)	Type of Sample (Matrix), LOD, LOQ	Extraction Method	Recovery (%), (RSD or SD)	Refs.
OP: parathion-methyl, fenitrothion, parathion, chlorfenvinphos, diazinon, ethion, fenchlorphos, chlorpyrifos, carbophenothion	C18 column (RES-ELUT, 150 mm × 4.6 mm, 5 μm) connected to a C18 guard column (20 mm × 2 mm)	Gradient programmed linearly for 20 min: 65:35 (v/v) methanol:water (start) to 100% methanol (final)	Bovine tissues, 0.04–0.25 μg/g	MSPD-*on-line*-SPE	91%–101% (≤12%)	[24]
Chlorpyrifos, chlorfenvinphos, diazinon, fenitrothion, parathion-methyl	C18 column (RES-ELUT, 150 mm, 4.6 mm, 5 μm) C18 column guard (13 mm, 4.6 mm)	Gradient programmed linearly for 10 min: 70:30 (v/v) methanol:water (start) to 100% methanol (final)	Bovine tissues, muscle LODs < 0.1 μg/g	MSPD	>94%, except for chlorfenvinphos in liver (55%) RSD: 15% in liver and 11.5% in muscle	[25]
Fluoroquinolones (FQs), organophosphorus (OP), and N-methyl carbamates (NMCs)	C18 column (Kromasil, 150 mm × 4.6 mm, 5 μm)	Gradient with: Solvent A (0.003 mol/L H_3PO_4), solvent B (acetonitrile): 13%–20% B (5 min), 20%–38% B (2 min), 38% B (8 min), 38%–64% B (5 min), 64%–13% B (2 min).	Porcine tissue LODs: 9–22 g/kg	MSPD	60.1%–107.7%	[26]
Fenarimol	C18 column (Symmetry, 2.1 × 150 mm, 3.5 μm)	Acetonitrile–water (60:40, v/v)	Liver, kidney, and gastric content samples. LOD 20 ng/g (liver samples), LOQ 60 ng/g (liver samples)	SPE	–	[27]
Isocarbophos, phosmet, parathion, parathion-methyl, fenitrothion, fonofos, phoxim	C18 column (Centurysil, 4.6 mm × 250 mm, 5.0 μm)	Methanol–water (70:30, v/v)	Water, LODs 0.1–0.3 ng mL^{-1}	UASEME	85%–99.6% (3.1–5.5)	[28]

Metolcarb, carbofuran, carbaryl, pirimicarb, isoprocarb, diethofencarb	C18 column (Centurysil, 4.6 mm × 250 mm, 5.0 μm)	Methanol–water (60:40, v/v)	Water LODs 0.1–0.3 ng mL^{-1}, LOQs 0.3–0.9 ng mL^{-1}	UASEME	81.0%–97.5% (3.2%–4.8%)	[29]
Urea, benzoylurea pesticides (benzthiazuron, metoxuron, monuron, fluometuron, isoproturon, diuron, linuron, chloroxuron, chlorbromuron, diflubenzuron, neburon, triflumuron, flucycloxuron)	SB-C18 column (Zorbax (DuPont), 250 mm × 4.6 mm, 5 μm)	Gradient: 0 min to 5 min: acetonitrile–water (10:90, v/v), 35 min: 90% acetonitrile–water (90:10, v/v). Water with 85% orthophosphoric acid pH of 2.5 and acetonitrile were used as mobile phase.	Drinking water, LODs (μg/L): Benzthiazuron, chlorbromuron, chloroxuron 0.001 μg/L; Monuron, isoproturon 0.002 μg/L; Metoxuron 0.003 μg/L; Neburon 0.004 μg/L; Fluometuron, flucycloxuron 0.005 μg/L; Triflumuron, diflubenzuron 0.006 μg/L; Diuron 0.007 μg/L	LLE	69% to 127% (SD ≤ 19)	[30]
Carbamate esters (aldicarb sulfone, oxamyl, methomyl) and other pesticides	C18 column (LiChrospher, 150 mm × 4.6 mm, 5 μm)	Gradient with acetonitrile–water (pH 5.8)	River water LODs in the range 0.02–0.1 μg/L. LODs (aliphatic carbamate esters): Oxamyl 0.5 μg/L, Methomyl 0.5 μg/L, Aldicarb sulfone 1.0 μg/L	SPE	14%–102%	[31]
Chlorfenvinphos, chlorothalonil, fenamiphos, iprodione, malathion, parathion-ethyl, parathion-methyl, procymidone, tebuconazole, triadimefon, triazophos, vinclozolin	C18 column (Hypersil, 100 × 0.46 mm, 5 μm)	Acetonitrile–water (60:40, v/v)	Groundwater LODs in the range 0.92–2.8 (μg mL^{-1})	–	–	[32]
Diazinon, fenitrothion, malathion, fenvalerate, phosalone, tridemorph	C18 column (4.6 mm × 150 mm, 5 μm)	Acetonitrile–water–methanol (20:55:25, v/v/v)	Human hair, water samples. LODs: in the range 0.004–0.095 ng/mL (aqueous matrices); 0.003–0.080 ng/mL (hair matrices)	HF-LPME	–	[33]

(Continued)

TABLE 14.3 (CONTINUED)
Examples of Application of HPLC-DAD in Pesticide Residue Analysis

Pesticides	Stationary Phase	Mobile Phase (v/v)	Type of Sample (Matrix), LOD, LOQ	Extraction Method	Recovery (%), (RSD or SD)	Refs.
Atrazine, bentazone, carbetamide, chloropropham, chlorothalonil, chlortoluron, desethylatrazine, desisopropylatrazine, dinoterb, diuron, ethofumesate, flusilazol, isoproturon, linuron, methabenzthiazuron, neburon, phenmedipham, prometryne, propyzamide, simazine, terbuthylazine, terbutryn	C18 polymer column (TSK ODS 80, 250 × 4.6 mm, 5 μm)	Gradient (acetonitrile–water): from 80:20 (v/v) at 0 min to 57:43 (v/v) at 30 min, 45:55 (v/v) at 55 min, 20:/80 (v/v) at 65 min and 100% of acetonitrile between 75 and 80 min.	River waters, LODs in the range 20–90 (ng L^{-1})	–	–	[34]
OP: isocarbophos, phosmet, parathion-methyl, triazophos, fonofos, phoxim	C18 column (Centurysil, 4.6 mm × 250 mm, 5.0 μm)	Methanol–water (70:30, v/v)	Water and watermelon, LODs in the range: 0.1–0.3 ng/mL (water), 1.0–1.5 ng/g (watermelon)	LLME, HF-MMSLPE MWCNT	Recoveries: 85.4%–100.8% (water samples at spiking levels 5.0 and 50.0 ng/mL), 82.6%–92.4% (watermelon samples at spiking levels of 5.0 and 50.0 ng/g)	[35]
Neonicotinoid pesticides	C18 column (Atlantis, 4.6 mm × 150 mm, 5 μm)	Acetonitrile–water (25:75, v/v)	Fruit juice and water samples. LODs 0.1–0.5 μg/L	VSLLME-SFO	85%–105%	[36]

Azoxystrobin, boscalid, bromopropylate, chlorfenapyr, chlorfluazuron, chromafenozide, cyazofamid, cyflufenamid, diethofencarb, diflubenzuron, etofenprox, famoxadone, fipronil, flubendiamide, flufenoxuron, hexythiazox, iprodione, isoxathion, kresoxim-methyl, lufenuron, myclobutanil, phenthoate, pyraclofos, pyridaben, pylidalyl, pyriproxyfen, teflubenzuron, trifloxystrobin	C18 column (SunFire, 250 mm × 4.6 mm, 5 μm); C18 guard column (SunFire, 20 mm × 4.6 mm, 5 μm).	Acetonitrile–water (70:30, v/v)	Tomatoes, green peppers, spinaches	LODs: 5 ng/mL (azoxystrobin, cyazofamid, diethofencarb, etofenprox, famoxadone, iprodione, isoxathion, pyraclofos, pyridaben, teflubenzuron); 10 ng/mL boscalid, bromopropylate, chromafenozide, cyflufenamid, hexythiazox, lufenuron, phenthoate, pylidalyl, pyriproxyfen, trifloxystrobin); 15 ng/mL (diflubenzuron, flubendiamide); 20 ng/mL (chlorfenapyr, chlorfluazuron, flufenoxuron, kresoxim-methyl, myclobutanil, fipronil)	70%–120%	[37]

(Continued)

TABLE 14.3 (CONTINUED)
Examples of Application of HPLC-DAD in Pesticide Residue Analysis

Pesticides	Stationary Phase	Mobile Phase (v/v)	Type of Sample (Matrix), LOD, LOQ	Extraction Method	Recovery (%), (RSD or SD)	Refs.
Folpet, chlorothalonil, quinomethionat, tetradifon, trifluralin	C18 column (Luna 250 mm × 4.6 mm, 5 μm) C18 guard column (4 mm × 3 mm)	Gradient of acetonitrile and water from 40% of acetonitrile to 90% in 20 min	Fruit juices, 0.5–1.0 (μg/kg)	SPE LODs: 0.5 μg/kg (folpet chlorothalonil quinomethionat); 1.0 μg/kg (tetradifon trifluralin)	93.8%–99.5%	[38]
Acetamiprid, azoxystrobin, cyprodinil, fenehexamid, fludioxonil, folpet, iprodione, metalaxyl, pirimicarb, tolyfluanid, phosmet	C18 column (Ultracarb ODS, Phenomenex, 250 mm × 4.6 mm, 5 μm)	Solvent A (0.1% of trifluoroacetic acid in water); Solvent B (methanol); The linear gradient program: 0–7.5 min, 20%–60% B in A, 7.5–11.5 min, 60% of B in A; 11.5–16.5 min, 60%–69% of B in A 16.5–25 min, 69%–85% B in A (25–45 min, column rinse and re-equilibration)	Lettuce (*Lactuca sativa*) 0.28–1.54 (μg/g)	SPME LODs (LOQs): Azoxystrobin 0.48 mg/kg (1.59 mg/kg); Cyprodinil 0.28 mg/kg (0.94 mg/kg); Fenhexamid 0.73 mg/kg (2.45 mg/kg); Fludioxonil 0.37 mg/kg (1.24 mg/kg); Folpet 0.47 mg/kg (1.57 mg/kg); Iprodione 1.54 mg/kg (5.14 mg/kg); Tolyfluanid 1.53 mg/kg (5.10 mg/kg)	97.1% ± 7.5%	[39]

Dithiocarbamate fungicides	C18 column (Kinetex, 50 × 4.6 mm, 2.6 μm), C18 guard column (50 × 4.6 mm)	Gradient conditions: start with water–acetonitrile (95:5, v/v) to 9 min water–acetonitrile (10:90, v/v)	Fruits and vegetables: apples, wine grapes, lettuces, peppers, tomatoes, strawberries. Quantification is based on external standard calibration curves made with dithiocarbamates-spiked blank-matrices (in mg dithiocarbamate/kg peel). LODs (LOQs) fruits: 0.11–0.31 (0.16–0.54) vegetables: 0.01–0.04 (0.02–0.12)	SLE, SPE	Mancozeb: 74%–139%, Maneb: 78%–115%, Propineb: 99%–128%	[40]
Thiophanate-methyl, carbofuran, carbaryl, tebuconazole, iprodione, oxyfluorfen, hexythiazox, fenazaquin		100% Milli-Q water (as mobile phase A) and 100% ACN (as mobile phase B)	Bananas LODs: Thiophanate-methyl 2.20 μg/L carbofuran 3.08 μg/L carbaryl 0.250 μg/L tebuconazole 3.86 μg/L	DDLME	69%–97% (except for thiophanate-methyl and carbofuran, which were 53%–63%) RSD ≤ 8.7%	[41]
Propoxur, carbaryl, carbendazim, thiabendazole, fuberidazole	C18 column (Agilent Sorbax SB, 150 mm × 4.6 mm, 5 μm)	Isocratic: water–methanol (50:50, v/v)	Tangerine, lemon, tomato, orange, and grapefruit juices. LODs: carbendazim 2.3 μg L^{-1}, thiabendazole 0.90 μg L^{-1}, propoxur 12 μg L^{-1}, fuberidazole 0.46 μg L^{-1}, carbaryl 0.32 μg L^{-1}	–	–	[42]
Atrazine, buturon, terbuthylazine, prometryn, bitertanol, procymidone, hexaflumuron, lufenuron, flufenoxuron, α-cypermethrin	XDB-C18 (ZORBAX Eclipse 150 mm × 4.6 mm column, 5 μm)	The gradients applied were, sequence: 30% B, 30 min linear to 76% B, 35 min linear to 100% B, and 35–45 min isocratic 100% B (A–H_2O, B–acetonitrile)	Wines, LODs and LOQs were in the ranges, respectively: 0.03 μg/mL–0.27 μg/mL and 0.10 μg/mL–0.81 μg/mL	SPE	69%–119%	[43]
Simazine, isoproturon, terbuthylazine, linuron, captan, terbutryn, procymidone, fenitrothion, clofentezine, bromopropylate	XDB-C18 (ZORBAX Eclipse 150 mm × 4.6 mm column, 5 μm)	The gradients applied were, sequence: 30% B, 30 min linear to 76% B, 35 min linear to 100% B, and 35–45 min isocratic 100% B (A–H_2O, B–acetonitrile)	Sunflower Seeds, LODs and LOQs were in the ranges, respectively: 0.01 μg/mL–0.57 μg/mL and 0.04 μg/mL–1.73 μg/mL	Four procedures: UAE/SPE (A), d-SPE (B), UAE/d-SPE (C), UAE/SPE/d-SPE (D)	Recoveries (A-D, four procedures) in the ranges: 9%–101% (A), 39%–75% (B), 8.5%–18.7% (C), 34%–107% (D)	[44]

(Continued)

TABLE 14.3 (CONTINUED)
Examples of Application of HPLC-DAD in Pesticide Residue Analysis

Pesticides	Stationary Phase	Mobile Phase (v/v)	Type of Sample (Matrix), LOD, LOQ	Extraction Method	Recovery (%), (RSD or SD)	Refs.
Fenuron, methabenzthiazuron, isoproturon, terbutryn, procymidone, fenitrothion, neburon, chlorfenvinphos, lufenuron, flufenoxuron, trifluralin, α-cypermethrin, monuron, fluometuron, dimethomorph, linuron, clofentezine, propazine, propachlor, terbuthylazine, bromopropylate	XDB-C18 (ZORBAX Eclipse 150 mm × 4.6 mm column, 5 μm)	The gradients applied were, sequence: 30% B, 30 min linear to 76% B, 35 min linear to 100% B, and 35–45 min isocratic 100% B (A–H_2O, B–acetonitrile)	Edible oils	SPE/QuEChERS (d-SPE with Z-Sep)	Recovery studies were performed at 75 ng g^{-1}, 125 ng g^{-1}, 250 ng g^{-1}, 500 ng g^{-1} and 1000 ng g^{-1} levels, yielding recovery rates in the range of 50%–130% for most of the analytes; RSD ≤ 15%	[45]

Note: DLLME, dispersive liquid–liquid microextraction; d-SPE, dispersive solid-phase extraction; HF-LPME, hollow fiber-based liquid phase microextraction; HF-MMSLPE, hollow fiber microporous membrane solid–liquid phase microextraction; LLE, liquid–liquid extraction; LLME, liquid–liquid microextraction; LOD, limit of detection; LOQ, limit of quantification; MSPD, matrix solid-phase dispersion; MWCNT, multiwalled carbon nanotube; OP, organophosphorus pesticide; QuEChERS, quick, easy, cheap, effective, rugged, and safe extraction method; RSD, relative standard deviations; SD, standard deviations; SLE, solid–liquid extraction; SPE, solid phase extraction; SPME, solid-phase microextraction; UASEME, ultrasound-assisted surfactant-enhanced emulsification microextraction; VSLLME-SFO, vortex-assisted surfactant-enhanced-emulsification liquid–liquid microextraction with solidification of floating organic droplet; Z-Sep, ZrO_2-based sorbent.

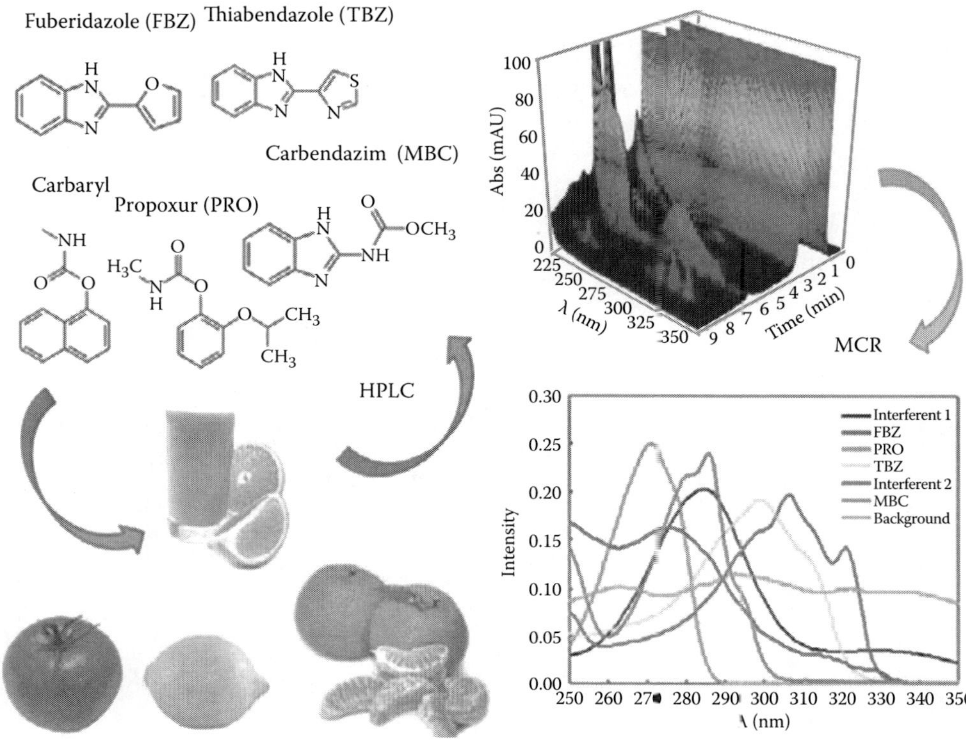

FIGURE 14.1 Five pesticides were determined in juice, fruit, and vegetable samples. Liquid chromatography was coupled to diode array detection. Chromatographic-spectral matrices were analyzed by multivariate curve resolution. (From Boeris, V. et al., *Analytica Chimica Acta*, 814, 23–30, 2014. With permission.)

nonlinearity and transform the discrete diode distances into a linear scale, a linear interpolation algorithm is applied, which utilizes a calibration table of wavelengths and real wavelength values obtained from emission lines of a deuterium lamp. The determination of peak purity is carried out using an interpolation algorithm, which takes into account a calibration table of wavelengths from the emission lines of the deuterium lamp at 486 and 656 nm.

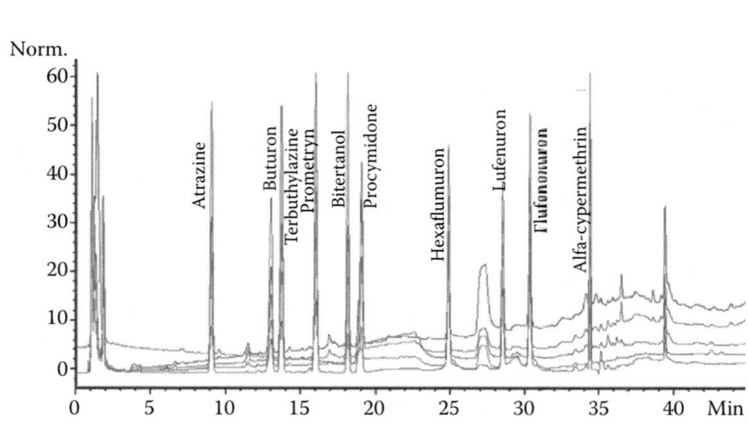

FIGURE 14.2 Chromatogram obtained by HPLC-DAD after SPE method from separated mixture of pesticides at a concentration level of 1 μg mL^{-1}. (From Tuzimski, T., *Journal of Liquid Chromatography and Related Technologies*, 35, 1415–1428, 2012. With permission.)

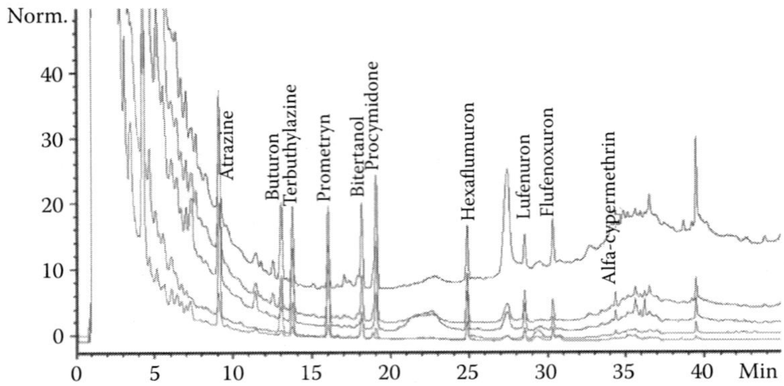

FIGURE 14.3 Chromatogram obtained by HPLC-DAD after SPE method from the spiked wine sample with a mixture of pesticides at a concentration level of 200 ng mL^{-1} of wine. (From Tuzimski, T., *Journal of Liquid Chromatography and Related Technologies*, 35, 1415–1428, 2012. With permission.)

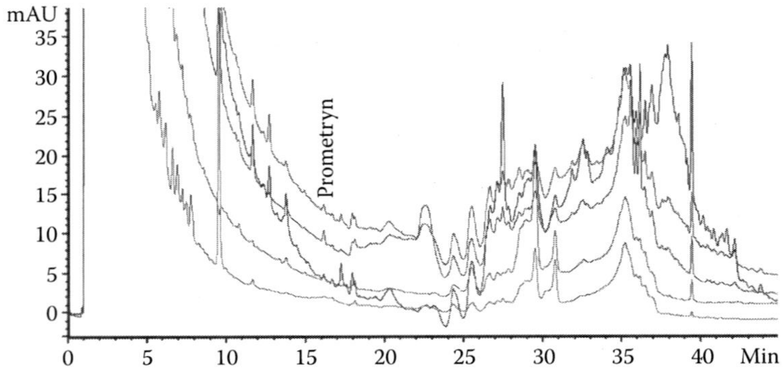

FIGURE 14.4 Chromatogram obtained by HPLC-DAD after SPE method from real wine sample with determination pesticide (prometryn). (From Tuzimski, T., *Journal of Liquid Chromatography and Related Technologies*, 35, 1415–1428, 2012. With permission.)

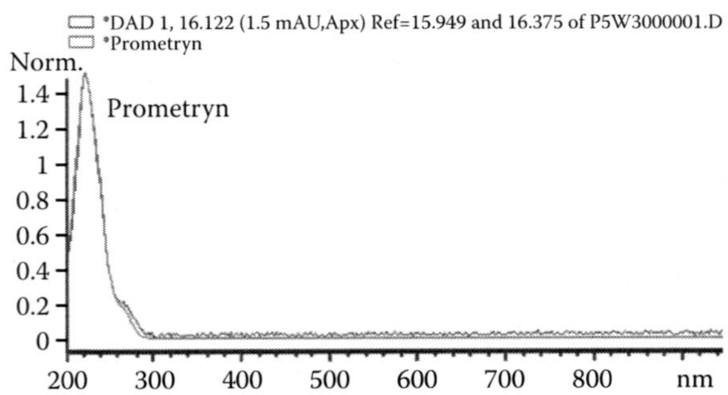

FIGURE 14.5 Comparisons of the UV spectrum of standard of prometryn (library) and spectrum found in wine samples by HPLC-DAD experiments after SPE on C18/SDB-1 cartridges. (From Tuzimski, T., *Journal of Liquid Chromatography and Related Technologies*, 35, 1415–1428, 2012. With permission.)

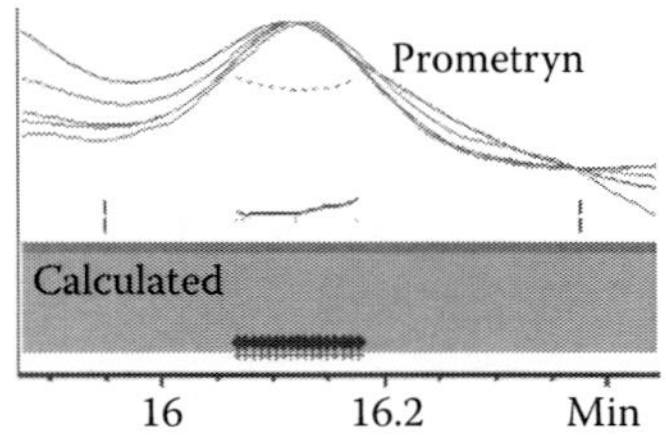

FIGURE 14.6 Purity of HPLC peak obtained for prometryn in wine sample after SPE on C18/SDB-1 cartridges. (From Tuzimski, T., *Journal of Liquid Chromatography and Related Technologies*, 35, 1415–1428, 2012. With permission.)

Application of HPLC-DAD can be useful for correct identification of pesticides in complicated mixtures and separation of analytes from the components of the matrix with a high content of lipids, for example, in sunflower seed samples [44]. Tuzimski and Rejczak [45] also described methodology relying on successive extractions of SPE/QuEChERS for the analysis of pesticide residues in grapeseed oil and extra virgin olive oils by HPLC-DAD. In the paper [45], the authors proposed the methodology of quantitative analysis of pesticides in food products (grapeseed oil and extra virgin olive oils) by HPLC-DAD after a SPE/QuEChERS procedure for their extraction. In the experiments, authors also evaluated the zirconium dioxide-based sorbent in the dispersive (d-SPE) step of the QuEChERS technique to decrease the matrix effect from samples with a high fat content [45]. Analysis of pesticide residues in food, especially containing complex matrices with a high content of fats, is a continuous challenge for analytical chemists. The proposed method allows obtaining the best optimal results for the correct identification and quantitative analysis of pesticide residues in food samples. Owing to the d-SPE cleanup step of samples and the use of the Z-Sep sorbent in the elaborated procedure, high recovery values for most pesticides were obtained [45]. The recovery values for 76% and 67% of studied pesticides for extra virgin olive oil samples and grapeseed oil samples ranged from 50% to 130%, in agreement with requirements of the European Union [Document (EU) N° SANCO/12571/2013] [45].

As illustrated in Figure 14.7 [45], the use of the zirconium dioxide-based sorbent (Z-Sep) enables purification of the sample and getting rid of interfering compounds included in the matrix. As a result, one obtains additional confirmation of the presence of pesticides in samples without adversely affecting the natural matrix components and a sufficiently high purity peak (Figure 14.8) [45].

Details on modes of extraction procedures with QuEChERS methodology applied to identification and quantitative analysis of pesticides can be found in Chapters 5 and 12.

Tuzimski confirmed the presence of analytes in natural samples by evaluation of purities of peaks of identified pesticides in both chromatographic methods: HPLC-DAD and thin-layer chromatography (TLC)-DAD [46–50]. In the HPLC-DAD technique, a match equal to or higher than 990 was fixed to confirm identification between both spectra for all the pesticides determined and purity of peaks. A TLC-DAD scanner allows scanning of the TLC plate along the track with simultaneous registration of the spectra in the wavelength range of UV-VIS (from 191 nm to 1033 nm) and allows evaluation of the purity of the peaks of the analytes. The peak purity index is a numerical index (in the range from 0 to 1) for the quality of the coincidence between two data sets. It is given by the least squares fit coefficient calculated for all intensity pairs in the two data sets under consideration. A peak purity index of 1 indicates that the compared spectra are identical, and the peaks of chromatographed compounds are pure. The analyte identification in water samples was accomplished by comparing pairs of spectra: the UV spectrum of the analyte and the UV spectrum of the reference compound. Purities indexes of identified pesticides were close to 1 [46–50].

Tuzimski described optimization of conditions for extraction of pesticides and their identification and quantification in environmental samples (samples of water from Łęczyńsko-Włodawskie Lake District [46–49] and Zemborzycki Reservoir [50]) by SPE and HPLC-DAD and TLC-DAD methods. Quantitative analysis of pesticides was obtained at optimal wavelength for each analyte on

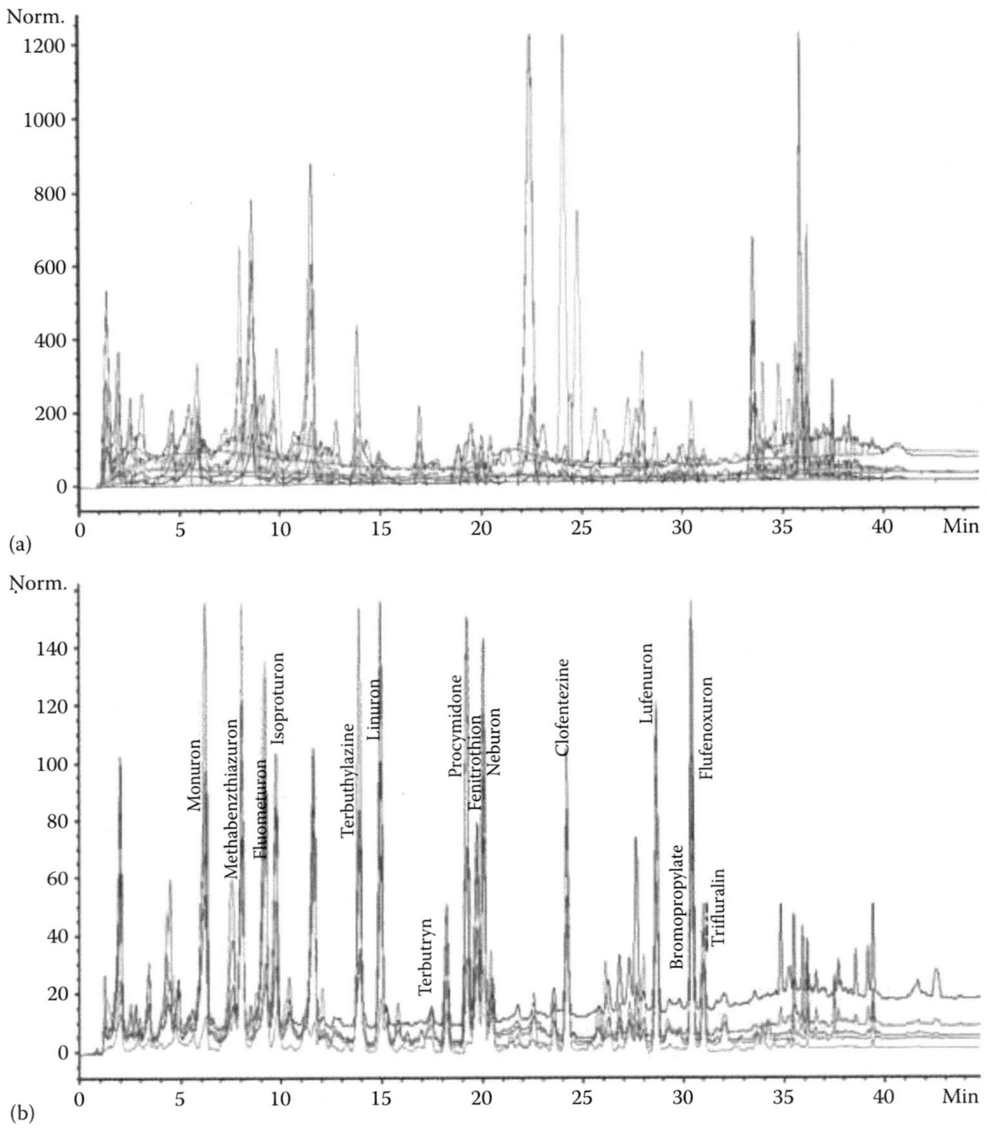

FIGURE 14.7 Chromatograms obtained by RP-HPLC-DAD following the SPE/QuEChERS method from spiked olive oil sample with a mixture of 17 pesticides at a concentration level of 1000 ng/g: (a) only after SPE step; (b) with Z-Sep sorbent for d-SPE cleanup step (propachlor and chlorfenvinphos were not detected). (From Tuzimski, T., Rejczak, T., Application of HPLC-DAD after SPE/QuEChERS with ZrO_2-based sorbent in d-SPE clean-up step for analysis of pesticides in edible oils, *Food Chemistry*, in press, 2015. With permission.)

the basis of calibration lines in both applied chromatographic techniques (HPLC-DAD and TLC-DAD). The LOD for most pesticides ranged from 0.01 to 0.46 μg/mL (HPLC-DAD) and from 0.04 to 0.65 μg/spot (TLC-DAD) [46–50]. The author compared values of recoveries for pesticides from different classes during SPE experiments by application of different types of sorbents and eluents. Four types of sorbents with different polarity—C18/SDB-1, C18, C18 Polar Plus, and CN—were used in SPE experiments. The samples of water (500 mL) were spiked with pesticides at a concentration level of 10 μg/L, which is characteristic of most environmental samples. The method of recovery was evaluated on the same type of SPE sorbent (on the same adsorbent) but with application of three different eluent solvents [methanol (MeOH), acetonitrile (MeCN), and tetrahydrofuran

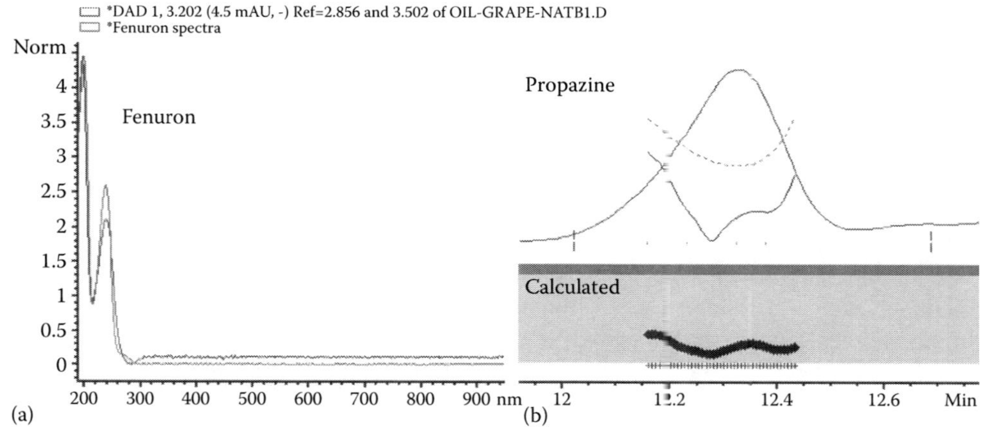

FIGURE 14.8 Examples of (a) comparison of the UV-VIS spectrum of standard of fenuron (library) and spectrum found in grapeseed oil sample by RP-HPLC-DAD experiments after SPE/QuEChERS (step B); (b) purity of propazine peak found in grapeseed oil sample analyzed by RP-HPLC-DAD method after the SPE/QuEChERS procedure (step C). (From Tuzimski, T., Rejczak, T., Application of HPLC-DAD after SPE/QuEChERS with ZrO_2-based sorbent in d-SPE clean-up step for analysis of pesticides in edible oils, *Food Chemistry*, in press, 2015. With permission.)

(THF)] [50]. Sorption of the modifier increases in the order MeOH, MeCN, and THF. Application of solvents with different properties, such as dipolar, proton donor, and proton acceptor, from different groups of the Snyder solvent selectivity triangle allows easy desorption of analytes from different classes of pesticides in small volumes of organic solvents. Recoveries from spiked samples were diverse and depended on the type of SPE sorbents applied and the type of solvents used for elution of target analytes. Recoveries for almost all pesticides were in the range of 60% to 115% [50].

In other publications [51–53], the same author optimized the conditions of extraction [ultrasound-assisted solvent extraction (UE or UAE) and SPE] and implemented them together with multidimensional planar chromatography in combination with scanning DAD (MDPC-DAD) and HPLC-DAD for quantitative analysis of pesticides in medical plant material and medical herbs (*Thymus vulgaris* L., *Melissa officinalis* L.). Heart-cut bands of the analyte and standard from the stationary phase (after elution in small volumes of solvents) were analyzed by HPLC-DAD on an octadecyl column (C18) in an aqueous system. The pure band of clofentezine was separated from the remaining components of the matrix. The spectra of detected pesticides in the extract and standard were ideally well fitting. Application of normal-phase (NP) systems in the stage part of the experiment of MDPC and RP systems in HPLC, in both steps with DAD detectors, permits more credible identification of analytes in samples of complicated natural origin [51]. In the other modes of the experiments [51], after optimization of extraction (UAE and SPE), the procedure was optimized for identification of pesticides, and then it was applied to identification and quantitative analysis of pesticides in *Thymus vulgaris* L. samples by MDPC in combination with diode-array scanning densitometry (MDPC-DAD) with NP and RP systems on dual-adsorbent plates with silica and octadecyl silica layers (Multi K SC5 and Multi K CS5 plates). Application of NP and RP systems in the two steps of MDPC-DAD permits separation of a pure spot of the detected pesticide, for example, clofentezine from remaining components of the matrix. Application of dual-adsorbent plates also permits concentration of the analyte and the disturbing components of the extract on the border between the two layers of adsorbents. Samples of *Thymus vulgaris* L. extract were spiked by clofentezine (at a concentration of 2.5 μg/g and 5 μg/g in the plant material), and recovery values were evaluated for different SPE sorbents and different elution solvents. Average recoveries from the spiked samples and the standard deviations (SD), were 80.1 ± 9.5% (SPE: C18, MeOH), 100.5 ± 9.5% (SPE: C18 Polar Plus, MeOH) at a concentration of 2.5 μg/g in plant material and 95.1 ± 8.5% (SPE: C18, MeCN) 5 μg/g in plant material (at $\lambda = 202$ nm) [51].

The SPE procedure on C18/SDB-1 cartridges (C18 500 mg on top + SDB copolymer 200 mg on bottom/6 mL) and elution with two solvents (with 5 mL MeOH and the next with 5 mL THF) was optimized for the identification of pesticides and was applied for identification and quantitative analysis of clofentezine in *Thymus vulgaris* L. samples by MDPC-DAD [52]. The samples of *Thymus vulgaris* L. extract were spiked by clofentezine at a concentration of 4.5 μg/g, 6 μg/g, 9 μg/g, and 12 μg/g in the plant material. Average recoveries from the spiked samples and the SDs were 55.8% ± 4.5 and 44.5% ± 6.5 (SPE: C18/SDB-1, THF eluates) by MDPC-DAD and HPLC-DAD, respectively. Application of elution with two solvents, in step-by-step mode, permits fractionation of samples and separation of clofentezine from the remaining components of the complicated matrix. The methanol eluates contained traces of clofentezine (<0.09%). SPE was used for partial purification of the sample (in the first step) and to elute the pesticides and their concentration and isolation from the remaining components of the complicated matrix (in the second step) [52].

Application of two-dimensional high-performance planar chromatography–diode array detector (2-D-HPTLC-DAD) and HPLC-DAD after UAE and SPE for identification and quantitative analysis of analytes in *Melissa officinalis* L. extracts was also described [53]. Dual-adsorbent Multi-K CS5 plates were developed with a RP system on octadecyl silica in the first direction and a NP system on silica in the second direction of 2-D experiments. 2-D-HPTLC-DAD was used to separate a seven-component mixture of pesticides (isoproturon, aziprotryne, hexazinone, flufenoxuron, methabenzthiazuron, procymidone, and α-cypermethrin) from other components of the sample (matrix). The Multi-K CS5 plates were scanned in the wavelength range of 200–400 nm. The purity indexes for compared spectra were always between 0.9911 and 0.9997 (the spots of analyte were pure). Heart-cut bands of analytes from the stationary phase (after 2-D-HPTLC-DAD experiments) were also injected on an octadecyl column (C18) and analyzed by RP-HPLC-DAD. The HPLC chromatograms obtained from the extract of *Melissa officinalis* L. after 2-D-HPTLC-DAD show that the purities of peaks of analytes separated from components of the matrix and that the compared spectra of the pairs—analyte and standard—are quite identical. The proposed procedure with application of 2-D-HPTLC-DAD and HPLC-DAD after SPE was proved correct for *Melissa officinalis* L.-spiked samples with seven pesticides at a concentration level of 10 μg/g in plant material after 1, 5, and 11 days. The method was characterized by good reproducibility. The number of examined samples of *Melissa officinalis* L. (*Labiatae*) was equal to 150 [53].

Afterward, SPE, HPLC-DAD, and TLC-DAD methods were applied for the determination of pesticides in wine samples [39]. SPE was used for isolation and concentration of pesticides in samples of five red wines from five countries: the Unites States, Bulgaria, Chile, Hungary, and France. In this series of SPE experiments, sorbents C18, C18 Polar Plus, and C18/SDB-1 were used. The quantities of the prometryn determined in the wine samples were in the range of 1.5–2.0 ng/mL of wine [39].

Application of LC connected with DAD is a cheaper solution than the use of LC-MS or LC-MS/MS. However, UV detection is limited to the compounds having chromophore groups (e.g., aromatic rings), and it is not suitable for the compounds that do not absorb in the UV range. The HPLC method is hardly suited for the structure elucidation of the unknown, even if PDAs are used. More structural information is available if HPLC is coupled with MS or MS/MS, which is now widely used for the analysis of pesticides in natural samples.

14.4.3 HPLC Coupled with Light-Scattering Detectors

The popular modes of light-scattering detectors are the following:

- Evaporative light-scattering detector (ELSD)
- Condensation nucleation light-scattering detector (CNLSD)
- Laser light-scattering detector (LLSD)

These detectors are universal detectors, which are very rarely applied in analysis of pesticides.

14.4.4 HPLC Coupled with Electrochemical Detectors

Electrochemical detectors are known for their high selectivity and sensitivity. In pesticide residue analysis, different types of these detectors [54,55] can be applied. The possibility of the amperometric detection of a number of pesticides, such as benomyl, thiram, linuron, metoxuron, desmedipham, dicuron, lenacil, and fludioxonil, widely used in agrochemical practice was described [54]. A microwave-assisted extraction (MAE) method was optimized (temperature, extraction time, and solvent volume) for determination of the herbicide simazine and the fungicide cymoxanil in soil samples by HPLC with reductive amperometrical detection [55]. HPLC coupled with electrochemical detectors is very rarely applied in the analysis of pesticides.

14.4.5 HPLC Coupled with Other Detectors

Fluorescence detectors are very sensitive and selective for analytes that fluoresce when excited by UV radiation. Sample components that do not fluoresce do not produce a detector signal, so sample cleanup may be simplified [5]. When a compound fluoresces, the desired emission wavelength is isolated with a filter or monochromator and directed to a photodetector, by which it is monitored and converted to an electronic signal for data processing. As fluorescence is emitted in all directions, it is common to monitor the emitted light at right angles to the incident light; this is simplified by the optics and reduced background noise. For many samples, the fluorescence detector is a hundredfold more sensitive than UV absorption and is one of the most sensitive HPLC detectors [5]. In pesticide residue analysis, it is rarely applied. HPLC with fluorescence detection (HPLC-FLD) has been developed for determination of pesticides in soil samples [56] and in water and fruit juice samples [57].

The corona-discharge detector, also called the charged-aerosol detector (CAD), is classified as a universal detector, which is sensitive to nearly all compounds that are sufficiently less volatile than the mobile phase, so they remain in the gas phase after the mobile phase is evaporated. In this CAD detector, as with other evaporative detectors, the mobile phase is restricted to volatile components (e.g., no phosphate buffer). It also requires particles that can be charged in the CAD detector. So far, this type of detector has not been used in the analysis of pesticides.

HPLC-MS and HPLC-MS/MS are powerful techniques because they typically provide highly precise quantitative results, very low detection limits, excellent instrument ruggedness, and an exceptional degree of selectivity for qualitative identification of target analytes.

However, in addition to high cost, HPLC-MS/MS has at least two other significant limitations:

- Unknown, nontargeted chemicals (analytes) are not readily detected or able to be identified.
- Ion suppression or an enhancement effect due to co-eluting matrix components can adversely affect the reality of quantitative results.

In addition, co-eluting matrix components can affect ionization efficiency through either signal suppression or signal enhancement. Methods to reduce the effects of the matrix on the signal include the use of

- Adjusting of HPLC or MS conditions
- Dilution
- Reduce or eliminate the co-eluting interferences by greater cleanup
- Isotopically labeled internal standards
- Matrix-matched calibration
- Continuous infusion of a marker compound
- The echo-peak technique

In Chapter 15 of this book, readers will gain useful information on avoiding some problems concerned with various modes of HPLC combined with MS experiments. Also, other details concerning application of these techniques for separation, detection, qualitative investigation of structures of analytes and quantitative determination of pesticides can be found in Chapter 15.

REFERENCES

1. Stalikas, C. D., Pilidis, G. A., Development of a method for the simultaneous determination of phosphoric and amino acid group containing pesticides by gas chromatography with mass-selective detection. Optimization of the derivatization procedure using an experimental design approach, *J. Chromatogr. A*, 872, 215–225, 2000.
2. Gerecke, A. C., Tixier, C., Bartels, T., Schwarzenbach, R. P., Müller, S. R., Determination of phenylurea herbicides in natural waters at 21 concentrations below 1 ng l^{-1} using solid-phase extraction, derivatization, and solid-phase microextraction–gas chromatography–mass spectrometry, *J. Chromatogr. A*, 930, 9–19, 2001.
3. Guo, L., Lee H. K., Low-density solvent based ultrasound-assisted emulsification microextraction and on-column derivatization combined with gas chromatography–mass spectrometry for the determination of carbamate pesticides in environmental water samples, *J. Chromatogr. A*, 1235 1–9, 2012.
4. Zhang, J., Lee H. K., Application of liquid-phase microextraction and on-column derivatization combined with gas chromatography–mass spectrometry to the determination of carbamate pesticides, *J. Chromatogr. A*, 1117, 31–37, 2006.
5. Snyder, L. R., Kirkland, J. J., Dolan, J. W., *Introduction to Modern Liquid Chromatography*, John Wiley & Sons, Inc., Hoboken, NJ, 2010.
6. Dong, M. W., *Modern HPLC for Practicing Scientists*, Wiley-Interscience, Hoboken, NJ, 2006.
7. Li, B. J., Signal-to-noise optimization in HPLC UV detection, *LCGC* 10, 856, 1992.
8. *High-Purity Solvent Guide*, Burdick & Jackson Laboratories, Muskegon, MI, 1980.
9. Riddick, J. A., Bunger, W. B., *Organic Solvents*, Wiley-Interscience, New York, 1970.
10. Snyder, L. R. in *High-Performance Liquid Chromatography: Advances and Perspectives*, Vol. 3, Horväth, C., (ed.), Academic Press, New York, 1983, p. 157.
11. Di Corcia, A., Marchetti, M., Multiresidue method for pesticides in drinking water using a graphitized carbon black cartridge extraction and liquid chromatographic analysis, *Anal. Chem.*, 63, 580–585, 1991.
12. Tran, A. T. K., Hyne, R. V., Doble, P., Determination of commonly used polar herbicides in agricultural drainage waters in Australia by HPLC, *Chemosphere*, 67, 944–953, 2007.
13. Heidari, H., Razmi, H., Multi-response optimization of magnetic solid phase extraction based on carbon coated Fe_3O_4 nanoparticles using desirability function approach for the determination of the organophosphorus pesticides in aquatic samples by HPLC–UV, *Talanta*, 99, 13–21, 2012.
14. Buonasera, K., D'Orazio, G., Fanali, S., Dugo, P., Mondello, L., Separation of organophosphorus pesticides by using nano-liquid chromatography. *J. Chromatogr. A*, 1216, 3970–3976, 2009.
15. Beale, D. J., Kaserzon, S. L., Porter, N. A., Roddick, F. A., Carpenter, P. D., Detection of s-triazine pesticides in natural waters by modified large-volume direct injection HPLC, *Talanta*, 82, 668–674, 2010.
16. Zhou, Q., Gao, Y., Bai, H., Xie, G., Preconcentration sensitive determination of pyrethroid insecticides in environmental water samples with solid phase extraction with SiO_2 microspheres cartridge prior to high performance liquid chromatography, *J. Chromatogr. A*, 1217, 5021–5025, 2010.
17. Qian, K., He, S., Tang, T., Shi, T., Li, J., Cao, Y., A rapid liquid chromatography method for determination of glufosinate residue in maize after derivatisation, *Food Chem.*, 127, 722–726, 2011.
18. Bratkowska, D., Fontanals, N., Borrull, F., Cormack, P. A. G., Sherrington, D. C., Marcé, R. M., Hydrophilic hypercrosslinked polymeric sorbents for the solid-phase extraction of polar contaminants from water, *J. Chromatogr. A*, 1217, 3238–3243, 2010.
19. He, L., Lou, X., Jiang, X., Qu, L., A new 1,3-dibutylimidazolium hexafluorophosphate ionic liquid-based dispersive liquid–liquid microextraction to determine organophosphorus pesticides in water and fruit samples by high-performance liquid chromatography, *J. Chromatogr. A*, 1217, 5013–5020, 2010.
20. He, L., Lou, X., Xie, H., Wang, C., Jiang, X., Lu, K., Ionic liquid-based dispersive liquid–liquid microextraction followed high-performance liquid chromatography for the determination of organophosphorus pesticides in water samples, *Anal. Chim. Acta*, 655, 52–59, 2009.
21. Basheer, C., Alnedhary, A. A., Rao, B. S. M., Lee, H. K., Determination of carbamate pesticides using micro-solid-phase extraction combined with high-performance liquid chromatography, *J. Chromatogr. A*, 1216, 211–216, 2009.

22. Nesterenko, E. P., Nesterenko, P. N., Connolly, D., Lacroix, F., Paull, B., Micro-bore titanium housed polymer monoliths for reversed-phase liquid chromatography of small molecules, *J. Chromatogr. A*, 1217, 2138–2146, 2010.
23. Tuzimski, T., Thin-layer chromatography (TLC) as pilot technique for HPLC. Utilization of retention database (R_F) vs. eluent composition of pesticides. *Chromatographia*, 56, 379–381, 2002.
24. Valencia, T. M. G., Llasera, M. P. G., Determination of organophosphorus pesticides in bovine tissue by an on-line coupled matrix solid-phase dispersion–solid phase extraction–high performance liquid chromatography with diode array detection method, *J. Chromatogr. A*, 1218, 6869–6877, 2011.
25. Llasera, M. P. G., Reyes-Reyes, M. L., A validated matrix solid-phase dispersion method for the extraction of organophosphorus pesticides from bovine samples, *Food Chem.*, 114, 1510–1516, 2009.
26. Wang, S., Mu, H., Bai, Y., Zhang, Y., Liu, H., Multiresidue determination of fluoroquinolones, organophosphorus and N-methyl carbamates simultaneously in porcine tissue using MSPD and HPLC–DAD, *J. Chromatogr. B*, 877, 2961–2966, 2009.
27. Proença, P., Marques, E. P., Teixeira, H., Castanheira, F., Barroso, M., Ávila, S., Vieira, D. N., A fatal forensic intoxication with fenarimol: Analysis by HPLC/DAD/MSD, *Forensic Sci. Int.*, 133, 95–100, 2003.
28. Wu, C., Liu, N., Wu, Q., Wang, C., Wang, Z., Application of ultrasound-assisted surfactant-enhanced emulsification microextraction for the determination of some organophosphorus pesticides in water samples, *Anal. Chim. Acta*, 679, 56–62, 2010.
29. Wu, Q., Chang, Q., Wu, C., Rao, H., Zeng, X., Wang, C., Wang, Z., Ultrasound-assisted surfactant-enhanced emulsification microextraction for the determination of carbamate pesticides in water samples by high performance liquid chromatography, *J. Chromatogr. A*, 1217, 1773–1778, 2010.
30. Dommarco, R., Santilio, A., Fornarelli, L., Rubbiani, M., Simultaneous quantitative determination of thirteen urea pesticides at sub-ppb levels on a Zorbax SB-C_{18} column, *J. Chromatogr. A*, 825, 200–204, 1998.
31. Papadopoulou-Mourkidou, E., Patsias, J., Development of a semi-automated high-performance liquid chromatographic-diode array detection system for screening pesticides at trace levels in aquatic systems of the Axios River basin, *J. Chromatogr. A*, 726, 99–113, 1996.
32. Rodríguez-Cuesta, M. J., Boqué, R., Rius, F. X., Martínez Vidal, J. L., Garrido Frenich, A., Development and validation of a method for determining pesticides in groundwater from complex overlapped HPLC signals and multivariate curve resolution, *Chemom. Intell. Lab. Syst.*, 77, 251–260, 2005.
33. Ebrahimi, M., Es'haghi, Z., Samadi, F., Hosseini, M.-S., Ionic liquid mediated sol–gel sorbents for hollow fiber solid-phase microextraction of pesticide residues in water and hair samples, *J. Chromatogr. A*, 1218, 8313–8321, 2011.
34. Irace-Guigand, S., Aaron, J. J., Scribe, P., Barcelo, D., A comparison of the environmental impact of pesticide multiresidues and their occurrence in river waters surveyed by liquid chromatography coupled in tandem with UV diode array detection and mass spectrometry, *Chemosphere*, 55, 973–981, 2004.
35. Wang, C., Wu, Q., Wu, C., Wang, Z., Determination of some organophosphorus pesticides in water and watermelon samples by microextraction prior to high performance liquid chromatography, *J. Sep. Sci.*, 34, 3231–3239, 2011.
36. Vichapong, J., Burakham, R., Srijaranai, S., Vortex-assisted surfactant-enhanced-emulsification liquid–liquid microextraction with solidification of floating organic droplet combined with HPLC for the determination of neonicotinoid pesticides, *Talanta*, 117, 221–228, 2013.
37. Watanabe, E., Kobara, Y., Baba, K., Eun, H., Aqueous acetonitrile extraction for pesticide residue analysis in agricultural products with HPLC-DAD, *Food Chem.*, 154, 7–12, 2014.
38. Topuz, S., Özhan, G., Alpertunga, B., Simultaneous determination of various pesticides in fruit juices by HPLC-DAD, *Food Control*, 16, 87–92, 2005.
39. Melo, A., Aguiar, A., Mansilha, C., Pinho, O., Ferreira, I. M. P. L. V. O., Optimisation of a solid-phase microextraction/HPLC/diode array method for multiple pesticide screening in lettuce, *Food Chem.*, 130, 1090–1097, 2012.
40. López-Fernández, O., Rial-Otero, R., González-Barreiro, C., Simal-Gándara, J., Surveillance of fungicidal dithiocarbamate residues in fruits and vegetables, *Food Chem.*, 134, 366–374, 2012.
41. Ravelo-Pérez, L. M., Hernández-Borges, J., Asensio-Ramos, M., Rodríguez-Delgado, M. Á., Ionic liquid based dispersive liquid–liquid microextraction for the extraction of pesticides from bananas, *J. Chromatogr. A*, 1216, 7336–7345, 2009.
42. Boeris, V., Arancibia, J. A., Olivieri, A. C., Determination of five pesticides in juice, fruit and vegetable samples by means of liquid chromatography combined with multivariate curve resolution, *Anal. Chim. Acta*, 814, 23–30, 2014.

43. Tuzimski, T., Determination of pesticides in wines samples by HPLC-DAD and HPTLC-DAD, *J. Liq. Chrom. Rel. Technol.*, 35, 1415–1428, 2012.
44. Tuzimski, T., Rejczak, T., Determination of pesticides in sunflower seeds by high-performance liquid chromatography coupled with a diode array detector, *J. AOAC Int.*, 97 (4), 1012–1020, 2014.
45. Tuzimski, T., Rejczak, T., Application of HPLC-DAD after SPE/QuEChERS with ZrO_2-based sorbent in d-SPE clean-up step for analysis of pesticides in edible oils, *Food Chem.*, in press, 2015.
46. Tuzimski, T., Application of SPE-HPLC-DAD and SPE-TLC-DAD to the determination of pesticides in real water samples, *J. Sep. Sci.*, 31, 3537–3542, 2008.
47. Tuzimski, T., Determination of pesticides in water samples from the Wieprz-Krzna Canal in the Łęczyńsko-Włodawskie Lake District of southeastern Poland by thin-layer chromatography with diode array scanning and high-performance column liquid chromatography with diode array detection, *J. AOAC Int.*, 91 (5), 1203–1209, 2008.
48. Tuzimski, T., Application of SPE–HPLC–DAD and SPE–HPTLC–DAD to the analysis of pesticides in lake water, *J. Planar Chromatogr.—Modern TLC*, 22 (4), 235–240, 2009.
49. Tuzimski, T., Sobczyński, J., Application of HPLC-DAD and TLC-DAD after SPE to the quantitative analysis of pesticides in water samples, *J. Liq. Chrom. Rel. Technol.*, 32 (9), 1241–1258, 2009.
50. Tuzimski, T., Application of HPLC and TLC with diode array detection after SPE to the determination of pesticides in water samples from the Zemborzycki reservoir (Lublin, Southeastern Poland), *J. AOAC Int.*, 93 (6), 1748–1756, 2010.
51. Tuzimski, T., New procedure for analysis of complex mixtures by use of multidimensional planar chromatography in combination with diode-array scanning densitometry and high-performance liquid chromatography coupled with diode-array detection, *J. Planar Chromatogr.—Modern TLC*, 23 (4), 184–189, 2010.
52. Tuzimski, T., Determination of clofentezine in medical herb extracts by chromatographic methods combined with diode array scanning densitometry, *J. Sep. Sci.*, 33, 1954–1958, 2010.
53. Tuzimski, T., Determination of analytes in medical herbs extracts by SPE coupled with two-dimensional planar chromatography in combination with diode array scanning densitometry and HPLC-diode array detector, *J. Sep. Sci.*, 34, 27–36, 2011.
54. Shapovalova, E. N., Yaroslavtseva, L. N., Merkulova, N. L., Yashin, A. Y., Shpigun, O. A., Separation of pesticides by high-performance liquid chromatography with amperometric detection, *J. Anal. Chem.*, 64 (2), 164–170, 2009.
55. Sabando, O. L., Balugera, Z. G., Goicolea, M. A., Rodriguez, E., Sarnpedro, M. C., Barrio, R. J., Determination of simazine and cymoxanil in soils by microwave-assisted solvent extraction and HPLC with reductive amperometrical detection, *Chromatographia*, 55 (11/12), 667–671, 2002.
56. Asensio-Ramos, M., Hernández-Borges, J., Borges-Miquel, T. M., Rodríguez-Delgado, M. Á., Ionic liquid-dispersive liquid–liquid microextraction for the simultaneous determination of pesticides and metabolites in soils using high-performance liquid chromatography and fluorescence detection, *J. Chromatogr. A*, 1218, 4808–4816, 2011.
57. Fu, L., Liu, X., Hu, J., Zhao, X., Wang, H., Wang, X., Application of dispersive liquid–liquid microextraction for the analysis of triazophos and carbaryl pesticides in water and fruit juice samples, *Anal. Chim. Acta*, 632, 289–295, 2009.

15 High-Performance Liquid Chromatography–Mass Spectrometry as a Method of Identification and Quantification of Pesticides

Ana Masiá and Yolanda Picó

CONTENTS

15.1 INTRODUCTION

A pesticide is any substance used to avoid pests in plants and animals. During the 1970s and 1980s, public and regulatory attention to these compounds increased, resulting in banned many traditional pesticides. Despite that, new groups of pesticides have appeared. They are called "currently used" or "modern" pesticides, a general term to describe a diverse group of chemicals. Pesticides are classified in different ways: One division is based on the chemical family they belong to, such as organophosphorus, triazines, carbamates, triazoles; another grouping approach is based on the type of organism they control, for instance herbicides (plants), insecticides (insects), and fungicides (fungi), etc. [1].

People are exposed to pesticide residues at low concentrations through their diets due to the widespread use of pesticides to control pests in crops in food production. Scientists are concerned about the health effects, but there has not been a clear understanding of them yet. Moreover, new alarming situations regarding the prevalence and effects of these compounds in the environment and concerns of synergies between them have recently appeared [1–3].

Analysis of pesticides is a difficult task due to the low concentration at which compounds usually occur [4]; consequently, it is necessary to develop new analytical methods with much lower detection limits than traditional referential ones. In this sense, sophisticated liquid chromatography (LC) and mass spectrometry (MS) instrumentation has experienced impressive progress over recent years, in terms of both technology development and application [5]. These methods offer a variety of platforms for sensitive detection of many types of molecules [6] and try to cover the high demand for residue analysis providing an increase in the productivity of laboratories as well as decreasing the cost of analyses [7].

With recent advances in LC-MS/MS instrumentation, this technique is quickly gaining acceptance for pesticide residue testing [8]. LC-MS/MS can also be used to simultaneously monitor hundreds of other potential contaminants, including those difficult to detect by gas chromatography (GC) and thermally labile, non-volatile, and high molecular weight species [9,10].

This chapter focuses on principles of MS and the developed advances in LC-MS, offering a general view of different mass analyzers and ionization sources to connect LC to MS as well as the advantages and disadvantages associated with the matrix effect. It also describes their application fields, covering relevant publications in the last three years.

15.2 PRINCIPLES AND INSTRUMENTATION

More than 80% of known organic species are amenable to separation by LC; therefore, LC separation is important to ensure the highest quality data. In addition, MS detectors are capable of providing structure, molecular weight, empirical formula, and quantitative information about specific pesticides and their specific and sensitive detection [6].

In LC analysis, sample extracts or solutions are injected onto a high-performance liquid chromatography (HPLC) column that, commonly, comprises a narrow stainless steel tube, which is packed with fine, chemically modified silica or polymeric particles. Compounds are separated on the basis of their relative interaction with the chemical coating of these particles (stationary phase) and the solvent eluting through the column (mobile phase). Components eluted from the chromatographic column are then introduced to the MS via a specialized interface [1,10–15].

For chromatographic separation, not only the analytical column but also the mobile phase is of great importance. Conventional LC platforms to separate pesticides by LC-MS are now very well established. Reversed-phase liquid chromatography (RP-LC) covers more than 95% of applications. Two types of eluent are typically used as the mobile phase: The most common is pure water (or high content water) eluent as the aqueous phase and methanol or acetonitrile as the organic phase. Mobile phases incorporate volatile buffers mostly by their influence in the signal intensity in positive ionization mode. When LC is coupled to MS, it is important that the additives are volatile. This is one of the reasons why buffers such as sodium phosphate are not used in LC-MS applications. Common buffers and additives used for LC-MS are formic, acetic, and trifluoracetic acids as well as ammonium formate, acetate, and carbonate [2,7,9,16–19].

Regarding MS analysis, a typical MS can be divided into three fundamental modules: ionization source, mass analyzer, and detector (see Figure 15.1).

The fundamentals of this technique are based on the production of a collection of gaseous ions from a sample in the ionization source. They travel to the mass analyzer in which they are separated according to their mass-to-charge (m/z) ratios by applying electromagnetic fields. The separated ions are detected and converted into an electrical signal, which, in turn, is converted into a digital response, which is sent to a data system in which the m/z ratios are stored together with their relative abundance for presentation in the format of m/z spectrum. In the most common instruments, the whole system must be maintained at very low pressure (high vacuum) as the ions cannot collide

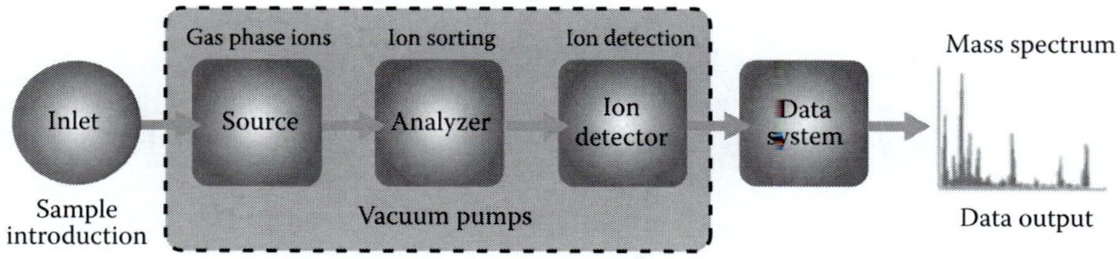

FIGURE 15.1 Diagram of the components of a mass spectrometer.

with any other forms of matter during the separation process. Therefore, only ions are detected in MS, and any nonionic particles that have no charge are removed from the MS by the continuous pumping that maintains the vacuum [2,20–23].

15.2.1 Miniaturization of Columns for Direct Coupling to MS

The column is the quintessence of an LC system. Traditional LC columns are 50–300 mm long and have internal diameters between 3 and 4.6 mm [3,7,19,24–35]. However, the combination of LC with MS has been a driver in the miniaturization of the LC columns as the atmospheric pressure ionization (API) interfaces require the combination of low flow rates (to form smaller droplets in the electrospray) and high efficiencies [36,37]. Disadvantages of working at very low flow rates include long run times that could be reduced at the same time that efficiency is maintained by miniaturization. The development of miniaturized chromatographic methods has attracted the interest of many researchers, providing new tools with higher sensitivity and peak resolution, increased efficiency, and shorter analysis times than those obtained with conventional LC [2,9,38]. Additional advantages of these systems are their economy of use, as smaller amounts of stationary-phase material are needed, and smaller volumes of mobile phase consumed. Alternative miniaturization strategies have been developed to obtain increased efficiency together with short analysis times. These strategies are mostly based on reducing the inner diameter of the LC columns or the particle size [39–41].

Among the advances within the former strategy, nowadays, commercially available LC columns have diameters that result in an impressive range of working flow rates. A typical classification grouping of chromatographic columns according to their diameters is presented in Table 15.1.

The use of analytical (3–4.6 mm i.d.) and narrow-bore (1–2 mm i.d.) columns to determine pesticide residues is the common practice. Medium-narrow LC columns (50 × 2.1 mm) and intermediate flow rates (ca. 0.2 mL/min) are often ideal [42]. The literature shows a clear shift toward smaller diameters in the use of normal-bore columns (from 4.6 to 2.1 mm, which offer a better electrospray ionization (ESI)-MS compatibility and lower mobile phase consumption) [11,12,21,38,40,41,43–58].

However, the use of micro- and nanobore columns within the field of pesticide residue analysis has not yet developed as a routine application. These systems are still used only for very special applications (limited sample volume, need for extreme peak capacity that can only be achieved by applying several thousand bars of pressure, use of rapid temperature pulses for trapping or selectivity enhancement, etc.) that commonly do not cover the field of pesticide residue determination [43].

TABLE 15.1
Classification of LC Columns by Their Internal Diameters

Column	Internal Diameter (mm)	Flow Rate (mL/min)
Normal bore	3–4.6	0.5–3
Narrow bore	1–2	0.02–0.3
Microbore	0.15–0.8	0.002–0.02
Nanobore	0.02–0.1	0.0001–0.001

According to the second strategy, the reduction of particle size, the emerging issue in the field of pesticide separation, is the introduction of new ultra-high performance liquid chromatography (UHPLC) platforms. The recent developments in stationary column phases, such as the use of smaller particle sizes (1.7 and 1.8 μm), have allowed improved peak resolution and, therefore, increased sensitivity in chromatographic separations. As an example, Ferrer et al. [59] evaluates a MS/MS methodology not only to screen but also quantitate and confirm 100 pesticides in a single analysis using a combination of the 1.8-μm LC columns (for maximum peak resolution) and two time segments with 100 transitions per segment in order to have both a quantifier ion and a qualifier ion, which satisfies the EU specifications for unequivocal identification and confirmation by MS.

UHPLC has been shown to give superior chromatographic resolution, reduce analysis time, and increase response. The low degree of band broadening in UHPLC also benefits MS detection, concentrating the analyte at the peak center and thereby increasing response. Thus, UHPLC coupled to MS offers improvements in performance for quantitative analysis over existing high LC-MS/MS methodologies. Pesticide residue analysis in food can also benefit from the high sample throughput obtained in UHPLC methods, keeping the high multiresidue capability required for such applications; this was recently demonstrated in a study comparing UHPLC and LC for pesticide residue analysis in baby foods. Leandro et al. [60] showed the advantages of the UHPLC-ESI-MS/MS method for the determination of 52 pesticides in cereal-based baby foods, oranges, and potatoes. The UHPLC method separates all of the pesticides, resolves structural isomers (e.g., butocarboxim sulfoxide and aldicarb sulfoxide), and has a short (7 min) cycle time.

Although UHPLC technology was launched only recently, many applications to determine pesticide residues can be found in the literature, which illustrates the interest in this technique [6,9, 22,50,61–66]. These applications have covered different aspects, for example, multiresidue analysis of pesticides in fruit and other commodities, confirmation of metabolites, and identification of unknown residues. An interesting example of the prospect of UHPLC separations is shown in Figure 15.2 [65]. The chromatogram shown in Figure 15.2a illustrates the conventional LC separation of a group of postharvest fungicides. On this account, a fast gradient started at time 0 min at 50% of methanol and increase linearly to 100% of methanol over 6 min, and then analytes were eluted under isocratic conditions (100% methanol), enabling relatively fast elution of the analytes. Using a UHPLC column, the application of the previous LC gradient resulted in a very fast elution of all analytes within 6 min (Figure 15.2b). The result of UHPLC separation tuning is shown in Figure 15.2c. Under optimum conditions, better separation in shorter analysis time is achieved.

The shortening of the analysis time is not so impressive as in conventional chromatography due to the slow flow rate [65].

Miniaturization of LC columns is also related to miniaturization of analytical devices to a lab-on-a-chip format, and it is one of the most popular topics in current analytical chemistry [67–69]. The microelectronic field has already been revolutionized by a number of new techniques. These techniques allow chromatographic microchannels with a high degree of accuracy. Furthermore, dozens of different miniaturized ESI/atmospheric pressure chemical ionization (APCI) and atmospheric pressure photoionization (APPI) sources have been published. Recently, an illustrative example shows the development of capillary liquid chromatography (capLC) with a microfabricated heated nebulizer chip for μAPPI–MS/MS to analyze selected carbamate pesticides in a tomato matrix. The limits of detection achieved with the capLC–μAPPI–MS/MS method in the positive ion mode were low, ranging from 0.25 ng mL^{-1} for pirimicarb to 5 ng mL^{-1} for oxamyl and methomyl, corresponding to 5 and 0.25 μg kg^{-1} for tomato samples, respectively, which are clearly below the maximum residue limits for them in fruit and vegetables. Figure 15.3 illustrates the selected reaction monitoring (SRM) chromatograms obtained with the optimized parameters from a tomato matrix sample spiked with the pesticide standard mixture at a concentration level of 0.5 μg mL^{-1} and a total ion chromatogram from a blank tomato sample [43].

Classical one-dimensional approaches do not always provide the resolving power and selectivity needed for the analysis of complex samples [70]. One solution is the use of multidimensional

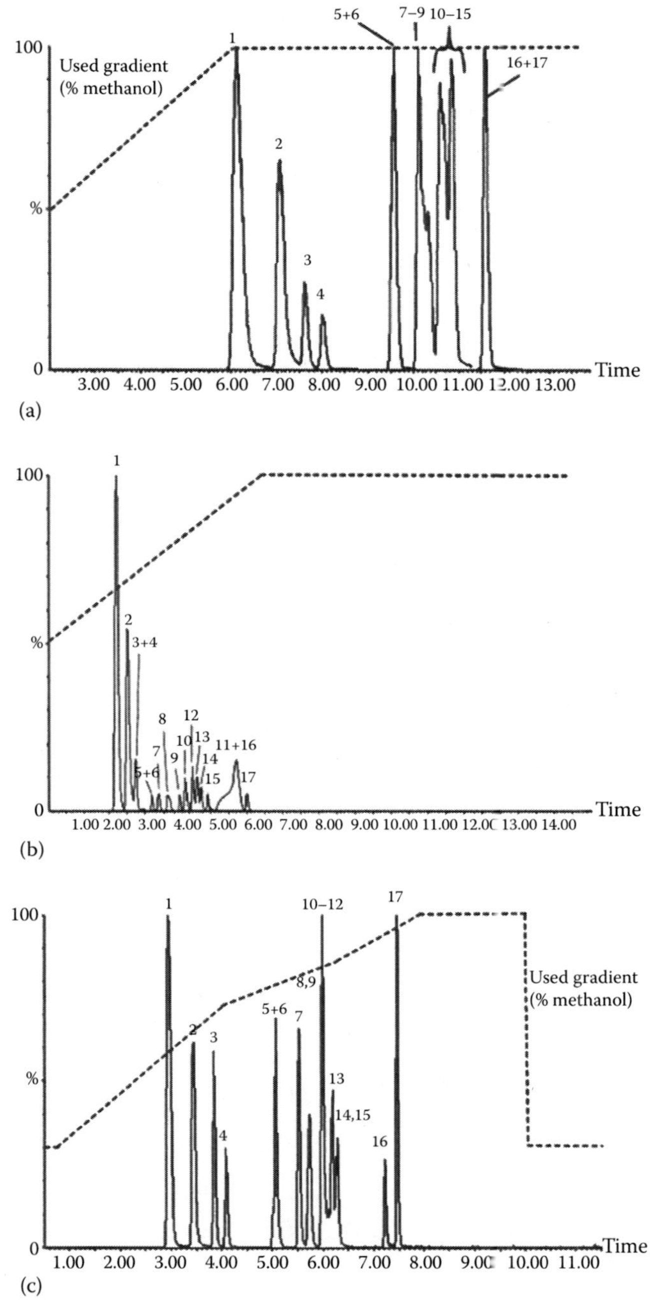

FIGURE 15.2 Chromatogram of apple crude extract spiked with 17 (semi)polar pesticides (conc. 0.02 mg kg^{-1} of each for [a] and [c] and 0.05 mg kg^{-1} of each for [b]) obtained by (a) HPLC–MS/MS, (b) UPLC–MS/MS when using same gradient as in HPLC, and (c) UPLC–MS/MS when using an optimized gradient. For illustration, used gradient (percentage of methanol) is shown. Peak identification: 1 = carbendazim, 2 = thiabendazole, 3 = carbofuran, 4 = carbaryl, 5 = linuron, 6 = methiocarb, 7 = epoxiconazole, 8 = flusilazole, 9 = diflubenzuron, 10 = tebuconazole, 11 = imazalil, 12 = propiconazole, 13 = triflumuron, 14 = bitertanol, 15 = prochloraz, 16 = teflubenzuron, 17 = flufenoxuron. (Reproduced from Kovalczuk, T. et al., *Anal. Chim. Acta*, 577, 8–17, 2006. With permission.)

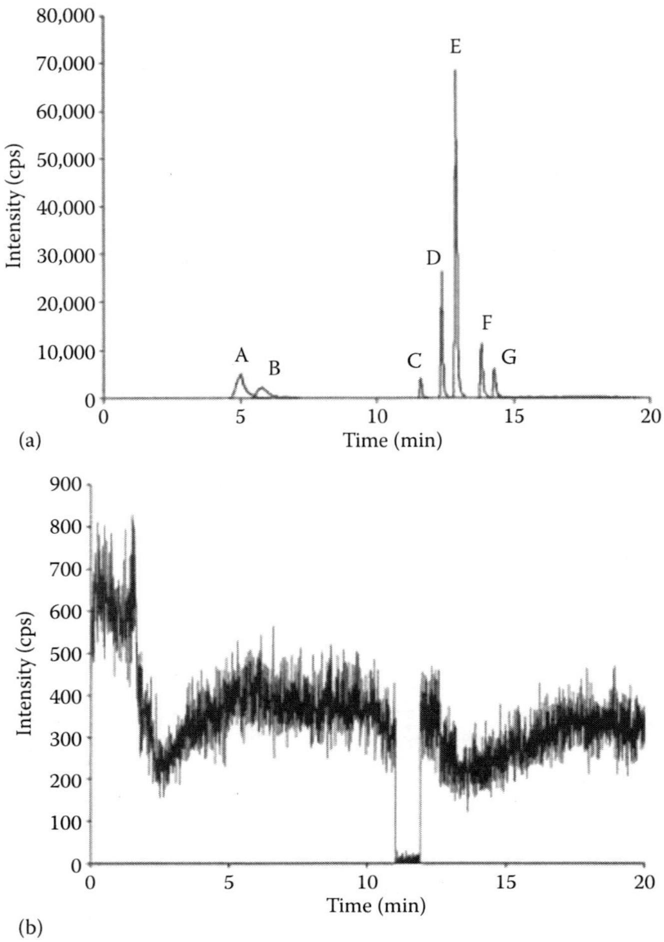

FIGURE 15.3 (a) SRM chromatograms of pesticides measured by capLC–μAPPI–MS/MS from a spiked tomato matrix sample at a concentration level of 0.5 μg mL^{-1}: (A) oxamyl, (B) methomyl, (C) aldicarb, (D) carbofuran, (E) pirimicarb, (F) methiocarb, and (G) ditalimfos. (b) A total ion chromatogram (sum of all SRM pairs) of a blank tomato sample. (Reproduced from Kruve, A. et al., *Anal. Chim. Acta*, 696, 77–83, 2011. With permission.)

chromatography. In the past few years, the field of LC has also clearly witnessed an unprecedented high interest in multidimensional separation. Different coupling techniques with various interface solutions have been developed. Recently, a fully automated system was developed for the determination of more than 300 different pesticides from various food commodities. The samples were extracted with acetonitrile prior to the injection into the two-dimensional LC system. No manual cleanup was needed. The separation of analytes and matrix compounds was carried out by a YMC-Pack Diol (2.1 mm × 100 mm; 5 μm; 120 Å) hydrophilic interaction liquid chromatography (HILIC) column in the first dimension. All analytes eluted within one small fraction at the beginning of the run. With a packed loop interface, this fraction was transferred to the analytical RP separation performed on an Agilent Poroshell 120 EC-C18 (2.1 mm × 100 mm; 2.7 μm; 120 Å). Some very polar compounds with a stronger retention on the HILIC column were measured directly. The method was validated for more than 300 pesticides in cucumber, lemon, wheat flour, rocket, and black tea. Figure 15.4 shows the principles and the achieved separation [52].

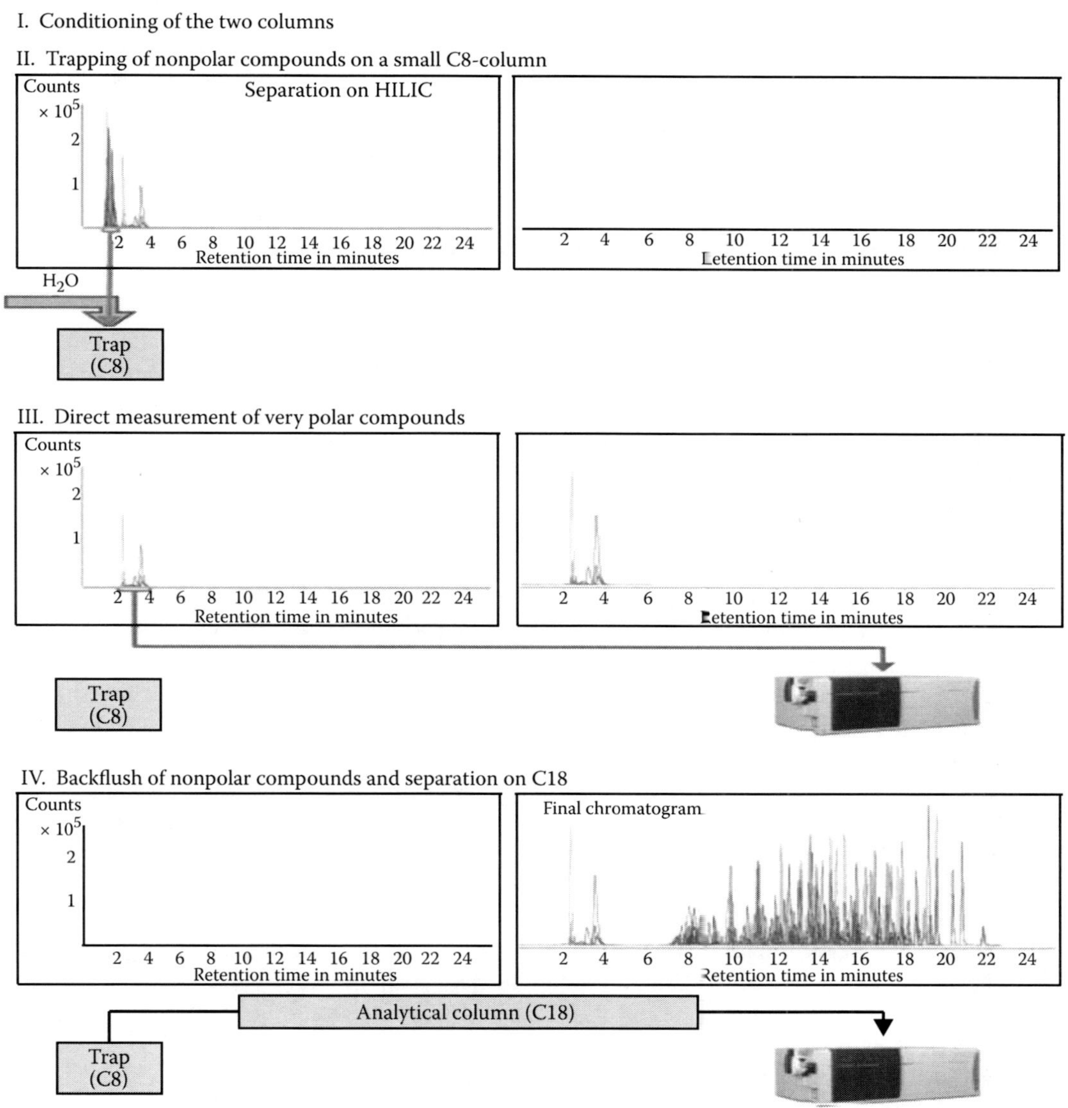

FIGURE 15.4 Principle of the two-dimensional method. (Reproduced from Kittlaus, S. et al., *J. Chromatogr. A*, 1283, 98–109, 2013. With permission.)

15.2.2 Online Hyphenation of HPLC-MS

Nowadays, the coupling of LC-MS is preferred over GC because of the polar nature of currently used pesticides. This situation was not the same 20 years ago when pesticide residue determination was almost restricted to GC. A recent comparison of GC and LC coupled with MS capabilities, regarding their scope and sensitivity, revealed that LC-MS showed better performance than GC for all pesticides except organochlorine [71].

Coupling LC to MS is an "odd couple" because the outlet of a chromatograph is at atmospheric pressure while MS operates at low pressure. Then, the liquid effluent must be vaporized to form volatized ions, and the mobile phase must be removed before introduction into the MS, because if the majority of the solvent was not eliminated, unacceptably high pressures would be developed in the ion source. This turns into a difficult task due to the high volumes that LC uses. A typical flow rate of 0.5–5.0 mL min^{-1} for the LC mobile phase becomes roughly 100–3000 mL min^{-1} of vapor [8,72]. To solve these handicaps, it has been necessary to develop interfaces able to produce a suitable and efficient transfer of the analytes from the solution phase to the gas phase.

The first experiments in coupling LC to MS date back to the late 1960s, but great development has taken place in the last 10 years [73,74]. The earliest interfaces, such as thermospray, particle beam, etc., were often difficult to use, presented limited sensitivity, and were not robust. They are now almost obsolete [73]. The introduction of API techniques was decisive for the increase of LC-MS applications due to a greatly expanded number of compounds that can be analyzed [72].

API is the general term used to name all ionization techniques in which analyte molecules are ionized first at atmospheric pressure, and then the ions generated are separated from neutral molecules mechanically and electrostatically. Ion losses are inevitable, and they occur mainly in the three parts of a MS: between the source and the analyzer, within the mass analyzer, and between the analyzer and detector. In this sense, API techniques present a particular challenge regarding to the efficient transfer of ions from the ion source to the mass analyzer [73].

Common API sources are the following:

- ESI is a soft ionization technique, in which analyte ions are generated in solution before reaching the MS. It is useful for polar and ionic solutes with high molecular weight (up to 600 u). There exist abundant literature in which the ESI mechanism is described in greater detail [5,6,9,10,26,38,44,61–63,75].

  LC eluent passes through a small-diameter capillary needle with a tip held at high potential and atmospheric pressure, at which a spray is formed because the liquid is charged electrically under the electrical field. Charged droplets go to the MS capillary in which a voltage is applied, and a dry gas is used to perform the desolvation process. Solvent from the initially formed droplets evaporates, becoming smaller in size, and their electric field densities convert to become more concentrated. This causes the like charges to repel one another, which increases the surface tension. When the charge droplets cannot support this surface tension, they explode by Coulomb repulsion into highly charged molecules capable of producing gas-phase ions [73]. The ions are sampled through a set of skimmer electrodes and finally analyzed in the mass analyzer (see Figure 15.5). Chemical properties and concentration of the additive, as well as pH, have a significant effect on the analyte response in ESI [5].
- APCI: Using APCI, the liquid flow from the LC is sprayed and rapidly evaporated by a coaxial nitrogen stream and by heating the nebulizer to a high temperature (350°C–500°C). Although these temperatures may degrade the analyte, the high flow rates and coaxial N_2 flow prevent breakdown of the molecules. The resulting gas phase solvent molecules are ionized by electrons discharged from a corona needle. The solvent ions then transfer charge to the analyte molecules through chemical reactions (chemical ionization) [73]. The analyte ions pass through a capillary sampling orifice into the mass analyzer (see Figure 15.5). Opposite to ESI, in APCI, the solvent-evaporation and ion-formation processes are separated. This allows the use of solvents that are unfavorable for ion formation. Another major difference can be found in the LC flow rates; ESI has its optimal performance at low flow rates (nL/min range), and APCI operates happily at high flow rates (1 mL/min and higher) because this interface easily evaporates the solvent [8,72,73]. APCI finds most of its applications in molecular weights below 1000 Da for medium- to low-polarity molecules [30,31].
- APPI is a relatively new technique [76]. As in APCI, a vaporizer converts the LC eluent to the gas phase. Photoions are formed by using a discharge lamp that generates vacuum-ultraviolet (VUV) photons and generates a current that flows through a collection electrode and forms the signal in the chromatogram (see Figure 15.6).

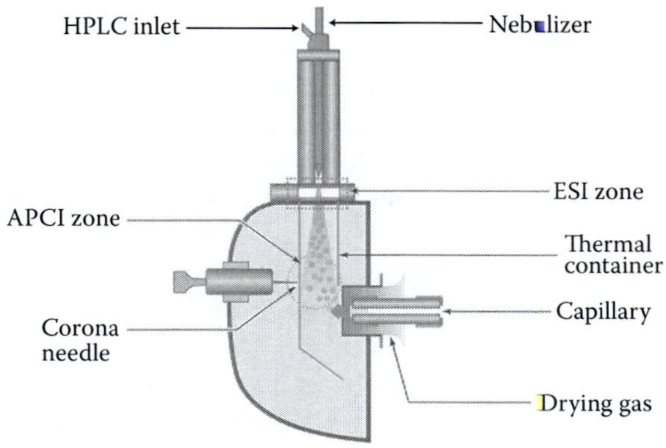

FIGURE 15.5 Diagram of ESI and APCI source.

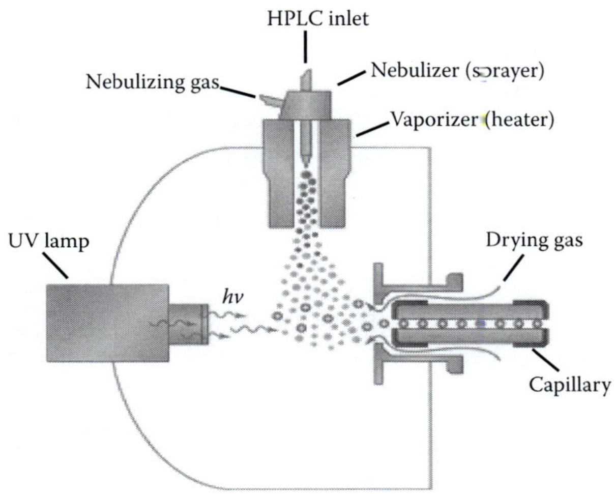

FIGURE 15.6 Scheme of the APPI source.

The photoions can be created by two reaction routes: the first one consists of single photon ionization. It occurs when the photons are absorbed by species that have ionization energy (IE) below the energy of the photons. The other one occurs when a carrier gas, such as nitrogen, is used, that strongly absorbs the VUV radiation. The addition of a large quantity of dopant, such as acetone or toluene, can greatly increase the ionization yield of the target compounds.

APPI is applicable to many of the same compounds that are typically analyzed by APCI. It seems particularly promising in two applications, highly nonpolar compounds and low flow rates (<100 μl/min), with which APCI sensitivity is sometimes reduced.

ESI and APCI are the sources employed for developing methods for determining pesticide residues [8]. It should be noted that in the last two years, the use of ESI [2,5,7,9,14,38,53,54,77,78] has been reported much more than the use of APCI [30,31]. APPI, the most recent incorporation into the API sources, was tested for a few classes of pesticides, and the results reported are promising as a way to minimize the matrix effect. The use of this interface is still quite restricted to a few applications [43,79–84].

Different studies performed by LC-MS/MS pointed out that the ESI interface was, on average, 20 times more sensitive compared to the APCI interface for a great number of pesticides than, depending on the study taken as a model, being as large as 200 [84,85].

15.2.3 Different Mass Analyzers

There are different types of mass analyzers that have been developed to cover the diverse set of applications. Even though the important features are common in the different designs, a brief description of each one is essential to understand accurately how a MS is formed.

The mass analyzer is used to separate ions according to their m/z ratio value, which is based on their characteristic behavior in the presence of an electric and/or magnetic field. The different sorts of mass analyzers are based on the way in which such fields are used to achieve an effective separation. All of them are characterized by parameters such as accuracy, resolution, mass range, tandem analysis capabilities, and scan speed. The accuracy is the capability of the mass analyzer to predict exactly the m/z ratio. This is related to the instrument stability and resolution, which means the ability to differentiate ions, that is, the ability to distinguish between ions of different m/z ratios. The mass range is related to the m/z range of work of the mass analyzer.

There are many types of mass analyzers, and each one presents advantages and disadvantages, depending on the requirements of the particular analysis. Triple quadrupole (QqQ), ion trap (IT), linear ion trap (LIT), time-of-flight (TOF), quadrupole time-of-flight (QTOF), and orbitrap analyzers are the most often used. There exist different classifications depending on their characteristics. They can be divided into those that provide nominal mass (QqQ, IT, and LIT) and those that provide accurate mass (TOF, QqTOF, and orbitrap). Some of them (QqQ, IT, QqLIT, QqTOF, and LIT-Orbitrap) are able to perform MS/MS. Among them, QqQ, QqTOF, and LIT-orbitrap belong to the instrumental group that performs tandem in space as they have different mass analyzers in physically different locations of the instruments. On the other hand, IT and LIT are tandem-in-time instruments because the various stages of MS are conducted in the same mass analyzer but at different times during the run [8,76].

15.2.3.1 QqQ

A quadrupole consists of four parallel rods arranged in a square, and each pair of opposite rods is connected electrically. This can work alone as single quadrupole (Q) or integrated in a QqQ system.

A single quadrupole system (Q) contains only one mass-filtering quadrupole. It can operate in full-scan mode and selected ion monitoring (SIM). In full-scan experiments, ions proceeding from the ion source pass through the quadrupole by applying DC (direct current) and RF (radio frequency) ramped voltage. This full-scan mode is little used to detect compounds at level concentrations, such as those of pesticides, due to low scan speed and low sensitivity. The sensitivity is higher in SIM mode, in which the DC and RF voltages applied on the quadrupole are constant, so only ions with a given m/z ratio pass through the rod assembly, and the others are expelled. First applications of LC-MS in pesticide residue analysis were carried out with this mass analyzer. However, it was not specific enough, mainly for complex matrices with a high number of interferents [8,59,72,86].

With the advent of commercially available MS/MS, the single quadrupole fell into disuse. Nowadays, QqQ is regarded as the most widely used technique for the routine multiresidue screening of pesticides in water and food [8,72,86]. It solves the limitations of the single quadrupole as it allows the performance of MS/MS by using three quadrupoles in series to obtain quantitative information for a variety of different analytes. Each quadrupole has a specific function: Q1 and Q2 perform such mass filter, and q2 acts as a collision cell (Figure 15.7).

Once the analytes have been ionized, they enter Q1. It scans across the m/z range and filter ions according to their m/z ratio by applying a combination of RF and DC voltages across the quadrupole. An electromagnetic field is generated in which ions follow complex trajectories. Ions with a given m/z ratio will have a stable trajectory to pass through the quadrupole, and all the others with unstable oscillations will collide with the rods and drift out of the quadrupole. The precursor ions selected in Q1 then pass to q2, which becomes a wide bandpass for the ions to Q3 when a RF voltage is applied to the quadrupole rods (no DC component). It also acts as collision cell (nonmass filtering)

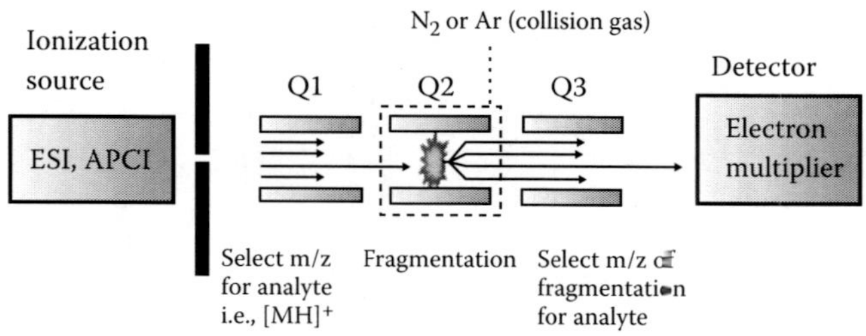

FIGURE 15.7 Diagram of QqQ system.

in which ions collide with a collision gas (N_2 or Ar), causing a collision-induced dissociation (CID) [72,73,86].

The product ions generated in q2 are filtered or scanned in Q3

It can operate in four different modes to obtain complementary qualitative information about the sample:

- Precursor ion mode: MS1 scans all ions that have been fragmented to obtain the product ion isolated in MS2.
- Product ion mode: The precursor ion is selected in MS1 and fragmented in the collision cell. All product ions generated are scanned in MS2.
- Neutral loss: MS1 and MS2 scan all the ions fragmented by the loss of a specific neutral mass.
- Multiple reaction monitoring (MRM): The common mode QqQ operation is SRM (single reaction monitoring) in which specific precursor-to-product ion transitions are selected and monitored by MS1 and MS2. It can also be called MRM (multiple reaction monitoring) in which multiple precursor and product ions can be monitored by quadrupoles with high sensitivity and selectivity. The disadvantage is that spectrum cannot be obtained.

Full scan acquisition with MS or MS/MS produces a great abundance of peaks that are difficult to identify because, besides the loss in sensitivity, the lack of libraries with LC-MS/MS spectra prevents identification of the unknown. Nowadays, a QqQ spectrometer is able to detect more than 100 pesticides simultaneously with enough sensitivity to determine residues at levels below 0.01 mg kg^{-1}. It is possible to use time window programs to improve the sensitivity when a very large number of compounds are determined simultaneously in the same run. MS/MS analysis requires two transitions, the first one quantitative and the second one confirmatory. Then, an increase in the limits of detection occurs. As an example of the good performance of QqQ, Figure 15.8 shows the chromatograms obtained in dynamic SRM mode corresponding to the LC-MS/MS analysis of wastewater spiked with 43 pesticides at 15 ng L^{-1}, and includes the signal to noise (S/N) ratios obtained for the pesticide transitions.

15.2.3.2 IT

The IT analyzer works on the same principles as the quadrupole mass analyzer. However, it consists of a circular ring electrode with two endcaps, forming a chamber altogether (Figure 15.9). A RF voltage is applied between the ring and endcap electrodes, and an electromagnetic field is generated, in which ions (that can be from inside the chamber or injected from an external source) are "trapped." Increasing the RF voltage gradually, ions are selectively ejected by the m/z ratio [72].

The primary advantage of IT as a non-scanning instrument is the high sensitivity of a full scan in the MS and MS/MS modes and its ability to perform multiple stages of MS (MS^n). A precursor ion is isolated, fragmented, and the product ions created are analyzed. This allows the elucidation

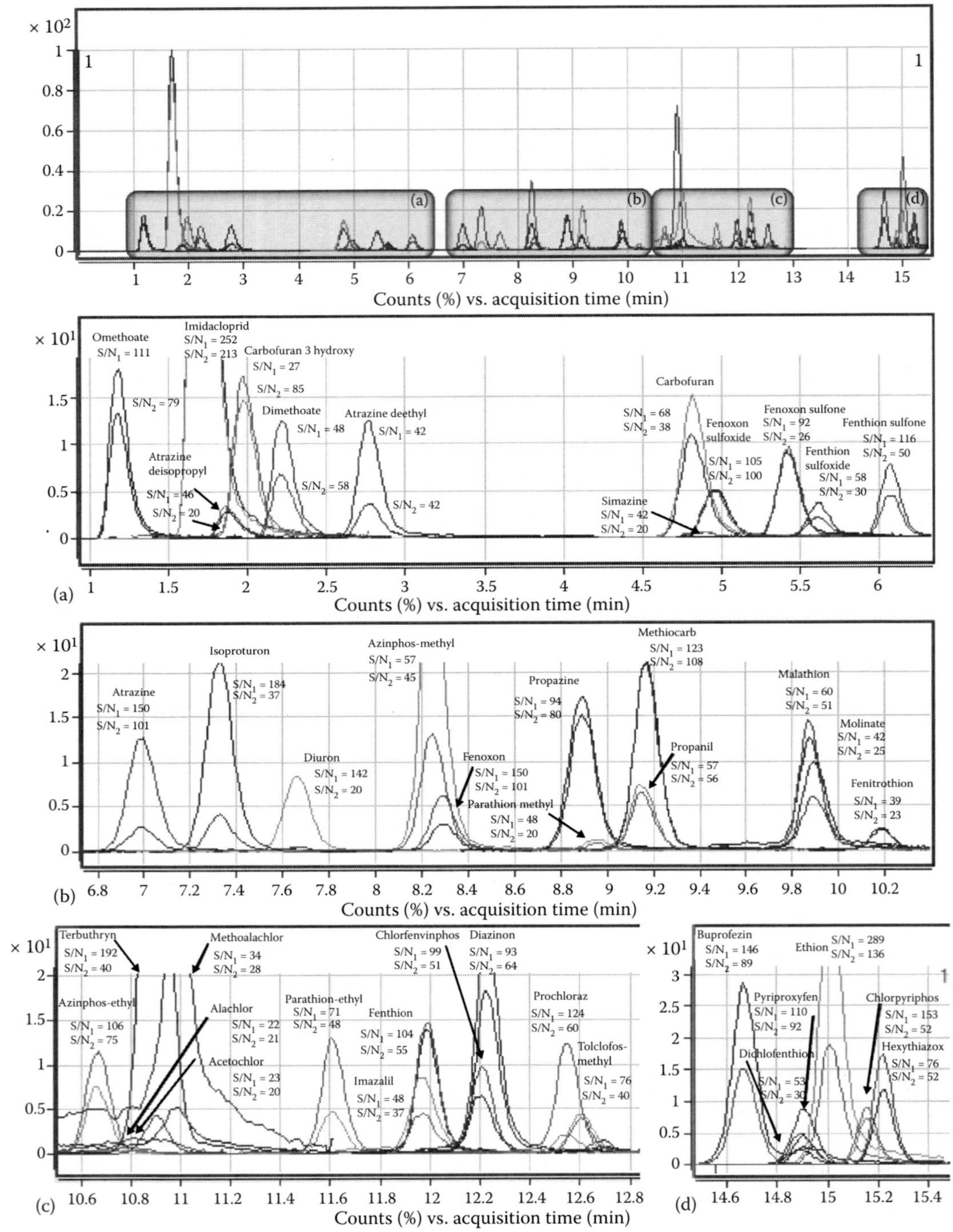

FIGURE 15.8 Chromatograms obtained in dynamic SRM mode corresponding to the LC–MS/MS analysis of wastewater spiked at 15 ng L^{-1} and includes the S/N ratios obtained for the pesticide transitions. The (a), (b), (c) and (d) letters in the shadow areas of the upper chromatogram identify the areas amplified below. There were important differences in the S/N ratios for the different pesticides and transitions, which are directly related to the LODs and LOQs. (Reproduced from Masia, A. et al., *Anal. Chim. Acta*, 761, 117–127, 2013. With permission.)

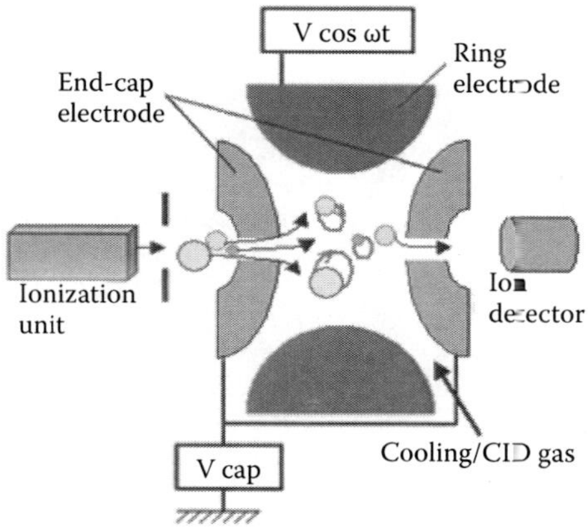

FIGURE 15.9 Diagram of QIT.

of structural information by the interpretation of the successive spectra obtained. However, spectra interpretation might be difficult when no well-known pesticide moieties and only C, H, O, N fragments are present. Furthermore, the presence of coextracted compounds can introduce extra difficulties in the correct selection of the diagnostic ions [74,87].

This process takes place in the same analyzer at different times. After trapping the ions, unwanted ions are ejected by ramping RF; then, the selected ions are fragmented using gas and supplementary frequencies applied to the endcap electrodes [73,88].

The limitations of this instrument are low resolution and mass shift, limited dynamic range because the space charge effects diminish performance, the inability to trap product ions below 50 m/z, and the existence of an upper limit on the ratio between the precursor mass and the lowest trapped fragment ion mass, which is approximately 0.3, depending on the q/z value, and the number of ions that can be simultaneously isolated and fragmented is limited [72–74].

In addition, the sensitivity of this detector is two orders of magnitude lower than that achievable using QqQ. However, there are some recent examples that determine ng of organophosphorus and carbamates [30]. This study achieves MS^3 determination as well as the change of operation between positive and negative ionization mode along the chromatographic analysis without any damage on peak intensity. The others use LC-IT-MS/MS to determine 10 multiclass fungicides in baby food attaining limits of detection (LODs) between 0.5 and 3.0 μg kg^{-1} within EU regulation [32], and to analyze benzoylurea insecticides in a variety of fruit, vegetables, and products of animal origin at levels below 10 μg kg^{-1} [31].

15.2.3.3 LIT

A LIT is basically a quadrupole. However, instead of stabilizing an ion trajectory, ions are confined radially by a two-dimensional RF axially by sloping potentials applied to end electrodes. The use of a quadrupole such as LIT significantly enhances IT performance while maintaining complete QqQ functionality [74].

The most common commercially available LIT instrument (Q TRAP, AB/MDS Sciex) is based on a QqQ platform on which the Q3 can be operated either in the normal RF/DC mode or in the LIT mode. In the LIT mode, the trapped ions are ejected axially in a mass-selective fashion using fringe field effects and detected by the standard detector of the system. Figure 15.10 shows the various modes of operation of the instrument. All specific scan functions of the QqQ are available. Moreover, it has what are called "enhanced" capabilities. The term "enhanced" is used when Q3 operates as a LIT [88].

As mentioned previously, the QqLIT system can operate in the classical QqQ mode with its particular strength of accurate and precise quantitation in the SRM mode. This has been widely

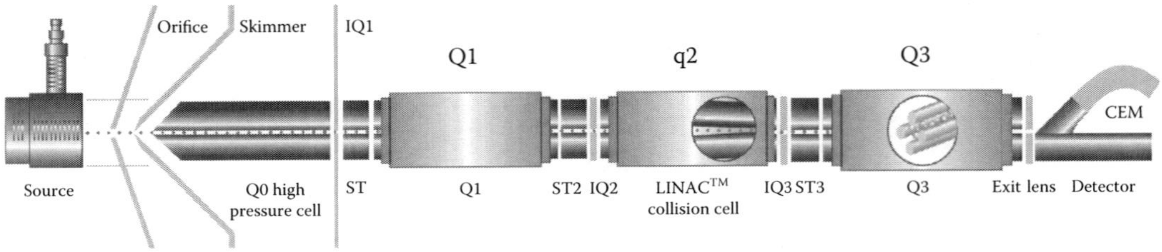

Mode of operation	Q1	q2	Q3
As a QqQ			
Q1 scan	Resolving scan	RF only	RF only
Q3 scan	RF only	RF only	Resolving scan
Product ion scan	Resolving (fixed)	Fragment	Resolving scan
Precursor ion scan	Resolving scan	Fragment	Resolving (fixed)
Neutral loss scan	Resolving scan	Fragment	Resolving (scan offset)
Selected reaction monitoring (SRM)	Resolving (fixed)	Fragment	Resolving (fixed)
As an ion trap			
Enhanced Q3 single MS	RF only	No fragment	Trap/scan
Enhanced product ion (EPI)	Resolving (fixed)	Fragment	Trap/scan
MS^3	Resolving (fixed)	Fragment	Isolation/frag. trap/scan
Time-delayed fragmentation	Resolving (fixed)	Trap/no frag.	Frag. trap/scan
Enhanced-resolution Q3 single MS	RF only	No fragment	Trap/scan
Enhanced multiple charge	RF only	No fragment	Trap/scan

FIGURE 15.10 Schematic of QqLIT (Q TRAP, ABSciex) and description of the various operation modes.

exploited in pesticide residue determination [33,34,41]. Of all the QqLIT modes reported in Figure 15.10, the most used in pesticide residue determination are the enhanced product ion mode and the MS^3 modes. With the QqLIT in enhanced product ion mode, the precursor ion selected in Q1 is fragmented in the collision cell q2, and fragment ions are trapped in Q3 operated in LIT mode. The MS^3 spectra are obtained by fragmenting the product ions trapped in Q3 by application of a single frequency of 85 Hz [8,88].

As an example of the additional capabilities of the QqLIT in comparison to QqQ, the data obtained from the information-dependent acquisition (IDA; EPI and MS^3) experiments performed on spinosyn A are presented in Figure 15.11 [89]. The spectra obtained in the EPI mode for spinosyn A shows two fragment ions (m/z 142 and 98) as a result of the fragmentation of the precursor ion m/z 732. However, the intensity of the fragment ion m/z 98 is approximately 10%. With the additional survey scan of MS^3 for spinosyn A, further structural information is obtained because the fragment ion m/z 142 provided another abundant fragment ion at m/z 98.0. Therefore, in this example, the use of this instrument enables one to unequivocally confirm the finding using a MS^3 step. In addition, the use of the EPI mode does not involve a remarkable sensitivity decrease, and the identification confirmation capabilities are better than a conventional QqQ instrument.

15.2.3.4 TOF

TOF MSs measure the time required for an ion to travel from the ion source to a detector crossing a field-free region, differing from quadrupole or IT systems that use an electric field to separate the ions with different m/z ratios [8,71,72]. A uniform electromagnetic force is applied to all ions at the same time, causing them to accelerate down through a field-free region. Consequently, kinetic energies are identical at the beginning of the flight tube, and their velocities depend on their m/z value. According to this, low m/z ions travel faster and arrive at the detector first, before high m/z ions [3,7,72].

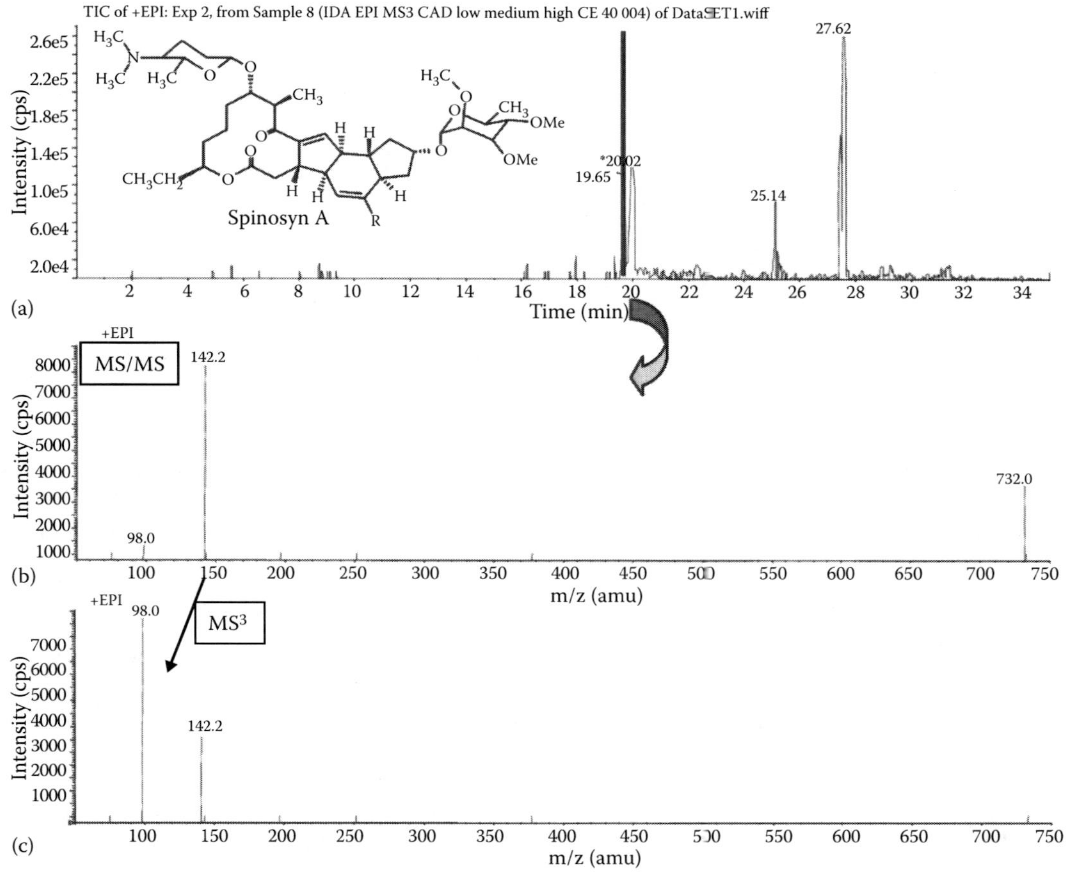

FIGURE 15.11 Application of LC-MS/MS for the confirmation of spinosyn A, a pesticide with low CID fragmentation in a QqQ, using a hybrid QqQ-LIT instrument: (a) total ion chromatogram; (b) enhanced product ion (EPI) scan mode; and (c) MS^3 mode. The combination of survey scans using the MRM, EPI, and MS^3 modes by means of IDA experiments is very useful for confirmatory purposes. (Reproduced from Garcia-Reyes, J. F. et al., *Anal. Chem.*, 79, 7308–7323, 2007. With permission.)

Several designs of TOF analyzers exist, and some of them employ reflectrons to improve resolution or the ability to distinguish two m/z ratios. This consists of an ion optic device placed at the end of the flight tube, which creates a decelerating and reflecting field that the ions penetrate (Figure 15.12). Depending upon their kinetic energy, they enter this field at different depths and then are reflected back into the flight tube, where they drift to the detector, which is placed close to the ion source. Faster ions travel further into the reflectrons, and slower ions travel less into the reflector. This way, both slow and fast ions of the same m/z value reach the detector at the same time rather than at different times, narrowing the bandwidth for the output signal [42,90].

This instrument is used in qualitative applications, such as the identification of nontargeted and/or unknown compounds, in which the acquisition of an entire mass spectrum is required. It is possible that TOF-MS can analyze high mass range, and it has a high acquisition speed reached by the quasisimultaneous detection of all ions, resulting in high full-scan spectral sensitivity. Additionally, it offers improved selectivity due to the high-resolution power (10,000 or more) expressed in terms of FWHM (full peak width at one half maximum). This feature linked to the capability to provide accurate mass measurements allows the obtaining of the elemental composition of parent and fragment ions and is used to identify unknown species. It also permits the identification of mass interferences with analytes having the same nominal mass and chromatographic retention time [3,35,42].

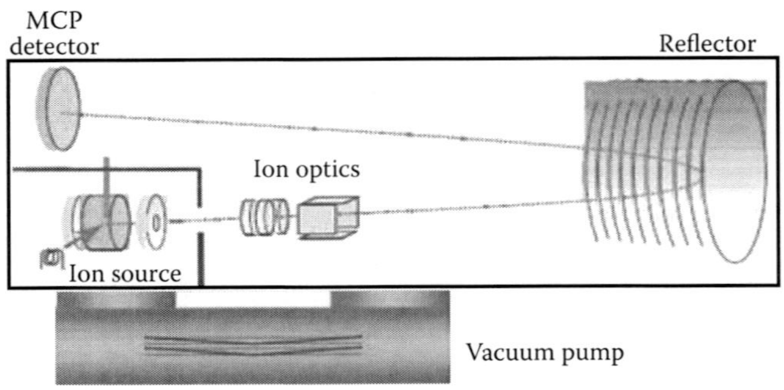

FIGURE 15.12 Diagram of QTOF.

The disadvantage is its limited dynamic range, that is, the ratio of the maximum to the minimum observable ion intensities or concentrations over which a linear response is obtained. This is the direct result of the types of detectors that are employed for handling the large number of high-resolution spectra, making the application of TOF instruments to quantitative analysis less attractive but not impossible [3,66,91].

The TOF analyzers have several advantages in the field of pesticide residue analysis as they can be used for target and nontarget analysis. The identification of target pesticides is based on accurate mass measurements of selected ions (protonated or deprotonated molecules ($[M + H]^+$ or $[M - H]^-$) and fragment ions) instead of in the precursor ion → product ion transitions [3,7,35,42,49,66,90,91]. The screening of pesticide residues could be accomplished at low concentration levels (10 μg kg^{-1}), and accuracy errors lower than two parts per million were obtained in most cases.

TOF MS has proven to be an attractive analytical tool for rapid detection and reliable identification of a large number of pesticides thanks to the full-spectrum acquisition at accurate mass with satisfactory sensitivity. This process is readily boosted when combined with specialized software packages, together with theoretical exact mass databases [49,66,91].

These mass analyzers also enable elucidating the most common degradation products (DPs) of pesticides, taking advantage of unique features of TOF (high sensitivity in full scan and accurate mass measurements), which allows acquiring a full-scan accurate mass spectrum of any peak in a chromatogram and represents a high value for the identification of the unknowns in the samples, because information on all the ions (molecular + fragments) generated by a specific compound is obtained. The empirical formula provided by the instrument and the searching in a specific chemical database could identify nontarget compounds. For example, Garcia-Reyes et al. [92] proposed a strategy for the identification of DPs, based on the use of "fragmentation–degradation" relationships. From a given parent species, the fragmentation patterns that occurred in-source (by collision induced dissociation or CID) could be used as a reference or model to predict possible DPs. Examples of this strategy have been shown for the identification of six DPs of amitraz and malathion on different food extracts, showing the unique potential of LC/TOF MS for the identification of unknown DPs in food without the use of standards a priori.

15.2.3.5 QTOF

QTOF can be described as a hybrid instrument consisting of a QqQ mass analyzer in which the third quadrupole has been replaced by a TOF analyzer. QTOF can operate in MS and MS/MS mode; they combine in both modes the high performance of TOF analysis. In the former, Q1 is operated in band pass mode, and the analysis is performed on the high-end TOF analyzer. For MS/MS, the first quadrupole is used for the isolation of precursor ions; the second one (q2) acts as a collision cell, in which the CID process is carried out; and the TOF analyzer provides accurate mass data for fragment ions formed in q2 (Figure 15.13).

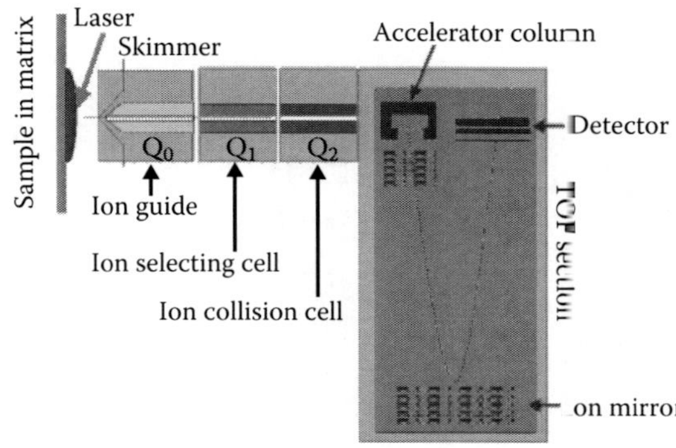

FIGURE 15.13 Hybrid QTOF configuration operating in MS/MS mode.

The hybrid configuration combines high sensitivity, mass resolution, and mass accuracy for both precursor (MS) and product ions (MS/MS). This constitutes a higher-order mass identification than those afforded by nominal mass measurements obtained by other types of mass analyzers [6,50,51,63,93]. QTOF also provides full-scan sensitivity in both modes because of the parallel feature detection, but it cannot be applied to the more specialized modes analogous of QqQ systems (precursor ion, neutral loss, and MRM scan). However, new techniques are emerging to address this limitation [8,51,72].

The instrument is considered to be suitable mostly for qualitative analysis, but it could be employed for quantitative analysis in some applications. The QqQ mass analyzer still achieves the best absolute sensitivity for targeted compounds; however, the higher resolution of QqTOF increases specificity, and consequently, it provides S/N benefit in some analytical situations [94].

This MS has been mostly used for the identification of nontarget, unexpected compounds, including transformation products of pesticides or even unknown compounds. As an interesting example of the application and the features of this mass analyzer within the field, Pico et al. [6] established the profiling of compounds and DPs from the postharvest treatment of apples and pears by a nontarget approach. Figure 15.14 shows the identification of ethoxyquin and imazalil and the mass spectrum of tentative DPs of ethoxyquin.

15.2.3.6 Orbitrap

The orbitrap mass analyzer consists of three electrodes: an inner spindle-like electrode (axial) connected with outer barrel-like (coaxial) electrodes placed opposite each other and electrically isolated. The central electrode holds the trap together and supports it via dielectric end spaces [72]. When ions are injected into the volume between the central and outer electrodes and a voltage is applied between them, a radial and axial electric field appears at the same time. The first one retains the ions on a nearly circular spiral inside the trap, and in the second, ions describe harmonic axial oscillations in the widest part of the trap, which produces a periodic signal detected on the outer electrodes as an image current that is converted into a frequency spectrum by means of a Fourier transform algorithm [95,96]. The frequency of these harmonic oscillations is independent of the ion velocity and is inversely proportional to the square root of the m/z ratio. A scheme of the orbitrap is shown in Figure 15.15.

The most important capabilities are high mass resolution and high mass accuracy, which are very useful in analysis of unknown samples; high resolving power; and a high dynamic range. The first commercial implementation was in a hybrid instrument that combines the orbitrap with LIT (LTQ orbitrap). Ions are generated in an API ionization source and trapped in the LTQ analyzer, where they are analyzed by means of MS and MS^n scan modes. The ions are ejected axially from the LTQ and collected in a C-shaped IT (C-Trap) before being transferred to the orbitrap analyzer, where they

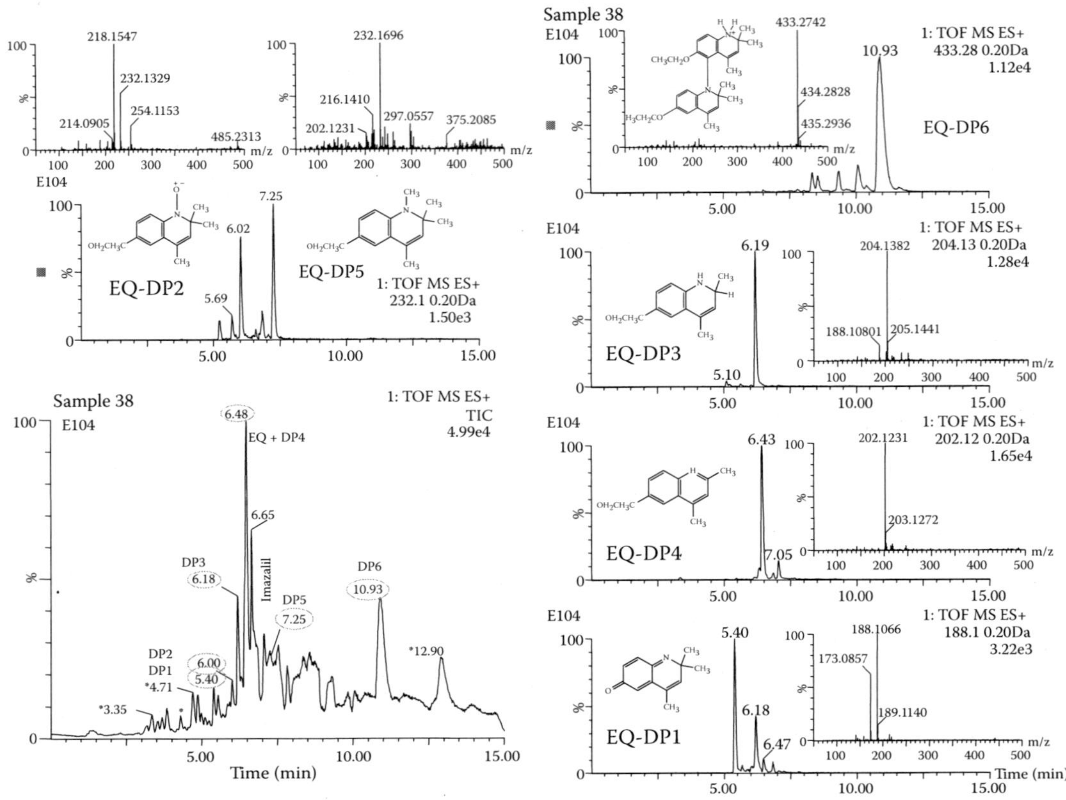

FIGURE 15.14 UPLC–ESI-QqTOF-MS chromatograms corresponding to an apple sample. Left bottom: total ion chromatogram; left upper and right: extracted ion chromatograms of EQ identified DPs, the mass spectrum and the proposed structure of which are shown as inserts. Peaks marked as (*) can be an unidentified contaminant. The other peaks of the chromatogram were also present in the extracts from non-postharvest treated apples and pears. (Reproduced from Pico, Y. et al., *Talanta*, 81, 281–293, 2010. With permission.)

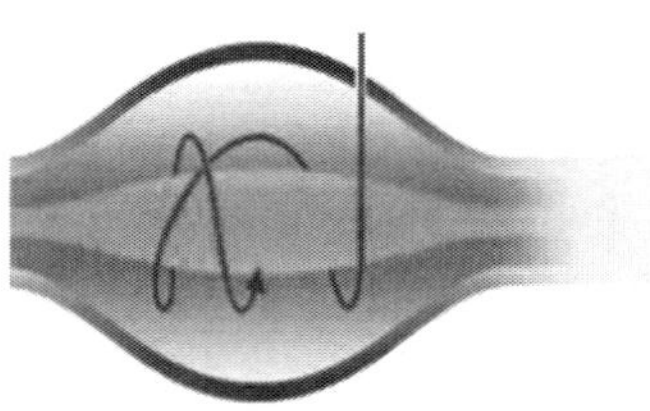

FIGURE 15.15 Orbitrap configuration.

are captured by rapidly increasing voltage on the center electrode. The trapped ions assume circular trajectories, and their axial oscillations are detected along the electrode.

The orbitrap or LIT-orbitrap is the latest addition to the extended family of mass analyzers (schematized in Figure 15.16). There are still few examples of the application of this mass analyzer. However, several studies [35,67,89] demonstrated its possibilities for screening and quantification of pesticide residues. In conclusion, each mass analyzer selects suitable analytical parameters, which confer them as features to identify and quantify pollutants in samples.

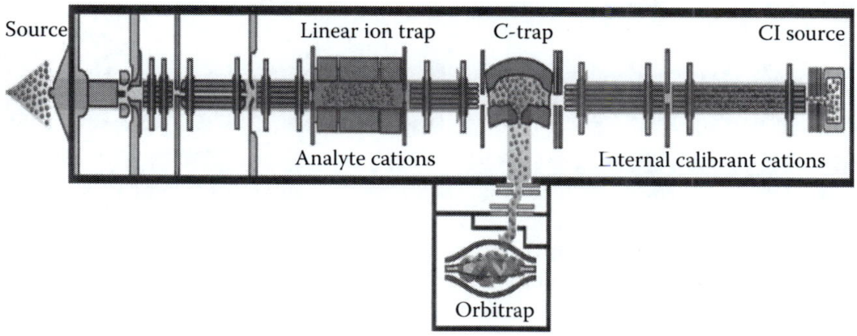

FIGURE 15.16 Diagram of LIT-orbitrap mass analyzer.

15.3 ANALYTICAL CHROMATOGRAPHY-MS (ADVANTAGES AND DISADVANTAGES)

LC coupled with MS detection (LC/MS) is one of the most powerful analytical tools for organic compound analysis. It provides high sensitivity, selectivity, specificity, and rapid analysis [8,72]. The sensitivity measures the change in instrument response, which corresponds to a change in analyte concentration. It is the capability of the analytical method to discriminate small differences in concentration or mass of the test analyte. Regarding the sensitivity, QqQ in SRM or Q and IT in SIM mode offer higher sensitivity than TOF analyzers, contrary to full-scan mode, in which TOF achieves more sensitivity than others. Comparing TOF and QTOF, both result in similar sensitivity. QTOF should obtain better S/N, but some ions are lost because the ion collection of the quadrupole filter does not have a 100% efficiency [43,72].

The terms "selectivity" and "specificity" are measures of the reliability of measurements in the presence of interferences. They are often used interchangeably, but they do not have exactly the same significance. Specificity generally refers to a method that produces a response for a single analyte only, that is, is the ability to assess unequivocally the analyte in the presence of components that may be expected to be present, including impurities, DPs, matrix, etc. Otherwise, selectivity refers to a method that provides responses for a number of chemical entities that may or may not be distinguished from each other. Then, selectivity studies should also assess interferences that may be caused by the matrix. Selectivity is related with tandem MS capabilities. QqQ and IT MS/MS present high selectivity. TOF is less selective than QTOF. In this, the selectivity of precursor ion scans is very high due to the high resolving power of the reflectron [6,97]. TOF analyzers provide much higher accuracy than any other instrument due to the excellent separation and detection in the flight tube. Then, TOF instruments make available high-accuracy fragment ions without comprising sensitivity.

Referring to dynamic range, QqQ shows one of three orders of magnitude, which allows the use of it for quantitative purposes, identifying and confirming the target compounds at very low concentrations (parts per trillion). Conversely, TOF and QTOF instruments have higher LOD and lower dynamic range of two orders of magnitude due to the ion saturation at the upper part of the concentration range. Due to that, quantification of target pesticides at ultra trace levels is a compromise in some cases with these mass analyzers [98]. Table 15.2 compares each type of mass analyzer according to the analytical parameters sensitivity, selectivity, and dynamic range.

Despite its enormous potential, LC-MS presents some disadvantages and limitations affecting both qualitative and quantitative determination, mainly the carryover and the matrix effects [94,99].

In an analytical method, the "carryover effect" is the presence of an analyte in a blank sample that follows the injection of a high concentration sample. This produces a memory effect caused by a residual analyte accumulating in the sample injection flow path that produces the unsuitable elution of the compound from the column [68,94].

TABLE 15.2
Comparison between LC-MS Systems

	Sensitivity in Full Scan	Selectivity	Accuracy	Dynamic Range	Unique Features for Pesticide Residue Determination
QqQ	Medium	High	Low, unit resolution	High	Highly sensitive SRM
IT-MS/MS	High	High	Low, unit resolution	Medium	MS^n, full ms range of precursor and product mass spectra with high sensitivity
QqLIT	High	High	Low, unit resolution		Combines features of QqQ and IT Highly sensitive SRM MS^3, full ms range of precursor and product mass spectra with high sensitivity
TOF-MS	High	Low	High	Medium	Accurate mass and sensitivity, high acquisition speed, resolution power >10,000 FWHM
QTOF-MS	High	High	High	Medium	Accurate mass and selectivity, high acquisition speed, resolution power >10,000 FWHM and MS/MS
Orbitrap	High	High	High	Medium	High speed scan, accuracy, and resolving power (>60,000 FWHM) and MS^n

Possible locations of an injection system in which a residual analyte can accumulate leading to carryover are the following:

- The injection syringe barrel and/or needle
- "Dead volume" spaces present within the injection valve
- The rotor of the injection valve
- The tubing material used within the injection system

Carryover denotes one type of systematic error that is derived from a preceding sample and introduced into the next sample, and in general, it causes a reduction in accuracy and precision of LC-MS/MS analyses. So minimizing this carryover effect during an analytical run is important for providing reproducible and reliable data. It is typically specified as a percentage of the lower limit of quantitation (LLOQ) of the analytical method being used, and it is desired to limit it to less than 20% of the LLOQ. Occasionally, carryover problems have simpler solutions, for instance, simply raising the needle height or minimizing the contact surface between the analyte and needle [37,94,99].

The "matrix effect" is an unexpected suppression or enhancement of the analyte response due to coeluting matrix constituents, affecting method performance parameters such as LOD, LOQ, linearity, accuracy, and precision. It occurs during the analyte ionization process, and it is the result of competition between nonvolatile matrix components and analyte ions for access to the droplet surface for transfer to the gas phase. It causes the loss of sensitivity and selectivity [72,99].

There are two types of matrix effect: the absolute matrix effect, which affects the accuracy of the method, and the relative matrix effect, which affects the precision of the method. Both cause the lack of selectivity and sensitivity, and they can be evaluated by several procedures [86,99,100].

The absolute matrix effect is the difference in response between the solvent sample and the postextraction spiked sample. It can be detected by postextraction addition and postcolumn infusion. Postextraction addition, allows the assessment of the matrix effect quantitatively by comparing the response of an analyte in neat solution to the response of the analyte spiked into a blank matrix sample that has been carried through the sample preparation process (matrix-matched standard) [99,100].

From the peak areas acquired, the percentage of matrix effect can be calculated (Equation 15.1):

$$\%ME = \frac{\text{Area of postextraction spiked}}{\text{Area of standard}} \times 100. \quad (15.1)$$

The recovery measures the efficiency of the analyte extraction process during sample pretreatment, and it allows the identification of whether the loss of sensitivity is due to poor recovery or to matrix suppression because both causes give the same result. However, it can only be determined adequately by means a method free from matrix effects (Equation 15.2).

$$\%RE = \frac{\text{Area of preextraction spiked}}{\text{Area of postextraction spiked}} \times 100. \quad (15.2)$$

It is also important to appreciate the difference between the recovery and the overall process efficiency. With the dates obtained above, it is possible to calculate the overall process efficiency (% PE) (Equation 15.3):

$$\%PE = \frac{\text{Area of preextraction spiked}}{\text{Area of standard}} \times 100 = \frac{(\%ME \times \%RE)}{100}. \quad (15.3)$$

In conclusion, the absolute matrix effect can be achieved by acquiring calibration plots with three sample sets. The first one consists of the analyte and the IS in mobile phase (standard solution), the second set consists of postextraction spiked samples (matrix-matched standard), and the last one preextraction spiked samples (fortified real samples) [90,99–101].

The postcolumn infusion method enables the evaluation of the absolute matrix effects of different sample pretreatment procedures. It identifies qualitatively chromatographic regions most likely to experience matrix effects. The extracted blank matrix is injected by an autosampler onto the analytical column to raise the background level so that the suppression matrix will show as negative peaks. If several compounds are determined in one method, all compounds should be infused separately to investigate possible matrix effects for every analyte [99,100].

The relative matrix effect is the difference in response between various lots of postextraction spiked sample. It can be evaluated by a comparison of the precision expressed as the percentage of relative standard deviation in repetitive injection of standards and postextraction spiked samples derived from various sample lots. The matrix effect in pesticide analysis by LC-MS is overcome by the use of calibration approaches: calibration using external standards prepared in sample matrix (matrix-matched standards) and calibration using internal standard. The advantage of using matrix-matched standards is the simplicity of application and economy; however, the need of obtaining a blank matrix for every sample type is mandatory, and it turns into an unpractical technique for routine analysis. Otherwise, calibration with internal standards is the best option if these compounds are available. An internal standard is an isotopically labeled compound with similar analytical behavior to the compounds of interest that is not expected to be found in the samples. It has a retention time identical or very close to the retention time of the analyte. It allows signal suppression to be corrected, as both labeled and native compounds suffer the same suppression effect [99,100].

15.4 APPLICATIONS OF HPLC-MS AND HPLC-MS/MS TO QUALITATIVE AND QUANTITATIVE ANALYSIS OF PESTICIDES

Table 15.3 provides the wide range of applications of LC-MS described in the literature between the years of 2010 and 2013. It summarizes key information on the analytical

TABLE 15.3
LC-MS Applications to Determine Pesticide Residues

			Separation		Determination		
Matrix	**Pesticide**	**Extraction**	**Column**	**Mobile Phase**	**Detection**	**Sensitivity**	**References**
			QqQ				
Surface water and wastewater	43 pesticides	Off-line SPE	LC-Narrow-bore column Luna C18 (15.0 cm × 0.21 cm, 3 μm)	Gradient MeOH-H_2O both 10 mM $HCOONH_4$ Flow rate: 0.4 mL/min	Agilent 6410 ESI in PI mode Dynamic MRM 2 precursor → product ion transition	LOD 0.04–2 ng/L	[9]
Red wine	18 pesticides	Comparison MASE and QuEChERS	LC-Narrow-bore column RP C18 Aqua column (50 × 2 mm, 5 μm, 125 Å)	Gradient: (A) 30% MeOH, 70% H_2O and (B) 90% MeOH, 10% H_2O; both with 2 mM $HCOONH_4$ Flow rate: 0.2 mL/min	AB Sciex, API 2000 ESI in PI mode MRM mode, 2 MS/MS transitions	LOQ 3 ng/L	[2]
Pond, river, and paddy water	OPPs	QuEChERS with AcN 1% acetic acid and SPE	LC-Narrow-bore column Hypersil BDS C8 column (100 × 2.1 mm, 2.4 mm)	(A) 0.2% acetic acid and 10 mM/L and NH_4OAc in H_2O; (B) 0.2% CH_3COOH in AcN Flow rate: 0.2 mL/min	Agilent 6410 ESI in PI compared with competitive indirect enzyme-linked immunosorbent assay (ciELISA) based on a monoclonal antibody (MAb)	LOQ 0.2–1 ng/mL	[38]
Fruit and vegetables	69 pesticides	QuEChERS	LC-Normal-bore and narrow-bore column Luna C18 (150 × 4.6 mm, 5 μm) Spherisorb ODS-2 (250 × 4.6 mm, 5 μm) Atlantis T3 (100 × 2.1 mm, 3 μm) Hypersil C18 (100 × 4 mm, 3 μm)	Gradient 0.1% HCOOH in H_2O and MeOH/AcN (1:1) with 0.1% HCOOH Flow rate: 200 μL/min	ABSciex, API 2000 SRM mode ESI in PI and NI mode 2 MS/MS transitions	LOQ < 40 μg/kg	[53]
Fruit and vegetables	150 pesticides	"Acetate buffered" QuEChERS	Comparison UHPLC and LC-Normal-bore column: Zorbax Eclipse C8 (150 × 4.6 mm, 5 μm) Zorbax C8 (100 × 2.1 mm, 1.8 μm)	LC: gradient AcN-H_2O with (A) 0.1% HCOOH, (B) 5 mM $HCOONH_4$, (C) 5 mM NH_4OAc Flow rate: 0.6 mL/min UHPLC: gradient H_2O (0.1% HCOOH) and AcN Flow rate: 300 μL/min	Agilent 6410 ESI in PI mode Dynamic MRM 2 most abundant MS/MS transitions	LOD < 0.5–10 μg/kg	[5]

Rice	4 herbicides, 9 fungicides, 2 insecticides	QuEChERS	LC-Normal-bore column Zorbax Eclipse XDB C8 (150 × 4.6 mm, 5 μm)	Gradient AcN-H_2O with 0.1% HCOOH Flow rate: 0.6 mL/min	Agilent 6410 ESI in PI and NI mode MRM, 2 most abundant MS/MS transitions	LOD < 10 μg/kg	[7]
Organic samples: 18 cereals, 13 vegetables, 9 wines	Pesticides Biopesticides Mycotoxines	QuEChERS	UHPLC column Acquity UHPLC BEH C18 (100 × 2.1 mm, 1.7 μm)	Gradient MeOH–H_2O 5 mM $HCOONH_4$ Flow rate: 0.45 mL/min	Waters, Acquity TQD ESI in PI mode SRM, 2 MS/MS transitions	LOD < 10 μg/kg	[77]
31 food samples	44 pesticides	QuEChERS	LC-Narrow-bore column SunFire C18 (2.1 × 150 mm, 3.5 μm)	Gradient: (A) distilled H_2O with 0.1% HCOOH, (B) AcN with 0.1% HCOOH Flow rate: 0.2 mL/min	Waters, Quattro Micro ESI in PI mode MRM, 2 MS/MS transitions	LOD 0.0003–0.0010 mg/kg LOQ 0.0078–0.0237 mg/kg	[54]
Cereals (wheat, rice, and corn)	6 sulfonylureas herbicides	SPE	LC-Normal-bore column Zorbax SB C18 (4.6 × 150 mm, 3.5 μm)	Gradient: (A) H_2O with 0.1% HCOOH, (B) AcN Flow rate: 0.2 mL/min	Agilent 6410\ ESI in PI mode MRM, 2 MS/MS transitions	LOD 0.043–0.23 μg/kg LOQ 0.14–0.77 μg/kg	[78]
Dried botanical dietary supplements	236 pesticides	QuEChERS-dSPE (clean-up)	UHPLC column Zorbax Eclipse Plus-C18 (2.1 × 100 mm, 1.8 μm)	Gradient of (A) 5 mM $HCOONH_4$ in H_2O with 0.01% HCOOH and (B) 0.01% HCOOH in AcN	Agilent 6460 ESI in PI mode Dynamic MRM, 2 precursor-to-product ion transitions	LOD < 5 μg/kg	[75]
Vegetables (pepper and tomato) and citrus fruit (orange and lemon)	54 pesticides	Extraction AcN and LLP aided by "salting out" with NaCl	LC-Normal-bore column Zorbax Eclipse XDB-C8 (150 × 4.6 mm, 5 μm)	Linear gradient: A–B, AcN 0.1% HCOOH. Flow rate: 0.6 mL/min	Agilent 6410 ESI in PI mode SRM, 2 MS/MS transitions	LOD 0.03–14.9 μg/kg	[102]
Meat products	188 OPPs and others	LE ethyl acetate-cyclohexane, GCP, and PSA/silica-gel SPE	LC-Normal-bore column Ascentis C18 (100 × 3.0 mm, 3 μm)	Gradient AcN and 10 mM NH_4OAc in H_2O Flow rate: 500 μL/min	Agilent MSD SL ESI in PI and NI mode	LOQ < 0.01 μg/g	[55]
Tea	Pesticide residues and OPPs and many other classes	QuEChERS	UHPLC column Acquity UPLC BEH Shield RP18 (150 × 2.1 mm, 1.7 μm)	Gradient: AcN/H_2O both with 0.02% HCOOH	Waters, Quattro Micro ESI in PI mode SRM, 2 MS/MS transitions	LOQ < 0.01 mg/kg	[61]
Total diet	73 OPPs and carbamates	QuEChERS	LC-Narrow-bore column Sinergi MAX-RP 80 A C_{12} (50 × 2 mm, 4 μm)	Gradient: MeOH/H_2O both with 5 mM $HCOONH_4$ Flow rate: 200 μL/min	ABSciex, API 4000 ESI in PI mode SRM, 2 MS/MS transitions	LOQ 10 μg/kg	[56]

(Continued)

TABLE 15.3 (CONTINUED)
LC-MS Applications to Determine Pesticide Residues

Matrix	Pesticide	Extraction	Separation		Determination		References
			Column	Mobile Phase	Detection	Sensitivity	
Grape	150 pesticides	QuEChERS	LC-Narrow-bore column Sinergi MAX-RP 80 A C_{12} (50 × 2 mm, 4 μm)	Gradient: MeOH/H_2O both with 5 mM $HCOONH_4$ Flow rate: 200 μL/min	Waters, Quattro Micro ESI in PI mode SRM, 2 MS/MS transitions	LOQ 10 μg/kg	[11]
Fruit and vegetables	9 OPPs	MSPD	UHPLC column Zorbax RRHT SB-C18 (100 × 2.1 mm, 1.8 μm)	Gradient of 0.1% HCOOH and 10 mM/L NH_4OAc, and AcN containing 0.1% HCOOH	Agilent 6410 ESI in PI mode SRM, 2 MS/MS transitions	LOD 0.06–0.15 μg/kg LOQ 0.2–0.5 μg/kg	[16]
Made tea, tea infusion, and spent leaves	42 pesticides	Ethyl acetate + cyclohexane (9:1; v/v) d-SPE with PSA + GCB + florisil	LC-Narrow-bore column Symmetry C18 (2.1 × 100 mm, 5 μm)	MeOH/H_2O 10/90 (v/v) with 5 mM NH_4OAc, and MeOH/H_2O 90/10 (v/v) with 5 mM NH_4OAc Flow rate: 200 μL/min	Waters, Quattro Micro ESI in PI and NI mode (only PI for OPPs) SRM	LOQ < 50 ng/mL	[57]
Milk	OPPs, other pesticides (multiresidue and mycotoxins)	Comparison of QuEChERS SPE "dilute-and-shoot" LLE SLE: ethyl acetate	UHPLC column Acquity BEH UPLC™ C18 (100 × 2.1 mm, 1.7 μm)	Gradient of MeOH and $HCOONH_4$ 5 mM in H_2O Flow rate: 350 μL/min	Waters Acquity TQD ESI in PI mode SRM, 2 transitions/ compound	LOD < 0.20 μg/kg LOQ < 0.67 μg/kg	[103]
Olives	OPPs and other pesticides	Two methods: QuEChERS MSPD with aminopropyl, Florisil and AcN	UHPLC column Zorbax Eclipse XDB-C18 (50 × 4.6 mm, 1.8 μm)	Gradient H_2O with 0.1% HCOOH and AcN Flow rate: 0.6 mL/min	Agilent 6410 Dynamic MRM 2 transitions	LOD < 10 μg/kg	[104]
Onion	Multiresidue pesticides	MSPD with C18 and AcN	LC-Normal-bore column Xterra (50 × 3 mm, 3.5 μm)	Isocratic H_2O and AcN both with 0.1% HCOOH (58:42) Flow rate: 0.6 mL/min	Waters, Quattro Micro ESI in PI mode 2 most abundant MS/MS transitions	LODs 0.003–0.03 mg/kg LOQs 0.01–0.1 mg/kg	[105]
Fruit and vegetables	11 OPPs	LE MeOH-water (80:20)	UHPLC column Acquity UPLC BEH C18 (50 × 2.1 mm, 1.7 μm)	Gradient MeOH–H_2O both 10 mM NH_4OAc Flow rate: 0.5 mL/min	Waters, TQD Premier ESI PI mode SRM, 2 MS/MS transitions	LOQ 2–26 pg	[62]
Fresh-cut vegetables	14 pesticides	QuEChERS	LC-Narrow-bore column Ascentis RPAmide (10 cm long, 2.1 mm i.d.)	Gradient: 5 mM $HCOONH_4$ and MeOH	ABSciex, API 3000 ESI in PI mode MRM, 2 transitions per compound	LOQ 0.01–0.05 mg/kg	[64]

Honey	30 pesticides	QuEChERS	LC-Normal-bore column Poroshell 120 EC-C18 (2.7 μm, 3 × 100 mm)	Gradient H_2O and MeOH, both with 10 mM NH_4OAc Flow rate: 0.4 mL/min	Agilent 6460 ESI in PI mode MRM, 2 transitions per compound	LOQ 2.73–75 ng/g	[17]
Plasma	OPPs adduct bound to proteins butyrylcholinesterase (BuChE) and albumin	Extraction with 1 mL *n*-hexane	UHPLC column Acquity HSS T3 (100 × 2.1 mm, 1.8 μm)	Gradient of 0.2% HCOOH in H_2O and 0.2% HCOOH in AcN. Flow rate: 0.1 mL/min	Thermo Scientific, TSQ Quantum Ultra triple quad MRM mode	–	[106]
Green tea	86 pesticides (insecticides, fungicides, and herbicides)	Comparison of three methods QuEChERS Ethyl acetate MiniLuke	LC-Normal-bore column Agilent Zorbax SB- C8 (4.6 × 150 mm, 5 μm)	Gradient of AcN and Milli-Q H_2O with 0.1% HCOOH Flow rate: 0.6 mL/min	Agilent 6410 ESI in PI and NI mode Dynamic MRM 2 transitions	LOD < 100 ng/g	[10]
Cereals	22 carbamate insecticides and 17 mycotoxins	QuEChERS	UHPLC column Acquity BEH C18 column (2.1 × 100 mm, 1.7 μm)	Gradient of H_2O (0.1% HCOOH, 0.50 mM NH_4OAc)/MeOH Flow rate: 0.2 mL/min	Waters Xevo™ TQ ESI in PI mode except for α-ZOL and β-ZOL in NI mode MRM mode	LOQ 0.20–29.7 μg/kg	[107]
Melon	Insecticides (dinotefuran and its metabolites, MNG, UF, and DN)	Modified QuEChERs acetate-buffered	LC-Normal-bore column YMC-Pack Pro C8 (150 × 4.6 mm, 3 μm)	Linear gradient (A) 5 mM NH_4OAc and 0.1% HCOOH in H_2O; (B) MeOH Flow rate: 0.5 mL/min	Agilent 6410 ESI in PI mode MRM with 2 mass transitions	LODs 0.02–0.05 mg/kg LOQs 0.06–0.16 mg/kg	[108]
Apple juice	12 pesticides	QuEChERS	LC-Normal-bore column Zorbax Eclipse XDB (150 × 4.6 mm, 5 μm)	Gradient of milli-Q H_2O with 0.1% HCOOH and AcN Flow rate: 0.6 mL/min	Agilent 6410 ESI in PI mode MRM with 2 mass transitions	LOQ 0.25–1 μg/kg	[109]
Fishpond water	Pentachlorophenol, niclosamide, and fenpropathrin	LLE with dichloromethane and acetone (4:1)	LC-Normal-bore column Acclaim 120-C18 column (3 × 150 mm, 3 μm)	Gradient of 20 mmol L^{-1} NH_4OAc of pH 4.5 and MeOH Flow rate: 0.3 mL/min	Thermo Finnigan LTQ-MS ESI in PI and NI mode SIM for pentachlorophenol MRM for niclosamide and fenpropathrin	LOD 0.02–1.2 ng/mL	[18]

(Continued)

TABLE 15.3 (CONTINUED)
LC-MS Applications to Determine Pesticide Residues

Matrix	Pesticide	Extraction	Separation		Determination		References
			Column	Mobile Phase	Detection	Sensitivity	
Infant formulas	Fungicides (genistein and dicarboximide)	Ultrasonic extraction with AcN/SPE	LC-Narrow-bore column LUNA C8 (50 × 2 mm, 5 μm)	Gradient of 0.05% acetic acid in MilliQ H_2O and AcN Flow rate: 200 μL/min	AB-SCIEX, API 3000 ESI MRM mode, 2 transitions	LOD 0.6–16.5 ng/g	[44]
Cosmetics and household products	Isothiazolinone biocides	MSPD	LC-Narrow-bore column Hypersil Gold aQ (100 × 2.1 mm, 3 μm)	Gradient of (A) H_2O/HCOOH (0.1%)/NH_4OAc (5 mM) and (B) MeOH/HCOOH (0.1%)/NH_4OAc (5 mM) Flow rate: 200 μL/min	Thermo Scientific, Quantum Access HESI in PI mode SRM with 2 mass transitions	LOD < ng/g LOQ < ng/g	[110]
Food samples (rice, orange, apple, spinach)	38 pesticides	QuEChERS	LC-Normal-bore column Prodigy ODS-3 (150 × 3 mm, 5 μm)	Gradient of 0.1% HCOOH in H_2O and 0.1% HCOOH in AcN Flow rate: 0.3 mL/min	ABSciex, API-3000 ESI in PI mode MRM, 2 transitions	–	[24]
Ground, surface, and wastewater	88 polar organic micropollutants (pesticides)	Online SPE	LC-Normal-bore column Atlantis T3 (3 × 150 mm, 3 μm)	Gradient of 5 mM NH_4OAc in H_2O and MeOH with 0.1% HCOOH Flow rate: 300 μL/min	Thermo Scientific, TSQ Quantum Ultra triple quadrupole ESI in PI and NI mode SRM, 2 transitions	LOQ < 0.1–87 ng/L (groundwater and surface water) LOQ < 1.5–206 ng/L (wastewater)	[25]
Green tea, red tea, black tea, and chamomile	86 pesticides (insecticides, fungicides, and herbicides)	QuEChERS (Clean-up: $CaCl_2$ instead of $MgSO_4$)	LC-Normal-bore column Zorbax SB (4.6 × 150 mm, 5 μm)	Gradient of AcN and milliQ H_2O with 0.1% HCOOH Flow rate: 0.6 mL/min	Agilent 6410 ESI in PI and NI mode Dynamic MRM	LODs 0.1–210 μg/kg	[26]
Grapes, baby food, and wheat flour	48 pesticides	Online Turbulent flow chromatography MCX-2 50 × 0.5 mm TurboFlow™ (TX) column	LC-Normal-bore column Hypersil BDS C18 (100 × 3 mm, 3 μm)	Gradient of H_2O and MeOH, both with 0.1% $HCOONH_4$ Flow rate: 1 mL/min	Thermo Fisher Scientific, TSQ Access Max triple quadrupole ESI in PI mode SRM	LOD < 0.8–6.0 ng/g (baby food) LOD < 0.8–10.3 ng/g (grapes and wheat flour)	[27]

Baby food	50 pesticides	QuEChERS	UHPLC column Zorbax Eclipse XDB-C18 (50 × 4.6 mm, 1.8 μm)	Gradient of MeOH and H_2O, both with 0.1% HCOOH Flow rate: 0.5 mL/min	Agilent 6410 ESI in PI mode MRM	LOQ < 10 μg/kg	[19]
Well water	Chlorinated phenols	—	UHPLC column Derivatization with densyl chloride Zorbax Eclipse Plus C18 Rapid Resolution (2.1 × 50 mm, 1.8 μm) Isomeric confirmation method: Zorbax Eclipse Plus C18 Rapid Resolution (2.1 × 100 mm, 1.8 μm)	Gradient of 0.1% HCOOH and MeOH Flow rate: 0.4 mL/min (underivatized and quantitation methods) Flow rate: 0.3 mL/min (isomeric confirmation method)	Agilent 6410 ESI in PI mode 2 transitions	LOD 0.01–1.0 μg/L	[39]
Vegetable (eggplant and lettuce) and fruit (strawberry)	Thiram (dithiocarbamate fungicide)	Extracted with sodium sulfate anhydrous, EDTA, and AcN	LC-Narrow-bore column Discovery C18 (50 × 2.1 mm, 5 μm)	H_2O–MeOH gradient at 0.1 mM NH_4OAc Flow rate: 300 μL/min	Waters, Quattro Micro ESI in PI mode Full scan acquisition 4 MS/MS transitions, one for quantification and 3 for confirmation	LOD < 0.0012 mg/kg	[40]
Drinking water samples	Organic contaminants (pharmaceuticals and pesticides)	SPE	LC-Normal-bore column XTerra MS C 18 (100 × 3 mm, 3.5 μm)	Gradient of (A) H_2O (acidified with 0.1% HCOOH) and (B) MeOH:H_2O (90:10 v/v) Flow rate: 0.30 mL/min	Water, Quattro Micro™ ESI in PI and NI mode MRM mode	LOQ < 0.006–0.208 mg/L	[28]
Infant milk formula	Pesticides	Microwave-assisted extraction (MAE) and solid phase extraction (SPE)	LC-Normal-bore column Zorbax Rx-SIL (4.6 × 250 mm, 5 μm)	Gradient of (A) 5 mM $HCOONH_4$ and 0.1% HCOOH in H_2O and (B) 5 mM $HCOONH_4$ and 0.1% HCOOH in AcN Flow rate: 1 mL/min	Agilent 6410 ESI in PI mode MRM, 2 transitions	LOD 0.12 to 2.53 μg/kg LOQ 0.41 to 8.42 μg/kg	[29]
Marine organisms (mainly *Mytilus edulis*)	Pharmaceuticals Perfluorinated compounds Pesticides	PLE and SPE	UHPLC column Nucleodur C18 Pyramid UHPLC (100 × 2 mm, 1.8 μm)	Gradient 2 mM of NH_4HCO_3 in H_2O and MeOH Flow rate: 0.3 mL/min	Thermo Electron, TSQ Vantage Triple-Stage Quadrupole ESI PI mode SRM mode, 2 MS/MS transitions	LOQ 0.1–10 ng/g	[22]
Minor crops (amaranth and parsley)	Spinetoram (insecticide) and its metabolites (demethyl and formyl)	QuEChERS	LC-Narrow-bore column Gemini C18 (50 × 2.0 mm, 3 μm)	Linear gradient of 10 mM NH_4OAc in H_2O and AcN Flow rate: 0.35 mL/min	Agilent 6410 ESI in PI mode MRM, 2 mass transitions	LOD 0.01 mg/kg LOQ 0.03 mg/kg	[45]

(*Continued*)

TABLE 15.3 (CONTINUED)
LC-MS Applications to Determine Pesticide Residues

Matrix	Pesticide	Extraction	Separation: Column	Separation: Mobile Phase	Determination: Detection	Determination: Sensitivity	References
IT							
Honey	12 Insecticides (OPPs and carbamates)	Comparison: QuEChERS SPE PLE SPME	LC-Normal-bore column Luna C18 (250 mm × 4.6 mm, 5 μm)	Gradient MeOH–H_2O Flow rate: 0.7 mL/min	Esquire 3000 (Brucker) APCI in both PI and NI modes Full-scan MS and MRM for MSn	CCα - 0.01–1.1155 μg/g	[30]
Variety of fruits and vegetables and products of animal origin	9 benzoylureas (BUs)	PLE	LC-Normal-bore column Luna C18 (150 × 4.60 mm, 5 μm)	Gradient of MeOH–H_2O Flow rate: 0.5 mL/min	Esquire 3000 (Brucker) APCI Full scan and MRM modes	LODs 0.7–3.4 μg/kg LOQs 2–10 μg/kg	[31]
Baby food	Fungicide residue	QuEChERS	LC-Normal-bore column Zorbax Eclipse XDB-C8 (50 × 4.6 mm, 5 μm)	H_2O with 0.1% HCOOH and AcN Flow rate: 0.6 mL/min	Esquire 6000 (Brucker) ESI in PI mode Full-scan and product ion scan MS/MS mode	LODs 0.5–3.0 μg/kg	[32]
LIT							
Lettuce (watery), maize grain (oily), wheat grain (dry), whole orange (acidic)	Insecticides (Chlorantraniliprole, Cyantraniliprole)	QuEChERS	LC-Narrow-bore column Aqua C18 (50 × 2 mm, 5 μm)	Gradient of (A) H_2O-MeOH (8:2 v/v) + 0.1% HCOOH + 5 mM NH_4OAc and (B) MeOH-H_2O (9:1 v/v) + 0.1% HCOOH + 5 mM NH_4OAc Flow rate: 0.25 mL/min	ABSciex, API 5000 ESI in PI mode, 2 transitions LCQ classic, DESI (Omnispray ionization), 3 stages of mass analysis, LTQ Orbitrap, ESI in PI mode, 2 transitions	LOQ 0.01 mg/kg	[41]
Porcine plasma and urine	OPPs (dimethoate and omethoate)	Deproteinization with AcN and dilution	LC-Normal-bore column Atlantis T3 C18 (150 mm × 4.6 mm, 5 μm)	(A) 0.1%, v/v, $HCOONH_4$ in H_2O and (B) ACN/H_2O 80:20, v/v; 0.1%, v/v, $HCOONH_4$	ABSciex, API 4000, ESI in PI mode, MRM, 2 transitions Comparison with flow injection ESI-MS/MS	LOD plasma 0.12–0.24 μg/mL LOQ plasma 0.24–0.49 μg/mL LOD urine 0.39–0.78 μg/mL LOQ urine 0.78–1.56 μg/mL	[33]

Fruit and vegetables	300 pesticides	QuEChERS	LC-Normal-bore column Reverse-phase C8 column (150 × 4.6 mm, 5 μm)	Gradient of deionized H_2O with 0.1% HCOOH, and AcN Flow rate: 0.6 mL/min	Applied Biosystems 3200 QTRAP ESI in PI mode EPI screening for 300 pesticides and MRM quantitation for 55 pesticides	LOD in MRM < 1-LOD: 20 ng/kg LOD on EPI < 1–40 ng/kg	[34]
Eggs	Phoxim and its phototransformation products	LE with AcN SPE with silica gel deactivated with trimethylamine	LC-Normal-bore column Zorbax SB-C18 (4.6 × 150 mm, 5 μm)	Gradient aqueous 0.1% (v/v) HCOOH and 0.05 M $HCOONH_4$ and MeOH Flow rate: 0.8 mL/min	Applied Biosystems 3200 QTRAP ESI in PI mode SRM. A total of six MS/MS transitions were monitored	CCα 0.0005–0.0044 mg/kg CCβ 0.0054–0.0224 mg/kg	[36]
Fruit and vegetables	191 pesticides	QuEChERS	LC-Narrow-bore column Ultra Aqueous, C18 (100 × 2.1 mm, 3 μm)	Gradient of MeOH-H_2O both 5 mM $HCOONH_4$, 0.1% HCOOH Flow rate: 300 μL/min	ABSciex 4000 Qtrap ESI in PI mode 2 most abundant MS/MS transitions	MDLs 0.5–5 μg/kg	[46]
Spices	Pesticides, dyes, and mycotoxins	LE with AcN	LC-Narrow-bore column Synergi Fusion-C18 (50 × 2 mm, 4 μm)	Gradient of MeOH-H_2O both 5 mM $HCOONH_4$ Flow rate: 200 μL/min	Applied Biosystems, 4000 Qtrap™ ESI in PI mode 2 most abundant MS/MS transitions	LOQ < 10 μg/kg	[12]
Fruit juices	53 pesticides (OPPs and others)	Centrifugation and mix 100 μL of juice with 900 μL of AcN	LC-Normal-bore column Reverse-phase C8 column (150 × 4.6 mm, 5 μm)	Gradient AcN and H_2O with 0.1% HCOOH Flow rate: 400 μL/min	ABSciex, 5500 QTRAP ESI in PI and NI mode SRM transitions with scheduled SRM mode 2 MS/MS transitions per compound	LOQ 0.1–5 μg/L	[37]
Fish muscle	13 pesticides	QuEChERS	UHPLC column Reverse phase Zorbax Eclipse C18 column (50 × 2.1 mm, 1.8 μm)	0.0125% (v/v, pH 4.04) CH_3COOH and 100% AcN	Applied Biosystems, 5500 QTRAP ESI in PI mode MRM, 2 transitions per compound	LOQ < 10 ng/g	[20]
River and underground waters	Chiral pesticides (enantiomers of mecoprop and dichlorprop)	SUSME and re-extraction in acetate buffer (pH 5.0)	LC-Normal-bore column Nucleodex α-PM (200 × 4.0 mm, 5 μm)	Isocratic: 65% MeOH and 35% 100 mM HCOOH/$HCOONH_4$ (pH 4.0) Flow rate: 0.5 mL/min	ABSciex 4000 Turbo Ion Spray (TIS) interface in NI mode SRM mode	LOQ 1–4 ng/L	[111]
Groundwaters and surface water	150 pesticide metabolites	Direct injection	LC-Narrow-bore column Synergy Fusion-RP 100A (50 × 2.0 mm, 2.5 μm)	Gradient of (A) 90/10 (v/v) H_2O/MeOH, (B) 10/90 (v/v) H_2O/MeOH each with 0.2% CH_3COOH. Flow rate: 250 μL/min	ABSciex, API 5500 ESI in PI and NI mode MRM with 2 mass transitions	LOQ 0.025–0.1 μg/L	[21]

(Continued)

TABLE 15.3 (CONTINUED)
LC-MS Applications to Determine Pesticide Residues

Matrix	Pesticide	Extraction	Separation: Column	Separation: Mobile Phase	Determination: Detection	Determination: Sensitivity	References
Fruits, cereals, spices, and oil seeds	288 pesticides, 38 mycotoxins	Comparison of 3 methods: QuEChERS Aqueous AcN extraction AcN extraction	UHPLC column Acquity UHPLC HSS T3 (100 × 2.1 mm, 1.8 μm)	ESI (+): 5 mM $HCOONH_4$ and 0.2% HCOOH in both Milli-Q H_2O and MeOH ESI (−): 5 mM NH_4OAc in Milli-Q H_2O and pure MeOH Gradient of flow rate: 0.3–0.7 mL/min	ABSciex, 5500 QTRAP TurboIon™ electrospray (ESI) in PI and NI mode MRM mode	LOQ (method A) < 10 μg/kg	[112]
Human whole blood and urine	Disulfoton and its oxidative metabolites	QuEChERS	LC-Narrow-bore column CAPCELL-PAK MG II column (35 × 2.0 mm, 5 μm)	Gradient of (A) 95% 10 mmol/L $HCOONH_4$ 5% MeOH and (B) 5% 10 mmol/L $HCOONH_4$ 95% MeOH Flow rate: 0.2 mL/min	ABsciex 3200 ESI (MRM-EPI) scan mode	LOD whole blood 0.90–1.15 ng/mL LOD urine 0.46–1.05 ng/mL LOQ whole blood and urine < 5 ng/mL	[47]
Green and black tea	Pesticide residues	Comparison of 3 methods: QuEChERS–dSPE, AcN–dSPE, AcN–HTpSPE	LC-Normal-bore column Chromolith Performance RP-18 end-capped (100 mm × 3.0 mm)	Gradient of AcN and 10 mM $HCOONH_4$ with 2% of MeOH	AB SCIEX, 5500 QTRAP ESI in PI mode MRM mode, 2 specific precursor-to-product ion transitions	LOQ < 0.002 mg/kg	[76]
Fruit juices	174 pesticides	QuEChERS	LC-Narrow-bore column Atlantis T3 octadecyl silica (C18) column (2.1 × 100 mm, 3 μm)	Buffer (A) 4 mM of $HCOONH_4$ and 0.1% HCOOH in H_2O and buffer (B) 4 mM of $HCOONH_4$ and 0.1% HCOOH in AcN Flow rate: 0.4 mL/min	AB Sciex 4000 ESI in PI mode MRM, 2 transitions	LOQ 0.04–0.5 μg/g	[48]

			TOF				
Rice	4 herbicides, 9 fungicides, 2 insecticides	QuEChERS	UHPLC column XDB-C18 (4.6 × 50 mm, 1.8 μm)	Gradient: (A) H_2O and acetonitrile (95/5) with 0.1% HCOOH, (B) AcN and H_2O (95/5) with HCOOH Flow rate: 0.6 mL/min	Agilent MSDTOF ESI in PI mode TIC in Full-scan	LOD < 10 μg/kg	[7]
Fruit-based soft drink	33 pesticides	SPE with Oasis HLB	UHPLC column Eclipse XDB-C18 (50 × 4.6 mm, 1.8 μm)	Gradient H_2O with 0.1% HCOOH and AcN Flow rate: 450 μL/min	Agilent MSD TOF ESI in PI mode In-source CID fragmentation	LOQ < 0.02–2 μg/L	[3]
Olive oil	OPPs and other pesticides	Two methods: QuEChERS MSPD with aminopropyl, Florisil and AcN	UHPLC column Eclipse XDB-C18 (50 × 4.6 mm, 1.8 μm)	Gradient H_2O with 0.1% HCOOH and AcN Flow rate: 0.6 mL/min	Agilent MSD TOF Dynamic MRM, 2 transitions	LOQ < 10 μg/kg	[35]
Surface water and soil samples	Pesticides and other pollutants	Waters: SPE Soils: ultrasonic bath with ethyl acetate	UHPLC column Acquity UHPLC BEH C18 (150 × 2.1 mm, 1.7 μm)	H_2O and MeOH, both acidified with 0.01% HCOOH Flow rate: 300 μL/min	Waters, Q-oaTOF Premier ESI in PI and NI mode Accurate-mass full-spectrum acquisition	–	[42]
Palm oil	Pesticides	QuEChERS	UHPLC column Zorbax Eclipse SB-C18 (50 × 2.1 mm, 1.8 μm)	Gradient AcN and H_2O with 0.1% HCOOH Flow rate: 0.25 mL/min	Agilent MSD TOF ESI in PI mode Accurate mass spectra	LOD < 5 ng/g LOQ < 9 ng/g	[90]
Wastewater effluent	84 pesticides and pharmaceuticals	SPE	UHPLC column ACQUITY UHPLC BEH C18 (100 × 2.1 mm, 1.7 μm)	Positive mode: 5 mM NH_4HCO_3, pH 9.5, pH adjusted with NH_3(aq) and 100% MeOH In negative mode: 0.05% CH_3COOH in H_2O and 0.05% CH_3COOH in MeOH Flow rate: 0.45 mL/min	Micromass LCT Premier XE ESI in PI and NI mode Resolution > 11,000 FHWM Mass range 100–1000	Positive polarity: ILD, 7.5 pg ILQ, 19 pg Negative polarity: ILD, 20 pg ILQ, 46 pg	[66]
Fruits, cereals, spices, and oil seeds	288 pesticides, 38 mycotoxins	Comparison of 3 methods: QuEChERS Aqueous AcN extraction AcN extraction	UHPLC column Acquity UHPLC HSS T3 (100 × 2.1 mm, 1.8 μm)	ESI (+): 5 mM $HCOONH_4$ and 0.2% HCOOH in both Milli-Q H_2O and MeOH ESI (–): 5 mM NH_4OAc in Milli-Q H_2O and pure MeOH Gradient of flow rate: 0.3–0.7 mL/min	Waters, LCT Premier XE (ESI) in PI and NI mode MRM mode	–	[91]

(*Continued*)

TABLE 15.3 (CONTINUED)
LC-MS Applications to Determine Pesticide Residues

			Separation		Determination		
Matrix	**Pesticide**	**Extraction**	**Column**	**Mobile Phase**	**Detection**	**Sensitivity**	**References**
Wine	15 fungicides	Sorptive extraction	LC-Narrow-bore column Zorbax Eclipse XDB-C18 (100 mm × 2.1 mm, 3.5 μm)	Gradient: Ultrapure H_2O and AcN, both with NH_4OAc 1 mM	Agilent 6520 Dual-Spray ESI source Operated in the 2-GHz Extended dynamic Range resolution mode	LOQs 0.1–2.2 ng/mL	[49]
			QTOF				
Apples and pears	Postharvest pesticides and metabolites of EQ and DPA	Ethyl acetate	UHPLC column Acquity C18 (15 cm × 2.1 mm, 1.7 μm)	Gradient MeOH–H_2O both 10 mM $HCOONH_4$ Flow rate: 200 μL/min	Water/Micromass ESI in PI mode	LOD 0.02–0.34 μg LOQ 0.05–0.60 μg	[6]
Fruits and vegetables	148 pesticides	QuEChERS	UHPLC column Acquity UHPLC BEH C18 (100 × 2.1 mm, 1.7 μm)	Gradient AcN and H_2O 10 mM $HCOONH_4$	Waters, QTOF Premier Full-scan MS and product ion full-scan MS/MS Mass range from *m/z* 50 to 950	LOQ < 10 μg/kg	[50]
Fruits and vegetables	148 pesticides	QuEChERS	LC-Narrow-bore column Atlantis dC18 (100 × 2.1 mm, 3 μm)	Gradient AcN and 10 mM NH_4OAc with 0.2% AcN Flow rate: 0.2 mL/min	Applied Biosystems, API 5000 ESI in PI mode 2 MS/MS transitions	LOQ < 5 μg/mL	[50]
Apple, strawberry, tomato, and spinach	212 pesticides	QuEChERS	UHPLC column Acquity UHPLC HSS T3 column (100 × 2.1 mm, 1.8 μm)	Gradient MeOH and 0.005 M $HCOONH_4$ in H_2O Flow rate: 200 μL/min	Waters LCT Premier XE ESI in PI and NI modes DRE and CID were used Target and nontarget screening	LOQ ≤ 10 μg/kg	[63]
Food	240 pesticides	QuEChERS	LC-Narrow-bore column Restek Ultra Aqueous C18 (100 × 2.1 mm, 3.0 μm)	Gradient H_2O 10 mM $HCOONH_4$/0.1% HCOOH and MeOH 10 mM $HCOONH_4$/0.1% HCOOH	ABSciex, 5600 Q-TOF ESI, PI mode MRM mode	–	[51]

Red pepper	Isobaric pesticides	Extraction with methano/water (80:20)	UHPLC column Zorbax C8 (4.6 × 150 mm, 1.8 and 3.5 μm) Zorbaz C18 (4.6 × 150 mm, 1.8 μm) Phenyl column (4.6 × 150 mm, 1.8 μm)	Gradient 90% H_2O 0.1% HCOOH and 10% AcN Flow rate: 0.6 mL/min	Agilent 6540 ESI in PI mode Single MS with full spectra and MS/MS to discriminate isobars	–	[93]
Surface water and wastewater	43 pesticides, 13 pharmaceuticals, 2 drugs	Offline SPE	UHPLC column Acquity UHPLC BEH C18 (100 × 2.1 mm, 1.7 μm)	Gradient MeOH-H_2O both 0.01% HCOOH Flow rate: 300 μL/min	Waters, Q-oaTOF Premier ESI interface in PI mode	–	[9]
Fruit and vegetables	97 pesticides	QuEChERS	UHPLC column XDB-C18 (4.6 × 50 mm, 1.8 μm)	Positive mode: 0.1% HCOOH and 5% MilliQ H_2O in AcN and 0.1% HCOOH in H_2O (pH 3.5) Flow rate: 0.6 mL/min Negative mode: 5% MilliQ H_2O in AcN and 5% AcN in HPLC-grade H_2O Flow rate: 0.6 mL/min	Agilent 6530 ESI in PI and NI mode Full-scan mode	–	[112]
Fruit and vegetables	53 pesticides	QuEChERS	LC-Normal-bore column Agilent Zorbax Eclipse XDB C8 (150 × 4.6 mm, 5 μm)	Gradient AcN and high-purity H_2O with 0.1% HCOOH Flow rate: 0.6 mL/min	Agilent 6530 ESI in PI mode 4 GHz High Resolution Mode Q-TOF-MS instrument was used as a TOF-MS system working in the MS mode under full-scan conditions	–	[94]
Apple and pear		Solvent extraction Ethyl acetate	UHPLC column Waters Acquity C18 (15 cm × 2.1 mm 1.7 μm)	H_2O and MeOH, both with 10 mM $HCOONH_4$ Flow rate: 200 μL/min	Waters Micromass ESI PI mode Resolution ~ 10,000 FHWM Scan MS and MS/MS mode	LOQ: 0.05–1 μg	[6]
Oranges and bananas	Multiclass pesticides	Solvent extraction MeOH/water (80:20, v/v)	UHPLC column Acquity C18 BEH (150 × 2.1 mm, 1.7 μm)	H_2O and MeOH, both with 0.01% HCOOH Flow rate: 0.3 mL/min	Waters, Premier ESI in PI and NI mode Resolution > 10,000 FHWM Scan MS and MS/MS mode	–	[98]

(Continued)

TABLE 15.3 (CONTINUED)
LC-MS Applications to Determine Pesticide Residues

Matrix	Pesticide	Extraction	Separation: Column	Separation: Mobile Phase	Determination: Detection	Determination: Sensitivity	References
ORBITRAP							
Fruit and vegetable peel (apples, pears, citrics)	Postharvest compounds and other xenobiotics	QuEChERS Direct peel monitoring	–	–	Thermo Fisher, LTQ Velos ESI in PI mode Full spectral acquisition mode	LOQ 1 ng	[95]
Fruit and vegetable peel (apples, pears, citrics)	Postharvest compounds and other xenobiotics	Ultrasound-assisted extraction	UHPLC column Kinetic C18 (100 × 2.1 mm, 1.7 μm)	Gradient MeOH-ultrapure H_2O both 10 mM $HCOONH_4$ Flow rate: 0.3 mL/min	Thermo Fisher, LTQ Velos ESI in PI mode Full spectral acquisition mode	LOQ 1 ng	[95]
Wastewater	Acidic pesticide and pharmaceutical contaminants	SPE	LC-Narrow-bore column Eclipse XDBC18 (2.1 × 150 mm, 5 μm)	Gradient of AcN/H_2O, both with 1 mM NH_4OAc	Thermo Scientific, LTQ Orbitrap Discovery HESI in NI mode Full-scan MS 30,000 FHWM resolution	LOQ 2.1–27 ng/L	[113]

methods and analytical conditions used in each case. Pesticides are analyzed by multiresidue methods that allow the determination of as many pesticides as possible. Several of the examples reported in Table 15.1 are able to analyze up to 300 pesticides [5,11,21,34,46,55] and even pesticides and other organic contaminants, such as mycotoxins, dyes, perfluorinated compounds, or pharmaceuticals [12,22,66,77,91,103,107,113,114].

Samples analyzed are mostly fruit and vegetables that are frequently monitored for legal/enforcement purposes that are aimed at establishing compliance with maximum residue limits (MRLs) [5,7,14,53]. Any food sample is susceptible to pesticide residue determination, including other agricultural products (e.g., wines, dried botanical dietary supplements, juices, baby food, etc.) [54,75,77,78], including products of animal origin, such as meat, milk, or honey [17,55,103]. There is a strict regulation in this sense to protect consumers from the harmful effects of pesticide residues, particularly some highly vulnerable or susceptible groups, such as children, pregnant woman, and older people.

Environmental samples are also frequently monitored for pesticide residue determination. To give just an idea of the importance of the problem, more than 98% of sprayed insecticides and 95% of herbicides reach a destination other than their target species, including nontarget species, air, water, bottom sediments, and soil. Consequently, all the environmental compartments can be contaminated by residues of these substances, which are a continuous threat to environmental safety. Among these samples, several biological fluids, such as urine, blood, plasma, or even tissues, have also been analyzed to determine pesticide levels. The analysis of biological matrices remains a challenging task due to sample complexity [3,33,61].

In all these fields, the advent of LC-MS has been responsible for a general trend toward the simplification of sample preparation to diminish the amount of sample needed, the number of off-line steps, and the amount of solvents employed. In this way, the sample pretreatments are based on extraction with an organic solvent miscible or not with water and off-line or online cleanup using solid phase extraction (SPE) cartridges. QuEChERs and SPE are the most used techniques, and they are replacing other traditional extraction techniques [8,72,74].

QuEChERs uses acetonitrile for extraction, anhydrous $MgSO_4$:NaCl to induce partitioning of acetonitrile extract from the water of the sample, and dispersive SPE to clean up. It has been validated for a large quantity of pesticides in food matrices mostly [1,2,7,8,11–15,22,38,54,62,64,77,106,107,115–121]. Some publications report a modified QuEChERs method, for instance, the use of acetate buffered to determine insecticides in melon [57] and the use of calcium chloride instead of magnesium sulfate in the clean-up step [36].

Off-line SPE has been applied to food and environmental samples to determine residue of pesticides and other organic pollutants, such as drugs and pharmaceuticals [9,14,114,122]. In other cases, the sample has been directly injected in the LC-MS/MS system without the necessity of analyte extraction [34,103].

Other authors have reported the comparison between several extraction techniques applied to different groups of pesticides and matrices [30,41,53,63,96,106] or the combination of two techniques in order to improve the extraction process, for instance, microwave-assisted extraction and SPE for the analysis of pesticides in infant milk formula [10]. Moreover, some alternative, less frequent procedures have been described, such as supramolecular solvent-based microextraction and re-extraction in acetate buffer (pH 5.0) [16], liquid extraction with dichloromethane and acetone (4:1) [3], matrix solid phase dispersion [50], turbulent flow chromatography [27], pressurized liquid extraction, [22], etc.

As it could also be observed, in Table 15.1, UHPLC separation has almost replaced the more conventional LC. Almost 50% of the reported applications used this system. It is indistinctly applied to determine pesticides in water [9], fruits and vegetables [5], dried botanical supplements [75], and tea [61]. This general use gives a good idea of the advantages and robustness of the technique.

A QqQ analyzer is the most used to determine pesticide residues, followed by the LIT and, more recently, TOF and QqTOF instruments; finally, few applications of the orbitrap mass analyzers have been reported, probably because this is the ultimate mass analyzer. Nowadays, the analysis of pesticide residue may be targeted or nontargeted. Target analysis is the conventional one, in which only those analytes selected a priori can be determined. This target analysis is mostly performed with QqQ and QLIT in multiple SRM but can also be performed by TOF, QqTOF, or orbitrap. With TOF, QqTOF, or orbitrap, several MS libraries can be self-constructed because, commonly, they are not commercially available. The full m/z chromatogram can be extracted, and target pesticides can be identified without the need of analytical standards. These systems also attain posttarget screening when unexpected pesticides, not searched for in the analysis, are suspected.

Another important issue on pesticide control to ensure food safety, which is still a challenge for the analyst, is the identification of nontarget pesticides and metabolites. The nontarget analysis offers the possibility of identifying unexpected pesticides, transformation products and/or impurities, or even untargeted compounds that can be toxic. This analysis is more complicated because it requires the identification of unknown compounds. There are some methods that were developed with the aim of identifying DPs of pesticides based on the information obtained by accurate MS, such as TOF and QTOF or orbitrap. There are already some interesting examples that report the identification of nontarget pesticides or even the identification of potential metabolites present in the sample. This last approach is very interesting to improve food safety.

15.5 CONCLUSIONS

LC-MS has clearly modified the strategy applied for the quantification of pesticide residues in food, biological, and environmental samples. MS is now playing a pivotal role in solving analytical problems concerning food and environmental safety. Because LC-MS and LC-MS/MS allow the achievement of high-throughput analysis with high sensitivity and selectivity, the sample preparation can be simplified and sped. The enhanced selectivity of MS/MS techniques is almost a must for complex sample matrices, such as food, sediments, blood, or soil.

LC-MS has achieved the sensitivity needed to meet European Union (EU) legislation for the analysis of pesticides in water (Directive 60/2000/EU, 2000) and food samples, which have been set through MRLs (European Commission, 1999; WHO, 2000) [98].

Certain limitations of the LC-MS technique are due to the matrix effect, and as a consequence, without appropriate sample preparation and additional chromatographic separation, the size of the error could be very high.

With the introduction of QqQ-MS, IT-MS^n, LIT, TOF, QqTOF, and orbitrap instruments, all major classes of pesticides can be detected, identified, and quantified satisfactorily. Currently, the first option is typically QqQ-MS or LIT-MS in multiple SRM for target pesticide analysis. However, the gradual introduction of the newly developed accurate MS, such as the QTOF or the orbitrap, with the distinctly enhanced selectivity, and the possibility of calculating elemental composition, will improve the performance of the analysis and has opened the door to the identification of nontarget and unknown pesticides as well as metabolites.

No doubt, in the next decade, LC-MS techniques and new analytical tools either for the characterization of a pesticide in parts per trillion levels or lower, and the detection of a pesticide with a very low threshold, will be among the leading options.

ACKNOWLEDGMENTS

This work has been supported by the Spanish Ministry of Economy and Competitiveness through the Program Consolider-Ingenio 2010 (Project No. CSD2009-00065) and Program on Fundamental Research (Projects No. CGL2011-29703-C02-00, CGL2011-29703-C02-01 and CGL2011-29703-C02-02).

REFERENCES

1. Andreu, V., Pico, Y., Determination of currently used pesticides in biota. *Anal. Bioanal. Chem.* 404: 2659–2681, 2012.
2. Moeder, M., Bauer, C., Popp, P., van Pinxteren, M., Reemtsma, T. Determination of pesticide residues in wine by membrane-assisted solvent extraction and high-performance liquid chromatography–tandem mass spectrometry. *Anal. Bioanal. Chem.* 403: 1731–1741, 2012.
3. Gilbert-Lopez, B., Garcia-Reyes, J. F., Mezcua, M., Ramos-Martos, N., Fernandez-Alba, A. R., Molina-Diaz, A., Multi-residue determination of pesticides in fruit-based soft drinks by fast liquid chromatography time-of-flight mass spectrometry. *Talanta* 81: 1310–1321, 2010.
4. Kumar, A., Malik, A. K., Pico, Y., Sample preparation methods for the determination of pesticides in foods using CE-UV/MS. *Electrophoresis* 31: 2115–2125, 2010.
5. Kmellar, B., Pareja, L., Ferrer, C., Fodor, P., Fernandez-Alba, A. R., Study of the effects of operational parameters on multiresidue pesticide analysis by LC-MS/MS. *Talanta* 84: 262–273, 2011.
6. Pico, Y., la Farre, M., Segarra, R., Barcelo, D., Profiling of compounds and degradation products from the postharvest treatment of pears and apples by ultra-high pressure liquid chromatography quadrupole-time-of-flight mass spectrometry. *Talanta* 81: 281–293, 2010.
7. Pareja, L., Colazzo, M., Perez-Parada, A., Besil, N., Heinzen, H., Boecking, B., Cesio, V., Fernandez-Alba, A. R., Occurrence and distribution study of residues from pesticides applied under controlled conditions in the field during rice processing. *J. Agric. Food Chem.* 60: 4440–4448, 2012.
8. Malik, A. K., Blasco, C., Pico, Y., Liquid chromatography-mass spectrometry in food safety. *J. Chromatogr. A* 1217: 4018–4040, 2010.
9. Masia, A., Ibanez, M., Blasco, C., Sancho, J. V., Pico, Y., Hernandez, F., Combined use of liquid chromatography triple quadrupole mass spectrometry and liquid chromatography quadrupole time-of-flight mass spectrometry in systematic screening of pesticides and other contaminants in water samples. *Anal. Chim. Acta* 761: 117–127, 2013.
10. Rajski, L., Lozano, A., Belmonte-Valles, N., Ucles, A., Ucles, S., Mezcua, M., Fernandez-Alba, A. R., Comparison of three multiresidue methods to analyse pesticides in green tea with liquid and gas chromatography/tandem mass spectrometry. *Analyst* 138: 921–931, 2013.
11. Afify, A. E.-M., Mohamed, M. A., El-Gammal, H. A., Attallah, E. R., Multiresidue method of analysis for determination of 150 pesticides in grapes using quick and easy method (QuEChERS) and LC-MS/MS determination. *J. Food Agric. Environ.* 8: 602–606, 2010.
12. Ferrer, A. C., Unterluggauer, H., Fischer, R., Fernandez-Alba, A., Masselter, S., Development and validation of a LC-MS/MS method for the simultaneous determination of aflatoxins, dyes and pesticides in spices. *Anal. Bioanal. Chem.* 397: 93–107, 2010.
13. Jia, Z., Mao, X., Chen, K., Wang, K., Ji, S., Comprehensive multiresidue method for the simultaneous determination of 74 pesticides and metabolites in traditional Chinese herbal medicines by accelerated solvent extraction with high-performance liquid chromatography/tandem mass spectrometry. *J. AOAC Int.* 93: 1570–1588, 2010.
14. Li, Y., Dong, F., Liu, X., Xu, J., Chen, X., Han, Y., Liang, X., Zheng, Y., Studies of enantiomeric degradation of the triazole fungicide hexaconazole in tomato, cucumber, and field soil by chiral liquid chromatography–tandem mass spectrometry. *Chirality* 25: 160–169, 2013.
15. Zhang, F., Huang, Z., Zhang, Y., Li, Z., Wang, M. Determination of 20 carbamate pesticide residues in food by high performance liquid chromatography–tandem mass spectrometry. *Se Pu = Chinese Journal of Chromatography/Zhongguo hua xue hui* 28: 348–355, 2010.
16. Guan, S. X., Yu, Z. G., Yu, H. N., Song, C. H., Song, Z. Q., Qin, Z., Multi-walled carbon nanotubes as matrix solid-phase dispersion extraction adsorbent for simultaneous analysis of residues of nine organophosphorus pesticides in fruit and vegetables by rapid resolution LC-MS-MS. *Chromatographia* 73: 33–41, 2011.
17. Barganska, Z., Slebioda, M., Namiesnik, J., Pesticide residues levels in honey from apiaries located of Northern Poland. *Food Control* 31: 196–201, 2013.
18. Jiang, H., Zhang, Y., Chen, X., Lv, J., Zou, J., Simultaneous determination of pentachlorophenol, niclosamide and fenpropathrin in fishpond water using an LC-MS/MS method for forensic investigation. *Anal. Meth.* 5: 111–115, 2013.
19. Vukovic, G., Shtereva, D., Bursic, V., Mladenova, R., Lazic, S., Application of GC-MSD and LC-MS/MS for determination of priority pesticides in baby foods in Serbian market. *Lwt Food Sci. Technol.* 49: 312–319, 2012.

20. Lazartigues, A., Wiest, L., Baudot, R., Thomas, M., Feidt, C., Cren-Olive, C., Multiresidue method to quantify pesticides in fish muscle by QuEChERS-based extraction and LC-MS/MS. *Anal. Bioanal. Chem.* 400: 2185–2193, 2011.
21. Reemtsma, T., Alder, L., Banasiak, U., A multimethod for the determination of 150 pesticide metabolites in surface water and groundwater using direct injection liquid chromatography–mass spectrometry. *J. Chromatogr. A* 1271: 95–104, 2013.
22. Wille, K., Kiebooms, J. A., Claessens, M., Rappe, K., Vanden Bussche, J., Noppe, H., Van Praet, N., De Wulf, E., Van Caeter, P., Janssen, C. R., De Brabander, H. F., Vanhaecke, L., Development of analytical strategies using U-HPLC-MS/MS and LC-ToF-MS for the quantification of micropollutants in marine organisms. *Anal. Bioanal. Chem.* 400: 1459–1472, 2011.
23. Zhou, Z., Zhang, M., Lin, S., Method establishment for the determination of six organophosphorus pesticides metabolites in human urine using LC-MS/MS. *Wei sheng yan jiu = Journal of Hygiene Research* 42: 122–126, 2013.
24. Kwon, H., Lehotay, S. J., Geis-Asteggiante, L., Variability of matrix effects in liquid and gas chromatography–mass spectrometry analysis of pesticide residues after QuEChERS sample preparation of different food crops. *J. Chromatogr. A* 1270: 235–245, 2012.
25. Huntscha, S., Singer, H. P., McArdell, C. S., Frank, C. E., Hollender, J., Multiresidue analysis of 88 polar organic micropollutants in ground, surface and wastewater using online mixed-bed multilayer solid-phase extraction coupled to high performance liquid chromatography–tandem mass spectrometry. *J. Chromatogr. A* 1268: 74–83, 2012.
26. Lozano, A., Rajski, L., Belmonte-Valles, N., Ucles, A., Ucles, S., Mezcua, M., Fernandez-Alba, A. R., Pesticide analysis in teas and chamomile by liquid chromatography and gas chromatography tandem mass spectrometry using a modified QuEChERS method: Validation and pilot survey in real samples. *J. Chromatogr. A* 1268: 109–122, 2012.
27. Hollosi, L., Mittendorf, K., Senyuva, H. Z., Coupled turbulent flow chromatography: LC-MS/MS method for the analysis of pesticide residues in grapes, baby food and wheat flour matrices. *Chromatographia* 75: 1377–1393, 2012.
28. Maldaner, L., Jardim, I. C., Determination of some organic contaminants in water samples by solid-phase extraction and liquid chromatography tandem mass spectrometry. *Talanta* 100: 38–44, 2012.
29. Fang, G., Lau, H. F., Law, W. S., Li, S. F. Y., Systematic optimisation of coupled microwave-assisted extraction-solid phase extraction for the determination of pesticides in infant milk formula via LC-MS/MS. *Food Chem.* 134: 2473–2480, 2012.
30. Blasco, C., Vazquez-Roig, P., Onghena, M., Masia, A., Pico, Y., Analysis of insecticides in honey by liquid chromatography-ion trap-mass spectrometry: Comparison of different extraction procedures. *J. Chromatogr. A* 1218: 4892–4901, 2011.
31. Brutti, M., Blasco, C., Pico, Y., Determination of benzoylurea insecticides in food by pressurized liquid extraction and LC-MS. *J. Sep. Sci.* 33: 1–10, 2010.
32. Gilbert-Lopez, B., Garcia-Reyes, J. F., Molina-Diaz, A., Determination of fungicide residues in baby food by liquid chromatography-ion trap tandem mass spectrometry. *Food Chem.* 135: 780–786, 2012.
33. John, H., Eddleston, M., Clutton, R., Worek, F., Thiermann, H., Simultaneous quantification of the organophosphorus pesticides dimethoate and omethoate in porcine plasma and urine by LC-ESI-MS/MS and flow-injection-ESI-MS/MS. *J. Chromatogr. B* 878: 1234–1245, 2010.
34. Kmellar, B., Abranko, L., Fodor, P., Lehotay, S., Routine approach to qualitatively screening 300 pesticides and quantification of those frequently detected in fruit and vegetables using liquid chromatography tandem mass spectrometry (LC-MS/MS). *Food Add. Contam.* 27: 1415–1430, 2010.
35. Gilbert-Lopez, B., Garcia-Reyes, J. F., Fernandez-Alba, A. R., Molina-Diaz, A., Evaluation of two sample treatment methodologies for large-scale pesticide residue analysis in olive oil by fast liquid chromatography–electrospray mass spectrometry. *J. Chromatogr. A* 1217: 3736–3747, 2010.
36. Lee, J. H., Park, S., Jeong, W. Y., Park, H. J., Kim, H. G., Lee, S. J., Shim, J. H., Kim, S. T., El-Aty, A., Im, M. H., Choi, O. J., Shin, S. C., Simultaneous determination of phoxim and its photo-transformation metabolite residues in eggs using liquid chromatography coupled with electrospray ionization tandem mass spectrometry. *Anal. Chim. Acta* 674: 64–70, 2010.
37. Ferrer, C., Martinez-Bueno, M., Lozano, A., Fernandez-Alba, A., Pesticide residue analysis of fruit juices by LC-MS/MS direct injection. One year pilot survey. *Talanta* 83: 1552–1561, 2011.
38. Xu, Z. L., Zeng, D. P., Yang, J. Y., Shen, Y. D., Beier, R. C., Lei, H. T., Wang, H., Sun, Y. M., Monoclonal antibody-based broad-specificity immunoassay for monitoring organophosphorus pesticides in environmental water samples. *J. Environ. Monit.* 13: 3040–3048, 2011.

39. Noestheden, M., Noot, D., Hindle, R., Fast, extraction-free analysis of chlorinated phenols in well water by high-performance liquid chromatography–tandem mass spectrometry. *J. Chromatogr. A* 1263: 68–73, 2012.
40. Peruga, A., Grimalt, S., Lopez, F. J., Sancho, J. V., Hernandez, F., Optimisation and validation of a specific analytical method for the determination of thiram residues in fruits and vegetables by LC-MS/MS. *Food Chem.* 135: 186–192, 2012.
41. Schwarz, T., Snow, T. A., Santee, C. J., Mulligan, C. C., Class T., Wadsley, M. P., Nanita, S. C., QuEChERS multiresidue method validation and mass spectrometric assessment for the novel anthranilic diamide insecticides chlorantraniliprole and cyantraniliprole. *J. Agric. Food Chem.* 59: 814–821, 2011.
42. Hernandez, F., Portoles, T., Ibanez, M., Bustos-Lopez, M., Diaz, R., Botero-Coy, A., Fuentes, C., Penuela, G., Use of time-of-flight mass spectrometry for large screening of organic pollutants in surface waters and soils from a rice production area in Colombia. *Sci. Total Environ.* 439: 249–259, 2012.
43. Kruve, A., Haapala, M., Saarela, V., Franssila, S., Kostiainen, R., Kotiaho, T. Ketola, R. A., Feasibility of capillary liquid chromatography-microchip-atmospheric pressure photoionization-mass spectrometry for pesticide analysis in tomato. *Anal. Chim. Acta* 696: 77–83, 2011.
44. Maggioni, S., Bagnati, R., Pandelova, M., Schramm, K. W., Benfenati, E. Genistein and dicarboximide fungicides in infant formulae from the EU market. *Food Chem.* 136: 116–119, 2013.
45. Park, K. H., Choi, J. H., Abd El-Aty, A., Cho, S. K., Park, J. H., Kim, B. M., Yang, A., Na, T. W., Rahman, M. M., Im, G. J., Shim, J. H., Determination of spinetoram and its metabolites in amaranth and parsley using QuEChERS-based extraction and liquid chromatography–tandem mass spectrometry. *Food Chem.* 134: 2552–2559, 2012.
46. Wong, J., Hao, C., Zhang, K., Yang, P., Banerjee, K., Hayward, D., Iftakhar, I., Schreiber, A., Tech, K., Sack, C., Smoker, M., Chen, X., Utture, S. C., Oulkar, D. P., Development and interlaboratory validation of a QuEChERS-based liquid chromatography–tandem mass spectrometry method for multiresidue pesticide analysis. *J. Agric. Food Chem.* 58: 5897–5903, 2010.
47. Usui, K., Hayashizaki, Y., Minagawa, T., Hashiyada, M., Nakano, A., Funayama, M., Rapid determination of disulfoton and its oxidative metabolites in human whole blood and urine using QuEChERS extraction and liquid chromatography–tandem mass spectrometry. *Legal Med.* 14: 309–316, 2012.
48. Tran, K., Eide, D., Nickols, S. M., Cromer, M. R., Sabaa-Srur, A., Smith, R. E., Finding of pesticides in fashionable fruit juices by LC-MS/MS and GC-MS/MS. *Food Chem.* 134: 2398–2405, 2012.
49. Fontana, A., Rodriguez, I., Ramil, M., Altamirano, J., Cela, R., Liquid chromatography time-of-flight mass spectrometry following sorptive microextraction for the determination of fungicide residues in wine. *Anal. Bioanal. Chem.* 401: 767–775, 2011.
50. Wang, J., Chow, W., Leung, D., Applications of LC/ESI-MS/MS and UHPLC QqTOF MS for the determination of 148 pesticides in fruits and vegetables. *Anal. Bioanal. Chem.* 396: 1513–1538, 2010.
51. Zhang, K., Wong, J. W., Yang, P., Hayward, D. G., Sakuma, T. Zou, Y., Schreiber, A., Borton, C., Tung-Vi, N., Kaushik, B., Oulkar, D., Protocol for an electrospray ionization tandem mass spectral production library: Development and application for identification of 240 pesticides in foods. *Anal. Chem.* 84: 5677–5684, 2012.
52. Kittlaus, S., Schimanke, J., Kempe, G., Speer, K., Development and validation of an efficient automated method for the analysis of 300 pesticides in foods using two-dimensional liquid chromatography–tandem mass spectrometry. *J. Chromatogr. A* 1283: 98–109, 2013.
53. Camino-Sanchez, F., Zafra-Gomez, A., Oliver-Rodriguez, B., Ballesteros, O., Navalon, A., Crovetto, G., Vilchez, J., UNE-EN ISO/IEC 17025:2005-accredited method for the determination of pesticide residues in fruit and vegetable samples by LC-MS/MS. *Food Add. Contam.* 27: 1532–1544, 2010.
54. Yang, A., Park, J. H., Abd El-Aty, A., Choi, J. H., Oh, J. H., Do, J. A., Kwon, K., Shim, K. H., Choi, O. J., Shim, J. H., Synergistic effect of washing and cooking on the removal of multi-classes of pesticides from various food samples. *Food Control* 28: 99–105, 2012.
55. Matsuoka, T., Akiyama, Y., Mitsuhashi, T., Validation of multi-residue method for determination of pesticides in meat products using official guideline of analytical methods in Japan. *J. Pest. Sci.* 36: 73–78, 2011.
56. Chung, S. W., Chan, B. T., Validation and use of a fast sample preparation method and liquid chromatography–tandem mass spectrometry in analysis of ultra-trace levels of 98 organophosphorus pesticide and carbamate residues in a total diet study involving diversified food types. *J. Chromatogr. A* 1217: 4815–4824, 2010.
57. Kanrar, B., Mandal, S., Bhattacharyya, A., Validation and uncertainty analysis of a multiresidue method for 42 pesticides in made tea, tea infusion and spent leaves using ethyl acetate extraction and liquid chromatography–tandem mass spectrometry. *J. Chromatogr. A* 1217: 1926–1933, 2010.

58. Li, Y., Dong, F., Liu, X., Xu, J., Chen, X., Han, Y., Liang, X., Zheng, Y., Development of a multi-residue enantiomeric analysis method for nine pesticides in soil and water by chiral liquid chromatography/tandem mass spectrometry. *J. Haz. Mat.* 250–251: 9–18, 2013.
59. Ferrer, I., Thurman, E. M., Zweigenbaum, J. A., Screening and confirmation of 100 pesticides in food samples by liquid chromatography/tandem mass spectrometry. *Rapid Commun. Mass Spectrom.* 21: 3869–3882, 2007.
60. Leandro, C. C., Hancock, P., Fussell, R. J., Keely, B. J., Comparison of ultra-performance liquid chromatography and high-performance liquid chromatography for the determination of priority pesticides in baby foods by tandem quadrupole mass spectrometry. *J. Chromatogr. A* 1103: 94–101, 2006.
61. Chen, G., Cao, P., Liu, R., A multi-residue method for fast determination of pesticides in tea by ultra performance liquid chromatography–electrospray tandem mass spectrometry combined with modified QuEChERS sample preparation procedure. *Food Chem.* 125: 1406–1411, 2011.
62. Grimalt, S., Sancho, J. V., Pozo, O. J., Hernandez, F., Quantification, confirmation and screening capability of UHPLC coupled to triple quadrupole and hybrid quadrupole time-of-flight mass spectrometry in pesticide residue analysis. *J. Mass Spectrom.* 45: 421–436, 2010.
63. Lacina, O., Urbanova, J., Poustka, J., Hajslova, J., Identification/quantification of multiple pesticide residues in food plants by ultra-high-performance liquid chromatography-time-of-flight mass spectrometry. *J. Chromatogr. A* 1217: 648–659, 2010.
64. Arienzo, M., Cataldo, D., Ferrara, L., Pesticide residues in fresh-cut vegetables from integrated pest management by ultra performance liquid chromatography coupled to tandem mass spectrometry. *Food Control* 31: 108–115, 2013.
65. Kovalczuk, T., Jech, M., Poustka, J., Hajslova. J., Ultra-performance liquid chromatography–tandem mass spectrometry: A novel challenge in multiresidue pesticide analysis in food. *Anal. Chim. Acta* 577: 8–17, 2006.
66. Nurmi, J., Pellinen, J., Multiresidue method for the analysis of emerging contaminants in wastewater by ultra performance liquid chromatography-time-of-flight mass spectrometry. *J. Chromatogr. A* 1218: 6712–6719, 2011.
67. Ríos, A., Zougagh, M., Avila, M. Miniaturization through lab-on-a-chip: Utopia or reality for routine laboratories? A review. *Anal. Chim. Acta* 740: 1–11, 2012.
68. De Stefano, L., Malecki, K., Rossi, A. M., Rotiroti, L., Corte, F. G. D., Moretti, L., Rendina, I., Integrated silicon-glass opto-chemical sensors for lab-on-chip applications. *Sensor Actuat. B-Chem.* 114: 625–630, 2006.
69. Sauter-Starace, F., Pudda, C., Delattre, C., Jeanson, H., Gillot, C., Sarrut, N., Constantin, O., Blanc, R., Lab on a chip: Advances in packaging for MEMS and lab on a chip. In: 4M 2006—Second International Conference on Multi-Material Micro Manufacture, Wolfgang, M., Dimov, S., Fillon, B. A., Bertrand, F., (Eds.), Oxford, Elsevier, 2006, pp. 3–7.
70. Purcaro, G., Moret, S., Conte, L., Sample pre-fractionation of environmental and food samples using LC-GC multidimensional techniques. *TrAC Trends Anal. Chem.* 43: 146–160, 2013.
71. Alder, L., Greulich, K., Kempe, G., Vieth, B., Residue analysis of 500 high priority pesticides: Better by GC-MS or LC-MS/MS? *Mass Spectrom. Rev.* 25: 838–865, 2006.
72. Pico, Y., Font, G., Ruiz, M. J., Fernandez, M., Control of pesticide residues by liquid chromatography-mass spectrometry to ensure food safety. *Mass Spectrom. Rev.* 25: 917–960, 2006.
73. Manisali, I., Chen, D. D. Y., Schneider, B. B., Electrospray ionization source geometry for mass spectrometry: Past, present, and future. *TrAC Trends Anal. Chem.* 25: 243–256, 2006.
74. Reemtsma, T., The use of liquid chromatography-atmospheric pressure ionization-mass spectrometry in water analysis- Part I: Achievements. *TrAC Trends Anal. Chem.* 20: 500–517, 2001.
75. Chen, Y., Al-Taher, F., Juskelis, R., Wong, J. W., Zhang, K., Hayward, D. G., Zweigenbaum, J., Stevens, J., Cappozzo, J., Multiresidue pesticide analysis of dried botanical dietary supplements using an automated dispersive SPE cleanup for QuEChERS and High-Performance Liquid Chromatography–Tandem Mass Spectrometry. *J. Agric. Food Chem.* 60: 9991–9999, 2012.
76. Oellig, C., Schwack, W., Planar solid phase extraction clean-up for pesticide residue analysis in tea by liquid chromatography–mass spectrometry. *J. Chromatogr. A* 1260: 42–53, 2012.
77. Romero-Gonzalez, R., Garrido Frenich, A., Martinez Vidal, J., Prestes, O., Grio, S., Simultaneous determination of pesticides, biopesticides and mycotoxins in organic products applying a quick, easy, cheap, effective, rugged and safe extraction procedure and ultra-high performance liquid chromatography–tandem mass spectrometry. *J. Chromatogr. A* 1218: 1477–1485, 2011.
78. Kang, S., Chang, N., Zhao, Y., Pan, C., Development of a method for the simultaneous determination of six sulfonylurea herbicides in wheat, rice, and corn by liquid chromatography–tandem mass spectrometry. *J. Agric. Food Chem.* 59: 9776–9781, 2011.

79. Chiaia-Hernandez, A. C., Krauss, M., Hollender, J., Screening of lake sediments for emerging contaminants by liquid chromatography atmospheric pressure photoionization and electrospray ionization coupled to high resolution mass spectrometry. *Environ. Sci. Technol.* 47: 976–986, 2013.
80. Yamamoto, A., Terao, T., Hisatomi, H., Kawasaki, H., Arakawa, R., Evaluation of river pollution of neonicotinoids in Osaka City (Japan) by LC/MS with dopant-assisted photoionisation. *J. Environ. Monit.* 14: 2189–2194, 2012.
81. Gu, C., Shamsi, S. A., CEC-atmospheric pressure ionization MS of pesticides using a surfactant-bound monolithic column. *Electrophoresis* 31: 1162–1174, 2010.
82. Itoh, N., Otake, T., Aoyagi, Y., Matsuo, M., Yarita, T., Application of pesticide quantification in unpolished rice by LC-dopant-assisted atmospheric pressure photoionization-MS. *Chromatographia* 70: 1073–1078, 2009.
83. Marchi, I., Rudaz, S., Veuthey, J. L., Atmospheric pressure photoionization for coupling liquid-chromatography to mass spectrometry: A review. *Talanta* 78: 1–18, 2009.
84. Wick, A., Fink, G., Ternes, T. A., Comparison of electrospray ionization and atmospheric pressure chemical ionization for multi-residue analysis of biocides, UV-filters and benzothiazoles in aqueous matrices and activated sludge by liquid chromatography–tandem mass spectrometry. *J. Chromatogr. A* 1217: 2088–2103, 2010.
85. Bester, K., Lamani, X., Determination of biocides as well as some biocide metabolites from facade run-off waters by solid phase extraction and high performance liquid chromatographic separation and tandem mass spectrometry detection. *J. Chromatogr. A* 1217: 5204–5214, 2010.
86. Soler, C., Manes, J., Pico, Y., The role of the liquid chromatography–mass spectrometry in pesticide residue determination in food. *Crit. Rev. Anal. Chem.* 38: 93–117, 2008.
87. Andreu, V., Pico, Y., Liquid chromatography-ion trap-mass spectrometry and its application to determine organic contaminants in the environment and food. *Curr. Anal. Chem.* 1: 241–265, 2005.
88. Rodriguez-Aller, M., Gurny, R., Veuthey, J. L., Guillarme, D., Coupling ultra high-pressure liquid chromatography with mass spectrometry: Constraints and possible applications. *J. Chromatogr. A* 1292: 2–18, 2013.
89. Garcia-Reyes, J. F., Hernando, M. D., Ferrer, C., Molina-Diaz, A., Fernandez-Alba, A. R., Large scale pesticide multiresidue methods in food combining liquid chromatography-time-of-flight mass spectrometry and tandem mass spectrometry. *Anal. Chem.* 79: 7308–7323, 2007.
90. Sobhanzadeh, E., Abu Bakar, N. K., Bin Abas, M. R., Nemati, K., A simple and efficient multi-residue method based on QuEChERS for pesticides determination in palm oil by liquid chromatography time-of-flight mass spectrometry. *Environ. Monit. Asses.* 184: 5821–5828, 2012.
91. Lacina, O., Zachariasova, M., Urbanova, J., Vaclavikova, M., Cajka, T., Hajslova, J., Critical assessment of extraction methods for the simultaneous determination of pesticide residues and mycotoxins in fruits, cereals, spices and oil seeds employing ultra-high performance liquid chromatography–tandem mass spectrometry. *J. Chromatogr. A* 1262: 8–18, 2012.
92. Garcia-Reyes, J. F., Molina-Diaz, A., Fernandez-Alba, A. R., Identification of pesticide transformation products in food by liquid chromatography/time-of-flight mass spectrometry via "fragmentation-degradation" relationships. *Anal. Chem.* 79: 307–321, 2007.
93. Thurman, E. M., Ferrer, I., Zavitsanos, P., Zweigenbaum, J. A., Analysis of Isobaric pesticides in pepper with high-resolution liquid chromatography and mass spectrometry: Complementary or redundant? *J. Agric. Food Chem.* 61: 2340–2347, 2013.
94. Ferrer, C., Lozano, A., Agueera, A., Giron, A., Fernandez-Alba, A., Overcoming matrix effects using the dilution approach in multiresidue methods for fruits and vegetables. *J. Chromatogr. A* 1218: 7634–7639, 2011.
95. Farre, M., Pico, Y., Barcelo, D., Direct peel monitoring of xenobiotics in fruit by direct analysis in real time coupled to a linear quadrupole ion trap-orbitrap mass spectrometer. *Anal. Chem.* 85: 2638–2644, 2013.
96. Ivanova, B., Spiteller, M., A novel UV-MALDI-MS analytical approach for determination of halogenated phenyl-containing pesticides. *Ecotoxicol. Environment. Saf.* 91: 86–95, 2013.
97. Diaz, R., Ibanez, M., Sancho, J. V., Hernandez, F., Target and non-target screening strategies for organic contaminants, residues and illicit substances in food, environmental and human biological samples by UHPLC-QTOF-MS. *Anal. Meth.* 4: 196–209, 2012.
98. Lacorte, S., Fernandez-Albaz, A. R., Time of flight mass spectrometry applied to the liquid chromatographic analysis of pesticides in water and food. *Mass Spectrom. Rev.* 25: 866–880, 2006.
99. Niessen, W. M. A., Manini, P., Andreoli, R., Matrix effects in quantitative pesticide analysis using liquid chromatography–mass spectrometry. *Mass Spectrom. Rev.* 25: 881–899, 2006.

100. Stahnke, H., Kittlaus, S., Kempe, G., Hemmerling, C., Alder, L., The influence of electrospray ion source design on matrix effects. *J. Mass Spectrom.* 47: 875–884, 2012.
101. Schuhn, B., Glauner, T., Kempe, G., Analytic of pesticides quantification and identification of residues in food by LC-MS/MS and triggered MRM. *Dtsch. Lebensm.* 108: 591–596, 2012.
102. Fenoll, J., Hellin, P., Martinez, C. M., Flores, P., Multiresidue analysis of pesticides in vegetables and citrus fruits by LC-MS-MS. *Chromatographia* 72: 857–866, 2010.
103. Aguilera-Luiz, M., Plaza-Bolanos, P., Romero-Gonzalez, R., Martinez Vidal, J., Garrido Frenich, A., Comparison of the efficiency of different extraction methods for the simultaneous determination of mycotoxins and pesticides in milk samples by ultra high-performance liquid chromatography–tandem mass spectrometry. *Anal. Bioanal. Chem.* 399: 2863–2875, 2011.
104. Gilbert-Lopez, B., Garcia-Reyes, J. F., Lozano, A., Fernandez-Alba, A. R., Molina-Diaz, A., Large-scale pesticide testing in olives by liquid chromatography–electrospray tandem mass spectrometry using two sample preparation methods based on matrix solid-phase dispersion and QuEChERS. *J. Chromatogr. A* 1217: 6022–6035, 2010.
105. Rodrigues, S. A., Caldas, S. S., Primel, E. G., A simple; efficient and environmentally friendly method for the extraction of pesticides from onion by matrix solid-phase dispersion with liquid chromatography–tandem mass spectrometric detection. *Anal. Chim. Acta* 678: 82–89, 2010.
106. van der Schans, M. J., Hulst, A. G., van der Riet-van Oeveren, Noort, D., Benschop, H. P., Dishovsky, C., New tools in diagnosis and biomonitoring of intoxications with organophosphorothioates: Case studies with chlorpyrifos and diazinon. *Chemico-Biological Interactions* 203: 96–102, 2013.
107. Zhang, J. M., Wu, Y. L., Lu, Y. B., Simultaneous determination of carbamate insecticides and mycotoxins in cereals by reversed phase liquid chromatography tandem mass spectrometry using a quick, easy, cheap, effective, rugged and safe extraction procedure. *J. Chromatogr. B* 915: 13–20, 2013.
108. Rahman, M. M., Park, J. H., Abd El-Aty, A., Choi, J. H., Yang, A., Park, K. H., Al Mahmud, M. N. U., Im, G. J., Shim, J. H., Feasibility and application of an HPLC/UVD to determine dinotefuran and its shorter wavelength metabolites residues in melon with tandem mass confirmation. *Food Chem.* 136: 1038–1046, 2013.
109. Martin, L., Mezcua, M., Ferrer, C., Gil, Garcia, M. D., Malato, O., Fernandez-Alba, A. R., Prediction of the processing factor for pesticides in apple juice by principal component analysis and multiple linear regression. *Food Add. Contam.* 30: 466–476, 2013.
110. Alvarez-Rivera, G., Dagnac, T., Lores, M., Garcia-Jares, C., Sanchez-Prado, L., Pablo Lamas, J., Llompart, M., Determination of isothiazolinone preservatives in cosmetics and household products by matrix solid-phase dispersion followed by high-performance liquid chromatography–tandem mass spectrometry. *J. Chromatogr. A* 1270: 41–50, 2012.
111. Caballo, C., Sicilia, M., Rubio, S., Stereoselective quantitation of mecoprop and dichlorprop in natural waters by supramolecular solvent-based microextraction, chiral liquid chromatography and tandem mass spectrometry. *Anal. Chim. Acta* 761: 102–108, 2013.
112. Malato, O., Lozano, A., Mezcua, M., Agueera, A., Fernandez-Alba, A. R., Benefits and pitfalls of the application of screening methods for the analysis of pesticide residues in fruits and vegetables. *J. Chromatogr. A* 1218: 7615–7626, 2011.
113. Cahill, M. G., Dineen, B. A., Stack, M. A., James, K. J., A critical evaluation of liquid chromatography with hybrid linear ion trap-orbitrap mass spectrometry for the determination of acidic contaminants in wastewater effluents. *J. Chromatogr. A* 1270: 88–95, 2012.
114. Chamkasem, N., Ollis, L. W., Harmon, T., Lee, S., Mercer, G., Analysis of 136 pesticides in avocado using a modified QuEChERS method with LC-MS/MS and GC-MS/MS. *J. Agric. Food Chem.* 61: 2315–2329, 2013.
115. Peng, J., Xiao, Y., Cao, H., Zhang, L., Tang, J., Determination of pirimicarb and paclobutrazol pesticide residues in food by HPLC-ESI-MS with a novel sample preparation method. *Anal. Lett.* 46: 35–47, 2013.
116. Riedel, M., Speer, K., Stuke, S., Schmeer, K., Simultaneous analysis of 70 pesticides using HPLC/MS/MS: A comparison of the multiresidue method of Klein and Alder and the QuEChERS method. *J. AOAC Int.* 93: 1972–1986, 2010.
117. Perez-Fernandez, V., Dominguez-Vega, E., Chankvetadze, B., Crego, A. L., Angeles Garcia, M., Luisa Marina, M., Evaluation of new cellulose-based chiral stationary phases Sepapak-2 and Sepapak-4 for the enantiomeric separation of pesticides by nano liquid chromatography and capillary electrochromatography. *J. Chromatogr. A* 1234: 22–31, 2012.
118. Flender, C., Leonhard, P., Wolf, C., Fritzsche, M., Karast, M., Analysis of boronic acids by nano liquid chromatography–direct electron ionization mass spectrometry. *Anal. Chem.* 82: 4194–4200, 2010.

119. Pico, Y., Ultrasound-assisted extraction for food and environmental samples. *TrAC Trends Anal. Chem.* 43: 84–99, 2013.
120. Juan-Garcia, A., Font, G., Juan, C., Pico, Y., Pressurised liquid extraction and capillary electrophoresis-mass spectrometry for the analysis of pesticide residues in fruits from Valencian markets, Spain. *Food Chem.* 120: 1242–1249, 2010.
121. Kim, J. H., Seo, J. S., Moon, J. K., Kim, J. H., Multi-residue method development of eight benzoylurea insecticides in mandarin and apple using high performance liquid chromatography and liquid chromatography-tandem mass spectrometry. *J. Kor. Soc. Appl. Biol. Chem.* 56: 47–54, 2013.
122. Polgar, L., Kmellar, B., Garcia-Reyes, J. F., Fodor, P., Comprehensive evaluation of the clean-up step in QuEChERS procedure for the multi-residue determination of pesticides in different vegetable oils using LC-MS/MS. *Anal. Meth.* 4: 1142–1148, 2012.

16 Multidimensional Liquid Chromatography Applied to the Analysis of Pesticides

Ahmed Mostafa, Heba Shaaban, and Tadeusz Górecki

CONTENTS

16.1 INTRODUCTION

A pesticide is any substance or mixture of substances intended for preventing, destroying, repelling, or mitigating any pest, such as fungi, insects, rodents, and weeds [1]. Pesticides are used intensively in modern agriculture. Because they are hazardous to human health and the environment, it is important to control pesticide residues after their application to food. Considerable attention is focused on regulating the allowable limits for pesticide residues in food and drinking water. For example, the European Union directives set the maximum admissible concentration of 0.1 μg/L for each pesticide in drinking water and 0.5 μg/L for the sum of pesticides [2,3]. Thus, analytical methods with detection limits as low as 0.02 μg/L [4] are required for pesticide monitoring in drinking water. This leads to the need for new methods of analysis to quantify the large number of pesticides that can be found at such low levels in water [5].

Single column (one-dimensional) chromatography analysis has been used for many years as a standard separation tool for analyzing compounds in a broad variety of fields, including pesticide analysis [6–9]. These separation techniques typically suffer from a general lack of resolving power because a uniform separation mechanism is utilized throughout. The ability of a chromatographic system to separate the individual analytes of a mixture is dependent on its peak capacity, which can be described as the maximum number of component peaks that can be placed, side by side, within the separation space with just enough resolution between neighbors [10]. Ideally, the peak capacity of a column far exceeds the number of individual analytes in the mixture, but this is rarely ever the case. In reality, many samples are too complex for conventional separations, and peak capacity is often far exceeded. This results in peak overlap, which decreases the quality of the analysis [11]. Therefore, this approach can be used successfully only for applications involving samples that are not very complex. One-dimensional chromatography typically cannot resolve all analytes of interest in the very complex matrices typically encountered in real-life samples, such as environmental, petroleum, pharmaceutical, food, and forensic samples. One of the most effective ways to enhance separation power and peak capacity is through multidimensional separation, a method introduced more than 50 years ago [12].

The development and the principles of multidimensional (MD) chromatographic separations were described by Giddings [13]. One of the oldest MD chromatographic separation techniques was termed "heart-cutting." These techniques involved sampling a fraction of the effluent from one column and subsequently injecting it into another column with a differing selectivity. Heart-cutting provided increased selectivity and peak capacity compared to one-dimensional separation by subjecting the sample effluent to two different separation mechanisms. However, this method proved to be effective only in target analysis, in which a limited number of heart-cuts requires additional second-dimension separation. If the entire sample (or at least a representative fraction of each sample component) requires analysis in two different dimensions, then comprehensive MD techniques should be used. Giddings [13,14] described the idea of comprehensive MD separation as a process in which the entire sample is subjected to all separation dimensions with preservation of the separations obtained in the previous dimensions. Although there is no inherent restriction to the number of independent separation methods used in a MD separation, practical constraints have limited the vast majority of the MD separations reported to date to two dimensions.

General aspects, including fundamentals, design, and applications of MD techniques, have been described in several interesting publications [10,15–21]. Therefore, only a brief description of each MD approach will be given in this chapter with the main focus on MD liquid chromatography (MDLC) and its applications in pesticide analysis. Readers interested in more specific instrumental or fundamental details are directed to the aforementioned review papers.

16.2 MDLC

16.2.1 Introduction to MDLC

The resolving power of high-performance liquid chromatography (HPLC) may be enhanced significantly by the introduction of MDLC, with which the two (or more) dimensions are based on different separation mechanisms. MD chromatography—also known as coupled-column chromatography (LC-LC coupling) or column switching—represents a powerful tool and an alternative procedure to classical one-dimensional HPLC methods. MDLC can be performed either off- or online and practically can be classified as off-line MDLC; online MDLC, including "heart-cutting" (LC-LC); and comprehensive MDLC (LC × LC). The difference between "heart-cutting" and comprehensive approaches is that the first enables reinjection of a limited number of the primary column effluent fractions to the secondary column, and in the second approach, the entire sample is subject to separation in both dimensions.

In off-line operation, the primary column effluent fractions are collected manually or by a fraction collector and then reinjected, either with or without concentration, into the secondary column. The advantages of this approach include simplicity, ease of operation, no need for switching valves, and no need for the mobile phase used in each column to be compatible [22]. On the other hand, this approach is often labor-intensive, time-consuming, and subject to sample loss or contamination, and recovery of analytes is often low [23]. To overcome these problems, online two-dimensional liquid chromatography (2D LC) based on valve switching was proposed [24,25]. A sample loop was used to collect the primary column heart-cuts, which were introduced into the secondary column, allowing only a few fractions of the primary column effluent to be analyzed in the second column. This approach was suitable for the characterization of specific fractions of the sample but not for complete ("comprehensive") characterization. Online 2D LC techniques are technically more complex and less straightforward to optimize but are easy to automate, thus improving reliability and sample throughput. They also shorten the analysis time and minimize sample loss or change because the analysis is performed in a closed-loop system [26].

The peak capacity enhancement of both the off-line and heart-cutting approaches is usually insufficient for full resolution of samples containing more than 100 to 200 relevant components. Comprehensive 2D LC (LC × LC) techniques are more suitable for this purpose [27–31]. In LC × LC, two columns are connected serially online via a switching valve interface. The main function of the interface is to collect narrow fractions of the primary column effluent for fast reinjection into the secondary column in multiple repeated alternating cycles (i.e., modulation periods) in real analysis time.

LC × LC setup is illustrated in Figure 16.1. A 10-port switching valve modulator was used with two identical-volume sampling loops (Figure 16.1a) or small trapping columns (Figure 16.1b). Loop 1 or trapping column 1 collects a fraction from the primary column effluent. The subsequent fractions are analyzed in the alternating valve operating cycles in which the two loops or trapping columns are regularly switched between the collection (1) and the elution (2) positions [22]. In this way, the entire sample is subjected to separation in both dimensions in alternating cycles.

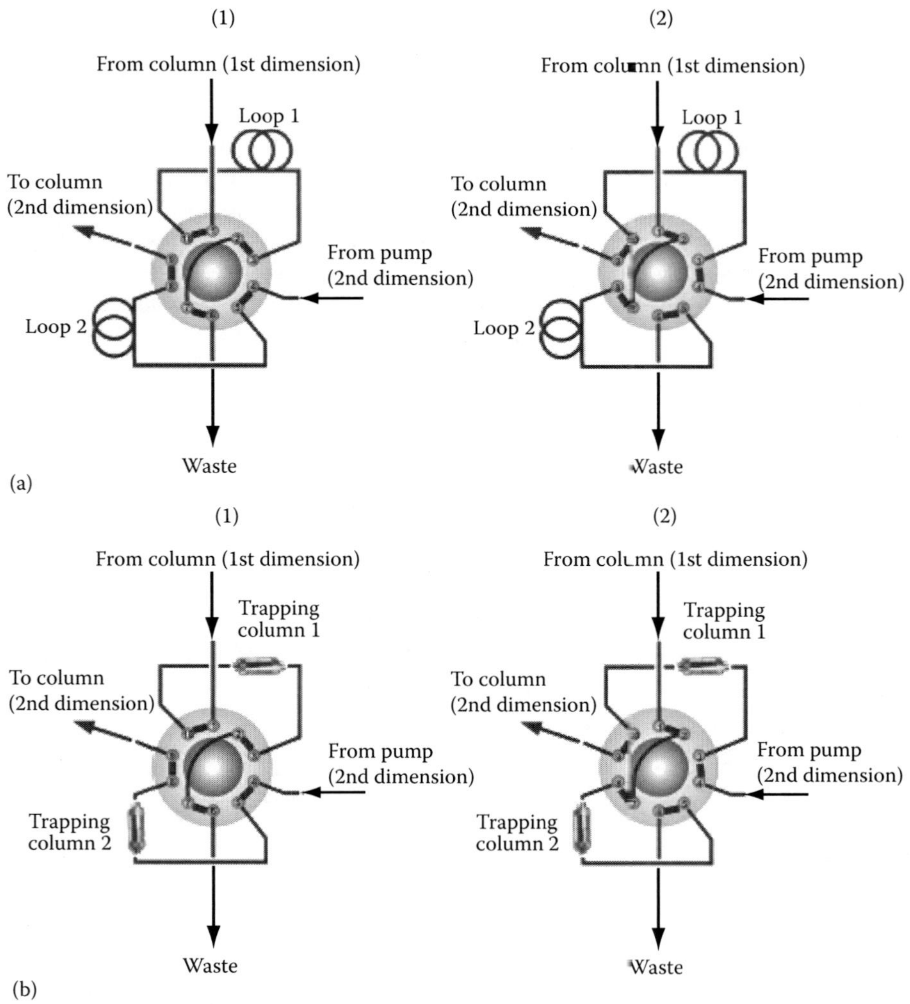

FIGURE 16.1 Standard comprehensive LC × LC setup with a 10-port valve modulator interface with two collecting loops, operating in alternating cycles (left to right) (a); two trapping columns substituting collecting loops, operating in alternating cycles (left to right) (b). (Jandera, P., Mondello, L.: *Multidimensional Liquid Chromatography: Theoretical Considerations. Comprehensive Chromatography in Combination with Mass Spectrometry*, Mondello, L. (Editor). 2011. Copyright Wiley-VCH Verlag GmbH & Co. KGaA. Reproduced with permission.)

The second-dimension separation in LC × LC has to be fast in order to be completed in a time shorter than or equal to the period of fraction collection, including the time necessary for the fraction transfer, usually 2 min or less. This constraint seriously limits the number of components that can be separated during a single second-dimension analysis cycle. This tradeoff can be solved using the stop-flow or stop-and-go technique, in which the primary and secondary columns are connected online via a six-port switching valve with no sampling loops [32]. In one position, the primary column is connected directly to the secondary column to which a fraction of the effluent is transferred. When the desired volume has passed onto the second column, the six-port valve is switched to the second position. Thus, the flow of the mobile phase flow is stopped in the first dimension, and the transferred fraction is separated in the second column. When the separation is finished, the valve is switched back to the original position, the mobile phase delivery onto the primary column is resumed and the entire procedure is repeated as many times as necessary (Figure 16.2) [33]. The stop-and-go technique allows the use of a longer second-dimension column with a higher plate number, thus increasing the number of compounds resolved at a cost of increased total analysis time per sample.

In LC × LC, switching valves are usually used as interfaces, and different combinations of the diverse LC separation modes can theoretically be coupled. The selectivity differences are based primarily on size, shape, polarity, hydrophobicity, degree of saturation, acidity/basicity, or charge. According to these properties, size exclusion chromatography, normal-phase liquid chromatography (NPLC), reversed-phase liquid chromatography (RPLC), hydrophilic interaction liquid chromatography (HILIC), ion exchange chromatography, or ion-pairing chromatography have been employed in the two dimensions of an LC × LC system, resulting in increased peak capacity [34–38]. Due to the significantly higher separation power of LC × LC compared to its one-dimensional counterpart, the technique has enjoyed significant interest in diverse fields and has been the subject of various reviews [15,16,39–47]. LC × LC is currently used for the analysis of complex environmental and petrochemical samples, pharmaceuticals, polymers, natural products, biological mixtures, and proteomics. Using RP mode in both dimensions is one of the most common approaches to LC × LC analysis. It has been demonstrated that orthogonality can be achieved by using either two different sets of mobile phases and one type of RP column, or the same mobile phase and two HPLC columns with different RP stationary phases [48].

Despite the great advantages associated with LC × LC, the technique has limitations related to (i) solvent immiscibility (because there is no evaporation step between the two separations, such as in off-line MDLC), (ii) limited analysis time in the second dimension, and (iii) transfer velocity (because the transfer has to be made quickly enough to avoid effluent losses from the first dimension). A basic requirement is that the first dimension separation has to be preserved in the second dimension [16]. The first-dimension peak should be sampled at least three times based on the

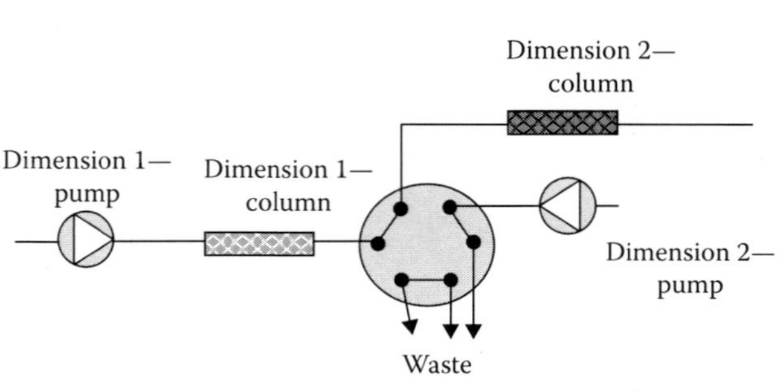

FIGURE 16.2 Comprehensive LC system with stop-and-go interface. During second-dimension analysis, the flow rate on the first dimension is interrupted with the primary flow path open to the laboratory atmosphere. (Francois, I., Sandra, K., Sandra, P.: *History, Evolution, and Optimization Aspects of Comprehensive Two-Dimensional Liquid Chromatography. Comprehensive Chromatography in Combination with Mass Spectrometry*, Mondello, L. (Editor). 2011. Copyright Wiley-VCH Verlag GmbH & Co. KGaA. Reproduced with permission.)

theoretical and experimental work of Murphy et al. [49]. However, in most current applications, the sampling rate has been lower than recommended.

The first LC × LC system was constructed by Erni and Frei [31] using a gel permeation (GP) column in the first dimension and a RP column in the second dimension. An eight-port switching valve equipped with two identical sampling loops was used to connect the two dimensions. The first dimension fractions were collected alternately in the sampling loops and reinjected into the second dimension for further separation. Even though this approach was not fully comprehensive because only seven fractions from the first dimension were transferred for further separation in the second dimension, it represented a significant step in separation science.

The first three-dimensional data obtained from an LC × LC system were presented in 1990 for a protein sample (Figure 16.3) [30]. The instrumental setup was the same as Erni and Frei's with a cation-exchange column in the first dimension and a size-exclusion column in the second dimension. This design was fully comprehensive as every first-dimension fraction was analyzed in the second dimension without stopped-flow operation.

The most common strategy in LC × LC is the employment of two columns connected in series through a switching valve with two identical injection loops or two trapping columns installed [50]. Gradient elution is often used in LC × LC applications, especially in the first dimension, as it provides considerably higher peak capacity in comparison to isocratic elution [51]. Cacciola et al. used two parallel columns working in alternating cycles in the second dimension [50]. The use of two alternative columns provided great differences in separation selectivity in each dimension and an almost orthogonal 2D system. In addition, the use of high temperatures in the second-dimension separation was explored using zirconia-based stationary phases. The system consisted of a RP column as the first dimension and two parallel zirconia-carbon columns working in alternating cycles in the second dimension (Figure 16.4).

A two-position 10-port valve was used as the interface connecting the two dimensions. Gradient elution was used in the first dimension, and isocratic high temperature (120°C) elution was employed in the second dimension. During each modulation period, a first-dimension effluent fraction was transferred for further separation onto one of the two zirconia columns, and the next first-dimension effluent fraction was trapped on the second column. Twelve phenolic standards were separated in 80 min (Figure 16.5). The results demonstrated that the use of high-temperature LC improved the resolution and speeded up the second-dimension separations.

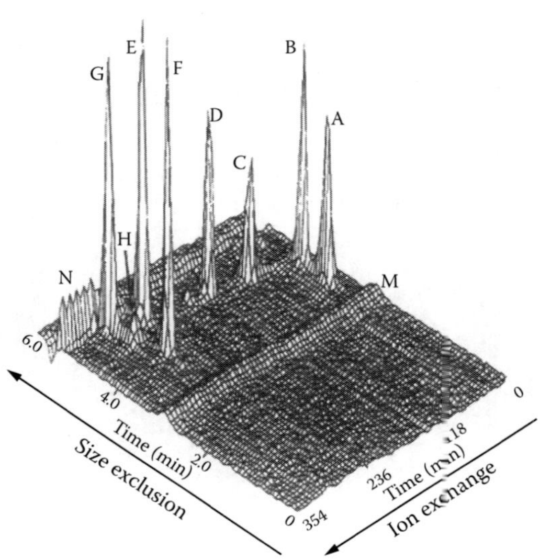

FIGURE 16.3 Three-dimensional plot of protein sample using a cation-exchange column and a size exclusion column in the first and second dimensions, respectively. (From Bushey, M. M., Jorgenson, J. W., *Anal. Chem.*, 62, 161–167, 1990. With permission.)

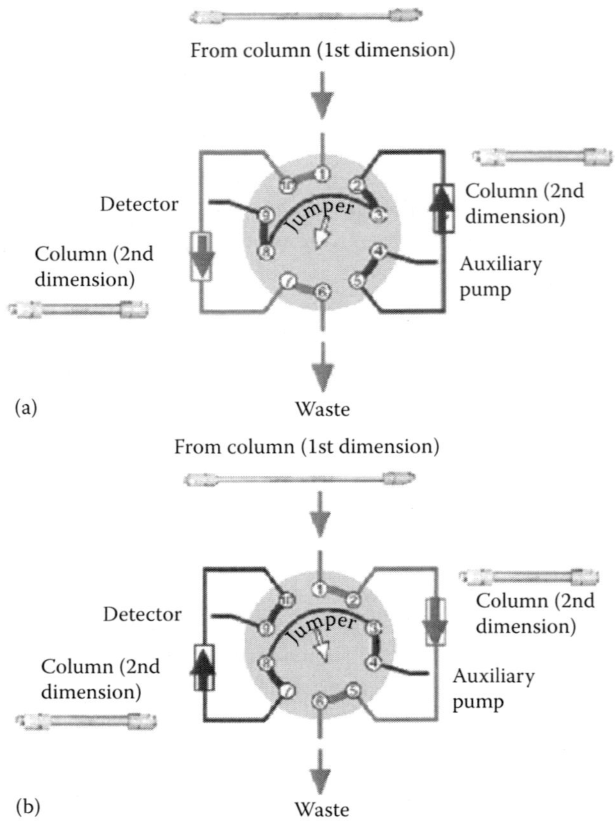

FIGURE 16.4 Two-dimensional LC × LC experimental setup with two alternating zirconia-carbon columns in the second dimension and a RP-18e column in the first dimension. (a) Loading of the first column effluent onto the zirconia-carbon column 1, separation of the previous fraction on the zirconia-carbon column 2; (b) loading of the next fraction of the first column effluent onto the zirconia-carbon column 2, separation of previously loaded fraction on the zirconia-carbon column 1. (From Cacciola, F. et al., *Chromatographia*, 66, 661–667, 2007. With permission.)

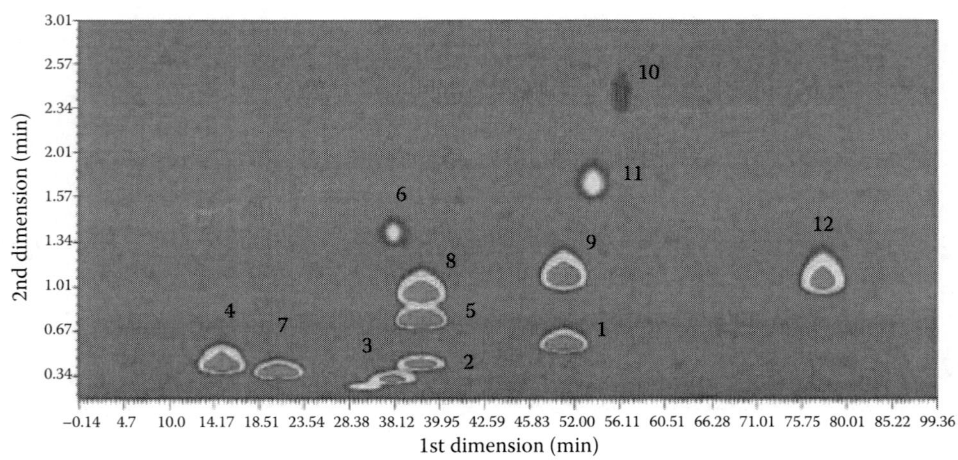

FIGURE 16.5 LC × LC 2D chromatogram of phenolic antioxidants on an RP column in the first dimension and on two zirconia-carbon columns working in alternating cycles in the second dimension. (From Cacciola, F. et al., *Chromatographia*, 66, 661–667, 2007. With permission.)

16.2.2 MDLC in Pesticide Analysis

The oldest and most common form of MDLC is the off-line approach. This technique requires knowledge of the retention of specific sample components before the fractionation can take place and is commonly used for the separation of specific components in a sample, that is, in target analysis. Numerous couplings of various LC modes for the off-line LC-LC separations have been used in different applications, such as food and pesticide analysis, as well as for the purification and preseparation of the sample prior to analytical separation using other techniques. For example, gel permeation chromatography (GPC) has been applied as an effective off-line technique for the cleanup of soil extracts prior to instrumental analysis for the trace analysis of several types of pesticides in soils [10,20,52,53]. The poor compatibility between the mobile phases and the relatively large elution volumes of the analyte-containing fractions hindered the online coupling of GPC and RPLC in this type of analysis.

Even though automation can solve some of the problems associated with off-line MDLC, such as poor reproducibility or the possibility of analyte degradation, coupling of different separation mechanisms is not a straightforward task. Solvent incompatibilities and immiscibility problems can arise, and the system is generally more difficult to operate. Nevertheless, this technique, commonly known as heart-cutting (LC-LC), is regarded as a powerful one. It has been successfully employed in different fields [54,55], and pesticide analysis is not an exception.

Online LC-LC methods can be used in profiling or heart-cutting modes. In the former mode, the aim is to fractionate and separate nearly all components of the sample matrix, and the latter mode has found more widespread use in targeted component analysis. This is a consequence of the fact that determination of individual analytes in complex samples is a problem that frequently requires some type of sample cleanup or preseparation as a prerequisite to the analytical measurement. In many applications, the first LC separation is used to eliminate components of the matrix that may interfere with analytes in the sample. Pesticide analysis in environmental samples usually implies a high degree of difficulty due to the complex matrices under study, and LC-LC can provide a step forward to solve the related problems. Numerous examples of LC-LC applications in the analysis of pesticides in foodstuffs (e.g., Refs. 56–59) and water samples (e.g., Refs. 54,60) have been published.

Off-line solid-phase extraction (SPE) in combination with coupled-column RPLC and UV detection was used as a powerful tool for the analysis of polar pesticides in water samples [61]. LC-LC offers efficient sample cleanup by the primary RP column, thus eliminating the early eluting interferences [60,61]. Meanwhile LC-LC in combination with UV detection is an attractive method for the analysis of polar pesticides because it is robust and rugged and allows the direct injection of aqueous samples without the need for extraction, derivatization, or other sample pretreatment. However, the main drawback is that the UV sensitivity is not sufficient for trace analysis. Therefore, sample preconcentration is a must for sensitive detection. Solid phase immunosorbents can be used before the LC-LC step with the aim to increase the selectivity of the LC-LC-UV system, but this step can be avoided by using the high selectivity and sensitivity of the MS detection [45].

The coextracted humic and fulvic acids usually severely interfere with trace analysis of acidic pesticides in environmental samples using RPLC-UV [62–65]. These types of interferences show up in the chromatogram as a broad "hump." Most analytes coelute on the steep slope or, even worse, on top of the "hump," making reliable quantification difficult or impossible. In the case of water samples, improved separation has been obtained by using selective SPE sorbents for off-line [62–65] or online [66,67] preconcentration. The LC-LC setup used a combination of C18 column in the first dimension with a C18-semipermeable surface (SPS) column in the second dimension. This column combination appeared to be more suitable than a two–C18 column combination for the analysis of acidic pesticides in water samples. The C18/SPS-C18 combination significantly decreased the baseline deviation caused by the coextracted humic/fulvic acid interferences [68].

An RP LC-LC method with UV detection was developed by Hidalgo et al. for the analysis of four triazine herbicides (simazine, atrazine, terbuthylazine, and terbutryn) in environmental and drinking water [54]. The RPLC-RPLC system used large volume injection of 2 mL directly injected into the chromatographic system. The mobile phase consisted of acetonitrile and water in both dimensions. Lower detection limits, between 0.01 and 0.05 μg/L, were obtained by preconcentration of 100 mL of water samples on C18 bonded phase SPE cartridges. The method was robust, selective, rapid (total analysis time of 7 min without the preconcentration step), and sensitive.

Van der Heeft et al. used a system composed of two different C18 columns connected through a switching valve and compared the performance of MS detection and UV detection coupled to RPLC-RPLC for the trace analysis of phenylurea herbicides in environmental water samples. The comparative study showed that LC-LC-MS was more selective and, in most cases, more sensitive than LC-LC-UV. The elution conditions were varied so that a desalting step could be performed in the first dimension to enhance the separation in the second dimension using MS for detection. Under these conditions, several phenylurea herbicides could be detected in water at levels below 0.01 μg/L in about 25 min total analysis time [60].

Pascoe et al. investigated the effect of LC-LC chromatographic separation on matrix-related signal suppression in electrospray ionization mass spectrometry [56]. A method incorporating online LC-LC-MS was developed to compensate for matrix effects and signal suppression in qualitative and quantitative analysis. The authors demonstrated that signal suppression could be induced by coelution of analytes and matrix components and/or column overload [69]. The online LC-LC system consisted of a binary and a quaternary pump and two columns, (C18, restricted access media or C8 as the first dimension, and C18 as the second dimension), connected through a six-port valve. An additional six-port valve was installed before the MS to divert the LC flow to the MS only during analyte elution. The system was applied for the determination of fenozide herbicides spiked in different matrices (e.g., wheat forage and pecan nut matrix). Absolute recoveries obtained with LC-LC-MS (MS) were, on average, greater than the desired value of at least 70%, therefore confirming that the LC-LC-MS (MS) method was effective for the simultaneous quantification of multiple compounds in a complex sample matrix with minimal sample cleanup. In addition, LC-LC-MS single-compound analysis was demonstrated to be more effective than multiple-compound analysis in reducing matrix-related signal suppression. The study of single- and multiple-compound analysis confirmed that column leaching from prior eluting compounds could augment signal suppression effects of later eluting analytes. Recovery for methoxy fenozide, the latest eluting compound, was increased by additional ~24% for single-compound LC-LC-MS (MS) analysis versus multiple-compound analyses.

Very recently, Kittlaus et al. developed a fully automated system for the determination of 300 different pesticides from various food commodities [70] without any manual sample cleanup. A YMC-Pack Diol HILIC column was used in the first dimension, and an Agilent Poroshell 120 EC-C18 column in the second dimension. Both columns were connected through a packed loop interface containing a C8 trapping column (Figure 16.6).

High orthogonality was obtained through the combination of HILIC and RP chromatography; thus, a good matrix separation was obtained. The HILIC column was used to replace the classical liquid–liquid extraction step during sample preparation. The first-dimension mobile phase was composed of water and acetonitrile containing ammonium formate and acetic acid, and the second-dimension mobile phase was composed of water and methanol with the same additives. To elute all pesticides within the first small fraction of the HILIC separation, formic acid was added to the mobile phase. At this time, most of the matrix compounds were still retained on the HILIC column. The packed loop interface had to collect all these analytes. A RP material had to be used for this purpose due to the nonpolar character of the second dimension. However, the high amount of acetonitrile in the first part of the HILIC separation was problematic for the retention of the analytes on this trap column. Thus, the eluate polarity had to be increased to increase the affinity of the nonpolar analytes to the stationary phase of the trap. This was achieved by adding water to the HILIC eluate. Very polar analytes with high retention on the HILIC column could be determined directly without any trapping or RP separation. Finally, trapped compounds were flushed back to the second-dimension RP column. The analytes were separated by gradient elution and detected

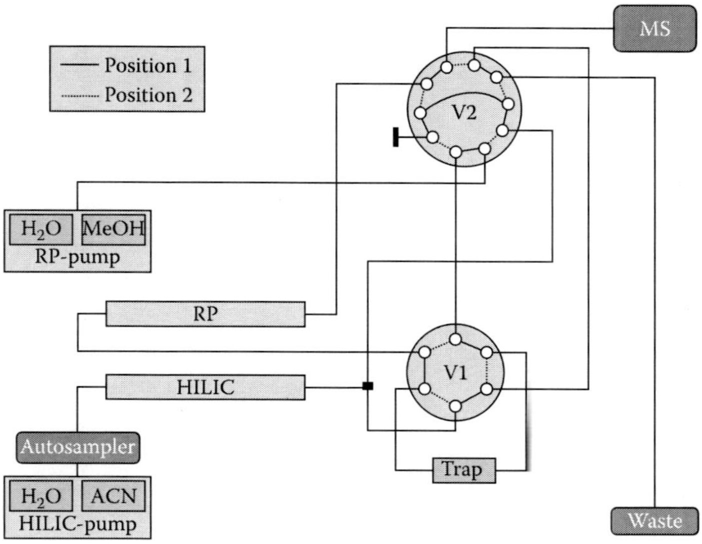

FIGURE 16.6 Kittlaus et al. 2D system configuration. (From Kittlaus, S. et al., *J. Chromatogr. A*, 1283, 98–109, 2013. With permission.)

with a triple quadrupole mass spectrometer working in the multiple reaction monitoring mode. This process is summarized in Figure 16.7.

The method was validated for more than 300 pesticides in cucumber, lemon, wheat flour, rocket, and black tea. The large majority of the compounds showed recoveries between 70% and 120%, and the relative standard deviations were under 20%. The limits of detection for nearly all compounds were at least at 0.01 mg/kg. Moreover, the method showed robust and accurate results even with "dirty" matrices, such as hops and tea.

16.3 OTHER MULTIDIMENSIONAL TECHNIQUES APPLIED TO ANALYSIS OF PESTICIDES

Combining two different forms of chromatography into a multidimensional system is not as popular as GC × GC or LC × LC. One of the main reasons is that the mobile phases exist in a different physical state, seemingly rendering interfacing more complex. Combinations such as LC (including SPE mode)-GC, supercritical fluid chromatography (SFC)-GC, or SFC-LC have proven to be very efficient in sample preparation and/or sample fractionation. For example, off-line LC separation before GC analysis can be used for the separation of chemical classes or targets out of a complex matrix that can then be analyzed by one-dimensional high-resolution GC. Off-line LC-GC is a two-dimensional technique that combines the primary LC column selectivity (often low efficiency) and the secondary GC column efficiency (often low selectivity). In the next sections, only multidimensional techniques applied for pesticide analysis are briefly discussed.

16.3.1 LC-GC

The first work describing coupling between LC and GC was published in 1980 [71]. However, the first automated system was not constructed until 1987 by Ramsteiner [72] for pesticide analysis in biological samples, and the first commercial instrument (Dualchrom 3000) was introduced in 1989 by Carlo Erba (Italy) [73].

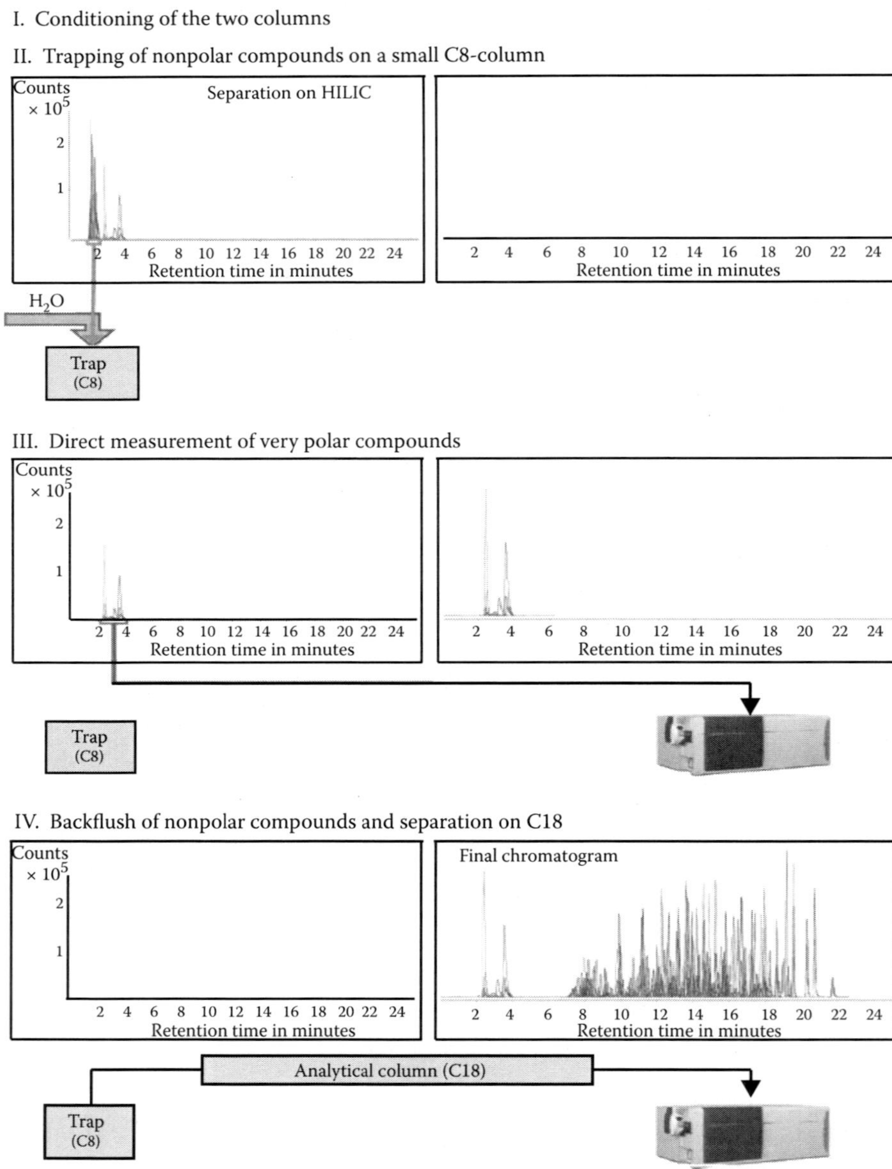

FIGURE 16.7 Principle of Kittlaus et al. 2D method. Phase I: conditioning for both dimensions. Phase II: The majority of the compounds eluted from the HILIC. After the addition of water, these compounds were trapped on the packed loop interface, and more polar matrix components were still on the HILIC column at this time. Phase III: water addition was stopped. The small and polar compounds from the HILIC column were eluted directly to the mass spectrometer without separation in the second dimension. Phase IV: The flow through the trap column was inverted. The trapped pesticides were eluted to the second dimension by an increasing methanol gradient. The matrix components with retention on the HILIC were flushed to waste by the mobile phase. (From Kittlaus, S. et al., *J. Chromatogr. A*, 1283, 98–109, 2013. With permission.)

One of the major problems of LC-GC is that a relatively large amount of liquid mobile phase must be eliminated before the GC analysis, which requires a special interface. Several designs have been developed in this respect. In addition, the target analytes must be volatile or semivolatile. For readers interested in a detailed description of LC-GC interfaces and instrumentation, more information is provided [19,73–75].

All interfaces used in LC-GC have advantages and disadvantages, and selection is based primarily on the analytical problem and on the LC conditions selected. Grob classified interfaces according to the LC mode used [53]. One of the most common interfaces in NPLC-GC is a wire interface. It was introduced for fully or partially concurrent effluent evaporation [76] and could be used with or without cosolvent trapping. The retention gap interface with different modifications is preferred for more volatile components [77,78]. The programmable temperature vaporizing (PTV) interface is used when large volumes of injection are needed. From a practical point of view, it is the easiest to use. The PTV interface has a split vent through which the solvent is removed while less volatile compounds remain in the liner, from which they can be thermally desorbed. PTV is the most commonly used interface in the comprehensive 2D LC × GC) field [79]. In RPLC-GC analysis, the mobile phase usually contains very polar solvents (e.g., methanol and water), which are difficult to remove effectively; hence special conditions and devices are needed for the transfer. One possibility is to use phase switching, akin to online liquid–liquid extraction of the mobile phase [80]. Another option is to employ vaporization of the RPLC effluent with hot injectors [53,81].

There are two modes of LC-GC, off-line and online. In the off-line mode, not all LC effluent is transferred onto the GC, and LC is employed as a prefractionation method before the GC analysis. It is useful when it is not possible to separate the compounds of interest in a single GC run. In the online approach, analytes are separated first in the LC and then the entire effluent or fractions of it are transferred to the capillary column of the GC via the selected interface for further separation. If only fractions of the LC effluent are transferred, the technique is called heart-cutting LC-GC, and if the entire effluent is transferred, it is called comprehensive MD LC-GC (LC × GC). The first LC × GC setup and application were described in 2000 by Quigley and coworkers [82].

Online LC-GC techniques are highly sensitive and selective, making possible the determination of pesticide residues in complex matrices, such as foods. The technique was used for the determination of the fungicide fenarimol in vegetables, such as cucumbers, tomatoes, and sweet peppers [83], using a loop-type interface with concurrent effluent evaporation. Two different detectors were used, an electron capture detector for the qualitative identification and a flame-ionization detector (FID) for the quantitative determination. On the other hand, Hyötyläinen et al. used another interesting interface for determining pesticides in red wine samples by RPLC-GC-FID [81]. A vaporizer/precolumn solvent split/gas discharge interface was used for the direct transfer of the aqueous effluent to the GC. Villén and coworkers applied their patented Through Oven Transfer Adsorption Desorption (TOTAD) interface in the analysis of pesticides in water [84,85] and olive oil samples [86,87].

The TOTAD interface is a modified PTV injector allowing solvent elimination in the same way as in PTV, but in this case, automation was possible (Figure 16.8). The PTV injector modifications included the pneumatics, sample introduction, and solvent elimination. The determination of pesticide residues in olive oil was achieved using a 50 mm × 4.6 mm column packed with modified silica (C4) and a nonpolar fused-silica column coated with 5% phenyl methyl silicone (30 mm × 0.32 mm × 0.25 μm). No sample pretreatment other than filtration was used. Methanol/water, 70/30 v/v, was used in the LC preseparation step. Good repeatability compensated for the low recovery values. Using a flame ionization detector, pesticide detection limits varied from 0.1 to 0.3 mg/L.

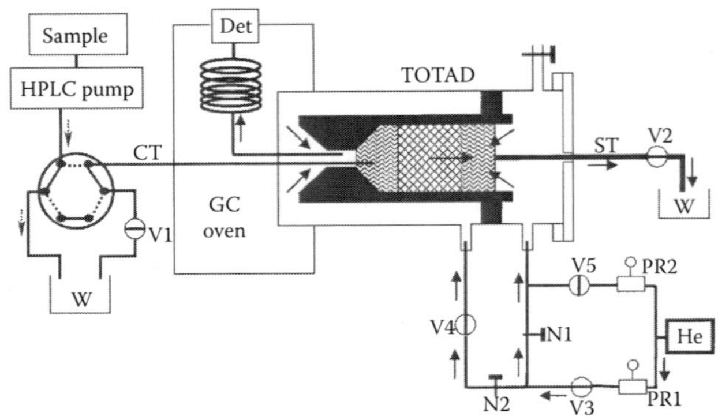

FIGURE 16.8 Scheme of the TOTAD interface used as an injector in Villén's study. Valves are positioned for LC separation, interface stabilization, and cleaning steps. (N) needle valve, (V) on–off valve, (PR) pressure regulator, (TT) stainless steel tubing used to transfer the effluent from LC to GC, (CT) silica capillary tubing, (ST) stainless steel tubing to allow for the exit of liquids and gases, and (W) waste. (From Alario, J. et al., *J. Chromatogr. Sci.*, 39, 65–69, 2001. With permission.)

16.3.2 Multidimensional Planar Chromatography

In planar chromatography (PC), the first-dimension fractions are not always transferred to another separation system, but rather a secondary separation is performed orthogonally on the same chromatographic plate. PC offers the possibility of MD separation by the use of the same layer and different eluent systems [88,89] or by the use of multiphase plates [90]. The following modes have most frequently been used for MD separations involving PC [91]:

i. Two-dimensional development on the same monolayer stationary phase with different mobile phases
ii. Two-dimensional development on the same bilayer stationary phase with the same or different mobile phases
iii. Multiple development in one, two, or three dimensions on the same monolayer stationary phase with different mobile phases
iv. Coupled layers with stationary phases with decreasing polarity developed with the same mobile phase
v. A combination of at least two of the abovementioned modes
vi. Coupled with another chromatographic technique, for example, GC or HPLC, which is used as the first dimension, while PC is used as the second dimension

Pesticides have been occasionally analyzed by MDPC. For example, 2D thin-layer chromatography (TLC) was used for the analysis of N-nitrosotriazine herbicides (cyanazine and terbutylazine and their reaction products) using different mobile phases for each dimension on silica gel plates [92]. In addition, large groups of pesticides were also separated on monolayer or bilayer stationary phases [93–95]. The success of 2D TLC separation depends on the difference between the selectivity of the two chromatographic systems used. The largest selectivity differences are obtained by combining NP and RP chromatography.

Graft TLC is a multiple system in which combined chromatographic plates with similar or different stationary phases are used. Analytes separated on the first chromatographic plate are transferred to the second plate by use of a strong mobile phase without scraping, extraction, or reapplication of the bands. It was first described in 1979 [96]. The technique has been used for pesticide analysis [97,98]. Complete separation of pesticide mixtures was also achieved by adsorbent-gradient 2D

TLC with NP development in the first dimension and RP development in the second dimension performed on HPTLC RP_{18} WF_{254S} plates or HPTLC CN F_{254S} plates.

16.4 CONCLUSIONS

In the last decade, new analytical needs have appeared and considerably modified the nature of separation sciences, widely expanding their fields of application and the variety of samples brought to the analysts. Pesticide analysis in environmental samples is very challenging due to the complexity of the matrices. In addition, this kind of analysis is becoming more complicated because we no longer need to identify a few analytes in a sea of unknowns. On the contrary, we need to identify and quantify most of these unknowns, including many at the trace level. Coupling two chromatographic separations using two different retention mechanisms, known as MD chromatography, is one of the most powerful separation techniques that can be very useful for the abovementioned kinds of analysis.

This chapter presents only a brief overview of some of the MD chromatographic techniques and their applications in pesticide analysis. These techniques enhance separation power and resolution, which makes them ideal for the analysis of pesticides in complex matrices, such as environmental or food samples. This can explain the reason why the number of papers devoted to MD chromatographic techniques has been growing steadily over the years. MDLC has exhibited a huge momentum since the 1990s. One can expect a fast increase in the number of new applications for various sample types when a new generation of 2D LC instrumentation with dedicated software for the direct conversion of raw data into 2D or 3D chromatograms becomes more widely available. Nevertheless, more research and innovation should be stimulated toward the development and design of new interfaces that allow combinations of highly orthogonal dimensions for optimum performance.

REFERENCES

1. U.S. Environmental Protection Agency, EPA 2013, 2012.
2. EEC Drinking Water Guideline, Council Directive 80/778/EEC of July 15, 1980 relating to the quality of water intended for human consumption, Official Journal L 229, 30/08/1980.
3. EEC Drinking Water Guideline, Council Directive 98/83/EC of November 3, 1998 on the quality of water intended for human consumption, Official Journal L 330, 05/12/1998.
4. Barcelo, D., Environmental Protection Agency and other methods for the determination of priority pesticides and their transformation products in water, *J. Chromatogr. A*, 643, 117–143, 1993.
5. Brouwer, E. R., Kofman, S., Brinkman, U. A. T., Selected procedures for the monitoring of polar pesticides and related microcontaminants in aquatic samples, *J. Chromatogr. A*, 703, 167–190, 1995.
6. Stanciu, G., Dobrinas, S., Birghila, S., Popescu, M., Determination of organic compounds from different types of coffee by HPLC and GC-ECD analysis, *Environ. Eng. Manag. J.*, 7, 661–666, 2008.
7. Hayward, D. G., Wong, J., Schenck, F. J., Zhang, K., Krynitsky, A. J., Begley, T., Multi-residue analysis of pesticides and POPs in fruits, vegetables and dried ginseng powders, *Organohalogen Comp.*, 70, 29–33, 2008.
8. Jaiswal, A. K., Determination of chloropyrifos in viscera using high performance thin layer chromatography plate, *J. Inst. Chem. (India)*, 81, 73–76, 2009.
9. Tran, K., Eide, D., Nickols, S. M., Cromer, M. R., Sabaa-Srur, A., Smith, R. E., Finding of pesticides in fashionable fruit juices by LC-MS/MS and GC-MS/MS, *Food Chem.*, 134, 2398–2405, 2012.
10. Bertsch, W., Two-dimensional gas chromatography. Concepts, instrumentation, and applications—Part 1: Fundamentals, conventional two-dimensional gas chromatography, selected applications, *J. High Resolut. Chromatogr.*, 22, 647–665, 1999.
11. Guiochon, G., Gonnord, M. F., Zakaria, M., Beaver, L. A., Siouffi, A. M., Chromatography with a two-dimensional column, *Chromatographia*, 17, 121–124, 1983.
12. Simmons, M. C., Snyder, L. R., Two-stage gas-liquid chromatography, *Anal. Chem.*, 30, 32–35, 1958.
13. Giddings, J. C., Two-dimensional separations: Concept and promise, *Anal. Chem.*, 56, 1258A–1270A, 1984.

14. Giddings, J. C., HRC CC, Concepts and comparisons in multidimensional separation, *J. High Resolut. Chromatogr. Chromatogr. Commun.*, 10, 319–323, 1987.
15. Sandra, K., Moshir, M., D'hondt, F., Tuytten, R., Verleysen, K., Kas, K., Francois, I., Sandra, P., Highly efficient peptide separations in proteomics. Part 2: bi- and multidimensional liquid-based separation techniques, *J. Chromatogr. B, Analyt. Technol. Biomed. Life Sci.*, 877, 1019–1039, 2009.
16. Herrero, M., Ibanez, E., Cifuentes, A., Bernal, J., Multidimensional *chromatography* in food analysis, *J. Chromatogr. A*, 1216, 7110–7129, 2009.
17. Edwards, M., Mostafa, A., Górecki, T., Modulation in comprehensive two-dimensional gas chromatography: 20 years of innovation, *Anal. Bioanal. Chem.*, 401, 2335–2349, 2011.
18. Mostafa, A., Edwards, M., Górecki, T., Optimization aspects of comprehensive two-dimensional gas chromatography, *J. Chromatogr. A*, 1255, 38–55, 2012.
19. Mondello, L., Dugo, P., Dugo, G., Lewis, A. C., Bartle, K. D., High-performance liquid chromatography coupled on-line with high resolution gas chromatography. State of the art, *J. Chromatogr. A*, 842, 373–390, 1999.
20. Bertsch, W., Two-dimensional gas chromatography. Concepts, instrumentation, and applications—Part 2: Comprehensive two-dimensional gas chromatography, *J. High Resolut. Chromatogr.*, 23, 167–181, 2000.
21. Marriott, P. J., Kinghorn, R. M., New operational modes for multidimensional and comprehensive gas chromatography by using cryogenic modulation, *J. Chromatogr. A*, 866, 203–2012, 2000.
22. Jandera, P., Mondello, L., in Mondello, L. (Editor), *Multidimensional Liquid Chromatography: Theoretical Considerations. Comprehensive Chromatography in Combination with Mass Spectrometry*, John Wiley & Sons, Inc., Hoboken, New Jersey, 2011.
23. van der Horst, A., Schoenmakers, P. J., Comprehensive two-dimensional liquid chromatography of polymers, *J. Chromatogr. A*, 1000, 693–709, 2000.
24. Janco, M., Hirano, T., Kitayama, T., Hatada, K., Berek, D., Discrimination of poly(ethyl methacrylate)s according to their molar mass and tacticity by coupling size exclusion chromatography and liquid chromatography at the critical adsorption point, *Macromolecules*, 33, 1710–1715, 2000.
25. Trathnigg, B., Rappel, C., Liquid exclusion–adsorption chromatography, a new technique for isocratic separation of nonionic surfactants: IV. Two-dimensional separation of fatty alcohol ethoxylates with focusing of fractions, *J. Chromatogr. A*, 952, 149–163, 2002.
26. Corradini, C., in Mondello, L., Lewis, A. C., Bartle, K. D. (Editors), *Coupled-Column Liquid Chromatography. Multidimensional Chromatography*, John Wiley & Sons, Ltd., Chichester, England, 2002.
27. Kohne, A. P., Welsch, T., Coupling of a microbore column with a column packed with non-porous particles for fast comprehensive two-dimensional high-performance liquid chromatography, *J. Chromatogr. A*, 845, 463–469, 1999.
28. Opiteck, G. J., Lewis, K. C., Jorgenson, J. W., Anderegg, R. J., Comprehensive on-line LC/LC/MS of proteins, *Anal. Chem.*, 69, 1518–1524, 1997.
29. Opiteck, G. J., Jorgenson, J. W., Anderegg, R. J., Two-dimensional SEC/RPLC coupled to mass spectrometry for the analysis of peptides, *Anal. Chem.*, 69, 2283–2291, 1997.
30. Bushey, M. M., Jorgenson, J. W., Automated instrumentation for comprehensive two-dimensional high-performance liquid chromatography of proteins, *Anal. Chem.*, 62, 161–167, 1990.
31. Erni, F., Frei, R. W., Two-dimensional column liquid chromatographic technique for resolution of complex mixtures, *J. Chromatogr. A*, 149, 561–569, 1978.
32. Blahova, E., Jandera, P., Cacciola, F., Mondello, L., Two-dimensional and serial column reversed-phase separation of phenolic antioxidants on octadecyl-, polyethyleneglycol-, and pentafluorophenylpropyl-silica columns, *J. Sep. Sci.*, 29, 555–566, 2006.
33. Francois, I., Sandra, K., Sandra, P., in Mondello, L. (Editor), *History, Evolution, and Optimization Aspects of Comprehensive Two-Dimensional Liquid Chromatography. Comprehensive Chromatography in Combination with Mass Spectrometry*, John Wiley & Sons, Inc., Hoboken, New Jersey, 2011.
34. Shellie, R. A., Haddad, P. R., Comprehensive two-dimensional liquid chromatography, *Anal. Bioanal. Chem.*, 386, 405–415, 2006.
35. Davis, J. M., Stoll, D. R., Carr, P. W., Dependence of effective peak capacity in comprehensive two-dimensional separations on the distribution of peak capacity between the two dimensions, *Anal. Chem.*, 80, 8122–8134, 2008.
36. Davis, J. M., Stoll, D. R., Carr, P. W., Effect of first-dimension undersampling on effective peak capacity in comprehensive two-dimensional separations, *Anal. Chem.*, 80, 461–473, 2008.
37. Li, X., Stoll, D. R., Carr, P. W., Equation for peak capacity estimation in two-dimensional liquid chromatography, *Anal. Chem.*, 81, 845–850, 2009.

38. Fairchild, J. N., Horvath, K., Guiochon, G., Approaches to comprehensive multidimensional liquid chromatography systems, *J. Chromatogr. A*, 1216, 1363–1371, 2009.
39. Tranchida, P. Q., Donato, P., Dugo, G., Mondello, L., Dugo, P., Comprehensive chromatographic methods for the analysis of lipids, *TrAC, Trends Anal. Chem.*, 26, 191–205, 2007.
40. Stoll, D. R., Li, X., Wang, X., Carr, P. W., Porter, S. E. G., Rutan, S. C., Fast, comprehensive two-dimensional liquid chromatography, *J. Chromatogr. A*, 1168, 3–43, 2001.
41. Jandera, P., Column selectivity for two-dimensional liquid chromatography, *J. Sep. Sci.*, 29, 1763–1783, 2006.
42. Guiochon, G., Marchetti, N., Mriziq, K., Shalliker, R. A., Implementations of two-dimensional liquid chromatography, *J. Chromatogr. A*, 1189, 109–168, 2008.
43. Francois, I., Sandra, P., Comprehensive supercritical fluid chromatography × reversed phase liquid chromatography for the analysis of the fatty acids in fish oil, *J. Chromatogr. A*, 1216, 4005–4012, 2009.
44. Evans, C. R., Jorgenson, J. W., Multidimensional LC-LC and LC-CE for high-resolution separations of biological molecules, *Anal. Bioanal. Chem.*, 378, 1952–1961, 2004.
45. Dugo, P., Kumm, T., Cacciola, F., Dugo, G., Mondello, L., Multidimensional liquid chromatographic separations applied to the analysis of food samples, *J. Liq. Chromatogr. Relat. Technol.*, 31 1758–1807, 2008.
46. Dugo, P., Cacciola, F., Kumm, T., Dugo, G., Mondello, L., Comprehensive multidimensional liquid chromatography: Theory and applications, *J. Chromatogr. A*, 1184, 353–368, 2008.
47. Berek, D., Two-dimensional liquid chromatography of synthetic polymers, *Anal. Bioanal. Chem.*, 396, 421–441, 2010.
48. Dugo, P., Mondello, L., Cacciola, F., Donato, P., in Mondello, L. (Editor), *Comprehensive Two-Dimensional Liquid Chromatography Applications. Comprehensive Chromatography in Combination with Mass Spectrometry*, John Wiley & Sons, Inc., Hoboken, New Jersey, 2011.
49. Murphy, R. E., Schure, M. R., Foley, J. P., Effect of sampling rate on resolution in comprehensive two-dimensional liquid chromatography, *Anal. Chem.*, 70, 1585–1594, 1998.
50. Cacciola, F., Jandera, P., Mondello, L., Comparison of high-temperature gradient heart-cutting and comprehensive LC × LC systems for the separation of phenolic antioxidants, *Chromatographia*, 66, 661–667, 2007.
51. Jandera, P., Can the theory of gradient liquid chromatography be useful in solving practical problems? *J. Chromatogr. A*, 1126, 195–218, 2006.
52. Shimmo, M., Hyötyläinen, T., Kallio, M., Anttila, P., Riekkola, M., Multidimensional and hyphenated techniques in aerosol analysis, *LC-GC Eur.*, 17, 640–645, 2004.
53. Grob, K., Efficiency through combining high-performance liquid chromatography and high resolution gas chromatography: Progress 1995–1999, *J. Chromatogr. A*, 892, 407–420, 2000.
54. Hidalgo, C., Sancho, J. V., Hernandez, F., Trace determination of triazine herbicides by means of coupled-column liquid chromatography and large volume injection, *Anal. Chim. Acta*, 338, 223–229, 1997.
55. Galera, M. M., Vazquez, P. P., Vidal, J. L. M., Fernandez, J. M., Gomez, J. L. P., Large-volume direct injection for determining naphthalene derivative pesticides in water using a restricted-access medium column in RPLC-LC with fluorescence detection, *Chromatographia*, 60, 517–522, 2004.
56. Pascoe, R., Foley, J. P., Gusev, A. I., Reduction in matrix-related signal suppression effects in electrospray ionization mass spectrometry using on-line two-dimensional liquid chromatography, *Anal. Chem.*, 73, 6014–6023, 2001.
57. Rule, G. S., Mordehai, A. V., Henion, J., Determination of carbofuran by online immunoaffinity chromatography with coupled-column liquid chromatography/mass spectrometry, *Anal. Chem.*, 66, 230–235, 1994.
58. Huber, J. F. K., Fogy, I., Fioresi, C., Residue analysis by multicolumn liquid chromatography—Herbicides in cereals, *Chromatographia*, 13, 408–412, 1980.
59. Tuinstra, L. G. M. T., Kienhuis, P. G. M., Automated two-dimensional HPLC residue procedure for glyphosate on cereals and vegetables with postcolumn fluoregenic labeling, *Chromatographia*, 24, 696–700, 1987.
60. Van der Heeft, E., Dijkman, E., Baumann, R. A., Hogendoorn, E. A., Comparison of various liquid chromatographic methods involving UV and atmospheric pressure chemical ionization mass spectrometric detection for the efficient trace analysis of phenylurea herbicides in various types of water samples, *J. Chromatogr. A*, 879, 39–50, 2000.
61. Hogendoorn, E. A., Hoogerbrugge, R., Baumann, R. A., Meiring, H. D., de Jong, A. P. J. M., van Zoonen, P., Screening and analysis of polar pesticides in environmental monitoring programmes by coupled-column liquid chromatography and gas chromatography–mass spectrometry, *J. Chromatogr. A*, 754, 49–60, 1996.

62. Pichon, V., Cau Dit Coumes, C., Chen, L., Guenu, S., Hennion, M.-C., Simple removal of humic and fulvic acid interferences using polymeric sorbents for the simultaneous solid-phase extraction of polar acidic, neutral and basic pesticides, *J. Chromatogr. A*, 737, 25–33, 1996.
63. Di Corcia, A., Crescenzi, C., Samperi, R., Scappaticcio, L., Trace analysis of sulfonylurea herbicides in water: Extraction and purification by a Carbograph 4 cartridge, followed by liquid chromatography with UV detection, and confirmatory analysis by an electrospray/mass detector, *Anal. Chem.*, 69, 2819–2826, 1997.
64. Barcelo, D., Hennion, M.-C., Sampling of polar pesticides from water matrices, *Anal. Chim. Acta*, 338, 3–18, 1997.
65. Sancho-Llopis, J. V., Hernandez-Hernandez, F., Hogendoorn, E. A., van Zoonen, P., Rapid method for the determination of eight chlorophenoxy acid residues in environmental water samples using off-line solid-phase extraction and on-line selective precolumn switching, *Anal. Chim. Acta*, 283, 287–296, 1993.
66. Hogenboom, A. C., Jagt, I., Vreuls, R. J. J., Brinkman, U. A. T., On-line trace level determination of polar organic microcontaminants in water using various pre-column–analytical column liquid chromatographic techniques with ultraviolet absorbance and mass spectrometric detection, *Analyst*, 122, 1371–1378, 1997.
67. Geerdink, R. B., van Tol-Wildenburg, S., Niessen, W. M. A., Brinkman, U. A. T., Determination of phenoxy acid herbicides from aqueous samples by improved clean-up on polymeric pre-columns at high pH, *Analyst*, 122, 889–893, 1997.
68. Hogendoorn, E. A., Westhuis, K., Dijkman, E., Heusinkveld, H. A. G., den Boer, A. C., Evers, E. A. I. M., Bauman, R. A., Semi-permeable surface analytical reversed-phase column for the improved trace analysis of acidic pesticides in water with coupled-column reversed-phase liquid chromatography with UV detection: Determination of bromoxynil and bentazone in surface water, *J. Chromatogr. A*, 858, 45–54, 1999.
69. Choi, B. K., Hercules, D. M., Gusev, A. I., Effect of liquid chromatography separation of complex matrices on liquid chromatography–tandem mass spectrometry signal suppression, *J. Chromatogr. A*, 907, 337–342, 2001.
70. Kittlaus, S., Schimanke, J., Kempe, G., Speer, K., Development and validation of an efficient automated method for the analysis of 300 pesticides in foods using two-dimensional liquid chromatography–tandem mass spectrometry, *J. Chromatogr. A*, 1283, 98–109, 2013.
71. Majors, R. E., Multidimensional high performance liquid chromatography, *J. Chromatogr. Sci.*, 18, 571–579, 1980.
72. Ramsteiner, K. A., On-line liquid chromatography–gas chromatography in residue analysis, *J. Chromatogr. A*, 393, 123–131, 1987.
73. Purcaro, G., Moret, S., Conte, L., Hyphenated liquid chromatography–gas chromatography technique: Recent evolution and applications, *J. Chromatogr. A*, 1255, 100–111, 2012.
74. Hyötyläinen, T., Riekkola, M. L., On-line coupled liquid chromatography–gas chromatography, *J. Chromatogr. A*, 1000, 357–384, 2003.
75. Cortes, H. J. (Editor), *Multidimensional Chromatography: Techniques and Applications*, Dekker, New York, NY, 1990.
76. Grob, K., Bronz, M., On-line LC-GC transfer via a hot vaporizing chamber and vapor discharge by overflow: Increased sensitivity for the determination of mineral oil in foods, *J. Microcolumn Sep.*, 7, 421–427, 1995.
77. Bosello, E., Grolimund, B., Grob, K., Lercker, G., Amado, R., Solvent trapping during large volume injection with an early vapor exit, Part 1: Description of the flooding process, *J. High Resolut. Chromatogr.*, 21, 355–362, 1998.
78. Hankemeier, T., van Leeuwen, S. P. J., Vreuls, R. J. J., Brinkman, U. A. T., Use of a presolvent to include volatile organic analytes in the application range of on-line solid-phase extraction-gas chromatography–mass spectrometry, *J. Chromatogr. A*, 811, 117–133, 1998.
79. Francois, I., Sandra, P., Sciarrone, D., Mondello, L., in Mondello, L. (Editor), *Other Comprehensive Chromatography Methods. Comprehensive Chromatography in Combination with Mass Spectrometry*, John Wiley & Sons, Inc., Hoboken, New Jersey, 2011.
80. Hyötyläinen, T., Keski-Hynnilae, H., Riekkola, M. L., Determination of morphine and its analogues in urine by on-line coupled reversed-phase liquid chromatography–gas chromatography with on-line derivatization, *J. Chromatogr. A*, 771, 360–365, 1997.
81. Hyötyläinen, T., Jauho, K., Riekkola, M. L., Analysis of pesticides in red wines by on-line coupled reversed-phase liquid chromatography–gas chromatography with vaporiser/precolumn solvent split/gas discharge interface, *J. Chromatogr. A*, 813, 113–119, 1998.

82. Quigley, W. W. C., Fraga, C. G., Synovec, R. E., Comprehensive LC × GC for enhanced headspace analysis, *J. Microcolumn Sep.*, 12, 160–166, 2000.
83. Rietveld, R., Quirijns, J., On-line liquid chromatography–gas chromatography for determination of fenarimol in fruiting vegetables, *J. Chromatogr. A*, 683, 151–155, 1994.
84. Alario, J., Perez, M., Vazquez, A., Villen, J., Very-large-volume sampling of water in gas chromatography using the Through Oven Transfer Adsorption Desorption (TOTAD) interface for pesticide-residue analysis, *J. Chromatogr. Sci.*, 39, 65–69, 2001.
85. Perez, M., Alario, J., Vazquez, A., Villen, J., Pesticide residue analysis by off-line SPE and on-line reversed-phase LC-GC using the through-oven-transfer adsorption/desorption interface, *Anal. Chem.*, 72, 846–852, 2000.
86. Sanchez, R., Vazquez, A., Andini, J. C., Villen, J., Automated multiresidue analysis of pesticides in olive oil by on-line reversed-phase liquid chromatography–gas chromatography using the through oven transfer adsorption-desorption interface, *J. Chromatogr. A*, 1029, 167–172, 2004.
87. Sanchez, R., Vazquez, A., Riquelme, D., Villen, J., Direct analysis of pesticide residues in olive oil by on-line reversed phase liquid chromatography–gas chromatography using an automated Through Oven Transfer Adsorption Desorption (TOTAD) interface, *J. Agric. Food Chem.*, 51, 6098–6102, 2003.
88. Guiochon, G., Beaver, L. A., Gonnord, M. F., Siouffi, A. M., Zakaria, M., Theoretical investigation of the potentialities of the use of a multidimensional column in chromatography, *J. Chromatogr. A*, 255, 415–437, 1983.
89. Zakaria, M., Gonnord, M. F., Guiochon, G., Applications of two-dimensional thin-layer chromatography, *J. Chromatogr. A*, 271, 127–192, 1983.
90. Johnson, E. K., Nurok, D., Computer simulation as an aid to optimizing continuous-development two-dimensional thin-layer chromatography, *J. Chromatogr. A*, 302, 135–147, 1984.
91. Nyiredy, Sz., Multidimensional planar chromatography, in Proceedings of the Dünnschicht-Chromatographie (in memoria Prof. Dr. Hellmut Jork), Kaiser, R. E., Günther, W. Gunz, H., Wulff, G. (Eds), InCom Sonderband, Düsseldorf, 1996.
92. Zwickenpflug, W., Weiss, H., Fuerst-Hunnius N., Richter E., Separation of the reaction products of cyanazine and terbuthylazine nitrosation by thin-layer chromatography, *Fresenius' J. Anal. Chem.*, 360, 679–682, 1998.
93. Tuzimski, T., Soczewiński, E., Use of a database of plots of pesticide retention (R_F) against mobile-phase composition. Part I. Correlation of pesticide retention data in normal- and reversed-phase systems and their use to separate a mixture of ten pesticides by 2D TLC, *Chromatographia*, 56, 219–223, 2002.
94. Tuzimski, T., Bartosiewicz, A., Correlation of retention parameters of pesticides in normal and RP systems and their utilization for the separation of a mixture of ten urea herbicides and fungicides by two-dimensional TLC on cyanopropyl-bonded polar stationary phase and two-adsorbent-layer multi-K plate, *Chromatographia*, 58, 781–788, 2003.
95. Tuzimski, T., Soczewiński, E., Correlation of retention parameters of pesticides in normal- and reversed-phase systems and their utilization for the separation of a mixture of 14 triazines and urea herbicides by means of two-dimensional thin-layer chromatography, *J. Chromatogr. A*, 961, 277–283, 2002.
96. Pandey, R. C., Misra, R., Rinehart, Jr., K. L., Graft thin-layer chromatography, *J. Chromatogr.*, 169, 129–139, 1979.
97. Tuzimski, T., Two-dimensional TLC with adsorbent gradients of the type silica-octadecyl silica and silica-cyanopropyl for separation of mixtures of pesticides, *J. Planar Chromatogr.—Mod. TLC*, 18, 349–357, 2005.
98. Tuzimski, T., Separation of multicomponent mixtures of pesticides by graft thin-layer chromatography on connected silica and octadecyl silica layers, *J. Planar Chromatogr.—Mod. TLC*, 20, 13–18, 2007.

17 Chiral Separation of Some Classes of Pesticides by High-Performance Liquid Chromatography Method

Imran Ali, Iqbal Hussain, Mohd Marsin Sanagi, and Hassan Y. Aboul-Enein

CONTENTS

17.1 INTRODUCTION

Pesticides are one of the notorious organic pollutants in our environment. They are classified according to the pests as insecticides, fungicides, herbicides, algaecides avicides, bactericides, miticides, molluscicides, nematicides, piscicides, and rodenticides. But major classes of pesticides are insecticides, herbicides, fungicides, and others that are grouped as miscellaneous pesticides. About 1693 pesticides are available in the world; most of the pesticides are organic chemicals, and some are inorganic and biological species [1]. These pesticides control insects by killing them or by changing the behavior of pests through a delivery system such as spraying, baits, slow-release diffusion, etc. The insecticides are classified as organochlorine compounds (biphenyl aliphatic, hexachlorocyclohexane, cyclodienes, and polychloro terpenes); organophosphate compounds–esters of phosphorus (phosphates, phosphonates, phosphorothioates, phosphorodithioates, phosphorothiolates, and phosphoramidates); organosulfur compounds, which contain two phenyl rings with a sulfur atom; carbamates; formamidines; dinitrophenols; organotins; pyrethroids (first-, second-, third-, and fourth-generation pyrethroids); nicotinoids; spinosyns; fiproles or phenylpyrazoles; pyrroles; pyrazoles; pyridazinones; quinazolines; benzoyl urea, botanicals (pyrethrum, nicotine, rotenone, limonene, or d-limonene and neem); synergists or activators; antibiotics; fumigants; insect repellents; inorganics; miscellaneous classes of insecticides (methoxyacrylates, naphthoquinones, nereistoxin analogues, pyridine azomethine, pyrimidinamines, tetronic acids); and some miscellaneous compounds of insecticides, which include etoxazole, pyridalyl, amidoflumet, pyriproxyfen, buprofezin, and tebufenozide [2].

Fungicides are useful for the control of fungal disease by specifically inhibiting or killing the fungus causing the disease. Different classes of fungicides on the basis of chemical compositions are benzimidazole, dicarboximide, imidazole, piperazine, triazole, phenylamide, oxathiin, anilino-pyrimidine, strobilurin, phenylpyrrole, chlorophenyls, chloronitrobenzene, triadiazole, cinnamic acid, hydroxyanilide, streptomyces, polyoxin, benzothiadiazole, phosphonate, dithiocarbamate, chloroalkythios, chloronitrile, phenylpyridin-amine, cyanoacetamide oxime, carbamate, aldehyde, mineral oils, and some inorganics [3]. Herbicides (weed killers) are useful for weed control and classified as phenoxy compounds, phenyl acetic acid, benzoic acid, phthalic acid, n-1-napthylpthalamic acid, aliphatic acid, substituted phenols, heterocyclic nitrogen derivatives, aliphatic organic nitrogen derivatives, carbamate, metal organic and inorganic salts, and hydrocarbons or oils [4].

Among these pesticides, about 28% (482) are chiral in nature, among which 149 are insecticides, 141 are herbicides, 97 are fungicides, and 95 are miscellaneous chiral pesticides [1]. These pesticides are useful for food production and decrease the rate of disease in crops, etc. But pesticides are also harmful for the environment and for human beings; therefore, the exercise to reduce the consumption of pesticides is necessary. On the basis of chirality, we can reduce the use of pesticides because only one enantiomer is biologically active toward target organisms, and the other enantiomer shows less effect but may have adverse effects on some nontarget organisms and serves as an unwanted burden to the environment [5,6]. Moreover, the enantio-selective behavior of the chiral pesticides and the physiological changes of plants may alter the food chain and, further, the ecological system. Besides, the degradation product of achiral pesticides may be chiral and toxic. The enantiomers can exhibit significant differences in biological activity and toxicity as well as in environmental behaviors [7–14]. Therefore, it is important to investigate and clarify the specific environmental fate of the chiral pesticides in the environment. In spite of this, most chiral pesticides are marketed and released into the environment as racemates.

Consequently, there is an urgent need to develop analytical methods to determine the optical purity and enantiomeric resolution of chiral pesticides. Thus, several analytical methods have been used to control the enantiomeric purity of different classes of pesticides. High-performance liquid chromatography (HPLC) has achieved a good reputation for chiral analysis of pesticides due to the availability of several chiral stationary phases (CSPs), high speed, sensitivity, and reproducibility. A variety of mobile phases, including normal (NP), reversed (RP), and new polar organic phases, are used in HPLC. About 80% enantiomeric separation of drugs, pharmaceuticals, and pesticides has been carried out by HPLC [15–22]. Due to the wide range of application of HPLC for the chiral resolution of pesticides, in this chapter, attempts have been made to explain the chiral resolution of pesticides by HPLC with NP and RP CSPs. Efforts have also been made to explain the mechanism of chiral separation of pesticides by HPLC.

17.2 MECHANISM OF CHIRAL SEPARATION

Some CSPs have been used for enantiomeric separation of pesticides. The most important chiral selectors are polysaccharides, cyclodextrins, macrocyclic glycopeptide antibiotics, proteins, crown ethers, ligand exchangers, Pirkle's types, and several others [15,23]. The enantiomeric recognition mechanism is one of the most important issues for chiral analytical scientists for applying CSPs properly and more precisely. But the chiral recognition mechanism at the molecular level on these CSPs is still not fully developed; however, in 1956, Pfeiffer [24] explained a three-point model for chiral recognition mechanisms. Further, this model was explained in more detail by Pirkle and Pochapsky [25]. According to this model, for chiral separation, a minimum of three interactions plays an important role between the CSP and at least one of the enantiomers, and at least one of these interactions is stereo-chemically dependent. This three-point model does not apply to every chiral species. After that, Groombridge et al. [26] postulated a four-point model for chiral recognition on some protein CSPs. Briefly, it is thought that diastereoisomeric complexes of the enantiomers are formed with CSPs, which have different physical and chemical properties, due to which these enantiomers get resolved with different retention times. Every CSP has different chiral recognition mechanisms. But the chiral recognition process on polysaccharides, cyclodextrins (CDs), macro cyclic glycopeptide antibiotics, proteins, and chiral crown ether (CCE)-based CSPs are more or less similar. The chiral grooves on polysaccharides, the cavities on CDs, the baskets on macrocyclic glycopeptide antibiotics, the bridges and loops on proteins, and the cavities on CCE-based CSPs provide a chiral environment for the enantiomers. Two enantiomers get fitted to different extents and, hence, elute at different retention times. The difference in the stabilities of the enantiomers on these CSPs are due to their different bonding and interactions, the most important of which are hydrogen bonding, dipole-induced dipole interactions, π–π complexation, inclusion complexation, anionic and cationic bonding, van der Waals forces, and so on [15–21]. Steric effects also play a crucial role in the chiral resolution of racemates. The different binding energies of the diastereoisomeric complexes are due to the various interactions mentioned above. Pirkle-type CSPs contain a chiral aromatic ring; therefore, the formation of a π–π charge transfer diastereoisomeric complex of the enantiomers (with the aromatic group) with a CSP is considered to be essential. In view of these facts, the π-acidic CSPs are suitable for chiral resolution of π donor solutes and vice versa. However, the newly developed CSPs that contain both π-acidic and π-basic groups are suitable for the chiral resolution of both types of solutes, that is, π-donor and π-acceptor analytes. Briefly, Pirkle-type CSPs contain a chiral moiety that provides a chiral environment for the enantiomers. Therefore, the enantiomers fit differently to this chiral moiety (due to their different spatial configurations). Accordingly, the two enantiomers form diastereoisomeric complexes that have different physical and chemical properties, along with different binding energies. Therefore, the two enantiomers elute at different retention times with the flow rate of the mobile phase, and hence, chiral separation occurs.

On ligand exchange CSPs, chiral resolution occurs due to the exchange of chiral ligands and enantiomers on specific metal ions through coordinate bonds. The two enantiomers have different exchange capacities because of the stereospecific nature of the ligand exchange process and, hence, chiral resolution takes place. Davankov et al. [27–29] suggested a theoretical model for the mechanisms of chiral resolution on these CSPs. In this model, the enantiomers are coordinated to metal ion in different ways, depending on their interactions with ligands bonded to the stationary phases, which act as a chiral selector. The authors explained that chiral resolution is due to different bonding along with the steric effects that result in the formation of diastereoisomeric complexes by the two enantiomers. These diastereoisomeric complexes are stabilized at different magnitudes by dipole–dipole interactions, hydrogen bonding, van der Waals forces, and steric effects and elute at different retention times. A general graphical representation of the chiral resolution mechanism of racemates on the abovementioned CSPs is shown in Figure 17.1.

17.3 COLUMNS AND ELUENTS

A variety of CSPs have been developed for the resolution of chiral pesticides. The most important classes of chiral selectors are explained above. These chiral selectors are available in the form of columns and capillaries and marketed with different trade names as shown in Table 17.1.

The development of CSPs enhanced the utility of the HPLC technique. These CSPs have been used frequently and successfully for the chiral resolution of many drugs, pharmaceuticals, and other environmental pollutants. Therefore, they may also be used for the enantiomeric separation of chiral pesticides, and some reports on the chiral separation of pesticides using the abovementioned CSPs are to be found in the literature. For significant resolution of chiral pesticides with these CSPs, a wide range of mobile phases as eluents have been used. Different composition of eluents for chiral separation depend on the CSP. For NP mode, various compositions of organic modifiers and, for RP, polar organic solvents with water and buffer can be employed. The different optimizing conditions for chiral resolution are composition of mobile phase, its pH, temperature, the amount injected on the HPLC machine, flow rate, detection, and so on.

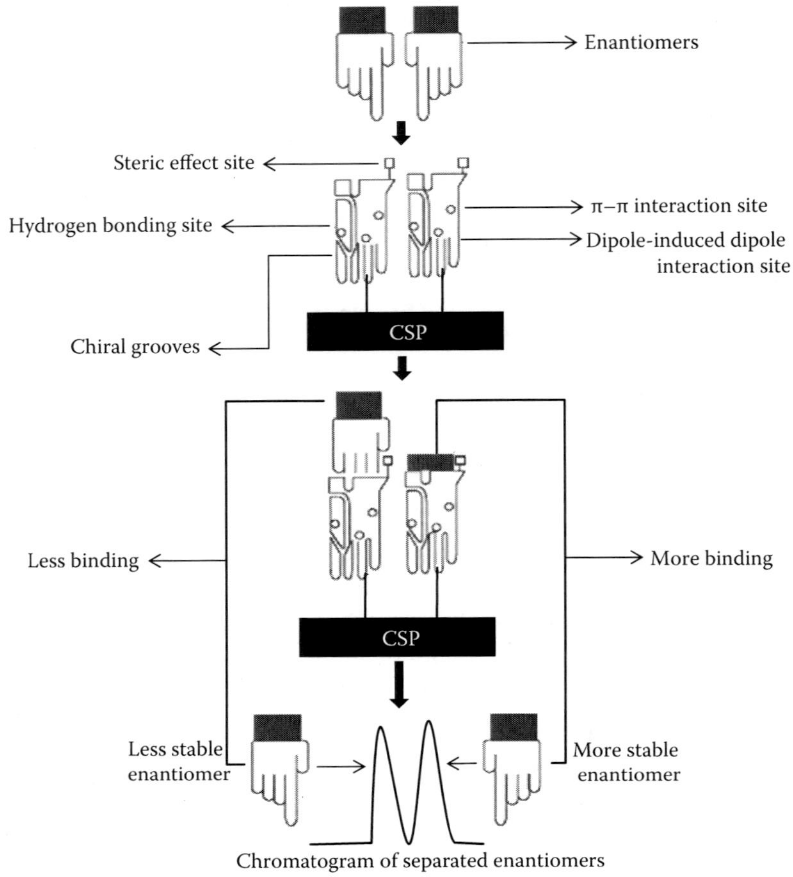

FIGURE 17.1 Schematic diagram of chiral resolution mechanism.

In the case of polysaccharide chiral columns, for NP chromatography, pure ethanol or 2-propanol as an eluent is recommended. To decrease the polarity of the mobile phase and increase the retention time of the enantiomers, hexane, cyclohexane, pentane, or heptane are used as one of the main constituents of the mobile phase. However, other alcohols are also used in the mobile phase. Normally, if pure ethanol or 2-propanol is not suitable as mobile phases, hexane, 2-propanol, or ethanol in the ratio of 80:20 are used as the mobile phase, and the change in mobile phase composition is carried out on the basis of observations. Finally, the optimization of the chiral resolution is carried out by adding small amounts of amines or acids (0.1%–1.0%). Similarly, chiral resolution on polysaccharide-based CSPs in RP mode is carried out by using aqueous mobile phases, and the selection of the mobile phase depends on the solubility and the properties of the pesticides to be analyzed. The choice of mobile phase in RP mode is very limited. Water is used as the main constituent of the mobile phases. The modifiers used are acetonitrile, methanol, and ethanol. Optimization of chiral resolution is carried out by adding small percentages of amines or acids (0.1%–1.0%). Some of the resolutions are pH-dependent and require a constant pH of mobile phase. Under such conditions, in general, the resolution is not reproducible when using a mobile phase such as water-acetonitrile or water–methanol, and therefore, a buffer with some organic modifiers (acetonitrile, methanol) have been used as the mobile phase. The optimization of the resolution is carried out by adjusting the pH values of buffers and the amounts of organic modifiers. The most commonly used buffers are perchlorate, acetate, and phosphate.

For CD-based columns, chiral resolutions have been carried out using aqueous mobile phases. Buffers of different concentrations and pH values have been developed and used for this purpose.

TABLE 17.1
Various NP and RP Chiral Columns and Their Trade Names

Columns	Trade Name
Polysaccharide-Based CSPs	
Chiralcel OB, Chiralcel OB-H, Chiralcel OJ, Chiralcel OJ-R, Chiralcel CMB, Chiralcel OC, Chiralcel OD, Chiralcel OD-H, Chiralcel OD-R, Chiralcel OF, Chiralcel OD-RH, Chiralcel OG, Chiralcel OA, Chiralcel CTA, Chiralcel OK, Chiralpak AD, Chiralpak AD-R, Chiralpak AR, Chiralpak AD-RH, and Chiralpak AS	Daicel Chemical Industries, Tokyo, Japan
CD-Based CSPs	
Cyclobond I, II, and III; Cyclobond AC, RN, Inc., SN; ApHpera ACD and BCD	Advanced Separation Technologies, Whippany, NJ, USA
Nucleodex β-OH, Nucleodex β-PM	Macherey-Nagel, Duren, Germany
ORpak CD-HQ and Orpak	
CDB-453 HQ, ORpak CDBS-453, Kanagawa, Japan	Showa Denko, Kanagawa, Japan
ORpak CDA-453 HQ and ORpak CDC-453 HQ	
Keystone β-OH and Keystone β-PM	Thermo Hypersil, Bellefonte, PA, USA
β-Cyclose-6-OH	Chiral Separations, La Frenaye, France
YMC Chiral CD BR, YMC Chiral NEA (R), and YMC Chiral NEA (S)	YMC, Kyoto, Japan
Macrocyclicglycopeptide Antibiotic–Based CSPs	
Chirobiotic R, Chirobiotic T, Chirobiotic V, Chirobiotic TAG, and Chirobiotic modified V	Advanced Separation Technologies, Inc., Whippany, NJ, USA
Protein-Based CSPs	
Chiral AGP, Chiral HSA, and Chiral CBH	Advanced Separation Technologies, Inc., Whippany, NJ, USA
Chiral AGP, Chiral CBH, and Chiral HAS	Chrom Tech, Ltd., Cheshire, UK
Resolvosil BSA-7 and Resolvosil BSA-7PX	Macherey-Nagel, Duren, Germany
Chiral AGP, Chiral CBH, and Chiral HAS	Regis Technologies, Morton Grove, IL, USA
AFpak ABA-894	Showa Denko, Kanagawa, Japan
Keystone HSA and Keystone BAS	Thermo Hypersil, Bellefonte, PA, USA
TSK gel Enantio L1 and TSK gel Enantio-OVM	Tosoh, Tokyo, Japan
EnantioPac	LKB Pharmacia, Bromma, Sweden
Bioptic AV-1	GL Sciences, Tokyo, Japan
Ultron ES-BSA, Ultron ES-OVM Column, Ultron ES-OGP Column, and Ultron ES-Pepsin	Shinwa Chemical Industries, Kyoto, Japan
Crown Ether–Based CSPs	
Crownpak CR	Chiral Technologies, Inc., Exton, PA, USA Separations Kasunigaseki-Chrome, Tokyo, Japan Daicel Chemical Industries, Tokyo, Japan
Opticrown RCA	USmac Corporation, Winnetka, Glenview, IL, USA
Chiralhyun-CR-1	K-MAC (Korea Materials & Analysis Corp.), South Korea
Chirosil CH RCA	Restech Corporation, Daedeok, Daejon, South Korea

(*Continued*)

TABLE 17.1 (CONTINUED)
Various NP and RP Chiral Columns and Their Trade Names

Columns	Trade Name
Ligand Exchange–Based CSPs	
Chirosolve	JPS Chemie, Switzerland
Chiralpak WH, Chiralpak WM, Chiralpak WE, and Chiralpak MA	Separations Kasunigaseki-Chrome, Tokyo, Japan
Nucleosil Chiral-1	Macherey-Nagel, Düren, Germany
Phenylglycine and leucine Chirex types	Regis Technologies, Morton Grove, IL, USA
Orpak CRX-853	Showa Denko, Kanagawa, Japan
Pirkle-Type CSPs	
Opticrown Chiralhyun-Leu-1 and Opticrown Chiralhyun-PG-1	Usmac Corporation, Glenview, IL, USA
Whelk-O 1, Whelk-O 2, Leucine, Phenylglycine, β-Gem 1, α-Burke 1, α-Burke 2, Pirkle 1-J, Naphthylleucine, Ulmo, and Dach	Regis Technologies, Morton Grove, IL, USA
Nucleosil Chiral-2	Macherey-Nagel, Düren, Germany
Sumichiral OA	Sumika Chemical Analysis Service, Konohana-ku Osaka, Japan
Kromasil Chiral TBB and Kromasil Chiral DMB	Eka Chemicals Separation Products, Bohus, Sweden
Chirex Type I	Phenomenex, Torrance, CA, USA
Chiris series	IRIS Technologies, Lawrence, KS, USA

Triethylammonium acetate (TEAA), phosphate, citrate, and acetate are among the most commonly used buffers [30–33]. Phosphate buffers, such as sodium, potassium, and ammonium phosphate, are commonly used. The stability constant of the complexes decreases due to the addition of organic solvents, and hence, the organic modifiers are used to optimize the chiral resolution. The most commonly used organic solvents are methanol and acetonitrile. Acetonitrile is a stronger organic modifier than methanol. Some other organic modifiers, such as ethanol, 2-propanol, 1-propanol, n-butanol, tetrahydrofuran (THF), triethylamine, and dimethylformamide, have also been used for the optimization of chiral resolution on CD-based CSPs [30,31,34,35]. The effect of the type and concentration of these organic modifiers varies from one analyte to another, and hence it is very difficult to predict the best strategy for their use as organic modifiers. Sometimes, the use of highly concentrated buffers under the RP mode decreases the lifetime and efficiency of the column. Therefore, the use of an alternative mobile phase, that is, a NP, is an advantage in chiral resolution on these phases. The most commonly used solvents in NP mode are hexane, cyclohexane, and heptane. However, some other solvents, such as such as dichloromethane, acetone, propanol, ethyl acetate, ethanol, and chloroform, have also been used as components of the mobile phase. The concentration of buffer is a very important aspect of chiral resolution on these phases under the RP mode. The addition of salts into the RP mobile phase has been found to improve the chiral resolution [33].

Antibiotic columns may be used in the NP, RP, and new polar ionic phase modes. Due to the complex structure of these antibiotics, most of them function equally well in RP, NP, and modified polar ionic phases. All three solvent modes generally show different selectivity with different analytes. In NP chromatography, the most commonly used solvents are typically hexane, ethanol, and methanol. The optimization of chiral resolution is achieved by adding some other organic acids and bases, such as acetic acid, THF, diethylamine (DEA), or triethylamine (TEA) [36,37]. In a RP system, buffers are mostly used as mobile phases with small amounts of organic modifiers. The use of buffers as mobile phases has increased the efficiency of the resolution. Ammonium nitrate, TEAA, and sodium citrate buffers have been used very successfully. A variety of organic modifiers have been used to alter selectivity [38–40]: acetonitrile, methanol, ethanol, 2-propanol, and THF have shown good selectivity for various analytes. In the RP mode, the amounts of organic modifiers are typically low, usually in the order of 10%–20%. The typical starting composition of the mobile phase is organic modifier–buffer (pH 4.0–7.0; 10:90). The use of alcohols as organic modifiers generally requires higher starting concentrations, for example, 20% for comparable retention when using acetonitrile or THF in a starting concentration of 10%. The effect of organic solvents on the enantio-selectivity also depends on the type of antibiotic. In fact, better recognition is obtained at lower buffer pH values or close to the isoelectric point of the antibiotics, especially for vancomycin. Using vancomycin, a low concentration of organic solvents did not significantly influence the separation, but enantio-resolution is improved for some compounds with ristocetin A and teicoplanin [41] even at low organic modifier concentrations. The effect of organic modifiers on chiral resolution varies from racemate to racemate [42]. A simplified approach has been proven to be very effective for the resolution of a broad spectrum of racemates. The first consideration in this direction is the structure of the analytes. If the compound has more than one functional group that is capable of interacting with the stationary phase and at least one of those groups is on or near the stereogenic center, then the first choice for the mobile phase would be the new polar ionic phase. Due to the strong polar groups present in macrocyclic peptides, it is possible to convert the original mobile phase concept to 100% methanol with the acid/base added to effect selectivity. The key factor in obtaining complete resolution is still the ratio of acid to base; the actual concentrations of the acid and base only affect the retention. Therefore, starting with a 1:1 ratio, some selectivity is typically observed, and then different ratios of 1:2 and 2:1 are applied to note the change in resolution indicating the trend. If the analyte is eluting too fast, the acid/base concentration is reduced. Conversely, if the analyte is too well retained, the acid/base concentration is increased. The parameters for the concentrations are between 1% and 0.00%. Above 1%, the analyte is too polar and indicates a typical RP system, and below 0.001%, it indicates a NP system. Both tri-fluoroacetic acid (TFA) and acetic acid have been used as the acid component with ammonium hydroxide and triethylamine as the base. For an analyte/pesticide that has only one functional group or for reasons of solubility, typical NP solvents (hexane/ethanol) or RP solvents (THF/buffer) are employed. The pH value is an important controlling factor for enantiomeric resolution in the NP, RP, and new polar ionic phases. In general, buffers are used as the mobile phases to control the pH in HPLC. The pH value of the buffers ranges from 4.0 to 7.0 in a RP system. It has been observed that, with an increasing pH value, the values of Rs, k, and α decrease. Therefore, the safest and most suitable pH values in RP systems vary from pH 4.0 to pH 7.0 [38,43].

Protein-based chiral columns were mostly used under RP, that is, aqueous mobile phases are frequently used. Buffers of differing concentrations and pH values are mostly used for chiral resolution on these CSPs. The most commonly used buffers were phosphate and borate, which were used in a concentration range of 20–100 mM with a 2.5–8.0 pH range. However, as with all silica-based CSPs, the prolonged use of an alkaline pH buffer (>8.5) is not suitable. On the other hand, at lower pH, irreversible changes are possible in cross-linked protein phases, and hence, the use of buffers with low pH values for long periods of time is not recommended. Therefore, a buffer that ranges from pH 3.0 to pH 7.0 should be chosen. A pH 4.5 ammonium acetate buffer may be useful. For the mobile phase development, any buffer (50 mM, pH 7.0) can be used, and the optimization is carried out by changing the concentration and the pH value. The successful use of organic solvents in the optimization of chiral resolution on these CSPs has been reported: the hydrophobic interactions are affected by the use of these solvents, the most important of which are methanol, ethanol, 1-propanol, 2-propanol, acetonitrile, and THF. These organic modifiers have been used in the range of 1%–10%. Care must be taken when using these organic modifiers as they can denature the protein. However, high concentrations of methanol and acetonitrile have been used on some of the cross-linked protein CSPs. The selection of these organic modifiers depends on the structure of the racemic compounds and the CSP used. In some cases, charged modifiers, such as hexanoic acid, octanoic acid, and quaternary ammonium compounds, have also been used for optimum chiral resolution [44,45].

Aqueous mobile phases containing organic modifiers and acids have been used on CCE-based CSPs. In all applications, aqueous and acidic mobile phases are used; the most commonly used mobile phases are aqueous perchloric acid and aqueous methanol containing sulfuric, TFA, or perchloric acid separately. Compounds with higher hydrophobicity, generally, have longer retention times on CCE-based CSPs, and therefore, organic modifiers are used to optimize the resolution [46]. This optimization is carried out by adjusting the amounts of methanol, sulfuric acid, and perchloric acid separately. In general, the separation is increased by an increase in methanol and a decrease in the acid concentrations. The other organic modifiers used are ethanol, acetonitrile, and THF, but methanol has been found to be the best one [47–49]. In addition to the composition of the mobile phase, the effects of other parameters, such as the temperature, the flow rate, the pH value, and the structure of the analytes, have also been studied, but only a few reports are available in the literature. It has been observed that, in general, lowering of the temperature results in better resolution. The flow rate may be used to optimize the chiral resolution of pesticides on these CSPs. Because all of the mobile phases are acidic in nature, the effect of the pH on chiral resolution is not significant.

Ligand exchange columns have been used for the chiral resolution of racemic compounds containing electron-donating atoms, and therefore, its application is confined. In most cases, buffers, sometimes containing organic modifiers, have been used as the mobile phase. Therefore, the optimization has been carried out by controlling the composition of the mobile phases. There are two strategies for the development and use of the mobile phases on these CSPs. With a CSP that has only a chiral ligand, an aqueous mobile phase containing a suitable concentration of metal ion is used, and in the case of a CSP that contains a metal ion complex as the chiral ligand, a mobile phase without a metal ion is used. In most applications, aqueous solutions of metal ions or buffers have been used as the mobile phases. The most commonly used buffers are ammonium acetate and phosphate. However, the use of a phosphate buffer is avoided if a metal ion is being used as the mobile phase additive to avoid complex formation between the metal ion and the phosphate, which may block the column. A literature search indicates that these buffers (20–50 mM) have frequently been used for successful chiral resolution, but in some instances, organic modifiers have also been used to improve the resolution. In general, acetonitrile has been used as the organic modifier [50]; however, some reports deal with the use of methanol, ethanol, and THF [51–54]. The concentrations of these modifiers vary from 10% to 30%. However, some reports have indicated the use of these organic modifiers by up to 75% [53,54]. In general, the chiral resolution of highly retained solutes is optimized by using organic modifiers. Basically, the organic modifiers reduce the hydrophobic interactions, resulting in an improved resolution. The pH of mobile phase has also been recognized as one of the most important controlling factors in chiral resolution on ligand exchange chiral phases [55]. The retention and the selectivity of enantiomeric resolution has also been investigated with respect to metal ion concentrations on these CSPs. Chiral resolution on CSPs containing only chiral ligands has been carried out using different concentrations of metal ions in the mobile phase.

NP mode has frequently been used for chiral resolution of racemic compounds on Pirkle-type CSPs. Hexane, heptane, and cyclohexane are the nonpolar solvents of choice on these chiral stationary phases. Aliphatic alcohols may be considered as hydrogen donors and acceptors and thus may interact at many points with the aromatic amide groups of CSPs, generating hydrogen bonds. Therefore, the addition of aliphatic alcohols improves the chiral resolution, and hence the alcohols are called organic modifiers. The most commonly used alcohol is 2-propanol; however, methanol, ethanol, 1-propanol, and n-butanol have also been used. Some reports have also indicated the use of dichloromethane and chloroform as organic modifiers with hexane. In addition to this, acidic and basic additives improve the chromatographic resolution. A small amount of acetic, formic, or TFA acid improves the peak shape and enantio-selectivity for acidic and basic solutes. Sometimes, there is a need to combine an acid and an organic amine (e.g., triethylamine) for strong basic racemic compounds. Some reports are also available for dealing with the RP eluents, but the prolonged use of a RP mobile phase is not recommended. With the development of more stable new CSPs, the use of the RP mobile phase mode became possible. Nowadays, both mobile phase modes are in use.

17.4 SEPARATION OF ENANTIOMERS OF PESTICIDES BY HPLC IN NP AND RP MODES

HPLC is the most popular and most widely applicable technology in the field of chiral analysis of a variety of pesticides due to the availability of a large number of chiral stationary phases in the form of NP and RP modes. Over the course of time, various types of LC approaches have been developed and used in this application, but HPLC remains the most suitable modality due to its various advantages, such as its high speed, sensitivity, and reproducibility. A variety of mobile phases, including NP, RP, and new polar organic phases, are used in HPLC. About 80% chiral resolution of pharmaceuticals, drugs, agrochemicals, and other compounds has been carried out using HPLC [15–20]. HPLC has also been used for the chiral separation of pesticides [21,22]. In spite of the variety of CSPs available for HPLC, it has not been used very frequently for the analysis of chiral environmental pollutants. This is due to the fact that some organochlorine pollutants are transparent to UV radiation, and hence, the very popular UV detector cannot be used in HPLC for the purpose of detecting such xenobiotics. However, HPLC can be coupled with MS, polarimetry, and other optical detection techniques for the chiral resolution of such types of pollutant. Apart from these points, a large number of reports are available for the chiral resolution of some UV-absorbing pesticides by HPLC in NP and RP modes.

17.4.1 Enantiomeric Separation of Pesticides in NP Mode

Enantiomeric separation of pesticides has been carried out by using various types of chiral stationary phases in the NP condition. Among all CSPs, polysaccharide-based CSPs are currently the most popular due to their versatility, durability, and loading capacity. They are effective under not only NP conditions, but also RP conditions using the appropriate mobile phases. The majority of polysaccharide-based CSPs employed were cellulose- and amylose-based polysaccharide columns [56,57]. Thus, Caccamese and Principato [58] separated the enantiomers of vincamine alkaloids using Chiralpak AD column with hexane–2-propanol and hexane-ethanol as mobile phases, separately. Ellington et al. [59] described the successful separations of the enantiomers of various organophosphorus pesticides (crotoxyphos, dialifor, fonofos, fenamiphos, fensulfothion, isofenphos, malathion, methamidophos, profenofos, crufomate, prothiophos, and trichloronat) using Chiralpak AD, Chiralpak AS, Chiralcel OD, Chiralcel OJ, and Chiralcel OG chiral columns with different mixtures of heptane and ethanol as an eluting solvent. They also studied the effect of the concentration of ethanol on the chiral resolution of organophosphorus pesticides and reported poor separation of enantiomers of fenamiphos, fensulfothion, isofenphos, profenofos, crufomate, and trichloronat pesticides at higher concentrations of ethanol. The effect of temperature on the chiral resolution of organophosphorus pesticides was also reported; the maximum chiral resolution was observed at low temperature. The effect of temperature (from 20°C to 60°C) on the chiral resolution of fensulfothion on Chiralcel OJ column is shown in Figure 17.2.

From this figure, it may be concluded that chiral resolution is improved at low temperature and becomes maximum at 20°C, but retention time increases with the decrease of temperature. Ali and Aboul-Enein [60] studied the effect of various polysaccharide CSPs on the chiral resolution of o,p-DDT and o,p-DDD on Chiralpak AD-R, Chiralcel OD-R, and Chiralcel OJ-R. The results of these findings are given in Table 17.2, which shows that the best resolution of these pesticides was obtained on Chiralpak AD-R under the NP mode.

Li et al. [61] developed a fast and precise HPLC method for the chiral resolution of phenthoate in soil samples. The authors used the Chiralcel OD column for chiral resolution with hexane-2-propanol (100:0.8, v/v) as the mobile phase. Aboul-Enein and Ali [62] have observed that chiral resolution on polysaccharide-based CSPs is pH-dependent under the NP mode. It was observed that only partial resolution of certain antifungal agents was achieved at lower pH, and the resolution was improved by increasing the pH using triethylamine on amylose and cellulose chiral columns.

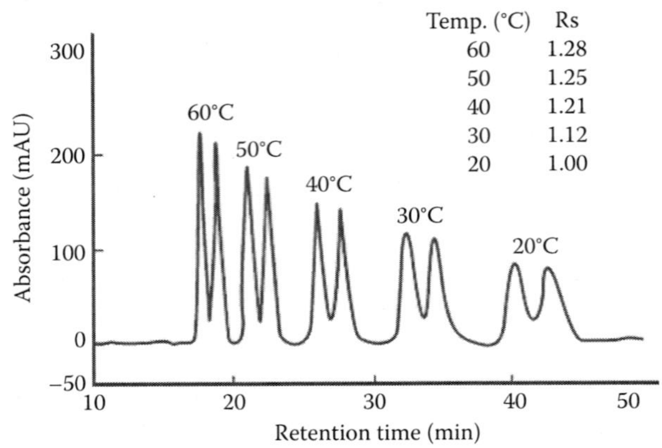

FIGURE 17.2 Effect of temperature on chiral separation of fensulfothion with Chiralcel OJ column. (From Ali, I., Aboul-Enein, H. Y., *Environ. Toxicol.*, 17, 329–333, 2002. With permission.)

Champion et al. [63] performed the enantiomeric separation of five polychlorinated compounds (*trans*-chlordane, *cis*-chlordane, heptachlor, heptachlor epoxide, and α-HCH) with different polysaccharide CSPs in NP mode, and the values of chromatographic parameters are shown in Table 17.2. From this table, it has been cleared that baseline separations were obtained for the enantiomers of *trans*-chlordane, *cis*-chlordane, and heptachlor on the Chiralcel OD column; α-HCH on the Chiralcel OJ column; and heptachlor epoxide on the Chiralpak AD column. The effect of the concentration of isopropanol (IPA) on the chiral resolution of *cis*-chlordane, *trans*-chlordane on the Chiralcel OD column is shown in Figure 17.3.

It may be concluded that 0% concentration of IPA is best as it gives the maximum resolution of *cis*- and *trans*-chlordane on the Chiralcel OD column. Wang et al. [64] observed the effect of alcohols (ethanol, n-propanol, IPA, isobutanol, n-butanol) for the resolution of fipronil, isocarbophos, and carfentrazone-ethyl pesticides on cellulose-tri (3, 5-dimethylphenylcarbamate) CSP. The concentration range for each alcohol was 2%–20%, but in the case of carfentrazone-ethyl, it was 0.1%–20% as shown in Table 17.2. The best results for the resolution of fipronil and isocarbophos were obtained with with 5% isobutanol and 5% IPA, respectively, although the best separation of carfentrazone-ethyl was obtained with 0.5% IPA and no resolution was found with ethanol, n-propanol, n-butanol, and isobutanol from 20% to 2%. Liu et al. [65] carried out enantiomeric resolution of synthetic pyrethroid (bifenthrin, permethrin, cypermethrin, and cyfluthrin) on sumichiral OA-2500-I and two chained Chirex 00G-3019-DO columns with hexane-1, 2-dichloroethane (500:1, v/v) and hexane-1,2-dichloroethane-ethanol (500:10:0.05, v/v/v), respectively. Lin et al. [66] performed the chiral separation of methamidophos on a Chiralcel OD column with a mobile phase of n-hexane-IPA (80:20, v/v) at 0.5 mL min^{-1} flow rate on OR and CD detectors and confirmed that R-(+)-methamidophos was eluted before S-(−)-methamidophos at 230 nm. Similarly, the same group of researchers studied the chiral separation of nematicide fosthiazate containing two stereogenic centers on a Chiralpak AD column with a mobile phase of n-hexane-ethanol (95:5, v/v) and 1.0 mL min^{-1} flow rate. The identification of peaks was confirmed with OR and CD detectors at 230 nm, and the values of capacity, separation, and resolution factors varies with the change of concentration of ethanol as shown in Table 17.3 [67].

Furthermore, Xu et al. [68] carried out the enantiomeric resolution of pyrethroid insecticides (λ-cyhalothrin) with a Chiralpak AD (amylase tris[3,5-dimethyl-phenyl carbamate]), Chiralpak AS (amylose tris[(S)-1-phenyl carbamate]), Chiralcel OD (cellulose tris[3,5-dimethylphenyl carbamate]), and Chiralcel OJ (cellulose tris[4-methyl benzoate]) CSPs. In this study, they concluded that all the CSPs are good for the resolution with a different ratio of eluting solvents as shown in Table 17.2 and confirmed that 5% ethanol is a good modifier in all the cases. They also studied the

TABLE 17.2
Chiral Separation of Pesticides on Different Columns with NP and RP Mobile Phase Conditions

Pesticides	Column/CSPs	Mobile Phase (v/v)	k_1	k_2	α	Rs	References
trans-Chlordane	CHIRALCEL-OD	Hexane/Iso-Propanol (99-1)	0.83	0.93	1.1	–	[63]
		Hexane (100)	2.4	2.9	1.2	–	
cis-Chlordane	CHIRALCEL-OD	Hexane/Iso-Propanol (99-1)	0.87	1.0	1.2	–	
		Hexane (100)	1.7	3.2	1.9	–	
Heptachlor	CHIRALCEL-OD	Hexane (100)	0.80	0.97	1.2	–	
α-HCH	CHIRALCEL-OJ	Hexane/Iso-Propanol (90-10)	1.0	1.4	1.4	–	
Heptachlor epoxide	CHIRALPAK-AD	MeOH	0.37	0.64	1.7	–	
Fipronil	Cellulose-tri (3,5-DMPC)	Hexane/Ethanol (80-20)	0.78	–	1.00	0.0	[64]
		Hexane/Ethanol (85-15)	0.99	–	1.12	0.25	
		Hexane/Ethanol (90-10)	1.69	–	1.14	0.63	
		Hexane/Ethanol (95-05)	4.27	–	1.18	1.01	
		Hexane/Ethanol (98-02)	10.70	–	1.22	1.27	
		Hexane/n-Propanol (80-20)	0.90	–	1.00	0.0	
		Hexane/n-Propanol (85-15)	1.40	–	1.00	0.0	
		Hexane/n-Propanol (90-10)	2.18	–	1.13	0.46	
		Hexane/n-Propanol (95-05)	4.94	–	1.19	1.06	
		Hexane/n-Propanol (98-02)	11.83	–	1.21	1.27	
		Hexane/Iso-Propanol (80-20)	0.90	–	1.26	0.0	
		Hexane/Iso-Propanol (85-15)	1.38		1.28	0.46	
		Hexane/Iso-Propanol (90-10)	2.54	–	1.31	0.91	
		Hexane/Iso-Propanol (95-05)	6.06	–	1.36	1.37	
		Hexane/Iso-Propanol (98-02)	15.56	–	1.37	1.69	
		Hexane/n-Butanol (80-20)	1.47	–	1.00	0.0	
		Hexane/n-Butanol (85-15)	1.38	–	1.00	0.0	
		Hexane/n-Butanol (90-10)	2.14	–	1.12	0.38	
		Hexane/n-Butanol (95-150)	4.85	–	1.17	1.05	
		Hexane/n-Butanol (98-02)	11.28	–	1.20	1.22	
		Hexane/Iso-Butanol (80-20)	1.16	–	1.29	0.0	
		Hexane/Iso-Butanol (85-15)	1.83	–	1.35	1.26	

(Continued)

TABLE 17.2 (CONTINUED)
Chiral Separation of Pesticides on Different Columns with NP and RP Mobile Phase Conditions

Pesticides	Column/CSPs	Mobile Phase (v/v)	k_1	k_2	α	Rs	References
		Hexane/Iso-Butanol (90-10)	3.15	–	1.37	1.47	
		Hexane/Iso-Butanol (95-05)	7.42	–	1.38	1.81	
		Hexane/Iso-Butanol (98-02)	–	–	–	–	
Isocarbophos	Cellulose-tri (3,5-DMPC)	Hexane/Ethanol (80-20)	0.77	–	1.17	0.30	
		Hexane/Ethanol (85-15)	0.97	–	1.23	0.51	
		Hexane/Ethanol (90-10)	1.21	–	1.34	0.78	
		Hexane/Ethanol (95-05)	2.13	–	1.25	1.32	
		Hexane/Ethanol (98-02)	3.69	–	1.27	1.61	
		Hexane/n-Propanol (80-20)	0.87	–	1.34	0.96	
		Hexane/n-Propanol (85-15)	1.05	–	1.33	1.14	
		Hexane/n-Propanol (90-10)	1.56	–	1.37	1.40	
		Hexane/n-Propanol (95-05)	2.61	–	1.40	1.81	
		Hexane/n-Propanol (98-02)	4.49	–	1.41	2.63	
		Hexane/Iso-Propanol (80-20)	0.96	–	1.52	1.47	
		Hexane/Iso-Propanol (85-15)	1.43	–	1.60	1.79	
		Hexane/Iso-Propanol (90-10)	1.90	–	1.67	2.18	
		Hexane/Iso-Propanol (95-05)	3.26	–	1.89	2.66	
		Hexane/Iso-Propanol (98-02)	5.93	–	1.66	2.42	
		Hexane/n-Butanol (80-20)	0.94	–	1.28	0.87	
		Hexane/n-Butanol (85-15)	1.09	–	1.28	0.91	
		Hexane/n-Butanol (90-10)	1.50	–	1.30	1.27	
		Hexane/n-Butanol (95-05)	2.42	–	1.32	1.38	
		Hexane/n-Butanol (98-02)	4.90	–	1.39	1.86	
		Hexane/Iso-Butanol (80-20)	1.01	–	1.40	1.23	
		Hexane/Iso-Butanol (85-15)	1.32	–	1.43	1.56	
		Hexane/Iso-Butanol (90-10)	1.65	–	1.43	1.61	
		Hexane/Iso-Butanol (95-05)	1.34	–	2.70	1.65	
		Hexane/Iso-Butanol (98-02)	6.27	–	1.68	2.56	
Carfentrazone-ethyl	Cellulose-tri (3,5-DMPC)	Hexane/Iso-Propanol (80-20)	0.78	–	1.00	0.0	
		Hexane/Iso-Propanol (85-15)	0.91	–	1.00	0.0	

		Hexane/Iso-Propanol (90-10)	1.08	–	1.10	0.24	
		Hexane/Iso-Propanol (95-05)	1.47	–	1.10	0.33	
		Hexane/Iso-Propanol (98-02)	2.49	–	1.10	0.45	
		Hexane/Iso-Propanol (99-01)	11.72	–	1.08	0.52	
		Hexane/Iso-Propanol (99.5-0.5)	5.45	–	1.07	0.58	
		Hexane/Iso-Propanol (99.9-0.1)	3.53	–	1.09	0.48	
Lambda-cyhalothrin	Chiralpak AD	Hexane/ethanol (95-05)	0.56	0.67	1.20	1.60	[68]
		Hexane/ethanol (96-04)	0.60	0.72	1.24	1.81	
		Hexane/ethanol (97-03)	0.63	0.81	1.28	2.06	
		Hexane/ethanol (99-01)	0.84	1.10	1.30	2.20	
		Hexane/ethanol (98-02)	1.10	1.35	1.32	2.35	
	Chiralpak AS	Hexane/1, 2-dichloroethane (95-05)	1.05	1.94	1.85	4.96	
		Hexane/1, 2-dichloroethane (96-04)	1.72	3.19	1.86	4.90	
		Hexane/1, 2-dichloroethane (97-03)	2.98	5.65	1.90	4.54	
		Hexane/1, 2-dichloroethane (98-02)	4.56	8.44	1.85	4.18	
		Hexane/1, 2-dichloroethane (99-01)	6.70	2.19	1.82	3.17	
	Chiralcel OD	Hexane/Iso-Propanol (95-05)	1.18	1.75	1.49	5.95	
		Hexane/Iso-Propanol (96-04)	1.25	1.92	1.53	6.54	
		Hexane/Iso-Propanol (97-03)	1.79	2.34	1.55	7.31	
		Hexane/Iso-Propanol (98-02)	2.24	3.59	1.60	8.41	
		Hexane/Iso-Propanol (99-01)	3.33	5.76	1.73	10.41	
	Chiralcel OJ	Hexane/ethanol (95-05)	1.56	1.99	1.27	3.22	
		Hexane/ethanol (96-04)	1.82	2.34	1.28	3.50	
		Hexane/ethanol (97-03)	2.24	2.90	1.30	3.82	
		Hexane/ethanol (98-02)	2.92	3.87	1.33	4.30	
		Hexane/ethanol (99-01)	4.52	6.13	1.36	4.79	
Salithion	Chiralcel OJ	Hexane/Iso-Propanol (95-05)	8.47	9.32	1.10	1.56	[69]
	Chiralcel OD	Hexane/Iso-Propanol (99.5-0.5)	6.76	7.40	1.09	1.42	
	Chiralpak AD	Hexane/Iso-Propanol (99.5-0.5)	4.15	5.03	1.21	4.14	
	Chiralpak OT (+)	Methanol (100)	2.02	2.52	1.25	2.28	
Neonicotinoid 1	Chiralcel OD-H	Hexane/ethanol (40-60)	1.40	1.97	1.41	2.74	[70]
		Hexane/ethanol (50-50)	1.92	2.65	1.38	2.75	
		Hexane/ethanol (60-40)	2.98	4.05	1.36	2.77	

(Continued)

TABLE 17.2 (CONTINUED)
Chiral Separation of Pesticides on Different Columns with NP and RP Mobile Phase Conditions

Pesticides	Column/CSPs	Mobile Phase (v/v)	k_1	k_2	α	Rs	References
		Hexane/ethanol (70-30)	5.25	7.01	1.34	2.87	
		Hexane/ethanol (80-20)	11.75	15.46	1.32	3.15	
Neonicotinoid 2	Chiralcel OD-H	Hexane/ethanol (40-60)	0.69	1.13	1.62	1.98	
		Hexane/ethanol (50-50)	1.02	1.67	1.64	2.15	
		Hexane/ethanol (60-40)	1.70	2.82	1.67	2.42	
		Hexane/ethanol (70-30)	3.26	5.57	1.71	2.84	
		Hexane/ethanol (80-20)	7.79	13.69	1.76	3.43	
Neonicotinoid 3	Chiralcel OD-H	Hexane/ethanol (50-50)	0.41	0.49	1.18	0.98	
		Hexane/ethanol (60-40)	0.64	0.75	1.17	1.09	
		Hexane/ethanol (70-30)	1.18	1.39	1.17	1.19	
		Hexane/ethanol (80-20)	2.50	2.90	1.16	1.27	
		Hexane/ethanol (90-10)	8.54	9.88	1.57	1.49	
Metalaxyl	Amylose tris-(S)-1-(PEC)	Hexane/Iso-Propanol (85-15)	3.01	4.66	1.55	2.51	[71]
		Hexane/Iso-Propanol (90-10)	4.28	6.68	1.56	2.80	
		Hexane/Iso-Propanol (95-05)	7.56	12.27	1.62	3.75	
Myclobutanil	Amylose tris-(S)-1-(PEC)	Hexane/Iso-Propanol (85-15)	5.08	5.96	1.17	0.86	
		Hexane/Iso-Propanol (90-10)	8.01	9.59	1.20	1.06	
		Hexane/Iso-Propanol (95-05)	17.49	21.74	1.24	1.49	
Imazalil	Amylose tris-(S)-1-(PEC)	Hexane/Iso-Propanol (85-15)	2.50	2.70	1.08	0.52	
		Hexane/Iso-Propanol (90-10)	3.71	4.08	1.10	0.70	
		Hexane/Iso-Propanol (95-05)	7.56	8.44	1.12	0.86	
Malathion	Amylose tris-(S)-1-(PEC)	Hexane/Iso-Propanol (90-10)	2.49	2.49	1.00	0.0	
		Hexane/Iso-Propanol (98-02)	3.19	3.48	1.09	0.77	
		Hexane/Iso-Propanol (95-05)	4.53	5.02	1.11	0.87	
Triadimefon	Amylose tris-(S)-1-(PEC)	Hexane/Iso-Propanol (85-15)	1.81	2.33	1.29	1.34	
		Hexane/Iso-Propanol (90-10)	2.48	3.27	1.32	1.43	
		Hexane/Iso-Propanol (95-05)	3.74	5.07	1.35	1.59	
		Hexane/Iso-Propanol (98-02)	5.97	8.47	1.42	1.84	
Ethofumesate	Amylose tris-(S)-1-(PEC)	Hexane/Iso-Propanol (90-10)	4.71	4.71	1.00	0.0	
		Hexane/Iso-Propanol (95-05)	6.66	6.92	1.04	0.40	

		Hexane/Iso-Propanol (98-02)	9.95	10.58	1.06	0.63	
Fipronil	Amylose tris-(S)-1-(PEC)	Hexane/Iso-Propanol (85-15)	2.83	3.45	1.22	1.03	
		Hexane/Iso-Propanol (90-10)	7.72	9.88	1.28	1.35	
		Hexane/Iso-Propanol (95-05)	14.63	19.42	1.33	2.03	
Napropamide	Amylose tris-(S)-1-(PEC)	Hexane/Iso-Propanol (85-15)	1.55	1.66	1.07	0.38	
		Hexane/Iso-Propanol (90-10)	2.04	2.25	1.10	0.55	
		Hexane/Iso-Propanol (95-05)	3.31	3.66	1.11	0.72	
		Hexane/Iso-Propanol (98-02)	7.65	8.76	1.14	1.14	
Paclobutrazol	Amylose tris-(S)-1-(PEC)	Hexane/Iso-Propanol (85-15)	2.03	2.36	1.16	0.81	
		Hexane/Iso-Propanol (90-10)	3.57	4.29	1.20	1.30	
		Hexane/Iso-Propanol (95-05)	6.38	7.74	1.21	1.66	
Metalaxyl	Amylopectin-tris-(PC)	Hexane/Iso-Propanol (80-20)	4.57	5.34	1.17	0.94	[72]
		Hexane/Iso-Propanol (85-15)	5.74	6.69	1.16	0.97	
		Hexane/Iso-Propanol (90-10)	9.11	10.77	1.18	1.37	
Hexaconazole	Amylopectin-tris-(PC)	Hexane/Iso-Propanol (80-20)	1.78	2.38	1.34	1.49	
		Hexane/Iso-Propanol (85-15)	2.33	3.19	1.36	1.83	
		Hexane/Iso-Propanol (90-10)	3.60	4.99	1.39	2.03	
		Hexane/Iso-Propanol (95-05)	7.53	10.60	1.41	2.45	
Myclobutanil	Amylopectin-tris-(PC)	Hexane/Iso-Propanol (70-30)	3.72	4.55	1.22	0.86	
		Hexane/Iso-Propanol (80-20)	5.63	6.93	1.23	1.10	
		Hexane/Iso-Propanol (85-15)	8.17	10.04	1.23	1.20	
		Hexane/Iso-Propanol (90-10)	13.11	16.27	1.24	1.47	
Tebuconazole	Amylopectin-tris-(PC)	Hexane/Iso-Propanol (80-20)	2.07	2.38	1.15	0.63	
		Hexane/Iso-Propanol (85-15)	2.80	3.32	1.19	0.92	
		Hexane/Iso-Propanol (90-10)	4.37	5.39	1.23	1.34	
		Hexane/Iso-Propanol (95-05)	9.84	12.01	1.22	1.54	
Uniconazole	Amylopectin-tris-(PC)	Hexane/Iso-Propanol (80-20)	2.11	2.13	1.01	0.05	
		Hexane/Iso-Propanol (85-15)	2.31	3.02	1.31	1.39	
		Hexane/Iso-Propanol (90-10)	3.86	5.17	1.34	1.48	
		Hexane/Iso-Propanol (95-05)	9.33	13.09	1.40	2.05	
Paclobutrazol	Amylopectin-tris-(PC)	Hexane/Iso-Propanol (85-15)	2.30	3.41	1.48	1.61	
		Hexane/Iso-Propanol (90-10)	3.79	5.88	1.55	2.19	
		Hexane/Iso-Propanol (95-05)	8.16	12.65	1.55	2.42	

(Continued)

TABLE 17.2 (CONTINUED)
Chiral Separation of Pesticides on Different Columns with NP and RP Mobile Phase Conditions

Pesticides	Column/CSPs	Mobile Phase (v/v)	k_1	k_2	α	Rs	References
Benalaxyl	Amylopectin-tris-(PC)	Hexane/Iso-Propanol (85-15)	2.59	2.91	1.12	0.66	
		Hexane/Iso-Propanol (90-10)	3.63	4.12	1.13	0.77	
		Hexane/Iso-Propanol (95-05)	4.75	5.59	1.18	0.83	
		Hexane/Iso-Propanol (97-03)	8.79	10.32	1.17	1.01	
EPN	Chiralpak AD	Hexane/Iso-Propanol (90-10)	0.86	0.96	1.11	1.13	[73]
		Hexane/Iso-Propanol (95-05)	0.90	1.04	1.16	1.41	
		Hexane/Iso-Propanol (96-04)	0.89	1.05	1.18	1.61	
		Hexane/Iso-Propanol (97-03)	0.94	1.08	1.15	2.31	
		Hexane/Iso-Propanol (98-02)	1.17	1.46	1.25	3.22	
		Hexane/Iso-Propanol (99-01)	1.86	2.61	1.40	5.39	
	Chiralpak AS	Hexane/Iso-Propanol (90-10)	1.23	1.45	1.18	1.89	
		Hexane/Iso-Propanol (95-05)	1.33	1.59	1.19	2.11	
		Hexane/Iso-Propanol (96-04)	1.56	1.87	1.19	2.28	
		Hexane/Iso-Propanol (97-03)	1.48	1.70	1.15	1.87	
		Hexane/Iso-Propanol (98-02)	1.93	2.24	1.16	2.13	
		Hexane/Iso-Propanol (99-01)	2.27	2.68	1.18	2.50	
o,p-DDT	Chiralpak AD-RH	Acetonitrile/Water (50-50)	15.41	19.77	1.24	2.47	[60]
o,p-DDD		Acetonitrile/Water (50-50)	–	–	–	–	
o,p-DDT	Chiralpak AD-RH	Acetonitrile/Iso-Propanol (50-50)	4.74	8.00	1.69	1.00	
o,p-DDD		Acetonitrile/Iso-Propanol (50-50)	3.26	4.11	1.26	0.60	
o,p-DDT	Chiralcel OD-RH	Acetonitrile/Water (50-50)	4.54	10.28	2.27	2.03	
o,p-DDD		Acetonitrile/Water (50-50)	–	–	–	–	
o,p-DDT	Chiralcel OJ-R	Acetonitrile/Iso-Propanol (50-50)	–	–	–	–	
o,p-DDD		Acetonitrile/Iso-Propanol (50-50)	–	–	–	–	
o,p-DDT		Acetonitrile/Water (50-50)	3.49	8.80	2.52	0.80	
o,p-DDD		Acetonitrile/Water (50-50)	–	–	–	–	
o,p-DDT		Acetonitrile/Iso-Propanol (50-50)	–	–	–	–	
o,p-DDD		Acetonitrile/Iso-Propanol (50-50)	–	–	–	–	
Epoxiconazole	Cellulose-tris-(3,5-DMPC)	Methanol/Water (75-25)	4.38	8.26	1.88	5.54	[74]
		Methanol/Water (80-20)	2.79	5.40	1.93	5.45	

		Methanol/Water (85-15)	1.74	3.34	1.92	5.32
		Methanol/Water (90-10)	1.16	2.22	1.91	5.05
		Methanol/Water (95-5)	0.90	1.75	1.96	5.00
		Methanol/Water (100-0)	1.01	1.62	1.60	3.41
		Acetonitrile/Water (50-50)	4.18	8.39	2.00	9.23
		Acetonitrile/Water (60-40)	1.85	3.86	2.08	7.71
		Acetonitrile/Water (70-30)	1.03	2.16	2.09	6.08
		Acetonitrile/Water (80-20)	0.65	1.39	2.12	5.14
		Acetonitrile/Water (90-10)	0.56	1.09	1.96	4.21
		Acetonitrile/Water (100-0)	0.95	1.75	1.84	4.83
Terallethrin	Cellulose-tris-(3,5-DMPC)	Methanol/Water (65-35)	15.09	15.91	1.05	0.66
		Methanol/Water (70-30)	8.18	8.58	1.05	0.54
		Methanol/Water (75-25)	4.52	4.74	1.05	0.53
		Methanol/Water (80-20)	2.63	2.74	1.04	0.42
		Methanol/Water (90-10)	1.01	1.01	1.00	–
		Methanol/Water (100-0)	0.70	0.70	1.00	–
		Acetonitrile/Water (50-50)	5.66	6.28	1.11	1.56
		Acetonitrile/Water (60-40)	2.18	2.43	1.12	1.22
		Acetonitrile/Water (70-30)	1.03	1.15	1.12	0.89
		Acetonitrile/Water (80-20)	0.53	0.60	1.13	0.66
		Acetonitrile/Water (90-10)	0.33	0.33	1.00	–
		Acetonitrile/Water (100-0)	0.51	0.51	1.00	–
Pyriproxyfen	Cellulose-tris-(3,5-DMPC)	Methanol/Water (80-20)	12.01	13.05	1.09	0.93
		Methanol/Water (85-15)	5.99	6.43	1.07	0.66
		Methanol/Water (90-10)	3.20	3.38	1.06	0.61
		Methanol/Water (95-05)	2.04	2.04	1.00	–
		Methanol/Water (100-0)	1.13	1.13	1.00	–
		Acetonitrile/Water (55-45)	12.38	13.09	1.06	0.91
		Acetonitrile/Water (60-40)	7.38	7.82	1.06	0.84
		Acetonitrile/Water (70-30)	3.20	3.39	1.06	0.72
		Acetonitrile/Water (80-20)	1.52	1.58	1.04	0.47
		Acetonitrile/Water (90-10)	0.75	0.75	1.00	–
		Acetonitrile/Water (100-0)	0.45	0.45	1.00	–

(Continued)

TABLE 17.2 (CONTINUED)
Chiral Separation of Pesticides on Different Columns with NP and RP Mobile Phase Conditions

Pesticides	Column/CSPs	Mobile Phase (v/v)	k_1	k_2	α	Rs	References
Benalaxyl	Cellulose-tris-(3,5-DMPC)	Methanol/Water (75-25)	4.02	4.82	1.20	1.59	
		Methanol/Water (85-15)	1.54	1.83	1.19	1.35	
		Methanol/Water (90-10)	1.04	1.22	1.17	1.13	
		Methanol/Water (95-05)	0.76	0.88	1.16	0.92	
		Methanol/Water (100:0)	0.64	0.72	1.14	0.76	
		Acetonitrile/Water (40-60)	14.20	14.20	1.00	–	
		Acetonitrile/Water (80-20)	0.54	0.54	1.00	–	
		Acetonitrile/Water (100-0)	0.38	0.38	1.00	–	
Lactofen	Cellulose-tris-(3,5-DMPC)	Methanol/Water (75-25)	14.71	6.26	1.11	1.07	
		Methanol/Water (80-20)	7.01	7.75	1.10	0.92	
		Methanol/Water (85-15)	3.13	3.46	1.11	0.83	
		Methanol/Water (90-10)	1.52	1.69	1.11	0.66	
		Methanol/Water (95-05)	0.79	0.88	1.11	0.63	
		Methanol/Water (100-0)	0.77	0.77	1.00	–	
		Acetonitrile/Water (50-50)	15.19	15.19	1.00	–	
		Acetonitrile/Water (80-20)	0.57	0.57	1.00	–	
		Acetonitrile/Water (100-0)	0.44	0.44	1.00	–	
Quizalofop-ethyl	Cellulose-tris-(3,5-DMPC)	Methanol/Water (75-25)	21.48	22.61	1.05	0.69	
		Methanol/Water (80-20)	11.40	12.02	1.05	0.59	
		Methanol/Water (85-15)	6.06	6.38	1.05	0.54	
		Methanol/Water (90-10)	3.40	3.56	1.05	0.45	
		Methanol/Water (95-05)	2.11	2.11	1.00	–	
		Methanol/Water (100-0)	1.20	1.20	1.00	–	
		Acetonitrile/Water (50-50)	13.04	13.04	1.00	–	
		Acetonitrile/Water (80-20)	1.02	1.02	1.00	–	
		Acetonitrile/Water (100-0)	0.45	0.45	1.00	–	
Diclofop-methyl	Cellulose-tris-(3,5-DMPC)	Methanol/Water (75-25)	14.43	14.43	1.00	–	
		Methanol/Water (85-15)	4.43	4.43	1.00	–	
		Methanol/Water (100-0)	0.77	0.77	1.00	–	
		Acetonitrile/Water (50-50)	12.73	13.95	1.10	1.53	

		Acetonitrile/Water (60-40)	4.35	4.76	1.09	1.22
		Acetonitrile/Water (70-30)	1.87	2.05	1.10	0.97
		Acetonitrile/Water (80-20)	0.87	0.95	1.10	0.72
		Acetonitrile/Water (90-10)	0.86	0.86	1.00	–
		Acetonitrile/Water (100-0)	0.34	0.34	1.00	–
Profenofos	Cellulose-tris-(3,5-DMPC)	Methanol/Water (70-30)	12.90	12.90	1.00	–
		Methanol/Water (80-20)	4.25	4.25	1.00	–
		Methanol/Water (100-0)	0.93	0.93	1.00	–
		Acetonitrile/Water (50-50)	8.47	8.47	1.00	–
		Acetonitrile/Water (70-30)	1.83	1.83	1.00	–
		Acetonitrile/Water (100-0)	0.62	0.62	1.00	–
Malathion	Cellulose-tris-(3,5-DMPC)	Methanol/Water (70-30)	6.90	6.90	1.00	–
		Methanol/Water (80-20)	2.41	2.41	1.00	–
		Methanol/Water (100-0)	0.77	0.77	1.00	–
		Acetonitrile/Water (70-30)	0.85	0.85	1.00	–
		Acetonitrile/Water (100-0)	0.49	0.49	1.00	–
Methamidophos	Cellulose-tris-(3,5-DMPC)	Methanol/Water (60-40)	1.28	1.28	1.00	–
		Methanol/Water (80-20)	1.82	1.82	1.00	–
		Methanol/Water (100-0)	0.76	0.76	1.00	–
		Acetonitrile/Water (40-60)	0.97	0.97	1.00	–
		Acetonitrile/Water (70-30)	2.35	2.35	1.00	–
		Acetonitrile/Water (100-0)	0.96	0.96	1.00	–
MCPA-isooctyl	Cellulose-tris-(3,5-DMPC)	Methanol/Water (80-20)	12.57	12.57	1.00	–
		Methanol/Water (90-10)	2.88	2.88	1.00	–
		Methanol/Water (100-0)	1.12	1.12	1.00	–
		Acetonitrile/Water (60-40)	10.37	10.37	1.00	–
		Acetonitrile/Water (80-20)	1.88	1.88	1.00	–
		Acetonitrile/Water (100-0)	0.79	0.79	1.00	–
Isofenphos-methyl	Cellulose-tris-(3,5-DMPC)	Methanol/Water (70-30)	5.49	5.49	1.00	–
		Methanol/Water (80-20)	1.99	1.99	1.00	–
		Methanol/Water (100-0)	0.65	0.65	1.00	–
		Acetonitrile/Water (50-50)	5.21	5.21	1.00	–
		Acetonitrile/Water (70-30)	1.07	1.07	1.00	–

(*Continued*)

TABLE 17.2 (CONTINUED)
Chiral Separation of Pesticides on Different Columns with NP and RP Mobile Phase Conditions

Pesticides	Column/CSPs	Mobile Phase (v/v)	k_1	k_2	α	Rs	References
		Acetonitrile/Water (100-0)	0.53	0.53	1.00	–	
Phenthoate	Cellulose-tris-(3,5-DMPC)	Methanol/Water (75-25)	7.54	7.54	1.00	–	
		Methanol/Water (90-10)	1.39	1.39	1.00	–	
		Methanol/Water (100-0)	0.87	0.87	1.00	–	
		Acetonitrile/Water (50-50)	6.57	6.57	1.00	–	
		Acetonitrile/Water (70-30)	1.26	1.26	1.00	–	
		Acetonitrile/Water (100-0)	0.57	0.57	1.00	–	
Fluroxypyr-meptyl	Cellulose-tris-(3,5-DMPC)	Methanol/Water (75-25)	14.85	14.85	1.00	–	
		Methanol/Water (90-10)	1.94	1.94	1.00	–	
		Methanol/Water (100-0)	0.80	0.80	1.00	–	
		Acetonitrile/Water (50-50)	17.75	17.75	1.00	–	
		Acetonitrile/Water (70-30)	2.33	2.33	1.00	–	
		Acetonitrile/Water (100-0)	0.65	0.65	1.00	–	
Acephate	Cellulose-tris-(3,5-DMPC)	Methanol/Water (60-40)	0.97	0.97	1.00	–	
		Methanol/Water (80-20)	0.88	0.88	1.00	–	
		Methanol/Water (100-0)	0.76	0.76	1.00	–	
		Acetonitrile/Water (50-50)	0.76	0.76	1.00	–	
		Acetonitrile/Water (80-20)	0.84	0.84	1.00	–	
		Acetonitrile/Water (100-0)	0.93	0.93	1.00	–	
Trichlorphon	Cellulose-tris-(3,5-DMPC)	Methanol/Water (60-40)	1.33	1.33	1.00	–	
		Methanol/Water (80-20)	0.87	0.87	1.00	–	
		Methanol/Water (100-0)	1.07	1.07	1.00	–	
		Acetonitrile/Water (80-20)	0.61	0.61	1.00	–	
		Acetonitrile/Water (100-0)	0.99	0.99	1.00	–	
2,4-D-ethylhexyl	Cellulose-tris-(3,5-DMPC)	Methanol/Water (80-20)	13.38	13.38	1.00	–	
		Methanol/Water (90-10)	3.13	3.13	1.00	–	
		Methanol/Water (100-0)	1.17	1.17	1.00	–	
		Acetonitrile/Water (60-40)	10.06	10.06	1.00	–	
		Acetonitrile/Water (80-20)	1.84	1.84	1.00	–	
		Acetonitrile/Water (100-0)	0.78	0.78	1.00	–	

Fenamiphos	Cellulose-tris-(3,5-DMPC)	Methanol/Water (70-30)	3.17	3.17	1.00	–	
		Methanol/Water (90-10)	0.85	0.85	1.00	–	
		Methanol/Water (100-0)	0.62	0.62	1.00	–	
		Acetonitrile/Water (40-60)	6.45	6.45	1.00	–	
		Acetonitrile/Water (70-30)	0.81	0.81	1.00	–	
		Acetonitrile/Water (100-0)	1.69	1.69	1.00	–	
Acetochlor	Cellulose-tris-(3,5-DMPC)	Methanol/Water (70-30)	4.33	4.33	1.00	–	
		Methanol/Water (90-10)	1.25	1.25	1.00	–	
		Methanol/Water (100-0)	0.82	0.82	1.00	–	
		Acetonitrile/Water (50-50)	4.22	4.22	1.00	–	
		Acetonitrile/Water (70-30)	1.13	1.13	1.00	–	
		Acetonitrile/Water (100-0)	0.66	0.66	1.00	–	
P-tefuryl Quizalofop-acid	Cellulose-tris-(3,5-DMPC)	Methanol/Water (85-15)	9.46	9.46	1.00	–	
		Methanol/Water (90-10)	1.75	1.75	1.00	–	
		Methanol/Water (100-0)	5.59	5.59	1.00	–	
		Acetonitrile/Water (60-40)	5.55	5.55	1.00	–	
		Acetonitrile/Water (70-30)	0.98	0.98	1.00	–	
		Acetonitrile/Water (100-0)	2.76	2.76	1.00	–	
Hexaconazole	Cellulose-tris-(3,5-DMPC) (5 μM)	Acetonitrile/Water (50-50)	3.20	–	1.15	3.94	[75]
		Acetonitrile/Water (70-30)	0.87	–	1.17	2.98	
		Acetonitrile/Water (90-10)	0.44	–	1.19	2.25	
	Cellulose-tris-(3,5-DMPC) (3 μM)	Acetonitrile/Water (50-50)	3.41	–	1.16	3.09	
		Acetonitrile/Water (70-30)	0.88	–	1.17	1.71	
		Acetonitrile/Water (90-10)	0.48	–	1.17	1.26	
	Cellulose-tris-(3,5-DMPC) (5 μM)	Methanol/Water (70-30)	5.45	–	1.14	2.59	
		Methanol/Water (80-20)	1.75	–	1.13	2.12	
		Methanol/Water (90-10)	0.66	–	1.13	1.44	
	Cellulose-tris-(3,5-DMPC) (3 μM)	Methanol/Water (70-30)	4.86	–	1.12	1.50	
		Methanol/Water (80-20)	1.71	–	1.09	0.76	
		Methanol/Water (90-10)	–	–	–	–	
Flutriafol	Cellulose-tris-(3,5-DMPC) (5 μM)	Acetonitrile/Water (50-50)	1.45	–	1.12	2.55	
		Acetonitrile/Water (70-30)	0.45	–	1.14	1.99	
		Acetonitrile/Water (90-10)	0.27	–	1.16	1.35	

(Continued)

TABLE 17.2 (CONTINUED)
Chiral Separation of Pesticides on Different Columns with NP and RP Mobile Phase Conditions

Pesticides	Column/CSPs	Mobile Phase (v/v)	k_1	k_2	α	Rs	References
	Cellulose-tris-(3,5-DMPC)	Acetonitrile/Water (50-50)	1.52	–	1.12	1.44	
	(3 µM)	Acetonitrile/Water (70-30)	0.49	–	1.07	0.50	
		Acetonitrile/Water (90-10)	–	–	–	–	
	Cellulose-tris-(3,5-DMPC)	Methanol/Water (70-30)	2.13	–	1.08	1.39	
	(5 µM)	Methanol/Water (80-20)	0.87	–	1.08	1.06	
		Methanol/Water (90-10)	–	–	–	–	
	Cellulose-tris-(3,5-DMPC)	Methanol/Water (70-30)	–	–	–	–	
	(3 µM)	Methanol/Water (80-20)	–	–	–	–	
		Methanol/Water (90-10)	–	–	–	–	
Diniconazole	Cellulose-tris-(3,5-DMPC)	Acetonitrile/Water (50-50)	3.71	–	1.11	3.09	
	(5 µM)	Acetonitrile/Water (70-30)	0.92	–	1.12	2.31	
		Acetonitrile/Water (90-10)	0.40	–	1.11	1.22	
	Cellulose-tris-(3,5-DMPC)	Acetonitrile/Water (50-50)	3.98	–	1.34	6.71	
	(3 µM)	Acetonitrile/Water (70-30)	0.93	–	1.13	1.30	
		Acetonitrile/Water (90-10)	–	–	–	–	
Tetraconazole	Cellulose-tris-(3,5-DMPC)	Acetonitrile/Water (50-50)	4.15	–	1.29	7.37	
	(5 µM)	Acetonitrile/Water (70-30)	0.86	–	1.31	5.04	
		Acetonitrile/Water (90-10)	0.30	–	1.39	3.35	
	Cellulose-tris-(3,5-DMPC)	Acetonitrile/Water (50-50)	4.49	–	1.29	5.79	
	(3 µM)	Acetonitrile/Water (70-30)	0.87	–	1.30	2.85	
		Acetonitrile/Water (90-10)	0.34	–	1.33	1.66	
	Cellulose-tris-(3,5-DMPC) (5 µM)	Methanol/Water (70-30)	–	–	–	–	
	Cellulose-tris-(3,5-DMPC) (3 µM)	Methanol/Water (70-30)	–	–	–	–	
Epoxiconazole	Cellulose-tris-(3,5-DMPC)	Acetonitrile/Water (50-50)	4.22	–	2.04	20.90	
	(5 µM)	Acetonitrile/Water (70-30)	1.03	–	2.10	16.89	
		Acetonitrile/Water (90-10)	0.42	–	2.27	12.30	
	Cellulose-tris-(3,5-DMPC)	Acetonitrile/Water (50-50)	4.57	–	2.00	16.71	

	(3 μM)	Acetonitrile/Water (70-30)	1.06	–	2.03	10.45
		Acetonitrile/Water (90-10)	0.47	–	2.08	7.06
	Cellulose-tris-(3,5-DMPC)	Methanol/Water (70-30)	8.71	–	1.29	5.86
	(5 μM)	Methanol/Water (80-20)	2.93	–	1.62	9.56
		Methanol/Water (90-10)	1.16	–	1.75	9.42
	Cellulose-tris-(3,5-DMPC)	Methanol/Water (70-30)	7.72	–	1.49	6.51
	(3 μM)	Methanol/Water (80-20)	2.78	–	1.55	5.90
		Methanol/Water (90-10)	1.19	–	1.67	5.19
Penconazole	Cellulose-tris-(3,5-DMPC)	Acetonitrile/Water (50-50)	4.20	–	1.22	7.58
	(5 μM)	Acetonitrile/Water (70-30)	1.34	–	1.05	1.18
		Acetonitrile/Water (90-10)	0.61	–	1.06	1.00
	Cellulose-tris-(3,5-DMPC)	Acetonitrile/Water (50-50)	5.29	–	1.05	1.12
	(3 μM)	Acetonitrile/Water (70-30)	–	–	–	–
		Acetonitrile/Water (90-10)	–	–	–	–
	Cellulose-tris-(3,5-DMPC)	Methanol/Water (70-30)	6.57	–	1.18	3.87
	(5 μM)	Methanol/Water (80-20)	2.31	–	1.17	3.20
		Methanol/Water (90-10)	0.95	–	1.17	2.29
	Cellulose-tris-(3,5-DMPC)	Methanol/Water (70-30)	5.95	–	1.15	2.54
	(3 μM)	Methanol/Water (80-20)	2.23	–	1.15	1.68
		Methanol/Water (90-10)	0.98	–	1.15	1.15
Myclobutanil	Cellulose-tris-(3,5-DMPC)	Acetonitrile/Water (50-50)	3.44	–	1.42	9.99
	(5 μM)	Acetonitrile/Water (70-30)	0.88	–	1.44	7.58
		Acetonitrile/Water (90-10)	0.38	–	1.51	5.10
	Cellulose-tris-(3,5-DMPC)	Acetonitrile/Water (50-50)	3.69	–	1.42	7.72
	(3 μM)	Acetonitrile/Water (70-30)	0.90	–	1.44	4.28
		Acetonitrile/Water (90-10)	0.42	–	1.45	2.85
	Cellulose-tris-(3,5-DMPC)	Methanol/Water (70-30)	5.07	–	1.26	4.47
	(5 μM)	Methanol/Water (80-20)	1.91	–	1.31	4.75
		Methanol/Water (90-10)	0.87	–	1.41	4.91
	Cellulose-tris-(3,5-DMPC)	Methanol/Water (70-30)	4.53	–	1.24	3.06
	(3 μM)	Methanol/Water (80-20)	1.83	–	1.27	2.43
		Methanol/Water (90-10)	0.89	–	1.37	2.25
Fenbuconazole	Cellulose-tris-(3,5-DMPC)	Acetonitrile/Water (50-50)	6.86	–	1.33	9.03

(Continued)

TABLE 17.2 (CONTINUED)
Chiral Separation of Pesticides on Different Columns with NP and RP Mobile Phase Conditions

Pesticides	Column/CSPs	Mobile Phase (v/v)	k_1	k_2	α	Rs	References
	(5 μM)	Acetonitrile/Water (70-30)	1.48	–	1.34	7.38	
		Acetonitrile/Water (90-10)	0.56	–	1.37	4.79	
	Cellulose-tris-(3,5-DMPC)	Acetonitrile/Water (50-50)	7.42	–	1.33	7.48	
	(3 μM)	Acetonitrile/Water (70-30)	1.49	–	1.34	4.46	
		Acetonitrile/Water (90-10)	0.59	–	1.34	2.80	
	Cellulose-tris-(3,5-DMPC)	Methanol/Water (70-30)	16.10	–	1.21	3.96	
	(5 μM)	Methanol/Water (80-20)	5.09	–	1.24	4.35	
		Methanol/Water (90-10)	1.97	–	1.29	4.80	
	Cellulose-tris-(3,5-DMPC)	Methanol/Water (70-30)	13.10	–	1.21	3.23	
	(3 μM)	Methanol/Water (80-20)	4.76	–	1.23	3.17	
		Methanol/Water (90-10)	1.96	–	1.28	2.89	
Triadimefon	Cellulose-tris-(3,5-DMPC)	Acetonitrile/Water (50-50)	2.72	–	1.15	3.82	
	(5 μM)	Acetonitrile/Water (70-30)	0.63	–	1.17	2.43	
		Acetonitrile/Water (90-10)	0.22	–	1.20	1.45	
	Cellulose-tris-(3,5-DMPC)	Acetonitrile/Water (50-50)	2.87	–	1.16	2.67	
	(3 μM)	Acetonitrile/Water (70-30)	0.65	–	1.17	1.26	
		Acetonitrile/Water (90-10)	–	–	–	–	
	Cellulose-tris-(3,5-DMPC)	Methanol/Water (70-30)	3.70	–	1.29	5.15	
	(5 μM)	Methanol/Water (80-20)	1.28	–	1.29	4.13	
		Methanol/Water (90-10)	0.51	–	1.28	2.73	
	Cellulose-tris-(3,5-DMPC)	Methanol/Water (70-30)	3.22	–	1.27	2.66	
	(3 μM)	Methanol/Water (80-20)	1.26	–	1.26	1.59	
		Methanol/Water (90-10)	0.55	–	1.23	1.00	

Note: DMPC, dimethylphenylcarbamate; PC, phenylcarbamate; PEC, phenylethylcarbamate.

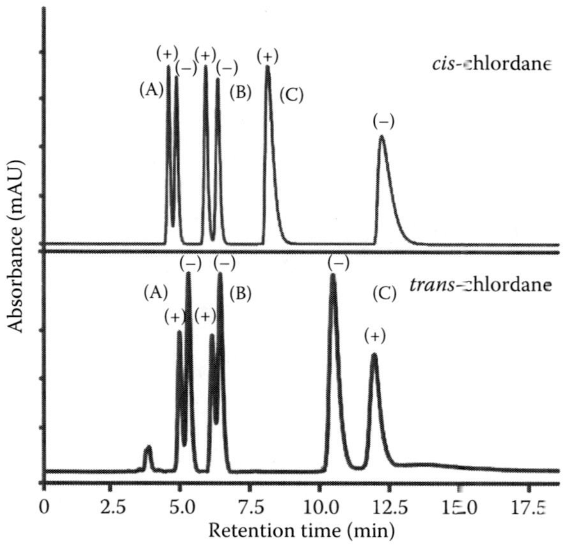

FIGURE 17.3 Effect of the concentration of IPA on chiral resolution of *cis*-chlordane, *trans*-chlordane on Chiralcel OD column. (A) Hexane-IPA (97:3, v/v), (B) Hexane-IPA (99:1, v/v), (C) Hexane (100%). (From Wang, P. et al., *Biomed. Chromatogr.*, 19, 454–458, 2005. With permission.)

TABLE 17.3
Effect of Concentration of Ethanol on the Chiral Resolution of R/S Methamidophos

Mobile Phase	Capacity Factor (k)				Separation Factor (α)						Resolution (Rs)		
(Hexane-Ethanol)	k_1	k_2	k_3	k_4	α_{12}	α_{23}	α_{34}	α_{14}	α_{13}	α_{24}	Rs_{12}	Rs_{23}	Rs_{34}
(95:05, v/v)	2.69	3.10	8.60	13.82	1.15	2.77	1.61	5.13	3.19	4.45	1.56	14.85	9.21
(90:10, v/v)	1.37	1.55	4.16	6.73	1.13	2.68	1.62	4.92	3.04	4.33	1.12	12.24	8.05
(85:15, v/v)	0.90	1.01	2.62	4.25	1.13	2.59	1.62	4.73	2.91	4.2	0.92	9.99	7.58

Source: Xu, C. et al., *Environ. Toxicol. and Chem.*, 27, 174–181, 2008.

effect of temperature on chiral resolution with two different columns as shown in Figure 17.4 and confirmed that 20°C is the optimum temperature for the baseline separation.

Li et al. [76] developed a method for chiral resolution of five organophosphorus compounds (Figure 17.5) on Chiralpak AD, Chiralpak AS, Chiralcel OD, and Chiralcel OJ columns.

The baseline separation of all compounds were obtained on a Chiralpak AD column by using a different ratio of hexane-ethanol and hexane-isopropanol as a mobile phase. Compound 1 was separated with hexane-ethanol (90:10, v/v); compounds 2, 3 and 4, 5 were eluted with hexane-isopropanol (90:10, v/v) and (95:5, v/v), respectively. Zhou et al. [69] also studied the enantiomeric resolution of salithion on different CSPs (Chiralcel OD, Chiralcel OJ, and Chiralpak AD) with different concentrations of isopropanol with hexaneat 1.0 mL min^{-1} flow rate, and the best separation was observed on the Chiralpak AD column with hexane-isopropanol (99.5/0.5, v/v) with various chromatographic values as shown in Table 17.2. The same group of workers resolute a long series of triazole fungicides (hexaconazole, triadimefon, tebuconazole, diniconazole, flutriafol, propiconazole, and difenoconazole) on Chrialcel OD and Chrialcel OJ columns. Authors observed that four compounds (hexaconazole, triadimefon, tebuconazole, diniconazole) were separated on the Chiralcel OD column although enantiomers of flutriafol were obtained by changing the mobile phase from hexane-2-propanol (90:10, v/v) to hexane-ethanol (90:10, v/v). In the case of propiconazole, only three

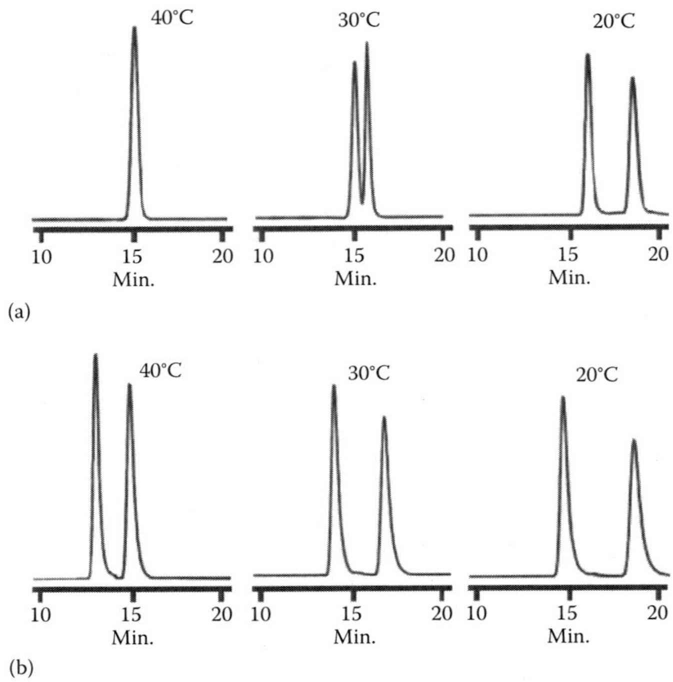

FIGURE 17.4 Effect of temperature on chiral separation of λ-cyhalothrin with two CSPs. (a) Chiralpak amylose tris-(3,5-dimethylphenyl-carbamate), n-hexane-ethanol (99:1, v/v), 0.40 mL min^{-1}; (b) Chiralpak amylose tris-([S]-α-methylbenzyl-carbamate), n-hexane-ethanol (97.5:2.5, v/v), 0.040 mL min^{-1}. (From Li, L. et al., *Chirality,* 20, 130–138, 2008. With permission.)

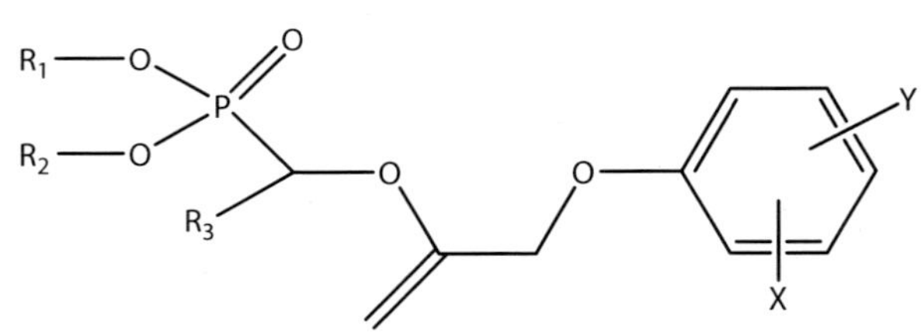

Organophosphorus compound	R_1	R_2	R_3	X	Y
1	C_2H_5	C_2H_5	CH_3	2-Cl	4-Cl
2	CH_3	CH_3	C_2H_5	2-Cl	4-Cl
3	CH_3	CH_3	CH_3	–	4-Cl
4	CH_3	CH_3	CH_3	2-Cl	–
5	CH_3	CH_3	CH_3	–	4-F

FIGURE 17.5 Chemical structure of organophosphorus compounds. (From Zhou, S. et al., *Chirality,* 21, 922–928, 2009. With permission.)

enantiomers could be separated on the Chiralcel OD column using hexane-2-propanol (90:10, v/v) with a flow rate of 0.6 mL min^{-1} at 15°C. A satisfactory resolution of difenoconazole was found on the Chiralel OJ column using hexane-ethanol (90:10, v/v) as the mobile phase. They also studied the effect of temperature with linear van't Hoff plot from 10°C to 35°C and concluded that the enantiomers of these triazole fungicides, except diniconazole and triadimefon, could be separated by differing the temperature [77]. Zang et al. [78] resolved the enantiomer of metalaxyl and metalaxyl acid on a Chiralcel OJ-H [cellulose-tris (4-methylbenzoate)] column with n-hexane-2-propanol-acetic acid (95:5:0.1, v/v/v) at 0.5 mL min^{-1} flow rate. Zhang et al. [78] carried out the chiral separation of three neonicotinoid insecticides with three different polysaccharide CSPs (Chiralcel OD-H, Chiralpak AD-H, and Chiralpak IB) by HPLC and supercritical fluid chromatography. In this study, workers also observed the effect of temperature and organic modifiers on the resolution of analytes and concluded that the best separation of all three analytes were obtained with a Chiralcel OD-H column using different ratios of hexane-ethanol as eluting solvents at a 1.0 mL min^{-1} flow rate as shown in Table 17.2. In the case of Chiralcel AD-H and Chiralpak IB columns, all the analytes were not resolved in good condition, and only two compounds were resolved on both CSPs. Emerick et al. [79] performed the enantioseparation of an organophosphorus compound (methamidophos) on four different polysaccharide CSPs: amylose tris-[(S)-1-phenylethylcarbamate] as CSP-1, amylose tris-(3,5-dimethylphenylcarbamate) as CSP-2, cellulose tris-(3,5-dimethylphenylcarbamate) as CSP-3, and a Chiralcel OD [cellulose tris-(3,5-dimethylphenylcarbamate)] as CSP-4 with different compositions of hexane-2 propanol as the mobile phase. A mixture of n-hexane-2-propanol (80:20, v/v) used initially and the chromatographic parameters k, α, and Rs obtained for CSP-1 and CSP-3 showed poor resolution with Rs less than 1.0. Furthermore, different ratios of 2-propanol (30%, 10%, 8%, and 5%) were also evaluated with CSP-1, CSP-3, or CSP-4; among the chromatographic conditions studied, the highest resolutions were 1.51 and 1.56 when n-hexane-2-propanol (95:5, v/v) and (98:2, v/v) were used with CSP-3 and CSP-4, respectively. The composition of ethanol in place of 2-propanol was also studied, and it was concluded that the use of ethanol at 10% to 20% reduced the retention factor with poor enantiomeric separation of methamidophos. Similarly, the effect of n-heptane in place of n-hexane with 1% acid additives was also studied, and there were no chromatographic parameters improved for the enantio-separation of methamidophos. Sun et al. [80] studied the chiral resolution of uniconazole and the enantio-selective effect on the growth of rice seedlings and cyanobacteria. Authors performed the chiral separation on a Chiralpak AD column by using n-hexane-IPA (85:15, v/v) as the mobile phase at 1.0 mL min^{-1} flow rate, and peaks were confirmed with CD and OR detector at 250 nm as shown in Figure 17.6.

Recently, Lao et al. [81] studied the enantio-selective degradation of warfarin in soil samples and resolute with a triproline CSP and fluoresence detector. The eluting solvent was hexane (0.1% TFA)-2-propanol in the ratio of (92:8, v/v) and (96:4, v/v), and it was observed that the baseline separation was obtained with letter one mobile phase composition, but peak broadening took place.

Wang et al. [71] synthesized the chiral selector [amylose tris-(S)-1-phenylethylcarbamate] for the enantiomeric resolution of 32 pesticides in which 10 pesticides showed the interaction with CSP and resolved with a different composition of hexane-IPA. They also studied the effect of temperature on resolution and concluded that, as the temperature increases, the capacity and separation factors decrease as shown in Table 17.4.

Similarly, Tan et al. [82] synthesized a new CSP [(S)-valine-(R)-1-phenyl-2-(4-methylphenyl) ethylamine] for the resolution of pyrethroid insecticides (fenpropathrin, fenvalerate, brofluthrinate, cypermethrin, and cyfluthrin). They also compared with the resolution on Pirkle type 1-A chiral stationary and got satisfactory baseline separation on a synthesized chiral selector. Oda et al. [45] studied the chiral resolution of warfarin on avidin and ovomucoid CSPs using methanol, ethanol, propanol, and acetonitrile organic modifiers. In general, the chiral recognition behavior of these modifiers on avidin and ovomucoid CSPs was in the order methanol > ethanol > propanol > acetonitrile. Recently, Pan et al. [83] compared the enantio-resolution of seven triazole fungicides (tebuconazole, hexaconazole, myclobutanil, diniconazole, uniconazole, paclobutrazol, and triadimenol) on a Pirkle type (S,S)-Whelk O1 chiral column and four different cellulose derivative columns, namely cellulose tribenzoate (CTB), cellulose tris-(4-methylbenzoate), cellulose triphenylcarbamate, and cellulose tris(3,5-dimethylphenyl carbamate) (CDMPC), in NP mode with ethanol, n-propanol, IPA, and n-butanol, respectively, as a polar modifier in hexane mobile phase and concluded that only two triazole fungicides (hexaconazole and triadimenol) were resolute with a different composition of alcohols in hexane. Among all these cellulose derivative columns, the best separation was achieved

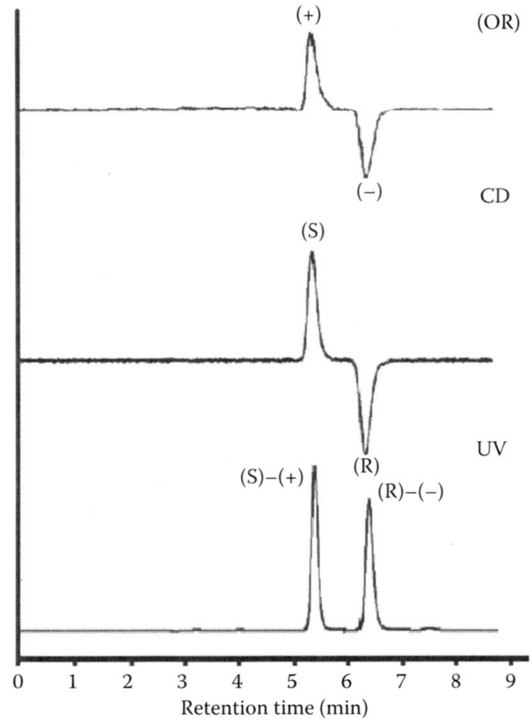

FIGURE 17.6 Chiral separation of uniconazole on Chiralpak AD column with CD and OR detectors. (From Lao, W., Gan, J., *Chirality*, 24, 54–59, 2012. With permission.)

on CDMPC, and there is no resolution achieved on CTB. Some other pesticides also resolved with different CSPs and mobile phases, and many workers compared the enantiomeric resolution with various organic mobile phases. Details of the chromatographic parameters are given in Table 17.2 [71–73].

17.4.2 Enantiomeric Separation of Pesticides in RP Mode

In RP chromatography, the enantiomeric resolution of pesticides were carried out by using aqueous mobile phases with different nonpolar CSPs. Among all CSPs, polysaccharide CSPs were used widely as shown in the literature. Most of the workers performed the chiral separation of pesticides on polysaccharide CSPs using polar organic modifiers with water or buffers. Ali and Aboul-Enein [60] determined the chiral resolution of o, p-DDT and o, p-DDD. The enantiomeric resolution of o, p-DDT and o, p-DDD has been achieved on the Chiralpak AD-RH, Chiralcel OD-RH, and Chiralcel OJ-R CSPs. The mobile phases used acetonitrile–water (50:50, v/v) acetonitrile–2-propanol (50:50, v/v) at a 1.0 mL min^{-1} flow rate. The detection was carried out at 220 nm for both pesticides. Tian et al. [74] developed an HPLC method for the enantiomeric resolution of 20 chiral pesticides on a CDMPC CSP under RP using a different composition of acetonitrile–water and methanol–water as eluents with an 0.8 mL min^{-1} flow rate, and the detection was at 210 and 230 nm, respectively. The values of these findings are also given in Table 17.2. Qiu et al. [75] carried out the chiral separation of nine triazole fungicides (hexaconazole, flutriafol, diniconazole, tetraconazole, epoxiconazole, penconazole, myclobutanil, fenbuconazole, and triadimefon) on Lux Cellulose-1 columns with different particle size (3.0 μm and 5.0 μm) with CDMPC. In this work, scientists used various ratios of acetonitrile and methanol with water as an organic modifier and reported the chromatographic parameters. Moreover, the effect of temperature on chiral resolution

TABLE 17.4
Effect of Temperature on Capacity, Separation, and Resolution Factors of Pesticides

		Capacity Factor			
Pesticides	**Temperature, °C**	k_1	k_2	**Separation Factor, α**	**Resolution, Rs**
Metalaxyl	0	4.04	6.78	1.68	1.67
	10	3.31	5.42	1.64	2.22
	20	3.01	4.66	1.55	2.51
	30	2.49	3.73	1.50	2.36
	40	2.22	3.19	1.43	2.22
Myclobutanil	0	24.53	31.53	1.29	1.27
	10	20.58	25.91	1.26	1.39
	20	17.49	21.47	1.24	1.49
	30	15.70	19.10	1.22	1.33
	40	14.06	16.85	1.20	1.18
Fenoxaprop-ethyl	0	9.42	12.29	1.30	0.83
	10	6.79	8.92	1.28	0.80
	20	6.07	7.53	1.24	1.02
	30	4.88	5.86	1.20	1.01
	40	4.34	5.15	1.19	1.04
Imazalil	0	9.97	11.25	1.30	0.81
	10	8.61	9.69	1.13	0.89
	20	7.56	8.44	1.12	0.86
	30	6.47	7.16	1.11	0.85
	40	5.59	6.12	1.09	0.76
Malathion	0	6.22	7.10	1.14	0.84
	10	5.05	5.70	1.13	0.80
	20	4.53	5.02	1.11	0.76
	30	3.72	4.03	1.08	0.58
	40	3.24	3.41	1.05	0.44
Triadimefon	0	8.29	12.32	1.49	1.63
	10	6.83	9.81	1.44	1.74
	20	5.97	8.47	1.42	1.84
	30	5.20	7.07	1.36	1.50
	40	4.83	6.40	1.33	1.36
Ethofumesate	0	14.01	15.02	1.07	0.73
	10	11.02	11.78	1.07	0.62
	20	9.95	10.58	1.06	0.63
	30	8.21	8.66	1.06	0.58
	40	7.21	7.54	1.05	0.46
Fipronil	0	10.11	13.09	1.29	0.84
	10	8.33	10.68	1.28	0.92
	20	7.72	9.88	1.28	1.35
	30	6.00	7.64	1.27	1.61
	40	4.97	6.28	1.26	1.81
Napropamide	0	11.16	13.38	1.20	0.97
	10	9.22	10.85	1.18	1.11
	20	7.65	8.76	1.14	1.14
	30	6.77	7.62	1.13	1.04
	40	5.54	6.18	1.12	1.07

(Continued)

TABLE 17.4 (CONTINUED)
Effect of Temperature on Capacity, Separation, and Resolution Factors of Pesticides

Pesticides	Temperature, °C	Capacity Factor k_1	k_2	Separation Factor, α	Resolution, Rs
Paclobutrazol	0	3.98	4.87	1.22	1.14
	10	3.81	4.64	1.22	1.26
	20	3.57	4.29	1.20	1.30
	30	3.29	3.88	1.18	1.34
	40	3.06	3.54	1.16	1.31
trans-permethrin	15	14.16	15.67	1.11	1.18
	22	12.01	14.60	1.21	1.26
	23	11.01	12.46	1.13	1.34
	25	9.67	11.19	1.16	1.32
	30	7.64	8.88	1.16	1.12
	35	6.64	7.64	1.15	1.02
	38	4.36	4.99	1.14	0.93
	43	3.27	3.74	1.14	0.78
cis-permethrin	15	23.91	26.34	1.10	2.02
	22	23.96	26.47	1.10	2.20
	23	20.57	22.76	1.11	2.21
	25	19.51	21.75	1.11	2.27
	30	17.78	20.27	1.14	2.25
	35	16.83	19.68	1.17	2.31
	38	11.38	14.14	1.24	2.00
	43	7.82	10.26	1.31	1.87

Source: Tan, X. et al., *J. Sep. Sci.*, 30, 1888–1892, 2007; Schneiderheinze, J. M. et al., *Chirality*, 11, 330–337, 1999.

was also studied. The best separation of all fungicides with different chromatographic conditions is shown in Figure 17.7. From this figure, it has been cleared that the polysaccharide CSP is good for the resolution of these fungicides with different ratios of acetonitrile and methanol with water.

Li et al. [84] also studied the enantiomeric separation of eight triazole fungicides (tetraconazole, fenbuconazole, epoxiconazole, diniconazole, hexaconazole, triadimefon, paclobutrazol, and myclobutanil) in soil samples on polysaccharide CSPs with an aqueous mobile phase, and a MS detector was used to distinguish the enantiomers. Authors concluded that the best chromatographic separation of all the 16 enantiomers of eight triazole fungicides were achieved with the Chiralcel OD-RH column with a mixture of acetonitrile-2 mM ammonium acetate in water (55:45, v/v) as a mobile phase. The baseline separation of all the peaks was observed with more than a 1.0 resolution factor. Authors also tested methanol as an organic modifier in place of acetonitrile in RPLC. However, satisfactory results were not achieved. Different concentrations of ammonium acetate buffer (0.5, 1, 2, 5, and 10 mM) was used to get better peaks, but a good response of the MS/MS detector was observed at 2 mM concentration. Moreover, other chromatographic conditions are also optimized, and the best results were obtained at a 0.45 mL min^{-1} flow rate at 25°C. Dong et al. [85] performed the enantio-selective analysis of myclobutanil (traizole fungicides) in cucumber and soil samples through the LC-MS method. In this technique, workers used a cellulose-based column (Chiralcel OD-RH, 150 mm × 4.6 mm i.d., 5 μm particle size) and two amylose-based columns (Chiralpak AD-RH and AS-RH, 150 mm × 4.6 mm i.d., 5 μm particle size) with a variety of RP mobile phase combinations, and the best chromatographic separation of two enantiomers of myclobutanil was achieved on a Chiralcel OD-RH column with acetonitrile–water (70:30, v/v) as a mobile phase and 0.5 mL min^{-1} flow rate at 40°C.

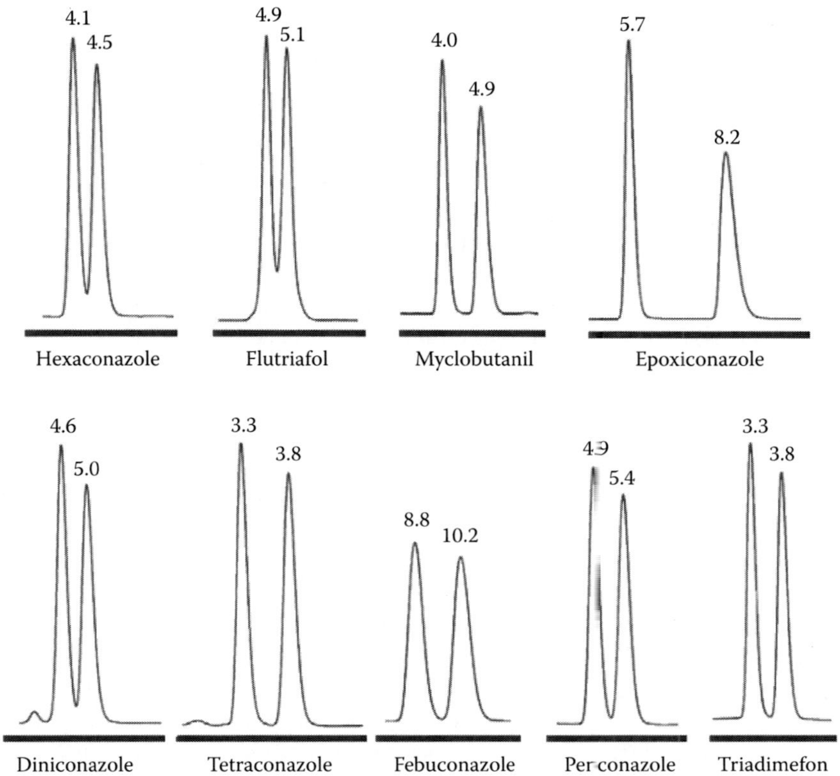

FIGURE 17.7 HPLC chromatogram of nine triazole fungicides. Mobile phases: methanol/water (75:25) on 3.0 μm column for flutriafol, tetraconazole, penconazole, myclobutanil, fenbuconazole, and triadimefon; acetonirile/water (60:40) on 3.0 μm column for epoxiconazole and acetonitrile/water (60:40) on 5.0 μm column for diniconazole. (From Li, Y. et al., *J. Chromatogr. A*, 1224, 51–60, 2012. With permission.)

Vetter et al. [86,87] described the separation of enantiomers of toxaphene on tert-butyldimethylsilylated-β-CD-based CSPs using HPLC. Blessington and Crabb [88] performed the chiral separation of aryloxypropionate herbicides on a Chiral AGP column. In this study, phosphate buffer (10 mM, pH 6)-2-propanol (94:4, v/v) as the mobile phase with detection at 240 nm was used. Recently, Shishovska et al. [89] carried out the chiral separation of permethrin enantiomers by using β-CD-based CSP and the separation of all enantiomers shown in chromatogram (Figure 17.8); retention times for *trans*-enatiomers were 19.8 and 22.7 min and for *cis*-enantiomers were 38.2 and 42.3 min, respectively. They also studied the effect of temperature on the chiral resolution and observed

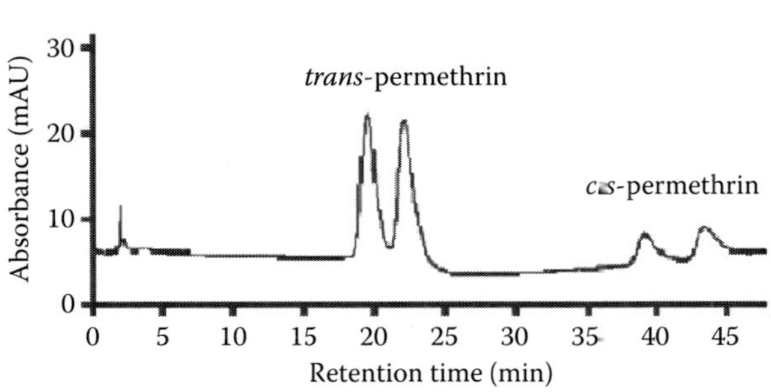

FIGURE 17.8 Chiral separation of permethrin enantiomers on β-CD-based CSP. (From Schneiderheinze, J. M. et al., *Chirality*, 11, 330–337, 1999. With permission.)

that, at high temperature, the values of chromatographic parameters k, α, and Rs decrease with an increase of temperature as shown in Table 17.4.

Schneiderheinze et al. [90] studied the chiral resolution of phenoxyalkanoic acid herbicides in plant and soil samples on Chirobiotic T CSPs with methanol-1% TEAA, pH 4.1 (60:40, v/v) as the mobile phase and detection was carried out by chiroptical detector. Möller et al. [91] resolved the enantiomers of α-HCH in brains of seals on a Chiraldex column by using methanol–water (75:25, v/v) as an eluting solvent. Dondi et al. [92] used a terguride-based CSP for the enantiomeric resolution of chrysanthemic acid [2,2-dimethyl-3-(2-methylpropenyl)-cyclopropanecarboxylic acid] and its halogen substituted analogues. In this work, an UV diode array and chiroptical detectors were used for the identification of the enantiomers and observed that isomers with (1R) configuration always eluted before those with (1S) configuration. The elution sequence of *cis*- and *trans*-isomers was strongly affected by the mobile phase pH whereas the enantio-selectivity remained the same.

Ludwig et al. [93] carried out the enantiomeric analysis of 2-(2,4-dichlorophenoxy) propionic acid (dichlorprop) and its degradation products in a marine microbial community on a Chiral AGP column with water-2-propanol-phosphate buffer (10 mM, pH 4.85) (94:4:2, v/v/v) mobile phase at 230 nm. Armstrong et al. [94] developed the method for the enantio-separation of warfarin, coumachlor, coumafuryl, bulan, crufomate, fonofos, anacymidol, napropamide, and 2-(3-chlorophenoxy)-propionamide pollutants on Chiral AGP and other cyclodextrin-based CSPs. Chu and Wainer [95] separated the enantiomers of warfarin in serum samples by using chiral HPLC. Weber et al. [96] used permethylated β-CD CSP for the chiral separation of phenoxypropionates. Recently, Malakova et al. [97] studied the chiral resolution of two enantiomers, (R)-(+)-warfarin and (S)-(−)-warfarin and their determination in the hepatoma HepG2 cell line by using a glycopeptide-based CSP with an acetonitrile–methanol–ammonium acetate buffer (10.0 mM, pH 4.1) (31:5:64, v/v) mobile phase at 1.2 mL min^{-1} flow rate, and the LOD was found to be 0.121 μmol L^{-1} for (S)-warfarin and 0.109 μmol L^{-1} for (R)-warfarin. Guillén-Casla et al. [98] analyzed the enantiomeric separation of a mixture of aryloxyphenoxypropionic herbicides (diclofop-acid and diclofop-methyl) with α_1-acid glycoprotein CSP by using one- and two-dimensional LC methods. In this method for achiral separation of samples containing diclofop-acid and diclofop-methyl racemate, mixtures were carried out by injecting 20.0 μL of sample on the C18-LUNA column. An isocratic mobile phase containing methanol-phosphate buffer (30 mM, pH 7) (73:23, v/v) at 1.0 mL min^{-1} was used. The chiral resolution of diclofopacid enantiomers and diclofop-methyl enantiomers were performed with phosphate buffer (70 mM, pH 7.0)-2-propanol (99.5:0.5, v/v) and phosphate buffer (30 mM, pH 7.0)-2-propanol (91:9, v/v), respectively, at 0.8 mL min^{-1} flow rate.

Recently, same group of workers carried out the chiral separation of aryloxyphenoxypropionic acid herbicides (fluazifop-butyl, quizalofop-ethyl, and mefenpyr-diethyl) on α_1-acid glycoprotein CSP. In this study, optimization of chromatographic conditions was performed through a factorial experimental design with phosphate buffer (pH 6.5–7.0) and propanol (5%–10%) at 15°C–25°C column temperature. Mathematical deconvolution has also been studied to determine peak areas and to calculate Rs, enantiomeric ratio, and enantiomeric fraction and concluded that peak deconvolution provides a simple, effective, and reproducible method for herbicide determination in soil matrices at the low levels of micrograms per gram [99]. Mano et al. [100] used a flavoprotein conjugated silica CSP for the chiral resolution of warfarin. They also studied the effect of pH on chiral resolution and observed that the best results were obtained at pH 4.0–4.8. The effect of different compositions of mobile phases and column particle size on capacity, separation, and resolution factors with RP mode by using different CSPs is also given in Table 17.2 [75].

17.5 CONCLUSION

The application of chiral pesticides into our environment is a serious issue as simple analyses do not provide the exact dose and toxicity of the pesticides. Besides, the different toxicities of the two enantiomers confused farmers for determining the lethal dose to the pests. Sometimes, the pest control remains unaffected due to an elusive dose of chiral pesticides. Besides, it is not possible to determine the fate of chiral pesticide degradation product without considering the chiral aspect of racemic pesticides. Chiral HPLC methods are available in the literature, and they should be used to study the fate of chiral pesticides before their application into our environment. The proper knowledge of chiral pesticides may design the exact dose and save our environment from the unnecessary load of pesticides. The agricultural scientists should think in terms of chiral pesticides and provide information to farmers.

REFERENCES

1. Ulrich, E. M., Morrison, C. N., Goldsmith, M. R., Foreman W. T., Chiral pesticides: Identification, description, and environmental implications, *Rev. of Environ. Conta. and Toxicol.*, 217, 1–74, 2012.
2. http://ipmworld.umn.edu/chapters/ware.htm.
3. http://www.apsnet.org/edcenter/intropp/topics/Pages/Fungicides.aspx.
4. http://www.agriinfo.in/default.aspx?page=topic&superid=1&topicid=804.
5. Richardson, S. D., Ternes, T. A., Water analysis: Emerging contaminants and current issues, *Anal. Chem.*, 77, 3807–3838, 2005.
6. Kurihara, N., Miyamot, J., Paulson, G. D., Zeeh, B., Skidmore, M. W., Hollingworth, R. M., Kuiper, H. A., Chirality in synthetic agro-chemicals: Bioactivity and safety consideration, *Pure Appl. Chem.*, 69, 2007–2025, 1997.
7. Ward, T. J., Hamburg, D. M., Chiral separations, *Anal. Chem.*, 76, 4635–4644, 2004.
8. Gasparrini, F., Misiti, D., Villani, C., High-performance liquid chromatography chiral stationary phases based on low-molecular-mass selectors, *J. Chromatogr. A*, 922, 35–50, 2001.
9. Garrison, A. W., Probing the enantioselectivity of chiral pesticides, *Environ. Sci. Technol.*, 40, 16–23, 2006.
10. Sekhon, B. S., Chiral pesticides, *J. Pestic. Sci.*, 34, 1–12, 2009.
11. Dong, F. S., Liu, X. G., Zheng, Y. Q., Cao, Q., Li, C. J., Stereoselective degradation of fungicide triadimenol in cucumber plants, *Chirality*, 22, 292–298, 2010.
12. Ye, J., Zhao, M. R., Liu, J., Liu, W. P., Enantioselectivity in environmental risk assessment of modern chiral pesticides, *Environ. Pollut.*, 158, 2371–2383, 2010.
13. Liu, W. P., Ye, J., Jin, M. Q., Enantioselective phytoeffects of chiral pesticides, *J. Agric. Food Chem.*, 57, 2087–2095, 2009.
14. Wang, Y., Tai, K., Yen, J., Separation, bioactivity, and dissipation of enantiomers of the organophosphorus insecticide fenamiphos, *Ecotoxicol Environ Saf.*, 57, 346–353, 2004.
15. Aboul-Enein, H. Y., Ali, I., *Chiral Separations by Liquid Chromatography and Related Technologies*, Dekker, New York, 2003.
16. Beesley, T. E., Scott, R. P. W., *Chiral Chromatography*, Wiley, New York, 1998.
17. Aboul-Enein, H. Y., Wainer, I. W., (eds), *The Impact of Stereochemistry on Drug Development and Use*, *Chemical Analysis*, vol. 142, Wiley, New York, 1997.
18. Subramanian, G. (ed.), *A Practical Approach to Chiral Separations by Liquid Chromatography*, VCH, Weinheim, 1994.
19. Allenmark, S., *Chromatographic Enantioseparation, Methods and Applications*, 2nd ed., Ellis Horwood, New York, 1991.
20. Krstulovic, A. M., *Chiral Separations by HPLC: Applications to Pharmaceutical Compounds*, Ellis Horwood, New York, 1989.
21. Huhnerfuss, H., Kallenborn, R., Chromatographic separation of marine pollutants, *J. Chromatogr. A*, 580, 191, 1992.
22. Kallenborn, R., Hühnerfuss, H., Chiral Environmental Pollutants, *Trace Analysis and Toxicology*, Springer-Verlag, *Berlin*, 2001.
23. Ali, I., Aboul-Enein, H. Y., *Chiral Pollutants: Distribution, Toxicity and Analysis by Chromatography and Capillary Electrophoresis*, John Wiley & Sons: Chichester, UK, 2004.
24. Pfeiffer, C. C., Optical isomerism and pharmacological action, a generalization, *Science*, 124, 29, 1956.
25. Pirkle, W. H., Pochapsky, T. C., Considerations of chiral recognition relevant to the liquid chromatography separation of enantiomers, *Chem. Rev.*, 89, 347–362, 1989.
26. Groombridge, J. J., Jones, C. G., Bruford, M. W., Nichols, R. A., Ghost alleles in the Mauritius kestrel, *Nature*, 403, 616, 2000.

27. Rogozhin, S. V., Davankov, V. A., German Patent 1932190, 1969 through *Chem. Abstr.* 72,90875c, 1970.
28. Rogozhin, S. V., Davankov, V. A., Ligand chromatography on asymmetric complex-forming sorbents as a new method for resolution of racemates, *J. Chem. Soc. Chem. Commun.*, 490, 1971.
29. Davankov, V. A., Chiral selectors with chelating properties in liquid chromatography: Fundamental reflections and selective review of recent developments, *J. Chromatogr. A*, 666, 55–76, 1994.
30. Han, S. M., Armstrong, D. W., Krstulovic, A. M., (ed.), *Chiral Separations by HPLC*, Ellis Horwood, Chichester, 208, 1989.
31. Stalcup, A. M., in G. Subramanian, (ed.), *A Practical Approach to Chiral Separations by Liquid Chromatography*, VCH, Weinheim, 95, 1994.
32. Feitsma, K. G., Bosmann, J., Drenth, B. F. H., De Zeeuw, R. A., A study of the separation of enantiomers of some aromatic carboxylic acids by high-performance liquid chromatography on a β-cyclodextrin-bonded stationary phase, *J. Chromatogr.*, 333, 59–68, 1985.
33. Han, S. M., Direct enantiomeric separations by high performance liquid chromatography using cyclodextrins, *Biomed. Chromatogr.*, 11, 259–271, 1997.
34. Piperaki, S., Kakoulidou-Tsantili, A., Parissi-Poulou, M., Solvent selectivity in chiral chromatography using a β-cyclodextrin-bonded phase, *Chirality*, 7, 257–266, 1995.
35. Hargitai, T., Okamoto, Y., Evaluation of 3,5-dimethylphenyl carbamoylated α-, β-, and γ-cyclodextrins as chiral stationary phases for HPLC. *J. Liq. Chromatogr.*, 16, 843–858, 1993.
36. Armstrong, D. W., Tang, Y., Chen, S., Zhou, Y., Bagwill, C., Chen, J. R., Macrocyclic antibiotics as a new class of chiral selectors for liquid-chromatography, *Anal. Chem.*, 66, 1473–1484, 1994.
37. *Chirobiotic Handbook, Guide to Using Macrocyclic Glycopeptide Bonded Phases for Chiral LC Separations*, 2nd ed., Advanced Separation Technologies, Inc., Whippany, NJ, 1999.
38. Armstrong, D. W., Rundlett, K. L. Chen, J. R., Evaluation of the macrocyclic antibiotic vancomycin as a chiral selector for capillary electrophoresis, *Chirality*, 6, 496–509, 1994.
39. Tesaova, E., Bosakova, Z., Pacakova, V., Comparison of enantioselective separation of N-tert.-butyloxycarbonyl amino acids and their non-blocked analogues on teicoplanin-based chiral stationary phase., *J. Chromatogr. A*, 838, 121–129, 1999.
40. Ekborg-Ott, K. H., Liu, Y., Armstrong, D. W., Highly enantioselective HPLC separations using the covalently bonded macrocyclic antibiotic, ristocetin A, chiral stationary phase, *Chirality*, 10, 434–483, 1998.
41. Armstrong, D. W., Liu, Y., Ekborg-Ott, K., A covalently bonded teicoplanin chiral stationary phase for HPLC enantioseparations, *Chirality*, 7, 474–497, 1995.
42. Peter, A., Torok, G., Armstrong, D. W., High-performance liquid chromatographic separation of enantiomers of unusual amino acids on a teicoplanin chiral stationary phase, *J. Chromatogr. A*, 793, 283–296, 1998.
43. Aboul-Enein, H. Y., Ali, I., Optimization strategies for HPLC enantioseparation of racemic drugs using polysaccharides and macrocyclicglycopeptide antibiotic chiral stationary phases, *IL Farmaco*, 57, 513–529, 2002.
44. Allenmark, S., in G., Subramanian, (ed.), *A Practical Approach to Chiral Separations by Liquid Chromatography*, Wiley VCH, Weinheim, 183–216, 1994.
45. Oda, Y., Mano, N., Asakawa, N., Yoshida, Y., Sato T. and Nakagawa T., Comparison of avidin and ovomucoid as chiral selectors for the resolution of drug enantiomers by high-performance liquid chromatography, *Anal. Sci.*, 9, 221, 1993.
46. Shinbo, T., Yamaguchi, T., Yanagishita, H., Kitamoto, D., Sakaik, K., Sugiura, M., Improved crown ether-based chiral stationary phase, *J. Chromatogr. A*, 625, 101–108, 1992.
47. Hyun, M. H., Jin, J. S., Lee, W., Liquid chromatographic resolution of racemic amino acids and their derivatives on a new chiral stationary phase based on crown ether, *J. Chromatogr. A*, 822, 155–161, 1998.
48. Hyun, M. H., Jin, J. S., Koo, H. J., Lee, W., Liquid chromatographic resolution of racemic amines and amino alcohols on a chiral stationary phase derived from crown ether, *J. Chromatogr. A*, 837, 75–82, 1999.
49. Hyun, M. H., Han, S. C., Jin, J. S., Lee, W., Separation of the stereoisomers of racemic fluoroquinolone antibacterial agents on a crown-ether-based chiral HPLC stationary phase, *Chromatographia*, 52, 473–476, 2000.
50. Feibush, B., Cohen, M. J., Karger, B. L., The role of bonded phase composition on the ligand-exchange chromatography of dansyl-D,L-amino acids, *J. Chromatogr.*, 282, 3–26, 1983.
51. Davankov, V. A., Bochkov, A. S., Belov, Y. P., Ligand-exchange chromatography of racemates: XV. Resolution of α-amino acids on reversed-phase silica gels coated with n-decyl-l-histidine, *J. Chromatogr.*, 218, 547–557, 1981.

52. Roumeliotis, P., Unger, K. K., Kurganov, A. A., Davankov, V. A., High-performance ligand-exchange chromatography of α-amino acid enantiomers–Studies on monomerically bonded 3-(L-prolyl)- and 3-(L-hydroxyprolyl)propyl silicas, *J. Chromatogr.*, 255, 51–66, 1983.
53. Roumeliotis, P., Kurganov, A. A., Davankov, V. A., Effect of the hydrophobic spacer in bonded [Cu(l-hydroxyprolyl)alkyl]+ silicas on retention and enantioselectivity of α-amino acids in high-performance liquid chromatography, *J. Chromatogr.*, 266, 439–450, 1983.
54. Shieh, C. H., Karger, B. L., Gelber, L. R., Feibush, B., Ligand exchange chromatography of amino alcohol enantiomers as Schiff bases, *J. Chromatogr.*, 406, 343–352, 1987.
55. Remelli, M., Fornasari, P., Dandi, F., Pulidori, F., Dynamic column-coating procedure for chiral ligand-exchange chromatography, *Chromatographia*, 37, 23–30, 1993.
56. Perez, S., Barcelo, D., Applications of LC-MS to quantitation and evaluation of the environmental fate of chiral drugs and their metabolites, *Trends Anal. Chem.*, 27, 836–846, 2008.
57. Ward, T. J., Ward, K. D., Chiral separations: Fundamental review 2010, *Anal. Chem.*, 82, 4712–4722, 2010.
58. Caccamese, S., Principato, G., Separation of the four pairs of enantiomers of vincamine alkaloids by enantioselective high-performance liquid chromatography, *J. Chromatogr. A*, 893, 47–54, 2000.
59. Ellington, J. J., Evans, J. J., Prickett, K. B., Champion, Jr., W. L., High-performance liquid chromatographic separation of the enantiomers of organophosphorus pesticides on polysaccharide chiral stationary phases, *J. Chromatogr. A*, 928, 145–154, 2001.
60. Ali, I., Aboul-Enein, H. Y., Determination of chiral ratio of o,p-DDT and o,p-DDD pesticides on polysaccharides CSPs by HPLC under reversed phase mode, *Environ. Toxicol.*, 17, 329–333, 2002.
61. Li, Z. Y., Zhang, Z. C., Zhou, Q. L., Gao, R. Y., Wang, Q. S., Fast and precise determination of phenthoate and its enantiomeric ratio in soil by the matrix solid-phase dispersion method and liquid chromatography, *J. Chromatogr. A*, 977, 17–25, 2002.
62. Aboul-Enein, H. Y., Ali, I., A comparison of chiral resolution of econazole, miconazole and sulconazole by HPLC using normal phase amylose CSPs, *Fresenius J. Anal. Chem.*, 370, 951–955, 2001.
63. Champion, Jr., W. L., Lee, J., Wayne, G. A., DiMarco, J. C., Matabe A., Prickett, K. B., Liquid chromatographic separation of the enantiomers of trans-chlordane, cis-chlordane, heptachlor, heptachlor epoxide and hexachlorocyclohexane with application to small-scale preparative separation, *J. Chromatogr. A*, 1024, 55–62, 2004.
64. Wang, P., Jiang, S., Liu, D., Jia, G., Wang, Q., Wang, P., Zhou, Z., Effect of alcohols and temperature on the direct chiral resolutions of fipronil, isocarbophos and carfentrazone-ethyl, *Biomed. Chromatogr.*, 19, 454–458, 2005.
65. Liu, W., Gan, J. J., Qin, S., Separation and aquatic toxicity of enantiomers of synthetic pyrethroid insecticides, *Chirality*, 17, S127–S133, 2005.
66. Lin, K., Zhou, S., Xu, C., Liu, W., Enantiomeric resolution and biotoxicity of methamidophos, *J. Agric. Food Chem.*, 54, 8134–8138, 2006.
67. Lin, K., Zhang, F., Zhou, S., Liu, W., Gan, J., Pan, Z., Stereoisomeric separation and toxicity of the nematicidefosthiazate, *Environ. Toxicol. and Chem.*, 26, 2339–2344, 2007.
68. Xu, C., Wang, J., Liu, W., Shengg, D., Tu, Y., Separation and aquatic toxicity of enantiomers of the pyrethroid insecticide lambda-cyhalothrin, *Environ. Toxicol. and Chem.*, 27, 174–181, 2008.
69. Zhou, S., Lin, K., Li, L., Jin, M., Ye, J., Liu, W., Separation and toxicity of salithion enantiomers, *Chirality*, 21, 922–928, 2009.
70. Zhang, C., Jin, L., Zhou, S., Zhang, Y., Feng, S., Zhou, Q., Chiral separation of neonicotinoid insecticides by polysaccharide-type stationary phases using high-performance liquid chromatography and supercritical fluid chromatography, *Chirality*, 23, 215–221, 2011.
71. Wang, P., Liu, Lei, X., Jiang, S., Zhou, Z., Enantiomeric separation of chiral pesticides by high-performance liquid chromatography on an amylose tris-(S)-1-phenylethyl carbamate chiral stationary phase, *J. Sep. Sci.*, 29, 265–271, 2006.
72. Wang, P., Liu, D., Jiang, S., Gu, X., Zhou, Z., The direct chiral separations of fungicide enantiomers on amylopectin based chiral stationary phase by HPLC. *Chirality*, 19, 114–119, 2007.
73. Sun, J., Liu, J., Tu, W., Xu, C., Separation and aquatic toxicity of enantiomers of the organophosphorus insecticide O-ethylO-4-nitrophenyl phenylphosphonothioate (EPN), *Chemosphere*, 81, 1308–1313, 2010.
74. Tian, Q., Lv, C., Wang, P., Ren, L., Qiu, J., Li, L., Zhou, Z., Enantiomeric separation of chiral pesticides by high performance liquid chromatography on cellulosetris-3,5-dimethyl carbamate stationary phase under reversed phase conditions, *J. Sep. Sci.*, 30, 310–321, 2007.

75. Qiu, J., Dai, S., Zheng, C., Yang, S., Chai, T., Bie, M., Enantiomeric separation of triazole fungicides with 3-lm and 5-lm particle chiral columns by reverse-phase high-performance liquid chromatography, *Chirality*, 23, 479–486, 2011.
76. Li, L., Zhou, S., Zhao, M., Zhang, A., Peng, H., Tan, X., Lin, C., He, A., Separation and aquatic toxicity of enantiomers of 1-(substituted phenoxyacetoxy) alkylphosphonate herbicides, *Chirality*, 20, 130–138, 2008.
77. Zhou, Y., Li, L., Lin, K., Zhu, X., Liu, W., Enantiomer separation of triazole fungicides by high-performance liquid chromatography, *Chirality*, 21, 421–427, 2009.
78. Zhang, X., Xia, T., Chen, J., Huang, L., Cai, X., Direct chiral resolution of metalaxyl and metabolite metalaxyl acid in aged mobile phases: The role of trace water, *J. Agric. Food Chem.*, 58, 5004–5010, 2010.
79. Emerick, G. L., Oliveira, R., Belaz, K. R. A., Goncalves, M., Deoliveira, G. H., Semi-preparative enantioseparation of methamidophos by HPLC-UV and preliminary in vitro study of butyryl cholinesterase inhibition, *Environ. Toxicol. and Chem.* 31, 239–245, 2012.
80. Sun, J., Zhang, A., Zhang, J., Xie, X., Liu, W., Enantiomeric resolution and growth-retardant activity in rice seedlings of uniconazole, *J. Agric. Food Chem.*, 60, 160–164, 2012.
81. Lao, W., Gan, J., Enantioselective degradation of warfarin in soils, *Chirality*, 24, 54–59, 2012.
82. Tan, X., Hou, S., Jiang, J., Wang, M., Preparation of a new chiral stationary phase for HPLC based on the (R)-1-phenyl-2-(4-methylphenyl)ethylamine amide derivative of (S)-valine and 2-chloro-3,5-dinitrobenzoic acid: Enantioseparation of amino acid derivatives and pyrethroid insecticides, *J. Sep. Sci.*, 30, 1888–1892, 2007.
83. Pan, C., Shen, B., Xu, B., Chen, J., Xu, X., Comparative enantioseparation of seven triazole fungicides on (S,S)-Whelk O1 and four different cellulose derivative columns, *J. Sep. Sci.*, 29, 2004–2011, 2006.
84. Li, Y., Dong, F., Liu, X., Xu, J., Li, J., Kong, Z., Chen, X., Liang, X., Zheng, Y., Simultaneous enantioselective determination of triazole fungicides in soil and water by chiral liquid chromatography/tandem mass spectrometry, *J. Chromatogr. A*, 1224, 51–60, 2012.
85. Dong, F., Cheng, L., Liu, X., Xu, J., Li, J., Li, Y., Kong, Z., Jian, Q., Zheng, Y., Enantioselective analysis of triazole fungicide myclobutanil in cucumber and soil under different application modes by chiral liquid chromatography/tandem mass spectrometry, *J. Agric. Food Chem.*, 60, 1929–1936, 2012.
86. Vetter, W., Klobes, U., Luckas, B., Hottinger, G., Enantiomeric resolution of persistent compounds of technical toxaphene (CTTs) on t-butyldimethylsilylated β-cyclodextrin phases, *Chromatographia*, 45, 255–262, 1997.
87. Vetter, W., Luckas, B., Enantioselective determination of persistent and partly degradable toxaphene congeners in high trophic level biota, *Chemosphere*, 41, 499–506, 2000.
88. Blessington, B., Crabb, N., Proposed primary reference methods for the determination of some commercially important chiral aryloxypropionate herbicides in both free acid and ester forms, *J. Chromatogr.*, 483, 349–358, 1989.
89. Shishovska, M., Trajkovska, V., HPLC-method for determination of permethrin enantiomers using chiral β-cyclodextrin-based stationary phase, *Chirality*, 22, 527–533, 2010.
90. Schneiderheinze, J. M., Armstrong, D. W., Berthod, A., Plant and soil enantioselective biodegradation of racemic phenoxyalkanoic herbicides, *Chirality*, 11, 330–337, 1999.
91. Möller, W. A., Bretzke, C., Hühnerfuss, H., Kallenborn, R., Kinkel, J. N., Kopf, J., Rimkus, G., The absolute configuration of (+)-α-1,2,3,4,5,6-hexachlorocyclohexane, and its permeation through the seal blood–brain barrier chromatographic separation of enantiomers of (Durchlässigkeit der Blut-Hirn-Schranke von Seehundenfür das Enantiomer (+)-α-1,2,3,4,5,6-Hexachlorcyclohexan und dessen absolute Konfiguration.), *Angew. Chem. Int. Ed. Engl.*, 33, 882–884, 1994.
92. Dondi, M., Flieger, M., Olsovska, J., Polcaro, C. M., Sinibaldi, M., High-performance liquid chromatography study of the enantiomer separation of chrysanthemic acid and its analogous compounds on a terguride-based stationary phase, *J. Chromatogr. A*, 859, 133–142, 1999.
93. Ludwig, P., Gunkel, W., Hühnerfuss, H., Chromatographic separation of the enantiomers of marine pollutants. Part 5: Enantioselective degradation of phenoxycarboxylic acid herbicides by marine microorganisms, *Chemosphere*, 24, 1423–1429, 1992.
94. Armstrong, D. W., Reid III, G. L., Hilton, M. L., Chang, C. D., Relevance of enantiomeric separations in environmental science, *Environ. Pollut.*, 79, 51–58, 1993.
95. Chu, Y. Q., Wainer, I. W., The measurement of warfarin enantiomers in serum using coupled achiral/chiral, high-performance liquid chromatography (HPLC), *Pharm. Res.*, 5, 680–683, 1988.

96. Weber, K., Kreuzig, R., Bahadir, M., On enantioselective separation of phenoxypropionates using permethylated β-cyclodextrin HPLC and GC columns, *Chemosphere*, 35, 13–20, 1997.
97. Malakova, J., Pavek, P., Svecova, L., Jokesova, I., Zivny, P., Palicka, V., New high-performance liquid chromatography method for the determination of (R)-warfarin and (S)-warfarin using chiral separation on a glycopeptide-based stationary phase, *J. Chromatogr., B*, 877, 3226–3230, 2009.
98. Guillén-Casla, V., Pérez-Arribas, L., One- and two-dimensional direct chiral liquid chromatographic determination of mixtures of diclofop-acid and diclofop-methyl herbicides, *J. Agric. Food Chem.*, 56, 2303–2309, 2008.
99. Guillén-Casla, V., Magro-Moral, J., Rosales-Conrado, N., Pérez-Arribas L. V., León-González, M. E., Polo-Díez, L. M., Direct chiral liquid chromatography determination of aryloxyphenoxy propionic herbicides in soil: Deconvolution tools for peak processing, *Anal. Bioanal. Chem.*, 400, 3547–3560, 2011.
100. Mano, N., Oda, Y., Asakawa, N., Yoshida, Y., Sato, T., Development of a flavoprotein column for chiral separation by high-performance liquid chromatography, *J. Chromatogr. A*, 623, 221–228, 1992.

18 Application in Pesticide Analysis

Liquid Chromatography—A Review of the State of Science for Biomarker Discovery and Identification

Peipei Pan and Jeanette M. Van Emon

CONTENTS

18.1 A BRIEF INTRODUCTION TO EXPOSURE BIOMARKERS FOR ENVIRONMENTAL CONTAMINANTS

18.1.1 ENVIRONMENTAL CONTAMINANTS

Throughout history, economic growth has led to an increased utilization of natural resources. This increased usage has manifested in the deposition of industrial wastewater, medical waste, and other waste residuals into the environment and the generation of polluted air. The widespread use of biohazardous pesticides, artificial food additives, pharmaceuticals, neutraceuticals, nanomaterials, and mineral-based beauty and skin care products also contribute to this environmental burden. The delicate balance between the growth of living organisms, resource utilization, and environmental stressors is being challenged. The persistence of environmental contaminants in many environmental compartments presents a potential risk to humans and ecosystems. Human exposure leading to adverse health outcomes can increase as the variety and quantity of contamination sources increase.

Environmental contaminants can enter into the body through different routes of exposure (e.g., dermal, inhalation, ingestion), leading to various outcomes. The harmful effect caused by a pollutant depends on the route of exposure as well as its innate toxicity, the effective exposure dose, and host factors, such as age, gender, nutritional state, and developmental stage [1].

18.1.2 Biomonitoring

The conventional approaches used for environmental monitoring make it difficult to determine causal connections between contaminants and biological responses. Depending on the chemical properties of the xenobiotic, exposures can be determined by measuring parent compound, metabolites, conjugates, or other biomarkers induced by the exposure. Biomonitoring is based on measuring the concentration of these biomarkers in biological fluids, such as blood, urine, saliva, sputum, sweat, and cerebrospinal fluid. The Centers for Disease Control and Prevention has stated that "biomonitoring measurements are the most health-relevant assessments of exposure because they measure the amount of the chemical that actually gets into people from all environmental sources, such as the air, soil, water, dust, or food combined" [2]. Biomonitoring is an effective approach for assessing human exposure to chemicals and to reduce uncertainties along the source-to-outcome continuum [3]. Environmental monitoring coupled with biomonitoring has become an increasingly important research area to study the interactions of contaminants with humans and ecosystems. Tracking the levels of environmental contaminants enables an assessment of exposure and the effects of environmental pollutants on an organism as well as the environment. Biomonitoring measurements aid in identifying contaminants of potential ecological or public health concern and provide data for risk assessments and effective risk management.

Biomonitoring has been used to monitor chemical exposure for decades; however, advances in analytical methodologies, such as with high-performance liquid chromatography (HPLC) in tandem with mass spectrometry (MS), enable the discovery and identification of new biomarkers in complex matrices as well as the measurement of more chemicals, in smaller concentrations, using smaller sample sizes [2]. Improvements in HPLC, such as the use of smaller particles (<2 μm i.d.) and ultra-high pressures up to 15,000 psi, enables the analysis of trace concentrations at low parts per trillion [4–6]. Through these advancements, biomonitoring has become more widely used for a variety of applications, including public health research, evaluation of environmental regulations, and risk assessment and management.

18.2 BIOMARKERS

Biomarkers have emerged as very powerful indicators of xenobiotic exposure and augment the traditional methods of environmental monitoring used to determine the presence and potential harmful effects of environmental contaminants. An increasing number of research papers from various fields, including exposure science, environmental epidemiology, toxicology, occupational and environmental medicine, analytical chemistry, public health, and pharmacology report on the utility of biomarkers [7]. Some commonly used biomarkers for pesticide biomonitoring have been identified by LC-MS (Table 18.1).

TABLE 18.1
Pesticide Biomonitoring Using LC with MS (Alphabetical by Separation Column)

Biomarker	Parent	Matrix	Column and Source	Analysis System	LOD (ng/mL)	References
Chlormequat		Urine	Atlantis Hydrophilic Interaction Chromatography (Waters)	LC-MS/MS	0.1	[8]
Para-nitrophenol	Parathion	Urine	BetaSil* phenyl (Thermo Scientific)	LC-MS/MS	0.1–1.5	[9,10]
3,5,6-Trichloro-2-pyridinol	Chlorpyrifos	Urine	BetaSil* phenyl (Thermo Scientific)	LC-MS/MS	0.1–1.5	[9,10]
2-Diethylamino-6-methyl pyrimidin-4-ol	Pirimiphos	Urine	BetaSil* phenyl (Thermo Scientific)	LC-MS/MS	0.1–1.5	[9,10]
5-Chloro-1-isopropyl-[3*H*]-1,2,4-triazol-3-one	Isazophos	Urine	BetaSil* phenyl (Thermo Scientific)	LC-MS/MS	0.1–1.5	[9,10]
3-Chloro-4-methyl-7-hydroxycoumarin	Coumaphos	Urine	BetaSil* phenyl (Thermo Scientific)	LC-MS/MS	0.1–1.5	[9,10]
2[(Dimethoxyphosphorothioyl) sulfanyl] succinic acid	Malathion	Urine	BetaSil* phenyl (Thermo Scientific)	LC-MS/MS	0.1–1.5	[9,10]
2-Isopropyl-6-methyl-4-pyrimidiol	Diazinon	Urine	BetaSil* phenyl (Thermo Scientific)	LC-MS/MS	0.1–1.5	[9,10]
3-Phenoxybenzoic acid	Several pyrethroids	Urine	BetaSil* phenyl (Thermo Scientific)	LC-MS/MS	0.1–1.5	[9,10]
cis- and *trans*-3-(2,2-Dichlorovinyl)-2,2-dimethylcyclopropane-1-carboxylic acids	Cyfluthrin, Permethrin, Cypermethrin	Urine	BetaSil* phenyl (Thermo Scientific)	LC-MS/MS	0.1–1.5	[9,10]
4-Fluoro-3-phenoxybenzoic acid	Cyfluthrin	Urine	BetaSil* phenyl (Thermo Scientific)	LC-MS/MS	0.1–1.5	[9,10]
cis-3-(2,2-Dibromovinyl)-2,2-dimethylcyclopropane-1-carboxylic acid	Deltamethrin	Urine	BetaSil* phenyl (Thermo Scientific)	LC-MS/MS	0.1–1.5	[9,10]
Atrazine mercapturate	Atrazine	Urine	BetaSil* phenyl (Thermo Scientific)	LC-MS/MS	0.1–1.5	[9,10]
Acetochlor mercapturate	Acetochlor	Urine	BetaSil* phenyl (Thermo Scientific)	LC-MS/MS	0.1–1.5	[9,10]
Alachlor mercapturate	Alachlor	Urine	BetaSil* phenyl (Thermo Scientific)	LC-MS/MS	0.1–1.5	[9,10]

(*Continued*)

TABLE 18.1 (CONTINUED)
Pesticide Biomonitoring Using LC with MS (Alphabetical by Separation Column)

Biomarker	Parent	Matrix	Column and Source	Analysis System	LOD (ng/mL)	References
Metolachlor mercapturate	Metolachlor	Urine	BetaSil* phenyl (Thermo Scientific)	LC-MS/MS	0.1–1.5	[9,10]
2,4,5-Trichlorophenoxyacetic acid	2,4,5-T	Urine	BetaSil* phenyl (Thermo Scientific)	LC-MS/MS	0.1–1.5	[9,10]
2,4-Dichlorophenoxyacetic	2,4-D	Urine	BetaSil* phenyl (Thermo Scientific)	LC-MS/MS	0.1–1.5	[9,10]
2-(Dimethylamino)-5,6-dimethylprimidin-4-ol; 5,6-Dimethyl-2-(methylamino)pyrim-idi-4-ol	Pirimicarb	Urine	Genesis C18 (Jones)	LC-MS	1.0	[11]
Dialkylphosphates	Organophosphate pesticides	Urine	Inertsil ODS3 C18 (Varian)	LC-MS/MS	2.0	[12]
Phosphorylated butyrylcholinesterase	Parathion and other OPs	Plasma	Jupiter C_{13} silica (Phenomenex); Immunoaffinity	LC-MS	NA	[13]
Organophosphorothioate albumin adducts	Organophosphate pesticides	Plasma	PepMap C18 (Dionex)	LC-MS/MS	NA	[14]
Butyrylcholinesterase pesticide adducts	Chlorpyrifos oxon; Aldicarb; Dichlorvos	Plasma	Vydac C18 polymeric reverse-phase nanocolumn (P.J. Cobert Assoc)	LC-MS/MS	NA	[15]
3,5-Dichloroaniline	Dicarboximide fungicides	Urine	Zorbax Eclipse XDB C18 (Agilent)	LC-MS/MS	0.1	[16]
Ethylenethiouria, propylenethiourea (bisdithiocarbamate fungicide metabolites)		Urine	Zorbax SB-C3 (Agilent)	LC-MS/MS	0.004–0.01	[17]
Ethylenethiourea;	Ethylene bisdithioicar-bamates;	Urine	Zorbax SB-C3 (Agilent)	LC-MS	0.001–0.282	[18]
Propylenethiourea	Propineb	Urine	Zorbax SB-C3 (Agilent)	LC-MS	0.001–0.282	[18]

The biomarker approach for environmental toxicology and ecological risk assessment and monitoring has been adopted from medical toxicology and pharmacology [19]. The term "biomarker" is used in a broad sense dependent on the research application and biological level of analysis [e.g., 20–24]. For environmental applications, a generally accepted definition of a biomarker is that it is a substance in a biological fluid or tissue that is measurable at the biochemical, cellular, physiological, or behavioral level that reflects exposure to an environmental agent resulting in cytotoxic or other biologic effects [20,21,23–25]. For example, a biomarker could be defined as an alteration in conformation and/or function induced by a genetic mutation caused by exposure to an environmental contaminant [26].

Exposure to a diversity of environmental contaminants may result in various cellular toxic response mechanisms that are mediated by different signal transduction pathways. Several studies have demonstrated that many pollutants can exert toxic effects on living organisms first by interacting with cellular macromolecules, such as proteins and nucleic acids, leading to DNA damage, apoptosis (programmed cell death), and heritable epigenetic mutations [27–35]. The toxic effects of pollutants can be further reflected at the tissue (histopathological), organ (malfunction), and organismal (deformity, carcinogenesis) levels as well as at the different levels of organization in an ecosystem (Table 18.2) [36–60].

TABLE 18.2
Biomarkers at Various Levels of Biological Organization

Biological Levels of Analysis	Cytotoxic Effect of Typical Biomarkers	Response Sensitivity	Response Specificity	Response Accuracy	Ecological Relevance
Macromolecules	Nucleic acids damage [36], enzyme conformation and activity alterations [37,38]				
Organelles and cells	Changes in cell membrane permeability and ion-channel function [39,40], motochondrial injury [36], chromosome fragility and variation [41,42], lysosomal damage [43]				
Tissues and organs	Tissue pathology [44], organ pathology [45]				
Individuals	Anatomical and physiological changes [46], behavioral alteration [47], infertility [48,49], mortality [50]				
Population	Productivity loss and size reduction [51], stress of competition [52]				
Community	Community restructuring [53] genetic diversity [54], instability reduction [55]				
Ecosystem	Dysfunction Diversity loss				

Note: Modified from Table 1 in Bucheli, T. D., Fent, K., *Environ. Sci. Technol.* 25: 201–268, 1995.

18.2.1 Biomarker Characteristics

Biomarkers can be used to measure the sensitive alterations in the state of biological systems before cellular damage advances to become a serious adverse condition. A list of criteria for an ideal biomarker includes the following [61]:

1. Sensitivity
2. Specificity
3. Reliability (able to measure accurately)
4. Extensively field-validated
5. Easy to obtain

18.2.2 Biomarker Classification

Biomarkers can be classified on the basis of different parameters for particular applications. Three main categories of exposure, effect (sometimes referred to as response), and susceptibility are typically used to classify biomarkers [20]. This classification is not exact and may require clarification, for example, DNA adducts could be considered to be a biomarker of both exposure and effect [62].

18.2.2.1 Biomarkers of Exposure

External exposure to environmental contaminants may result in absorption, leading to an internal dose, followed by distribution, metabolism, and excretion. The U.S. National Academy of Sciences, Institute of Medicine, defines a biomarker of exposure as "a constituent or metabolite that is measured in a biological fluid or tissue that has the potential to interact with a biological macromolecule." Biomarkers of exposure can be divided into three subgroups: potential dose or external dose, internal or absorbed dose, and biologically effective dose [63]. This type of biomarker is an effective tool to assess exposure to environmental contaminants. Compared to conventional methods that measure trace amounts of xenobiotics in biological media, biomarkers are able to detect and quantify the metabolic products of xenobiotic metabolism as well as the interaction between xenobiotics and their internal targets, such as macromolecules or cells [64].

18.2.2.2 Biomarkers of Effect

Ideally, a biomarker of effect is a quantitative measurable alteration that, depending on its magnitude, can be directly linked with a pathological change [25]. These biomarkers are biological indicators of the body's response to external exposures and may indicate the presence of early subclinical problems or the impairment of normal physiological function, which may lead to an adverse health outcome.

Biomarkers of effect can track early biological effects, including mild biochemical modifications in target tissues, which may develop into nonreversible, severe functional, and/or structural damage [63,65], such as chromosome aberrations [66] and targeted gene mutations [67]. For example, micronuclei formation detected in cultured peripheral lymphocytes has been used as a biomarker to assess the genotoxic potential of exposure to high concentrations of organic solvents, such as toluene [68].

18.2.2.3 Biomarkers of Susceptibility

Biomarkers of susceptibility are indicators of an inherent or acquired ability of an organism to respond to the challenge of being exposed to a foreign substance [25]. Biomarkers of susceptibility can be inherited or induced and may indicate differences between individuals or populations affecting their response sensitivity [20,69,70]. Genetic polymorphism, preexisting conditions (e.g., obesity, diabetes, diminished organ function), genotypic characteristics, differences in metabolic rate, variations in serum immunoglobulin concentrations, and cellular regeneration rate from environmental insults are examples of biomarkers of susceptibility [69,71–75].

Genetic biomarkers provide an estimate of how genetic variations influence an individual's susceptibility to an environmental agent. Following a genomics approach, inherited genetic susceptibility has proven to be an important factor in the toxicological response manifested, leading to a new focus on identifying biomarkers of genetic variations in xenobiotic metabolism genes [70,76–78].

The metabolism of xenobiotic agents is typically a process to convert toxic molecules into more readily excretable hydrophilic compounds via the renal and biliary systems [79,80]. Polymorphisms in genes that encode xenobiotic-metabolizing enzymes have been extensively used as biomarkers for the evaluation of increased susceptibility to environmental toxins [71,81].

18.2.3 "-Omic" Approaches for Biomarker Discovery

Several "-omic" approaches are being employed to determine biomarkers for both clinical and environmental exposure applications. HPLC-MS-based proteomics is a rapidly developing technique suitable for both qualitative and quantitative assessment of proteins, particularly for monitoring protein profiles expressed by cell cultures that have been exposed to xenobiotics when stable isotope labeling is used. Figure 18.1 provides the overall approach for determining biomarkers using cell cultures and HPLC-MS/MS. Stable isotope labeling with amino acids in cell culture (SILAC) is a simple and accurate approach that depends on the incorporation of amino acids containing substituted stable isotope nuclei (e.g., ^{2}H, ^{13}C, ^{15}N) into proteins in living cells [82]. In addition to genomic and proteomic platforms, other "-omic" techniques, such as metabolomics, lipidomics, glycomics, and secretomics, are also being used, and many require HPLC-MS capability (Table 18.3).

18.3 A DESCRIPTION OF LC TECHNIQUES USED FOR BIOMARKER STUDIES

In the mid 1970s, Csaba Horváth first introduced the acronym "HPLC" standing for high-pressure liquid chromatography in his Pittcon presentation. The continued development of chromatographic performance has allowed the use of very small particles, small column diameters, and very high fluid pressures to achieve enhanced separations in shorter periods of time. Thus, the acronym HPLC has become known as high-performance liquid chromatography, rather than "high pressure."

Table 18.4 lists several commercially available detectors for use in HPLC for environmental and biological monitoring. Each detector has its own merits and distinctions for application to different target chemicals. A data acquisition system collects the data information from the HPLC and presents it as a chromatogram, providing qualitative data (retention time) and quantitative data (area under curve).

Traditional HPLC detectors include ultraviolet-visible, fluorescence, refractive index, electrochemical, and MS, as shown in Table 18.4. MS detection provides the most sensitivity and specificity. MS data provides valuable information about the molecular weight, structure, identity, and quantity of the target analyte.

18.3.1 Components of MS

A MS consists of four main components: inlet system, ion source, mass analyzer, and detector. An ion source is an electromagnetic device to generate charged ions. These ions are then transferred by electromagnetic fields to a mass analyzer [109]. Improvements in ion source techniques, such as atmospheric pressure ionization (API), have greatly expanded the number of compounds that can be successfully analyzed by HPLC-MS [110]. Sample molecules are ionized under atmospheric pressure, and the ions are then mechanically and electrostatically separated from neutral molecules. A mass analyzer sorts and identifies ions by their mass/charge (m/z) ratios [102]. Four basic types of mass analyzers are most often used in HPLC-MS: quadrupole,

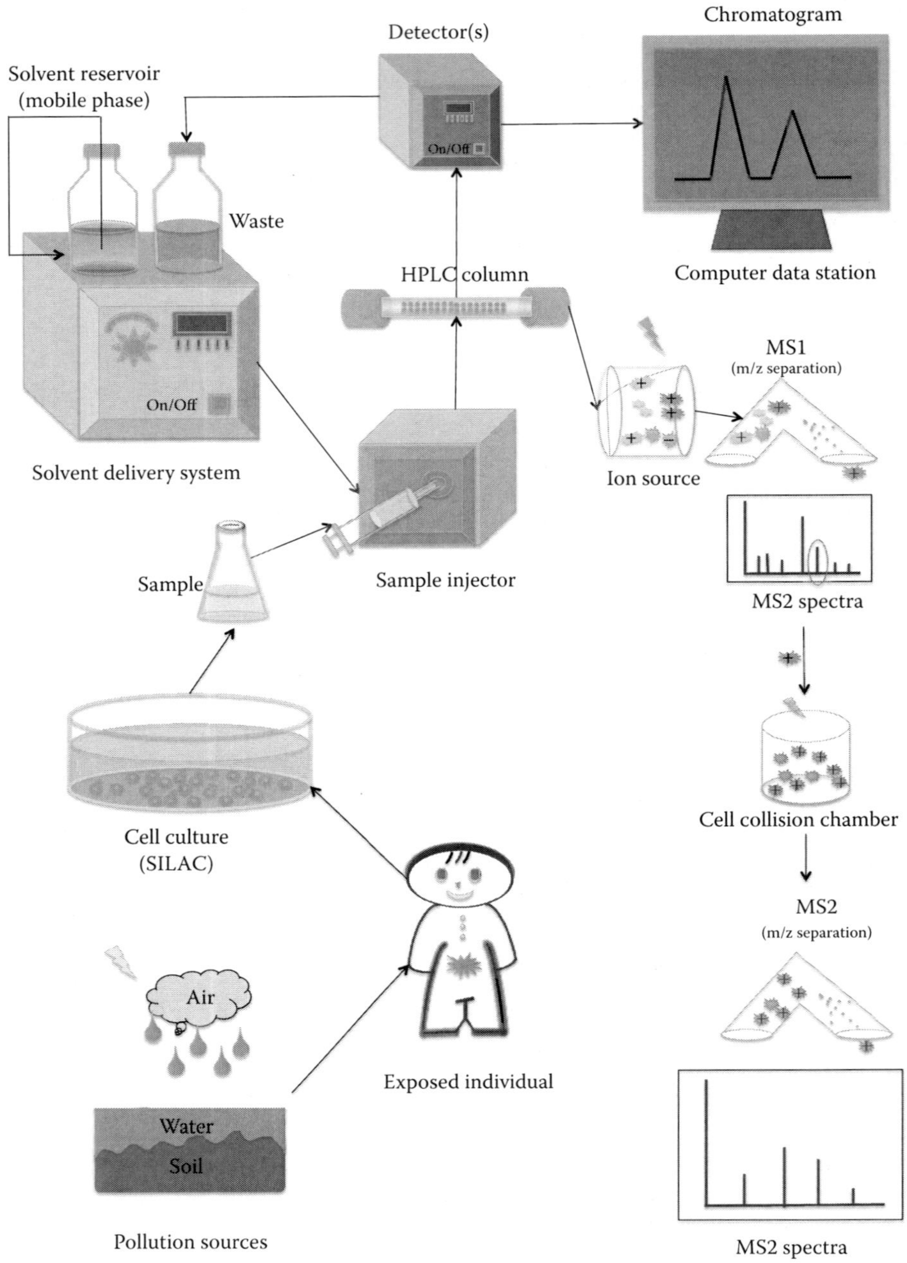

FIGURE 18.1 A biomarker approach for linking exposures to adverse outcome pathways using SILAC, human cells (primary cells from tissue specimens or cell lines from commercial sources), and HPLC-MS/MS.

ion trap, time-of-flight (TOF), and Fourier transform-ion cyclotron resonance [110]. An MS detector can generate both two-dimensional abundance data and three-dimensional MS data to determine the molecular weight and structure [110]. A full description of all types of ion sources, detectors, and their underlying mechanisms is outside the scope of this chapter. We present a brief overview of MS techniques used in tandem with HPLC for developing biomonitoring methods.

TABLE 18.3
High Throughput "-Omic" Approaches for Biomarkers Discovery

Various Approaches	Biomarker Discovery Techniques	References
Genomic approach	Northern blot SAGE library DNA microarray	[83–86]
Proteomic approach	2D-PAGE LC-MS SELDI-TOF (or MALDI-TOF) Antibody array Tissue microarray Immunoassays (e.g., iTRAQ, RIA, ELISA, and USERA) Synchronous fluorescence spectroscopy ^{32}P-postlabeling assay	[87–93]
Metabolomics approach	Serotonin production pathway activated in an alcoholic person	[93]
Lipidomics approach	Mass spectrometry, chromatography, nuclear magnetic resonance	[94–96]
Imaging biomarkers	Cardiac imaging (e.g., coronary angiography, magnetic resonance imaging, optical coherence tomography, near infrared spectroscopy, positron emission tomography) Molecular imaging (e.g., magnetic resonance imaging)	[97]

18.3.2 Basic Principles of MS Detection

A MS weighs molecules electronically by ionizing molecules and then sorting and identifying the ions according to their m/z ratios [111]. The sample is introduced into the MS and vaporized, forming ions. The ions of charge z (where z is the elementary charges carried by the ions) are produced by an ion source within the instrument. All of the ions are then accelerated by an electric field so that they have similar kinetic energy. The ions with different m/z ratios are separated in various ways, such as by a quadruple mass filter. The ions at each m/z value are detected in proportion to their abundance producing a mass spectrum (a plot of ion abundance versus m/z ratios) [112]. The ions generated from the sample provide information concerning the nature and the structure of their precursor molecule [113]. The resulting mass spectrum provides information to determine the molecular weight and structure of the sample compounds and the relative abundance of a specific compound in the sample matrix [114].

Biomarker analyses are frequently performed by HPLC in tandem with two mass analyzers (HPLC-MS/MS). The first mass analyzer is a nondestructive mass analyzer, such as a quadrupole or an ion trap, that initially separates the sample components into parent ions and sequentially releases parent ions of known m/z ratios. In a collision cell, the parent ions interact and collide with molecules of an inert gas (helium, neon, argon, or nitrogen) to break into pieces of daughter ions [115]. These daughter ions then move into a second mass detector, such as a TOF analyzer, providing a full spectrum of mass data for the daughter ions to complement the parent ion MS data. MS/MS can detect trace amounts of the target in complex samples with exceptional sensitivity.

HPLC-MS/MS has been used as a very powerful analytical technique for separating and quantifying trace quantities of biomarkers from complex mixtures. For the quantification of very complex samples, MS/MS has the advantage over conventional MS to select and direct particular ions of interest to the second mass detector for further investigation. Because the matrix-induced background signal is greatly reduced, sample cleanup prior to chromatographic separation becomes much less critical.

TABLE 18.4
Different Types of Detectors Available for HPLC

Types of Detectors	LOD	Principle of Operation	References
Absorbance detector (ultraviolet-visible-photodiode array)	~pg level	Measures the ability of a sample to absorb light at one or several wavelengths; PDA measures a spectrum of wavelength simultaneously, but limited with Hg lamp.	[98]
Fluorescence detector	~fg level	Measures the ability of a compound to absorb and then emit fluorescent signal. The fluorescence intensity is monitored to quantify the compounds concentration. Each compound has a characteristic fluorescence.	[99]
Refractive-index detector	~ng level	Measures the change in refractive index or the ability of sample molecules to refract light. Light proceeds through a bi-modular flow-cell to a photodetector. Detection occurs when the light is bent due to samples eluting from the column, and this is read as disparity between the two channels of the flow-cell.	[100]
Evaporative light scatting detector	~ng level	Measures the light scattered by analyte in the eluent, which is nebulized and then evaporated to form fine particles.	[101]
Mass spectrometer	~pg level	Measures the m/z ratios of ionized samples to elucidate their chemical structures.	[102]
Nuclear magnetic resonance detectors	~ng level	Measures the nuclear magnetic resonance signals of samples irradiated in an external magnetic field to determine their structure.	[103]
Near-infrared detectors	~μg level	Measures the stretching and bending vibrations of particular chemical bonds of samples at a certain wavelength.	[104]
Electrochemical detectors	~fg level	Measures gain or loss of electrons from migrating samples as they pass between electrodes at a given electrical potential.	[105]
Radiochemical detectors	~ng level	Measures the fluorescence associated with beta-particle ionization, involving use of radiolabeled material.	[106]
Conductivity detectors	~μg level	Measures the change of electric current of samples as they pass between electrodes imposed with a constant voltage. No salts or buffers in mobile phase.	[107]
Element specific detector (e.g., ICP detector)	~pg level	Measures all elements of a separated sample simultaneously and calculate the total molecular formula of sample. The mobile phase must be aqueous and must not contain any element of interest.	[108]

18.3.3 Isotope Dilution MS Coupled with HPLC

Stable isotope dilution mass spectrometry (SIDMS) is a quantitative technique providing high accuracy and precision. It is able to determine ultra-trace concentrations (<pg/g) of compounds in biological and environmental matrices [116]. The term "SID" refers to the use of a known concentration of a stable isotope-labeled internal standard spiked into the sample. The stable isotope-labeled analog is identical to the endogenous target analyte with the exception of having a heavier mass [117]. The isotope-labeled internal standard has identical physicochemical properties as the target analyte and can act as a surrogate during the extraction and analysis steps, preventing loss of the target analyte [117,118]. Thus, the internal standard can verify the presence of the compound and normalize experimental variables, such as matrix effects and system instabilities.

SIDMS in combination with HPLC (LC-SIDMS) can provide a very high degree of specificity for quantification. LC-MS/MS is applicable to the analysis of a wide range of biomarkers and is now widely used in the discovery and validation of exposure biomarkers [117].

Lindh et al. have studied the plant growth regulator chlormequat in human urine as a biomarker of exposure by HPLC-MS using an Atlantis hydrophilic interaction chromatography (HILIC) column (3 μm particle size, 150 mm × 2.1 mm) and 0.05 M acetic acid/ammonium acetate buffer (pH 3.75) in water and acetonitrile as the mobile phase. The experiment was carried out using the selected reaction monitoring in the positive ion mode. [$^2H_4$] labeled chlormequat was used as the internal standard for quantification. The limit of detection (LOD) was determined to be 0.1 ng/ml with a reproducibility of 3%–6% [8].

Norrgran et al. have proposed a method for quantification of six herbicide metabolites (atrazine mercapturate, acetochlor mercapturate, metolachlor mercapturate, 2,4,5-trichlorophenoxyacetic acid, 2,4-dichlorophenoxyacetic acid, alachlor mercapturate) by HPLC-MS using a Betasil Hexylphenyl column (3 μm particle size, 4.6 mm × 100 mm) and automated liquid delivery of internal standards and acetate buffer. Isotope dilution calibration was used for quantification of all analytes. The mobile phase was acetic acid in water or acetonitrile. The limit of detection ranged from 0.036 to 0.075 ng/ml [119].

Lindh et al. developed a method to quantify 3,5-dichloroaniline (3,5-DCA) as a biomarker of the fungicides vinclozolin and iprodione in human urine by LC-MS/MS using a Rapid Resolution Zorbax Eclipse XDB C18 column (1.8 μm particle size, 2.1 mm × 50 mm). The mobile phase consisted of water and methanol with 0.5% acetic acid. The urine samples were treated by basic hydrolysis to degrade the fungicides, their metabolites, and conjugates to 3,5-DCA. The 3,5-DCA was then extracted using toluene and derivatized using pentafluoropropionic anhydride. Analysis of the derivative was carried out using selected reaction monitoring in the negative ion mode. Quantification of the derivative was performed using [$^{13}C_6$]-labeled 3,4-DCA as an internal standard with good precision and linearity in the range of 0.1–200 ng/ml urine. The limit of detection was determined to be 0.1 ng/ml [16].

18.3.4 LC/Electrospray Ionization-Tandem MS

Electrospray ionization (ESI) is a common API technique [110]. ESI uses electrical energy to transfer sample compounds into gaseous phase ions for analysis. Column effluent is sprayed into a chamber at atmospheric pressure in the presence of a strong electrostatic field and a heated drying gas. The electrostatic field further breaks down the compounds to form droplets. The heated drying gas causes the solvent in the droplets to evaporate. As the droplets shrink, the charge concentration in the droplet increases and ions are ejected into the gas phase and exported into the mass analyzer [110].

HPLC coupled with ESI-tandem MS (HPLC/ESI-MS/MS) is a very powerful technique capable of analyzing both small and large molecules of various polarities in a complex sample. With the additional separation capacities of MS/MS, sample purification prior to HPLC may not be necessary [120].

Sams et al. developed a method to quantify two major metabolites [2-(dimethylamino)-5,6-dimethylpyrimidin-4-ol (DDHP) and 5,6-dimethyl-2-(methylamino)pyrimidin-4-ol (MDHP)] of the carbamate insecticide pirimicarb in human urine by HPLC-MS. HPLC-MS was carried out on an Agilent 1100 chromatograph interfaced to an ion trap MS. A Genesis C18 column (3 μm particle size, 250 mm × 2.1 mm) was used for chromatographic separation of the analytes. The MS was operated using positive ESI. Nitrogen was used as the nebulizer gas at 15 psi and as the drying gas at 6l/min, 350°C. Metabolites were detected at the following m/z ratios: MDHP (154), DDHP (168), and internal standard (168) [121].

Li et al. developed a method to detect pesticide adducts in tryptic digests of butyrylcholinesterase (BChE) as biomarkers of pesticide exposure in human plasma from patients poisoned by pesticides. BChE was purified from 2 ml serum by ion exchange chromatography at pH 4, followed by procainamide affinity chromatography at pH 7. A 5-ml aliquot of HPLC-purified, tryptic BChE peptides was injected onto a Vydac C18 polymeric reverse-phase nano-column for HPLC separation. Peptides were separated with a 90-min linear gradient of acetonitrile (0%–60%) and formic acid (0.1%) and then electrosprayed through a fused silica emitter directly into a hybrid quadrupole linear ion trap mass spectrometer (QTRAP 4000) [122].

18.3.5 Ultra-HPLC with MS

The evolution of chromatographic methods has been, in part, due to the reduction in the particle size of the column packing materials. However, a decrease in the particle diameter leads to an increase in the column backpressure. A column backpressure exceeding 10,000 psi (~700 bar) is referred to as ultra-HPLC (UHPLC) [123]. UHPLC has a higher separation and throughput for the rapid discovery and monitoring of biomarkers via the utilization of sub-2-micron column particle sizes at high linear velocities of 3.5–6 mm/s [124]. Columns packed with sub-2-micron particles are generally divided into two categories: (1) short columns (less than 5 cm); and (2) capillary columns (inner diameter less than 100 μm). Recent studies have shown that the efficiency of 1.0 mm i.d. columns (15 cm long) packed with 1.5 μm packing materials is approximately twice as high as 150 mm analytical columns packed with 3 μm materials [123].

Most UHPLC biomarker analyses are performed using reversed phase columns, such as the Acquity High Strength Silica T3 and the Acquity C18 with a bridged ethylsiloxane-silica hybrid adsorbent [125–129]. HILIC is sometimes used, but it is applicable only for the separation of polar metabolites [125]. MS is normally the detector for UHPLC separation using positive or negative electrospray [130]. Heat-assisted ESI is also used when it is necessary to eliminate potential interferences and improve sensitivity [131]. Triple quadrupole, quadrupole-TOF, and orbitrap instrumentation have all been used as mass analyzers [125,132–135].

Chen et al. described a method to determine nine environmental phenols (bisphenol A; 2,3,4-trichlorophenol; 2,4,5-trichlorophenol; pentachlorophenol; triclosan; 4-tert-octylphenol; 4-n-octylphenol; 4-n-nolyphenol; and benzophenone-3) in human urine by UHPLC-ESI-MS/MS. A [$^{13}C_6$]-labeled internal standard was added to the samples before the analytes were extracted and preconcentrated with solid-phase extraction. The chromatographic separation was carried out on an Acquity UPLC BEH, C18 column (1.7 μm particle size, 2.1 mm × 100 mm) maintained at 35°C equipped with a filter (Frit, 0.2 μm, 2.1 mm) and a Van Guard BEH, C18 precolumn (1.7 μm). The mobile phases were methanol and water. The ions were detected by a Waters Quattro Premier MS using an ESI probe in the negative ion mode and with a multiple reaction monitoring mode. The flow rate was 0.25 mL/min (11 min run time). The LOD for all nine compounds ranged from 0.02 to 0.90 ng/mL [136].

Alwis et al. have developed a sensitive and high throughput method to simultaneously measure 28 metabolites as biomarkers of exposure to volatile organic compounds in human urine using reverse-phase UHPLC coupled with ESI-MS/MS. The chromatographic separation was performed using an Acquity UPLC HSS T3 column (1.8 μm particle size, 2.1 mm × 150 mm). The mobile phase was 15 mM ammonium acetate pH 6.8 and acetonitrile. The eluent from the column was ionized using an electrospray interface (−4000 v), and the MS was operated in the multiple reaction monitoring mode for negative ions. The ion source temperature was 650°C. The LOD for all 28 metabolites ranged from 0.5 to 20 ng/mL [137].

Hsiao et al. have described a sensitive and high throughput method to detect the conjugate 2,5-dichlorophenol glucuronide as a biomarker of exposure to 1,4-dichlorobenzene in human urine using solid-phase extraction for sample preparation and UHPLC-MS/MS with negative ESI for detection. The mobile phase was ammonium acetate buffer and methanol [138].

18.3.6 Nano-HPLC

Nano-LC was first introduced by Karlsson and Novotny in 1988 as a complementary separation method to conventional HPLC [139]. Nano-HPLC has several advantages, such as higher efficiency, ability to work with minute sample sizes, lower consumption of mobile phases (reducing the use of organic solvents), and better compatibility with MS [139,140]. Nano-HPLC is highly compatible with MS due to the relatively low flow rate (40–600 nL/min) that allows the transfer of the entire effluent from the column [140,141].

The chromatographic separation is performed in capillary columns of i.d.s in the range between 10 and 100 μm. Capillary columns are typically fused silica or polyetheretherketone (PEEK™). Several HPLC stationary phases are applicable to both silica and PEEK™. The development of smaller particle sizes (3–5 μm d_p) with uniform pore sizes improves efficiency, resolution, and selectivity, all with a shorter analysis time. A drawback to this technique is the increase in backpressure. More recently, particles of 1.5–1.8 μm d_p have been successfully employed in ultra-performance LC [140].

MS can be easily coupled with nano-LC instrumentation through the different nanospray interfaces available. Normally, the capillary column used for the chromatographic separation is connected to the emitter tip through a zero dead volume union attached to a power supply (voltages between 1000 and 2500 V) [140–143].

Nano-LC has been applied to the separation of a wide number of compounds in different areas, such as proteomic and pharmaceutical research [140]. Nano-LC has not been widely used for the analysis of compounds of environmental interest up to now although HPLC is one of the major techniques for the analysis of pollutants and their metabolites. Very few studies report the application of nano-LC in environmental analysis [140]. One study reported by Cappiello et al. described the use of a new nano-LC gradient generator coupled to a modified direct electron ionization LC-MS interface for the analysis of pesticides, nitropolynuclear aromatic hydrocarbons, and hormones [144]. Rosales-Conrado et al. proposed the enantiomeric separation of phenoxy acid herbicides mecoprop, dichlorprop, and fenoprop in their acid form by nano-LC using a 75 μm i.d. capillary column packed with vancomycin-modified silica particles of 5 μm. This separation capability is important as the (R) isomers of the phenoxy acid herbicides show much higher herbicide activity and a different metabolism than the (S) isomers [145]. Aryal et al. described an approach for the detection and quantification of phosphorylated BChE activity as an exposure biomarker of organophosphates and nerve agents by coupling magnetic bead-based immune-affinity purification with LC-MS/MS. They purified BChE protein by biotinylated anti-BChE polyclonal antibodies conjugated to streptavidin magnetic beads. The peptide samples were analyzed using an automated nano-flow, metal-free nano-LC system with an i.d. capillary column (40 cm × 50 μm) packed with 3 μm Jupiter C18 silica. The heated capillary was maintained at 200°C and the ESI voltage was held at 2.2 kV [146].

18.4 CONCLUSIONS

Exposure to environmental contaminants may result in human health effects and may also negatively impact the balance within an ecosystem. Environmental monitoring is needed to provide environmental fate and transport information to determine the deposition of contaminants and identify potential exposure sources and routes. Biomonitoring is critical to detect exposures and early biological changes caused by environmental contaminants that may lead to an adverse outcome. Both monitoring approaches require reliable analytical methods to provide data of known quality.

Biomarkers are effective tools for biomonitoring studies to assess exposure to environmental contaminants. Biomarkers are vital to understanding the relationships between exposure and adverse outcomes in humans and the environment. LC, particularly in combination with MS, has been extensively utilized for the discovery and identification of new biomarkers to determine metabolites, biotransformation enzymes, biotransformation products, stress proteins, and other markers. The continued advancements in LC methods will undoubtedly keep pace with the increasing variety and chemical composition of environmental contaminants and enable the discovery and identification of new biomarkers to safeguard human health and the environment.

Sensitive and high throughput analytical methods are essential for the detection of various biomarkers of exposure to support biomonitoring studies, such as the National Biomonitoring Program conducted by the Centers for Disease Control and Prevention [147] and U.S. EPA environmental monitoring studies. The combination of data from both types of monitoring studies provides insight to the complete exposure scenario from source to outcome.

REFERENCES

1. NLM (Content Source); Monosson, E. (Topic Editor) "Toxicity". In: *Encyclopedia of Earth.* Cleveland, C. J. (eds.) (Washington, D.C.: Environmental Information Coalition, National Council for Science and the Environment). *Encyclopedia of Earth*, 2010.
2. Stephenson, J., Testimony: Before the Subcommittee on Superfund, Toxics and Environmental Health, Committee on Environment and Public Works, U.S. Senate: United States Government Accountability Office. Biomonitoring: EPA Could Make Better Use of Biomonitoring Data. 2010.
3. Sobus, J. R., Tan, Y. M., Pleil, J. D., Sheldon, L. S., A biomonitoring framework to support exposure and risk assessments. *Sci. Total Environ.* 409 (22), 4875–4884, 2011.
4. MacNair, J. E., Lewis, K. C., Jorgenson, J. W., Ultrahigh-pressure reversed-phase liquid chromatography in packed capillary columns. *Anal. Chem.* 69, 983–989, 1997.
5. Mazzeo, J. R., Neue, U. D., Kele, M., Plumb, R. S., Advancing LC performance with smaller particles and higher pressure. *Anal. Chem.* 77, 460A–467A, 2005.
6. Wu, N., Lippert, J. A., Lee, M. L., Practical aspects of ultrahigh pressure capillary liquid chromatography. *J. Chromatogr. A* 911, 1–12, 2001.
7. Calafat, A., Background paper on BPA biomonitoring and biomarker studies. World Health Organization, 2011.
8. Lindh, C. H., Littorin, M., Johannesson, G., Jönsson, B. A., Analysis of chlormequat in human urine as a biomarker of exposure using liquid chromatography triple quadrupole mass spectrometry. *J. Chromatogr. B* 879 (19), 1551–1556, 2011.
9. Panuwet, P., Prapamontol, T., Chantara, S., Barr, D. B., Urinary pesticide metabolites in school students for northern Thailand. *J. Hyg. Environ. Health* 212, 288–297, 2009.
10. Olsson, A. O., Baker, S. E., Nguyen, J. V., Romanoff, L. C., Udunka, S. O., Walker, R. D., Flemmen, K. L., Barr, D. B., A liquid chromatography–tandem mass spectrometry multiresidue method for quantification of specific metabolites of organophosphorus pesticides, synthetic pyrethroids, selected herbicides, and DEET in human urine. *Anal. Chem.* 76, 2453–2461, 2004.
11. Sams, C., Patel, K., Jones, K., Biological monitoring for exposure to pirimicarb: Method development and a human oral dosing study. *Toxicol. Lett.* 192: 56–60, 2010.
12. Dulaurent, S., Saint-Marcoux, F., Marquet, P., Lachatre, G., Simultaneous determination of six dialkylphosphates in urine by liquid chromatography tandem mass spectrometry. *J. Chromatogr. B* 831, 223–229, 2006.
13. Aryal, U. K., Lin, C., Kim, J., Heibeck, T. H., Wang, J., Qian, W., Lin, Y., Identification of phosphorylated butyrylcholinesterase in human plasma using immunoaffinity purification and mass spectrometry. *Anal. Chim. Acta.* 723, 68–75, 2012.
14. Noort, D., Holst, A. G., van Zuylen, A., van Rijssel, E., van der Schans, M. J., Covalent binding of organophosphorothioates to albumin: A new perspective for OP-pesticide biomonitoring? *Springer-Verlag,* 2009.
15. Li, B., Ricordel, I., Schopfer, L. M., Baud, F., Mégarbane, B., Masson, P., Lockridge, O., Dichlorvos, chlorpyrifos oxon and Aldicarb adducts of butyrylcholinesterase, detected by mass spectrometry in human plasma following deliberate overdose. *J. Appl. Toxicol.* 30 (6), 559–565, 2010.
16. Lindh, C. H., Littorin, M., Amilon, A., Jönsson, B. A., Analysis of 3,5-dichloroaniline as a biomarker of vinclozolin and iprodione in human urine using liquid chromatography/triple quadrupole mass spectrometry. *Rapid. Commun. Mass Spectrom.* 21 (4), 536–542, 2007.
17. Jayatilaka, N. K., Montesano, M. A., Whitehead, R. D., Schloth, S. J., Needham, L. L., Barr, D. B., High-throughput sample preparation for the quantitation of acephate, methamidophos, omethoate, dimethoate, ethylenethiourea, and propylenethiourea in human urine using 96-well-plate automated extraction and high-performance liquid chromatography–tandem mass spectrometry. *Arch. Environ. Contam. Toxicol.* 61 (1), 59–67, 2011.
18. Montesano, A., Olsson, A. O., Kuklenyik, P., Needham, L. L., Bradman, A., Barr, D. B., Method for determination of acephate, methamidophos, omethoate, dimethoate, ethylenethiourea and propylenethiourea in human urine using high-performance liquid chromatography–atmospheric pressure chemical ionization tandem mass spectrometry. *J. Expo. Sci. Environ. Epidemiol.* 17, 321–330, 2007.
19. McCarty, L. S., Power, M., Munkittrick, K. R., Bioindicators versus biomarkers in ecological risk assessment. *Hum. Ecol. Risk Assess.* 8 (1), 159–164, 2002.
20. Committee on Biological Markers of the National Research Council. Biological markers in environmental health research. *Environ. Health Perspect.* 74: 3–9, 1987.
21. Biomarkers Definitions Working Group. Biomarkers and surrogate endpoints: Preferred definitions and conceptual framework. *Clin. Pharmacol. Ther.* 69 (3), 89–95, 2001.

22. Benson, W. H., DiGiulio, R. T., Biomarkers in hazard assessment of contaminated sediments. In: Burton, G. A. ed., *Sediment Toxicity Assessment.* Lewis Publishers, Boca Raton, FL, 1992.
23. Depledge, M. H., Amaral-Mendes, J. J., Daniel, B., Halbrook, R. S., Loepper-Sams, P., Moore, M. N., Peakall, D. B., The conceptual basis of the biomarker approach. In: Peakall, D. B., Shugart, L. R. (eds.), *Biomarkers: Research and Application in the Assessment of Environmental Health*, NATO ASI Series H: Cell Biology. Springer Verlag, Berlin, 68: 15–29, 1993.
24. Walker, C. H., Sibly, R. M., Hopkin, S. P., Peakall, D. B., *Principles of Ecotoxicology.* Taylor & Francis Ltd., London, 1996.
25. World Health Organization. Biomarkers and Risk Assessment: Concepts and Principles Environmental Health Criteria, No. 155. Geneva, Switzerland, 1993.
26. Bucheli, T. D., Fent, K., Induction of cytochrome P450 as a biomarker for environmental contamination in aquatic ecosystems. *Environ. Sci. Technol.* 25: 201–268, 1995.
27. Pasco, N., Hay, J., Webber, J., Biosensors: MICREDOX—A new biosensor technique for rapid measurement of BOD and toxicity. *Biomarkers* 6, 83–89, 2000.
28. Bhabra, G., Sood, A., Fisher, B., Cartwright, L., Saunders, M., Evans, W. H., Surprenant, A., Lopez-Castejon, G., Mann, S., Davis, S. A., Hails, L. A., Ingham, E., Verkade, P., Lane, J., Heesom, K., Newson, R., Case, C. P., Nanoparticles can cause DNA damage across a cellular barrier. *Nat. Nanotechnol.* 12: 876–883, 2009.
29. Yen, C. C., Ho, T. J., Wu, C. C., Change, C. F., Su, C. C., Chen, Y. W., Jinn, T. R., Lu, T. H., Cheng, P. W., Su, Y. C., Liu, S. H., Huang, C. F., Inorganic arsenic causes cell apoptosis in mouse cerebrum through an oxidative stress-regulated signaling pathway. *Arch. Toxicol.* 85 (6), 565–575, 2011.
30. Song, C., Gao, H. W., Wu, L. L., Transmembrane transport of microcystin to Danio rerio zygotes: Insights into the developmental toxicity of environmental contaminants. *Toxicol. Sci.* 122 (2), 395–405, 2011.
31. Cao, J., Chen, J., Wang, J., Jia, R., Xue, W., Luo, Y., Gan, X., Effects of fluoride on liver apoptosis and Bcl-2, Bax protein expression in freshwater teleost, Cyprinus carpio. *Chemosphere* 91 (8), 1203–1212, 2013.
32. Chen, K. M., Guttenplan, J. B., Zhang, S. M., Aliaga, C., Cooper, T. K., Sun, Y. W., Deltondo, J., Kosinska, W., Sharma, A. K., Jiang, K., Bruggeman, R., Ahn, K., Amin, S., El-Bayoumy, K., Mechanisms of oral carcinogenesis induced by dibenzo[a,l]pyrene: An environmental pollutant and a tobacco smoke constituent. *Int. J. Cancer* 133 (6), 1300–1309, 2013.
33. Komissarova, E. V., Rossman, T. G., Arsenite induced poly(ADP-ribosyl)ation of tumor suppressor P53 in human skin keratinocytes as a possible mechanism for carcinogenesis associated with arsenic exposure. *Toxicol. Appl. Pharmacol.* 243 (3), 399–404, 2010.
34. Xu, Z., Zhang, Y. L., Song, C., Wu, L. L., Gao, H. W., Interactions of hydroxyapatite with proteins and its toxicological effect to zebrafish embryos development. *PLoS One* 7(4): e32818, 2012.
35. Collotta, M., Bertazzi, P. A., Bollati, V., Epigenetics and pesticides. *Toxicology* 307: 35–41, 2013.
36. Fujii, Y., Tomita, K., Sano, H., Yamasaki, A., Hitsuda, Y., Adcock, I. M., Shimizu, E., Dissociation of DNA damage and mitochondrial injury caused by hydrogen peroxide in SV-40 transformed lung epithelial cells. *Cancer Cell. Int.* 2 (1), 16, 2002.
37. Chen, Q. X., Zhang, R. Q., Xue, X. Z., Yang, P. Z., Hen, S. L., Zhou, H. M., Effect of methanol on the activity and conformation of acid phosphatase from the prawn Penaeus penicillatus. *Biochemistry (Mosc.).* 65 (4), 452–456, 2000.
38. Rajalakshmi, S., Mohandas, A., Copper-induced changes in tissue enzyme activity in a freshwater mussel. *Ecotoxicol. Environ. Saf.* 62 (1), 140–143, 2005.
39. Atchison, W. D. Effects of toxic environmental contaminants on voltage-gated calcium channel function: From past to present. *J. Bioenerg. Biomembr.* 35 (6), 507–532, 2003.
40. Hu, W. Y., Jones, P. D., DeCoen, W., King, L., Fraker, P., Newsted, J., Giesy, J. P., Alterations in cell membrane properties caused by perfluorinated compounds. *Comp. Biochem. Physiol. C. Toxicol. Pharmacol.* 135 (1), 77–88, 2003.
41. Genualdo, V., Perucatti, A., Iannuzzi, A., Di Meo, G. P., Spagnuolo, S. M., Caputi-Jambrenghi, A., Coletta, A., Vonghia, G., Iannuzzi, L., Chromosome fragility in river buffalo cows exposed to dioxins. *J. Appl. Genet.* 53 (2), 221–226, 2012.
42. Pierce, B. L., Kibriya, M. G., Tong, L., Jasmine, F., Argos, M., Roy, S., Paul-Brutus, R., Rahaman, R., Rakibuz-Zaman, M., Parvez, F., Ahmed, A., Quasem, I., Hore, S. K., Alam, S., Islam, T., Slavkovich, V., Gamble, M. V., Yunus, M., Rahman, M., Baron, J. A., Graziano J. H., Ahsan, H., Genome-wide association study identifies chromosome 10q24.32 variants associated with arsenic metabolism and toxicity phenotypes in Bangladesh. *PLoS Genet.* 8(2): e1002522, 2012.
43. Lowe, D. M., Pipe, R. K., Contaminant induced lysosomal membrane damage in marine mussel digestive cells: An in vitro study. *Aquat. Toxicol.* 30 (4), 357–365, 1994.

44. Satpute, R. M., Lomash, V., Hariharakrishnan, J., Rao, P., Singh, P., Gujar, N. L., Bhattacharya, R., Oxidative stress and tissue pathology caused by subacute exposure to ammonium acetate in rats and their response to treatments with alpha-ketoglutarate and N-acetyl cysteine. *Toxicol. Ind. Health*. [Epub ahead of print], 2012.
45. Calderón-Garcidueñas, L., Reed, W., Maronpot, R. R., Henríquez-Roldán, C., Delgado-Chavez, R., Calderón-Garcidueñas, A., Dragustinovis, I., Franco-Lira, M., Aragón-Flores, M., Solt, A. C., Altenburg, M., Torres-Jardón, R., Swenberg, J. A., Brain inflammation and Alzheimer's-like pathology in individuals exposed to severe air pollution. *Toxicol. Pathol.* 32 (6), 650–658, 2004.
46. Maruthi Sridhar, B. B., Diehl, S. V., Han, F. X., Monts, D. L., Su, Y., Anatomical changes due to uptake and accumulation of Zn and Cd in Indian mustard (*Brassica juncea*). *Environ. Exp. Bot.* 54 (2), 131–141, 2005.
47. Schultz, M. M., Bartell, S. E., Schoenfuss, H. L., Effects of triclosan and triclocarban, two ubiquitous environmental contaminants, on anatomy, physiology, and behavior of the fathead minnow (Pimephales promelas). *Arch. Environ. Contam. Toxicol.* 63 (1), 114–124, 2012.
48. Lewis, C., Ford, A. T., Infertility in male aquatic invertebrates: A review. *Aquat. Toxicol.* 120–121, 79–89, 2012.
49. Latini, G., Del Vecchio, A., Massaro, M., Verrotti, A., De Felice, C., Phthalate exposure and male infertility. *Toxicology* 226 (2–3), 90–98, 2006.
50. Norwood, W. P., Borgmann, U., Dixon, D. G., An effects addition model based on bioaccumulation of metals from exposure to mixtures of metals can predict chronic mortality in the aquatic invertebrate hyalella azteca. *Environ. Toxicol. Chem.* 32 (7), 1672–1681, 2013.
51. Clark, K. E., Niles, L. J., Stansley, W., Environmental contaminants associated with reproductive failure in bald eagle (Haliaeetus leucocephalus) eggs in New Jersey. *Bull. Environ. Contam. Toxicol.* 61 (2), 247–254, 1998.
52. Boone, M. D., Semlitsch, R. D., Interactions of an insecticide with competition and pond drying in amphibian communities. *Ecol. Appl.* 12 (1), 307–316, 2002.
53. Masson, S., Desrosiers, M., Pinel-Alloul, B., Martel L., Relating macroinvertebrate community structure to environmental characteristics and sediment contamination at the scale of the St. Lawrence River. *Hydrobiologia* 647, 35–50, 2010.
54. Bickham, J. W., Sandhu, S., Hebert. P. D., Chikhi, L., Athwal, R., Effects of chemical contaminants on genetic diversity in natural populations: Implications for biomonitoring and ecotoxicology. *Mutat. Res.* 463 (1), 33–51, 2000.
55. Nunes, A. C., Auffray, J. C., Mathias, M. L., Developmental instability in a riparian population of the Algerian mouse (Mus spretus) associated with a heavy metal-polluted area in central Portugal. *Arch. Environ. Contam. Toxicol.* 41 (4), 515–521, 2001.
56. Ghio, A. J., Carraway, M. S., Madden, M. C., Composition of air pollution particles and oxidative stress in cells, tissues, and living systems. *J. Toxicol. Environ. Health B Crit. Rev.* 15 (1), 1–21, 2012.
57. Kupryianchyk, D., Rakowska, M. I., Roessink, I., Reichman, E. P., Grotenhuis, T. J., Koelmans, A. A., In situ treatment with activated carbon reduces bioaccumulation in aquatic food chains. *Environ. Sci. Technol.* 47 (9), 4563–4571, 2013.
58. Liu, J., Zhao, M., Zhuang, S., Yang, Y., Yang, Y., Liu, W., Low concentrations of o,p'-DDT inhibit gene expression and prostaglandin synthesis by estrogen receptor-independent mechanism in rat ovarian cells. *PLoS One* 7 (11), e49916, 2012.
59. Monteiro, C., Santos, C., Pinho, S., Oliveira, H., Pedrosa, T., Dias, M. C., Cadmium-induced cyto- and genotoxicity are organ-dependent in lettuce. *Chem. Res. Toxicol.* 25 (7), 1423–1434, 2012.
60. Dom, N., Vergauwen, L., Vandenbrouck, T., Jansen, M., Blust, R., Knapen, D., Physiological and molecular effect assessment versus physico-chemistry based mode of action schemes: Daphnia magna exposed to narcotics and polar narcotics. *Environ. Sci. Technol.* 46 (1), 10–18, 2012.
61. Stratton, K., Shetty, P., Wallace, R., Bondurant, S., (eds.) *Clearing the Smoke: Assessing the Science Base for Tobacco Harm Reduction*. National Academy Press, Washington, DC, 636 pp., 2001.
62. Pfau, W., DNA adducts in marine and freshwater fish as biomarkers of environmental contamination. *Biomarkers* 2: 145–151, 1997.
63. Araoud, M., *Biological Markers of Human Exposure to Pesticides, Pesticides in the Modern World—Pests Control and Pesticides Exposure and Toxicity Assessment.* Stoytcheva, M., (eds.) ISBN: 978-953-307-457-3, InTech, Available from: http://www.intechopen.com/books/pesticides-in-the-modern-world-pests-control-and-pesticides-exposure-and-toxicity-assessment/biological-markers-of-human-exposure-to-pesticides, 2011.

64. van der Oost, R., Beyer, J., Vermeulen, N. P., Fish bioaccumulation and biomarkers in environmental risk assessment: A review. *Environ. Toxicol. Pharmacol.* 13 (2), 57–149, 2003.
65. Tsigou, E., Psallida, V., Demponeras, C., Boutzouka, E., Baltopoulos, G., Role of new biomarkers: Functional and structural damage. *Crit. Care Res. Pract.* 2013, 361078, 2013.
66. Ballarini, F., Ottolenghi, A., Chromosome aberrations as biomarkers of radiation exposure: Modelling basic mechanisms. *Adv. Space. Res.* 31 (6), 1557–1568, 2003.
67. Robles, A. I., Harris, C. C., Clinical outcomes and correlates of TP53 mutations and cancer. *Cold Spring Harb. Perspect. Biol.* 2 (3), a001016, 2010.
68. Pitarque, M., Vaglenov, A., Nosko, M., Pavlova, S., Petkova, V., Hirvonen, A., Creus, A., Norppa, H., Marcos, R., Sister chromatid exchanges and micronuclei in peripheral lymphocytes of shoe factory workers exposed to solvents. *Environ. Health Perspect.* 110 (4), 399–404 2002.
69. Zeiger, M., Biomarkers: The clues to genetic susceptibility. *Environ. Health Perspect.* 102 (1), 50–57, 1994.
70. Stellman, J. M., (eds.) *Encyclopaedia of Occupational Health and Safety*. International Labour Organization, 1998.
71. Hong, J. Y., Yang, C. S., Genetic polymorphism of cytochrome P450 as a biomarker of susceptibility to environmental toxicity. *Environ. Health Perspect.* 4, 759–762, 1997
72. Simkhovich, B. Z., Kleinman, M. T., Kloner, R. A., Air pollution and cardiovascular injury epidemiology, toxicology, and mechanisms. *J. Am. Coll. Cardiol.* 52 (9), 719–726, 2008.
73. Lewalter, J., Leng, G., Consideration of individual susceptibility in adverse pesticide effects. *Toxicol. Lett.* 107 (1–3), 131–44, 1999.
74. Zhang, L., Bassig, B. A., Mora, J. L., Vermeulen, R., Ge, Y., Curry, J. D., Hu, W., Shen, M., Qiu, C., Ji, Z., Reiss, B., McHale, C. M., Liu, S., Guo, W., Purdue, M. P., Yue, F., Li, L., Smith, M. T., Huang, H., Tang, X., Rothman, N., Lan, Q., Alterations in serum immunoglobulin levels in workers occupationally exposed to trichloroethylene. *Carcinogenesis* 34 (4), 799–802, 2013.
75. Ghadially, R., Brown, B. E., Sequeira-Martin, S. M., Feingold, K. R., Elias, P. M., The aged epidermal permeability barrier. Structural, functional, and lipid biochemical abnormalities in humans and a senescent murine model. *J. Clin. Invest.* 95 (5), 2281–2290, 1995.
76. Chokkalingam, A. P., Metayer, C., Scelo, G. A., Chang, J. S., Urayama, K. Y., Aldrich, M. C., Guha, N., Hansen, H. M., Dahl, G. V., Barcellos, L. F., Wiencke, J. K., Wiemels, J. L., Buffler, P. A., Variation in xenobiotic transport and metabolism genes, household chemical exposures, and risk of childhood acute lymphoblastic leukemia. *Cancer. Causes. Control.* 23 (8), 1367–1375, 2012.
77. Luo, Y. P., Chen, H. C., Khan, M. A., Chen, F. Z., Wan, X. X., Tan, B., Ou-Yang, F. D., Zhang, D. Z., Genetic polymorphisms of metabolic enzymes-CYP1A1, CYP2D6, GSTM1, and GSTT1, and gastric carcinoma susceptibility. *Tumour Biol.* 32 (1), 215–222, 2011.
78. Rodriguez, H., O'Connell, C., Barker, P. E., Atha, D. H., Jaruga, P., Birincioglu, M., Marino, M., McAndrew, P., Dizdaroglu, M., Measurement of DNA biomarkers for the safety of tissue-engineered medical products, using artificial skin as a model. *Tissue Eng.* 10 (9–10), 1332–1345, 2004.
79. Caldwell, J., Gardner, I., Swales, N., An introduction to drug disposition: The basic principles of absorption, distribution, metabolism, and excretion. *Toxicol. Pathol.* 23 (2), 102–114, 1995.
80. Xu, C., Li, C. Y., Kong, A. N., Induction of phase I, II and III drug metabolism/transport by xenobiotics. *Arch. Pharm. Res.* 28 (3), 249–268, 2005.
81. Dougherty, D., Garte, S., Barchowsky, A., Zmuda, J., Taioli, E., NQO1, MPO, CYP2E1, GSTT1 and GSTM1 polymorphisms and biological effects of benzene exposure—A literature review. *Toxicol. Lett.* 182 (1–3), 7–17, 2008.
82. Amanchy, R., Kalume, D. E., Pandey, A., Stable isotope labeling with amino acids in cell culture (SILAC) for studying dynamics of protein abundance and posttranslational modifications. *Sci. STKE* 2005 (267), pl2, 2005.
83. Ali, A., Krone, P. H., Pearson, D. S., Heikkila, J. J., Evaluation of stress-inducible hsp90 gene expression as a potential molecular biomarker in Xenopus laevis. *Cell Stress Chaperon.* 1 (1), 62–69, 1996.
84. Kneller, J. M., Ehlen, T., Matisic, J. P., Miller, D., Van Niekerk, D., Lam, W. L., Marra, M., Richards-Kortum, R., Follen, M., Macaulay, C., Jones, S. J., Using LongSAGE to detect biomarkers of cervical cancer potentially amenable to optical contrast agent labelling. *Biomark. Insights* 2, 447–461, 2007.
85. Loukopoulos, P., Shibata, T., Katoh, H., Kokubu, A., Sakamoto, M., Yamazaki, K., Kosuge, T., Kanai, Y., Hosoda, F., Imoto, I., Ohki, M., Inazawa, J., Hirohashi, S., Genome-wide array-based comparative genomic hybridization analysis of pancreatic adenocarcinoma: Identification of genetic indicators that predict patient outcome. *Cancer Sci.* 98 (3), 392–400, 2007.

86. He, Y. D., Genomic approach to biomarker identification and its recent applications. *Cancer Biomark.* 2 (3–4), 103–133, 2006.
87. González-González, M., Garcia, J. G., Montero, J. A., Fernandez, L. M., Bengoechea, O., Muñez, O. B., Orfao, A., Sayagues, J. M., Fuentes, M., Genomics and proteomics approaches for biomarker discovery in sporadic colorectal cancer with metastasis. *Cancer Genomics Proteomics* 10 (1), 19–25, 2013.
88. Seibert, V., Ebert, M. P., Buschmann, T., Advances in clinical cancer proteomics: SELDI-ToF-mass spectrometry and biomarker discovery. *Brief. Funct. Genomic Proteomic* 4 (1), 16–26, 2005.
89. Datta, M. W., True, L. D., Nelson, P. S., Amin, M. B., The role of tissue microarrays in prostate cancer biomarker discovery. *Adv. Anat. Pathol.* 14 (6), 408–418, 2007.
90. Perera, F. P., Poirier, M. C., Yuspa, S. H., Nakayama, J., Jaretzki, A., Curnen, M. M., Knowles, D. M., Weinstein, I. B., A pilot project in molecular cancer epidemiology: Determination of benzo[a]pyrene—DNA adducts in animal and human tissues by immunoassays. *Carcinogenesis* 3 (12), 1405–1410, 1982.
91. Li, Y. Q., Sui, W., Wu, C., Yu, L. J., Derivative matrix isopotential synchronous fluorescence spectroscopy for the direct determination of 1-hydroxypyrene as a urinary biomarker of exposure to polycyclic aromatic hydrocarbons. *Anal. Sci.* 17 (1), 167–170, 2001.
92. Kato, S., Petruzzelli, S., Bowman, E. D., Turteltaub, K. W., Blomeke, B., Weston, A., Shields, P. G., 7-Alkyldeoxyguanosine adduct detection by two-step HPLC and the 32P-postlabeling assay. *Carcinogenesis* 14 (4), 545–550, 1993.
93. Seneviratne, C., Johnson, B. A., Serotonin transporter genomic biomarker for quantitative assessment of ondansetron treatment response in alcoholics. *Front. Psychiatry* 3, 23, 2012.
94. Hu, C., van der Heijden, R., Wang, M., van der Greef, J., Hankemeier, T., Xu, G., Analytical strategies in lipidomics and applications in disease biomarker discovery. *J. Chromatogr. B. Analyt. Technol. Biomed. Life Sci.* 877 (26), 2836–2846, 2009.
95. Min, H. K., Lim, S., Chung, B. C., Moon, M. H., Shotgun lipidomics for candidate biomarkers of urinary phospholipids in prostate cancer. *Anal. Bioanal. Chem.* 399 (2), 823–830, 2011.
96. Suhre, K., Römisch-Margl, W., de Angelis, M. H., Adamski, J., Luippold, G., Augustin, R., Identification of a potential biomarker for FABP4 inhibition: The power of lipidomics in preclinical drug testing. *J. Biomol. Screen.* 16 (5), 467–475, 2011.
97. Xue, S., Qiao, J., Pu, F., Cameron, M., Yang, J. J., Design of a novel class of protein-based magnetic resonance imaging contrast agents for the molecular imaging of cancer biomarkers. *Wiley Interdiscip. Rev. Nanomed. Nanobiotechnol.* 5 (2), 163–179, 2013.
98. Mesmer, M. Z., Flurer, R. A., Determination of bromethalin in commercial rodenticides found in consumer product samples by HPLC-UV-vis spectrophotometry and HPLC-negative-ion APCI-MS. *J. Chromatogr. Sci.* 39 (2), 49–53, 2001.
99. Kaddoumi, A., Mori, T., Nakashima, M. N., Wada, M., Nakashima, K., High performance liquid chromatography with fluorescence detection for the determination of phenylpropanolamine in human plasma and rat's blood and brain microdialysates using DIB-Cl as a label. *J. Pharm. Biomed. Anal.* 34 (3), 643–650, 2004.
100. Nelofar, A., Laghari, A. H., Yasmin, A., Validated HPLC-RI method for the determination of lactulose and its process related impurities in syrup. *Indian J. Pharm. Sci.* 72 (2), 255–258, 2010.
101. Peng, C. A., Ferreira, J. F., Wood, A. J., Direct analysis of artemisinin from Artemisia annual using high-performance liquid chromatography with evaporative light scattering detector, and gas chromatography with flame ionization detector. *J. Chromatogr. A* 1133 (1–2), 254–258, 2006.
102. Sparkman, O. D., *Mass spectrometry desk reference*. Pittsburgh: Global View Pub, 2000.
103. Lin, Y., Schiavo, S., Orjala, J., Vouros, P., Kautz, R., Microscale LC-MS-NMR platform applied to the identification of active cyanobacterial metabolites. *Anal. Chem.* 80 (21), 8045–8054, 2008.
104. Hellgeth, J. W., Taylor, L. T., FTIR detection of liquid chromatographically separated species. *J. Chromatogr. Sci.* 24, 519–528, 1986.
105. Donker, M. G., Reinhoud, N. J., van Valkenburg, C. F. M., Attomole detection limits in micro HPLC-ECD. *Monitoring Molecules in Neuroscience*, 13–14, 2001.
106. Domoradzki, J. Y., Pottenger, L. H., Thornton, C. M., Hansen, S. C., Card, T. L., Markham, D. A., Dryzga, M. D., Shiotsuka, R. N., Waechter, J. M. Jr., Metabolism and pharmacokinetics of bisphenol A (BPA) and the embryo-fetal distribution of BPA and BPA-monoglucuronide in CD Sprague-Dawley rats at three gestational stages. *Toxicol. Sci.* 76 (1), 21–34, 2003.
107. Bhushan, B., Halasz, A., Spain, J. C., Hawari, J., Initial reaction(s) in biotransformation of CL-20 is catalyzed by salicylate 1-monooxygenase from Pseudomonas sp. strain ATCC 29352. *Appl. Environ. Microbiol.* 70 (7), 4040–4047, 2004.

108. Knapp, G., Leitner, E., Michaelis, M., Platzer, B., Schalk, A., Element specific GC-detection by plasma atomic emission spectroscopy-A powerful tool in environmental analysis. *Int. J. Environ. Anal. Chem.* 38, 369–378, 1990.

109. Dass, C., *Fundamentals of Contemporary Mass Spectrometry*. John Wiley & Sons. p. 5. 2007.

110. Food Safety Applications in Mass Spectrometry: A practical reference for applying current developments in Agilent MS technologies to food analysis Reference. Agilent Technologies, 5989-0916EN, www.agilent.com/chem, 2005.

111. Patrick, A., Combined liquid chromatography mass spectrometry. Section III. Applications of thermospray. *Mass Spectrometry Reviews* 11, 3, 1992.

112. IUPAC, Compendium of Chemical Terminology, 2nd ed. (the "Gold Book"), 1997. Online corrected version: "mass spectrum," 2006.

113. van Heuveln, F., Meijering, H., Wieling, J., Inductively coupled plasma-MS in drug development: Bioanalytical aspects and applications. *Bioanalysis* 4 (15), 1933–1955, 2012.

114. http://media.rsc.org/Modern%20chemical%20techniques/MCT1%20Mass%20spec.pdf.

115. Petrović, M., Hernando, M. D., Díaz-Cruz, M. S., Barceló, D., Liquid chromatography–tandem mass spectrometry for the analysis of pharmaceutical residues in environmental samples: A review. *J. Chromatogr. A* 1067 (1–2), 1–14, 2005.

116. Heumann, K. G., Isotope dilution mass spectrometry. *Int. J. Mass Spectrom.* 118, 575–592, 1992.

117. Ciccimaro, E., Blair, I. A., Stable-isotope dilution LC–MS for quantitative biomarker analysis. *Bioanalysis* 2 (2), 311–341, 2010.

118. Oe, T., Ackermann, B. L., Inoue, K., Berna, M. J., Garner, C. O., Gelfanova, V., Dean, R. A., Siemers, E. R., Holtzman, D. M., Farlow, M. R., Blair, I. A., Quantitative analysis of amyloid beta peptides in cerebrospinal fluid of Alzheimer's disease patients by immunoaffinity purification and stable isotope dilution liquid chromatography/negative electrospray ionization tandem mass spectrometry. *Rapid Commun. Mass Spectrom.* 20 (24), 3723–3735, 2006.

119. Norrgran, J., Bravo, R., Bishop, A. M., Restrepo, P., Whitehead, R. D., Needham, L. L., Barr, D. B., Quantification of six herbicide metabolites in human urine, *J. Chromatogr. B* 830, 185–195, 2006.

120. Ho, C. S., Lam, C. W., Chan, M. H., Cheung, R. C., Law, L. K., Lit, L. C., Ng, K. F., Suen, M. W., Tai, H. L., Electrospray ionisation mass spectrometry: Principles and clinical applications. *Clin. Biochem. Rev.* 24 (1), 3–12, 2003.

121. Sams, C., Patel, K., Jones, K., Biological monitoring for exposure to pirimicarb: Method development and a human oral dosing study. *Toxicol. Lett.* 192, 56–60, 2010.

122. Li, B., Ricordel, I., Schopfer, L. M., Baud, F., Megarbane, B., Masson, P., Lockridge, O., Dichlorvos, chlorpyrifos oxon and aldicarb adducts of butyrylcholinesterase, detected by mass spectrometry in human plasma following deliberate overdose. *J. Appl. Toxicol.* 30, 559–565, 2010.

123. Anspach, J. A., Maloney, T. D., Colón, L. A., Ultrahigh-pressure liquid chromatography using a 1-mm id column packed with 1.5-microm porous particles. *J. Sep. Sci.* 30 (8), 1207–1213, 2007.

124. Kofman, J., Zhao, Y., Maloney, T., Baumgartner, T., Bujalski, R., Ultra-high performance liquid chromatography: Hope or hype? *American Pharmaceutical Review*, 2006

125. Denoroy, L., Zimmer, L., Renaud, B., Parrot, S., Ultra high performance liquid chromatography as a tool for the discovery and the analysis of biomarkers of diseases: A review. *J. Chromatogr. B. Analyt. Technol. Biomed. Life Sci.* 927, 37–53, 2013.

126. Horvath, T. D., Matthews, N. I., Stratton, S. L., Mock, D. M., Boysen, G., Measurement of 3-hydroxyisovaleric acid in urine from marginally biotin-deficient humans by UPLC-MS/MS. *Anal. Bioanal. Chem.* 401 (9), 2805–2810, 2011.

127. van der Ham, M., Albersen, M., de Koning, T. J., Visser, G., Middendorp, A., Bosma, M., Verhoeven-Duif, N. M., de Sain-van der Velden, M. G., Quantification of vitamin B6 vitamers in human cerebrospinal fluid by ultra performance liquid chromatography–tandem mass spectrometry. *Anal. Chim. Acta* 712, 108–114, 2012.

128. Kawanishi, H., Toyo'oka, T., Ito, K., Maeda, M., Hamada, T., Fukushima, T., Kato, M., Inagaki, S., Rapid determination of histamine and its metabolites in mice hair by ultra-performance liquid chromatography with time-of-flight mass spectrometry. *J. Chromatogr. A* 1132 (1–2), 148–156, 2006.

129. Sugiura, K., Min, J. Z., Toyo'oka, T., Inagaki, S., Rapid, sensitive and simultaneous determination of fluorescence-labeled polyamines in human hair by high-pressure liquid chromatography coupled with electrospray-ionization time-of-flight mass spectrometry. *J. Chromatogr. A* 1205 (1–2), 94–102, 2008.

130. Ackermans, M. T., Kettelarij-Haas, Y., Boelen, A., Endert, E., Determination of thyroid hormones and their metabolites in tissue using SPE UPLC-tandem MS. *Biomed. Chromatogr.* 26 (4), 485–490, 2012.

131. Boysen, G., Collins, L. B., Liao, S., Luke, A. M., Pachkowski, B. F., Watters, J. L., Swenberg, J. A., Analysis of 8-oxo-7,8-dihydro-2′-deoxyguanosine by ultra high pressure liquid chromatography-heat assisted electrospray ionization-tandem mass spectrometry. *J. Chromatogr. B* 878 (3–4), 375–380, 2010.
132. Lame, M. E., Chambers, E. E., Blatnik, M., Quantitation of amyloid beta peptides Aβ(1-38), Aβ(1-40), and Aβ(1-42) in human cerebrospinal fluid by ultra-performance liquid chromatography–tandem mass spectrometry. *Anal. Biochem.* 419 (2), 133–139, 2011.
133. Mattarucchi, E., Guillou, C., Critical aspects of urine profiling for the selection of potential biomarkers using UPLC-TOF-MS. *Biomed. Chromatogr.* 26 (4), 512–517, 2012.
134. Plumb, R. S., Johnson, K. A., Rainville, P., Smith, B. W., Wilson, I. D., Castro-Perez, J. M., Nicholson, J. K. UPLC/MS(E); a new approach for generating molecular fragment information for biomarker structure elucidation. *Rapid Commun. Mass Spectrom.* 20 (13), 1989–1994, 2006.
135. Dunn, W. B., Broadhurst, D., Brown, M., Baker, P. N., Redman, C. W., Kenny, L. C., Kell, D. B., Metabolic profiling of serum using Ultra Performance Liquid Chromatography and the LTQ-Orbitrap mass spectrometry system. *J. Chromatogr. B* 871 (2), 288–298, 2008.
136. Chen, M., Zhu, P., Xu, B., Zhao, R., Qiao, S., Chen, X., Tang, R., Wu, D., Song, L., Wang, S., Xia, Y., Wang, X., Determination of nine environmental phenols in urine by ultra-high-performance liquid chromatography-tandem mass spectrometry. *J. Anal. Toxicol.* 36 (9), 608–615, 2012.
137. Alwis, K. U., Blount, B. C., Britt, A. S., Patel, D., Ashley, D. L., Simultaneous analysis of 28 urinary VOC metabolites using ultra high performance liquid chromatography coupled with electrospray ionization tandem mass spectrometry (UPLC-ESI/MSMS). *Anal. Chim. Acta* 750, 152–160, 2012.
138. Hsiao, P. K., Shih, T. S., Chen, C. Y., Chiung, Y. M., Lin, Y. C., Evaluation of 1,4-dichlorobenzene exposure and associated health effects: Hematologic, kidney, and liver functions in moth repellent workers. *Epidemiology* 22 (1), p. S33, 2011.
139. Karlsson, K. E., Novotny, M., Separation efficiency of slurry-packed liquid chromatography microcolumns with very small inner diameters. *Anal. Chem.* 60 (17), 1662–1665, 1988.
140. Hernández-Borges, J., Aturki, Z., Rocco, A., Fanali, S., Recent applications in nanoliquid chromatography. *J. Sep. Sci.* 30 (11), 1589–1610, 2007.
141. Tomer, V., Separations combined with mass spectrometry. *Chem. Rev.* 101, 297–328, 2001.
142. Edwards, E., Thomas-Oates, J., Hyphenating liquid phase separation techniques with mass spectrometry: On-line or off-line. *Analyst.* 130, 13–17, 2005.
143. Abian, J., Oosterkamp, A. J., Gelpi, E. J., Comparison of conventional, narrow-bore and capillary liquid chromatography/mass spectrometry for electrospray ionization mass spectrometry: Practical considerations. *Mass Spectrom.* 34, 244–254, 1999.
144. Cappiello, A., Famiglini, G., Mangani, F., Palma, P., Siviero, A., Nano-high-performance liquid chromatography–electron ionization mass spectrometry approach for environmental analysis. *Anal. Chim. Acta* 493 (2), 125–136, 2003.
145. Rosales-Conrado, N., León-González, M. E., D'Orazio, G., Fanali, S., Enantiomeric separation of chlorophenoxy acid herbicides by nano liquid chromatography–UV detection on a vancomycin-based chiral stationary phase. *J. Sep. Sci.* 27 (15–16), 1303–1308, 2004.
146. Aryal, U. K., Lin, C., Kim, J., Heibeck, T. H., Wang, J., Qian, W., Lin, Y., Identification of phosphorylated butyrylcholinesterase in human plasma using immunoaffinity purification and mass spectrometry. *Anal. Chim. Acta* 723, 68–75, 2012.
147. http://www.cdc.gov/biomonitoring/Endrin_BiomonitoringSummary.html.

19 Ultra-Performance Liquid Chromatography Applied to Analysis of Pesticides

María de los Ángeles Herrera Abdo, José Luis Martínez Vidal, and Antonia Garrido Frenich

CONTENTS

19.1 INTRODUCTION

The use of plant-protection products, such as pesticides, is widespread worldwide. They are widely used in agricultural practice [1], both at cultivation and postharvest steps, allowing an increase in crop productivity and improved product quality. Other uses include public health to control vector-borne diseases, such as malaria or dengue; disinfection of livestock and pets; homes; gardens; maintenance of reservoirs of water; or industry [1].

The main consequence of their use is the presence of pesticide residues in both food and environmental matrices. Consequently, several international organizations have established stringent regulatory controls on pesticide use in order to minimize exposure for the general population to pesticide residues in food. These controls have allowed the establishment of maximum residue limits for a wide variety of pesticide/commodity combinations [2,3]. In addition, pesticides have been listed as priority pollutants in environmental samples by the United States Environmental Protection Agency [4,5] and the European Union [6].

As a result of the widespread application of pesticides, their control and monitoring of residual levels in food and environmental matrices is highly necessary, not only in order to meet regulatory requirements, but especially to protect the consumer and the environment. To this aim, analytical methodologies applied must be adequate to identify and accurately quantify the concentration of any pesticide residue detected, usually at very low levels, as well as being able to determine as many pesticide residues as possible from the hundreds of compounds commonly used. Multiresidue/multiclass methods based on chromatographic techniques, mainly gas (GC) and high-performance liquid chromatography (HPLC) coupled to mass spectrometry (MS) analyzers, have usually been applied to the determination of pesticide residues worldwide. However, recently, more and more methods are using HPLC because there has been a clear trend to use more polar pesticides of low volatility or thermal lability, which are not amenable GC compounds or require derivatization before analysis by this technique [7].

In routine pesticide monitoring laboratories that analyze hundreds of samples per day, in addition to achieving high-quality results with wide analytical scope, analysis time is an important issue to consider in the choice of an analytical method in order to increase sample throughput. Four approaches have been mainly applied for fast separations in LC [8–10]: (i) operate at high temperatures, (ii) use of monolithic columns, (iii) use of fused core columns, and (iv) use of columns with porous sub-2-μm particles. The first approach has as its main weaknesses the scarce number of stationary phases that can operate at high temperatures (>60°C) as well as the potential degradation of both temperature-sensitive analytes and column packing stationary phases. The use of monolithic columns allows increasing the flow rate and shortening the column length although mixtures with a high number of compounds could not be separated adequately. Other limiting factors are the small number of commercially available column dimensions and stationary phases. The third option, based on fused core columns, allows increasing the speed of analysis and separation efficiency using superficially porous particles. However, this approach has as its major disadvantages loading capacity and retention slightly lower than conventional HPLC. Finally, the use of columns with porous sub-2-μm particles offers high-speed analysis with high efficiency and, in consequence, an improvement in resolution and sensitivity. The use of small columns with fine particles also reduces solvent consumption. However, the drawback is a higher column backpressure that is not acceptable

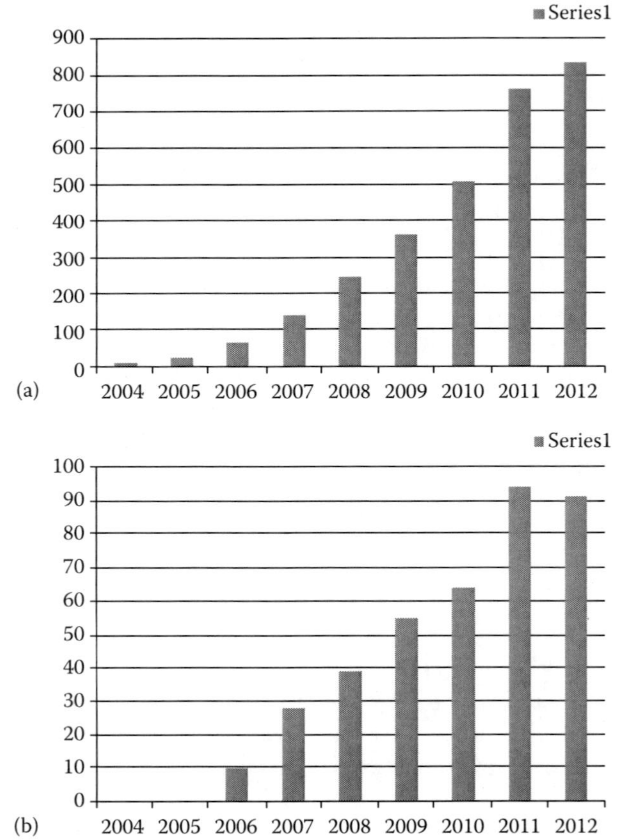

FIGURE 19.1 Number of articles or reviews published each year in the field of UPLC since 2004, (a) with keywords UPLC or UHPLC and, (b) with an additional filter (keyword pesticide). (From Scopus database. Available at http://www.scopus.com/home.url. Date of information gathering: December 2012.)

for conventional HPLC systems, and therefore, it was necessary to develop new LC systems capable of supporting these high pressures.

This last approach has been the most widely used since 2004, the year in which a new generation of stationary phases that can withstand high pressures (to about 15,000 psi), along with the development of compatible LC systems, were introduced in the market under the trade name of ultra-performance LC (UPLC). The first commercially available UPLC instrument was introduced by the Waters company under the ACQUITY UPLC™ system, and since then, many suppliers have also commercialized these systems [11–13]. Generally, these UPLC systems differ in some of their specifications, such as maximum backpressure and acquisition rate, flow rate and injection volume range, dead volume, or injection time. Although, in some cases, different vendors use the term ultra-high-performance liquid chromatography (UHPLC), this chapter will use UPLC to refer to LC systems that support pressures greater than 6000 psi.

The demands of high-throughput analyses have led to a growing interest in UPLC in different fields. Figure 19.1 shows the exponential growth in the use of UPLC with five related articles published in 2004 to around a thousand publications in 2012. Main applications of UPLC include pharmaceutical analyses (drug discovery and development, bioequivalence studies, quality control of drugs, etc.), metabolomics/metabonomics, proteomics, and chiral separations [9–11], and UPLC has also been used in multiresidue pesticide analysis in environmental, food, and drink matrices [9–11,14]. The first publications related to UPLC in the field of pesticide residues appeared in 2006 with significant growth since then until 2011, which was maintained in 2012 (Figure 19.1).

The actual trend in pesticide residue analysis focuses on the use of UPLC combined with MS detection using electrospray ionization (ESI). In consequence, this chapter is focused on UPLC-MS applied to analysis of pesticides, which, at the moment, is a widely accepted technique used for purposes of monitoring pesticides as well as for regulatory issues in food and beverages [15–45] as well as in environmental samples [46–58]. In addition, this chapter reviews the basic principles of the UPLC technique as well as a comparison between HPLC and UPLC, including the main mobile and stationary phases used.

19.2 BASIC PRINCIPLES

To increase the efficiency of separations and thus increase the resolution, there has been a trend throughout the evolution of HPLC toward the use of stationary phases with smaller particle sizes (from 5 to 2 μm) [59]. The introduction of UPLC has meant an improvement in this field, allowing the use of sub-2-μm particle sizes. However, the UPLC is based on the chromatographic principles of HPLC.

According to the Van Deemter equation, which is an empirical formula that describes the relationship between the height equivalent of a theoretical plate (HETP or column efficiency) and linear velocity (flow rate), a decrease in HETP is predicted with the use of smaller particle sizes [10,60] (Equation 19.1):

$$\text{HETP} = A + B/\nu + C\nu \tag{19.1}$$

where A, B, and C are factor characteristics of each column, and ν is the average linear velocity of the mobile phase; the optimum ν is inversely proportional to the particle diameter (Equation 19.2):

$$\nu_{opt} = \sqrt{\frac{B}{C}} \approx \frac{3Dm}{dp} \tag{19.2}$$

where Dm is the diffusion coefficient of an analyte in the mobile phase, and dp is the diameter of the packing material.

The factors of the Van Deemter equation contribute to band broadening from the following:

- Eddy diffusion: The *A* factor describes the peak broadening due to the presence of stationary phase particles in the column. It is smallest when the packed chromatographic column particles are small. This term is independent of the mobile phase velocity.
- Longitudinal diffusion: The *B* constant represents the natural diffusion tendency of each analyte in the mobile phase along the longitudinal direction of a chromatographic column. This constant affects peak broadening only at low flow rates (below the minimum HETP). In consequence, this term is reduced at high flow rates, and for that, it is divided by ν.
- Resistance to mass transfer: The *C* constant represents the kinetic resistance to equilibrium in the chromatographic separation process. This resistance is the time lag involved in moving from the packing stationary phase and back again. The greater the flow of gas, the more a molecule on the packing tends to lag behind molecules in the mobile phase. So the *C* term is proportional to ν. Resistance to mass transfer in the mobile phase can be decreased significantly using very small particle sizes.

The hyperbolic form of the van Deemter equation (Figure 19.2) predicts that there is an optimum velocity at which HETP achieves a minimum value and hence a maximum separation efficiency [59]. For lower speeds than the optimum, the longitudinal diffusion causes a widening of the band and thus an increase of HETP. For speeds greater than optimum, the difficulty to reach equilibrium between the phases creates band spreading.

The higher optimum linear velocity can be used for smaller chromatographic particle sizes. Figure 19.2 shows the performance of various particle diameters in the van Deemter plot [59]. Sub-2-μm particle sizes demonstrate a significant gain in efficiency, which does not diminish at increased linear velocities. Optimal separations can be carried out at higher velocities and over a wider range of velocities. In consequence, small particle sizes should be used for both high efficiency and speed in LC (Figure 19.3) [61].

The problem is that the smaller the filler particles, the greater the resistance that opposes the flow of the mobile phase, and thus greater pressures are generated. According to Darcy's law, the

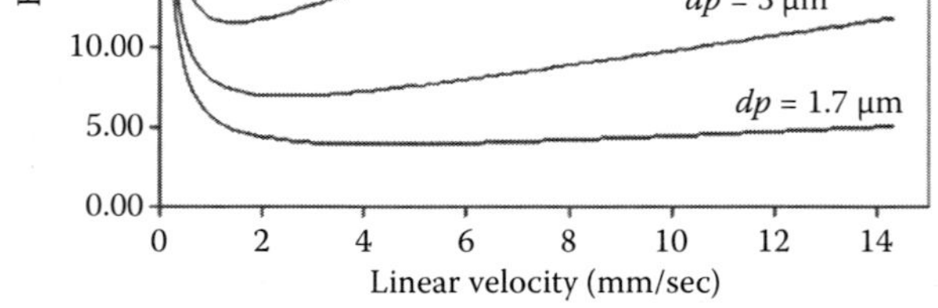

FIGURE 19.2 Van Deemter curves for different particle sizes (10 μm, 5 μm, 3 μm, 1.7 μm). (Reprinted from Nováková, L. et al., *J. Sep. Sci.*, 29, 2433–2443, 2006. With permission.)

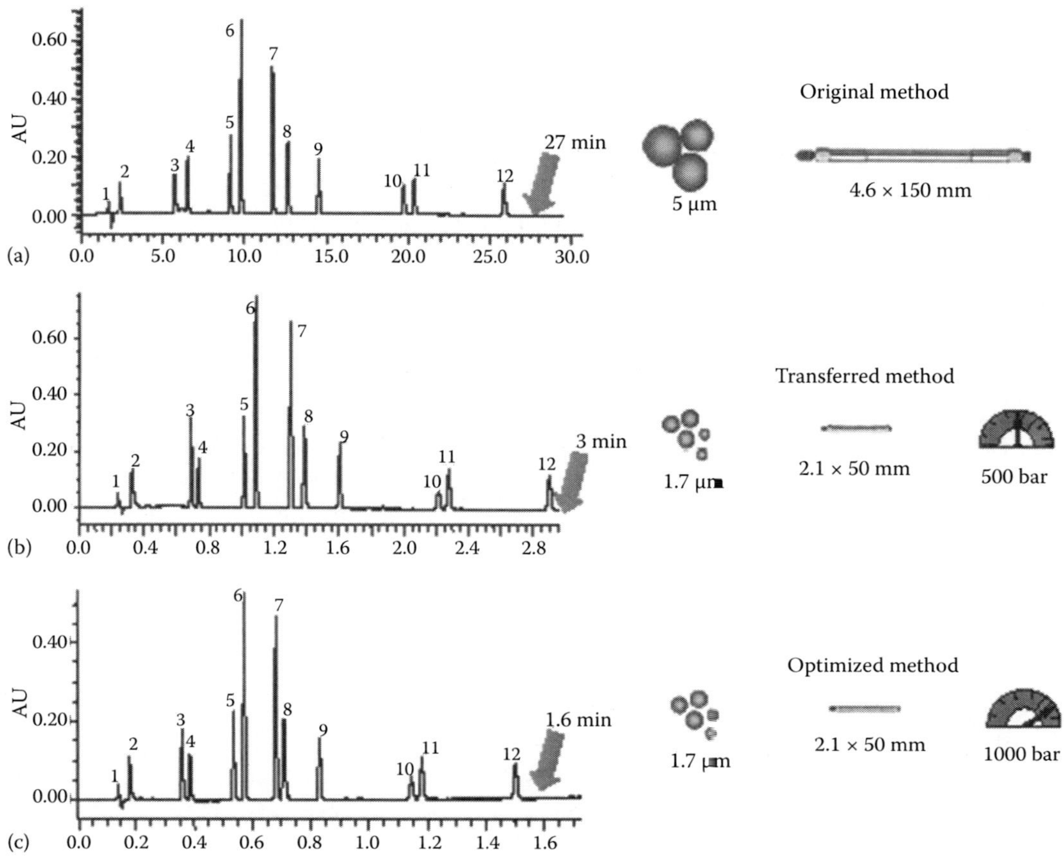

FIGURE 19.3 Method transfer from conventional HPLC to UHPLC. Separation of a pharmaceutical formulation containing the main product (6) and 11 impurities in gradient mode with HPLC and UHPLC systems: (a) original HPLC method: column XBridge C18 150 × 4.6 mm, 5 μm; flow rate, 1000 μL min^{-1}; injected volume, 20 μL; total gradient time, 45 min. (b) Transferred UHPLC method: column Acquity BEH C18 50 × 2.1 mm, 1.7 μm; flow rate, 610 μL min^{-1}; injected volume, 1.4 μL; total gradient time, 5.1 min. (c) Transferred and optimized UHPLC method: column Acquity BEH C18 50 × 2.1 mm, 1.7 μm; flow rate, 1000 μL min^{-1}; injected volume, 1.4 μL; total gradient time, 3.1 min. (Reprinted from Guillarme, D. et al., *Anal. Bioanal. Chem.*, 397, 1069–1082, 2010. With permission.)

pressure drop across a column, ΔP, is inversely proportional to the square of *dp* at the optimum ν [10] (Equation 19.3):

$$\Delta P = \frac{\phi \eta L \nu}{dp^2} \tag{19.3}$$

where ϕ is the flow resistance factor, η is the mobile phase viscosity, and L is the length of the packed bed.

The use of sub-2-μm particle sizes allows also the obtaining of better resolution and reduced analysis time in LC. In the fundamental resolution (*Rs*) equation (Equation 19.4),

$$Rs = \frac{\sqrt{N}}{4}\left(\frac{\alpha - 1}{\alpha}\right)\left(\frac{k}{k+1}\right) \tag{19.4}$$

Rs is related to the number of theoretical plates (*N*) in the column, the selectivity factor (α), and the retention factor (*k*). To obtain high resolution, the three terms must be maximized. An increase in

N by lengthening the column leads to an increase in retention time and increased band broadening, which may not be desirable. But because N is inversely proportional to dp (Equation 19.5),

$$N \propto \frac{1}{dp} \tag{19.5}$$

To increase the N, the HETP can be reduced by reducing the size of the stationary phase particles.

Separations can be greatly improved by controlling the capacity factor, k. This can be achieved by changing the composition of the mobile phase. The α can also be manipulated to improve separations. When α is close to unity, optimizing k and increasing N is not enough to give good separation in a reasonable time. In these cases, k is optimized first, and then α is increased generally by changing the mobile phase composition, column temperature, or composition of stationary phase.

As the particle size is lowered by a factor of three, from, for example, 5 μm (HPLC-scale) to 1.7 μm (UPLC-scale), N is increased by three, and Rs by the square root of three or 1.7. N is also inversely proportional to the square of the peak width (w). In addition, as the dp decreases to increase N and subsequently Rs, an increase in sensitivity is obtained. Also, peak height (H) is inversely proportional to the peak width (w) (Equation 19.6):

$$H \propto \frac{1}{w} \tag{19.6}$$

In consequence, an increase in sensitivity is obtained in UPLC because narrower peaks are taller peaks. Narrower peaks also mean more peak capacity per unit time in gradient separations, desirable for many applications.

As mentioned earlier, efficiency is proportional to column length and inversely proportional to the dp; therefore, the column can be shortened by the same factor as the particle size without loss of Rs. Using a flow rate three times higher due to the smaller particles and shortening the column by one third (again due to the smaller particles), the separation is completed in one ninth the time while maintaining Rs (Figure 19.3). Therefore, it is possible to increase throughput and the speed of analysis without affecting the chromatographic performance [61].

19.3 COMPARISON AND FUNDAMENTAL DIFFERENCES: HPLC AND UPLC

UPLC has demanded the development of new instrumentation, which can take advantage of the separation performance (by reducing dead volume) and is consistent with the pressures (up to 15,000 psi, compared with up to 5000 psi in HPLC). Table 19.1 shows a comparison of certain characteristics of HPLC and UPLC [60,62].

UPLC instruments must have certain characteristics in order to take advantage of sub-2-μm particles, such as injection valves that must withstand high pressures and operate at high speeds, low system volume and dwell volume to reduce dispersion at lower flow rates, and low injection volume to minimize band spreading. These features can improve ionization efficiencies when UPLC is coupled to MS, and in consequence, a higher sensitivity is obtained.

Undoubtedly, one of the main differences between HPLC and UPLC is the analytical column: particle size, diameter, and length. The use of smaller particle size provides higher column efficiency, resolution, and speed. The reduction of column length also improves the run time. The use of smaller-diameter columns in UPLC allows the minimization of the effects of frictional heating of the mobile phase and, as a consequence, the loss of performance.

19.3.1 Stationary Phases Used in UPLC

The major chromatographic mode to determine pesticides by UPLC is reversed phase (RP). In this mode, medium polarity, polar, and even ionic pesticides are separated by partitioning between an

TABLE 19.1
Comparison between HPLC and UPLC

Characteristic	HPLC	UPLC
Maximum pressure	5000 psi	15,000 psi
Injector	Normal speed	High speed
Injection volume	More	Less
Flow rate	More	Less
Dwell volume	High	Low
Particle size	3 to 5 μm	<2 μm
Diameter	3.0–4.6	1.0–2.1
Length	10–25 cm	5–10 cm
Solvent consumption	More	Less
Run time	Long	Short
Sensitivity	Less	More

apolar stationary phase and a polar mobile phase. Most stationary phases are based on silica particles packed in capillary columns.

At the beginning of UPLC, nonporous 1- to 1.5-μm silica particles were used. This technique provides high efficiency, but the low surface area produces poor capacity and retention and high pressures. In order to overcome these limitations, porous silica particles or polymeric stationary phases were applied, but they have their own disadvantages [13]. The introduction in 2000 of a new generation of hybrid materials intended to overcome the above limitations but failed in mechanical stability, which was improved with the bridged ethyl hybrid (BEH) technology [60].

Different bonded phases are available for UPLC separations based on BEH technology [60]: (i) BEH C18 and C8; (ii) BEH Shield RP_{18} (embedded polar group column), (iii) BEH phenyl, and (iv) BEH hydrophilic interaction chromatography. As can be seen in Tables 19.2 and 19.3, BEH C18 columns are the most stationary phases used for the pesticide determination by UPLC. Also, high strength silica (HSS) particles are applied for determining polar pesticides in RP-UPLC [26,28,35,48,54,55]. Available chemistries for UPLC separations based on HSS particles are (i) T3; (ii) C18, and (iii) C18 SB. HSS particles provide superior polar pesticide retention compared to BEH technology.

Typically, 5- or 10-cm columns packed with 1.7- to 1.8-μm particles and with an internal diameter of 2.1 mm are used in UPLC (Tables 19.2 and 19.3). The use of smaller particles provides higher column efficiency and requires a shorter column length for a given separation. Also, the reduction of column length in UPLC improves the run time because the retention of the analytes decreases.

19.3.2 Mobile Phases Used in UPLC

The selection of mobile phase strongly depends on the column type used for separation. As noted in the above section, for pesticide residue analysis, most often RP phases are applied. To separate polar and semipolar pesticides in one run, gradient elution is used at a flow rate varying from 0.2 mL/min [16,33,39] to 0.6 mL/min [35,43]. An appropriate way is to work with binary mixtures of polar solvents. The most common binary solvent systems are based on acetonitrile–water or methanol–water. Mixtures with acetonitrile offers lower viscosity, lower column pressure, and sharper peaks. However, mixtures with methanol generally offer slightly better efficiency than acetonitrile and are sometimes preferred for economical and toxicological reasons. In general, mixtures of methanol and water are mostly used for UPLC separation of pesticides in food and beverages (Table 19.2) and environmental matrices (Table 19.3).

TABLE 19.2
Overview of Stationary Phases and Mobile Phases Used in Literature for UPLC Separation of Pesticides in Food and Beverages

Matrix	No. Pesticides	Stationary Phase	Mobile Phases	Extraction	LOD (µg/kg)	Analysis	Refs.
Orange (fruit, peel, pulp, oil, juice)	5	BEH C18 (50 × 2.1 mm i.d., 1.7 µm)	Gradient mode (0.25 mL/min): phase A, MeOH; phase B, 1 mmol/L NH_4Ac in water	QuEChERS (MeOH)	0.03–3	UHPLC-QqQ-MS/MS (MRM)	[15]
Cucumber	7	BEH C18 (50 × 2.1 mm i.d., 1.7 µm)	Gradient mode (0.2 mL/min): phase A, ACN; phase B, 0.2% formic acid in water	HF-LPME (chloroform)	0.01–0.31	UHPLC-QqQ-MS/MS (MRM)	[16]
Fruits and vegetables	Multiresidue (71)	BEH C18 (100 × 2.1 mm i.d., 1.7 µm)	Gradient mode (0.45 mL/min): phase A, 5% MeOH and 95% 2 mM NH_4Ac; phase B, 95% MeOH and 5% 2 mM NH_4Ac	SE (TFE and toluene)	0.12–2.16	UHPLC-QqQ-MS/MS (MRM)	[17]
Tomato	1	BEH C18 (50 x 2.1 mm i.d.,1.7 µm)	Gradient mode (0.3 mL/min): phase A, MeOH; phase B, 0.2% formic acid in water	SE (ACN)	0.02	UHPLC-QqQ-MS/MS (MRM)	[18]
Honey	Multiclass (>350)	Hypersil GOLD aQ C18 (100 × 2.1 mm i.d., 1.7 µm)	Gradient mode (0.3 mL/min): phase A, 0.1% formic acid and NH_4Ac 4 mM in water; phase B, 0.1% formic acid and NH_4Ac 4 mM in MeOH	SE (ACN)	1–50	UPLC-orbitrap-MS (Full scan)	[19]
Orange, banana, corn	Multiresidue (>230)	BEH C18 (100 × 2.1 mm i.d., 1.7 µm)	Gradient mode (0.3 mL/min): phase A, ACN with 0.01% formic acid; phase B, MeOH with 0.01% formic acid	SE (Water: MeOH, 20:80 v/v) SE (ACN: water, 80:20 v/v, with 0.1% formic acid)—Corn	Not reported	UPLC–QTOF-MS/MS (product ion scan)	[20]
Cucumber, tomato	2 (parent and metabolite)	BEH C18 (100 × 2.1 mm i.d., 1.7 µm)	Gradient mode (0.5 mL/min): phase A, 2 mM NH_4Ac in water; phase B, ACN	QuEChERS (ACN + water)	2–4	UHPLC-QqQ-MS/MS (MRM)	[21]
Cereals, vegetables, wine	Multiclass (>90)	BEH C18 (100 × 2.1 mm i.d., 1.7 µm)	Gradient mode (0.45 mL/min): phase A, MEOH; phase B, 5 mM ammonium formate in water	QuEChERS (ACN; ACN + water, cereals)	<10	UPLC-QqQ-MS/MS (MRM)	[22]

Fruits, vegetables	4	BEH C18 (100 × 2.1 mm i.d., 1.7 μm)	Gradient mode (0.3 mL/min): phase A, ACN; phase B, 0.1% formic acid in water	QuEChERS (ACN)	0.06–6	UPLC-QqQ-MS/MS (MRM)	[23]
Pear, apple	Identification of unknown compounds	BEH C18 (15 × 2.1 mm i.d., 1.7 μm)	Gradient mode (0.2 mL/min): phase A, 10 mM ammonium formate in water; phase B, MEOH with 10 mM ammonium formate	SE (ethyl acetate)	Not reported	UPLC–QTOF-MS (full scan)	[24]
Fruits, vegetables	11	BEH C18 (50 × 2.1 mm i.d., 1.7 μm)	Gradient mode (0.3 mL/min): phase A, 0.5 mM NH_4Ac in MeOH; phase B, 5 mM NH_4Ac in water	SE (MeOH:H_2O, 80:20 v/v)	0.1–1.3 pg (QqQ) 0.44–8.75 pg (TOF) 0.31–12.5 pg (QTOF)	UPLC-QqQ-MS/MS (SRM) UPLC–QTOF-MS (full acquisition, product ion scan)	[25]
Food plants	Multiresidue (212)	HSS T3 C18 (100 × 2.1 mm i.d., 1.8 μm)	Gradient mode (0.3–0.6 mL/min): phase A, MeOH; phase B, 0.5 mM ammonium formate in water	QuEChERS (ACN)	Not reported	UPLC–TOF-MS (full scan)	[26]
Cereal grains	Multiresidue (64)	BEH C18 (50 × 2.1 mm i.d., 1.7 μm)	Gradient mode (0.45 mL/min): phase A, 10 mM ammonium formate in water (pH 3, adjusted using formic acid); phase B, 10 mM ammonium formate in MeOH	QuEChERS (MeOH + water)	Not reported	UPLC-QqQ-MS/MS (MRM)	[27]
Apple, potato, cabbage	7	HSS T3 C18 (100 × 2.1 mm i.d., 1.8 μm)	Gradient mode (0.3 mL/min): phase A, 0.1% formic acid in water; phase B, ACN	ACN + SPE Oasis HLB	Not reported	UPLC-QqQ-MS/MS (MRM)	[28]
Berries	Multiresidue (148)	BEH C18 (100 × 2.1 mm i.d., 1.7 μm)	Gradient mode (0.4 mL/min): phase A, ACN; phase B, 10 mM NH_4Ac with 2% ACN in water	QuEChERS (ACN)	≤5	UPLC–QTOF-MS (full scan product ion scan)	[29]
Tea	6	BEH C18 (100 × 2.1 mm i.d., 1.7 μm)	Gradient mode (0.3 mL/min): phase A, ACN; phase B, 0.2% formic acid in water	SE (ACN) + SPE florisil	1–9	UPLC-QqQ-MS/MS (MRM)	[30]
Fruits, vegetables	Multiresidue (148)	BEH C18 (100 × 2.1 mm i.d., 1.7 μm)	Gradient mode (0.4 mL/min): phase A, ACN; phase B, 10 mM NH_4Ac in water	QuEChERS (ACN)	≤5	UPLC–QTOF-MS (full scan product ion scan)	[31]

(*Continued*)

TABLE 19.2 (CONTINUED)
Overview of Stationary Phases and Mobile Phases Used in Literature for UPLC Separation of Pesticides in Food and Beverages

Matrix	No. Pesticides	Stationary Phase	Mobile Phases	Extraction	LOD (μg/kg)	Analysis	Refs.
Wine	Multiresidue (72)	BEH C18 (100 × 2.1 mm i.d., 1.7 μm)	Gradient mode (0.2 mL/min): phase A, ACN; phase B, 10 mM NH_4Ac in water	Modified QuEChERS (ACN)	≈1 μg/l	UPLC-QqQ-MS/MS (MRM)	[33]
Infant foods	Multiresidue (138)	BEH C18 (100 × 2.1 mm i.d., 1.7 μm)	Gradient mode (0.4 mL/min): phase A, ACN; phase B, 10 mM NH_4Ac in water	QuEChERS (ACN)	1	UPLC–QTOF-MS (full scan product ion scan)	[34]
Milk	10	HSS T3 C18 (100 × 2.1 mm i.d., 1.8 μm)	Gradient mode (0.6 mL/min): phase A, 0.01% acetic acid in water:ACN (900:100, v/v); phase B, 5 mM ammonium formate in MeOH:ACN (750:250, v/v)	LLE (ethyl acetate)	5	UPLC-QqQ-MS/MS (MRM)	[35]
Fruit juices	Multiresidue (90)	BEH C18 (100 × 2.1 mm i.d., 1.7 μm)	Gradient mode (0.35 mL/min): phase A, MeOH; phase B, 0.01% formic acid in water	QuEChERS (ACN)	≤0.7 μg/L	UPLC-QqQ-MS/MS (MRM)	[36]
Strawberry	Multiresidue (100)	BEH C18 (50 × 2.1 mm i.d., 1.7 μm)	Gradient mode (0.48 mL/min): phase A, 5 mM NH_4Ac in water:MeOH 95:5 v/v; phase B, 5 mM NH_4Ac in MEOH	SE (ethyl acetate)	Not reported	UPLC-QqQ-MS/MS (MRM) UPLC–QTOF-MS (full scan)	[37]
Beverages	Multiresidue (>50)	BEH C18 (100 × 2.1 mm i.d., 1.7 μm)	Gradient mode (0.35 mL/min): phase A, MeOH; phase B, 0.01% formic acid in water	HF-LPME (1-octanol)	0.01–2 μg/L	UPLC-QqQ-MS/MS (MRM)	[38]
Rice	13	BEH C18 (100 × 2.1 mm i.d., 1.7 μm)	Gradient mode (0.20 mL/min): phase A, 0.1% (v/v) formic acid in water; phase B, 0.1% (v/v) formic acid in MeOH	QuEChERS (ACN + water)	0.5	UPLC-QqQ-MS/MS (MRM)	[39]

Vegetables	Multiresidue (53)	BEH C18 (100 × 2.1 mm i.d., 1.7 μm)	Gradient mode (0.35 mL/min): phase A, 0.01% formic acid in water; phase B, MeOH	QuEChERS (ACN)	<3	UPLC-QqQ-MS/MS (MRM)	[40]
Maize, meat, milk, egg, honey	Multiclass (172)	BEH C18 (100 × 2.1 mm i.d., 1.7 μm)	Gradient mode (0.4 mL/min): phase A, 100% water containing 1 mM ammonium formate and 20 μl/L formic acid; phase B, water:MeOH, 5:95 v/v, containing 1 mM ammonium formate and 20 μl/L formic acid	Generic extraction LLE: (water/acetone-1% formic acid, for milk and honey) SE: (water-ACN-1% formic acid, rest of matrices)	<10–50	UPLC-QqQ-MS/MS (MRM) UPLC–QTOF-MS (full scan)	[41]
Fruits	Identification of unknown compounds	BEH C18 (50 × 2.1 mm i.d., 1.7 μm)	Gradient mode (0.4 mL/min): phase A, 10 mM NH_4Ac in water; phase B, MeOH	SE (ethyl acetate)	0.4	UPLC–QqTOF-MS (full scan)	[42]
Potato, orange, baby food	Multiresidue (52)	BEH C18 (50 × 2.1 mm i.d., 1.7 μm)	Gradient mode (0.6 mL/min): phase A, 17.5 mM acetic acid in MeOH; phase B, 17.5 mM acetic acid in water	SE (ACN:acetic acid, 99:1 v/v),	Not reported	UPLC-QqQ-MS/MS (MRM)	[43]
Baby food	17	BEH C18 (100 × 2.1 mm i.d., 1.7 μm)	Gradient mode (0.3 mL/min): phase A, 90% water, 10% MeOH, and 20 mM NH_4Ac; phase B: 10% water, 90% MeOH, and 20 mM NH_4Ac	SE (ACN)	1	UPLC-QqQ-MS/MS (MRM)	[44]
Baby food	17	BEH C18 (100 × 2.1 mm i.d., 1.7 μm)	Gradient mode (0.3 mL/min): phase A, water; phase B: MeOH	SE (ACN)	Not reported	UPLC-QqQ-MS/MS (MRM)	[45]

Note: ACN, acetonitrile, BEH, bridged ethyl hybrid; HF-LPME, hollow fiber-liquid-phase microextraction; HSS, high-strength silica; LLE, liquid–liquid extraction; MeOH, methanol; MRM, multireaction monitoring; MS/MS, tandem mode; NH_4Ac, ammonium acetate; QqQ, triple quadrupole; QTOF, quadrupole-time-of-flight; QuEChERS, quick, easy, cheap, effective, rugged, and safe; SE, solid extraction; SPE, solid phase extraction; TFE, tetrafluoroethane.

TABLE 19.3
Overview of Stationary Phases and Mobile Phases Used in Literature for UPLC Separation of Pesticides in Environmental and Biological Matrices

Matrix	No. Pesticides	Stationary Phase Composition	Mobile Phase	Extraction	LOD (μg/L)	Analysis	Refs.
Wastewater	Multiresidue (>230)	BEH C18 (100 × 2.1 mm i.d., 1.7 μm)	Gradient mode (0.3 mL/min): phase A, ACN with 0.01% formic acid; phase B, MeOH with 0.01% formic acid	SPE (Oasis HLB)	Not reported	UPLC–QTOF-MS (product ion scan)	[20]
Soil	2 (parent and metabolite)	BEH C18 (100 × 2.1 mm i.d., 1.7 μm)	Gradient mode (0.5 mL/min): phase A, 2 mM NH_4Ac in water; phase B, ACN	QuEChERS (ACN + water)	2–4 μg/kg	UHPLC-QqQ-MS/MS (MRM)	[21]
Surface, drinking, groundwater	2 degradation products	BEH ShieldRP$_{18}$ C18 (50 × 2.1 mm i.d., 1.7 μm)	Gradient mode (0.36 mL/min): phase A, water; phase B: CAN	Direct large volume injection	0.01	UPLC-QqQ-MS/MS (MRM)	[46]
Soil	15	BEH C18 (100 × 2.1 mm i.d., 1.7 μm)	Gradient mode (0.3 mL/min): phase A, MeOH; phase B: 5 mM ammonium formate in water	QuEChERS approach (water + ACN)	1–5 μg/kg	UPLC-QqQ-MS/MS (MRM)	[47]
Surface, groundwater	2	HSS T3 C18 (50 × 2.1 mm i.d., 1.8 μm)	Gradient mode (0.3 mL/min): phase A, MeOH; phase B: water	Direct injection	0.1	UPLC-QqQ-MS/MS (MRM)	[48]
Wastewater	Multiresidue (84)	BEH C18 (100 × 2.1 mm i.d., 1.7 μm)	Gradient mode (0.45 mL/min): phase A: 5 mM ammonium bicarbonate, pH 9.5, in water; phase B: MeOH (ESI +) phase A: 0.05% acetic acid in water; phase B: 0.05% acetic acid in MeOH (ESI –)	SPE (Oasis MCX + Strata-X)	0.015–0.026	UPLC-TOF-MS (full scan)	[49]
Groundwater	1 parent + 4 degradation products	BEH C18 (100 × 2.1 mm i.d., 1.7 μm)	Gradient mode (0.3 mL/min): phase A, 0.01% formic acid in water; phase B: 0.01% formic acid in MeOH	Direct injection	Not reported	UPLC–QTOF-MS (product ion scan)	[50]
Wastewater	Multiresidue (39)	BEH C18 (50 × 2.1 mm i.d., 1.7 μm)	Gradient mode (0.3 mL/min): phase A, MeOH; phase B: 0.01% formic acid in water	SPE (C18)	0.01–0.5	UPLC-QqQ-MS/MS (MRM)	[51]

Drinking water	8 parent + degradation products	BEH C18 (50 × 2.1 mm i.d., 1.7 μm)	Gradient mode (0.35 mL/min): phase A, ACN; phase B: water	Online SPE (polymeric cartridges PLRP-s)	Not reported	UPLC–QTOF-MS (full scan product ion scan)	[52]
Surface, drinking, groundwater	1 degradation product	HSS T3 C18 (50 × 2.1 mm i.d., 1.8 μm)	Gradient mode (0.36 mL/min): phase A, ACN; phase B: 0.01% formic acid in water	Direct injection	0.01	UPLC-QqQ-MS/MS (MRM)	[53]
Surface, ground, wastewater	Multiresidue (37)	HSS T3 C18 (50 × 2.1 mm i.d., 1.8 μm)	Gradient mode (0.3 mL/min): phase A, 0.01 mM NH_4Ac in water; phase B: 0.01 mM NH_4Ac in MeOH	SPE (Oasis HLB)	0.05–1	UPLC-QqQ-MS/MS (MRM)	[54]
Surface water	Multiresidue (31)	BEH C18 (100 × 2.1 mm i.d., 1.7 μm)	Gradient mode (0.4 mL/min): phase A, 0.1% formic acid in ACN; phase B, water:ACN, 9:1 with 0.1% formic acid	SPE (Oasis HLB)	0.001–0.020	UPLC-QqQ-MS/MS (MRM)	[55]
Wastewater	Degradation products of two parent	BEH C18 (100 × 2.1 mm i.d., 1.7 μm)	Gradient mode (0.4 mL/min): phase A, ACN; phase B, ACN (10%), 0.1% formic acid in water	SPE (Oasis HLB)	Not reported	UPLC-QqQ-MS/MS (MRM)	[56]
Water	Multi-residue (>40)	BEH C18 (100 × 2.1 mm i.d., 1.7 μm)	Gradient mode (0.3 mL/min): phase A, MeOH; phase B, 0.01% formic acid in water	SPE (Oasis HLB)	<0.02	UPLC-QqQ-MS/MS (MRM)	[57]
Water	9	BEH C18 (100 × 2.1 mm i.d., 1.7 μm)	Gradient mode (0.5 mL/min): phase A, 0.1% formic acid in water; phase B, ACN	SPE (Oasis HLB)	0.0001–0.020	UPLC-QqQ-MS/MS (MRM)	[58]

Note: ACN, acetonitrile; BEH, bridged ethyl hybrid; ESI, electrospray ionization; HSS, high strength silica; MeOH, methanol; MRM, multireaction monitoring; MS/MS, tandem mode; NH_4Ac, ammonium acetate; QqQ, triple quadrupole; QTOF, quadrupole-time-of-flight; QuEChERS, quick, easy, cheap, effective, rugged, and safe; SPE, solid phase extraction; TOF, time-of-flight.

Despite the mobile phase being important to obtain a good chromatographic separation, it also affects the analyte ionization and the sensitivity of the MS. So pesticide charge should be suppressed by modification of the mobile phase pH for optimum retention, but this can have a negative influence on MS response. On the contrary, for optimized ESI, the pH should be adjusted to promote the charged state of the pesticide over its neutral species as ionization takes place in the liquid phase. In ESI, the total ion current, and therewith sensitivity, is influenced mainly by two parameters. The first is the electrospray voltage; a higher voltage results in a higher ion current. The second way to improve the ion current is the enhancement of the eluent's conductivity by addition of buffers. Typical buffers used in pesticide residue analysis are ammonium acetate, ammonium formate, formic acid, and acetic acid in concentrations between 0.5 mmol/L [32] and 20 mmol/L [44] (Tables 19.2 and 19.3). Less common has been the use of ammonium bicarbonate in a multiresidue method [49]. In general, as it can be observed in Tables 19.2 and 19.3, the aqueous phase is modified by the addition of a buffer. However, in other applications, both aqueous and organic phases contain the same [17,19,20,24,25,27,32,37,39,41,43,44,50,54] or different [35,49] buffers. There are only four applications that do not use buffers [45,46,48,52]; two of them use mixtures of methanol and water [45,48], and the other two use [46,52] acetonitrile and water.

It is also important to note that the quality of solvents has influence on the results. In general, solvent quality for HPLC-MS is necessary although volatile buffers must be used. On the contrary, residues of the nonvolatile buffers will be accumulated in the mass analyzer, which may inhibit the correct function of the system.

19.4 UPLC APPLIED TO ANALYSIS OF PESTICIDES

Significant UPLC–MS methods in pesticide residue analysis covered by this chapter in food and environmental samples are summarized in Tables 19.2 and 19.3. Two main steps in the analytical methodology applied for pesticide residues are sample preparation and chromatographic analysis; this section follows this differentiation.

On the one hand, sample preparation for pesticides in these matrices usually involves extraction, cleanup, and concentration steps prior to UPLC–MS. Simple matrices, such as surface, drinking, and groundwater, can sometimes be directly analyzed without pretreatment [46,48,50,53]. However, most food and water samples require some pretreatment before the analysis to isolate the target pesticides, to eliminate interferences, and to concentrate the sample. In the optimization of multiresidue methods, a compromise is required in order to get acceptable recoveries for the simultaneous extraction of as many pesticides as possible. In addition, sample extraction methods for multiresidue analysis should be as simple as possible in order to achieve high sample throughput.

For solid food samples, solvent extraction (SE) is the most widely used extraction technique in pesticide analysis. Toward this aim, solvents of medium polarity are commonly selected for SE, such as acetonitrile [18,19,28,30,43–45], methanol [20,25,32], ethyl acetate [24,25,37,42], or acetone [41]. Extraction conditions, particularly pH, must usually be adjusted to facilitate analyte extraction using formic [16,18–20,23,27,28,30,36,38–41] or acetic acid [25,43]. Solid phase extraction (SPE), using as sorbents Oasis HLB [28] and florisil [30], is selected as an additional cleanup step following SE for complex matrices.

In a recent study [17], a pressurized liquid solvent containing a mixture of 1,1,1,2-tetrafluoroethane and toluene as solvents was selected for the extraction of pesticide residues. Both solvents are immiscible with water and cannot dissolve matrix components (sugars and glycerides) from the target fruits and vegetables; therefore, a cleaning step is not required.

For liquid food samples, liquid–liquid extraction (LLE) has been applied as an extraction procedure for pesticides from several matrices. Acetonitrile saturated in n-hexane is used for the extraction of phosmet and phosmet metabolites from treated olive oil samples [32]. Ethyl acetate and extraction with water–acetone with formic acid were found to be the default solvents of choice for benzimidazole carbamate residues in milk [35] and multiclass pesticides in milk and honey samples

[41], respectively. However, new trends in pesticide residue analysis have been focused on the miniaturization of the sample preparation methodology, moving to the development of straightforward, faster, cost-effective, and environmentally friendly procedures. In this sense, an alternative to LLE is hollow-fiber liquid-phase microextraction (HF-LPME), which is a miniaturized technique of LLE that greatly reduces the amount of organic solvent required. The utility of HF-LPME has been recently demonstrated for the determination of seven pesticides in cucumber using chloroform as the acceptor phase and a mixture of methanol:water 1:1, v/v, as the desorption phase [17]. Also, HF-LPME was employed for the multiresidue extraction of more than 50 pesticides in alcoholic beverages using 1-octanol as the acceptor phase and methanol as the desorption phase [37].

In recent years, extraction procedures based on QuEChERS (quick, easy, cheap, effective, rugged, and safe) methodology have been used. The original QuEChERS method is based on SE using acetonitrile as an extractant in the presence of magnesium sulfate and sodium chloride, followed by a dispersive SPE (dSPE) cleanup step with primary–secondary amine. Several modifications of the method have been introduced, ranging from the modification of the used organic solvents (methanol, ethyl acetate), salts (sodium acetate), and dSPE sorbents (C18, graphitized carbon black), depending on the target pesticides and nature of the matrix. Acetonitrile is the preferred organic solvent in the QuEChERS method to extract pesticide residues in a variety of food matrices, such as fruit and vegetables [15,21–23,26,29,31,40], juice [36], or infant foods [34] although methanol is also used as an extraction solvent, for example, for the multiresidue extraction of pesticides from cereal grains [27] and wine [33]. To improve the extraction efficiency of low moisture containing samples, such as cereals [22,27] or rice [40], the addition of water is carried out.

In relation to environmental samples, pesticide analysis has mainly been carried out in water [20,46,48–58] and, to a lesser extent, in soil [21,47] samples. Sample treatment methods for water samples are usually based on SPE using polymeric sorbents [20,49,52,54–58] although C18 [51,59] has also been reported. Extraction of pesticides from soil matrices has been performed by QuEChERS, using acetonitrile as an extractant and adding water to improve the extraction efficiency.

On the other hand, in relation to the chromatographic analysis, the majority of the UPLC methods make use of the triple quadrupole (QqQ) mass analyzer, using the multiple-reaction monitoring (MRM) mode (Tables 19.2 and 19.3). UPLC-QqQ operating in tandem mode (MS/MS) has become, so far, the most widely used technique for the quantitation of target pesticides because a high sensitivity and selectivity are achieved. Even though the majority of UPLC-QqQ-MS/MS methods proposed focus on the development of large multiresidues (>30 pesticides) or even multiclass methods, for instance, pesticides and veterinary drugs [19,49]; pesticides and mycotoxins [22]; or pesticides, mycotoxins, and plant toxins [41], several deal with short multiresidue methods (<30 pesticides) or a unique pesticide [18]. Nowadays, UPLC-QqQ-MS/MS methods are capable of analyzing approximately 100 pesticides simultaneously, for instance, in wine [33] and fruit juices [36] with sufficient sensitivity for detection at μg/L level.

A major limitation of the QqQ analyzers working in MRM mode is that they are limited on the number of transitions that can be monitored, and so in their ability to analyze a large number of pesticides in a single chromatographic run. In addition, UPLC-QqQ-MS/MS methods are only focused on target analysis, and unknown or nontarget compounds are missed, thus making it difficult to detect the presence of metabolites or transformation products of pesticides. Therefore, other mass analyzers, such as time-of-flight (TOF) MS, hybrid quadrupole (Q)TOF, or orbitrap, have been introduced to solve these problems. These mass analyzers allow the screening of a virtually unlimited number of pesticides, including both target and nontarget compounds.

Thus, the coupling of UPLC-TOF provides an excellent analytical tool working in full scan mode for the identification/quantification of 212 pesticides in food plants [26] and 84 pesticides and pharmaceuticals in wastewater [49] at trace levels. Compared to UPLC-MS/MS, UPLC-TOF-MS showed an adequate quantification of 100 pesticides in strawberries [37]. The QTOF analyzer can be simply operated as a TOF analyzer (QTOF-MS) in full-scan mode or as a tandem mass

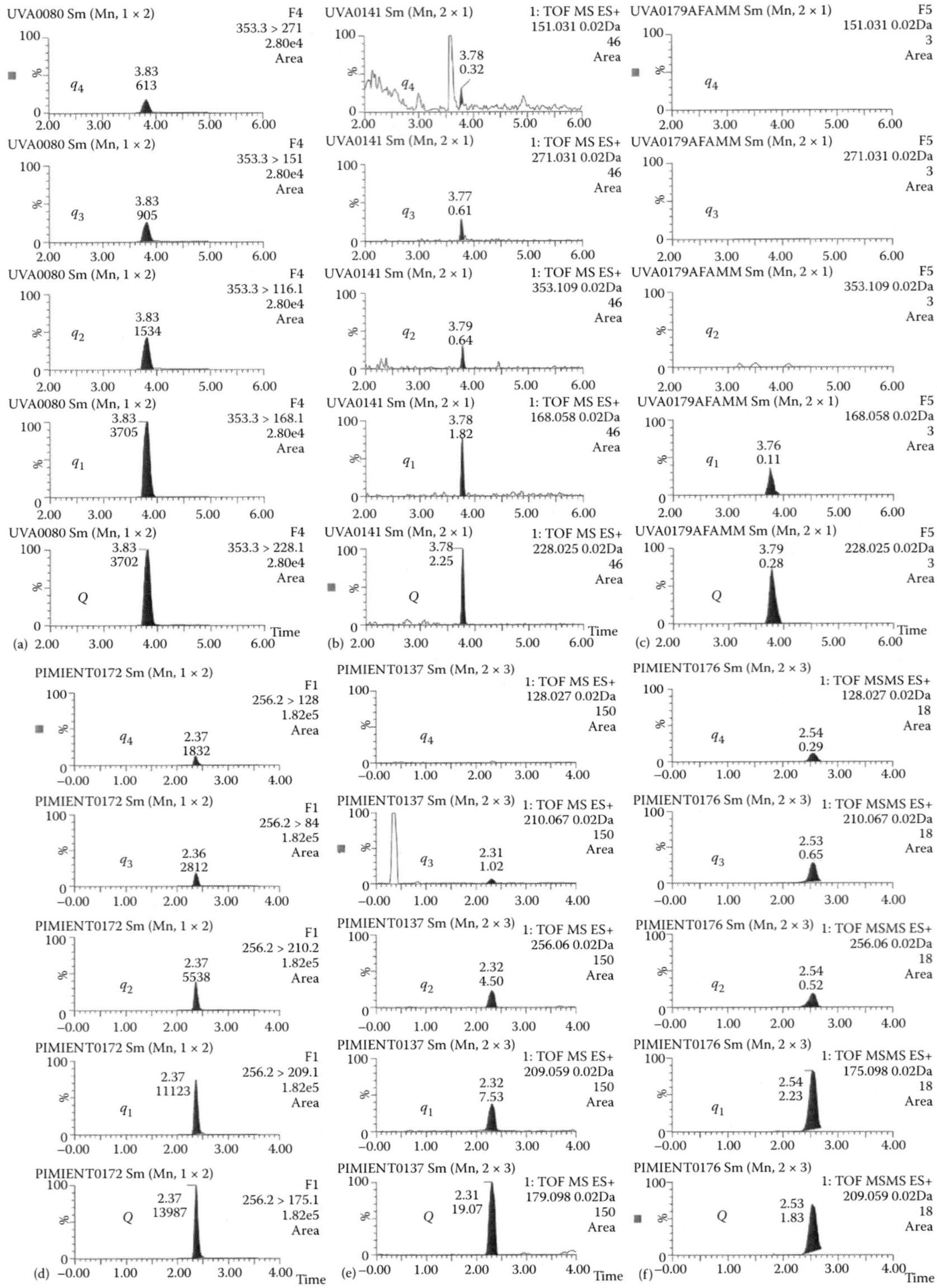

FIGURE 19.4 Chromatograms for grape sample positive to hexythiazox (a, b, and c) and pepper sample positive to imidacloprid (d, e, and f) obtained by QqQ, (left), TOF (center), and QTOF (right). *Q*, quantification ion; *q*i, first, second, third, or fourth confirmation ions. (Reprinted from Grimalt, S. et al., *J. Mass Spectrom.*, 45, 421–436, 2010. With permission.)

spectrometer (QTOF-MS/MS) in the product-ion scan mode. Both capabilities have been evaluated in the target and nontarget screening for multiclass compounds [20]. The nontargeted screening presents important drawbacks at low compound concentrations although an interesting advantage associated with TOF-MS-based approaches concerns the possibility of performing retrospective analysis [20]. However, compared to LC-MS/MS methods, UHPLC-QTOF-MS/MS showed a poor repeatability and large measurement uncertainty, but it was an ideal tool for posttarget screening and confirmation of pesticides in fruits and vegetables [29,31]. Also, compared to LC-MS/MS, UPLC-QTOF-MS/MS showed poor repeatability and large measurement uncertainty for quantification, but it was adequate for screening of many pesticides as possible in a single analysis and confirming the identity of pesticides based on accurate mass measurement at low level [34].

The utility of UPLC-TOF-MS and UPLC-QTOF-MS/MS for the detection of pesticides and their transformation products in food and water samples has been evaluated. Picó et al. used an UPLC-QTOF-MS method for the identification of pesticide residues and degradation products from the postharvest treatment of pears and apples [24]. The levels of the metabolites found exceeded several times those of the parent compounds [24]. The application of UPLC-QTOF-MS for the successful identification of three pesticides (imazalil, carbendazim, and ethoxyquin) in a pear extract has been reported [42]. The potential of the UPLC-QTOF-MS and UPLC-QTOF-MS/MS techniques as a quantification tool was also discussed, taking imazalil as an example, concluding that UPLC-QTOF-MS might become a powerful analytical tool for both identification of unknown pesticides and quantification of target pesticides [42]. Another study shows the use of UPLC-QTOF-MS to discover the presence of pesticide metabolites in food samples [32] as well as the advantages of UPLC-QTOF-MS/MS to elucidate and/or confirm the presence of detected metabolites [32].

The identification of transformation products of the herbicide bromacil after groundwater chlorination has been investigated using UPLC-TOF-MS [50]. Also, the identification of the four main degradation products from disinfection by-products of selected triazines in drinking water was carried out by UPLC-QTOF-MS/MS [52].

A comparison of the three most used mass analyzers (QqQ, TOF, and QTOF) coupled to UPLC in the field of pesticide analysis, taking 11 pesticides and nine vegetable matrices as a model, was carried out. It was concluded that the QqQ analyzer was the most satisfactory for quantification purposes; the TOF platform was the most adequate for screening purposes, and the QTOF analyzer was the most powerful for confirmation purposes [25]. Figure 19.4 shows the UPLC chromatograms confirming the presence of hexythiazox in a grape sample by using the QqQ, TOF, and QTOF analyzers. Finally, a remarkable multiclass method for the screening of >350 compounds (pesticides and veterinary drugs) in honey with UPLC-orbitrap-MS has been recently reported [19]. UHPLC–orbitrap-MS can also provide adequate quantification of target compounds [19].

In general, it is worth noting that the developed UPLC methods showed good quantitative results with detection values lower than the maximum levels established by the EU in food or water matrices.

19.5 SUMMARY AND CONCLUSIONS

In pesticide monitoring laboratories, analysis time is an important issue to consider in the choice of an analytical method in order to increase sample throughput. Toward this aim, UPLC using columns with porous sub-2-μm particles has been the most widely used approach since 2004. The use of smaller particle sizes provides higher column efficiency, resolution, and speed analysis, and the use of small columns with fine particles also reduces solvent consumption and improves the run time.

RP-UPLC is the most applied chromatographic mode to separate pesticides by partitioning between an apolar stationary phase and a polar mobile phase. Most stationary phases are based on silica that has been chemically modified with octadecyl (C18) using BEH technology. The most common binary solvent systems used as a mobile phase are based on acetonitrile–water or methanol–water mixtures. In general, an aqueous phase is modified by the addition of a volatile buffer in UPLC-MS in order to improve the pesticide ionization and therefore the sensitivity.

Finally, some general conclusions regarding the analysis of pesticides by UPLC-MS in food and environmental samples can be outlined. First, it can be indicated that in the last few years, extraction procedures based on QuEChERS methodology have been mainly employed for food samples due to this approach being quicker and easier than other methodologies previously used; however, SPE is the most used technique for water samples. Second, as for the MS analyzers used, the QqQ analyzer has been the most widely applied although the use of TOF and QTOF analyzers has increased in recent years, and orbitrap applications are still scarce. In general, the QqQ platform was the most satisfactory analyzer for quantification purposes, the TOF analyzer was the most adequate for screening purposes, and the QTOF analyzer was the most powerful for confirmation purposes.

ACKNOWLEDGMENTS

We gratefully acknowledge the Spanish Ministry of Economy and Competitiveness and FEDER for financial support through project AGL2010-21370.

REFERENCES

1. Ramírez, J. A., Lacasaña, M., Plaguicidas: Clasificación, uso, toxicología y medición de la exposición, *Arch. Prev. Riesgos Labor.*, 4, 67–75, 2001.
2. U.S. Department of Agriculture, Foreign Agricultural Service, Maximum Residue Limit Database. www.fas.usda.gov/htp/MRL.asp (last accessed July 2013).
3. EC Regulation No. 396/2005, Off. J. Eur. Commun. L70/1-16, http://ec.europa.eu/food/plant/protection/pesticides/community_legislation_en.htm (last accessed July 2013).
4. http://www.epa.gov/waterscience/methods/pollutants.html (last accessed July 2013).
5. http://www.epa.gov/safewater/contaminants/index.html (last accessed July 2013).
6. Directive 2008/105/EC. Off. J. Eur. Communities L 348/84, 24.12.2008.
7. Garrido Frenich, A., Martínez Vidal, J. L., Pastor-Montoro, E., Romero-González, R., High-throughput determination of pesticide residues in food commodities by use of ultra-performance liquid-chromatography-tandem mass spectrometry, *Anal. Bioanal. Chem.*, 390, 947–959, 2008.
8. Wnag, Y., Ai, F., Ng, S.-C., Yang Tan, T. T., Sub-2 µm porous silica materials for enhanced separation performance in liquid chromatography, *J. Chromatogr. A*, 1228, 99–109, 2012.
9. Guilarme, D., Schappler, J., Rudaz, S., Veuthey, J. L., Coupling ultra-high-performance liquid chromatography with mass spectrometry, *Trends Anal. Chem.*, 29, 15–27, 2010.
10. Wu, N., Clausen, A. M., Fundamental and practical aspects of ultrahigh pressure liquid chromatography for fast separations, *J. Sep. Sci.*, 30, 1167–1182, 2007.
11. Novakavá, L., Vickovâ, H., A review of current trends and advances in modern bio-analytical methods: Chromatography and sample preparation, *Anal. Chim. Acta*, 656, 8–35, 2009.
12. Nguyen, D. T. T., Guillarme, D., Henisch, S., Barrioulet, M.-P., Rocca, J. L., Rudaz, Veuthey, J. L., High throughput liquid chromatography with sub-2 µm particles at high pressure and high temperature, *J. Chromatogr. A*, 1167, 76–84, 2007.
13. Swartz, M. E., UHPLCTM: An introduction and review, *J. Liquid Chromatogr. and Rel. Technol.*, 28, 1253–1263, 2005.
14. Botitsi, H. V., Garbis, S. D., Economou, A., Tsipi, D. F., Current mass spectrometry strategies for the analysis of pesticides and their metabolites in food and water matrices, *Mass Spectrom. Rev.*, 30, 907–939, 2011.
15. Li, Y., Jiao, B., Zhao, Q., Wang, C., Gong, Y., Zhang, Y., Chen, W., Effect of commercial processing on pesticide residues in orange products, *Eur. Food Res. Technol.*, 234, 449–456, 2012.
16. Wang, J., Du, Z., Yu, W., Qu, S., Detection of seven pesticides in cucumbers using hollow fibre-based liquid-phase microextraction and ultra-high pressure liquid chromatography coupled to tandem mass spectrometry, *J. Chromatogr. A*, 1247, 10–17, 2012.
17. Turkoz Bakırc, G., Hısil, Y., Fast and simple extraction of pesticide residues in selected fruits and vegetables using tetrafluoroethane and toluene followed by ultrahigh-performance liquid chromatography/tandem mass spectrometry, *Food Chem.*, 135, 1901–1913, 2012.

18. Kong, Z., Dong, F., Xu, J., Liu, X., Zhang, C., Li, J., Li, Y., Chen, X., Shan, W., Zheng, Y., Determination of difenoconazole residue in tomato during home canning by UPLC-MS/MS, *Food Control*, 23, 542–546, 2012.
19. Gómez-Pérez, M. L., Plaza-Bolaños, P., Romero-Gonzalez, R., Martínez Vidal, J. L., Garrido Frenich, A., Comprehensive qualitative and quantitative determination of pesticides and veterinary drugs in honey using liquid chromatography–Orbitrap high resolution mass spectrometry, *J. Chromatogr. A*, 1248, 130–138, 2012.
20. Díaz, R., Ibañez, M., Sancho, J. V., Hernández, F., Target and non-target screening strategies for organic contaminants, residues and illicit substances in food, environmental and human biological samples by UHPLC-QTOF-MS, *Anal. Methods*, 4, 196–209, 2012.
21. Dong, F., Liu, X., Xu, J., Li, J., Li, Y., Shan, W., Song, W., Zheng, Y., Determination of cyantraniliprole and its major metabolite residues in vegetable and soil using ultra-performance liquid chromatography/tandem mass spectrometry, *Biomed. Chromatogr.*, 2, 377–383, 2012
22. Romero-Gonzalez, R., Garrido Frenich, A. Martínez Vidal, J. L., Prestes, O. D., Grio, S. L., Simultaneous determination of pesticides, biopesticides and mycotoxins in organic products applying a quick, easy, cheap, effective, rugged and safe extraction procedure and ultra-high performance liquid chromatography–tandem mass spectrometry, *J. Chromatogr. A*, 1218, 1477–1485, 2011.
23. Liu, X., Xu, J., Dong, F., Li, Y., Song, W., Zheng, Y., Residue analysis of four diacylhydrazine insecticides in fruits and vegetables by quick, easy, cheap, effective, rugged, and safe (QuEChERS) method using ultra-performance liquid chromatography coupled to tandem mass spectrometry, *Anal. Bioanal. Chem.*, 401, 1051–1058, 2011.
24. Picó, Y., la Farré, M., Segarra, R., Barceló, D., Profiling of compounds and degradation products from the postharvest treatment of pears and apples by ultra-high pressure liquid chromatography quadrupole-time-of-flight mass spectrometry, *Talanta*, 81, 281–293, 2010.
25. Grimalt, S., Sancho, J. V., Pozo, O. J., Hernández, F., Quantification, confirmation and screening capability of UHPLC coupled to triple quadrupole and hybrid quadrupole time-of-flight mass spectrometry in pesticide residue analysis, *J. Mass Spectrom.*, 45, 421–436, 2010.
26. Lacina, O., Urbanova, J., Poustka, J., Hajslova, J., Identification/quantification of multiple pesticide residues in food plants by ultra-high-performance liquid chromatography-time-of-flight mass spectrometry, *J. Chromatogr. A*, 1216, 648–659, 2010.
27. Mastovska, K., Dorweiler, K. J., Lehotay, S. J., Wegscheid, J. S., Szpylka, K. A., Pesticide multiresidue analysis in cereal grains using modified QuEChERS method combined with automated direct sample introduction GC-TOFMS and UPLC-MS/MS techniques, *J. Agric. Food Chem.*, 58, 5959–5972, 2010.
28. Liu, S., Zheng, Z., Wei, F., Ren, Y., Gui, W., Wu, H., Zhu, G., Simultaneous determination of seven neonicotinoid pesticide residues in food by ultraperformance liquid chromatography tandem mass spectrometry, *J. Agric. Food Chem.*, 58, 3271–3278, 2010.
29. Wang, J., Leung, D., Cho, W., Applications of LC/ESI-MS/MS and UHPLC QqTOF MS for the determination of 148 pesticides in berries, *J. Agric. Food Chem.*, 58, 5904–5925, 2010.
30. Lu, C., Liu, X., Dong, F., Xu, J., Song, W., Zhang, C., Li Y., Zheng, Y., Simultaneous determination of pyrethrins residues in teas by ultra-performance liquid chromatography/tandem mass spectrometry, *Anal. Chim. Acta*, 678, 56–62, 2010.
31. Wang, J., Chow, W., Leung, D., Applications of LC/ESI-MS/MS and UHPLC QqTOF MS for the determination of 148 pesticides in fruits and vegetables, *Anal. Bioanal. Chem.*, 396, 1513–1538, 2010.
32. Hernández, F., Grimalt, S., Pozo, O. J., Sancho, J. V., Use of ultra-high-pressure liquid chromatography–quadrupole time-of-flight MS to discover the presence of pesticide metabolites in food samples, *J. Sep. Sci.*, 32, 2245–2261, 2009.
33. Zhang, K., Wong, J. W., Hayward, D. G., Sheladia, P., Krynitsky A. J., Schenck, F. J., Webster, M. G., Ammann, J. A., Ebeler, S. E., Multiresidue pesticide analysis of wines by dispersive solid-phase extraction and ultra high-performance liquid chromatography-tandem mass spectrometry, *J. Agric. Food Chem.*, 57, 4019–4029, 2009.
34. Wang, J., Leung, D., Applications of ultra-performance liquid chromatography electrospray ionization quadrupole time-of-flight mass spectrometry on analysis of 138 pesticides in fruit- and vegetable-based infant foods, *J. Agric. Food Chem.*, 57, 2162–2173, 2009.
35. Keegan, J., Whelan, M., Danaher, M., Crooks, S., Sayers, R. Anastasio, A., Elliott, C., Brandon, D., Furey, A., O'Kennedy, R., Benzimidazole carbamate residues in milk: Detection by surface plasmon resonance-biosensor, using a modified QuEChERS (quick, easy, cheap, effective, rugged and safe) method for extraction, *Anal. Chim. Acta*, 654, 111–119, 2009.

36. Romero-González, R., Garrido Frenich, A., Martínez Vidal, J. L., Multiresidue method for fast determination of pesticides in fruit juices by ultra performance liquid chromatography coupled to tandem mass spectrometry, *Talanta*, 76, 211–225, 2008.
37. Taylor, M. J., Keenan, G. A., Reid, K. B., Uría Fernández, D., The utility of ultra-performance liquid chromatography/electrospray ionisation time-of-flight mass spectrometry for multi-residue determination of pesticides in strawberry, *Rapid Commun. Mass Spectrom.*, 222, 2731–2746, 2008.
38. Plaza-Bolaños, P., Romero-González, R., Garrido Frenich, A., Martínez Vidal, J. L., Application of hollow fibre liquid phase microextraction for the multiresidue determination of pesticides in alcoholic beverages by ultra-high pressure liquid chromatography coupled to tandem mass spectrometry, *J. Chromatogr. A*, 1208, 16–24, 2008.
39. Koesukwiwata, U., Sanguankaewa, K., Leepipatpiboona, N., Rapid determination of phenoxy acid residues in rice by modified QuEChERS extraction and liquid chromatography–tandem mass spectrometry, *Anal. Chim. Acta*, 626, 10–20, 2008.
40. Garrido Frenich, A., Martínez Vidal, J. L., Pastor-Montoro, E., Romero-González, R., High-throughput determination of pesticide residues in food commodities by use of ultra-performance liquid chromatography–tandem mass spectrometry, *Anal. Bioanal. Chem.*, 390, 947–959, 2008.
41. Mol, H. G. J., Plaza-Bolaños, P., Zomer, P., de Rijk, T. C., Stolker, A. A. M., Mulder, P. P. J., Toward a generic extraction method for simultaneous determination of pesticides, mycotoxins, plant toxins, and veterinary drugs in feed and food matrixes, *Anal. Chem.*, 80, 9450–9459, 2008.
42. Picó, Y., la Farré, M., Soler, C., Barceló, D., Identification of unknown pesticides in fruits using ultra-performance liquid chromatography–quadrupole time-of-flight mass spectrometry imazalil as a case study of quantification, *J. Chromatogr. A*, 1176, 123–134, 2007.
43. Leandro, C. C., Hancock, P., Fussell, R. J., Keely, B. J., Ultra-performance liquid chromatography for the determination of pesticide residues in foods by tandem quadrupole mass spectrometry with polarity switching, *J. Chromatogr. A*, 1144, 161–169, 2007.
44. Leandro, C. C., Hancock, P., Fussell, R. J., Keely, B. J., Comparison of ultra-performance liquid chromatography and high-performance liquid chromatography for the determination of priority pesticides in baby foods by tandem quadrupole mass spectrometry, *J. Chromatogr. A*, 1103, 94–101, 2006.
45. Kovalczuk, T., Jech, M., Poustka, J., Hajslová, J., Ultra-performance liquid chromatography–tandem mass spectrometry: A novel challenge in multiresidue pesticide analysis in food, *Anal. Chim. Acta*, 577, 8–17, 2006.
46. Kowal, S., Balsaa, P., Werres, F., Schmidt, T. C., Reduction of matrix effects and improvement of sensitivity during determination of two chloridazon degradation products in aqueous matrices by using UPLC-ESI-MS/MS, *Anal. Bioanal. Chem.*, 403, 1707–1717, 2012.
47. Prestes, O. D., Padilla-Sánchez, J. A., Romero-González, R., López-Grio, S., Garrido Frenich, A., Martínez Vidal, J. L., Comparison of several extraction procedures for the determination of biopesticides in soil samples by ultrahigh pressure LC-MS/MS, *J. Sep. Sci.*, 35, 861–868, 2012.
48. Ripollés, C., Sancho, J. V., López, F. J., Hernández, F., Liquid chromatography coupled to tandem mass spectrometry for the residue determination of ethylenethiourea (ETU) and propylenethiourea (PTU) in water, *J. Chromatogr. A*, 1243, 53–61, 2012.
49. Nurmi, J., Pellinen, J., Multiresidue method for the analysis of emerging contaminants in wastewater by ultra performance liquid chromatography–time-of-flight mass spectrometry, *J. Chromatogr. A*, 1218, 6712– 6719, 2011.
50. Ibañez, M., Sancho, J. V., Pozo, O. J., Hernández, F., Use of quadrupole time-of-flight mass spectrometry to determine proposed structures of transformation products of the herbicide bromacil after water chlorination, *Rapid Commun. Mass Spectrom.*, 222, 2731–2746, 2008.
51. Barco-Bonilla, N., Romero-González, R., Plaza-Bolaños, P., Garrido Frenich, A., Martínez Vidal, J. L., Analysis and study of the distribution of polar and non-polar pesticides in wastewater effluents from modern and conventional treatments, *J. Chromatogr. A*, 1217, 7817–7825, 2010.
52. Brix, R., Bahi, N., Lopez de Alda, M. J., Farré, M., Fernández, J. M., Barceló, D., Identification of disinfection by-products of selected triazines in drinking water by LC-Q-ToF-MS/MS and evaluation of their toxicity, *J. Mass Spectrom.*, 44, 330–337, 2009.
53. Kowal, S., Balsaa, P., Werres, F., Schmidt, T. C., Determination of the polar pesticide degradation product N,N-dimethylsulfamide in aqueous matrices by UPLC–MS/MS, *Anal. Bioanal. Chem.*, 395, 1787–1794, 2009.
54. Marín, J. M., Gracia-Lor, E., Sancho, J. V., López, F. J., Hernández, F., Application of ultra-high-pressure liquid chromatography–tandem mass spectrometry to the determination of multi-class pesticides in environmental and wastewater samples. Study of matrix effects, *J. Chromatogr. A*, 1216, 1410–1420, 2009.

55. Gervais, G., Brosillon, S. Laplanche, A., Helen, C., Ultra-pressure liquid chromatography–electrospray tandem mass spectrometry for multiresidue determination of pesticides in water, *J. Chromatogr. A*, 1202,163–172, 2008.
56. Farré, M. J., Brosillon, S., Domenech, J., Peral, J., Evaluation of the intermediates generated during the degradation of diuron and linuron herbicides by the photo-Fenton reaction, *J. Photochem. Photobiol. A: Chem.*, 189, 364–373, 2007.
57. Pastor Montoro, E., Romero-González, R., Garrido Frenich, A., Hernández Torres, M. E., Martínez Vidal, J. L., Fast determination of herbicides in waters by ultra-performance liquid chromatography/tandem mass spectrometry, *Rapid Commun. Mass Spectrom.*, 21, 3585–3592, 2007.
58. Mezcua, M., Agüera, A., Lliberia, J. L., Cortés, M. A, Bagó, B., Fernández-Alba, A. R., Application of ultra performance liquid chromatography–tandem mass spectrometry to the analysis of priority pesticides in groundwater, *J. Chromatogr. A*, 1109, 222–227, 2006.
59. Nováková, L., Sollchová, D., Sollch, P., Advantages of ultra performance liquid chromatography over high-performance liquid chromatography: Comparison of different analytical approaches during analysis of diclofenac gel, *J. Sep. Sci.*, 29, 2433–2443, 2006.
60. Kalyan, J., Pekamwar, S. S., Ultra performance liquid chromatography: A recent development in HPLC, *Int. J. Pharm. Tech.*, 4, 1800–1821, 2012.
61. Guillarme, D., Ruta, J., Rudaz, S., Veuthey, J.-L., New trends in fast and high-resolution liquid chromatography: A critical comparison of existing approaches, *Anal. Bioanal. Chem.*, 397, 1069–1082, 2010.
62. Patil, V. P., Tathe, R. D., Debed, S. J., Angadi, S. S., Kale, S. H., Ultra performance liquid chromatography: A review, *Int. Res. J. Pharm.*, 2, 39–44, 2011.

20 High-Performance Liquid Chromatography versus Other Modern Analytical Methods for Determination of Pesticides

Tomasz Tuzimski and Joseph Sherma

CONTENTS

20.1 INTRODUCTION

High-performance liquid chromatography (HPLC) is one of several chromatographic methods used for identification, separation, and determination of multicomponent chemical mixtures. In 2010, the third edition of a book titled *Introduction to Modern Liquid Chromatography* edited by Lloyd R. Snyder, Joseph J. Kirkland, and John W. Dolan was published [1]; this is an excellent book for all practitioners of chromatography. Our book focuses on the HPLC of pesticides but also covers some of the more general aspects of HPLC.

Knowledge of physicochemical properties of pesticides, as presented in Chapter 2, allows the fate and behavior of such chemicals in the environment to be predicted. This knowledge aids the choice of the optimal conditions for determination of pesticides by HPLC (Chapters 3 and 6–9) in samples prepared for analysis (Chapter 12). Selection of sample preparation techniques depends mainly on the type and properties of the analytes (Chapter 4) and nature and properties of the sample (matrix) in which they are situated (Chapter 5).

On the other hand, the choice of the type of HPLC (Chapters 14–16 and 19) depends not only on the nature of the sample, but frequently on the availability of a good sample preparation technique. The detection and measurement associated with the chromatographic process (Chapters 14–16 and 19) has major impact on sample preparation. Also, the choice of a specific detection system (Chapters 14, 15, and 18) may be determined by the availability of a specific sample preparation procedure. These interactions between sample preparation, chromatographic separation, and the detection system play an important role in choosing the operations in a pesticide analysis, especially for quantitative analysis (Chapter 13), in chiral separation of analytes (Chapter 17), kinetics studies (Chapter 10), and photochemical degradation (Chapter 11) of pesticides in the environment.

20.2 PROS AND CONS OF HPLC VERSUS OTHER MODERN ANALYTICAL METHODS FOR DETERMINATION OF PESTICIDES

Pesticide residue analysis has developed largely by adapting techniques and instrumentation to the unique problems of ultra-low-level analysis in complex matrices. Residues in fruits and vegetables, cereals, processed baby food, and foodstuffs of animal origin are controlled through a system of statutory maximum residue limits (MRLs). The MRL is defined as "The maximum concentration of pesticide residue [expressed as milligrams of residue per kilogram of commodity (mg/kg)] likely to occur in or on food commodities and animal feeds after the use of pesticides according to good agricultural practice" (Proposed Pan American Health Organization/World Health Organization Plan of Action for Technical Cooperation in Food Safety, 2006–2007). MRLs vary ordinarily within the interval 0.0008–50 mg/kg (The Applicant Guide: Maximum Residue Levels, The Pesticides Safety Directorate, York, United Kingdom) and typically between 0.01 and 10 mg/kg for the adult population. Lower values of MRLs are set for baby food—the European Community specified an MRL of 0.010 mg/kg (Pesticides and the Environment, A Strategy for the Sustainable Use of Plant Protection Products and Strategy Action Plans, London, United Kingdom); the lowest levels are set for particular special residues (Status of Active Substances under European Union [EU] review [doc. 3010]; Commission Directive 2003/13 and /14; Council Directive 98/83/).

The search for the optimal steps of methods used for the analysis of pesticides is continually ongoing. With the help of a relatively uncommon technique called electron monochromator-mass spectrometry (MS), Dane et al. detected three nitro pesticides—flumetralin, pendimethalin, and trifluralin—in cigarette smoke for the first time [2]. The results suggest that approximately 10% of pesticide residues on tobacco survive the combustion process. Pendimethalin and trifluralin are considered "possible human carcinogens" by the U.S. Environmental Protection Agency, and all three pesticides are suspected endocrine-disrupting compounds [2]. Flumetralin has been banned for use on tobacco in the EU.

There is a strong interrelation between sample preparation and the analytical chromatographic process. Selection of sample preparation procedures depends mainly on the properties of analytes and matrices (see Chapters 5 and 12). The most efficient approach to pesticide analysis involves the use of chromatographic methods. In HPLC, sample preparation (cleanup) is usually required prior to injection in order to remove components that can damage the column or interfere with the separation and/or detection of the analytes. The purity of samples must usually be greater than for thin-layer chromatography (TLC) because the stationary phase (layer) is used only once as opposed to a column onto which multiple samples are injected. Especially trace analysis, as in the determination of compound impurities, may impose additional requirements on both sample preparation and the detector.

One of the most widely developed and used sample preparation techniques is QuEChERS (pronounced "catchers," an acronym for quick, easy, cheap, effective, rugged, and safe), first published in 2003 by Anastassiades et al. [3]. The QuEChERS method was originally designed for the analysis of fruits and vegetables, but it continues to undergo modification for application to a broad array of analytes, for example, pesticides (especially herbicides and fungicides) and drugs (antibiotics and other compounds throughout the entire food supply) in a vast array of matrices, for example, animal products (meat, fish, kidney, chicken, milk, and honey), cereals and grain products, and food and beverages (wine, juice, fruit, and vegetables) [4–24]. Now it is considered an "approach" rather than a "method" with great power for quick sample extraction and cleanup, especially when coupled with

HPLC-MS/MS or gas chromatography (GC)-MS methods. QuEChERS involves three steps that are presented in Figure 20.1 [24]:

- Liquid microextraction
- Solid-phase cleanup
- HPLC-MS/MS or GC-MS analysis

Readers can find details in Chapter 5, Section 5.2.

The sample treatment applied depends heavily on the complexity of the matrix. In general, in pesticide analysis of crops, one distinguishes the following four types of matrices: high water content (e.g., tomato), high acidic content (e.g., citrus), high sugar content (i.e., raisins), and high-fat content (olives or avocado). In all cases, it is often necessary to apply cleanup stages to remove nondesirable components of the matrix, for example, pigments. There are some applied procedures for pesticide multiresidue analysis in matrices with high water, acid, or sugar content (Figure 20.2) [25].

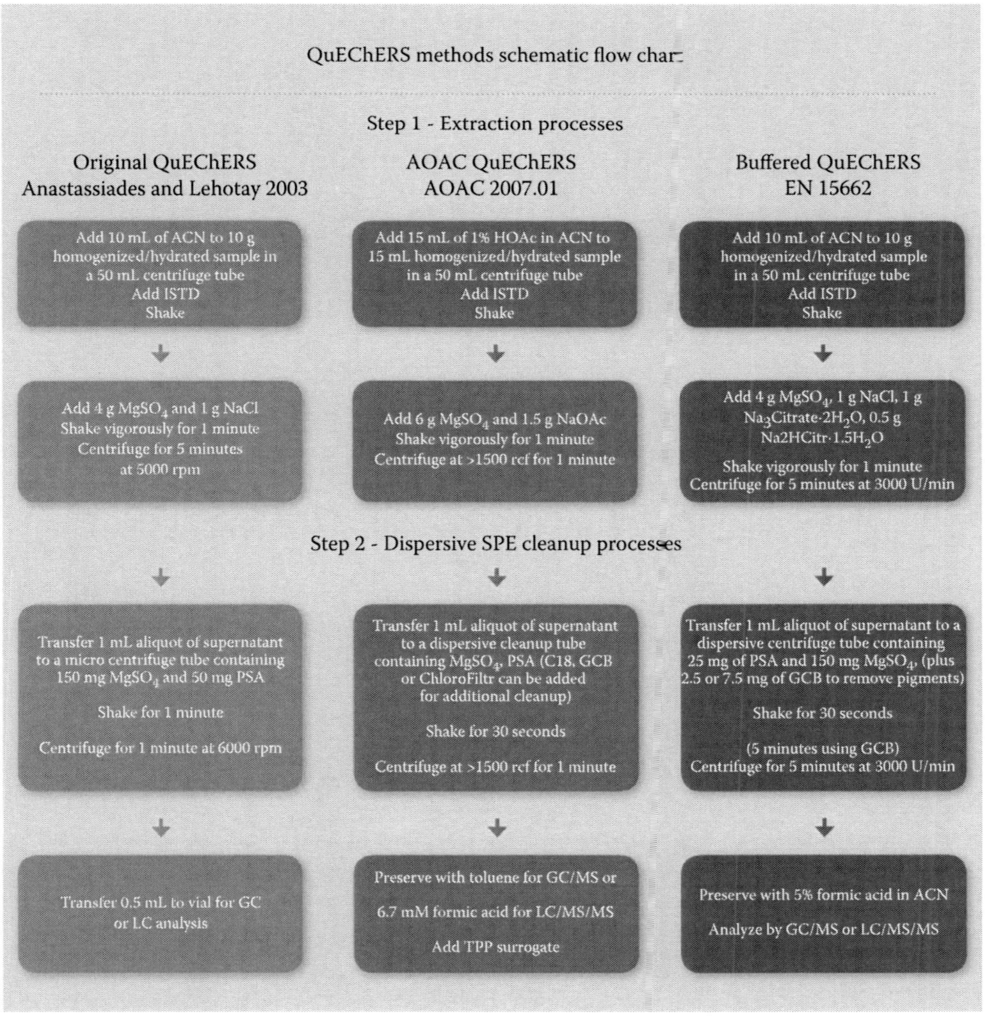

FIGURE 20.1 Three general procedures of QuEChERS methodology. (From Agilent Technologies, Inc., 5990-5324EN QuEChERS Poster, 2011, U.S.A., August 9, 2011. With permission.)

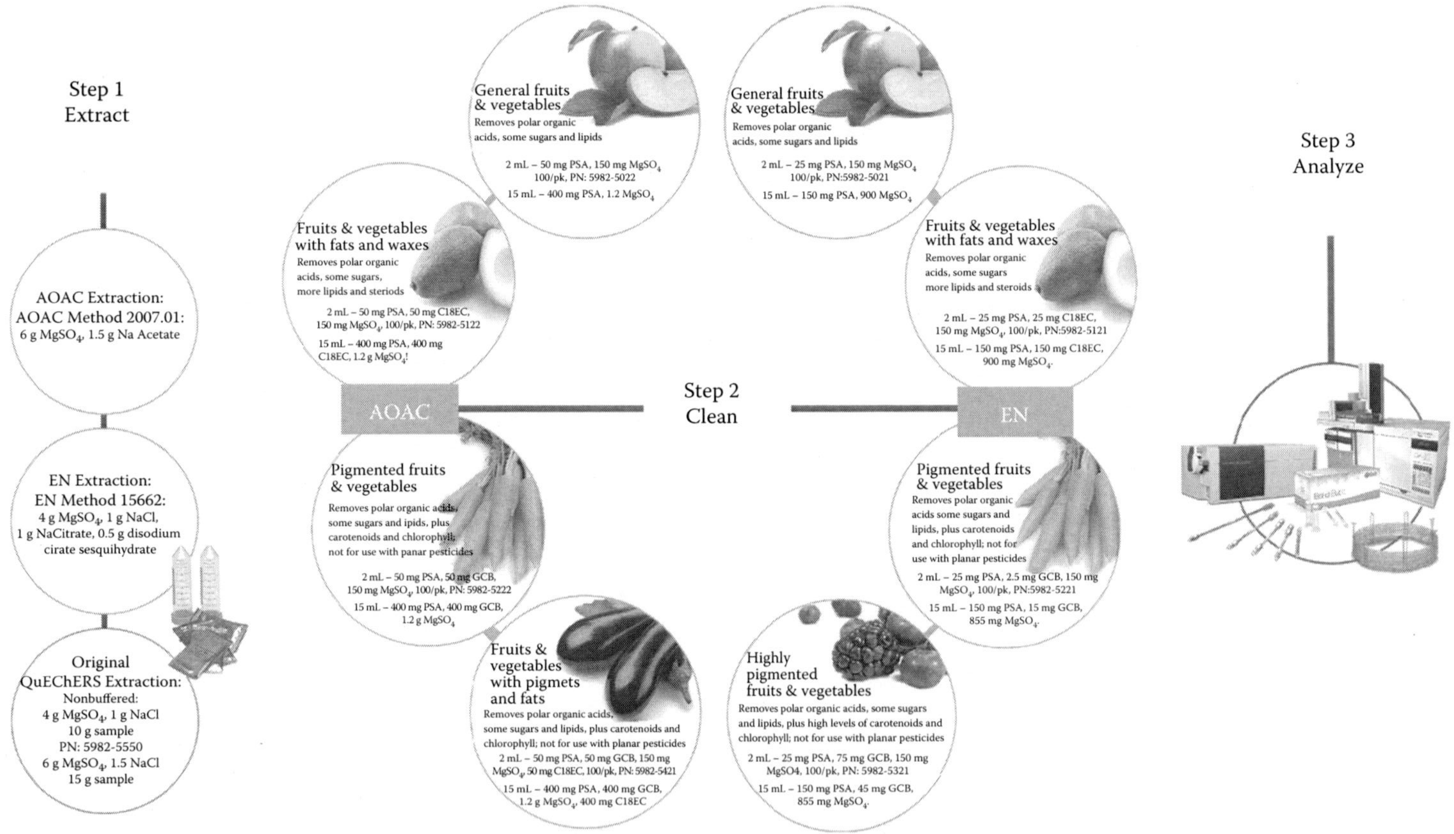

FIGURE 20.2 Procedures of the QuEChERS technique for pesticide multiresidue analysis in matrices with high water content (e.g., tomato), high acidic content (e.g., citrus), high sugar content (i.e., raisins), and high fat content (olives or avocado). (With permission from UCT, LLC (PA, USA) QuEChERS Information Booklet. Pesticide Residue Analysis. Available at www.unitedchem.com.)

About 75% of all known compounds cannot be determined by GC but can by HPLC. Although GC is considerably more efficient than HPLC (higher values of the plate number, *N*), which means faster and/or better separations are possible, GC is limited to samples that are volatile below 300°C. Therefore, HPLC is more widely applied in pesticide analysis.

When analyzing environmental samples, often we do not know what we will find in the sample. Then, we usually choose reversed-phase (RP)-HPLC for identification of unknown analytes. RP-HPLC is usually a first choice for separation of both neutral and ionic compounds using a less polar bonded phase column, such as octylsilyl (C8) or octadecylsilyl (C18), and a mobile phase that is, in most cases, a mixture of water and an organic modifier (acetonitrile, methanol, tetrahydrofuran, dioxane, or isopropanol). Methanol and acetonitrile are most frequently used. Compared to other forms of HPLC (for example, normal-phase [NP] and ion-exchange chromatography), RP-HPLC methods are usually more convenient, robust, and versatile. A common problem in RP-HPLC is poor retention of very polar compounds. Different problems may be encountered during the development and subsequent routine use of an HPLC procedure. Some of them can be anticipated in advance, allowing experiments to be carried out that will minimize the likelihood of their occurrence [1]:

- Poor retention of very polar samples
- Overlooked peaks
- Poor batch-to-batch reproducibility of the column
- Nonrobust separation conditions
- Variations in equipment

Different ways to solve some of the problems occurring in the analysis of pesticides (troubleshooting) are presented in Table 20.1 [1].

20.2.1 HPLC as the Basic Method for Analysis of Pesticides

Compared to other separation techniques, such as GC, TLC, supercritical fluid chromatography (SFC), capillary electrophoresis (CE), and capillary electrochromatography (CEC), HPLC is exceptional in terms of the following characteristics [1]:

- Almost universal applicability; few samples are excluded from the possibility of HPLC separation.
- Remarkable assay precision (relative standard deviation, RSD, ±0.5% or better in many cases).
- A wide range of equipment, columns, and other materials is commercially available, allowing the use of HPLC for almost every application.
- Most laboratories that deal with a need for analyzing chemical mixtures are equipped for HPLC; it is often the first choice technique for pesticide analysis.

In the past decade, HPLC has emerged as a technique for the separation of environmental and other types of complex samples because of its outstanding chromatographic resolving power, the possibilities to automate the analysis, and its compatibility with MS detection using an electrospray ionization (ESI) source. The advent of ESI-based LC-MS/MS resulted in the development of many multiresidue methods for the analysis of polar and thermally labile pesticides. Using one [26] or two [27,28] multiple reaction monitoring (MRM) stages [26], these methods are capable of analyzing >200 pesticides in a single LC-MS/MS analysis.

To obtain high selectivity for target analytes, direct analysis in real time (DART) MS has been combined with a highly sophisticated mass analyzer, such as time-of-flight (TOF) or orbital [29,30]. Especially HPLC-DART-TOF-MS is a very valuable technique for analysis of pesticide residues.

TABLE 20.1
Potential Problems and Troubleshooting

Problems	Reasons	Troubleshooting		Example in Pesticide Analysis
Poor retention of very polar samples	Very polar analytes are difficult to separate with nonpolar sorbents because of weak retention. Appropriate mobile phase selection is restricted because most solvents show strong elution strength for these analytes.	For nonionized analytes	Necessary to switch to normal-phase chromatography, which retains polar solutes strongly	For details please see Chapters 3 and 7
		For weakly retained solutes that are ionized	Ion-pair chromatography	For details please see Chapter 8
			Ion-exchange chromatography	For details please see Chapter 8
Poor retention of very nonpolar samples	Very nonpolar analytes are difficult to separate in systems with a nonpolar sorbent because of very strong retention. Mobile phase selection is limited because most solvents show too low strength for these separations.			For details, please see Chapters 7–9
Overlooked peaks	Poor detection sensitivity	Poor detection sensitivity often can be dealt with by the complementary use of a nonspecific detector, which is advisable when using UV detection for samples whose composition is not fully known at the start of method development.		For details, please see Chapters 7 and 14

Solvent not compatible with UV detection	When also used in a sample preparation step or as a component of the mobile phase that is nontransparent at the lowest possible wavelength for UV detection (e.g., acetone).	Change solvent(s) used in extraction technique step or/and as component of mobile phase for one that is more compatible with UV detection at wavelengths below 230 nm for increased detection sensitivity. Another possibility is a more thorough removal of the solvent after the stage of sample preparation and before chromatography.	For details, please see Chapters 7 and 14
Failure of the chromatographic procedure to separate two adjacent peaks (overlapping peaks)		An alternative approach, following the apparent separation of all peaks in the sample by a "primary" procedure, is the development of an orthogonal separation whose selectivity is very different from that of primary method.	For details, please see Chapters 3, 7–9, 14
One peak is much larger than the other		An orthogonal separation should be able to move a missing peak to another part of the chromatogram where it is more noticeable. During the method development, additional "hidden" peaks can be appear or they may be "covered" by another peak(s) in the chromatogram.	For details, please see Chapter 16 and Ref. [1]
In M3, which is able to deconvolute overlapping peaks			For details, please see Chapter(s) 15 (7 and 14)

(*Continued*)

TABLE 20.1 (CONTINUED)
Potential Problems and Troubleshooting

Problems	Reasons	Troubleshooting	Example in Pesticide Analysis
Nonrobust separation conditions	Can result in a loss in resolution from small, inadvertent changes in one or more of the separation conditions. For example, small variations in mobile-phase pH are difficult to avoid during normal laboratory operation, yet they can result in significant changes in resolution when ionizable compounds are present in the sample.	To confirm that the final method is robust, the effect on resolution of small changes in each separation condition should be determined. It is usually possible to modify separation conditions so as to improve method robustness.	For details, please see Chapters 3, 7–9, and 14
Poor batch-to-batch reproducibility of the column	This is more likely to arise for complex samples when the chromatogram is crowded and many peaks have marginal or barely adequate resolution. Small, unintended changes in column selectivity can result in decrease in resolution for one or more peaks.	After the conditions for the final method are selected, several different manufacturing lots of the column are tested to confirm equivalent performance. Usually all tested column lots will provide adequate separation, and this helps eliminate concern about column reproducibility in the future.	For details, please see Chapter 6 and Ref. [1]
Variations in equipment and their effect on the separation	Should be addressed during method development.	The most important requirement is the development of standard test procedures that will guarantee satisfactory performance of the equipment. The hold up or dwell volume of equipment used for gradient elution often varies from system to system, and this can lead to failure of the method as a result of consequent changes in relative retention.	For details, please see Chapter 13 and Ref. [1]

Source: Snyder, L. R., Kirkland, J. J., Dolan, J. W.: *Introduction to Modern Liquid Chromatography.* 2010. Copyright Wiley-VCH Verlag GmbH & Co. KGaA. Reproduced with permission; and our own experience.

The study of xenobiotics present in fruit peel by exposing it (without any pretreatment) to DART coupled to a high-resolution orbitrap mass spectrometer was reported for the first time by Farre et al. [31]. When comparing overall time and workload demands (sample preparation and instrumental analysis), DART-TOF-MS represents an excellent option thanks to the straightforward sample examination (no chromatographic separation or sophisticated extraction steps are required), and measurement in real time on the fruit peel was achieved. Furthermore, the limit of detection (LOD) values were quite acceptable, and quantification after analysis was possible. A comparison of the results obtained using the direct peel screening DART-based method was made with those obtained by DART analysis of solvent extracts as well as those obtained analyzing these extracts by ultra-HPLC MS (UHPLC-Orbitrap) [31]. Different LC-MS systems are compared in Chapter 15 (see Table 15.2). (Note: UPLC is usually used in the literature for ultra-performance liquid chromatography [column stationary phase particles <2 um] carried out with a Waters Corp. ACQUITY system, and UHPLC is usually used for systems from other manufacturers.)

Modern determination techniques for pesticides must yield identification quickly with high confidence for timely enforcement of tolerances. A protocol for the collection of LC ESI–quadruple linear ion trap (Q-LIT) MS library spectra after QuEChERS was developed by Zhang et al. [32]. Following their protocol, an enhanced product ion (EPI) library of 240 pesticides was developed by use of spectra collected from two laboratories. An LC-Q-LIT-MS workflow using scheduled MRM (sMRM) survey scanning, information-dependent acquisition triggered collection of EPI spectra, and library searching was developed and tested to identify the 240 target pesticides in one single LC-Q-LIT-MS analysis. By use of LC retention time, one sMRM survey scan transition, and a library search, 75%–87% of the 240 pesticides were identified in a single LC-MS analysis at fortified concentrations of 10 ng/g in 18 different foods [32].

20.2.2 Other Modern Analytical (Chromatographic) Methods as Alternatives Used to Determine Pesticides

20.2.2.1 GC

Until the late 1990s, GC-MS had been the main choice for pesticide analysis with the exception of polar and/or thermally labile pesticides that are not suitable for GC procedures [33]. GC-MS and GC-tandem MS (MS/MS) are now the most common techniques used for the qualitative and quantitative analysis of volatile pesticides. The lack of volatility or thermal instability of many analytes is the main limiting factor for the use of this technique. GC is, therefore, preferred to HPLC for gases, most low boiling compounds, and many higher boiling compounds that are thermally stable under the conditions of separation. GC is not applicable for very high boiling or nonvolatile materials. Sometimes, analytes must be derivatized for GC analysis [34–38].

GCs have available several very sensitive and/or element-specific detectors (e.g., flame photometric, thermionic, and electrochemical) that permit very low detection limits. GC with element-selective detectors or an electron capture detector provides analyte LODs of 10^{-9}–10^{-12} g (1 ng–1 pg) [34]. Hyphenated techniques, such as GC-MS, GC-MS/MS, and HPLC-MS also give analyte detectabilities of 10^{-9}–10^{-12} g but with exceptional, often single analyte, selectivity [39,40]. The excellent detectability (LODs < 20 fg) when using atmospheric pressure chemical ionization (APCI) combined with state-of-the-art MS/MS was demonstrated for real samples [41].

Optimal preparation of samples for analysis is a key step. Microwave accelerated selective Soxhlet extraction (MA-SSE) is a novel technique investigated by Zhou et al. [42]. A Soxhlet extraction system containing a glass filter was designed as an extractor. During the procedure of MA-SSE, both the target analytes and the interfering components were extracted from the sample into the extraction solvent enhanced by microwave irradiation. After the solvent flowed through the sorbent, the interfering components were sorbed, and the target analytes remaining in the solvent were collected in the extraction bottle. No cleanup or filtration was required after extraction. The efficiency of the MA-SSE approach was demonstrated by the determination of organophosphorus and carbamate pesticide residues in ginseng by GC-MS [42].

20.2.2.2 TLC

TLC, although less efficient and sensitive than, for example, GC-MS or LC-MS, has several advantages. High-performance TLC (HPTLC) is most effective for the low-cost analysis of samples requiring minimal sample cleanup. TLC and HPTLC are included in the method classification planar chromatography (PC). The techniques are selected for pesticide analysis because of the following [43,44]:

- Single use of stationary phase minimizes sample preparation requirements.
- Parallel separation of numerous samples provides high throughput.
- Ease of postchromatographic derivatization improves method selectivity.
- Detection and/or quantification steps can easily be repeated under different conditions.
- All chromatographic information is stored on the plate and can be reevaluated if required.
- Several screening protocols for different analytes can be carried out simultaneously.
- Selective derivatizing reagents can be used for individual or group identification of the analytes.
- Detection of the separated zones with specific and sensitive color-forming reagents.
- Visual detection of ultraviolet (UV)-absorbing compounds is possible in field analyses by use of a UV lamp and phosphor-containing (F) layers.
- Detection of radioactive compounds by contact with x-ray film, digital bio- and autoradiography, and even quantitative assay by use of enzymes is possible.
- Plates can be documented by videoscans or photographs.
- TLC combined with modern video-scanning and slit-scanning densitometry enables quantitative analysis.
- TLC coupled with densitometry enables detection of the zones through scanning of the chromatograms with UV–Vis light in the transmission, reflectance, or fluorescence mode.
- With multiwavelength scanning of the chromatograms, spectral data of the analytes can be directly acquired from the TLC plates and can be compared with the spectra of the analytes from a software library.
- Additional information for structural elucidation can be obtained by TLC combined with MS (fast atom bombardment, secondary ion MS, and, especially, ESI using the CAMAG TLC-MS interface).
- The whole procedure of chromatographic development can be followed visually, so any distortion of the solvent front, etc., can be observed directly.
- The chromatogram can be developed simply by dipping the plate into a mobile phase.
- The possibility of 2D development with a single sorbent.
- The possibility of 2D development on, for example, silica–C18 silica coupled layers (Whatman Multi-K SC5 and CS5 dual phase). (Note: The Whatman TLC plate product line was sold to GE Healthcare, which discontinued it in the Spring of 2013. Analtech now offers many plates comparable to former Whatman plates, including CS5 and SC5.)
- TLC is the easiest technique that performs multidimensional (MD) separation (e.g., by graft chromatography or MD chromatography).

Biennial review papers on TLC in pesticide analysis written by Sherma have appeared in the literature [45–55]. There are some fundamental books [56–59] and book chapters [43,44,60] on the theory, techniques, instrumentation, and applications of TLC to guide workers in the field.

20.2.2.2.1 TLC as a Pilot Technique for HPLC

PC as a pilot technique for HPLC has been described for analysis of pesticides [61]. Retention value (R_F) versus mobile phase composition relationships for nearly 100 pesticides were determined for

TLC on various sorbents, for example, C18; these relationships constitute a retention database [62] and can be used to choose optimal conditions for experiments in HPLC. TLC R_M can be calculated using the following equations:

$$R_M = \log \frac{(1 - R_F)}{R_F} \tag{20.1}$$

For HPLC, log k is calculated from

$$k = \frac{(t_r - t_0)}{t_0} \tag{20.2}$$

where t_0 is the hold-up volume determined using a nonretained solute (uracyl), and t_r is the retention time. Results were given as log $k_{(HPLC)}$ versus $R_{M(TLC)}$ and show for most pesticides an insignificant dispersion of points and relatively high correlation coefficient r (0.9195–0.9936), which permits TLC to qualify as a pilot technique for HPLC [61,62]. Because several screening protocols for different analytes can be carried out simultaneously on the same plate (ca., 40), TLC can be applied as a pilot technique for HPLC and screening of a large number of samples in environmental analysis.

20.2.2.2.2 TLC Coupled with a Diode Array Detector

Hyphenated techniques, such as HPTLC coupled to UV diode array detection (DAD) and to MS, provide extensive online structural information on the separated compounds. TLC combined with modern scanning densitometry provides the possibility of quantitative analysis [63–67]. The method offers a simple and economical alternative to other chromatographic techniques, especially column HPLC. Application of a modern fiber-optic TLC scanner with a DAD has several advantages [43,44,60], for example,

- The scanner can measure TLC plates simultaneously at different wavelengths without destroying the plate surface and permits parallel recording of chromatograms and in situ UV spectra in the range of 191–1033 nm; therefore, it is possible to obtain doubly credible correct identification of the compounds in a chromatogram.
- The TLC scanner permits determination of each compound at its optimum wavelength, thus offering optimum sensitivity for detection and quantification of each component.
- The TLC–DAD scanner permits measurement of a 3D chromatogram, $A = f(\lambda, t)$, with absorbance as a function of wavelength and distance.
- The TLC–DAD scanner can compare parallel UV spectra of an unknown compound and a standard from a library of spectra.
- Software is available that allows the user access to all common parameters used in HPLC–DAD: peak purity, resolution, identification via spectral library match, etc.

The TLC–DAD scanner is especially useful for correct identification of components of difficult, complicated mixtures, such as in plant extract and toxicological analysis [43,44]. Examples of TLC-DAD analysis of pesticides are presented in the following [68–75].

Application of solid phase extraction (SPE) and TLC-DAD for identification and quantitative analysis of fenitrothion in fresh apple juice was demonstrated [68]. Figure 20.3 shows an example of the 3D plot (scanning range × trace distance × absorbency) obtained from an apple extract [68]. Identification was achieved by comparing the UV spectrum obtained from the extract and a fenitrothion standard. Figure 20.4 shows UV spectra obtained from fenitrothion standards at eight concentrations (100–1000 μg mL^{-1}) and the UV spectrum obtained (TLC–DAD) from fenitrothion in an extract from freshly squeezed apple juice [68].

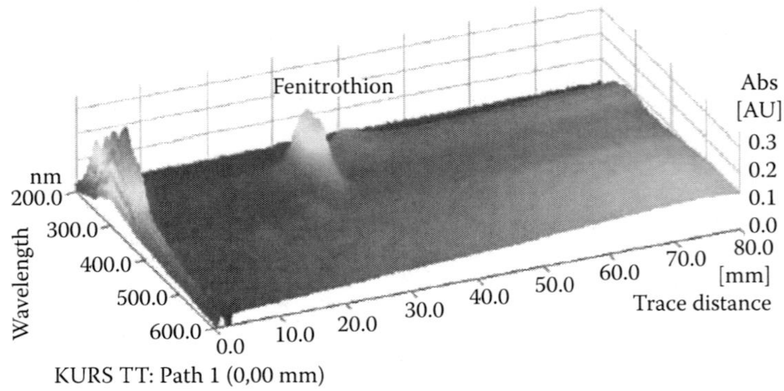

FIGURE 20.3 Three-dimensional plot obtained from an apple extract containing fenitrothion. (From Tuzimski, T., *J. Planar Chromatogr. Mod.—TLC*, 18, 419, 2005. With permission.)

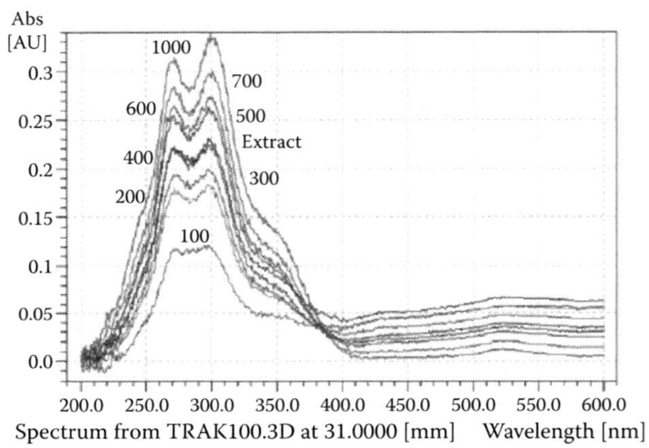

FIGURE 20.4 UV spectra obtained from fenitrothion standards at eight concentrations (100–1000 μg mL^{-1}) and from an extract of freshly squeezed apple juice containing fenitrothion. (From Tuzimski, T., *J. Planar Chromatogr.—Mod. TLC*, 18, 419, 2005. With permission.)

The peak purity index, *P*, is a numerical index for the quality of the coincidence between two data sets. It is given by the least-squares fit coefficient calculated for all intensity pairs in the two data sets under consideration. The following equation is applied:

$$P = \frac{\sum_i (s_i - \bar{s})(r_i - \bar{r})}{\sqrt{\sum_i (s_i - \bar{s})^2 \sum (r_i - \bar{r})^2}} \tag{20.3}$$

where s_i and r_i are the respective intensities for the same abscissa value, *i* is the number of data points, and $\bar{s}$ and $\bar{r}$ are the average intensities of the first and second data set. A peak purity index has values in the range from 0 to 1. A peak-purity index of 1 indicates that the compared spectra are identical.

The components of two mixtures of pesticides that were separated by 2D-TLC with sorbent gradients of the type $C18_W$ bonded silica (w = water wettable) or silica-cyanopropyl (CN) were identified by R_F in both chromatographic systems and by comparison of UV spectra [69]. In other papers, applications of fiber optical scanning densitometry in analysis of water samples from nine lakes and from Wieprz-Krzna Canal from Łęczyńsko-Włodawskie Lake District (Southeast Poland) were demonstrated [70–73]. Atrazine, clofentezine, chlorfenvinphos, hexaflumuron, terbuthylazine, lenacyl, neburon, bitertanol, and metamitron were enriched from canal water samples by SPE on C18/SDB-1, C18, C18 Polar Plus, and CN cartridges. The recovery rates were high for all extraction materials except CN, for which the values were lower. SPE was used not only for the preconcentration of analytes but also for their fractionation. The analytes were eluted first with methanol and next with dichloromethane [70–73]. The method was validated for precision, repeatability, and accuracy. The calibration plots were linear between 0.1 and 50.0 μg mL^{-1} for all pesticides, and the correlation coefficient, r, values were between 0.9994 and 1.000 as determined by HPLC–DAD. Calibration plots were linear between 0.1 and 1.5 mg zone^{-1} for all pesticides, and the r values were between 0.9899 and 0.9987 determined by TLC–DAD. The LOD was between 0.04 and 0.23 mg zone^{-1} (TLC–DAD) and 0.02 and 0.45 μg mL^{-1} (HPLC–DAD) [68].

In another paper [74], in the SPE experiments, the analytes were eluted with methanol, acetonitrile, or tetrahydrofuran. Next, the eluates were analyzed by HPLC–DAD (Figures 20.5 through 20.7) and by TLC–DAD (Figures 20.8 and 20.9) [74]. The identities of the peaks/bands of

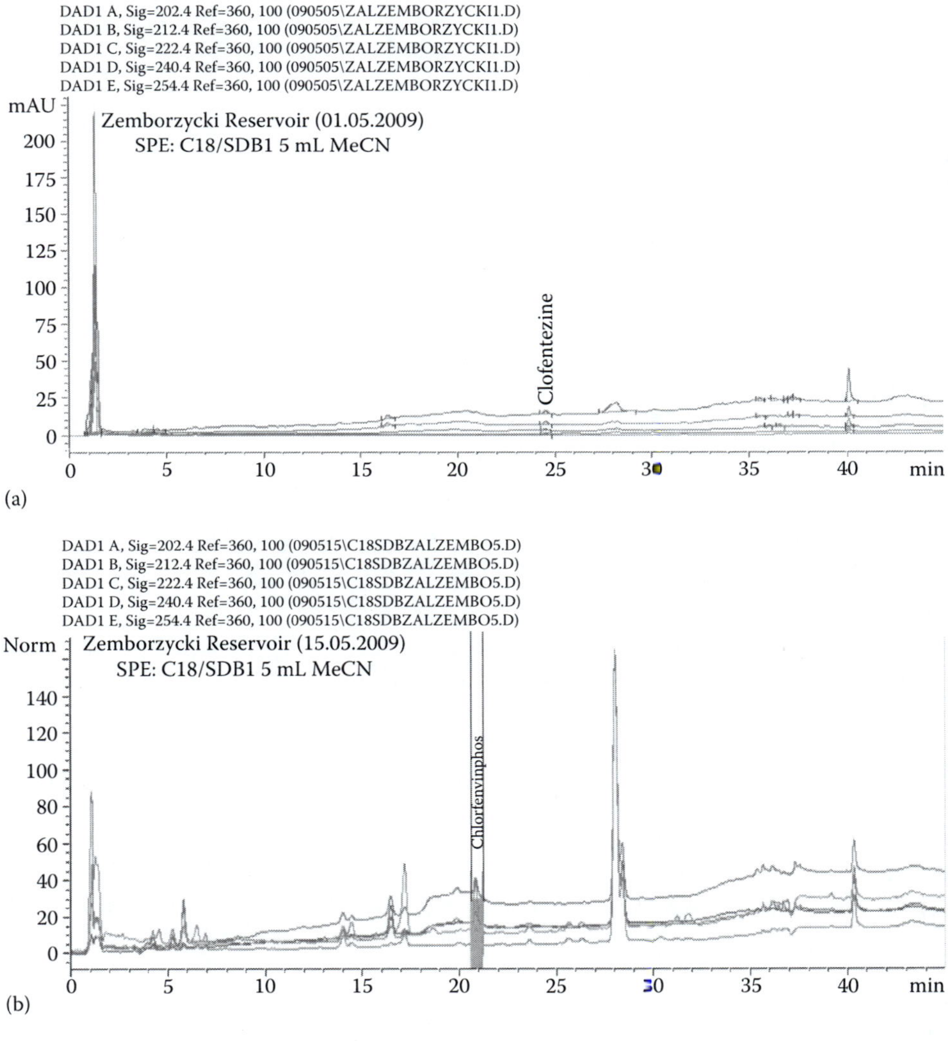

FIGURE 20.5 Chromatograms of water obtained from the Zemborzycki Reservoir showing detection and quantification by HPLC-DAD of (a) clofentezine (May 1, 2009) and (b) chlorfenvinphos (May 15, 2009). (From Tuzimski, T., *J. AOAC Int.*, 93, 1748, 2010. With permission.)

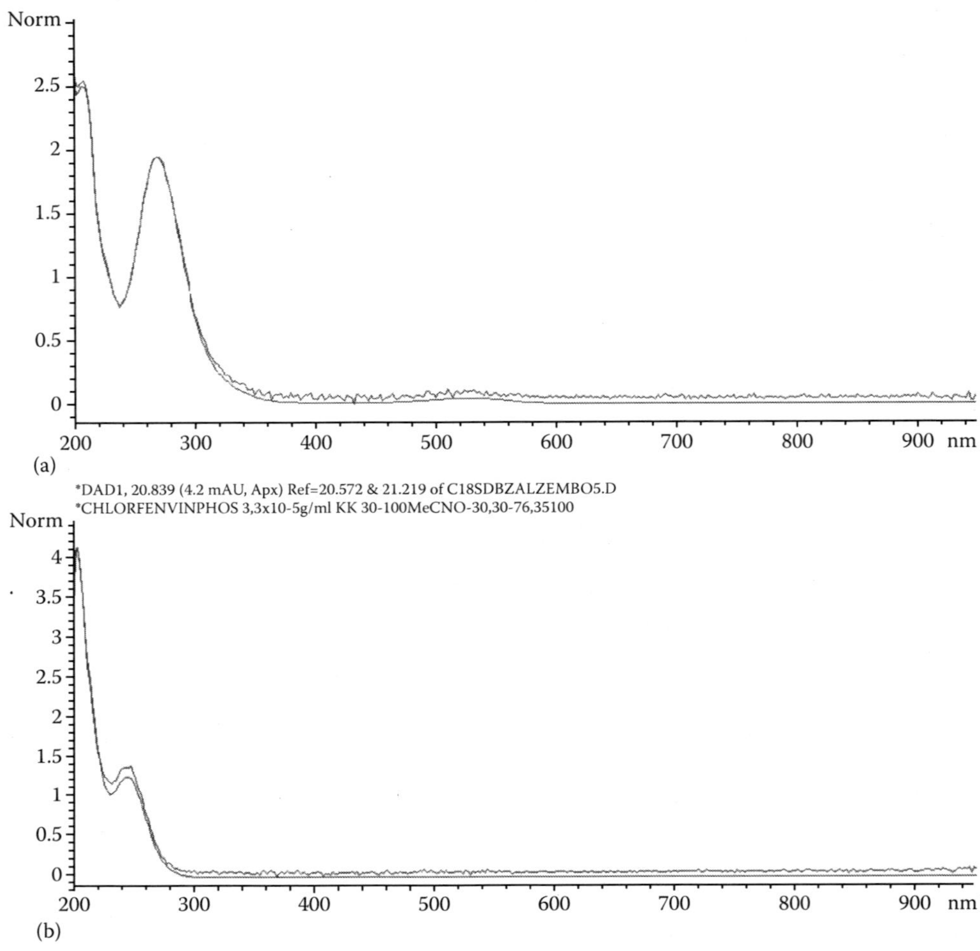

FIGURE 20.6 Comparison of spectra of analytes found in surface waters from the Zemborzycki Reservoir with spectra of pesticide standards (library) (by HPLC-DAD method): (a) clofentezine and (b) chlorfenvinphos. (From Tuzimski, T., *J. AOAC Int.*, 93, 1748, 2010. With permission.)

analytes in the water samples were confirmed by overlaying their UV absorption spectra with those of the standards of these compounds in both methods: HPLC-DAD (Figure 20.6) and TLC-DAD (Figure 20.8). A match equal to or higher than 0.990 (99%) or values of purity index higher than 0.9500 (95%) were defined as confirming the identification between both spectra for analytes determined by HPLC-DAD and TLC-DAD. Figure 20.7 shows the purity of peaks of the investigated pesticides by HPLC-DAD. The least-squares fit values of the spectrum from a fortified sample of water (Figure 20.9a) and the Zemborzycki Reservoir (Figure 20.9b) and a spectrum from the clofentezine standard are also presented (obtained by the TLC-DAD method).

Applications of SPE, ultrasound-assisted extraction, and HPLC–DAD and/or TLC–DAD to the determination of pesticides in medical herbal samples were also described [75–77].

20.2.2.2.3 TLC Coupled with MS

The key to successful adoption of coupled TLC and MS is producing a viable and useful interface, which will be a simple device that transforms the distribution of samples on an *xy* plane (the layer) into a sequence of sample molecules in a gas or liquid stream, mimicking column LC-MS or GC-MS analysis. Now, the very convenient and universal CAMAG TLC-MS interface is available, which

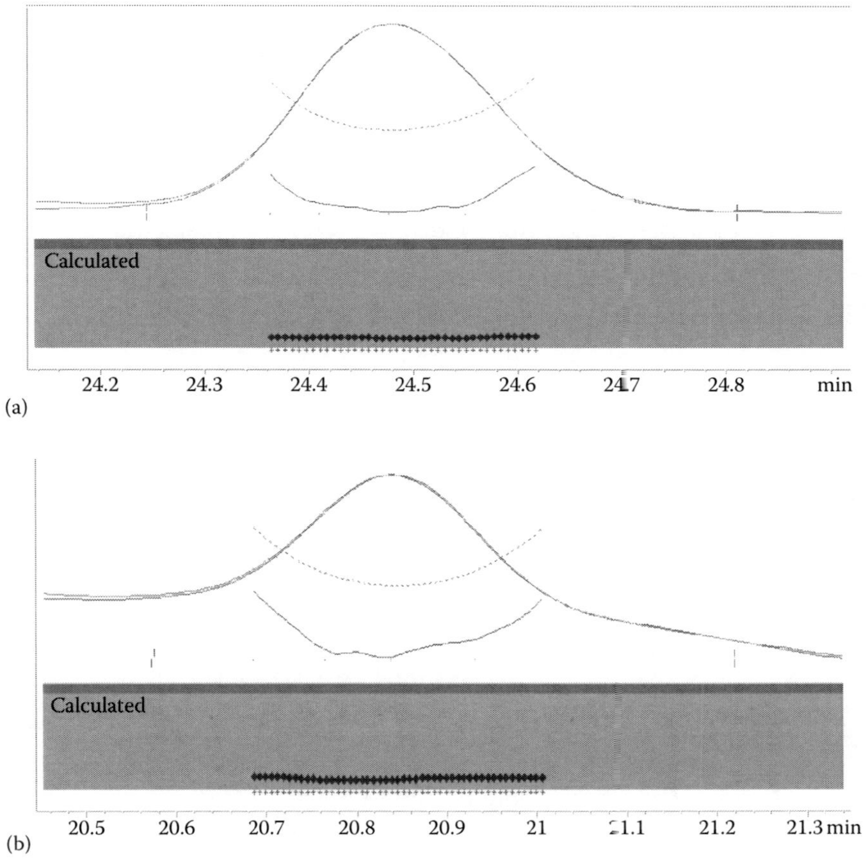

FIGURE 20.7 Purity of peaks of analytes found in surface waters from the Zemborzycki Reservoir: (a) clofentezine and (b) chlorfenvinphos (by HPLC-DAD method). (From Tuzimski, T., *J. AOAC Int.*, 93, 1748, 2010. With permission.)

can semiautomatically extract zones of interest and direct them online into any brand of HPLC-MS system. Application of this TLC-MS interface has several advantages, for example [43,78–86],

- It is compatible with all common HPLC-MS systems and types of TLC/HPTLC plates.
- It is quickly and easily connected (by two fittings) to any LC-coupled MS without adjustments or MS modifications.
- It permits rapid and convenient extraction (elution) directly into most types of MS systems (APCI-MS, atmospheric pressure photoionization–MS, or ESI-MS) but has been used online exclusively with the latter so far.
- It is especially useful for identification of unknown substances.
- It gives the possibility of confirmation of target compounds.
- It is the solution to an ongoing analysis issue, that is, traditional TLC/HPLC methods are unable to definitively identify unknown compounds.
- It eliminates scraping off the plate and gives the possibility of extraction into vials for offline nuclear magnetic resonance (NMR) or attenuated total reflectance–Fourier transform infrared (FTIR) spectrometry or for offline static nanospray, direct inlet electron impact (EI), and matrix-assisted laser desorption/ionization (MALDI) MS.
- It gives the possibility of automatic cleaning of the piston (elution head) between the extractions.

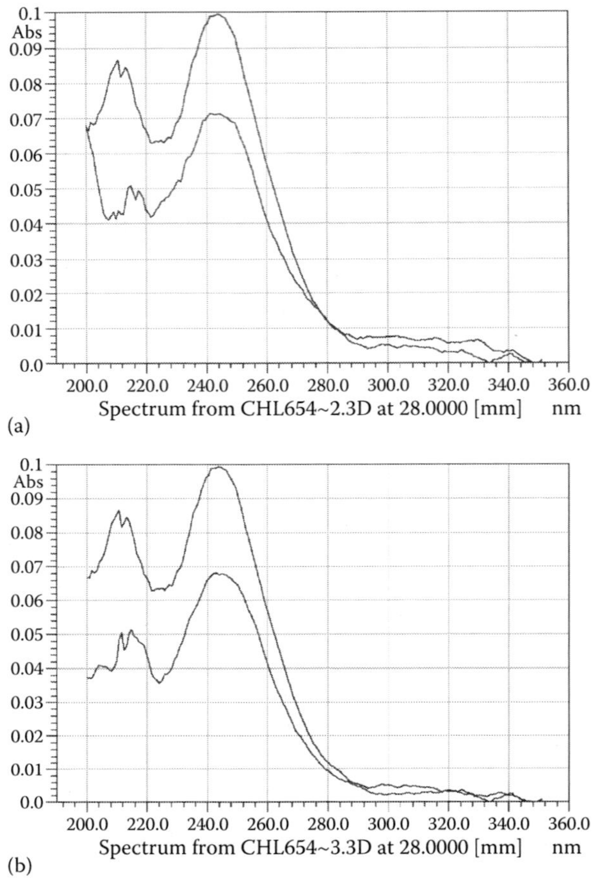

FIGURE 20.8 Comparison of a spectrum of chlorfenvinphos found in (a) fortified surface water and (b) the Zemborzycki Reservoir with a spectrum of chlorfenvinphos standard (library) (by TLC-DAD method). (From Tuzimski, T., *J. AOAC Int.*, 93, 1748, 2010. With permission.)

According to CAMAG, the TLC-MS interface enables substances to be directly extracted from a TLC/HPTLC plate, and sensitive MS signals are obtained within a minute substance zone^{-1}. The interface extracts the complete substance zone with its depth profile and thus allows detection comparable to HPLC down to pg zone^{-1} levels. The interface has been proven to be the most reliable, versatile, and widely applied interface for TLC/HPTLC-MS coupling [43,78–86]. Other currently available commercial interfaces include those by IonSense for DART MS, Bruker Daltonics for MALDI, and Advion, Inc., for liquid surface analysis coupled to chip-based nanospray high-resolution (HR) MS.

Oellig and Schwack [87] described pesticide residue screening by planar SPE cleanup and microliter flow injection TOF MS (HTpSPE-uL-FIA-TOF-MS). Sample extraction was performed according to the QuEChERS method, and the raw extracts were applied to an amino (NH_2) bonded silica gel 60 F_{254s} aluminum-backed plate. After twofold development, the target zones were eluted by the CAMAG TLC-MS interface into autosampler vials (Figure 20.10a), and uL-FIA-TOF-MS was then carried out without a column for chromatographic separation (Figure 20.10b). Without chromatographic separation, the complete information from an injected sample was focused in a single FIA peak from which a single MS was extracted from the full scan data covering the entire

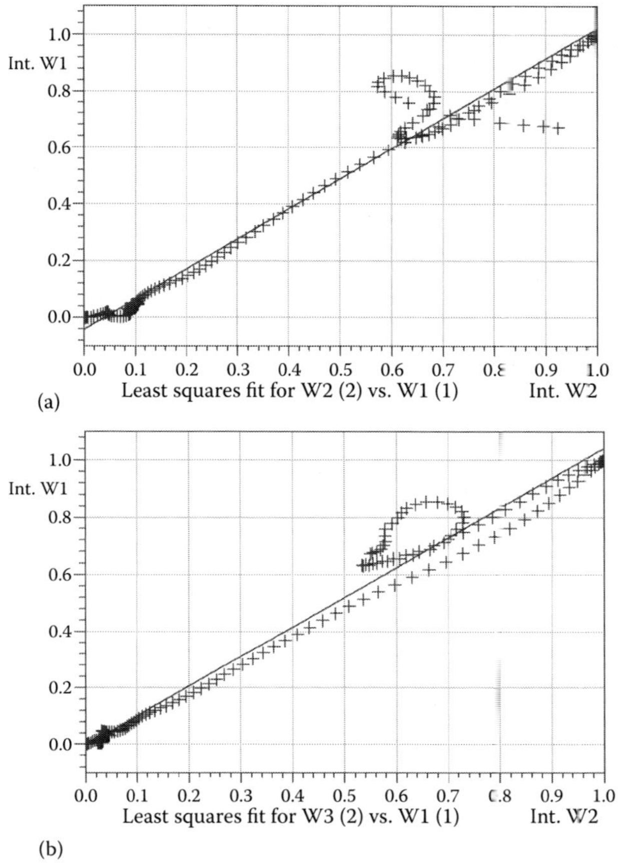

FIGURE 20.9 Correlation curve of peak purity of spectra of chlorfenvinphos found in (a) fortified surface water and (b) the Zemborzycki Reservoir and of chlorfenvinphos standard (library) (by TLC-DAD method). (From Tuzimski, T., *J. AOAC Int.*, 93, 1748, 2010. With permission.)

sample peak (Figure 20.10c), enabling a rapid screening process. Comparison of μL-FIA-TOF-MS of cucumber blank extracts and extracts spiked with a pesticide mixture using different cleanup methods is presented in Figure 20.11 [87].

20.2.2.2.4 TLC Coupled with NMR and FTIR Spectroscopy

Other hyphenated techniques, such as TLC coupled to NMR and FTIR spectroscopy, were applied to pesticide-specific detection in the past, but now they are used less and less.

20.2.2.3 Automated Multiple Development Modes of Pesticide Separations

A special device for automated multiple development (AMD) of a chromatogram was described by Perry et al. [88] followed by a programmable setup constructed by Burger [89], and an instrument is now produced by CAMAG designated as the AMD 2. A full separation process comprising 20–25 steps takes a long time. However, this is compensated by simultaneous separation of many samples on one chromatographic plate and the ability to use the system outside of working hours without worker intervention. Therefore, the final analysis is characterized by a relatively high throughput. This throughput can be increased by reduction of the number of steps of the AMD procedure. Application of special software for the simulation of the PC process can additionally enhance this procedure [90,91].

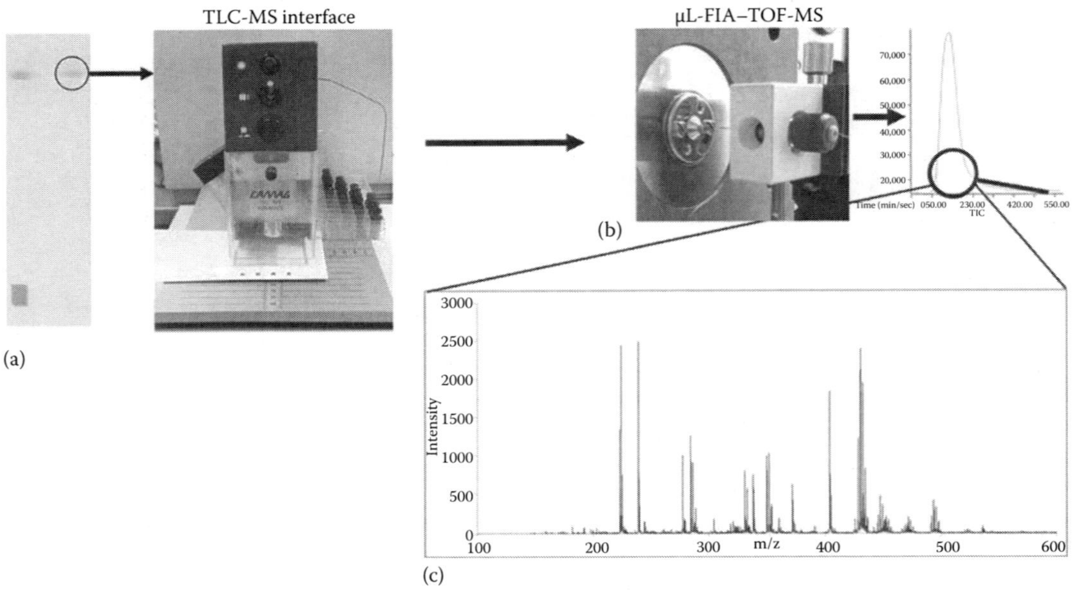

FIGURE 20.10 Flow chart of HTpSPE–μL-FIA–TOF-MS screening analysis: (a) elution of pesticides from the target analyte zone, using the CAMAG TLC-MS interface, after the twofold development (planar solid phase extraction cleanup); (b) injection of extracts into the μL-FIA–TOF-MS system; (c) the obtained full-scan mass spectrum extracted from the entire FIA peak. (From Oellig, C., Schwack W., *J. Chromatogr. A*, 1351, 1, 2014. With permission.)

Advantages of the AMD instrument are the following [43,92]:

- The mobile phase for each development is prepared automatically by mixing appropriate portions of solvents from up to five different reservoirs.
- Gradient development can be accomplished with a similar number of the mobile-phase components.
- Chromatography is monitored, and the run stops when the selected developing distance is reached.
- The chromatographic plate (usually an HPTLC plate) is developed repeatedly in the same direction.
- Each step of the chromatogram development follows complete evaporation of the mobile phase from the chromatographic plate and is performed over a longer migration distance of the mobile phase front than the one before.
- Each step of the chromatogram development uses a solvent of lower elution strength than the one used in the preceding run; this means that a complete separation process proceeds under conditions of gradient elution.
- A focusing effect of the solute bands takes place during the separation process, which leads to very narrow component zones and high efficiency of the chromatographic system comparable to HPLC.
- AMD is highly reproducible.
- AMD has major applications for separation of components spanning a wide polarity range or that are similar. In the first case, a steep gradient provides best results. In the second case, the focusing effect of multiple developments combined with a shallow gradient, that is, small increases of developing distance, and a large number of steps gives the best separation.
- If the experimentation is realized with the regulation of temperature, all of the conditions combine to increase the reproducibility in the analysis.

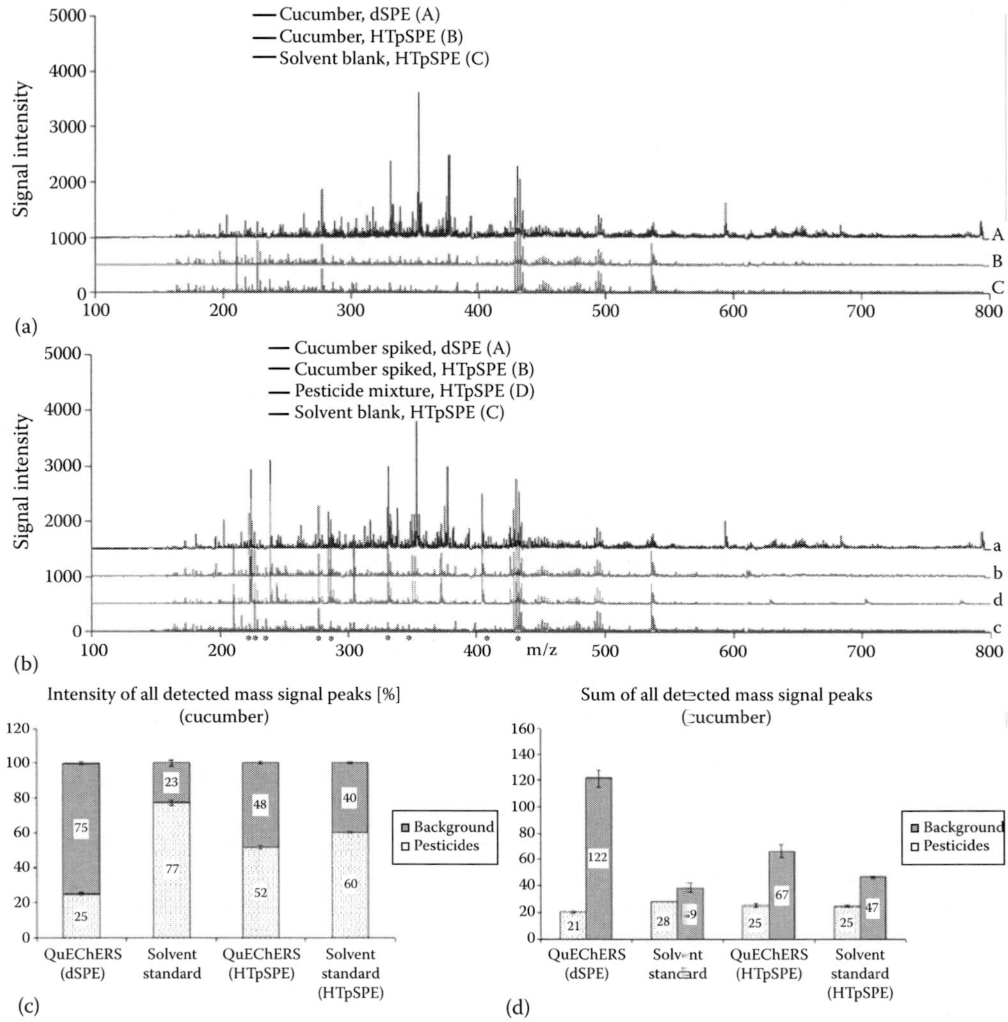

FIGURE 20.11 μL-FIA–TOFMS mass spectra covering the entire sample peak of cucumber blank extracts including the internal standards Sudan II and tris(1,3-dichloro-2-propyl)phosphate (TDCPP) (a) and extracts spiked with a pesticide mixture (b); sample concentration 0.25 g/mL, QuEChERS raw extracts after dispersive SPE (dSPE) (A) and after HTpSPE (B). Track D refers to a pesticide (*) solvent standard mixture of acetamiprid, mepanipyrim, pirimicarb, Sudan II, penconazole, fenarimol, chlorpyrifos, azoxystrobin, and TDCPP after HTpSPE (ordered by increasing m/z). Track C indicates a solvent blank after HTpSPE, including Sudan II and TDCPP, showing the TLC plate background signals. The relative S/N intensity values of all detected mass signals in the corresponding mass spectra (S/N > 100) are presented in (c), and (d) depicts the absolute numbers of all obtained mass signals from the related mass spectra (S/N > 100). (From Oellig, C., Schwack W., *J. Chromatogr. A*, 1351, 1, 2014. With permission.)

- The fully automated development of the plates (preconditioning time, automated mobile phase gradient, and drying time) also leads to good precision of the analysis, and the appropriate different mobile phase used for each step permits a sharper separation in well-defined experimental conditions with no spot diffusion in the layer sorbent and also reproducible R_F values.
- AMD permits the analysis of very small quantities and produces sharper separations because of the absence of diffusion in the sorbent at the upper R_F values, which is very favorable for quantitative densitometry.

Poole and Belay [93] reviewed the essential methods and parameters of multiple development techniques in PC (including AMD). Evaluation of parameters such as change in the zone width versus number of developments, zone separation versus number of developments through AMD, and several typical applications of AMD were described. The chromatogram is developed several times on the same plate, and each step of the development follows the complete evaporation of the mobile phase from the chromatographic plate of the previous development. On the basis of the development distance and the composition of the mobile phase used for consecutive development steps, multiple development techniques are classified into four categories [43,94]:

- Unidimensional multiple developments (UMD), in which each step of chromatogram development is performed with the same mobile phase and the same migration distance of the mobile phase front.
- Incremental multiple development (IMD), in which the same mobile phase but an increasing development distance in each subsequent step is applied.
- Gradient multiple development (GMD), in which the same development distance but a different composition of the mobile phase in each step is applied.
- Bivariant multiple development (BMD), in which the mobile phase composition and distance are varied in each step of the plate development.

These modes of chromatogram development are mainly applied for analytical separations due to their good efficiency, which is comparable to HPLC. In a typical isocratic AMD run, the development distance is increased during consecutive development steps whereas the mobile-phase strength is constant. In the initial stage of the AMD gradient procedure, the solvent of the highest strength is used (e.g., methanol, acetonitrile, or acetone); in the next stages, an intermediate or base solvent of medium strength (e.g., chlorinated hydrocarbons, ethers, esters, or ketones); and in the final stage, a nonpolar solvent (e.g., heptane, hexane) [95].

Several parameters must be considered to obtain the best separation in AMD: choice of solvents, gradient profile of solvents, and number of steps. All modes of multiple development can be easily performed using chambers for automatic development, which are manufactured by some firms. However, these devices are relatively expensive. Typical horizontal chambers for PC should be considered for application in multiple developments in spite of more manual operations in comparison with the automatic chromatogram development.

Especially, the Chromdes horizontal DS chamber could be considered for separations with multiple developments. This chamber can be easily operated due to its convenient maintenance, including cleaning the mobile phase reservoir. For the separation of a more complicated sample mixture, a computer simulation could be used to enhance the efficiency of the optimization procedure [43,90,91,96–99].

The AMD technique is suitable for the analysis of very complex mixtures and compounds with similar structures, and HPTLC-AMD can be a timesaving analytical technique compared to GC or HPLC. It is possible to apply up to 16 samples and two standard mixtures containing a certain spectrum of pesticides of interest together on one plate. This means that 16 samples can be analyzed in approximately 3 h (time for the multiple development process), a period of time in which considerably fewer samples can be analyzed with the other chromatographic techniques.

A technique reciprocal to AMD was introduced by Matysik et al. [100–102] called multiple gradient development: the plate is developed several times (with evaporation of mobile phase after each), starting with a weak mobile phase over the full distance, then with mobile phases of increasing strengths over decreasing distances. This process retains separation of zones with higher R_F values achieved in the preceding developments.

HPTLC-AMD was used to screen water samples for pesticides [103]. A universal gradient based on dichloromethane was employed to check for the presence of pesticides from different classes, such as phenylureas, carbamates, triazines, phenoxycarboxylic acids, and others. In total, 283 pesticides were analyzed applying this gradient [103]. A TLC method using AMD was developed for the determination of six phenylurea herbicides in food [104]. The herbicides were extracted with acetone and purified by SPE, the extract was evaporated, and the residue was dissolved in acetone, applied onto a silica gel plate, and chromatographed by AMD. A 25-step gradient composed of acetonitrile, dichloromethane, acetic acid, toluene, and hexane was used. Quantification was done by UV measurement. The LOD was 0.01 ppm [104].

Summing up, TLC is one of the principal separation techniques that plays a key role in pesticide analysis. The very strong points of (HP)TLC are the following:

- Its ability to analyze multiple samples simultaneously on a single plate.
- Its ability to be used as a pilot technique for HPLC.
- Application to pesticides of different chemical classes in a wide variety of sample types because of the great array of layer types; development techniques; detection methods based physical, chemical, and biological methods; and instrumental techniques for qualitative and quantitative analysis, for example, TLC-densitometry and TLC-MS.

In the next part of the chapter, the reader will gain useful information about the new method of PC–MD chromatography (MDPC), applications of different modes of MDPC, and its combination with the DAD (MDPC–DAD) and HPLC–DAD for qualitative and quantitative determination of pesticides in environmental and other types of samples.

20.2.2.4 MD Chromatography

The coupling of chromatographic techniques is clearly attractive for the analysis of multicomponent mixtures of analytes. Truly comprehensive 2D hyphenation is generally achieved by frequent sampling from a first column into a second, which is a very rapid analytical approach. In this section are presented different modes of MD chromatographic separation techniques, including MDGC, MDLC, and MDPC, applied to analysis of pesticides. Reviews of MD chromatography in pesticide analysis can be found in chapters published by Tuzimski [105,106]. Some fundamental and detailed information on the topic are included in a book titled *Multidimensional Chromatography* [107] and some chapters [108–113].

20.2.2.4.1 MDGC

To overcome the problem of coelution, analysis of pesticides is generally performed by heart-cut MDGC [105,114–117] and comprehensive 2D GC (GC × GC) [105,114–117]. In the heart-cut MDGC technique, two independent GC systems are coupled so that one or more unresolved fractions are transferred directly (online) from a first column (first dimension) to a second column (second dimension) where separation of the compounds will occur. In comprehensive GC × GC, the entire sample is separated very quickly on two different columns [105,114–117].

20.2.2.4.2 MDLC

MDLC has been applied to analysis of nonvolatile pesticides, but presently it is used less often than MDGC. MDLC (also known as coupled-column chromatography [LC-LC coupling] or column switching) represents a powerful tool and an alternative procedure to classical 1D HPLC. MDLC separation has been defined as a technique that is mainly characterized by two distinct criteria: (i) the first criterion for an MD system is that sample components must be displaced by two or more separation techniques involving orthogonal separation mechanisms; and (ii) the second criterion is that components that are separated by any single separation dimension must not be recombined in any further separation dimension [105,118,119].

MDLC can be performed as in MDGC either in the online or offline mode [105]. The online mode of MDLC has the advantage of automation by using pneumatic or electronically controlled valving, which switches the column effluent directly from the primary column into the secondary column. The online technique is more reproducible, and no loss of sample or contamination occurs. Automation improves reliability and sample throughput and shortens the analysis time as well as minimizing sample loss or change because the analysis is performed in a closed-loop system [105,109].

What characterizes LC-LC coupling when compared to conventional multistep chromatography is the requirement that the whole chromatographic process be carried out online. The transferred volume of the mobile phase from the first column to the second column (from 1D to 2D) can correspond to a group of peaks, a single peak, or a fraction of a peak, so that different parts of the sample may follow different paths through the LC-LC configuration [105,109]. Details on modes of MDLC applied to identification and quantitative analysis of pesticides can be found in Chapter 16.

20.2.2.4.3 MDPC

Details on modes of MDPC applied to identification and quantitative analysis of pesticides can be found in published chapters (also online with free access) by Tuzimski [105,106].

Giddings defined MD chromatography as a technique that includes two criteria [110]:

- The components of the mixture are subjected to two or more separation steps in which their migration depends on different factors.
- When two components are separated in any single step, they always remain separated until completion of the separation.

Nyiredy divided MDPC techniques as follows [59,113,120]:

- Comprehensive 2D planar chromatography (PC × PC)—multidimensional development on the same monolayer stationary phase and two developments with different mobile phases or using a bilayer stationary phase and two developments with the same or different mobile phases.
- Targeted or selective 2D PC (PC + PC)—in which, following the first development from the stationary phase, a heart-cut spot is applied to a second stationary phase for subsequent analysis to separate the compounds of interest.
- Targeted or selective 2D PC (PC + PC), second mode—in which, following the first development, which is finished and the plate dried, two lines must be scraped into the layer perpendicular to the first development and the plate developed with another mobile phase to separate the compounds that are between the two lines. For the analysis of multicomponent mixtures containing more than one fraction, separation of components of the next fractions should be performed with suitable mobile phases.
- Modulated 2D PC (nPC)—in which, on the same stationary phase, mobile phases of decreasing solvent strengths and different selectivities are used.
- Coupled-layer PC (PC-PC)—in which two plates with different stationary phases are turned face to face (one stationary phase to second stationary phase) and pressed together so that a narrow zone of the layers overlaps and the compounds from the first stationary phase are transferred to the second plate and separated with a different mobile phase.
- Combination of MDPC methods—in which the best separation of multicomponent mixtures is realized by parallel combination of stationary and mobile phases, which are changed simultaneously.

By use of this technique, for example, after separation of compounds in the first dimension with changed mobile phases, the plate is dried and the separation process is continued in a perpendicular direction by use of the grafted technique with a changed mobile phase (based on the idea of coupled TLC plates, denoted as graft TLC in 1979) [121].

In this chapter, readers can find a general framework with ways and possibilities for separation of pesticides by MDPC. High separation efficiency can be obtained using modern PC techniques that comprise 2D development, chromatographic plates with different properties, a variety of solvent combinations for mobile phase preparation, and various forced-flow techniques and multiple development modes. By combination of these possibilities, MDPC can be performed in various ways.

20.2.2.4.3.1 Comprehensive 2D Chromatography on One Sorbent Separation of components of mixtures can be realized by comprehensive 2D chromatography on one sorbent. One of the powerful tools for obtaining optimal separation of substances by 2D TLC is the use of graphical correlation plots of retention data for two chromatographic systems that differ with regard to modifiers and/or sorbents [122]. The interpretation of plots is illustrated in Figure 20.12. The plots in Figure 20.12 directly indicate the positions of zones on a 2D chromatogram (2D-TLC). As shown in Figure 20.12f, the best separation of complex mixtures by 2D-TLC is possible with differentiated R_F values in both systems; then the correlation plots of retention parameters for two chromatographic systems are poor [62]. Good separation can be achieved when the zones are spread over the whole chromatographic plate area [123–127]. The largest differences were obtained by combination of NP and RP systems with the same chromatographic layer, for example, CN [126]. An example of this type of 2D development is illustrated in Figure 20.13d.

20.2.2.4.3.2 2D-TLC In 2D development, the mixtures can be simultaneously applied at each corner of the chromatographic plate so that the number of separated samples can be higher in comparison to the classical 2D development in which only one initial zone is applied [128]. An example of this type of 2D development is illustrated in Figure 20.13a through 20.13d, which shows a video

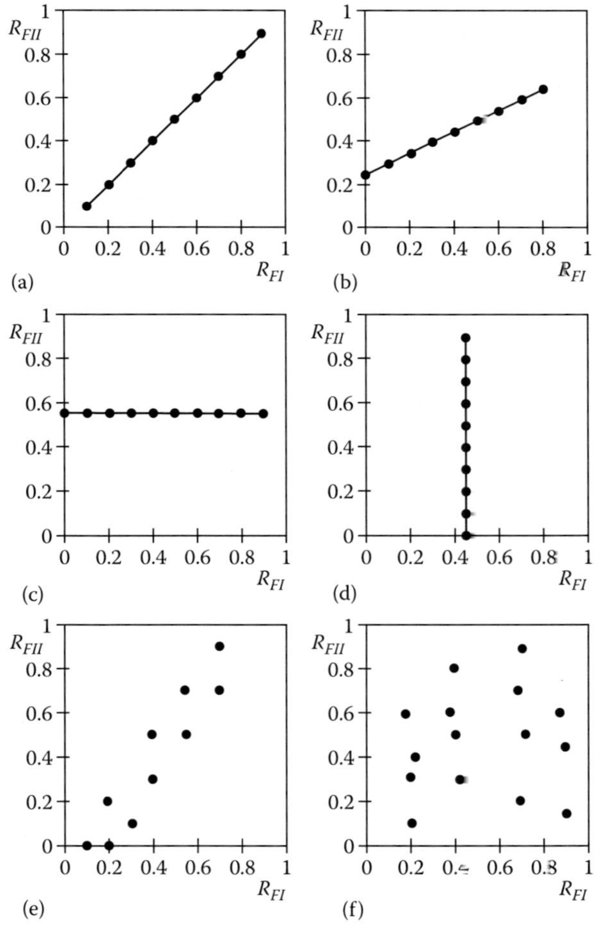

FIGURE 20.12 Characteristic correlations of R_{FII} versus R_{FI} (retardation factors [R_F] for two compared eluent/adsorbent systems [I and II]). (From Tuzimski, T., Soczewiński, E., Retention and selectivity of liquid-solid chromatographic systems for the analysis of pesticides (Retention database), in: *Problems of Science, Teaching and Therapy*, Medical University of Lublin, Poland, No. 12, Lublin, October 2002, 219 pp. With permission.)

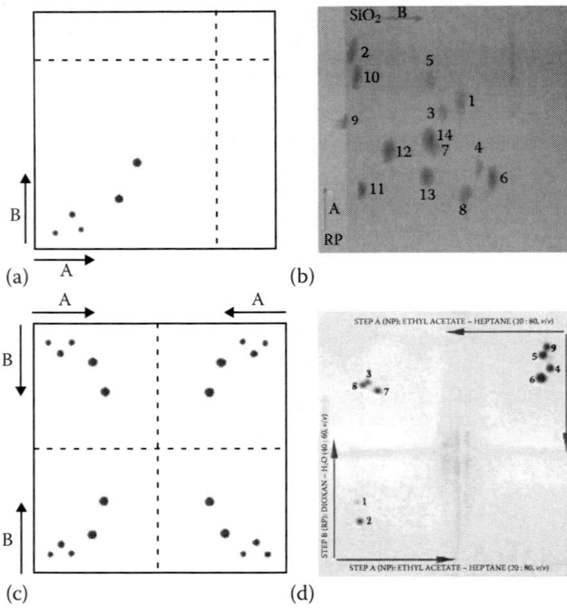

FIGURE 20.13 Two-dimensional development. (a) Schematic presentation of 2D chromatogram. (With kind permission from Springer Science+Business Media: *Planar Chromatography, a Retrospective for the Third Millennium*, Nyiredy, Sz. (Ed.), 2001, 69–87, Dzido, T. H.) (b) 2D chromatogram of the 14-component mixture of pesticides (1—metribuzin, 2—metamitron, 3—simazine, 4—propazine, 5—cyanazine, 6—aziprotryne, 7—desmetryn, 8—terbutryn, 9—hexazinone, 10—metoxuron, 11—chloroxuron, 12—methabenzthiazuron, 13—chlorbromuron, 14—metobromuron) presented as a video scan of dual-phase Multi-K CS5 plate in systems: A (first direction): methanol–water (60:40, v/v) on C18 silica sorbent, B (second direction): tetrahydrofuran-*n*-heptane (20:80, v/v) on silica gel. (From Tuzimski, T., and Soczewiński, E., *J. Chromatogr. A*, 961, 277–283. With permission.) (c) Schematic presentation of 2D chromatogram of four samples simultaneously separated on the plate. (With kind permission from Springer Science+Business Media: *Planar Chromatography, a Retrospective for the Third Millennium*, Nyiredy, Sz. (Ed.), 2001, 69–87, Dzido, T. H.) (d) 2D chromatograms of three fractions of the mixture of nine pesticides (1—fenuron, 2—monuron, 3—fluometuron, 4—buturon, 5—neburon, 6—monolinuron, 7—methabenzthiazuron, 8—chlorotoluron, 9—pencycuron) presented as a video scan of the HPTLC plate (CN) in systems with A (first direction): ethyl acetate-*n*-heptane (20:80, v/v), B (second direction): dioxane–water (40:60, v/v). (From Tuzimski, T., Soczewiński, E., *Chromatographia*, 59, 121–128, 2004. With permission.)

scan of the plate with separation of three fractions of the mixture of nine pesticides by 2D PC with NP/RP systems on a chemically bonded CN layer.

Nyiredy [59,130] described the technique of joining two different sorbent layers to form a single plate. Large differences were obtained by combination of NP systems of the type silica/nonaqueous mobile phase and RP systems of the type C18 silica/water + organic modifier (methanol, acetonitrile, dioxane) on Whatman multiphase plates with a narrow zone of silica gel and a wide zone of C18 (or vice versa) (Multi K SC5 or CS5 plates) [62,123–127]. Tuzimski and Soczewiński [62,123–125] and Tuzimski and Bartosiewicz [127] first used bilayer Multi K plates for separation of complex mixtures (Figure 20.13b).

Method development for 2D-TLC of complex mixtures can be formulated as follows [62]:

- Determine R_F versus percentage of modifier plots for polar sorbent and nonaqueous eluents composed of heptane (or hexane) and two to three polar modifiers; choose compositions of mobile phases for optimal differentiated retention of the components (in the range of 0.05–0.70).
- Determine R_F versus percentage of modifier concentration plots for aqueous RP systems (C18, CN, or other sorbents) for methanol and acetonitrile modifiers and choose the optimal concentration of modifier.

- Correlate the R_F values for NP/RP combinations and choose that corresponding to optimal spacing of spots on the plate area.
- Use the optimal combination of NP/RP mobile phases for a bilayer or monolayer plate (silica, CN, etc.).

20.2.2.4.3.3 Graft-TLC Graft-TLC, a novel multiplate system with layers of the same or different sorbents for isolation of the components of natural and synthetic mixtures on a preparative scale, was first described by Pandey et al. [121]. The procedure of performing reproducible graft-TLC analysis was described in detail by Tuzimski [131] and is presented in Figure 20.14. An example of this technique is demonstrated in Figures 20.14 and 20.15 [69,131].

In graft-TLC experiments with connected sorbent layers, several mixtures can be applied as spots 1 cm from the edge of the first sorbent, for example, a silica gel plate. Several samples can be developed at the same time in the first direction on the first sorbent (up to 10 with 20 cm × 10 cm plates). After drying, the plate used in the first run is cut into 2 cm × 10 cm strips. The cut strips should have smooth edges without irregularities resulting from partial loss of sorbent because such irregularities may lead to deformation of the zones during the transfer to the second sorbent layer. If the sorbent edge is uneven, it should be smoothed before attachment to the second sorbent [131]. Then, individual strips are clamped to other plates, and compounds are transferred. Individual strips should be connected (2 mm overlap) to 10 cm × 10 cm HPTLC plates along the longer (10 cm) side of the strip. It is essential that two of the plates are in close contact but without disintegration of the overlapping layers. To achieve this, the HPTLC plates are placed between thicker glass plates pressed together with screw clamps. The transfer of analyzed compounds is performed in a vertical glass chamber as the joined plates are difficult to develop in horizontal chambers. The most important issue in graft-TLC is to choose an appropriate mobile phase to transfer compounds from the first sorbent to another. The choice of this mobile phase depends on the choice of the first and second sorbents and the character of the transferred substances (whether polar or nonpolar). Methanol is usually applied for transferring compounds from the first sorbent to another layer [131]. If the analyzed compounds are strongly sorbed on the first layer, the addition of organic acids, and also water, to the transferring mobile phase, is advised so that all sample compounds have $R_F \approx 1.0$.

If, after transfer from, for example, the silica layer, the spots are spread along the 1 cm transfer distance, the second HPTLC plate can be developed to a distance of 1 cm with a strong mobile phase to improve their shapes. The application of a narrow strip of the first sorbent may also play the same role as the preconcentrating zone in the case of a multiphase (bilayer) plate.

The sample components are not only separated in the first step of a graft-TLC experiment, but also concentrated, and as such developed in the second direction. The concentration is also performed during the transfer as the strong mobile phase used in this procedure transfers the analyzed substances to another sorbent as very thin bands. Graft-TLC separations (2D PC on connected layers) of three mixtures of pesticides were described [69,131]. An example of this technique is demonstrated in Figure 20.15 [69,131].

Complete separation of the components of a pesticide mixture was also achieved by sorbent-gradient 2D TLC when, first, NP development was performed on silica gel and, second, RP development was performed on HPTLC F_{254S} plates or HPTLC CN F_{254S} plates (Figure 20.15a). Figure 20.15b shows the video scan, and Figure 20.15c shows the densitogram; complete separation of the components of the mixture of pesticides is apparent.

20.2.2.4.3.4 Combination of MDPC Very difficult separations of multicomponent mixtures of compounds require the application of MDPC combining different separation systems. A new procedure for separation of complex mixtures by combination of different modes of MDPC was described [132,133]. By this new procedure, 14 or 22 compounds from complex mixtures were separated on 10 cm × 10 cm TLC and HPTLC plates [132,133]. Figure 20.16 shows an example of this procedure step by step for separation of 22 compounds from a complex mixture on a TLC plate [133]. After

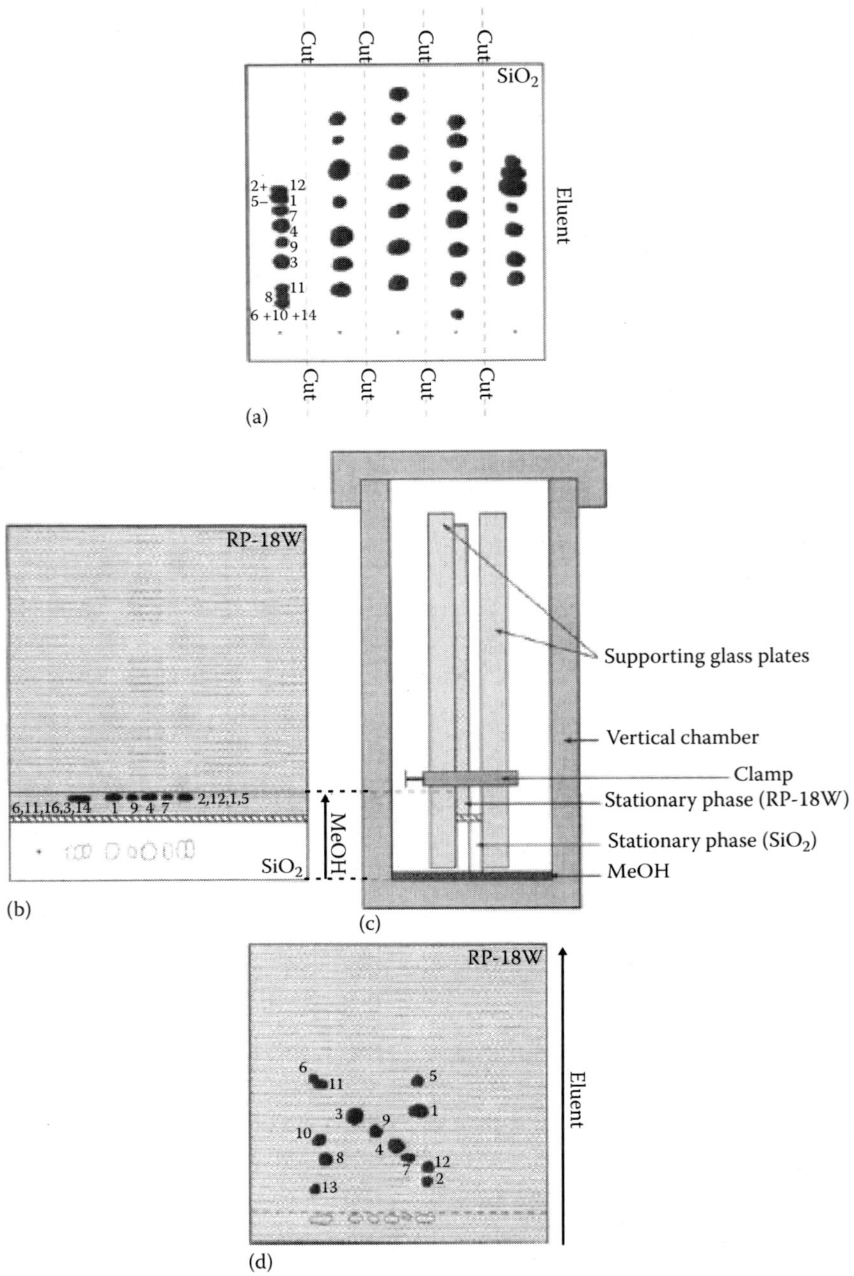

FIGURE 20.14 Transfer of a mixture of pesticides from the first plate to the second one. (a) First development with partly separated mixtures of pesticides on silica plate (mobile phase: ethyl-acetate-*n*-heptane 20 + 80 or 30 + 70 (v/v)). After development, the silica plate was dried and cut along the dashed lines into 2 cm × 10 cm strips. (b) A narrow strip (2 cm × 10 cm) was connected (2 mm overlap–hatched area) to a 10 cm × 10 cm HPTLC C_{18}W plate along the longer (10 cm) side of the strip. The partly separated mixture of pesticides was transferred in a vertical chamber to the second plate, using methanol as a strong eluent to a distance of about 1 cm. (c) Schematic diagram of cross section of two connected sorbent layers. (d) The HPTLC C_{18}W plate was developed in the second dimension with organic—water mobile phase (methanol–water 60 + 40 or 75 + 25, (v/v)) in a Chromdes DS chamber (1—aziprotryne, 2—fenvalerate, 3—desmetryn, 4—terbutryn, 5—pyriproxyfen, 6—benzthiazuron, 7—fluroglycofen-ethyl, 8—bensultap, 9—benalaxyl, 10—thiabendazole, 11—metalaxyl, 12—tetramethrin, 13—atrazine). (From Tuzimski, T., *J. Planar Chromatogr.—Mod. TLC*, 20, 13–18, 2007. With permission.)

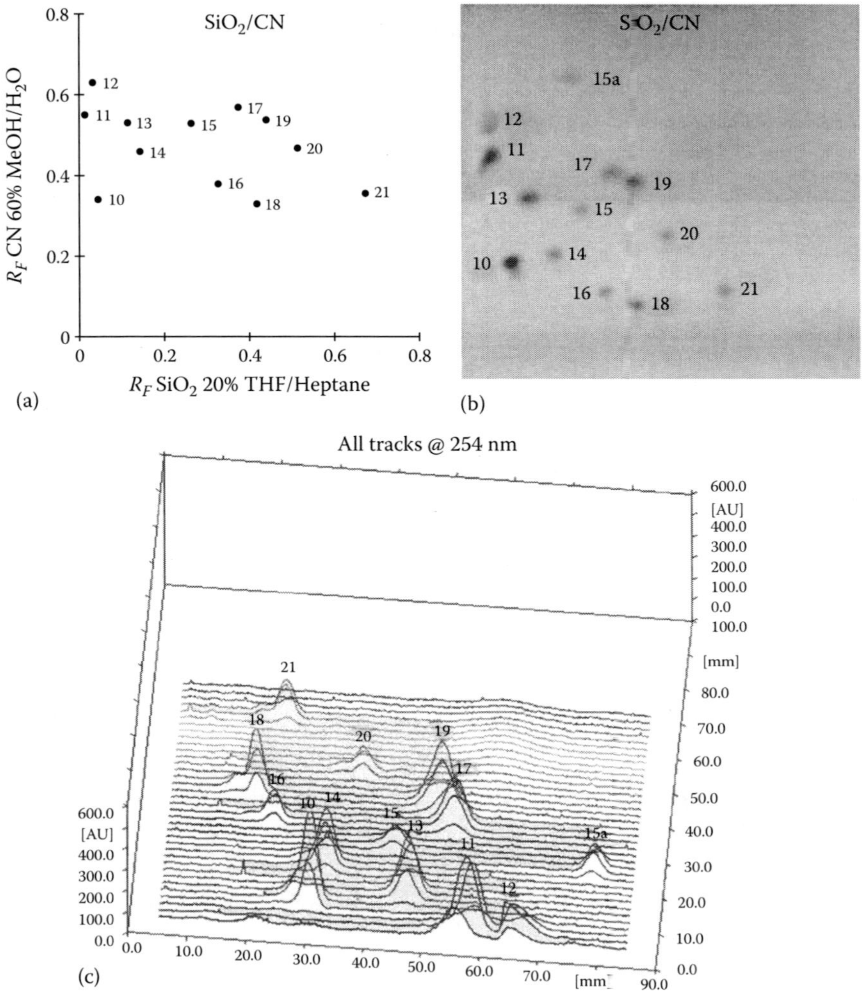

FIGURE 20.15 (a) Correlation between NP R_F values obtained with tetrahydrofuran–n-heptane (20:80, v/v) mobile phase on silica and RP R_F values obtained with methanol–water (60:40, v/v) mobile phase on a CN layer. This pair of NP and RP systems was chosen for 2D TLC with a sorbent gradient. The video scan (b) and densitogram (c) of the plate show the separation achieved for the 13-component pesticide mixture (10—bitertanol, 11—hexazinone, 12—chlorsulfuron, 13—methabenzthiazuron 14—phenmedipham, 15 and 15a—metiokarb, 16—dichlofluanid, 17—propachlor, 18—procymidone, 19—terbuthylazine, 20—propyzamide, 21—tri-allate). (From Tuzimski, T., *J. Planar Chromatogr.—Mod. TLC* 18, 354, 2005.)

optimization of the mobile phases for separation of the components of all groups of pesticides based on the solvent classification by Snyder [134] and the "Prisma" method described by Nyiredy and coworkers [135–137], these mobile phases were used for MDPC (Figure 20.16a through 20.16f).

Mixtures of pesticides were applied as spots 1 cm from the bottom and 3.5 cm from the left edge of the plate. TLC plates were developed in the first dimension (step I) with ethyl acetate–n-heptane (40:60, v/v), as an NP mobile phase. HPTLC plates were developed in the first dimension (step I) with ethyl acetate–n-heptane, 50:50 (v/v) as an NP mobile phase. After drying in air for 20 min, the plates were turned by 90° (so that the partly separated components of the complex mixture of compounds were on the start line [origin] of the next step). Next, one or two lines (approximately 1 mm wide) were scraped in the sorbent layer perpendicular to the direction of the first development, so the zone(s) of the target compounds were between the lines. Some of the sorbent layer (approximately 5 mm wide)

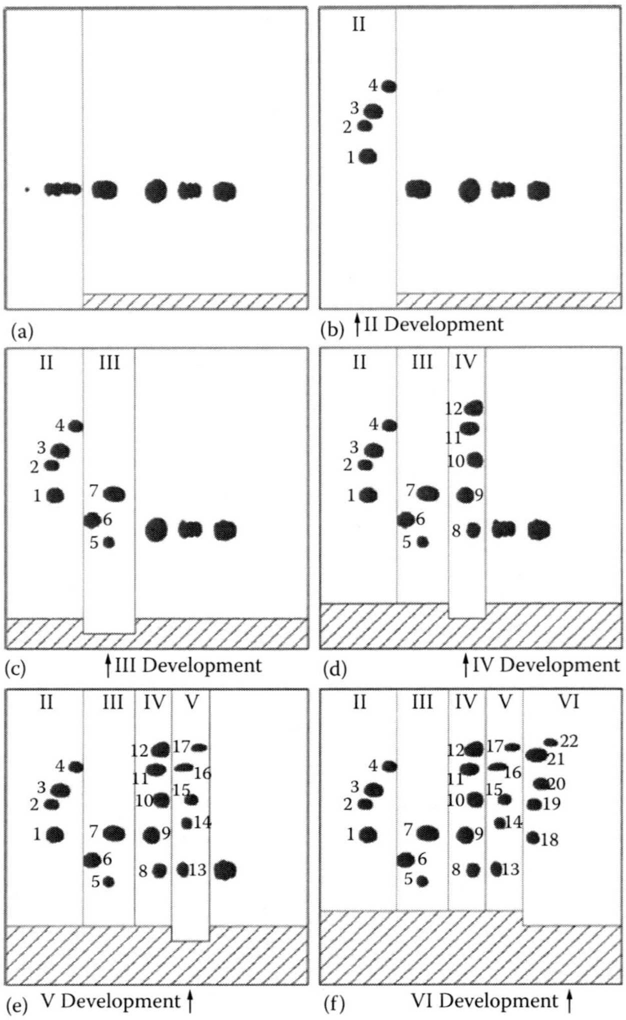

FIGURE 20.16 Illustration of step-by-step selective MDPC separation. (a) The dried plate after the first separation (first development) prepared for separation of the first group of compounds. One line (approximately 1 mm thick) was scraped in the stationary phase (continued) perpendicular to the first development in such a way that the spot(s) of the target compounds are between the line and the edge of the plate. For separation of the first group of compounds, another 5 mm wide region of the silica gel layer was removed from the bottom of the plate (the hatched lines indicate the stationary phase removed) so that, in the next step, the mobile phase runs only up a narrow strip of sorbent. (b) The dried plate after separation of the first group of pesticides (1–4) by use of acetonitrile–chloroform (15:85, v/v) mobile phase in the second development. (c) The prepared and dried plate after separation of the components of the second group of pesticides (5–7) by development with 100% chloroform twice over the same distance (UMD). (d) The prepared and dried plate after separation of the five components of the third group of pesticides (8–12) with nitromethane–dichloromethane (5:95, v/v) mobile phase in the fourth development. (e) The prepared and dried plate after separation of the five components of next group of pesticides (13–17) with nitromethane–chloroform (5:95, v/v) mobile phase in the fifth development. (f) The prepared and dried plate after separation of the five components of the last group of pesticides (18–22) with toluene–*n*-heptane (70:30, v/v) mobile phase in the sixth development (1—metamitron, 2—prochloraz, 3—metoxuron, 4—chloroxuron, 5—metalaxyl, 6—chlorotoluron, 7—methabenzthiazuron, 8—desmetryn, 9—napropamide, 10—tetrachlorvinphos, 11—metobromuron, 12—chlorbromuron, 13—atrazine, 14—flamprop-M-isopropyl, 15—benomyl, 16—captan, 17—procymidone, 18—bromopropylate, 19—vinclozolin, 20—fenvalerate, 21—methoxychlor, 22—fenchlorphos). (From Tuzimski, T., *J. Planar Chromatogr.—Mod. TLC*, 21, 49–54, 2008. With permission.)

must be removed to ensure that the mobile phase of the second development (step II) develops only the spot(s) of the target compounds between the two lines. Next, all of the plate and stationary phase except the part to be developed was covered by glass plates that were fixed with clamps. This procedure was repeated in subsequent steps (steps III–VI). Before each of steps III to VI, a region of the sorbent layer must again be removed from the plate to ensure that the mobile phase only develops the zones of the group of compounds of interest. (The regions removed before each development are shown by the hatched lines in Figure 20.16). The plates were developed in an unsaturated vertical chamber in MDPC experiments (Figure 20.16a through 20.16f). To prevent the mobile phase from migration on the sorbent with constituents that are not supposed to be chromatographed during the particular step, the part of the plate from which the sorbent was removed can be covered with a lipophilic substance (wax). In this case, the plates can be developed in a horizontal chamber in MDPC experiments.

The compounds from the first group (1–4) were chromatographed with acetonitrile–chloroform (15:85, v/v) as the mobile phase (Figure 20.16b). The pesticides in the second group (5–7) were chromatographed twice with chloroform as the mobile phase over the same distance (Figure 20.16c). The plate was dried for approximately 5 min between the two steps. The plate was then dried after the second separation of compounds of this group. Another portion of the stationary phase (the next 5 mm; hatched lines in Figure 20.16d) was then removed to ensure that the mobile phase used for development IV (Figure 20.16d) affected only the zones of the next group of compounds (8–12) between the two lines. Separation of components 13–17 with nitromethane–chloroform (5:95, v/v) mobile phase in the next step on the TLC plate is depicted in Figure 20.16e. Separation of pesticides of the last group (18–22) with toluene–*n*-heptane (70:30, v/v) mobile phase is depicted in Figure 20.16f. Separation of 22 components from a complex mixture by developments I–VI by MDPC was also achieved on a silica HPTLC plate. The best results were obtained with ethyl acetate–*n*-heptane (50:50, v/v) mobile phase in the first direction, which separated the pesticides into five groups (1–4, 5–7, 8–12, 13–17, and 18–22). As an example, a video scan of the MD separation of the 22 components on a silica HPTLC plate is shown in Figure 20.17 [133]. The separation can be characterized as PC × (PC + nPC + PC + PC + PC).

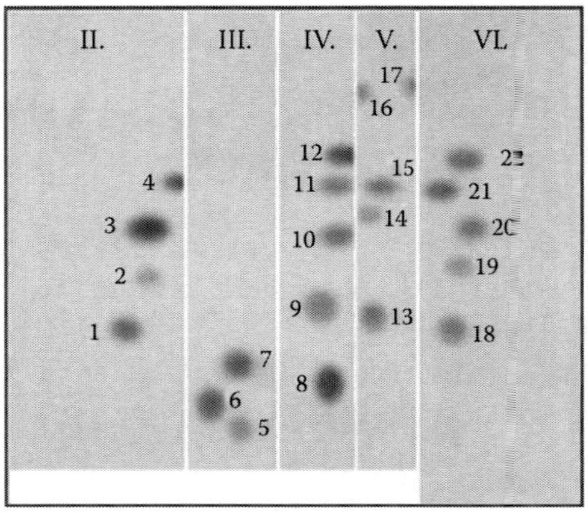

FIGURE 20.17 Video scan at 254 nm of a silica HPTLC plate showing separation of the 22 components of a mixture by developments I–VI in MDPC (names of pesticides the same as in Figure 20.16: 1—metamitron, 2—prochloraz, 3—metoxuron, 4—chloroxuron, 5—metalaxyl, 6—chlorotoluron, 7—methabenzthiazuron, 8—desmetryn, 9—napropamide, 10—tetrachlorvinphos, 11—metobromuron, 12—chlorbromuron, 13—atrazine, 14—flamprop-M-isopropyl, 15—benomyl, 16—captan, 17—procymidone, 18—bromopropylate, 19—vinclozolin, 20—fenvalerate, 21—methoxychlor, 22—fenchlorphos). (From Tuzimski, T., *J. Planar Chromatogr.—Mod. TLC*, 21, 49–54, 2008. With permission.)

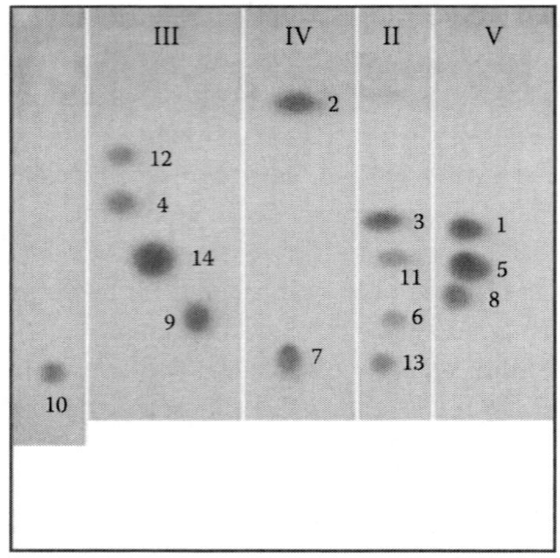

FIGURE 20.18 Video scan at 254 nm showing complete separation of 14 pesticides by MDPC (1—diuron, 2—dinoseb, 3—monolinuron, 4—trifluralin, 5—isoproturon, 6—propoxur, 7—alachlor, 8—lenacil, 9—tetradifon, 10—hexachlorobenzene, 11—carbaryl, 12—p,p′-DDT, 13—simazine, 14—4,4′-dibromobenzophenone). (From Tuzimski, T., *J. Sep. Sci.*, 30, 964–970, 2007. With permission.)

Also, a video scan is shown (Figure 20.18) of a complete separation of the next 14-component mixture of pesticides in the Ist–Vth developments by MDPC [PC × (nPC + PC + PC + PC)] [132].

The best combination for MDPC is the parallel combination of stationary and mobile phases. Separations of multicomponent mixtures were realized on multiphase plates [75,76]. The largest differences were obtained by combination of NP systems of the type silica/nonaqueous mobile phase in the first step of MDPC and RP systems of the type C18/water + organic modifier (methanol, acetonitrile, dioxane) in the next steps of MDPC on multiphase plates, for example, with a narrow zone of SiO_2 and a wide zone of C18 (or vice versa, Whatman Multi K SC5 or CS5 plates) [75]. This type of MDPC is performed by applying the multicomponent mixture in the corner of a square chromatographic plate (20 cm × 20 cm) and by development in the first direction with the first mobile phase on the narrow zone of SiO_2 (Multi K SC5 plate) or C18 silica (Multi K CS5 plate) [75,76].

20.2.3 SFC

SFC [138] has been seldom reported for pesticide residue analysis. However, one example is the high throughput determination of low concentrations of 17 pesticides with a wide range of polarities and molecular weights within 11 min using SFC-MS/MS with a polar imbedded RP column [139]. Diquat dibromide, cypermethrin, and tralomethrin were detected at ng/L levels in the presence of various other pesticides using a single mobile phase.

20.2.4 CEC

CEC is a method in which the mobile phase is driven across the stationary phase by use of electroosmosis instead of pressure as in column LC [140]. Again, relatively few papers have reported the use of this method for pesticide residue analysis with the following two as examples.

Zhao et al. [141] prepared a novel molecularly imprinted capillary monolithic column using trichlorfon as the template molecule by combining a nonhydrolytic sol-gel process with a molecular imprinting technique. Using this column, trace trichlorfon residues were determined in vegetables (cucumber, cauliflower, and leek) with an LOD (S/N = 3) of 92.5 ug kg^{-1} and method quantification limit of 305 ug kg^{-1}. The method was successfully validated for linearity, precision, and accuracy.

Perez-Fernandez et al. [142] used two novel polysaccharide-based chiral stationary phases for chiral separation of 16 pesticides, including herbicides, insecticides, and fungicides. CEC gave higher efficiency and chiral resolution in a comparative study with nano-LC. The CEC method was evaluated for LOD, limit of quantification, precision, and accuracy, and its applications to quantification of metalaxyl and its enantiomeric impurity (metalaxyl-S) in a commercial fungicide product marketed as enantiomerically pure (metalaxyl-M) and residues in soil and tap water after SPE were described.

20.3 NONCHROMATOGRAPHIC METHODS USED FOR PESTICIDE ANALYSIS

Liu et al. proposed a highly sensitive, rhodamine B-covered gold nanoparticle (RB AuNP)-based assay with dual readouts (colorimetric and fluorometric) for detecting organophosphorus and carbamate pesticides in complex solutions [143]. The detection mechanism is based on the fact that these pesticides can inhibit the activity of acetylcholinesterase (AChE), thus preventing the generation of thiocholine (which turns the RB-AuNP solutions blue and unquenches the fluorescence of RB simultaneously). The color of the RB-AuNP solution remains red, and the fluorescence of RB remains quenched. By use of this dual-readout assay, the lowest detectable concentrations for several kinds of pesticides including carbaryl, diazinon, malathion, and phorate were measured to be 0.1, 0.1, 0.3, and 1 μg/L, respectively, all of which are much lower than the MRLs as reported in the EU pesticides database as well as those from the U.S. Department of Agriculture (USDA). This assay allows detection of pesticides in real samples, such as agricultural products and river water. The results in detecting pesticide residues collected from food samples via this method agreed well with those from HPLC. This simple assay is, therefore, suitable for sensing pesticides in complex samples, especially in combination with other portable platforms [143].

CE is a nonchromatographic method because there is no mobile phase [138]; mobility of the compounds to be separated is obtained by application of voltage. Its use in pesticide analysis has been reported in a relatively limited number of papers, apparently because of its comparatively low sensitivity. For example, online concentration techniques necessary to increase sample size and meet LOD levels for determination of agrochemical residues by CE were reviewed [144], and molecularly imprinted polymer SPE was used prior to CE for determination of trichorfon residues in lettuce and radish vegetables with an enrichment factor of 160 [145].

Immunoassays have been widely used in pesticide residue analysis and can serve as an important complementary technique to HPLC. A book on pesticide immunoassays [146] and reviews on multi-analyte immunoassays [147]; enzyme-linked immunosorbent assays (ELISAs) for quantitative, rapid, and simple detection and quantification of neonicotinoid insecticides in food or environmental matrices [148]; and detection of mycotoxins, pesticides, and veterinary drugs in agricultural products and foodstuffs [149] were published recently. Immunoassays tend to be simple, rapid, and sensitive, but most are single analyte methods. Twenty-one direct competitive ELISA test kits were developed for 21 kinds of insecticide and fungicide residues for use in inspection of farm products by farmers and regulators; preparation methods of antibodies for the pesticides, development of the ELISAs, commercialization of kits, and applicability to analysis of samples were discussed [150]. An ELISA kit and HPLC were compared in terms of LOD, RSD, and percentage recovery, and matrix interference for determination of residues of chlorpyrifos ethyl in water and sediment samples. Results were comparable, and it was suggested that ELISA could be used in regular pesticide monitoring programs, particularly in developing countries where HPLC is not a commonly available analytical technique [151].

20.4 CONCLUSIONS

In our opinion, key strategies and future development of the HPLC methods for analysis of major classes of pesticides and their residues are associated with high sensitivity and selectivity of analytical techniques and with increasingly lower values of MRLs, and they are the following:

- Developing new sensitive and selective HPLC/UHPLC-MS techniques (QqQ-MS, IT-MS^n, LIT-MS, TOF, QqTOF, and orbitrap instruments).
- Simplified preparation steps especially for high-throughput analysis, such as use of automated devices. Dr. Alex Krynitsky, Section Editor for Residues and Trace Elements of the *Journal of AOAC International*, attended the North American Chemical Residue Workshop, St. Pete Beach, FL, July 20–23, 2014 (formerly the Florida Pesticide Residue Workshop), and noted that many investigators are trying to shorten the extraction/cleanup by using more automation, that is, Gerstel devices on the front end of an LC-MS/MS system [152]. Gerstel offers devices for automated SPE, automated disposable pipet extraction (DPX), and automated DPX-QuEChERS according to its website.
- These new sample preparation methods will reduce the impact of matrix effects. However, some analysts, knowing the sensitivity and selectivity of the newer MS systems, are relying on "dilute and shoot" methods [152] without cleanup.
- Wider application of the QuEChERS sample preparation technique and further modifications to include more analytes and sample matrices; this procedure requires only a few milliliters of solvent and is capable of generating recoveries of 90%–110% with RSDs <5% for a wide range of LC or GC amenable compounds.
- Replacing hazardous chemicals with safer reagents, reducing the number of sample pretreatment steps, and developing novel techniques for direct detection are typical approaches for developing green analytical methodologies.
- Further development of MD techniques.
- Further development of miniaturization.
- Compared to HPLC/UHPLC-MS and -MS/MS and GC-MS and -MS/MS, other chromatographic and nonchromatographic methods will be used only in analyses for certain types of analyte/sample matrix combinations when these methods would offer characteristic advantages in particular situations, for example, sample complexity, field analysis, cost, speed, experience of the analysts, and availability of required instruments in the analytical laboratory.

From the perspective of food safety control, especially for baby food, developing sensitive and broad-spectrum screening tools to detect lethal levels of various pesticides in foods will have significant practical applications.

REFERENCES

1. Snyder, L. R., Kirkland, J. J., Dolan, J. W. *Introduction to Modern Liquid Chromatography,* John Wiley & Sons, Inc., Hoboken, NJ, 2010.
2. Dane, A. J., Havey, C. D., Voorhees. K. J., The detection of nitro pesticides in mainstream and sidestream cigarette smoke using electron monochromator–mass spectrometry, *Anal. Chem.*, 78, 3227–3233, 2006.
3. Anastassiades, M., Lehotay, S. J., Stajnbaher, D., Schenck, F. J., Fast and easy multiresidue method employing acetonitrile extraction/partitioning and "dispersive solid-phase extraction" for the determination of pesticide residuesin produce. *J. AOAC Int.* 86, 412–431, 2003.
4. Pesticide residues in foods by MeCN extraction and partitioning with magnesium sulfate. Official Methods of Analysis of AOAC International; AOAC International: Gaithersburg, MD, 2007; Method 2007.1.
5. EN 15662 (2008). Foods of plant origin—Determination of pesticide residues using GC-MS and/or LC-MS/MS following acetonitrile extraction/partitioning and cleanup by dispersive SPE–QuEChERS-method; http://esearch.cen.eu/esearch/ (accessed Feb 23, 2013).

6. Lehotay, S. J., Mastovska, K., Yun, S. J., Evaluation of two fast and easy methods for pesticide residue analysis in fatty food matrixes. *J. AOAC Int.* 88 (2), 630–638, 2005.
7. Lehotay, S. J., de Kok, A., Hiemstra, M., van Bodegraven, P., Validation of a fast and easy method for the determination of residues from 229 pesticides in fruits and vegetables using gas and liquid chromatography and mass spectrometric detection. *J. AOAC Int.* 88, 595–614, 2005.
8. Nguyen, T. D., Lee, B. S., Lee, B. R., Lee, D. M., Lee, G. H., A multiresidue method for the determination of 109 pesticides in rice using Quick Easy Cheap Effective Rugged and Safe (QuEChERS) sample preparation method and gas chromatography/mass spectrometry with temperature control and vacuum concentration, *Rapid Commun. Mass Spectrom.* 18, 3115–3122, 2007.
9. Lehotay, S. J., Determination of pesticide residues in foods by acetonitrile extraction and partitioning with magnesium sulfate: Collaborative study. *J. AOAC Int.* 90, 485–520, 2007.
10. Nguyen, T. D., Yu, J. E., Lee, D. M., Lee, G. H., A multiresidue method for the determination of 107 pesticides in cabbage and radish using QuEChERS sample preparation method and gas chromatography–mass spectrometry. *Food Chem.* 110, 207–213, 2008.
11. Lee, J.-M., Park, J.-W., Jang, G.-C., Hwang, K.-J., Comparative study of pesticide multi-residue extraction in tobacco for gas chromatography–triple quadrupole mass spectrometry. *J. Chromatogr. A* 1187, 25–33, 2008.
12. Lehotay, S. J., Anastassiades, M., Majors, R. E., The QuEChERS revolution. *LC-GC Europe* 23, 418–429, 2010.
13. Lin, Y., Chamkasem, N., Zhang, K., Wong, J., Inter-laboratory validation of a modified QuEChERS and LC-MS/MS method for analysis of about 250 pesticide residues in six fatty food products. Proceedings of the 58th ASMS Conference on Mass Spectrometry and Allied Topics, Salt Lake City, UT, May 23–27; ASMS: Santa Fe, NM, 2010.
14. Lin, Y., Laboratory Information Bulletin 4472, Analysis of twenty-six pesticide residues in fatty food products using a modified QuEChERS method and LC-MS/MS; U.S. Food and Drug Administration, Division of Field Science: Rockville, MD, 2011.
15. Chamkasem, N., Ollis, L. W., Harmon, T., Lee S., Mercer G., Analysis of 136 pesticides in avocado using a modified QuEChERS method with LC-MS/MS and GC-MS/MS, *J. Agric. Food Chem.* 61, 2315–2329, 2013.
16. Wang, J., Chow, W., Leung, D., Chang, J., Application of ultrahigh-performance liquid chromatography and electrospray ionization quadrupole orbitrap high-resolution mass spectrometry for determination of 166 pesticides in fruits and vegetables. *J. Agric. Food Chem.* 60, 12088–12104, 2012.
17. Wang, J., Leung, D., Applications of ultra-performance liquid chromatography electrospray ionization quadrupole time-of-flight mass spectrometry for analysis of 138 pesticides in fruit- and vegetable-based infant foods. *J. Agric. Food Chem.* 57, 2162–2173, 2009.
18. Chen, X., Bian, Z., Hou, H., Yang, F., Liu, S., Tang, G., Hu, Q., Development and validation of a method for the determination of 159 pesticide residues in tobacco by gas chromatography–tandem mass spectrometry. *J. Agric. Food Chem.* 61, 5746–5757, 2013.
19. Guan W., Li, Z., Zhang, H., Hong, H., Rebeyev, N., Ye Y., Ma, Y., Amine modified graphene as reversed-dispersive solid phase extraction materials combined with liquid chromatography–tandem mass spectrometry for pesticide multi-residue analysis in oil crops. *J. Chromatogr. A* 1286, 1–8, 2013.
20. Mastovska, K., Dorweiler, K. J., Lehotay, S. J., Wegscheid, J. S., Szpylka, K. A., *J. Agric. Food Chem.* 58, 5959, 2010.
21. Lacina, O., Zachariasova, M., Urbanova, J., Vaclavikova, M., Cajka, T., Hajslova, J., Critical assessment of extraction methods for the simultaneous determination of pesticide residues and mycotoxins in fruits, cereals, spices and oil seeds employing ultra-high performance liquid chromatography–tandem mass spectrometry. *J. Chromatogr. A* 1262, 8–18, 2012.
22. Cajka, T., Sandy, C., Bachanova, V., Drabova, L., Kalachova, K., Pulkrabova, J., Hajslova, J. Streamlining sample preparation and gas chromatography–tandem mass spectrometry analysis of multiple pesticide residues in tea. *Anal. Chim. Acta* 742, 51–60, 2012.
23. Lehotay, S. J., Son, K., Kwon, H., Koesukwiwat, U., Fu, W., Mastovska, K., Hoh, E., Leepipatpiboon, N., Comparison of QuEChERS sample preparation methods for the analysis of pesticide residues in fruits and vegetables. *J. Chromatogr. A* 1217, 2548–2560, 2010.
24. UCT, LLC (PA, USA) *QuEChERS Information Booklet. Pesticide Residue Analysis.* www.unitedchem .com.
25. 5990-5324EN QuEChERS Poster, Agilent Technologies, Inc. 2011, U.S.A., August 9, 2011.
26. Ortelli, D., Edder, P., Corvi, C., Multiresidue analysis of 74 pesticides in fruits and vegetables by liquid chromatography–electrospray–tandem mass spectrometry. *Anal. Chim. Acta* 520, 33–45, 2004.

27. Wong, J., Hao, C. Y., Zhang, K., Yang, P., Banerjee, K., Hayward, D., Iftakhar, I., Schreiber, A., Tech, K., Sack, C., Smoker, M., Chen, X. R., Utture, S. G., Oulkar, D. P., Development and interlaboratory validation of a QuEChERS-based liquid chromatography-tandem mass spectrometry method for multiresidue analysis. *J. Agric. Food Chem.* 58, 5897–5903, 2010.
28. Zhang, K., Wong, J., Yang, P., Tech, K., DiBenedetto, A. L., Lee, N. S., Hayward, D. G., Makovi, C. M., Krynitsky, A. J., Banerjee, K., Jao, L., Dasgupta, S., Smoker, M. S., Simonds, R., Schreiber, A., Multiresidue pesticide analysis of agricultural commodities using acetonitrile salt-out extraction, dispersive solid-phase sample cleanup, and high-performance liquid chromatography–tandem mass spectrometry. *J. Agric. Food Chem.* 59, 7636–7646, 2011.
29. Hajslova, J., Cajka, T., Vaclavik, L., Challenging applications offered by direct analysis in real time (DART) in food quality and safety analysis. *Trends Anal. Chem.* 30 (2), 204–218, 2011.
30. Edison, S. E., Lin, L. A., Parrales, L., Practical considerations for the rapid screening of pesticides using ambient pressure desorption ionisation with high resolution mass spectrometry. *Food Addit. Contam., Part A* 28 (10), 1393–1404, 2011.
31. Farré, M., Picó, Y., Barcelo, D., Direct peel monitoring of xenobiotics in fruit by direct analysis in real time coupled to a linear quadrupole ion trap–orbitrap mass spectrometer. *Anal. Chem.* 85 (5), 2638–2644, 2013.
32. Zhang, K., Wong, J. W., Yang, P., Hayward, D. G., Sakuma, T., Zou, Y., Schreiber, A., Borton, C., Nguyen, T.-V., Kaushik, B., Oulkar, D., Protocol for an electrospray ionization tandem mass spectral product ion library: Development and application for identification of 240 pesticides in foods. *Anal. Chem.* 84, 5677–5684, 2012.
33. Alder, L., Greulich, K., Kempe, G., Vieth, B., Residue analysis of 500 high priority pesticides: Better by GC-MS or LC-MS/MS. *Mass Spectrom. Rev.* 25, 838–865, 2006.
34. Vassilakis, I., Tsipi, D., Scoullos, M., Determination of a variety of chemical classes of pesticides in surface and ground waters by off-line solid-phase extraction, gas chromatography with electron-capture and nitrogen–phosphorus detection, and high-performance liquid chromatography with postcolumn derivatization and fluorescence detection. *J. Chromatogr. A* 823, 49–58, 1998.
35. Stalikas, C. D., Pilidis, G. A., Development of a method for the simultaneous determination of phosphoric and amino acid group containing pesticides by gas chromatography with mass-selective detection. Optimization of the derivatization procedure using an experimental design approach. *J. Chromatogr. A* 872, 215–225, 2000.
36. Gerecke, A. C., Tixier, C., Bartels, T., Schwarzenbach, R. P., Müller, S. R., Determination of phenylurea herbicides in natural waters at concentrations below 1 ng l^{-1} using solid-phase extraction, derivatization, and solid-phase microextraction–gas chromatography–mass spectrometry. *J. Chromatogr. A* 930, 9–19, 2001.
37. De Alwis, G. K. H., Needham, L. L., Barr, D. B., Measurement of human urinary organophosphate pesticide metabolites by automated solid-phase extraction, post extraction derivatization, and gas chromatography–tandem mass spectrometry. *J. Chromatogr. B* 843, 34–41, 2006.
38. Guo, L., Lee, H. K., Low-density solvent based ultrasound-assisted emulsification microextraction and on-column derivatization combined with gas chromatography–mass spectrometry for the determination of carbamate pesticides in environmental water samples. *J. Chromatogr. A* 1235, 1–9, 2013.
39. Mastovska, K., Lehotay, S. J., Anastassiades, M., Combination of analyte protectants to overcome matrix effects in routine GC analysis of pesticide residues in food matrixes. *Anal. Chem.* 77, 8129–8137, 2005.
40. Pang, G. F., Fan, C.-L., Zhang, F., Li, Y., Chang, Q.-Y., Cao, Y.-Z., Liu, Y.-M., Li, Z.-Y., Wang, Q.-J., Hu, X.-Y., Liang, P., High-throughput GC/MS and HPLC/MS/MS techniques for the multiclass, multiresidue determination of 653 pesticides and chemical pollutants in tea. *J. AOAC Int.* 94 (4), 1253–1296, 2011.
41. Portolés, T., Mol, J. G. J., Sancho, J. V., Hernández, F., Advantages of atmospheric pressure chemical ionization in gas chromatography tandem mass spectrometry: Pyrethroid insecticides as a case study. *Anal. Chem.* 84, 9802–9810, 2012.
42. Zhou, T., Xiao, X., Li, G., Microwave accelerated selective soxhlet extraction for the determination of organophosphorus and carbamate pesticides in ginseng with gas chromatography/mass spectrometry. *Anal. Chem.* 84, 5816–5822, 2012.
43. Tuzimski, T., Basic principles of planar chromatography and its potential for hyphenated techniques, in: *High-Performance Thin-Layer Chromatography (HPTLC)*, Srivastava, M. M. (Ed.), Springer-Verlag Berlin Heidelberg, Chapter 14, pp. 247–310, 2011.
44. Tuzimski, T., Use of planar chromatography in pesticide residue analysis, in: *Handbook of Pesticides: Methods of Pesticide Residue Analysis*, Nollet, L. M. L., Rathore, H. S. (Eds.), CRC Press/Taylor & Francis Group, Boca Raton, FL, Chapter 9, pp. 187–264, 2010.

45. Sherma, J., Review of advances in the thin layer chromatography of pesticides: 2010–2012. *J. Environ. Sci. Health B* 48, 417–430, 2013.
46. Sherma, J., Review of advances in the thin layer chromatography of pesticides: 2008–2010. *J. Environ. Sci. Health B* 46, 557–568, 2011.
47. Sherma, J., Review of advances in the thin layer chromatography of pesticides: 2006–2008. *J. Environ. Sci. Health B* 44, 193–203, 2009.
48. Sherma, J., Review of advances in the thin layer chromatography of pesticides: 2004–2006. *J. Environ. Sci. Health B* 42, 429–440, 2007.
49. Sherma, J., Thin layer chromatography of pesticides—A review of applications for 2002–2004. *Acta Chromatogr.* 15, 5–30, 2005.
50. Sherma, J., Recent advances in the thin layer chromatography of pesticides: A review. *J. AOAC Int.* 86, 602–611, 2003.
51. Sherma, J., Recent advances in the thin layer chromatography of pesticides. *J. AOAC Int.* 84, 993–999, 2001.
52. Sherma, J., Recent advances in thin layer chromatography of pesticides. *J. AOAC Int.* 82, 48–53, 1999.
53. Sherma, J., Review: Determination of pesticides by thin layer chromatography. *J. Planar Chromatogr.—Mod. TLC* 10, 80–89, 1997.
54. Sherma, J., Biennial review of pesticide analysis. *Anal. Chem.* 67, 1R–20R, 1995.
55. Sherma, J., Biennial review of pesticide analysis. *Anal. Chem.* 65, 40R–54R, 1993.
56. Wall, P. E, *Thin-Layer Chromatography—A Modern Approach*, The Royal Society of Chemistry, Cambridge, UK, 2005.
57. Fried, B., Sherma, J. (Eds.), *Handbook of Thin-Layer Chromatography,* 3rd edition, revised and expanded, Marcel Dekker Inc., New York, 2003.
58. Srivastava, M. M., *High Performance Thin Layer Chromatography (HPTLC)*, Springer, Heidelberg, Germany, 2011.
59. Nyiredy, Sz., *Planar Chromatography, a Retrospective View for the Third Millennium*, Springer Scientific Publisher, Budapest, Hungary, 2001.
60. Dzido, T. H., Tuzimski, T., Chambers, sample application, and chromatogram development, in: *Thin-Layer Chromatography in Phytochemistry*, Waksmundzka-Hajnos, M., Sherma, J., Kowalska, T. (Eds.), CRC Press/Taylor & Francis Group, Boca Raton, Chapter 7, pp. 119–174, 2008.
61. Tuzimski, T., Thin-layer chromatography (TLC) as pilot technique for HPLC. Utilization of retention database (R_F) vs. eluent composition of pesticides. *Chromatographia* 56, 379–381, 2002.
62. Tuzimski, T., Soczewiński, E., Retention and selectivity of liquid-solid chromatographic systems for the analysis of pesticides (Retention database), in: *Problems of Science, Teaching and Therapy*, Medical University of Lublin, Poland, No 12, Lublin, October 2002, 219 pp. (limited number available on request: tomasz.tuzimski@umlub.pl).
63. Spangenberg, B., Does the Kubelka–Munk theory describe TLC evaluations correctly? *J. Planar Chromatogr.—Mod. TLC* 19 (111) 332–341, 2006.
64. Spangenberg, B., Klein, K.-F., Fibre optical scanning with high resolution in thin-layer chromatography. *J. Chromatogr. A* 898, 265–269, 2000.
65. Spangenberg, B., Klein, K.-F., New evaluation algorithm in diode-array thin-layer chromatography. *J. Planar Chromatogr.—Modern TLC* 14 (4) 260–265, 2001.
66. Spangenberg, B., Lorenz, K., Nasterlack, S., Fluorescence enhancement of pyrene measured by thin layer chromatography with diode-array detection. *J. Planar Chromatogr.—Mod. TLC* 16 (5) 331–337, 2003.
67. Ahrens, B., Blankenhorn, D., Spangenberg, B., Advanced fiber optical scanning in thin-layer chromatography for drug identification. *J. Chromatogr. B* 772, 11–18, 2002.
68. Tuzimski, T., Use of thin-layer chromatography in combination with diode array scanning densitometry for identification of fenitrothion in apples. *J. Planar Chromatogr.—Mod. TLC* 18 (106) 419–422, 2005.
69. Tuzimski, T., Two-dimensional TLC with adsorbent gradients of the type silica-octadecyl silica and silica-cyanopropyl for separation of mixtures of pesticides. *J. Planar Chromatogr—Mod. TLC* 18 (105), 349–357, 2005.
70. Tuzimski, T., Determination of pesticides in water samples from the Wieprz-Krzna Canal in the Łęczyńsko-Włodawskie Lake District of Southeastern Poland by thin-layer chromatography with diode array scanning and high performance column liquid chromatography with diode array detection. *J. AOAC Int.* 91 (5), 1203–1209, 2008.
71. Tuzimski, T., Application of SPE–HPLC–DAD and SPE–TLC–DAD to the determination of pesticides in real water samples. *J. Sep. Sci.* 31 (20), 3537–3542, 2008.

72. Tuzimski, T., Application of SPE–HPLC–DAD and SPE–HPTLC–DAD to the analysis of pesticides in lake water. *J. Planar Chromatogr.—Mod. TLC* 22 (4), 235–240, 2009.
73. Tuzimski, T., Sobczyński, J., Application of HPLC–DAD and TLC–DAD after SPE to the quantitative analysis of pesticides in water samples. *J. Liq. Chromatogr. Relat. Technol.* 32 (9) 1241–1258, 2009.
74. Tuzimski, T., Application of HPLC and TLC with diode array detection after SPE to the determination of pesticides in water samples from the Zemborzycki Reservoir in (Lublin, southeastern Poland). *J. AOAC Int.* 93 (6) 1748–1756, 2010.
75. Tuzimski, T., New procedure for determination of analytes in complex mixtures by multidimensional planar chromatography in combination with diode array scanning densitometry (MDPC–DAD) and high performance liquid chromatography coupled with DAD detector (HPLC–DAD). *J. Planar Chromatogr.—Mod. TLC.* 23 (3), 184–189, 2010.
76. Tuzimski, T., Determination of clofentezine in medical herbs extracts by chromatographic methods combined with diode array scanning densitometry. *J. Sep Sci.* 33, 1954–1958, 2010.
77. Tuzimski, T., Determination of analytes in medical herbs extracts by SPE coupled with two-dimensional planar chromatography in combination with diode array scanning densitometry and HPLC-diode array detector. *J. Sep Sci.* 34, 27–36, 2011.
78. Jautz, U., Morlock, G., Efficacy of planar chromatography coupled to (tandem) mass spectrometry for employment in trace analysis. *J. Chromatogr. A* 1128 (1–2), 244–250, 2006.
79. Jautz, U., Morlock, G., Validation of a new planar chromatographic method for quantification of the heterocyclic aromatic amines most frequently found in meat. *Anal. Bioanal. Chem.* 387 (3) 1083–1093, 2007.
80. Morlock, G., Ueda, Y., New coupling of planar chromatography with direct analysis in real time mass spectrometry. *J. Chromatogr. A* 1143 (1–2), 243–251, 2007.
81. Morlock, G., Jautz, U., Comparison of two different plunger geometries for HPTLC-MS coupling via an extractor-based interface. *J. Planar Chromatogr.—Mod. TLC* 21 (5) 367–371, 2008.
82. Aranda, M., Morlock, G., Simultaneous determination of riboflavin, pyridoxine, nicotinamide, caffeine and taurine in energy drinks by planar chromatography-multiple detection with confirmation by electrospray ionization mass spectrometry. *J. Chromatogr. A* 1131 (1–2) 253–260, 2006.
83. Morlock, G., Schwack, W., Determination of isopropylthioxanthone (ITX) in milk, yoghurt, and fat by HPTLC-FLD, HPTLC-ESI/MS and HPTLC-DART/MS. *Anal. Bioanal. Chem.* 385 (3) 586–595, 2006.
84. Alpmann, A., Morlock, G., Improved online coupling of planar chromatography with electrospray mass spectrometry: Extraction of zones from glass plates. *Anal. Bioanal. Chem.* 386 (5) 1543–1551, 2006.
85. Alpmann, A., Morlock, G., Rapid and sensitive determination of acrylamide in drinking water by planar chromatography and fluorescence detection after derivatization with dansulfinic acid. *J. Sep. Sci.* 31 (1) 71–77, 2008.
86. Morlock, G., Dytkiewitz, E., Analytical strategy for rapid identification and quantification of lubricant additives in mineral oil by high-performance thin-layer chromatography with UV absorption and fluorescence detection combined with mass spectrometry and infrared spectroscopy. *J. AOAC Int.* 91 (5), 1237–1244, 2008.
87. Oellig, C., Schwack, W., Planar solid phase extraction clean-up and microliter-flow injection analysis–time-of-flight mass spectrometry for multi-residue screening of pesticides in food. *J. Chromatogr. A* 1351, 1–11, 2014.
88. Perry, J. A., Haag, K. W., Glunz, L. J., Programmed multiple development in thin-layer chromatography. *J. Chromatogr Sci.* 11, 447–453, 1973.
89. Burger, K., TLC-PMD, thin layer chromatography with gradient elution in comparison to HPLC. *Fresenius' Z. Anal. Chem.* 318, 228–233, 1984.
90. Markowski, W., Computer-aided optimization of gradient multiple development thin-layer chromatography. III. Multi-stage development over a constant distance. *J. Chromatogr. A* 726, 185–192, 1996.
91. Markowski, W., Gradient development in TLC, in: *Encyclopedia of Chromatography*, Cazes, J., (Ed), Marcel Dekker, Inc., New York, pp. 699–713, 2005.
92. CAMAG: Instruments, Tools and Concepts for HPTLC. Available at http://www.camag.com.
93. Poole, C. F., Belay, M. T., Progress in automated multiple development. *J. Planar Chromatogr.—Mod. TLC* 4, 345–359, 1991.
94. Szabady, B., Nyiredy, Sz., *Dünnschicht-Chromatographie in Memorian Professor Dr. Hellmut Jork*, Kaiser, R. E., Gunther, W., Gunz, H., Wulff, G., (Eds), InCom Sonderband, Dusseldorf, Germany, pp. 212–224, 1996.
95. Ebel, S., Volkl, S., *Dtsch. Apoth. Ztg.* 130, 2162–2169, 1990.

96. Markowski, W., Soczewiński, E., Computer-aided optimization of gradient multiple-development thin layer Chromatography. I. Two-stage development. *J. Chromatogr.* 623, 139–147, 1992.
97. Markowski, W., Soczewiński, E., Computer-aided optimization of stepwise gradient and multiple development thin-layer chromatography. *Chromatographia* 36, 330–336, 1993.
98. Markowski, W., Computer-aided optimization of gradient multiple-development thin-layer chromatography. I. Multi-stage development. *J. Chromatogr.* 635, 283–289, 1993.
99. Markowski, W., Czapińska, K. L., Błaszczak, M., Determination of the constants of the Snyder-Soczewiński equation by means of gradient multiple development. *J. Liq. Chromatogr.* 17, 999–1009, 1994.
100. Markowski, W., Matysik, G., Analysis of plant extracts by multiple development thin-layer chromatography. *J. Chromatogr.* 646, 434–438, 1993.
101. Matysik, G., Modified programmed multiple gradient development (MGD) in the analysis of complex plant extracts. *Chromatographia* 43, 39–43, 1996.
102. Matysik, G., Soczewiński, E., Stepwise gradient development in thin-layer chromatography. IV. Miniaturized generators for continuous and stepwise gradients. *J. Chromatogr.* 446, 275–282, 1988.
103. Butz, S., Stan, H. J., Screening of 265 pesticides in water by thin-layer chromatography with automated multiple development. *Anal. Chem.* 67, 620–630, 1995.
104. Lautie, J.-P., Stankovic, V., Automated multiple development TLC of phenylurea herbicides in plants, *J. Planar Chromatogr.—Mod. TLC* 9, 113–115, 1996.
105. Tuzimski, T., Multidimensional chromatography in pesticides analysis, in: *Pesticides, Strategies for Pesticides Analysis*, Stoytcheva, M. (Ed.), InTech Rijeka, 2011, Chapter 7, pp. 155–196.
106. Tuzimski, T., Determination of pesticides in complex samples by one dimensional (1D-), two-dimensional (2D-) and multidimensional chromatography, in: *Pesticides in the Modern World—Trends in Pesticides Analysis*, Stoytcheva, M. (Ed.), InTech Rijeka, 2011, Chapter 12, pp. 281–318.
107. Mondello, L., Lewis, A. C., Bartle, K. D. (Eds.), *Multidimensional Chromatography*, John Wiley & Sons, Ltd., Chichester, England.
108. Bertsch, W., Multidimensional gas chromatography, in: *Multidimensional Chromatography. Techniques and Applications*, Cortes, H. J. (Ed.), Marcel Dekker, New York, 1990, pp. 74–144.
109. Corradini, C., Coupled-column liquid chromatography, in: *Multidimensional Chromatography*, Mondello, L., Lewis, A. C., Bartle, K. D. (Eds.), 2002, pp. 109–134, John Wiley & Sons, Ltd., Chichester, England, 2002, pp. 109–134.
110. Giddings, J. C., Use of multiple dimensions in analytical separations, in: *Multidimensional Chromatography*, Cortes, H. J. (Ed.), Marcel Dekker, New York, 1990, pp. 251–299.
111. Mariott, P. J., Orthogonal GC-GC, in: *Multidimensional Chromatography*, Mondello, L.; Lewis, A. C.; Bartle, K. D. (Eds.), John Wiley & Sons, Ltd., Chichester, England. 2002, pp. 77–108.
112. Lewis, A. C., Multidimensional high resolution gas chromatography, in: *Multidimensional Chromatography*, Mondello, L., Lewis, A. C., Bartle, K. D. (Eds.), John Wiley & Sons, Ltd., Chichester, England, 2002, pp. 47–75.
113. Nyiredy, Sz., Multidimensional planar chromatography, in: *Multidimensional Chromatography*, Mondello, L., Lewis, A. C., Bartle, K. D. (Eds.), pp. 171–196, John Wiley & Sons, Ltd., Chichester, England, 2002, pp. 171–196.
114. Liu, Z., Phillips, J. B., Comprehensive two-dimensional gas chromatography using an on-column thermal modulator interface. *J. Chromatogr. Sci.* 29, 227–231, 1991.
115. Phillips, J. B., Xu, J., Comprehensive multidimensional gas chromatography. *J. Chromatogr.,* 703, 327–334, 1995.
116. Phillips, J. B., Ledford, E. B., Thermal modulation: A chemical instrumentation component of potential value in improving portability. *Field Anal. Chem. Technol.* 1, 23–29, 1996.
117. Phillips, J. B., Beens, J., Comprehensive two-dimensional gas chromatography: A hyphenated method with strong coupling between the two dimensions. *J. Chromatogr. A* 856, 331–347, 1999.
118. Giddings, J. C., Two-dimensional separation: Concepts and promise. *Anal. Chem.* 56, 1258A–1270A, 1984.
119. Giddings, J. C., Concepts and comparison in multidimensional separation. *J. High Resolut. Chromatogr. Chromatogr. Commun.* 10, 319–323, 1987.
120. Nyiredy, Sz., Multidimensional planar chromatography. *LC GC Eur.* 16, 52–59, 2003.
121. Pandey, R. C., Misra, R., Rinehart, Jr., K. L., Graft thin-layer chromatography. *J. Chromatogr.* 169, 129–139, 1979.
122. De Spiegeleer, B., Van den Bossche, W., De Moerlose, P., Massart, D., A strategy for two-dimensional, high-performance thin-layer chromatography, applied to local anesthetics. *Chromatographia* 23, 407–411, 1987.

123. Tuzimski, T., Soczewiński, E., Chemometric characterization of the R_F values of pesticides in thin-layer chromatography on silica with mobile phases comprising a weakly polar diluent and a polar modifier. Part V. *J. Planar Chromatogr.—Mod. TLC* 15, 164–168, 2002.
124. Tuzimski, T., Soczewiński, E., Correlation of retention parameters of pesticides in normal- and reversed-phase systems and their utilization for the separation of a mixture of 14 triazines and urea herbicides by means of two-dimensional thin-layer chromatography. *J. Chromatogr. A* 961, 277–283, 2002.
125. Tuzimski, T., Soczewiński, E., Use of database of plots of pesticide retention (R_F) against mobile-phase composition. Part I. Correlation of pesticide retention data in normal- and reversed-phase systems and their use to separate a mixture of ten pesticides by 2D-TLC. *Chromatographia* 56, 219–223, 2002.
126. Tuzimski, T., Soczewiński, E., Use of database of plots of pesticide retention (R_F) against mobile-phase compositions for fractionation of a mixture of pesticides micropreparative thin-layer chromatography. *Chromatographia* 59, 121–128, 2004.
127. Tuzimski, T., Bartosiewicz, A., Correlation of retention parameters of pesticides in normal and RP systems and their utilization for the separation of a mixture of ten urea herbicides and fungicides by two-dimensional TLC on cyanopropyl-bonded polar stationary phase and two-adsorbent-layer Multi-K plate. *Chromatographia* 58, 781–788, 2003.
128. Tuzimski, T., Separation of a mixture of eighteen pesticides by two-dimensional thin layer chromatography on a cyanopropyl-bonded polar stationary phase. *J. Planar Chromatogr.—Mod. TLC* 17, 328–334, 2004.
129. Dzido, T. H., Modern TLC chambers, in: *Planar Chromatography, a Retrospective View for the Third Millennium*, Nyiredy, Sz. (Ed.), Springer Scientific Publisher, Budapest, Hungary, 2001, pp. 69–87.
130. Szabady, B., Nyiredy, Sz., The versatility of multiple development, in: *Dünnschicht-Chromatographie in Memorian Professor Dr. Hellmut Jork*, Kaiser, R. E., Günther, W., Gunz, H., Wulff, G. (Eds.), InCom Sonderband, Düsseldorf, Germany, 1996, pp. 212–224.
131. Tuzimski, T., Separation of multicomponent mixtures of pesticides by graft thin layer chromatography on connected silica and octadecyl silica layers. *J. Planar Chromatogr.—Mod. TLC* 20, 13–18, 2007.
132. Tuzimski, T., A new procedure for separation of complex mixtures of pesticides by multidimensional planar chromatography. *J. Sep. Sci.* 30, 964–970, 2007.
133. Tuzimski, T., Strategy for separation of complex mixtures by multidimensional planar chromatography. *J. Planar Chromatogr.—Mod. TLC* 21, 49–54, 2008.
134. Snyder, L. R., Classification of solvent properties of common liquids. *J. Chromatogr. Sci.* 16, 223–231, 1978.
135. Dallenbach-Tölke, K., Nyiredy, Sz., Meier, B., Sticher, O., Optimization of overpressured layer chromatography of polar, naturally-occurring compounds by the Prisma model. *J. Chromatogr.* 365, 63–72, 1986.
136. Nyiredy, Sz., Dallenbach-Tölke, K., Sticher, O., Correlation and prediction of the K′ values for mobile phase optimization in HPLC. *J. Liq. Chromatogr. Relat. Technol.* 12, 95–116, 1989.
137. Nyiredy, Sz., Fatér, Z., Automatic mobile phase optimization, using the "PRISMA" model, for the TLC separation of apolar compounds. *J. Planar Chromatogr.—Mod. TLC* 8, 341–345, 1995.
138. Sherma, J., A field guide to instrumentation: Supercritical fluid chromatography. *J. AOAC Int.* 87, 89A–94A, 2004.
139. Ishibashi, M., Ando, T., Sakai, M., Matsubara, A., Uchikata, T., Fukusaki, E., Bamba, T., High-throughput simultaneous analysis of pesticides by supercritical fluid chromatography/tandem mass spectrometry. *J. Chromatogr. A* 1266, 143–148, 2012.
140. Sherma, J., A field guide to instrumentation: Capillary electrophoresis. *Inside Laboratory Management* 6 (6), 36–40, 2002.
141. Zhao, T., Wang, Q. Q., Li, J., Qiao, X. G., Xu, Z. X., Study on an electrochromatography method based on organic–inorganic hybrid molecularly imprinted monolith for determination of trichlorfon in vegetables. *J. Sci. Food Agric.* 94, 1974–1980, 2014.
142. Perez-Fernandez, V., Dominguez-Vega, E., Chankvetadze, B., Crego, A. L., Garcia, M. A., Marina, M. L., Evaluation of new cellulose based chiral stationary phases Sepapak-2 and Sepapak-4 for the enantiometric separation of pesticides by nano liquid chromatography and capillary electrochromatography. *J. Chromatogr. A* 1234, 22–31, 2012.
143. Liu, D., Chen, W., Wei, J., Li, X., Wang, Z., Jiang, X., A highly sensitive, dual-readout assay based on gold nanoparticles for organophosphorus and carbamate pesticides. *Anal. Chem.* 84, 4185–4191, 2012.
144. Fang, R., Yi, L. X., Shao, Y. X., Zhang, L., Chen, G. H., Online preconcentration in capillary electrophoresis for analysis of agrochemical residues. *J. Liq. Chromatogr. Relat. Technol.* 37, 1465–1497, 2014.

145. Zhao, T., Gao, H. J., Wang, X. L., Zhang, L. M., Qiao, X. D. G., Xu, Z. X., Study on a molecularly imprinted solid phase extraction coupled to capillary electrophoresis method for the determination of trace trichlorfon in vegetables. *Food Anal. Method.* 7, 1159–1185, 2014.
146. Malgorzata, J., Pesticides immunoassay, in: *Pesticides—Strategies for Pesticide Analysis,* Stoycheva, M. (Ed.), Intech, open access: http://www.intechopen.com/books/pesticides-strategies-for-pesticides-analysis, DOI: 10.5772/565 (accessed on July 21, 2014).
147. Yan, X., Li, H. X., Yan, Y., Su, X. G., Developments in pesticide analysis by multi-analyte immunoassays: A review. *Anal. Method.* 6, 3543–3554, 2014.
148. Watanabe, E., Miyake, S., Yogo, Y., Review of enzyme linked immunosorbent assays (ELISAs) for analysis of neonicotinoid insecticides in agro-environments. *J. Agric. Food Chem.* 81, 12459–12471, 2013.
149. Dzantiev, B. B., Byzova, N. A., Urusov, A. E., Zherdev, A. V., Immunochromatographic methods in food analysis. *Trends Anal. Chem.* 55, 81–93, 2014.
150. Miyake, S., Development of immunoassay test kit “SmartAssay Series” for pesticide analysis. *Horiba Technical Report*, Fukuoka, Japan, May 31, 2012.
151. Otieno, P. O., Owour, P. O., Lelah, J. O., Pfister, G., Schramm, K.-W., Comparative evaluation of ELISA kit and HPLC DAD for the determination of chlorpyrifos ethyl residues in water and sediments. *Talanta* 117, 250–257, 2013.
152. Personal communication to JS, Dr. Alex Krynitsky, U.S. FDA–CFSAN, College Park, MD, July 22, 2014.

Index

Page numbers followed by f and t indicate figures and tables, respectively.

B

C

D

E

F

G

I

N

O

P

S

U

V

W

X

Z